First-Level Graduate Semester Cours[...]
Some Experience Using Lagrange's E[...]

[1]Chapter 7 is included here because it is conventional to address field equations. However, coverage of the material in any of the later chapters, especially Section 9.1 on substructuring, is a reasonable alternative.

FIRST EDITION

MECHANICAL AND STRUCTURAL VIBRATIONS
Theory and Applications

JERRY H. GINSBERG
Georgia Institute of Technology

JOHN WILEY & SONS, INC.
New York / Chichester / Weinheim / Brisbane / Singapore / Toronto
http://www.wiley.com/college

To my family

Rona, Mitch, Danny, Tracie, Leah, Beth, Abby, and Tundra

Acquisitions Editor   *Joseph Hayton*
Marketing Manager   *Katherine Hepburn*
Senior Production Editor   *Michael Farley*
Senior Designer   *Kevin Murphy*
Production Management Services   *Publication Services*

This book is printed on acid-free paper.

ISBN 0-471-12808-2 (cloth: alk. paper)

10 9 8 7 6 5 4 3 2 1

PREFACE

I have used several textbooks to teach undergraduate and graduate vibrations courses, but never with complete satisfaction. None shared my vision of the subject, and few treated all the concepts that I consider to be important to engineering practice and research. Nevertheless, I did not believe that it was a worthwhile endeavor to add to the multitude of available texts. The usage of Georgia Tech as the Olympic Village in 1996 caused me to change my view. Shortening of the academic quarters in order to fit the Olympic schedule forced me to rethink how to cover the usual topics in an undergraduate course. I developed a way of introducing students to vibrations of beams without necessitating the solution of partial differential equations. As I prepared each lecture on the topic, I realized that the concept was actually quite powerful and broad. That led me to recognize that much of what I did in class was different from what is contained in most texts. What has evolved is a work that I believe substantially alters the standard approach, in a manner that offers numerous benefits to both the student and the instructor.

A fundamental aspect of linear vibration theory is its unified nature, reflecting the fact that the physics of various system are alike. Emphasizing this underlying theme, rather than the mathematical tools required to analyze a specific conceptual model, brings recognition of the connection between topics and the fact that features of one type of model system might be extremely relevant to others. From this, the student comes to develop an expertise in which knowledge of the phenomena associated with each type of system drives the modeling process. In addition, the capabilities of modern software have drastically changed the range of examples and homework exercises that can be addressed. Attention needs to be given to the usage of this software. One of the features addressed here is the way in which algorithms for efficient use of software alter the way we perceive formal mathematical relations.

What one will find inside is a development that assembles many standard topics in a different order and adds material not often found in a basic text. I am unaware of any single book that treats all of the topics covered here. The subjects are addressed as comprehensively as possible, given the size limitations of a general textbook. The imperative to avoid superficiality was my primary reason for omitting three important topics: plates and shells, random vibrations, and nonlinear vibrations, each of which has been addressed in a number of fine textbooks.

Friends have asked me whether I have written an undergraduate- or graduate-level text, to which I respond that it is aimed at everyone. It should not be regarded as a purely graduate text because much of the development is accessible to anyone with fundamental skills. At the same time, I do not wish to oversimplify the field, because graduating students must be prepared for the complexities of real engineering practice. Indeed, one of my fundamental precepts is that a good textbook should make a complicated world accessible to any student, rather than oversimplifying reality until it lacks relevance.

I have attempted to address the needs of undergraduates and graduates alike by suitably structuring Chapters 2 to 6. Core concepts and techniques required to pursue topics in later chapters occupy the leading sections, while nonessential topics at a higher level appear later in these chapters. Within each section one will find numerous examples and homework exercises covering a range of difficulty levels, such that some are available for instruction at every level. Pete Rogers, my colleague, remarked that most texts become progressively more difficult, whereas this one has a series of peaks and valleys. A more descriptive picture would be a sawtooth-like distribution of difficulty, in the sense that the early chapters each build to a peak and then the following chapter returns to a fundamental level based on minimal coverage of preceding chapters. The following overview surveys the key features.

- Chapter 1 addresses modeling discrete systems of particles and rigid bodies. Newton-Euler equations of motion are discussed for simple systems. More complicated systems are addressed by a power balance formulation that is valid solely for linear time-invariant systems. The approach closely parallels Lagrange's equations but is much more accessible to undergraduate students. Students already knowledgeable in Lagrange's equations can focus on the linearization techniques, which are discussed quite comprehensively.

- Chapter 2 is a fairly conventional treatment of free and transient forced vibration of one-degree-of-freedom systems. The primary innovation is analysis of forced response combining the superposition principle, the concept of time delay, and tabulated standard solutions. This approach, which came to me when I realized that students rely on tables when they employ Laplace transforms, enhances a fundamental understanding of the manner in which systems respond to excitation. It also is well suited to mathematical software.

- Chapter 3 is devoted to harmonic excitation. One of the thrusts here is to acquaint students with FFT technology. Many students upon graduation are assigned the task of carrying out experiments and interpreting data, yet they are unaware of the fundamental concepts and associated issues, such as aliasing and leakage. Furthermore, digital signal processing of audio and visual data has had a profound technological impact, which makes it imperative that engineers have some familiarity with these techniques. Toward this goal complex variables are employed to evaluate harmonic steady-state response. Treatment of complex Fourier series for periodic excitation follows naturally. After discussion of the formal mathematical issues, including the Gibbs phenomenon, the discrete Fourier transform (DFT) is developed as a way to evaluate series coefficients numerically, and the Cooley-Tukey FFT algorithm is explained. This leads to a discussion of how to use numerical software to analyze the periodic response of one-degree-of-freedom systems, which serves as a springboard to an extensive development of the proper use of FFTs to analyze transient vibration.

- Chapter 4 is devoted to modal analysis of multi-degree-of-freedom systems. Two- and three-degree-of-freedom systems are analyzed manually in order to illustrate the formal development of the general eigenvalue problem leading to vibrational modes. Once the fundamental concepts have been well ingrained, computational methods are discussed. Many vibrations texts have focused on the numerical eigenvalue algorithms. However, the procedures have become embedded in the standard mathematical software, as well as the scientific subroutine libraries for most programming languages. Hence, the usage of such capabilities, as well as recognition of potential errors, is the primary concern. The remainder of this chapter is a fairly conventional treatment of modal equations, but I think the reader will find the examples quite interesting.

- Chapter 5 addresses the usage of transfer functions to study the response of multi-degree-of-freedom systems to harmonic excitation. The directness offered by the method enables one to focus on fundamental phenomena, such as those encountered in a tuned vibration absorber. The concept of a frequency domain transfer function naturally leads to experimental modal analysis concepts, which is another topic that I think has not been treated sufficiently at a fundamental level.

- Chapter 6 is the most innovative. It uses Ritz series, commonly called the method of assumed modes, as the analytical tool for modeling vibration of bars. The concept is introduced as an approximate technique leading to discretized equations of motion, whose form is like those governing multi-degree-of-freedom systems. This eases the students into the phenomena of continuum vibrations without simultaneously transitioning to the much greater sophistication entailed in solving partial differential equations. Another advantage of the approach is that attachment of masses, springs, and dashpots at arbitrary locations is described directly. This enhances the realism of the systems that can be analyzed, as well as recognition that continua are not intrinsically different from lumped element models. The discussion then turns to modal analysis. When I began down this path, I was quite surprised to realize that one can not only evaluate mode functions, but also derive mathematical orthogonality relations that are more general than those usually derived from Sturm-Liouville theory for the field equations, in the sense that they allow for the presence of attached masses and springs at arbitrary locations. Questions regarding convergence are addressed by studying the Rayleigh ratio and its implications. A particularly interesting phenomenon stemming from ill-conditioning is encountered in an example.

- Chapter 7 uses field equations to study continuum vibrations. It begins with separation of variables. The analysis incorporates arbitrary mixed boundary conditions, but otherwise is fairly standard. Its purpose is to verify the Ritz series formulation and to bring to the fore some fundamental physical phenomena. For example, a section on the wavelike nature of modal vibrations leads to identification of asymptotic approximations required to evaluate high-frequency mode functions. I have not found a corresponding treatment of this representation in other texts. After forced response using analytical mode functions is developed, a wave description leading to transfer matrices is developed. The closing section uses Timoshenko beam theory to examine the validity of classical beam theory for wave propagation and modal analysis. A new feature here is a discussion of how to evaluate Timoshenko beam modes for arbitrary boundary conditions.

- Chapter 8 develops finite element analysis for beams. Its genesis was a set of notes I had semiseriously called "Everything You Need to Know about Finite Elements, unless You Wish to Run an FE Program." My objective here is to give an appreciation for each of the fundamental operations carried out in those programs, as well as to illustrate some of the enhanced capabilities for vibration analysis afforded by such programs.

- Chapter 9 discusses techniques for merging substructures. The concept is introduced with a restricted formulation using Lagrange multipliers. After the basic approach is developed, Hurty's formulation of component mode synthesis is adapted to the Ritz series description, in order to give an appreciation of how the Lagrange multiplier formulation may be generalized. The examples were selected to typify the operations and, equally important, to illustrate some fundamental vibrational phenomena, such as coupling of different types of displacement, and mode localization.

- Chapter 10 is another one that is new from the standpoint of fundamental texts. Arbitrary damping properties invalidate the standard techniques of modal analysis, but they can be addressed in the state space. Some have chosen to address the issue in a manner that leads to asymmetric system matrices, which causes the topic to be incorporated into analyses of gyroscopic systems. However, the formulation presented here yields a symmetric eigenvalue problem. This leads to an ability to perform analyses in the state space without drastically altering the procedures and algorithms that have become ingrained.

- Chapter 11 extends the state-space development to include gyroscopic and Coriolis inertial effects, and follower and feedback forces. The matrix eigenvalue problem in this case is inherently nonsymmetric, and therefore not self-adjoint. Exposure to the state-space formulation in the preceding chapter considerably expedites extension to such systems. Here again, the examples have been chosen to illustrate both the general procedures as well as specific phenomena. In particular, the examples of transverse vibration of a translating cable and stability of a pipe with internal flow were research issues not too long ago.

- Chapter 12 applies the general development in the preceding chapter to a number of canonical models for rotating machinery. I believe it is important to modern engineering practice to understand the problems that can be encountered in high-speed machinery. Each model is selected to illustrate the diversity of dynamic phenomena, including some counterintuitive effects of damping and stiffness. The intention here is to provide a working knowledge of the fundamental issues, as well as to stimulate some readers to explore the subject in greater detail.

I have endeavored to endow each chapter with certain common attributes. Solved examples were selected to illustrate the general concepts and procedures. Another criterion applied to most examples is that they introduce and discuss interesting and important vibrational phenomena. The objective is to enable students to explain why something happens. A minor feature in this regard is that I preface every solution with a few remarks indicating what the student should get out of the development.

I have attempted to let the various topics drive the selection of mathematical tools. Not much prior knowledge is required in this regard. Some fundamental capabilities in the algebra of complex numbers, matrix algebra, and the method of undetermined coefficients for differential equations are the most important math prerequisites. Clearly, availability of mathematical software, such as MATLAB and Mathcad, is essential.

The issue of how mathematical software should be incorporated into a text is a difficult one. There are a variety of packages available, with strong advocates for each. In such an environment, I believe it would be a mistake to focus on one. My experience, especially based on probing students in my classes, is that MATLAB and Mathcad are best suited for the *study* of vibrations. On the other hand, some research-type issues might be handled better by symbolic languages such as Maple, while large computational models for some engineering applications might best be treated with a compiled language such as FORTRAN. Where appropriate, the text material discusses general issues pertaining to the implementation of various procedures. The examples describe specific algorithms and program fragments for MATLAB (Version 5.2) and Mathcad, which are intended to provide students with a starting point for their own work. However, the student and instructor are not constrained to employ any specific software, because MATLAB and Mathcad cover the basic paradigms (programmatic versus worksheet) employed in most alternatives.

It is my desire that this book be one that the instructor, as well as the student, feels an urgency to read in its entirety. I hope that you will share my enthusiasm for the approach advocated here.

ACKNOWLEDGMENTS

The people who have most influenced the material in this book are the numerous undergraduate and graduate students at Georgia Tech whom I have taught using some form of the manuscript. Regardless of whether they directly provided comments, questions, and corrections, or indirectly gave me feedback through their work and classroom participation, they all provided valuable measures by which I could judge how well their needs were being met. Of the students, those deserving specific mention are Harry Garner, Ryan Rye, and Mike Swinson, who were quite thorough in their inspections. Two of my present doctoral research assistants were especially valuable. Nicole L. Zirkelback, while serving as a teaching intern under my direction, had many pertinent observations regarding the perspective of the undergraduate students. She also was an information source and "sounding board" for the development of rotordynamics, and she provided useful criticism in the treatment of substructuring concepts. Michael V. Drexel focused on the state-space modal concepts and experimental modal analysis.

My colleagues were a great source of support. Aldo Ferri was a valuable resource for this book, as he was for my textbook on advanced dynamics. I long ago lost track of the number of excellent insights and suggestions he provided. Also, by sharing his experiences with the use of a draft manuscript as the textbook in an undergraduate class, he had a profound effect on the initial chapters. I must thank Ward O. Winer, the Chair of the Woodruff School of Mechanical Engineering, for understanding the scope of this project. That understanding was sorely tested when I was late for, or completely forgot, appointments.

The efforts of Luis San Andres of Texas A&M and Universiteit Twente are noteworthy. He learned of the manuscript from Nicole Zirkelback, who is a Texas A&M graduate, and requested permission to use it for the course he was teaching. He provided detailed feedback regarding his experience, as well as solutions for some of the homework exercises. In addition, he did a great job in soliciting comments and criticisms from the students. This information was vital to the final organization of the opening chapters. All the reviewers, J. Gregory McDaniel of Boston University, S. J. Siavoshani of Oakland University, Serge Abrate of Southern Illinois University, Raul Longoria of the University of Texas at Austin, George Flowers of Auburn University, Zhikun Hou of Worchester Polytechnic Institute, and E. Harry Law of Clemson University, through their insightful criticisms, both positive and negative, were important to ensuring that I remained true to my objective of meeting a broad range of needs. I look forward to hearing from them, as well as you, the reader, in the future.

A special recognition goes to my editor at John Wiley & Sons, Joseph Hayton. His commitment to this project often was a vital impetus in keeping me focused on finishing the book.

I have saved recognition of the contributions of my family for last, because their role is without end. The grandchildren, Leah Morgan Ginsberg, Elizabeth Rachel Ginsberg, and our newest addition, Abigail Rose Ginsberg, help me just by being wondrous. I cannot thank their parents, Mitchell Robert and Tracie Sears Ginsberg, enough for bringing them into my life, as well as for understanding my preoccupation with this project. My son, Daniel Brian Ginsberg, by sharing my drive toward intellectual achievement, albeit in an entirely different endeavor, gives me a special sense of accomplishment. Finally, I would like to express my love and admiration for my wife, Rona Axelrod Ginsberg. She, more than anyone, bore the brunt of my efforts to write this book. I cannot count the number of times I was late and neglected her in other ways, as a result of concentrating on this project. However, she is more than my devoted companion. She is an editor without equal. With her, I had no need for a style manual.

CONTENTS

CHAPTER 1

EQUATIONS OF MOTION FOR DISCRETE SYSTEMS

1.1   THE STUDY OF VIBRATIONS: AN OVERVIEW

If you have driven over a pothole, or experienced an earthquake, or sat in a noisy airplane, or had to rebalance a washing machine during the spin cycle, or watched a tree shake in a gusty wind, you have experienced a vibration phenomenon. This subject is a specialized area of dynamics, which concerns the motion of physical systems. The distinguishing characteristic of a vibration is that the motion consists of an oscillation relative to a reference state. In most cases this reference state is the system's static equilibrium position, as would be the case for a building. The reference state might also be a steady motion—for example, the spinning of a satellite. The motion is induced by a disturbance force, which is one of four basic effects that influence a vibratory response. Movement away from the reference state leads to generation of elastic restoring forces. Movement of the system requires that there be an acceleration, which changes the system's momentum. The inertial resistance affects the interplay between the elastic and disturbing forces. In addition, a variety of friction-type phenomena act in opposition to movement of the system.

The general objectives of a vibrations study are to predict, explain, and/or control the oscillatory response. In this text we will develop analytical tools for predicting the way in which a system will vibrate. We also will explore how measured response data for one type of excitation can be used to predict a system's response to a different excitation.

Concern for vibrational phenomena can have various motivations. Safety considerations require that we identify situations that cause excessive motions, for they will result in the generation of large stresses. The oscillatory nature of vibratory response means that large stresses are likely to cause significant fatigue damage. Even if a vibration does not damage a system, it might be a source of physical discomfort. A common situation illustrating this point is an automobile whose tires are out of balance. Vibration is also likely to be a source of noise generation and therefore an annoyance to anyone who is exposed to the system. The general public often takes quietness to be a mark of quality, as in the case of automobiles. Indeed, such regard is probably well warranted, because vibration and noise criteria represent some of the most demanding design and manufacturing requirements.

The key element that determines the level of difficulty for a vibrations study is the conceptual model we create to represent the system. The model must account for the four basic phenomena—excitation, elasticity, inertia, and dissipation—with sufficient faithfulness to reality that we will obtain a good correlation between the model's prediction and the actual system response. At the same time, we need to avoid making the model overly sophisticated by capturing more details of the system than are necessary. An excessively complicated model will require far more analytical, computational, and experimental effort than necessary. Indeed, if we are not careful, we could construct a model that cannot be fully solved. We will begin in this chapter to develop our modeling expertise. Further study, in which we see what phenomena are captured by each type of model, will enhance our capability.

One way of categorizing a vibration study is according to the nature of its model. The simplest types of models feature one or several rigid bodies whose mass resists acceleration. Springs attached to the rigid bodies represent restorative forces that tend to return a system to its undisturbed state. Resistance to movement, regardless of its nature, is provided by dashpot devices, which are typified by the hydraulic pistons in a typical automotive shock absorber. A prediction of the response consists of a determination of the manner in which the position variables change as a function of elapsed time. From such knowledge, we may evaluate internal forces for the purpose of design. We also could create a computer animation that pictorially displays the response. Because these models use a finite number of mechanical elements, whose action is described by a finite number of position variables, we say that they are *discrete system models.*

The alternative to a discrete model is a *continuum model,* in which the elastic components of the system also have inertia. Such a model is the dynamic analog of the types of systems one encounters in the study of mechanics of materials and elasticity. A continuous system might consist of only one body, such as a beam representing a simple bridge, or it might be an intricate network of structural elements, such as the assembly of panels and stiffeners that form an aircraft fuselage. The equations of motion associated with a continuum model are significantly more complicated than those of a discrete system.

Other categorizations are often applied to vibrations problems. One concerns the nature of the disturbance. A *free vibration* study examines situations where a system has already been disturbed at the instant we begin to study it. Such a disturbance is associated with initial conditions. The system will vibrate even if no further disturbing force is applied. In contrast, for a *forced vibration* study, the effects of forces that act during the observation interval are of primary interest. We will assign equal importance to free and forced vibration phenomena.

Within the class of forced vibrations, a finer categorization pertains to whether the excitation forces are deterministic or random. All excitations have some degree of randomness associated with them, but such variability can often be ignored. For example, we will see that any rotating body whose center of mass does not coincide with the axis of rotation will generate an oscillatory excitation that depends on the rotation rate. It is not possible to hold a rotation rate absolutely constant, but it is usually permissible to ignore such variability. Common sources of random excitations are fluid turbulence in wind gusts and the number and distribution of vehicles on a bridge. In such cases we would not be able to make a reasonable estimate of the most likely response unless we examined statistical behavior. A different reason for a probabilistic model arises when the system parameters cannot be assigned with reasonable certainty, which might be the case with some manufacturing processes. The techniques for random vibration analysis are beyond the scope of this text. However, comprehension of the deterministic phenomena we encounter here is vital to the study of random vibrations.

A different categorization pertains to approximations and restrictions to be imposed on the model. Anyone who has studied ordinary differential equations knows that linear equations are much easier to solve than nonlinear equations. It is quite likely that even the simplest model we create for a system will have several features that could lead to nonlinear equations. We will circumvent this difficulty by requiring that the motion of the system be small in some sense. In most cases, this limitation still leads to a highly useful model, with which we can identify circumstances where the motion would be sufficiently large to be dangerous or undesirable. However, the restriction to linear phenomena means that certain inherently nonlinear behaviors, such as chaotic response, will not be accessible. One reason we shall not address nonlinear vibration phenomena is that the tools for their study are founded in large part on the linear vibration techniques we shall develop.

The creation of a conceptual model involves the discipline of mechanics. A fundamental dynamics course that treats the kinematics and dynamics of a rigid body in planar motion is sufficient for studying vibrations of discrete systems. In addition, knowledge of the core concepts of mechanics of materials, such as stress-strain laws and the relations governing internal forces in a bar, will be useful for the treatment of continuum models.

A modern approach to the study of vibrations requires implementation of computational capabilities as well as formal mathematical tools. The mathematical tools we shall use in the beginning chapters require a degree of familiarity with linear algebra, such as matrix operations and the solution of simultaneous equations. The algebra of complex numbers (polar vs. rectangular representations, addition, multiplication, etc.) will also play a prominent role. We will see that ordinary differential equations govern the motion of discrete systems. The method of undetermined coefficients will be sufficient to handle all situations of interest. It is not until Chapter 7, which concerns the vibration of fundamental continuum models, that we will encounter partial differential equations. However, by that point we will have already derived an alternative analytical description for continua. It will provide essentially the same insight as that afforded by approaches leading to partial differential equations, but the alternative is easier to implement and more generally applicable.

Before we initiate our study of vibrations, a word is in order regarding the computational software that will be needed. A variety of packages are in wide use, and their advocates feel strongly about them. These include MATLAB, Mathcad, Maple, Mathematica, and Macsyma. The author has used a few of these, and received his education at a time when FORTRAN was the norm. Now some individuals prefer C++. The important aspect for our work is that familiarity with any *one* of these environments is sufficient. Any particular mathematical operation might be implemented in different ways, depending on which package one chooses to use. In order to be cognizant of, and thereby avoid, potential sources of error, we will examine these issues in the appropriate locations. Particular attention in the solution of several exercises will be focused on usage of MATLAB and Mathcad. These developments may be readily adapted to other software. However, the reader with little computational experience who has not yet selected software to use is forewarned that there are significant differences in the learning curves for the various programs.

1.2 SYSTEM ELEMENTS

As noted, discrete systems are constituted from three types of components representing different effects: inertia, elasticity, and resistance to movement. We will review here the fundamental aspects of each. Our task in the following sections and chapters is to describe how the various system components work together.

Rigid Bodies: Inertial Objects A rigid body's inertia is responsible for the resistance to acceleration of a system. Rigid bodies also transfer forces between locations. There is no need to consider the particle model separately because the physical laws for a particle are a subset of those for a rigid body.

Rigid bodies can execute three types of motion. In a translation, lines in the body maintain a constant orientation, so that all points in the body have the same velocity and acceleration. A classic example of a system of translating bodies is a railroad train moving along a straight track. The instantaneous position of a translating body in planar motion may be described in terms of the distance any point in the body has moved. It should be noted, however, that the definition of translation does not require that the path followed by points in the body be straight.

In a pure rotation, at least one point in the body remains stationary. Our concern until Chapters 11 and 12 will be with systems in plane motion. Pure rotation in that case requires that there be a fixed axis of rotation. A gear mounted on rigid shaft is an important example of a body in pure rotation. We may define the instantaneous position of a rotating body in terms of the angle by which any line in the body rotates.

General motion combines translation and pure rotation. A basic theorem of kinematics, which we will explore later, states that a general motion may be pictured as a translation following any point in the body, onto which is superposed a pure rotation about the selected point. General motion is extremely common, with examples ranging from a simple rolling wheel to linkages, vehicles on a rough road, and airplane flight. Because it is necessary to account simultaneously for translational and rotational aspects, the development of system models for bodies in general motion can be quite challenging.

Springs: Restorative Effects A restorative force originates from any source that tends to return a system to its undisturbed state. Such forces are most commonly associated with elastic deformation of a body, but they also can arise from the pendulum-like effect of gravity. The fundamental component we use to represent restorative forces is a coil spring whose ends are pulled apart. The axial force in a spring depends on the elongation Δ, which is the difference between the spring's current length L and unstretched length ℓ_0,

$$\boxed{\Delta = L - \ell_0} \tag{1.2.1}$$

The force relation is taken to be linear,

$$\boxed{F = k\Delta} \tag{1.2.2}$$

The proportionality factor k is the *spring stiffness*, whose units are N/m in the SI system. Linearity is an approximation that is valid if the strain of elastic bodies is sufficiently small. Also, note that the preceding is also taken to apply for compression, in which case Δ is negative, corresponding to decreased length.

For simplified studies, we may use a spring to represent any system component that exerts a force tending to return a system to its reference state. Figure 1.1 lists a number of ways in which an elastic bar may be used as a spring component. If we neglect inertial effects, based on the bar's mass being much less than that of other system components, then application of a force at some point will result in a proportional displacement of that point, as described in Figure 1.1. Thus the mathematical model of the bar's effect will be the same as that obtained if the bar were replaced by a coil spring aligned in the direction of that displacement and having the corresponding k.

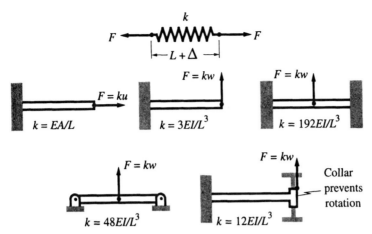

FIGURE 1.1 Equivalent spring constants for elastic bars. *EA* is the extensional rigidity, and *EI* is the flexural rigidity.

A system may also contain a *torsional spring*, which exerts a couple (moment) that is proportional to the rotational deformation it experiences, so that

$$M = k_T \theta \qquad (1.2.3)$$

The SI units used to describe k_T are N-m/rad. As depicted in Figure 1.2, a torsional spring element is usually depicted as a spiral whose ends are fastened to other bodies. The figure shows that a coil spring twisted about its axis can be considered to be a torsional spring, as can elastic bars, when θ is taken as the angle by which the bar rotates at the location where the couple is applied. The texts by Seirig (1969) and Cochin and Plass (1990) provide compact lists of spring constants for a range of common system components.

It often happens that spring-like system components are interconnected. Figure 1.3 depicts springs that are connected in parallel, in series, and in a combination of the two basic types.

A parallel connection is one in which the springs undergo the same deformation and the forces exerted by each add. Because Δ_1 is the amount by which both parallel springs are elongated, the forces within the springs are $k_1\Delta_1$ and $k_2\Delta_1$. The force F

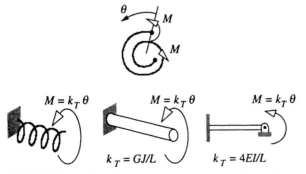

FIGURE 1.2 Coil spring and elastic bars as equivalent torsional springs. *GJ* is the torsional rigidity, and *EI* is the flexural rigidity.

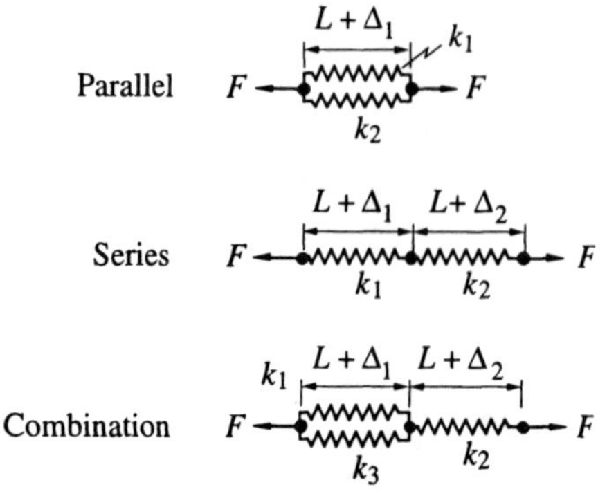

FIGURE 1.3 Different types of connections between springs.

exerted by the pair is therefore $F = k_1\Delta_1 + k_2\Delta_1$. The *equivalent stiffness* k of the pair is such that it gives $F = k\Delta_1$, from which it follows that

$$\boxed{k = k_1 + k_2} \tag{1.2.4}$$

If more than two springs are connected in parallel, we merely add the additional stiffnesses to the ones already included.

In a series connection, the increase in the distance between the ends is the sum of the amount by which each spring is lengthened. In this case the force applied is transmitted equally through the springs. In other words, the elongations add, and the force is the same. The amount by which each spring is lengthened is related to the corresponding spring stiffness by $\Delta_1 = F/k_1$ and $\Delta_2 = F/k_2$. The total elongation is $\Delta = \Delta_1 + \Delta_2$, so we have

$$\Delta = \frac{F}{k_1} + \frac{F}{k_2} \tag{1.2.5}$$

If we were to replace this with an equivalent single spring, we would have $\Delta = F/k$, from which it follows that

$$\boxed{\frac{1}{k} = \frac{1}{k_1} + \frac{1}{k_2} \quad \Rightarrow \quad k = \frac{k_1 k_2}{k_1 + k_2}} \tag{1.2.6}$$

Situations where more than two springs are connected in series are handled by adding additional reciprocals to the preceding.

Connecting springs in series yields a lower stiffness value than either individual element, because the overall deformation is obtained by adding the individual contributions. In contrast, a parallel connection of springs raises the equivalent spring stiffness, because in that case the individual forces, rather than the extensions, add.

The way to obtain the equivalent stiffness for springs connected in an arbitrary manner is to apply successively the rules for parallel and then series connections. In the case of Figure 1.3, we begin by replacing k_1 and k_3, which act in parallel, by a single equivalent spring whose stiffness is $k_{1,3} = k_1 + k_3$. This operation yields a series connection consisting of k_2 and $k_{1,3}$. We obtain the equivalent stiffness by equating the reciprocal of the equivalent stiffness to the sum of the reciprocals of the series-connected springs. In the case of Figure 1.3, this step leads to

$$\frac{1}{k} = \frac{1}{k_{1,3}} + \frac{1}{k_2} = \frac{1}{k_1 + k_3} + \frac{1}{k_2} \quad \Rightarrow \quad k = \frac{(k_1 + k_3)k_2}{k_1 + k_2 + k_3}$$

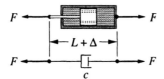

$F \longleftarrow \boxed{} \longrightarrow F$

$\longleftarrow L + \Delta \longrightarrow$

$F \longleftarrow \boxed{} \longrightarrow F$ **FIGURE 1.4** Physical and schematic representations of a
   c dashpot.

An important aspect of combining springs into a single equivalent system is recognizing when it is permissible to do so. The analyses require that the springs be directly connected. Specifically, *no bodies having mass may be present at the connection.* In the next sections we will develop techniques to account for the way springs and masses work together.

Dashpots: Resistance to Velocity Forces that oppose velocity originate from many sources. A device that is designed to provide such resistance is a *dashpot*, also sometimes referred to as a *damper*. A simplified picture of such a device appears in Figure 1.4, where a piston moves within a sealed cylinder that is filled with a fluid. The resistance is obtained from the viscous forces generated when the fluid flows through orifices in the piston. The lower part of the figure is the simplified schematic form in which dashpots are usually depicted. We take the axial force F to be proportional to the velocity of the piston relative to the cylinder, so that

$$ F = c\dot{\Delta} \tag{1.2.7} $$

The factor of proportionality c is the *dashpot constant.* Its SI units are N-s/m.

A dashpot may also be torsional. Such a device exerts a torque that is proportional to the rate at which the rotational deformation changes:

$$ M = c_T\dot{\theta} \tag{1.2.8} $$

where the SI units of the *torsional dashpot constant* c_T are N-m-s. A torsional dashpot may be obtained by attaching blades to a shaft in order to get fluid resistance to rotation. The fluid coupling in an automotive automatic transmission acts like a torsional dashpot for the drive train.

The presence of dashpot in a system represents a design decision to introduce resistive forces. Such forces also arise from a number of sources that are not under our control, including fluid resistance, acoustic radiation, friction between surfaces and at joints, and internal dissipation. We often approximate these effects by including a dashpot in our model. However, the actual processes that are being approximated are seldom well understood. In such situations, the dashpot constants are likely to be selected on the basis of experimental measurements, previous experience, or an educated guess.

The relations for force and moment dashpots are analogous to those for springs. The similarity extends to the analysis and rules for replacing parallel- and series-connected dashpots by a single equivalent dashpot. Specifically, we add the constants to obtain the equivalent constant for parallel-connected dashpots, and equate the sum of the reciprocals of the constants to the equivalent reciprocal for series-connected dashpots.

EXAMPLE 1.1

An elastic beam, whose flexural rigidity is *EI*, is attached to two springs as shown. Determine the static displacement of the block resulting from application of the static force *F*.

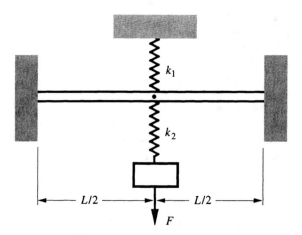

Solution The purpose of this example is to emphasize that the primary consideration in replacing a set of springs with an equivalent single spring is the way the springs act, rather than their visual appearance. In particular, although springs k_1 and k_2 are arranged collinearly, they do not act exactly as though they are series connected. The analysis must account for the spring-like action of the beam. From Figure 1.1, the stiffness at midspan of a beam whose ends are clamped is $k_3 = 192EI/L^3$. We next observe that k_1 and k_3 undergo the same displacement. This is the condition for a parallel connection, so the single spring equivalent to k_1 and the beam is

$$k_{1,\text{beam}} = k_1 + k_3 = k_1 + \frac{192EI}{L^3}$$

We now proceed to consider k_2. The displacement of the block is the sum of the displacement at the beam and the amount by which k_2 is stretched. Furthermore, the force F applied to the block is transmitted through k_2 and then applied to $k_{1,\text{beam}}$. Both observations are characteristics of a series connection, so we have

$$\frac{1}{k} = \frac{1}{k_2} + \frac{1}{k_{1,\text{beam}}} = \frac{1}{k_2} + \frac{1}{k_1 + \dfrac{192EI}{L^3}}$$

Taking the reciprocal of this expression leads to

$$k = \frac{k_2(k_1 L^3 + 192EI)}{k_1 L^3 + k_2 L^3 + 192EI}$$

1.3 GENERALIZED COORDINATES

The primary step in mathematically modeling a physical system is selection of the variables that describe the system's instantaneous position. By making this selection, we implicitly are selecting the parts that must be included in the model. For example, if we decide that a certain object is sufficiently small to consider it to be a particle, we have made a modeling assumption that its rotational inertia effects are unimportant. Similarly, if we approximate a part as being a rigid body, we are assuming that deformational effects are unimportant to its overall motion.

Our initial efforts will concern systems that may be represented as a collection of particles and rigid bodies. We begin by asking what geometric information is required to draw a sketch of the system's location at an arbitrary instant. The variables we use to locate each body may be coordinates of points relative to a set of orthogonal Cartesian axes; or curvilinear coordinates, such as polar coordinates; or angles of lines relative to some reference orientation. Furthermore, position variables describing a specific object may be measured from fixed reference positions, or they may be measured relative to other objects contained within the system. Regardless of how they are defined, the variables we select are called the system's *generalized coordinates*. We shall use the generic symbol q_n to refer to these variables.

Most of the systems we will encounter undergo planar motion, which means that their parts all move in a common plane. To understand the nature of generalized coordinates, let us consider a fundamental system—specifically, two particles that are interconnected by a spring and are free to move anywhere on a plane. Figure 1.5 shows six generalized coordinates for this system. Although six variables are defined, four are sufficient to locate both particles. The set $q_1 = X_A$, $q_2 = Y_A$, $q_3 = X_B$, $q_4 = Y_B$ locate both particles in terms of distances along the reference directions. Alternatively, we can use $q_1 = X_A$, $q_2 = Y_A$ to locate particle A, and then use polar coordinates $q_3 = r$, $q_4 = \theta$ centered on particle A to locate particle B. The minimum number of independent variables required to locate a system at any instant are the system's *number of degrees of freedom*, which we will denote by the symbol N.

Although the system in Figure 1.5 has four degrees of freedom, we are not free to select any four variables as generalized coordinates. For example, setting $q_1 = X_A$, $q_2 = X_B$, $q_3 = r$, $q_4 = \theta$ would be unacceptable for two reasons. First, it is apparent that the overall position in the Y direction would not be defined, because r and θ only set the distance in the Y direction from particle A to particle B. Second, the aforementioned four variables are not independent because $r \cos(\theta) = X_B - X_A$.

Geometric equations relating generalized coordinates will also exist if we select more generalized coordinates than the number of degrees of freedom. For example, if we use X_A, Y_A, X_B, Y_B, and r as generalized coordinates, we have $r^2 = (X_B - X_A)^2 + (Y_B - Y_A)^2$. These observations lead us to the concept of *unconstrained generalized coordinates*, which are any N position variables that are not related by geometrical or other kinematical conditions.

Engineering systems feature a variety of guides and supports that prevent displacement of points in certain directions. These restrictions are kinematical in nature, and reduce the number of degrees of freedom relative to what a system would have if

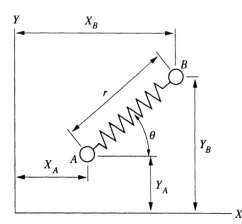

FIGURE 1.5 Two particles on a horizontal plane connected by a spring.

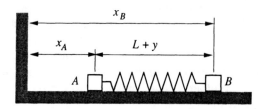

FIGURE 1.6 Two particles joined by a spring and moving horizontally.

each of its parts could move freely. Figure 1.6 shows one way in which the two-particle system could be further constrained. The particles are represented there as blocks that slide over a rigid horizontal surface. Three position variables, x_A, x_B, and y, are defined in the figure, but only two are independent, so $N = 2$ for this system.

The vertical positions of the blocks in Figure 1.6 are not variable because of the normal force exerted between each block and the floor. We say that these forces are *constraint forces*, which is synonymous with *reactions*, because they enforce the constraint condition that the blocks cannot penetrate the surface. The magnitude of constraint forces is unknown and unrestricted, for these forces will be as large as necessary to impose the associated constraint condition. (The present normal forces may only push upward because the surfaces cannot prevent the blocks from moving upward.) Compare the systems in Figures 1.5 and 1.6. In the first, there are no constraint forces; this leaves all position coordinates as unknowns that may be determined only by considering the inertial equations of motion. For the system in Figure 1.6, the normal forces are unknown constraint forces, and two position variables that previously were free now are system properties.

Figure 1.7 connects the two particles in Figure 1.5 by a rigid bar. This system may be located at any instant by specifying the values of X_A, Y_A, and θ, so $N = 3$. Replacing the spring by a rigid, massless bar is equivalent to considering the stiffness of the spring to be infinite, so the relation between spring deformation and spring force is undefined. The spring force now becomes an axial constraint force F imposing the condition that the distance r between the particles cannot change. Here again, the constraint force has the effect of converting a geometric variable into a system property.

A different viewpoint for the system in Figure 1.7 is to consider it to be a rigid body whose mass is concentrated at two locations. A fundamental principle of kinematics is Chasle's theorem (see any basic text on dynamics of rigid bodies), according to which the motion of a rigid body is the superposition of a translation following any point in the body and a rotation of the body about that point. For planar motion, we

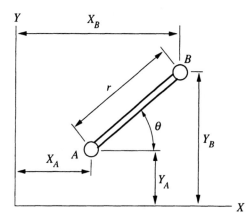

FIGURE 1.7 Two particles on a horizontal plane joined by a rigid bar.

need two variables to describe the motion of a point and one variable to describe the rotation of the body, which occurs about an axis perpendicular to the plane. It follows that any rigid body that moves in a plane without constraint has three degrees of freedom. In this viewpoint, using $q_1 = X_A$, $q_2 = Y_A$, $q_3 = \theta$ for the system in Figure 1.7 is equivalent to applying Chasle's theorem with point A chosen for the translational motion. Other choices obviously are possible, but in any case three geometric variables are required to locate the system's instantaneous position.

Because any rigid body moving freely in a plane has three degrees of freedom, a system having M rigid bodies undergoing planar motion will have no more than $3M$ degrees of freedom. This number is reduced by connections between the bodies and to the ground. In Figure 1.8, two bars are joined by a pin. We may locate the system by giving the coordinates X_B and Y_B, and the two angles of elevation θ_1 and θ_2, so $N = 4$. The pin imposes the constraint condition that the connection points on each body must have the same position coordinates X_B and Y_B, so there are two motion constraints imposed on the two bodies. The number of bodies is $M = 2$ and there are two constraint conditions, and $N = 3M - 2$ is consistent with our identification of $N = 4$ based on constructing the position. The constraint forces associated with the pin connection are the components of the force exerted between the two bars at pin B.

If we further constrain the two bars in Figure 1.8 by pinning the free end of one and attaching the end of the other to a collar on a guide, we obtain a "four-bar linkage" like the one in Figure 1.9. In comparison with the system in Figure 1.8, the linkage has three additional motion restrictions—specifically, that the position coordinates of pin A are constant and that end C must be situated somewhere along the guide for the collar. Hence, this system has only one degree of freedom. A suitable generalized coordinate is the angular position of either bar. The associated constraint forces are the horizontal and vertical forces exerted by pin A, and the normal force exerted on collar C by the guide.

It is apparent that unconstrained generalized coordinates represent the manner in which a system can move consistently with the kinematical constraints imposed in it. An important corollary is that the values of unconstrained generalized coordinates at any instant depend solely on the forces causing the system to move. We say that such forces are the *excitation*. It follows that a mathematical model of a system will contain N equations relating unconstrained generalized coordinates for the system. The (unknown) constraint forces do not appear in these equations,

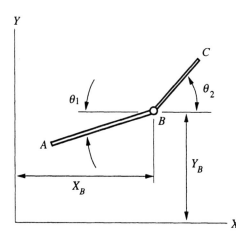

FIGURE 1.8 Two pinned bars executing a planar motion.

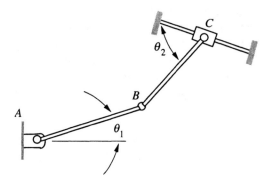

FIGURE 1.9 A four-bar linkage.

which are called *equations of motion*. We will see that these are differential equations. Solution of the equations of motion constitutes the system's response.

For any system of particles and rigid bodies, we identify a set of generalized coordinates by inspection. We do so by considering what position variables must be known in order to draw a sketch locating the system's instantaneous position. If the full set of variables may be selected freely, without violating any of the kinematical constraint conditions, the variables constitute a set of unconstrained generalized coordinates for the system. If there are geometrical relations between the generalized coordinates, we use these equations to eliminate the excess variables. The system's number of degrees of freedom is the number of unconstrained generalized coordinates.

Each set of generalized coordinates we have considered so far have been defined relative to a stationary location. Usually, if a system has a static equilibrium position, we will use that position to define the generalized coordinates. We say that a system is *time invariant* if its generalized coordinates describe the system's displacement relative to a stationary reference position.

It is not always desirable to measure generalized coordinates relative to a fixed location. For example, suppose we were interested in the in-flight vibrational properties of an aircraft turbine engine. We would locate the position of the engine's parts relative to the aircraft by defining a set of generalized coordinates relative to the airframe. In that case, locating any engine part with respect to the Earth would require that we know both the instantaneous values of the generalized coordinates and the instantaneous position of the aircraft. This is a typical case of a *time-dependent system*, which describes any situation where the generalized coordinates are defined with respect to a reference frame that is itself moving.

1.4 NEWTON-EULER MOTION EQUATIONS

Once we have decided how to represent the model elements and selected a suitable set of unconstrained generalized coordinates, our task is to obtain a set of differential equations of motion. As noted, the constraint forces should not appear in these equations. A number of different approaches have been developed to derive such equations. One uses the laws of Newton and Euler, which are based on linear and angular momentum. In general, we will derive equations of motion from the Newton-Euler laws only if the system fits the following conditions:

- The system undergoes planar motion.
- All rigid bodies undergo either translation or pure rotation.
- All forces acting on the bodies either have constant orientation or are oriented parallel to the direction in which the point of application moves.
- Rigid bodies are not directly connected to each other, although they may be connected via springs and dashpots.

If these four conditions are not met, the level of complexity of the system will be such that it is preferable to derive the equations of motion using methods we shall develop later. The advantage of the Newton-Euler formulation in the circumstances we consider here is that it will provide more physical insight into the interplay between forces.

The decision to limit our initial studies to systems in planar motion is partially motivated by the fact that the dynamic principles for a body in planar motion are much simpler than their counterparts for three-dimensional motion. In a planar motion, all points in a system move in parallel planes. We denote one such plane as XY. The angular velocity of a body must be perpendicular to the plane, $\bar{\omega} = \omega \bar{K}$, so that only the rotation rate ω is variable.

The Newton-Euler kinetics laws for a rigid body are derived by considering the momentum of a system of particles. Examination of the rate of change of such a system's linear momentum leads to the equivalent of Newton's second law for a particle:

$$\boxed{\sum \bar{F} = m\bar{a}_G} \qquad (1.4.1)$$

where $\sum \bar{F}$ is the resultant of the external forces (that is, forces applied to the rigid body by other bodies), m is the total mass, and $\bar{a}_G$ is the acceleration of the center of mass G.

The principle governing rotation, which is associated with Euler, emerges from an evaluation of angular momentum. For planar motion of a rigid body, the angular momentum about the center of mass G is $I_G\,\bar{\omega}$. The quantity I_G is the centroidal mass moment of inertia of the body about an axis perpendicular to the plane of motion through point G. Its dimensions are mass $\times$ length^2, which is kg-m^2 in SI units. The quantity $\bar{\omega}$ is expressed as a vector by the right-hand rule. The moment of the external forces about the center of mass is the rate at which this angular momentum changes. When XY is the plane in which motion occurs, we have

$$\sum M_{GZ} = I_G \dot{\omega} \qquad (1.4.2)$$

The notation $\sum M_{GZ}$ denotes that the moment sum is about the Z axis that intersects the center of mass. Positive moments, as well as ω, are defined in the sense of the right-hand rule relative to this axis.

A particularly useful alternative form of the moment equation applies to the case of pure rotation. With point O denoted as the point in the body that is stationary, the resultant moment equation is

$$\boxed{\sum M_{OZ} = I_O \dot{\omega}} \qquad (1.4.3)$$

where I_O is the mass moment of inertia about the axis of rotation. It is preferable to use eq. (1.4.3) to describe pure rotation because preventing point O from moving requires imposition of constraint forces at that point. By taking the moment equation about point O, we avoid the appearance of these unknown forces in the equations of motion.

The restrictions imposed on the situations to be addressed here lead to the simplification that each body in the system will be governed by a single kinetics equation.

- Translating bodies are governed by eq. (1.4.1). The position of a translating body is defined by the distance by which it has been displaced, so we use that distance as a generalized coordinate q_j, where j is the number assigned to the body. If $\bar{e}_t$ denotes the direction in which the translation occurs, the acceleration will be $\bar{a}_G = \ddot{q}_j \bar{e}_t$. Recall that we seek equations of motion that do not contain constraint forces. Such forces act perpendicularly to $\bar{e}_t$, as exemplified by the normal force exerted on a block sliding along the ground. It follows that the only force equation for translating bodies that will not contain constraint forces is the one obtained from the tangential direction. Thus, the kinetics law we form for a translating body is

$$\Sigma \bar{F} \cdot \bar{e}_t = m\ddot{q}_j \qquad (1.4.4)$$

- Rotating bodies are governed by eq. (1.4.3). The rotation of the body is defined in terms of the angle to some line in the body measured from some fixed reference line, such as horizontal or vertical. We define this angle to be the generalized coordinate q_j, so the angular velocity is $\bar{\omega} = \dot{q}_j \bar{K}$. Because the reactions required to keep point O stationary give no contribution to a moment sum about that point, the kinetics law we form for a body in pure rotation is

$$\Sigma M_{OZ} = I_O \ddot{q}_j \qquad (1.4.5)$$

A resultant force or moment equation applies to each rigid body. Forming each equation requires that we describe the force or moment resultant in terms of the generalized coordinates. An aid that has evolved to ensure that all forces are considered is the *free body diagram* (FBD). Each body is isolated from the rest of the system in these drawings, and all forces exerted on the body are shown as vectors. The coordinate directions to be used for force components should also be shown in the FBD. Any force exerted between bodies must be consistent with Newton's Third Law. Hence, if a force exerted between bodies 1 and 2 appears in the FBD for body 1 with a certain set of components, the force appearing in the FBD for body 2 should have the opposite set of components. Furthermore, the lines of action must be the same.

With eqs. (1.4.4) and (1.4.5), we have characterized the inertial aspects of the system model in terms of the generalized coordinates. It also is necessary to characterize in terms of the generalized coordinates the spring and dashpot forces that contribute to force or moment sums. Let us first consider a typical system featuring translational motion.

In Figure 1.10(a), two collars that slide over a horizontal guide are connected by spring k_2 and dashpot c_2. As a result, application of the excitation force F to the right collar results in movement of both collars. The generalized coordinates defined in the figure are the displacements of the blocks, so $q_1 = x_1$ and $q_2 = x_2$. We define the reference position, at which $x_1 = x_2 = 0$, to be the "as-built" configuration, which is the static equilibrium position in the absence of the excitation force. Because x_1 and x_2 define the location of each block, and they are not related kinematically, $N = 2$ for the system.

Figure 1.10(b) shows the free body diagram for each block. (The normal and gravity forces, which equilibrate, are omitted from these diagrams.) The spring and dashpot forces are drawn as tensile effects, corresponding to positive values of their elongation Δ and elongation rate $\dot{\Delta}$, respectively. In part (a) of the figure L_1 and L_2 are

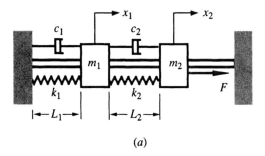

(a)

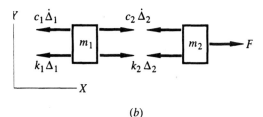

(b)

FIGURE 1.10 A two-degree-of-freedom vibratory system.

the lengths of the springs when the blocks are at their reference position. These lengths become $L_1 + x_1$ and $L_2 + x_2 - x$, respectively, when the blocks move. (Note that positive x_2 lengthens spring 2, whereas positive x_1 shortens it.) We denote as ℓ_1 and ℓ_2 the respective lengths in the undeformed state. It follows that the elongations of the springs in the displaced state are given by

$$\Delta_1 = L_1 + x_1 - \ell_1, \qquad \Delta_2 = L_2 + x_2 - x_1 - \ell_2$$

Differentiation yields expressions for the elongation rates in terms of the generalized coordinates,

$$\dot{\Delta}_1 = \dot{x}_1, \qquad \dot{\Delta}_2 = \dot{x}_1 - \dot{x}_1$$

The preceding expressions enable us to describe the spring and dashpot forces, so we proceed to form the motion equations. The tangential direction for both blocks is X, so we write

Collar 1: $\sum F_X = -k_1\Delta_1 - c_1\dot{\Delta}_1 + k_2\Delta_2 + c_2\dot{\Delta}_2 = m_1\ddot{x}_1$

Collar 2: $\sum F_X = F - k_2\Delta_2 - c_2\dot{\Delta}_2 = m_2\ddot{x}_2$
$$(1.4.6)$$

Substitution of the expressions for the elongations leads to two differential equations of motion, as required for a two-degree-of-freedom system:

$$m_1\ddot{x}_1 + (c_1 + c_2)\dot{x}_1 - c_2\dot{x}_2 + (k_1 + k_2)x_1 - k_2x_2 + k_1(L_1 - \ell_1) - k_2(L_2 - \ell_2) = 0$$

$$m_2\ddot{x}_2 + c_2\dot{x}_2 - c_2\dot{x}_1 + k_2x_2 - k_2x_1 + k_2(L_2 - \ell_2) = F \qquad (1.4.7)$$

The terms containing spring lengths are constants, so they represent static forces. The equations of motion reduce to conditions for static equilibrium if we set $F, x_1, x_1, \dot{x}_2$, and $\dot{x}_2$ to zero. This leads to $k_1(L_1 - \ell_1) = k_2(L_2 - \ell_2)$ and $k_1(L_2 - \ell_2) = 0$; in other words, $L_1 = \ell_1$ and $L_2 = \ell_2$. We can discern both conditions without writing equations of motion by merely recognizing that there is no external force to balance the force in spring 2 in the static equilibrium state. Application of these conditions on the spring lengths to eq. (1.4.7) eliminates static effects.

A compact way in which to write linear equations of motion is to use matrix notation, with the generalized coordinates assembled as a column. This leads to

$$\begin{bmatrix} m_1 & 0 \\ 0 & m_2 \end{bmatrix} \begin{Bmatrix} \ddot{x}_1 \\ \ddot{x}_2 \end{Bmatrix} + \begin{bmatrix} (c_1 + c_2) & -c_2 \\ -c_2 & c_2 \end{bmatrix} \begin{Bmatrix} x_1 \\ x_2 \end{Bmatrix} + \begin{bmatrix} (k_1 + k_2) & -k_2 \\ -k_2 & k_2 \end{bmatrix} \begin{Bmatrix} x_1 \\ x_2 \end{Bmatrix} = \begin{Bmatrix} 0 \\ F \end{Bmatrix}$$

(1.4.8)

We will see in subsequent sections that all time-invariant systems have linearized equations of motion that fit this form, where coefficient matrices multiply columns of generalized accelerations, generalized velocities, and generalized coordinates. These are referred to as the *inertia, damping,* and *stiffness matrices,* respectively. The force column will contain the generalized forces.

Let us now turn our attention to bodies that execute a pure rotation by considering Figure 1.11(a), where a bar rotates in the vertical plane about pin O, with a spring attached at one end. The position depicted in Figure 1.11(a), in which the bar is vertical, corresponds to the static equilibrium position. In this position, the spring is horizontal and unstretched. We define the angle of rotation θ as the generalized coordinate for this one-degree-of-freedom system.

The forces that act when the system has moved away from the static equilibrium position are depicted in the free body diagram, Figure 1.11(b). The moment equation about the fixed point O eliminates the constraint forces holding that point fixed. One of the terms in that equation is the moment of the spring force about the pivot, whose evaluation requires consideration of geometrical features. As depicted in Figure 1.11(b), the attachment point moves to position B'. The angle between the spring's line of action and the bar is β, so the moment is $-(k\Delta)L \sin \beta$. The value of Δ is the amount by which the spring's length is reduced, $\Delta = \ell_0 - |\vec{r}_{B'/C}|$. It is possible to express both $|\vec{r}_{B'/C}|$ and β in terms of θ, but the evaluation is complicated.

This spring force certainly does not fit the restricted class we imposed earlier, because it is neither fixed in direction nor parallel to the direction in which the end of the bar moves. However, by introducing a fundamental assumption about the response, we fit the spring force to the latter restriction and thereby considerably simplify the analysis. We impose the *small displacement approximation*. This essentially is an assumption that the displacement of any system is so small that, at any instant,

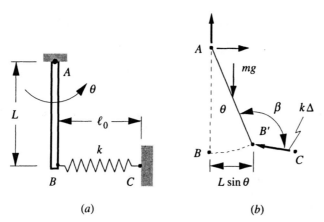

(a) (b)

FIGURE 1.11 (a) Spring attached to a bar. (b) Evaluation of the moment of the spring force.

all parts of the system seem to have the orientation with which they started. Here this means that we consider the rotation of the bar to be very small, $|\theta| \ll 1$. In that case, point B essentially moves horizontally. The amount of this displacement is obtained by letting $\sin \theta \approx \theta$, so the horizontal distance between the original and current positions of point B is the arc length $L\theta$. This is approximately how much the spring's length is reduced, so we have $\Delta \approx L\theta$. Furthermore, when θ is small, the spring is essentially horizontal, so we have $\beta \approx \pi/2$. Thus, the moment is approximately $-L(kL\theta)$.

The small displacement approximation also applies to the moment of the gravity force. The lever arm of this force relative to point A is $(L/2) \sin \theta$. Because θ is small, this simplifies to $(L/2)\theta$, so the gravitational moment is $-mg(L/2)\theta$. The moment of inertia of a thin bar about its end is $\frac{1}{3}mL^2$, so the equation of motion is

$$\Sigma M_A = -kL^2\theta - \tfrac{1}{2}mgL\theta = (\tfrac{1}{3}mL^2)\ddot{\theta} \quad \Rightarrow \quad \ddot{\theta} + \left(\frac{3k}{m} + \frac{3g}{2L}\right)\theta = 0 \qquad (1.4.9)$$

It is important that the differential equation is linear in θ. Linearity, which results from the small displacement approximation, is highly desirable from the standpoint of solving the equations of motion. However, this simplification comes with a penalty, in that there is a limit to the magnitude of θ for which the solution will be accurate.

The restrictions imposed on the types of systems for which we would employ the Newton-Euler equations (planar translation or pure rotation, forces that are constant or parallel to displacement, no interconnected bodies) are selected to avoid complications in deriving linearized approximations of all effects. These restrictions also are such that we can readily prevent the appearance of constraint forces in the force or moment sums. Under these conditions, the Newton-Euler approach is quite useful.

EXAMPLE 1.2

The sketch depicts a model of a two-story building. Four columns (only two are visible), each having flexural rigidity EI, are used to support a floor. The floors are considered to be rigid. The columns are welded at their ends. In combination with the rigidity of the floors, this leads to a model in which the ends of the columns can displace, but not rotate. Each floor has mass m. Derive the differential equations of motion governing the horizontal displacements u_1 and u_2 of each floor.

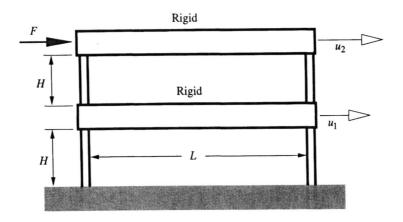

Solution This example will demonstrate the use of equivalent stiffnesses in the derivation of equations of motion. We are concerned with horizontal displacement, and the first floor is connected to the ground by the four lower columns. The deformation of these columns is measured by u_1. The four upper columns act like horizontal springs connecting the two floors, and their deformation is measured by $u_2 - u_1$. Thus, the building model is equivalent to a two-degree-of-freedom system like the one in Figure 1.10(a), with $x_1 = u_1$ and $x_2 = u_2$. The stiffness of the springs connecting m_1 to the ground, and m_2 to m_1, is $4k$, where k is the stiffness of a single beam that is clamped at one end and free to displace, but not rotate, at the other end. From Figure 1.1, we find that $k = 12EI/H^3$, where H is the length of a column. We therefore replace k_1 and k_2 in Figure 1.10(b) with $48EI/L$, and omit the dashpot forces. The resulting equations of motion are

$$m\ddot{u}_1 + \frac{48EI}{H^3}[u_1 - (u_2 - u_1)] = 0$$

$$m\ddot{u}_2 + \frac{48EI}{H^3}(u_2 - u_1) = F$$

EXAMPLE 1.3

The springs in the diagram represent elasticity of a cable wrapped around the flywheel. There is no slippage of the cable relative to the flywheel, and the moment of inertia of the flywheel about its pivot is I_O. Derive the differential equations of motion for the system.

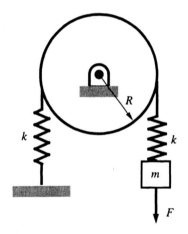

Solution This exercise will give us practice in working with the kinematics of rotational motion, as well as treating systems where bodies execute different types of motion. The rotation of the flywheel may be described by the angle of rotation θ of any radial line. We take θ to be counterclockwise, based on the right-hand rule for a z axis out of the plane. The vertical displacement of the suspended block is defined by u, which we measure positive downward from the static equilibrium position. The free body diagrams of each body are as shown, where $k\Delta_1$ and $k\Delta_2$ are the forces exerted by the springs on either side of the flywheel, based on Δ_1 and Δ_2 being elongations.

To eliminate the reactions at the bearing, the useful motion equation for the flywheel is

$$\sum M_{OZ} = I_O\ddot{\theta} = R(k\Delta_1) - R(k\Delta_2)$$

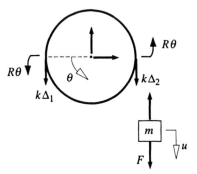

The appropriate force equation for the block is

$$\sum F_t = m\ddot{u} = -k\Delta_2 + F$$

Note in this equation that we have omitted the weight of the block, because we take Δ_2 to be the effect of movement away from static equilibrium, and the force in the spring at static equilibrium is canceled by the weight.

The elongations are related to the generalized coordinates. We must be careful to properly account for signs in describing these relations. To do so, we consider the influence on the length of each spring if all generalized coordinates are positive. Points on the perimeter of the flywheel displace circumferentially by $R\theta$. Thus, positive θ decreases the length of the left spring, such that $\Delta_1 = -R\theta$. Positive θ has a tensile effect for the right spring, corresponding to pulling the spring upward by $R\theta$, and $u > 0$ also elongates the spring by pulling the lower end downward. Thus, we have $\Delta_2 = R\theta + u$. Substitution of the expressions for the elongations into the preceding force and moment equations yields

$$I_O\ddot{\theta} + Rk(R\theta) + Rk(R\theta + u) = 0$$

$$m\ddot{u} + k(R\theta + u) = F$$

1.5 POWER BALANCE METHOD FOR TIME-INVARIANT SYSTEMS

We shall now develop a method for deriving equations of motion that is more efficient and reliable than the Newton-Euler formulation, especially when the system has bodies that execute general motion. The method requires characterization of energy and work quantities. This is a fundamental difference from the Newton-Euler approach, which is based on momentum considerations. Another difference is that energy-based techniques usually allow us to avoid decomposing multibody systems into their individual parts. Regardless of which approach we employ, small displacement approximations are usually necessary to obtain linear equations. However, the approximation tends to be easier to implement in energy-based methods.

The primary method we shall employ is based on the power balance law, which is the time derivative of the work-energy principle for a system. The derivation of the method is somewhat heuristic, but it is justifiable in light of the Lagrange equation approach derived in Appendix A. The two methods give the same result for linearized time-invariant systems, as proven in Section A.2. A formulation based on the power balance law has the advantage that the concept, and all of the quantities that arise in its formulation, should be familiar to anyone with a modest background in dynamics.

However, such a formulation is more limited in applicability than Lagrange's equations, which are generally valid.

Readers already familiar with Lagrange's equations should focus on derivation of the expressions for kinetic and potential energy, power dissipation, and power input. Such considerations are required for both approaches, and are treated in greater detail than found in standard dynamics texts. For those individuals, the principle of power balance, and its implications for equations of motion, merely provide an interesting alternative perspective. Section A.2 is especially useful in this regard.

1.5.1 Power Balance Law

To invoke the power balance formulation, we require that a few conditions be met:

- The system must have a static equilibrium position.
- The system must consist of a specified collection of bodies. Specifically, variable mass systems, such as those associated with control volumes used to study flow effects, are excluded.
- The generalized coordinates should be defined such that they are zero when the system occupies its static equilibrium position:

$$\text{Static equilibrium} \quad \Leftrightarrow \quad q_1 = q_2 = \cdots = q_N \equiv 0 \qquad (1.5.1)$$

The first two conditions ensure that the system is time invariant. The third condition is not absolutely necessary, but its imposition leads to certain standard forms that are conducive to vibration studies.

The general work-energy principle for any system states that the increase in mechanical energy, which is defined as the sum of the kinetic energy T and potential energy V, equals the amount of work done on the system by all nonconservative forces, W_{nc}. We use the symbol Δ to denote an increment resulting from moving a system from one state to another, so

$$\Delta T + \Delta V = \Delta W_{nc} \qquad (1.5.2)$$

We now consider the system states to be separated by a small time interval Δt. We divide the relation by Δt and take the limit as $\Delta t \longrightarrow 0$. In this manner, the incremental quantities become rates of change. Furthermore, the definition of the power $\mathcal{P}$ associated with a force is that it is the rate at which the force does work. We thereby obtain the power form of the work-energy principle,

$$\dot{T} + \dot{V} = \mathcal{P} \qquad (1.5.3)$$

To obtain the final form of the power law, we recognize that the nonconservative forces contributing to $\mathcal{P}$ are either external excitations or dissipation effects. The power input by forces external to a system is $\mathcal{P}_{in}$. Dissipation mechanisms may conceptually be visualized as dashpot devices. Changing the length of a dashpot requires that work be done on it. This work is dissipated within the dashpot as nonmechanical forms of energy. Instantaneously, this is manifested as power input to the dashpot and therefore depleted from the system. Hence, we denote the effect as $-\mathcal{P}_{dis}$. These considerations lead us to the *power balance law* for a vibratory system,

$$\boxed{\dot{T} + \dot{V} = \mathcal{P}_{in} - \mathcal{P}_{dis}} \qquad (1.5.4)$$

This relation is quite general, but it represents only one equation. There are N generalized coordinates to determine, which leads to the question, How can we obtain

N equations of motion from the single scalar power law? The answer lies in accounting for the specific forms of kinetic energy, potential energy, power dissipation, and power input.

1.5.2 Linearized Energy and Power

We begin by considering the kinetic energy of the system in Figure 1.10. If the generalized coordinates are the displacements x_1 and x_2 of each collar from its static equilibrium position, then the kinetic energy of that system is

$$T = \tfrac{1}{2}(m_1\dot{x}_1^2 + m_2\dot{x}_2^2) \tag{1.5.5}$$

However, other choices for the generalized coordinates locating the collars are possible. For example, we could locate collar 2 relative to collar 1 by letting y_2 be the increase in the distance separating the two collars, with $y_2 = 0$ at the static equilibrium position. Then the displacement of collar 2 is $x_1 + y_2$. The corresponding velocities are $v_1 = \dot{x}_1$ and $v_2 = \dot{x}_1 + \dot{y}_2$, which leads to the kinetic energy being

$$T = \tfrac{1}{2}[m_1\dot{x}_1^2 + m_2(\dot{x}_1 + \dot{y}_2)^2] = \tfrac{1}{2}[(m_1 + m_2)\dot{x}_1^2 + 2m_2\dot{x}_1\dot{y}_2 + m_2\dot{y}_2^2] \tag{1.5.6}$$

The first form of T for this system consists of a sum of squares, while the second form has all possible products of the two generalized velocities $\dot{x}_1$ and $\dot{x}_2$. The coefficients in both forms are constants. We say that both forms of T constitute *quadratic sums*. (*Quadratic* here refers to second-degree products.) The variables in these quadratic sums are time derivatives of generalized coordinates. We refer to such quantities as *generalized velocities* and denote them as $\dot{q}_j$.

We have restricted our attention to cases where the generalized coordinates locate a system relative to its static equilibrium condition. Furthermore, we seek linearized equations of motion corresponding to the small displacement approximation. Under such conditions, a system's kinetic energy expression will always reduce to a quadratic sum whose variables are the generalized velocities. If N represents the number of degrees of freedom, and therefore the required number of generalized coordinates, T must contain products of $\dot{q}_1$ with every $\dot{q}_n$ ranging from $n = 1$ to $n = N$, plus products of $\dot{q}_2$ with every $\dot{q}_n$ ranging from $n = 1$ to $n = N$, and so on. We shall let M_{jn} denote the constant factor of a product of $\dot{q}_j$ and $\dot{q}_n$, so a general quadratic sum for T will have the form

$$\begin{aligned} T = \tfrac{1}{2}[&\dot{q}_1(M_{11}\dot{q}_1 + M_{12}\dot{q}_2 + \cdots + M_{1N}\dot{q}_N) \\ &+ \dot{q}_2(M_{21}\dot{q}_1 + M_{22}\dot{q}_2 + \cdots + M_{2N}\dot{q}_N) + \cdots \\ &+ q_N(M_{N1}\dot{q}_1 + M_{N2}\dot{q}_2 + \cdots + M_{NN}\dot{q}_N)] \end{aligned} \tag{1.5.7}$$

First, we observe that squares of each generalized velocity occur once, while products of different generalized velocities occur twice. It is irrelevant which of the two generalized velocities is the first factor in a mixed product. Hence, we merge the coefficients of products of different generalized velocities. This is done by multiplying one of the coefficients by 2 and considering the two coefficients to be equal; that is, $M_{nj} = M_{jn}$. Thus, we rewrite the preceding as

$$\boxed{\begin{aligned} T = \tfrac{1}{2}[&(M_{11}\dot{q}_1^2 + M_{22}\dot{q}_2^2 + \cdots) + (2M_{12}\dot{q}_1\dot{q}_2 + 2M_{13}\dot{q}_1\dot{q}_3 + \cdots) \\ &+ (2M_{23}\dot{q}_2\dot{q}_3 + 2M_{24}\dot{q}_2\dot{q}_4 + \cdots) + \cdots] \end{aligned}} \tag{1.5.8}$$

This form will be useful when we derive the equations for a specific system. A more compact notation uses the summation symbol. We ensure inclusion of every possible product of two generalized velocities by using two summations with different indices. Thus, an alternative description of eq. (1.5.8) is

$$T = \frac{1}{2} \sum_{j=1}^{N} \sum_{n=1}^{N} M_{jn} \dot{q}_j \dot{q}_n, \qquad M_{nj} = M_{jn} \qquad (1.5.9)$$

The factors M_{jn} are (constant) *inertia coefficients*. They form an $N \times N$ symmetric matrix. Matrix notation will be especially useful when we solve the equations of motion, which leads us to another alternative way of writing eq. (1.5.8). Let $\{q\}$ be a column vector of the generalized coordinates,

$$\{q\} = [q_1 \ q_2 \ \cdots \ q_N]^{\mathrm{T}} \qquad (1.5.10)$$

The sum over n in eq. (1.5.9) involves the column number for the elements of $[M]$, so it forms the elements of $[M]\{\dot{q}\}$. The sum over j involves the row number of the $[M]$ elements, so that sum is equivalent to forming $\{\dot{q}\}^{\mathrm{T}}[M]$; in other words, the matrix representation of the general quadratic sum for kinetic energy is

$$T = \tfrac{1}{2}\{\dot{q}\}^{\mathrm{T}}[M]\{\dot{q}\}, \qquad [M] = [M]^{\mathrm{T}} \qquad (1.5.11)$$

We will see in the next section that the power dissipation is formed by squaring the extension rate of dashpots. Consequently, $\mathcal{P}_{\mathrm{dis}}$ also will be a quadratic sum containing all possible products of two generalized velocities. It follows that the general form of this quantity will be like T, except that there is no common $\frac{1}{2}$ factor. We let C_{jn} denote the constant factors of the products in $\mathcal{P}_{\mathrm{dis}}$, so the alternative ways of representing this quantity are the expanded form,

$$\mathcal{P}_{\mathrm{dis}} = [(C_{11}\dot{q}_1^2 + C_{22}\dot{q}_2^2 + \cdots) + (2C_{12}\dot{q}_1\dot{q}_2 + 2C_{13}\dot{q}_1\dot{q}_3 + \cdots)$$
$$+ (2C_{23}\dot{q}_2\dot{q}_3 + 2C_{24}\dot{q}_2\dot{q}_4 + \cdots) + \cdots] \qquad (1.5.12)$$

the summation form,

$$\mathcal{P}_{\mathrm{dis}} = \sum_{j=1}^{N} \sum_{n=1}^{N} C_{jn} \dot{q}_j \dot{q}_n, \qquad C_{nj} = C_{jn} \qquad (1.5.13)$$

and the matrix form,

$$\mathcal{P}_{\mathrm{dis}} = \{\dot{q}\}^{\mathrm{T}}[C]\{\dot{q}\}, \qquad [C]^{\mathrm{T}} = [C] \qquad (1.5.14)$$

Because dashpots are the prototypical device for dissipating energy, we refer to C_{jn} as *damping coefficients*.

The potential energy V has a somewhat similar representation. A potential energy function may be derived by considering the work done by conservative forces that depend only on position. Consequently, the potential energy depends only on the instantaneous position of the system, which means that the potential energy depends only on the values of the generalized coordinates. We indicate such dependence by writing $V = V(q_j)$, where q_j denotes the entire set of variables.

Any analytical function of generalized coordinates may be expanded in a Taylor series relative to reference values. Let us use $q_j = 0$ as the reference for such an expansion. Note that the kinetic energy in eq. (1.5.9) contains only quadratic sums. Correspondingly, we stop the Taylor series expansion of V at quadratic terms. If the system has only one degree of freedom, so that $V = V(q_1)$, the Taylor series expansion is

$$V(q_1) = V_0 + \left(\frac{dV}{dq_1}\right)_0 q_1 + \frac{1}{2}\left(\frac{d^2 V}{dq_1^2}\right)_0 (q_1)^2 \qquad (1.5.15)$$

where a zero subscript indicates that the associated quantity should be evaluated at $q_1 = 0$. Taylor series for functions that depend on several independent variables have a similar form. One difference is that the total derivative becomes a sequence of partial derivatives. Another modification is that every possible combination of first and second derivatives must be included. An expanded version of the series is

$$V(q_1, q_2, \ldots) = V_0 + \left[\left(\frac{\partial V}{\partial q_1}\right)_0 q_1 + \left(\frac{\partial V}{\partial q_2}\right)_0 q_2 + \cdots\right]$$
$$+ \frac{1}{2}\left[\left(\frac{\partial^2 V}{\partial q_1^2}\right)_0 (q_1)^2 + 2\left(\frac{\partial^2 V}{\partial q_1 \partial q_2}\right)_0 q_1 q_2 + \left(\frac{\partial^2 V}{\partial q_2^2}\right)_0 (q_2)^2 + \cdots\right]$$

$$(1.5.16)$$

This series may be written more compactly in summation form. Let us denote the first-derivative coefficients as $(F_0)_j$, and let K_{jn} denote the second-derivative coefficients. This leads to

$$V = V_0 + \sum_{j=1}^{N} (F_0)_j q_j + \frac{1}{2}\sum_{j=1}^{N}\sum_{n=1}^{N} K_{jn} q_j q_n, \qquad K_{nj} = K_{jn} \qquad (1.5.17)$$

where

$$K_{jn} = \left(\frac{\partial^2 V}{\partial q_j \partial q_n}\right)_0, \qquad (F_0)_j = \left(\frac{\partial V}{\partial q_n}\right)_0 \qquad (1.5.18)$$

The quantities K_{jn} are *stiffness coefficients*. We will see that the $(F_0)_j$ represent static forces. An explicit expansion of the summation representation of V is

$$V = V_0 + [(F_0)_1 q_1 + (F_0)_2 q_2 + \cdots] + \frac{1}{2}[(K_{11}q_1^2 + K_{22}q_2^2 + \cdots)$$
$$+ (2K_{12}q_1q_2 + 2K_{13}q_1q_3 + \cdots)$$
$$+ (2K_{23}q_2q_3 + 2K_{24}q_2q_4 + \cdots) + \cdots] \qquad (1.5.19)$$

This expression may also be written in matrix form as

$$V = V_0 + \{q\}^T\{F_0\} + \frac{1}{2}\{q\}^T[K]\{q\} \qquad (1.5.20)$$

The last step in our investigation of standard mathematical representations is examination of the power being input to the system by forces external to the system. Perhaps the most important aspect of this step comes from observations in Section 1.3. We saw there that constraint forces (i.e., reactions) cannot affect unconstrained generalized coordinates. Correspondingly,

> Constraint forces do not contribute to the power input.

(This property does not apply to time-dependent systems, which is why the power balance formulation is not useful for such systems.)

The forces of concern for the equations of motion are excitations. If one such force, $\bar{F}_k$, is applied to a point whose velocity is $\bar{v}_k$, the associated power is $\mathcal{P} = \bar{F}_k \cdot \bar{v}_k$. The total power input is the sum of contributions from all forces, so the external power input is

$$\mathcal{P}_{\text{in}} = \Sigma \bar{F}_k \cdot \bar{v}_k \qquad (1.5.21)$$

The velocity of any point might depend on more than one generalized velocity, but linearization based on small displacement approximations requires that we approximate that dependence as linear. Thus, $\mathcal{P}_{\text{in}}$ will contain force components multiplying generalized velocities. In other words, the explicit form of the power input is

$$\mathcal{P}_{\text{in}} = Q_1 \dot{q}_1 + Q_2 \dot{q}_2 + \cdots \qquad (1.5.22)$$

The coefficients Q_j are time dependent, because they are formed from the external forces. We say that the Q_j constitute a set of *generalized forces*. The corresponding summation form of the power input is

$$\mathcal{P}_{\text{in}} = \sum_{j=1}^{N} Q_j \dot{q}_j \qquad (1.5.23)$$

To write this in matrix form, we define a vector of generalized forces,

$$\{Q\} = [Q_1 \; Q_2 \; \cdots]^{\text{T}} \qquad (1.5.24)$$

The resulting expression for the power input by the excitations is

$$\mathcal{P}_{\text{in}} = \{\dot{q}\}^{\text{T}}\{Q\} \qquad (1.5.25)$$

We will employ the explicit forms of T, V, $\mathcal{P}_{\text{dis}}$, and $\mathcal{P}_{\text{in}}$ when we construct the actual equations of motion. The matrix form will be useful when we solve the equations of motion, as well as for derivations of the standard form of the equations of motion, which is the next topic.

1.5.3 Linearized Equations of Motion

Our task now is to substitute the standard representations of the energy and power terms into the power balance law. This requires that we evaluate the time derivative of T and V, for which we use the matrix forms. The process of differentiating T in eq. (1.5.11) follows the manner in which products of scalar quantities are differentiated.

Specifically, because $[M]$ is an array of constants, we have

$$\dot{T} = \frac{1}{2}\frac{d}{dt}(\{\dot{q}\}^{\mathrm{T}}[M]\{\dot{q}\})$$

$$= \frac{1}{2}\left(\frac{d}{dt}\{\dot{q}\}^{\mathrm{T}}\right)[M]\{\dot{q}\} + \frac{1}{2}\{\dot{q}\}^{\mathrm{T}}[M]\left(\frac{d}{dt}\{\dot{q}\}\right) \qquad (1.5.26)$$

$$= \frac{1}{2}\{\ddot{q}\}^{\mathrm{T}}[M]\{\dot{q}\} + \frac{1}{2}\{\dot{q}\}^{\mathrm{T}}[M]\{\ddot{q}\}$$

The last two terms actually are the same. To prove this, we use an identity for the transpose of a "triple" product, which states that $[[a]\,[b]\,[c]]^{\mathrm{T}} \equiv [c]^{\mathrm{T}}[b]^{\mathrm{T}}[a]^{\mathrm{T}}$. Applying this identity to the first term gives

$$\{\ddot{q}\}^{\mathrm{T}}[M]\{\dot{q}\} \equiv \{\dot{q}\}^{\mathrm{T}}[M]^{\mathrm{T}}\{\ddot{q}\} \qquad (1.5.27)$$

Because $[M]$ is a symmetric matrix, $[M]^{\mathrm{T}} \equiv [M]$, both terms in $\dot{T}$ have identical values, which reduces the expression to

$$\dot{T} = \{\dot{q}\}^{\mathrm{T}}[M]\{\ddot{q}\} \qquad (1.5.28)$$

Because the last term of V appearing in eq. (1.5.20) has a form similar to T, and the stiffness matrix $[K]$ is symmetric, differentiating the last term in eq. (1.5.20) yields a term that resembles the above expression for $\dot{T}$. Furthermore, both V_0 and $\{F_0\}$ are constants, so we find that

$$\dot{V} = \{\dot{q}\}^{\mathrm{T}}\{F_0\} + \{\dot{q}\}^{\mathrm{T}}[K]\{q\} \qquad (1.5.29)$$

We are now ready to assemble the power balance law. When we substitute $\dot{T}$ from eq. (1.5.28), $\dot{V}$ from eq. (1.5.29), $\mathcal{P}_{\mathrm{dis}}$ from eq. (1.5.14), and $\mathcal{P}_{\mathrm{in}}$ from eq. (1.5.25) into eq. (1.5.4), we obtain

$$\{\dot{q}\}^{\mathrm{T}}[M]\{\ddot{q}\} + \{\dot{q}\}^{\mathrm{T}}\{F_0\} + \{\dot{q}\}^{\mathrm{T}}[K]\{q\} = \{\dot{q}\}^{\mathrm{T}}\{Q\} - \{\dot{q}\}^{\mathrm{T}}[C]\{\dot{q}\} \qquad (1.5.30)$$

The casual observer might look at this expression, note that $\{\dot{q}\}^{\mathrm{T}}$ is a common premultiplying factor for each term, and proceed to cancel it out. Indeed, that is what we shall do, but only after we justify the operation. The questionable aspect of canceling the common factor stems from the fact that the power balance law is a scalar, so the preceding relation is a single equation. If we cancel the $\{\dot{q}\}^{\mathrm{T}}$ factor, each of the remaining terms in that equation is a vector. Collecting like elements from each vector leads to a set of N scalar equations—it seems as though we are trying to extract more information than the original power balance law contained. A mathematical reason to be concerned about canceling the common factor stems from the observation that it is not difficult to construct a variety of nonzero vectors $\{a\}$ for which $\{a\}^{\mathrm{T}}\{x\} = \{a\}^{\mathrm{T}}\{y\}$, even though $\{x\}$ and $\{y\}$ are arbitrary unequal vectors.

The operation of canceling the $\{\dot{q}\}^{\mathrm{T}}$ factor may be justified on logical grounds. We could bring all terms in eq. (1.5.30) to the left side of the equality sign, in which case the form of the equation is $\{\dot{q}\}^{\mathrm{T}}\{Z\} = 0$. We would then argue that the differential equations of motion should not be altered in form if different excitations are applied to the system. However, because $\{Q\}$ constitutes the excitation, each value of $\{Q\}$ will generate a different response, so almost any $\{\dot{q}\}^{\mathrm{T}}$ can be obtained at any instant. In the face of such multiple possibilities, satisfying $\{\dot{q}\}^{\mathrm{T}}\{Z\} = 0$ cannot stem from selecting the elements of $\{\dot{q}\}^{\mathrm{T}}$ in a special manner. We must therefore conclude that $\{Z\} = \{0\}$.

For anyone who is uncomfortable with this argument, rigorous justification of the cancellation operation comes from the fact that the resulting equations of motion are identical to what is obtained when the Lagrange equation method is applied to a time-invariant system. This is proven in Section A.2 of Appendix A. Upon cancellation of the common factor $\{\dot{q}\}^T$, the terms remaining in eq. (1.5.30) are

$$[M]\{\ddot{q}\} + [C]\{\dot{q}\} + [K]\{q\} = \{Q\} - \{F_0\} \qquad (1.5.31)$$

This matrix equation is an abbreviated notation for N scalar equations obtained by equating like elements. Thus, we have obtained N differential equations of motion.

The applicability of eq. (1.5.31) is limited by the restrictions stated earlier. We cannot use the formulation to describe small movements that occur relative to a steady motion. In addition, it is required that $\{q\}$ be measured from the static equilibrium position. The static equilibrium condition is characterized by $\{q\} = \{\dot{q}\} = \{\ddot{q}\} = \{0\}$. Let us examine the implications of this definition on the equations of motion. Strictly speaking, the power balance law does not apply to static systems, because all rates of change and power quantities are identically zero. However, we may consider a static situation to be the limit of a dynamic process that evolves so slowly that the effects of $\{\dot{q}\}$ and $\{\ddot{q}\}$ are negligible. Because we have defined $\{q\} = \{0\}$ as the static equilibrium position, the state of static equilibrium corresponds to

$$\boxed{\{F_0\} = \{Q_{\text{static}}\}} \qquad (1.5.32)$$

where $\{Q_{\text{static}}\}$ represents the values of the generalized forces required to hold the system at rest in its static equilibrium position.

The experienced reader will recognize eq. (1.5.32) as a special form of the static equilibrium principle of virtual work and potential energy (Shames and Dym, 1985). The elements of $\{F_0\}$ are defined in eq. (1.5.17) as the first-order coefficients of a Taylor series expansion of V. By imposing eq. (1.5.32) on the coefficients $(F_0)_j$ that arise in such an expansion, we can ascertain parameters necessary for static equilibrium. In many situations the static system is conservative, by which we mean that the forces that hold the system in equilibrium are all conservative. Then if V contains the potential energy of all of these forces, it must be that $\{F_0\} = \{0\}$. Furthermore, if the static system is not conservative, we may use eq. (1.5.32) to replace $\{F_0\}$ by the set of nonconservative forces, $\{Q_{\text{static}}\}$, acting on the system when it is in static equilibrium. When we make this substitution, the equations of motion appear as

$$\boxed{[M]\{\ddot{q}\} + [C]\{\dot{q}\} + [K]\{q\} = \{\hat{Q}\}, \quad \text{where } \{\hat{Q}\} = \{Q\} - \{Q_{\text{static}}\}} \qquad (1.5.33)$$

In this perspective we interpret $\{\hat{Q}\}$ as the increment in the generalized forces causing the system to move away from its static equilibrium position. Another name for such quantities is *disturbance forces*. Note that we will use the caret notation for disturbance generalized forces only if doing so is necessary to prevent confusion.

We know that a linearized analysis relative to the static equilibrium position of a system must lead to the standard differential equations described by eq. (1.5.33). It follows that determining the equations of motion for such systems reduces to the task of identifying $[M]$, $[K]$, $[C]$, and $\{Q\}$ corresponding to our choice for the generalized coordinates. The basic concept for doing this is common to each quantity and relies on matching the general representations of T, V, $\mathscr{P}_{\text{dis}}$, and $\mathscr{P}_{\text{in}}$ to the specific forms describing the energy and power terms for the system of interest. Hence, our attention now turns to the task of describing these effects in terms of the generalized coordinates we have selected.

1.5.4 Implementation

Kinematics Constructing the kinetic energy and power terms requires that we describe the velocity of various points in terms of the generalized coordinates, while constructing the potential energy requires a description of displacements. We obtain such expressions by using the standard kinematical tools. One particularly useful principle relates the velocity of two points in a rigid body. For motion in the XY plane, the angular velocity is $\bar{\omega} = \dot{\theta}\bar{K}$. The general kinematical relation between the velocity of any two points in a rigid body executing a planar motion is

$$\bar{v}_B = \bar{v}_A + \dot{\theta}\bar{K} \times \bar{r}_{B/A} \tag{1.5.34}$$

where $\bar{r}_{B/A}$ is the position vector extending from point A to point B. A typical situation is shown in Figure 1.12, where A_0 and B_0 mark the positions of the respective points when the generalized coordinates are all zero.

Suppose the body in Figure 1.12 is connected only to springs and dashpots, so that its movement is not constrained by other bodies. Then a suitable set of generalized coordinates would be the displacement of point A in the X and Y directions, whose components are denoted as u_A and w_A, and the rotation θ of the body from its equilibrium position. We define ψ to be the angle of the position vector $\bar{r}_{B/A}$, measured relative to the X axis, when the body is at the equilibrium position. When the body has rotated by θ from this position, the position vector will be $\bar{r}_{B/A} = \ell[\cos(\psi + \theta)\bar{I} + \sin(\psi + \theta)\bar{J}]$, and the velocity of point A is $\bar{v}_A = \dot{u}_A\bar{I} + \dot{w}_A\bar{J}$. Thus, eq. (1.5.34) states that the velocity of point B is $\bar{v}_B = \dot{u}_A\bar{I} + \dot{w}_A\bar{J} + \dot{\theta}\ell[\cos(\psi + \theta)\bar{J} - \sin(\psi + \theta)\bar{I}]$.

Observe that the generalized velocities $\dot{u}_A$, $\dot{w}_A$, and $\dot{\theta}$ occur linearly in $\bar{v}_B$. If $\bar{v}_B$ is to be a linear description, the coefficients of the generalized velocities must be zero, but the coefficient of $\dot{\theta}$ contains a sine and cosine of $\psi + \theta$, which is not constant. There are no contradictions here, for we have not accounted for the small displacement approximation associated with linearization. Because θ is limited to small values in that approximation, we set $\theta = 0$ in the expression for $\bar{v}_B$. The resulting relation is equivalent to $\bar{v}_B = \dot{u}_A\bar{I} + \dot{w}_A\bar{J} + \dot{\theta}\bar{K} \times (\bar{r}_{B/A})$, where the position vector describes the relation between the two points when the system is at its equilibrium position.

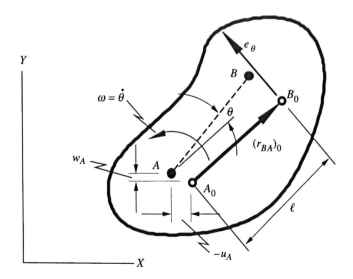

FIGURE 1.12 Linearized description of the relationship between the velocities of points in a rigid body.

This simple example is at the heart of almost every task involving the description of motion. Specifically,

Linearized expressions for the motion of points and the rotation of bodies in time-invariant systems will result from using the static equilibrium position to describe the location of points and the orientation of lines.

An example of the foregoing pertains to the analysis of linkages. In such systems, we often have one part that slides along another. A typical case is depicted in Figure 1.13, where collar C slides over bar AB. Because the collar is attached to bar CD, by characterizing the velocity of the collar we will be able to relate the motion of the two bars. If we know $\bar{v}_A$ and $\dot{\theta}$ for arm AB, the relative velocity equation gives

$$\bar{v}_C = \bar{v}_A + (\bar{v}_C)_{AB} + \dot{\theta}\bar{K} \times \bar{r}_{C/A} \tag{1.5.35}$$

where $(\bar{v}_C)_{AB}$ is the relative velocity of collar C, by which we mean the velocity that would be seen by an observer on rotating bar AB. To linearize this relation, we use the coordinates of the points at the equilibrium position to describe $\bar{r}_{C/A}$. We also note that $(\bar{v}_C)_{AB}$ is tangent to bar AB. Because the rotation θ of this bar is considered to be very small, this tangent direction will be essentially the same as the orientation of the bar at the equilibrium position, which we denote by a unit vector $(\bar{e}_{B/A})_0 \equiv \bar{r}_{B/A}/|\bar{r}_{B/A}|$. Correspondingly, we write

$$\boxed{\bar{v}_B = \bar{v}_A + v_{\text{rel}}(\bar{e}_{B/A})_0 + \dot{\theta}\bar{K} \times (\bar{r}_{C/A})_0} \tag{1.5.36}$$

where v_{rel} is the rate at which the collar slides over bar AB.

Cables and rigid massless links provide a way of interconnecting bodies. Figure 1.14 depicts a common situation where a block is suspended from a bar by a cable that passes over a small pulley. Suppose that we have chosen θ to be the generalized coordinate for the position of the bar, and that the position depicted in the figure corresponds to static equilibrium, at which location $\theta = 0$. Because the bar executes a pure rotation about its pivot, the velocity of points along the bar will be perpendicular to the bar. At the reference position, it follows that the velocity of point A, where the cable is fastened, is $\bar{v}_A = \ell\dot{\theta}$ upward. From the perspective of the cable, this velocity may be viewed as having two components. The component of $\bar{v}_A$ perpendicular to the cable swings the cable around. In contrast, the component of $\bar{v}_A$ parallel to the cable, which is $\bar{v}_A \cdot \bar{e}_{\text{cable}}$, decreases the distance between point A and the tangent point on the pulley. The speed of the block matches the rate at which this distance is decreasing, because a cable is idealized as being inextensible. Hence, we find that $v_{\text{block}} =$

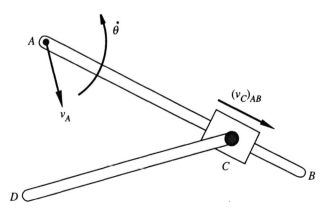

FIGURE 1.13 The relative velocity relation.

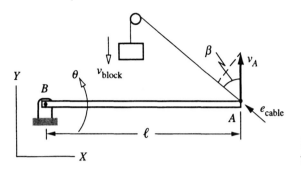

FIGURE 1.14 Kinematics of a
typical cable.

$\bar{v}_A \cdot \bar{e}_{\text{cable}}$. For the sense in which the vectors are defined, a positive dot product will correspond to downward motion of the block. Finally, to linearize the analysis we use the equilibrium position to describe the orientation of the cable. From $\bar{v}_A = \ell\dot{\theta}\,\bar{J}$, we find that $v_{\text{block}} = \bar{v}_A \cdot (\bar{e}_{\text{cable}})_0 = \ell\dot{\theta}\,\cos(\beta)$.

The insights we gain from the analysis of the block may be extended to treat displacements, which we denote as $\Delta\bar{r}$ in order to emphasize that they are *changes* in position. In Figure 1.15 the block has been replaced with a spring. Point A follows a circular path, but for small rotation angles we may consider the arc of the circle to be vertical, so we have $\Delta\bar{r}_A = \ell\theta$ upward. Also, because θ is very small, the orientation of the cable will be changed little by this displacement. In that case, the length of the cable from point A to the pulley will be decreased by an amount equaling the component of $\Delta\bar{r}_A$ measured along the cable. This results in shortening the spring by $\Delta = -\Delta\bar{r}_A \cdot (\bar{e}_{\text{cable}})_0 = -\ell\theta\,\cos\beta$.

The specific forms of the preceding displacements are less important than their relationship to the corresponding velocity relation. For example, consider $\bar{v}_A$ and $\Delta\bar{r}_A$ in Figures 1.14 and 1.15. Displacement is the time integral of velocity, so $\Delta\bar{r}_A = \int \bar{v}_A\,dt$. The only quantity that depends on time in $\bar{v}_A$ is θ, so we have $\Delta\bar{r}_A = \int(\ell\dot{\theta}\bar{J})\,dt = \ell\theta\bar{J}$. A similar analysis applies to the speed of the block compared with the elongation of the spring. That is, $\Delta = \int(-v_{\text{block}})\,dt$. Because $\dot{\theta}$ is the only variable in $\bar{v}_{\text{block}}$, we find that $\Delta = \int[-\ell\dot{\theta}\,\cos\beta]\,dt = -\ell\theta\,\cos\beta$. The ability to obtain an expression for displacement by a simple manipulation of a velocity expression is a prime simplification afforded by linearization. In general,

A linearized description of the displacement of a point may be obtained directly from a linearized description of that point's velocity. To do so, we merely replace all generalized velocities $\dot{q}_j$ with the corresponding generalized coordinate q_j.

We express this mathematically as

$$\boxed{\bar{v}_p = \bar{d}_1\dot{q}_1 + \bar{d}_2\dot{q}_2 + \cdots \qquad \Rightarrow \qquad \Delta\bar{r}_p = \bar{d}_1 q_1 + \bar{d}_2 q_2 + \cdots} \qquad (1.5.37)$$

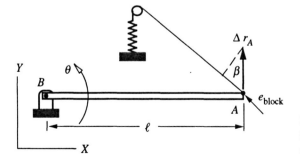

FIGURE 1.15 Linearized
kinematical analysis of deformation
of a spring.

The vectors $\overline{d}_j$ represent geometrical properties evaluated at the equilibrium position. For the sake of simplifying the notation, the use of a subscript 0 to indicate such an evaluation will be suppressed.

Kinetic Energy The kinetic energy of a rigid body in planar motion is

$$T = \tfrac{1}{2}m\,\overline{v}_G \cdot \overline{v}_G + \tfrac{1}{2}I_G\omega^2 \tag{1.5.38}$$

where point G is the center of mass, and I_G is the moment of inertia about the axis perpendicular to the plane of motion and intersecting point G. In the special case where the body is in pure rotation, with point O being the stationary point, an alternative form of the kinetic energy is

$$T = \tfrac{1}{2}I_O\omega^2 \tag{1.5.39}$$

The kinetic energy of the system is formed by adding the (scalar) contributions of each body contained in the system.

To illustrate how we derive inertia coefficients from general expressions such as these, consider a two-degree-of-freedom system that consists of a body in general motion. Two generalized coordinates, q_1 and q_2, locate the position of this body relative to its equilibrium position. We perform a kinematical analysis in order to express $\overline{v}_G$ and ω in terms of $\dot{q}_1$ and $\dot{q}_2$. Equation (1.5.37) indicates that these expressions will be of the form

$$\overline{v}_G = \overline{d}_1\dot{q}_1 + \overline{d}_2\dot{q}_2, \quad \omega = s_1\dot{q}_1 + s_2\dot{q}_2 \tag{1.5.40}$$

The actual values of each $\overline{d}_n$ and s_n will depend on the nature of the system. Neither $\overline{v}_G$ nor ω is a generalized velocity, so they may not appear explicitly in the expression for T. We therefore substitute the expressions for $\overline{v}_G$ and ω into eq. (1.5.38) and collect the coefficients of each product $\dot{q}_j\dot{q}_n$, as follows:

$$\begin{aligned}
T &= \tfrac{1}{2}m(\overline{d}_1\dot{q}_1 + \overline{d}_2\dot{q}_2) \cdot (\overline{d}_1\dot{q}_1 + \overline{d}_2\dot{q}_2) + \tfrac{1}{2}I_G(s_1\dot{q}_1 + s_2\dot{q}_2)^2 \\
&= \tfrac{1}{2}[(m\overline{d}_1 \cdot \overline{d}_1 + I_G s_1^2)\dot{q}_1^2 + (m\overline{d}_2 \cdot \overline{d}_2 + I_G s_2^2)\dot{q}_2^2 + 2(m\overline{d}_1 \cdot \overline{d}_2 + I_G s_1 s_2)\dot{q}_1\dot{q}_2]
\end{aligned} \tag{1.5.41}$$

The explicit standard form of the kinetic energy, eq. (1.5.8), for any two-degree-of-freedom system is

$$T = \tfrac{1}{2}(M_{11}\dot{q}_1^2 + 2M_{12}\dot{q}_1\dot{q}_2 + M_{22}\dot{q}_2^2) \tag{1.5.42}$$

This general description must give the same value of T as the specific expression for the system, eq. (1.5.41), regardless of the actual values of the generalized velocities. Such equivalence is possible only if the coefficients of corresponding terms in eqs. (1.5.41) and (1.5.42) match. This leads to

$$M_{11} = m\overline{d}_1 \cdot \overline{d}_1 + I_G s_1^2, \quad M_{22} = m\overline{d}_2 \cdot \overline{d}_2 + I_G s_2^2$$
$$M_{12} = M_{21} = m\overline{d}_1 \cdot \overline{d}_2 + I_G s_1 s_2 \tag{1.5.43}$$

The matching process itself is straightforward. Thus, most of the effort entailed in deriving the coefficients for a system lies in the kinematical operations relating the velocity of each center of mass and the angular velocity of each body to the generalized velocities.

EXAMPLE 1.4

The sketch shows a model of an automobile chassis and its suspension. The center of mass G and the spring attachment points A and B are essentially at the same elevation, so only vertical motion displacement of these points needs to be considered. The mass of the chassis is $m = 1400$ kg and the radius of gyration about the center of mass G is 1.1 m.

(a) Derive the inertia matrix corresponding to generalized coordinates y_G and θ that describe the motion relative to the center of mass.

(b) Derive the inertia matrix corresponding to the generalized coordinates y_A and y_B, which are the upward displacements of the points where the springs are attached.

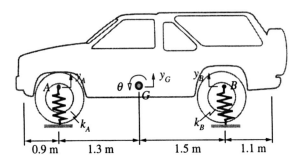

Solution This exercise will demonstrate that the inertia matrix values are intrinsically tied to the definition of the generalized coordinates. The vehicle is executing a general motion, so we form the kinetic energy relative to the center of mass. The moment of inertia is related to the radius of gyration r by

$$I_G = mr^2 = 1400(1.1)^2 = 1694 \text{ kg-m}^2$$

The kinetic energy is the sum of translational and rotational parts,

$$T = \tfrac{1}{2}mv_G^2 + \tfrac{1}{2}I_G\omega^2$$

The task now is to express the motion variables in terms of the generalized velocities.

In the first case, we have $q_1 = y_G$, $q_2 = \theta$. The velocity of the center of mass is $v_G = \dot{y}$ upward, and the rotation rate of the chassis is $\omega = \dot{\theta}$. Expressed in terms of the generalized coordinates, the kinetic energy is

$$T = \tfrac{1}{2}m\dot{y}_G^2 + \tfrac{1}{2}I_G\dot{\theta}^2$$

On the other hand, the standard quadratic sum for T in expanded form for this system is

$$T = \tfrac{1}{2}(M_{11}\dot{y}_G^2 + 2M_{12}\dot{y}_G\dot{\theta} + M_{22}\dot{\theta}^2)$$

Matching like coefficients leads to

$$M_{11} = m = 1400 \text{ kg}, \qquad M_{22} = I_G = 1694, \qquad M_{12} = 0$$

The second set of generalized coordinates use $q_1 = y_A$ and $q_2 = y_B$. In this case we need to express v_G and ω in terms of $\dot{y}_A$ and $\dot{y}_B$. We find these relations by using the fact that points A, B, and G belong to the same body, so that

$$\bar{v}_A = \bar{v}_G + \bar{\omega} \times \bar{r}_{A/G}, \qquad \bar{v}_B = \bar{v}_G + \bar{\omega} \times \bar{r}_{B/G}$$

We use the static equilibrium configuration to describe each term in the preceding expression. In this position $\bar{r}_{A/G}$ and $\bar{r}_{B/G}$ are both horizontal. Considering $\bar{\omega} \times \bar{r}$ to represent circular motion relative to point G, with counterclockwise $\dot{\theta}$ being positive, leads to $\bar{\omega} \times \bar{r}_{A/G}$ being $-1.3\dot{\theta}$ downward, while $\bar{\omega} \times \bar{r}_{B/G}$ is $1.5\dot{\theta}$ upward. We set $\bar{v}_G = \dot{y}_G$ upward, so all velocities are in the vertical direction only; specifically,

$$\dot{y}_A = \dot{y}_G - 1.3\dot{\theta}, \qquad \dot{y}_B = \dot{y}_G + 1.5\dot{\theta}$$

We solve these equations for $\dot{y}_G$ and $\dot{\theta}$, which leads to

$$\dot{y}_G = \frac{1}{2.8}(1.3\dot{y}_B + 1.5\dot{y}_A), \qquad \dot{\theta} = \frac{1}{2.8}(\dot{y}_B - \dot{y}_A)$$

The kinetic energy of the system is

$$T = \tfrac{1}{2}m\dot{y}_G^2 + \tfrac{1}{2}I_G\dot{\theta}^2 = \tfrac{1}{2}m(\dot{y}_G^2 + 1.21\,\dot{\theta}^2)$$

Elimination of $\dot{y}_G$ and $\dot{\theta}$ leads to

$$T = \frac{1}{2}m\left[\left(\frac{1.3\dot{y}_B + 1.5\dot{y}_A}{2.8}\right)^2 + 1.21\left(\frac{\dot{y}_B - \dot{y}_A}{2.8}\right)^2\right]$$

We expand the squares and collect like coefficients of the generalized velocities. The result of these operations is

$$T = \tfrac{1}{2}m(0.4413\dot{y}_A^2 + 0.3699\dot{y}_B^2 + 2(0.0944)\dot{y}_A\dot{y}_B)$$

In order to avoid confusing $[M]$ for the two formulations, we use a prime to denote the matrices associated with $q_1 = y_A$ and $q_2 = y_B$, so that

$$T = \tfrac{1}{2}(M'_{11}\dot{y}_A^2 + 2M'_{12}\dot{y}_A\dot{y}_B + M'_{22}\dot{y}_B^2)$$

Matching the two descriptions of T yields

$$M'_{11} = 0.4413m, \qquad M'_{22} = 0.39923m, \qquad M'_{12} = M'_{21} = 0.0944m$$

Clearly, the derivation of the inertia matrix in this formulation is more tedious than it was for the first choice of generalized coordinates. However, if we were to follow the later developments in order to determine $[K]$, we would find that the second set of generalized coordinates substantially simplifies the evaluation of the potential energy.

A comparison of the value of $[M]$ for the alternative sets of generalized coordinates reveals that in the first case the inertia matrix is diagonal. The matrix notation is a shorthand way of writing individual differential equations. Off-diagonal elements of either $[M]$, $[K]$, or $[C]$ cause more than one generalized coordinate to appear in these scalar equations, which means that the generalized coordinates are coupled. When $[M]$ is not diagonal, we say that the equations of motion are *inertially coupled*. Similarly, if $[K]$ is not diagonal, we say that the equations of motion are *inertially coupled*. The issue of coupling is addressed in more detail in Section 4.3.1.

Power Dissipation   The determination of the damping coefficients follows essentially the same procedure as that used for the inertia coefficients. Implementing that procedure requires a fundamental relation for the power dissipated in a single dashpot. Figure 1.16 shows a dashpot whose ends are fastened to different bodies. In this sketch $\bar{e}_{B/A}$ is a unit vector oriented toward point B from point A. Because we seek a linearized description, we use the initial relative orientation of the ends to construct this direction. Like the case of a cable, at each end it is the velocity component parallel to the dashpot that results in changing the distance between the ends. These components are

$$\dot{u}_A = \bar{v}_A \cdot \bar{e}_{B/A}, \qquad \dot{u}_B = \bar{v}_B \cdot \bar{e}_{B/A} \tag{1.5.44}$$

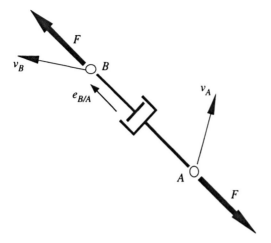

FIGURE 1.16 Power input to a dashpot.

(Note that by using an overdot on the u quantities, we are implying that they are displacements. This view will be very useful when we examine the potential energy stored in springs.) Because both velocity components are taken with respect to the same direction, which extends from end A to end B, positive $\dot{u}_B$ causes the dashpot to elongate and positive $\dot{u}_A$ causes it to shorten, as illustrated in Figure 1.16. The rate at which the dashpot is being elongated is the difference between these velocity components,

$$\dot{\Delta} = \dot{u}_B - \dot{u}_A \qquad (1.5.45)$$

The tensile force in the dashpot is $F = c\dot{\Delta}$. At end B, $\overline{F}$ is in the sense of $\dot{u}_B$, so the power input to the dashpot at that end is positive, being given by $F\dot{u}_B$. At the right end, $\overline{F}$ is opposite the sense of $\dot{u}_A$, so the associated power is $-F\dot{u}_A$. The net power input to the dashpot is therefore

$$\mathscr{P} = F\dot{u}_B - F\dot{u}_A = c\dot{\Delta}(\dot{u}_B - \dot{u}_A) = c\dot{\Delta}^2 \qquad (1.5.46)$$

The power input to the dashpot is converted to nonmechanical energy, so it represents dissipation of the system's mechanical energy. If a system has more than one dashpot, the power dissipation at any instant is the sum of the power dissipated in each dashpot, so we have

$$\mathscr{P}_{\mathrm{dis}} = \sum_j c_j \dot{\Delta}_j^2 \qquad (1.5.47)$$

This formula for $\mathscr{P}_{\mathrm{dis}}$ is comparable to the standard formulas relating kinetic energy to the center of mass velocity and angular velocity of a body. This similarity leads us to the procedure by which we may identify the damping matrix $[C]$. The first step is to describe the velocity of each end of each dashpot as a linear function of the generalized velocities. We obtain such descriptions by using the methods outlined in the beginning of this section. We then form the corresponding velocity components parallel to the respective dashpots by projecting each velocity onto the orientation of the dashpot at the equilibrium position, as described by eq. (1.5.44). Equation (1.5.45) then gives Δ as a linear combination of generalized velocities. For example, in the case of a two-degree-of-freedom system, such a description will have the form $\dot{\Delta}_1 = r_1 \dot{q}_1 + r_2 \dot{q}_2$ where r_1 and r_2 are constants that depend on the features of the system.

We derive expressions like the preceding one for each dashpot, which we then use to form eq. (1.5.47). We expand the squares and collect like coefficients of each product of generalized velocities. In the case of a two-degree-of-freedom system with two dashpots, this step would appear as follows:

$$\mathcal{P}_{dis} = c_1(r_1\dot{q}_1 + r_2\dot{q}_2)^2 + c_2(s_1\dot{q}_1 + s_2\dot{q}_2)^2$$
$$= (c_1 r_1^2 + c_2 s_1^2)\dot{q}_1^2 + (c_1 r_2^2 + c_2 s_2^2)\dot{q}_2^2 + 2(c_1 r_1 r_2 + c_2 s_1 s_2)\dot{q}_1\dot{q}_2 \tag{1.5.48}$$

This quadratic sum, which is specific to the system of interest, must give the same result as the general description of $\mathcal{P}_{dis}$ given explicitly by eq. (1.5.12). This equivalence must apply for any set of generalized velocities, which means that the coefficients must match. In the case of the example system, the result is

$$C_{11} = c_1 r_1^2 + c_2 s_1^2, \qquad C_{22} = c_1 r_2^2 + c_2 s_2^2$$
$$C_{12} = C_{21} = c_1 r_1 r_2 + c_2 s_1 s_2 \tag{1.5.49}$$

EXAMPLE 1.5

The columns supporting the second floor of a building are braced by a diagonal dashpot, as shown. Assume that the first and second floors only translate horizontally, so that u_1 and u_2 serve as generalized coordinates. Determine the corresponding damping matrix.

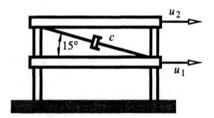

Solution This exercise is a straightforward demonstration of the matching process leading to the damping matrix. We define the unit vector parallel to the dashpot to extend from the upper left to the lower right corner. The displacement at the left end is u_2 horizontally, so the velocity component at this end parallel to the dashpot is $\dot{u}_2 \cos 15°$. By the same reasoning, the velocity component at the right end parallel to the dashpot is $\dot{u}_1 \cos 15°$. Positive u_1 elongates the dashpot while positive u_2 shortens it, so the elongation rate is $\dot{\Delta} = \dot{u}_1 \cos 15° - \dot{u}_2 \cos 15° = 0.9659(\dot{u}_1 - \dot{u}_2)$. Thus, the power dissipation is

$$\mathcal{P}_{dis} = c\dot{\Delta}^2 = 0.9330c(\dot{u}_1^2 - 2\dot{u}_1\dot{u}_2 + \dot{u}_2^2)$$

The expanded standard form of $\mathcal{P}_{dis}$ is

$$\mathcal{P}_{dis} = C_{11}\dot{u}^2 + C_{11}\dot{u}^2 + 2C_{12}\dot{u}_1\dot{u}_2$$

Matching the coefficient of each $\dot{u}$ product between the two descriptions leads to

$$C_{11} = C_{22} = 0.9330c, \qquad C_{12} = C_{21} = -0.9330c$$

Potential Energy Forces that are describable by a potential energy function are conservative. (The conservative property means that when work is done to change the state of the body or device associated with that force, the work is stored for later use.) The most common elements storing potential energy are springs and gravity. In most systems the mass of springs is much less than the mass of the bodies to which they are attached, so we idealize them by ignoring their inertial effects. (There are some situations, notably the valve train in internal combustion engines, in which the mass of the springs is substantial. In such situations, it is possible to introduce a kinetic energy correction that accounts for springs. This matter is described in the context of eq. (6.1.20).) To determine the stiffness coefficients for a system, we need to describe the system's potential energy in terms of the generalized coordinates. In the following discussion we shall examine alternative strategies for obtaining stiffness coefficients for each type of force.

The potential energy of a spring is

$$\boxed{V_{\text{spring}} = \tfrac{1}{2}k\Delta^2}\qquad(1.5.50)$$

where the elongation Δ is defined in eq. (1.2.1) to be the increase of the spring's length relative to its unstretched length ℓ_0. To explore the construction of V, consider the situation in Figure 1.17, where one end of a spring is stationary. Point B_0 is the position of the moving end when the generalized coordinates are zero, which is defined to be the static equilibrium position. From the Pythagorean theorem, we may find the deformed length of the spring in terms of the coordinates of the end points:

$$L = [(x_B - x_A)^2 - (y_B - y_A)^2]^{1/2}\qquad(1.5.51)$$

We form $\Delta = L - \ell_0$, which leads us to

$$V_{\text{spring}} = \tfrac{1}{2}k(L - \ell_0)^2 = \tfrac{1}{2}k(L^2 - 2\ell_0 L + \ell_0^2)\qquad(1.5.52)$$

By definition, the position of point B depends on the generalized coordinates, so the deformed length L depends on these variables. Consequently, forming the preceding expression for a specific system leads to the description of V_{spring} as a function of the generalized coordinates.

To see how eq. (1.5.18) may be used to find the stiffness coefficients associated with V_{spring}, consider the spring supporting the bar in Figure 1.18. The position illustrated there corresponds to static equilibrium. Let θ be the generalized coordinate. The length of the spring when θ is nonzero may be found from the law of cosines:

$$L = \left[\left(\frac{b}{2}\right)^2 + b^2 - 2\left(\frac{b}{2}\right)b\cos\left(\frac{\pi}{2} - \theta\right)\right]^{1/2} \equiv b\left[\frac{5}{4} - \sin(\theta)\right]^{1/2}\qquad(1.5.53)$$

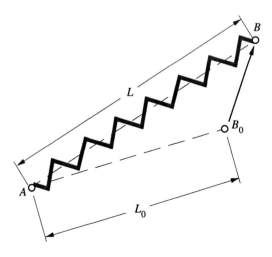

FIGURE 1.17 Deformation of a spring.

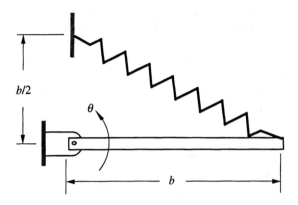

FIGURE 1.18 A typical elastic spring.

According to eq. (1.5.52), the potential energy stored in the spring is

$$V = \frac{1}{2}k\left\{ b^2\left[\frac{5}{4} - \sin(\theta)\right] - 2\ell_0 b\left[\frac{5}{4} - \sin(\theta)\right]^{1/2} + \ell_0^2 \right\} \qquad (1.5.54)$$

When we use eq. (1.5.18) to evaluate the stiffness coefficient associated with θ, we have

$$K_{11} = \left(\frac{\partial^2 V}{\partial \theta^2}\right)_{\theta=0} = \frac{1}{2}k\frac{d}{d\theta}\left\{ -b^2\cos(\theta) - \ell_0 b\left[\frac{5}{4} - \sin(\theta)\right]^{-1/2}[-\cos(\theta)] \right\}\Bigg|_{\theta=0}$$

$$= \frac{1}{2}k\left\{ b^2\sin(\theta) + \frac{1}{2}\ell_0 b\left[\frac{5}{4} - \sin(\theta)\right]^{-3/2}[-\cos(\theta)]^2 \right.$$

$$\left. -\ell_0 b\left[\frac{5}{4} - \sin(\theta)\right]^{-1/2}[+\sin(\theta)]\right\}\Bigg|_{\theta=0} = \frac{2}{5\sqrt{5}}k\ell_0 b \qquad (1.5.55)$$

It is always correct to employ the first of eqs. (1.5.18) to determine the stiffness coefficients K_{jn}. Furthermore, doing so is a logical extension of using the second of those equations to determine the coefficients $(F_0)_j$ from which equilibrium conditions may be established. However, even if we were to use a symbolic program such as Maple to evaluate the derivatives, it is obvious that constructing the potential energy function for springs can be quite tedious.

In most situations, the stiffness coefficients associated with springs may be obtained by a simpler, alternative procedure. One of the attractive features of this alternative is that its steps are directly analogous to the procedure for determining $[M]$ and $[C]$. The simplification, which is called the *stiff spring approximation*, relies on the fact that if the spring constant k is very large, then the spring will be deformed little when the system is in the static equilibrium position.

We begin by examining the displacement of the end of a spring. Figure 1.19 repeats the situation in Figure 1.17, except that the displacement $\Delta \bar{r}_B$ is represented in terms of two components. The unit vector $\bar{e}_{B/A}$ marks the initial orientation of the spring: u_B is the axial component and w_B is the transverse component. We use the Pythagorean theorem to describe the current length:

$$L = [(L_0 + u_B)^2 + w_B^2]^{1/2} = (L_0^2 + 2L_0 u_B + u_B^2 + w_B^2)^{1/2} \qquad (1.5.56)$$

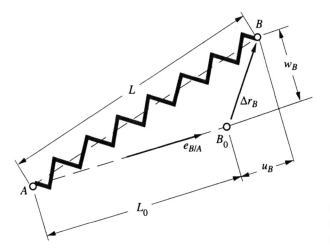

FIGURE 1.19 Deformation analysis for a linear spring's potential energy.

Let us expand L in a series, based on u_B and w_B being much less than L_0. We truncate this series at the first order, so that

$$L = L_0 \left[1 + \left(2\frac{u_B}{L_0} + \frac{u_B^2}{L_0^2} + \frac{w_B^2}{L_0^2} \right) \right]^{1/2} \approx L_0 \left(1 + \frac{u_B}{L_0} \right) \tag{1.5.57}$$

where the terms that are omitted are quadratic or higher degree.

Recall that ℓ_0 is the length of the spring when it is unstressed, while L_0 is the length when the spring is at the static equilibrium position. The stiff spring approximation involves setting $L_0 = \ell_0$. Strictly speaking, this approximation requires that the spring be unstressed in the static equilibrium position. This is likely to be the case if the system moves in the horizontal plane, for then the spring need not exert any force to counter the effect of gravity. Even if motion occurs in the vertical plane, it usually is possible to consider $L_0 \approx \ell_0$. Let F_0 denote the axial force (positive if it is tensile) in the spring required to hold the system in static equilibrium. Because the elongation in that state is $L_0 - \ell_0 = F_0/k$, we will find that $L_0 \approx \ell_0$ if $|F_0|/(kL_0) \ll 1$. This is the case of a relatively stiff spring, which is typical of engineering applications.

When we set $L_0 = \ell_0$ in eq. (1.5.57), we find that $L \approx \ell_0 + u_B$. Thus, the potential energy in the stiff spring approximation is

$$V_{\text{spring}} = \tfrac{1}{2} k (L - \ell_0)^2 = \tfrac{1}{2} k u_B^2 \tag{1.5.58}$$

From a practical standpoint, the approximation should be quite good if $|F_0|/(kL_0) < 0.01$, which corresponds to a static spring length L_0 that is within 1% of the unstretched length ℓ_0.

Although we considered the case where only one end of the spring is movable, we could repeat the derivation to prove that it applies if both ends move. Specifically, the total elongation will be the difference between the displacement component of each end parallel to $\bar{e}_{B/A}$. Thus, we are led to the general description of the potential energy stored in springs,

$$\boxed{V_{\text{springs}} = \frac{1}{2} \sum_j k_j \Delta_j^2, \quad \Delta_j = u_{Bj} - u_{Aj}, \quad \text{if } \frac{|F_{0j}|}{k_j L_{0j}} \ll 1} \tag{1.5.59}$$

The significance of this result is that u_A and u_B represent the component of displacement at each end parallel to the original orientation of the spring. According to the discussion of kinematics, the linearized displacement of a point may be obtained directly from the point's velocity. To do so, we merely replace any $\dot{q}_j$ variable with the corresponding q_j.

These observations lead to the general procedure for obtaining the contribution of springs to the stiffness coefficients. We begin by performing a kinematical velocity analysis. We do this in order to express, in terms of the $\dot{q}_j$ variables, the velocity component of each end of a spring parallel to the spring's orientation at static equilibrium. We convert these velocity expressions to displacement components by replacing each generalized velocity with the corresponding generalized coordinate. We use the displacement components to describe the elongation of each spring, and then use that elongation to form V according to eq. (1.5.59). Because the displacements will be linear in the q_j variables, the potential energy V that results will be a quadratic sum. We then collect the coefficients of each unique combination $q_j q_n$. The last step is to identify the stiffness coefficients by matching this quadratic sum for V to the standard one, eq. (1.5.19). This procedure is a close parallel to that by which we obtain $[C]$ from $\mathcal{P}_{\text{dis}}$.

As an illustration of this procedure, let us redo the analysis of K_{11} for this system in Figure 1.18. The velocity of the right end of the bar in that figure is $b\dot{\theta}$ upward. The angle between the spring and vertical is β, where the Pythagorean theorem indicates that $\cos\beta = 1/\sqrt{5}$. The extension rate of the spring is $\dot{\Delta} = -b\dot{\theta}\cos\beta$ (the minus sign indicates that upward movement of the end of the bar results in shortening of the spring), so the change in the spring's length is $\Delta = -b\theta/\sqrt{5}$. The corresponding potential energy is $V_{\text{sp}} = \frac{1}{2}k\Delta^2 = \frac{1}{2}(kb^2/5)\theta^2$, which leads to $K_{11} = kb^2/5$. The stiffness coefficient we obtained previously was $K_{11} = 2k\ell_0 b/(5\sqrt{5})$. If the stiff spring approximation applies, then $\ell_0 \approx \sqrt{5}/4b$, which leads to the expression for K_{11} we derived here.

The other conservative force we commonly encounter is gravity. The potential energy of this force is

$$\boxed{V_{\text{gravity}} = mgH} \qquad (1.5.60)$$

where H denotes the height of the center of mass relative to a reference elevation called the datum. A body's weight tends to affect a system differently depending on whether the body rotates and how large its contribution is relative to springs. Also, when motion occurs in the horizontal plane, gravity effects are irrelevant to the motion. As we did for springs, let us explore the issues by considering a specific example. The system in Figure 1.20 consists of a bar that is connected to a collar that may slide along the vertical guide bar. The springs have been selected such that the system is in static equilibrium in the position shown, and the spring constants are sufficiently large to warrant using the stiff spring approximation.

The movement of the system away from the static equilibrium position is defined by the vertical displacement y of the collar and the rotation θ, which we take to be the generalized coordinates. We form the potential energy of gravity by adding mgH terms for each body. It is wise to use some fixed location in the system as the datum, so we use the guide bar's lower support for this purpose. The collar and bar each have mass, so

$$V_{\text{grav}} = m_1 g(L_0 + y) + m_2 g\left[L_0 + y - \frac{L}{2}\cos(\theta + \beta)\right] \qquad (1.5.61)$$

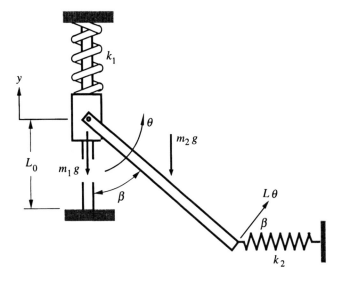

FIGURE 1.20 Typical two-degree-of-freedom system.

To determine the corresponding contribution of gravity to the stiffness coefficients, we employ the first of eqs. (1.5.18). This entails evaluating partial derivatives, such that

$$\left(K_{11}\right)_{\text{grav}} = \left(\frac{\partial^2 V}{\partial y^2}\right)_{y=\theta=0} = 0$$

$$\left(K_{22}\right)_{\text{grav}} = \left(\frac{\partial^2 V}{\partial \theta^2}\right)_{y=\theta=0} = m_2 g \frac{L}{2} \cos(\beta) \tag{1.5.62}$$

$$\left(K_{12}\right)_{\text{grav}} = K_{21} = \left(\frac{(\partial)^2 V}{\partial y \partial \theta}\right)_{y=\theta=0} = 0$$

Several aspects of these expressions are representative of general properties. We see that gravity does not affect stiffness coefficients pertaining to y, which represents a translational motion. The explanation for this lies in the nature of the displacement associated with y. If y increases and θ remains constant, then the entire system translates upward. In such a motion, the component of each weight force in the direction of motion does not change, which means that gravity has a constant effect. Such an effect influences the static equilibrium position, but not movement away from that position.

Only K_{22} for our sample system contains a nonzero contribution from gravity. If the static equilibrium position corresponds to $\beta = \pi/2$, even this term goes away. The reason for this is that when $\beta = \pi/2$, changing θ results in vertical displacement of the center of mass, which is parallel to the gravity force. In this case, the component of the gravity force in the direction of the displacement remains constant, so gravity has a static effect. In contrast, if $\beta = 0$ at static equilibrium, we have $K_{22} = m_2 g L/2$. This situation resembles a pendulum, in which the effect of gravity is to exert a moment $m_2 g(L/2) \sin \theta \approx m_2 g(L/2) \theta$ about the pivot that tends to return the system back to $\theta = 0$. This restoring moment may be thought of as the product of θ and an equivalent spring constant K_{22}. For an arbitrary β, $K_{22} = m_2 g(L/2) \cos \beta$ represents this variable effect of the weight.

There are a number of circumstances under which it is permissible to ignore gravity in the determination of stiffness coefficients. Three that we have seen already are motion in the horizontal plane, translation along a straight line, and rotational motion in which the center of mass displaces in the vertical direction. Even when none of these conditions is met, it might be that gravity is unimportant in comparison with the effect of springs. From a qualitative viewpoint, the stiffness coefficient for gravity in a rotational motion can be expected to be proportional to $mg\ell$, where ℓ is a characteristic length for the system. By the same token, stiffness coefficients associated with springs when a system rotates will be proportional to $k\ell^2$. Thus, it might be possible to ignore gravity of $k\ell \gg mg$.

EXAMPLE 1.6

The small sphere in the sketch is connected to three unevenly arranged springs situated in the vertical plane. The angles θ_1, θ_2, and θ_3 give the orientation of the springs at the equilibrium position, which corresponds to $x = y = 0$. Determine the stiffness matrix associated with x and y.

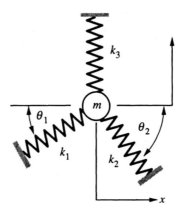

Solution This exercise will provide practice in constructing spring elongations, as well as in matching terms to identify the stiffness matrix. We begin by drawing a free body diagram, which helps us to account for all forces.

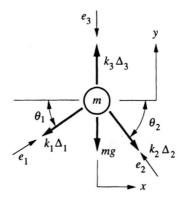

Unit vectors parallel to each spring, extending from the fixed end to the sphere, are shown in the free body diagram. The displacements x and y define the position of the sphere at any instant, so we set $q_1 = x$, $q_2 = y$.

The motion occurs in the vertical plane, so there is gravitational potential energy. There is no obvious fixed point in this system, so we define the datum to be the elevation of the sphere at static equilibrium. The potential energy of gravity is $V_{gr} = mgy$. Because this is linear in the generalized coordinates, eq. (1.5.18) indicates that gravity does not contribute to the stiffness coefficients.

The potential energy of the springs is the sum of their individual contributions,

$$V_{sp} = \tfrac{1}{2}k_1\Delta_1^2 + \tfrac{1}{2}k_2\Delta_2^2 + \tfrac{1}{2}k_3\Delta_3^2$$

It is not stated otherwise, so we assume that the stiff spring approximation is applicable. Then each elongation is the displacement component in the direction of the respective spring. The displacement of the disk from the reference position is $\Delta\bar{r} = x\bar{i} + y\bar{j}$. We may obtain Δ_1 pictorially by projecting the x and y components onto $\bar{e}_1$, as shown in the following sketch.

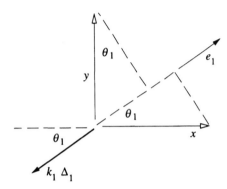

This leads to $\Delta_1 = x\cos\theta_1 + y\sin\theta_1$. Alternatively, we can obtain Δ_1 mathematically by using a dot product to project $\Delta\bar{r}$ onto $\bar{e}_1$, which leads to $\Delta_1 = \Delta\bar{r} \cdot \bar{e}_1 = (x\bar{i} + y\bar{j}) \cdot (\cos\theta_1\bar{i} + \sin\theta_1\bar{j})$. Let us use the latter procedure for the other springs. The unit vector extending from the fixed end to the moving end of spring 2 is $-\cos\theta_2\bar{i} + \sin\theta_2\bar{j}$, so the corresponding elongation is $\Delta_2 = \Delta\bar{r} \cdot \bar{e}_2 = (x\bar{i} + y\bar{j}) \cdot (-\cos\theta_2\bar{i} + \sin\theta_2\bar{j}) = -x\cos\theta_2 + y\sin\theta_2$. For spring 3, we have $\bar{e}_3 = -\bar{j}$, so that $\Delta_3 = \Delta\bar{r} \cdot \bar{e}_3 = (x\bar{i} - y\bar{j}) \cdot (-\bar{j}) = -y$. Thus, the potential energy is

$$V_{sp} = \tfrac{1}{2}k_1(x\cos\theta_1 + y\sin\theta_1)^2 + \tfrac{1}{2}k_2(-x\cos\theta_2 + y\sin\theta_2)^2 + \tfrac{1}{2}k_3(-y)^2$$

$$= \tfrac{1}{2}[(k_1\cos^2\theta_1 + k_2\cos^2\theta_2)x^2 + 2(k_1\cos\theta_1\sin\theta_1 - k_2\cos\theta_2\sin\theta_2)xy$$

$$+ (k_1\sin^2\theta_1 + k_2^2\sin^2\theta_2 + k_3)y^2]$$

The quadratic terms in the standard form of V for a two-degree-of-freedom system are

$$V = \tfrac{1}{2}(K_{11}y^2 + 2K_{12}xy + K_{22}y^2)$$

There is no contribution of gravity to the stiffness coefficients, so matching the standard representation to the preceding description of V_{sp} yields

$$K_{11} = k_1\cos^2\theta_1 + k_2\cos^2\theta_2, \qquad K_{22} = k_1\sin^2\theta_1 + k_2\sin^2\theta_2 + k_3$$

$$K_{12} = K_{21} = k_1\cos\theta_1\sin\theta_1 - k_2\cos\theta_2\sin\theta_2$$

EXAMPLE 1.7

Generalized coordinates for the system in Figure 1.20 are the vertical displacement y of the collar and the rotation θ of the bar. Determine the corresponding $[K]$.

Solution We will see in this exercise how to combine the contributions of springs and gravity to the stiffness matrix. For the springs, we assume that the stiff spring approximation is applicable. The potential energy stored in the springs is

$$V_{sp} = \tfrac{1}{2}k_1\Delta_1^2 + \tfrac{1}{2}k_2\Delta_2^2$$

Positive y shortens spring k_1, so that $\Delta_1 = -y$. For spring k_2, we must describe the displacement at the lower end of the bar in terms of y and θ, which we accomplish using the analogy with velocity. Relating the velocity at both ends of the bar gives

$$\bar{v}_B = \bar{v}_A + \dot{\theta}\bar{k} \times \bar{r}_{B/A}$$

where we designate the upper end as point A and the lower end as point B. Also, the x direction is to the right and the y direction is upward, so that $\bar{k}$ is out of the plane of the diagram. We describe $\bar{r}_{B/A}$ as its value at the static equilibrium position, so that $\bar{r}_{B/A} = L\sin\beta\bar{i} - L\cos\beta\bar{j}$. Also, we know that $\bar{v}_A = \dot{y}\bar{j}$, so we have

$$\bar{v}_B = \dot{y}\bar{j} + L\dot{\theta}\sin\beta\bar{j} + L\dot{\theta}\cos\beta\bar{i}$$

To convert this to displacement, we employ the linearization analogy with velocity, which allows us to cancel the time derivatives:

$$\Delta\bar{r}_B = y\bar{J} + L\theta\sin\beta\bar{j} + L\theta\cos\beta\bar{i}$$

Spring k_2 is horizontal, so the unit vector extending from its fixed end to its moving end is $\bar{e}_2 = -\bar{i}$. We then find the component of displacement at point B parallel to the spring to be $\Delta_2 = \Delta\bar{r}_B \cdot \bar{e}_2 = -L\theta\cos\beta$.

It follows from the description of Δ_1 and Δ_2 that

$$V_{sp} = \tfrac{1}{2}k_1 y^2 + \tfrac{1}{2}k_2(L\theta\cos\beta)^2$$

The quadratic terms in the expanded general form of potential energy are

$$V_{sp} = \tfrac{1}{2}[(K_{11})_{sp}y^2 + (K_{22})_{sp}\theta^2 + 2(K_{12})_{sp}y\theta]$$

so matching the two descriptions yields

$$(K_{11})_{sp} = k_1, \qquad (K_{22})_{sp} = k_2 L^2\cos^2\beta, \qquad (K_{12})_{sp} = (K_{21})_{sp} = 0$$

We now focus on the effect of gravity. The contribution of gravity to the stiffness coefficients was found in eqs. (1.5.62). We add the spring and gravity portions, which leads to

$$K_{11} = (K_{11})_{sp} + (K_{11})_{gr} = k_1$$

$$K_{22} = (K_{22})_{sp} + (K_{22})_{gr} = k_2 L^2\cos^2\beta + m_2 g\frac{L}{2} + k_2 L^2\cos^2\beta$$

$$K_{12} = (K_{12})_{sp} + (K_{12})_{gr} = 0 = (K_{21})_{sp}$$

Power Input The technique for identifying the generalized forces also employs a matching procedure. If force $\bar{F}_p$ acts at a point whose velocity is $\bar{v}_p$, the power supplied by the source of that force is $\bar{F}_p \cdot \bar{v}_p$. In planar motion, the power input to body b rotating at rate ω_b by a torque Γ_b is $\Gamma_b\omega_b$, where the contribution is positive if Γ and ω are in the same sense. The total power input to a system is the sum of the individual contributions, so

$$\boxed{\mathscr{P}_{in} = \sum_p \bar{F}_p \cdot \bar{v}_p + \sum_b \Gamma_b\omega_b} \tag{1.5.63}$$

This expression is the starting point for identification of the generalized forces. We perform a linearized kinematical analysis using the techniques in Section 1.5.4 in order to describe each $\bar{v}_p$ and ω_b as linear combinations of the generalized velocities in the manner of eq. (1.5.40). In the case of a single force and a single torque, the result of using such expressions is

$$\mathcal{P}_{in} = \bar{F} \cdot \bar{v} + \Gamma \omega = \bar{F} \cdot (\bar{d}_1 \dot{q}_1 + \bar{d}_2 \dot{q}_2) + \Gamma(s_1 \dot{q}_1 + s_2 \dot{q}_2)$$
$$= (\bar{F} \cdot \bar{d}_1 + \Gamma s_1)\dot{q}_1 + (\bar{F} \cdot \bar{d}_2 + \Gamma s_2)\dot{q}_2 \qquad (1.5.64)$$

Matching this to the standard description of $\mathcal{P}_{in}$ in eq. (1.5.22) yields the generalized forces:

$$Q_1 = \bar{F} \cdot \bar{d}_1 + \Gamma s_1, \qquad Q_2 = \bar{F} \cdot \bar{d}_2 + \Gamma s_2 \qquad (1.5.65)$$

As noted earlier, constraint forces do not input power to time-invariant systems. Consequently, we may ignore such forces when we determine the Q_j quantities.

An important class of excitations is associated with motion of the ground. Two examples are the excitation applied to a building when its foundation moves during an earthquake, and the excitation applied to an automobile as it moves over a rough road. To account for the force within a spring or dashpot connected to the ground in this situation, we use a superposition of effects. In eq. (1.5.59) for the potential energy of a spring and eq. (1.5.47) for power dissipated in a dashpot, Δ_j represents the difference between the displacements at the two ends. Let u_A be the displacement component at the connection of the device to one of the system's bodies, and let u_g be the displacement at the grounded end. This is illustrated in Figure 1.21, where a shock absorber consisting of a parallel spring and dashpot supports a body, in order to isolate it from ground motion $z(t)$ in the vertical direction.

The displacement u_g is the component of z parallel to the spring, while u_A is the component of the body's displacement parallel to the spring. The total elongation is $u_A - u_g$, so the force exerted by the spring and dashpot on the body is $k(u_A - u_g) + c(\dot{u}_A - \dot{u}_g)$. Because the motion of the body is defined by the generalized coordinates, the relation between u_A and the q_n variables is unaffected by the ground motion. We therefore separate the force exerted on the body into two parts, as shown in Figure 1.21. In one part, we consider the deformation to be $\Delta = u_A$, as though the ground were fixed. The corresponding force is $k\Delta + c\dot{\Delta}$, which we show as tensile because positive u_A elongates the spring and dashpot. This part is accounted for in the usual way, by adding $\frac{1}{2}k\Delta^2$ to the potential energy and $c\dot{\Delta}^2$ to the power dissipation.

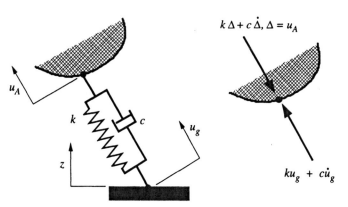

FIGURE 1.21 Effect of ground motion on the forces exerted by springs and dashpots.

The other part of the force exerted by the spring and dashpot is $ku_g + c\dot{u}_g$, which represents the force induced by the ground motion. This force is compressive because positive u_g shortens the devices. Because u_g is the component of z parallel to the spring, and we know z as a function of t, it follows that $ku_g + c\dot{u}_g$ represents a known force that acts to move the system. We use the power input to account for this force. At the point where the force is applied to the system, the velocity parallel to the force is $\dot{u}_A$. Correspondingly, the term to add to the power input is

$$\mathscr{P}_{in} = (ku_g + c\dot{u}_g)\dot{u}_A \qquad (1.5.66)$$

The consequence of handling ground motion in this manner is that the expressions for V and $\mathscr{P}_{dis}$ will be the same as they would be if the ground were stationary. Thus, *the $[K]$ and $[C]$ matrices will be the same as they would be if the ground were stationary*. The only effect of the motion is to add terms to $\mathscr{P}_{in}$, which show up as contributions to the generalized forces.

EXAMPLE 1.8

A piece of electronic equipment is isolated from ground motion by a pair of shock absorbers. Assume that the equipment translates only in the vertical direction, with y being the upward displacement relative to the static equilibrium position. What is the generalized force corresponding to the ground translating upward at $z(t)$?

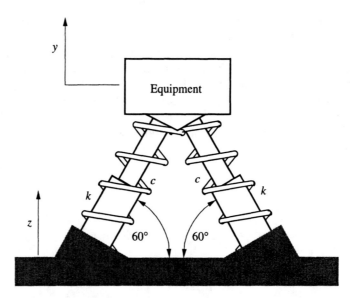

Solution Demonstration of the procedure for incorporating the effects of ground motion into the generalized forces is the objective of this exercise. We draw a free body diagram, in which the spring and dashpot forces are broken into two parts.

Terms containing Δ_1 or Δ_2 are associated with the upward displacement y. These forces are accounted for in the potential energy and power dissipation, which lead to $[K]$ and $[C]$. These quantities were not requested in the problem statement, so we proceed to the portion of the forces that depend on the ground displacement. The displacement at ground points A and B is z upward, and the angle from vertical to either device is $30°$. Thus, $(u_g)_A = (u_g)_B = z\cos 30°$. Positive z results in shortening each shock absorber, so the respective forces are compressive.

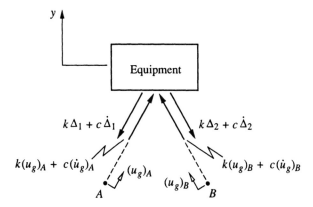

This is consistent with the manner in which the forces are depicted in the free body diagram. Thus, the forces exerted by the dashpot as a result of the ground motion are $(F_g)_A = (F_g)_B = k(0.866z) + c(0.866\dot{z})$.

The velocity of the equipment package is $\dot{y}$ upward. Each shock absorber force is at an angle of 30° relative to this velocity. Hence, the power input is

$$\mathscr{P}_{\text{in}} = [(F_g)_A \cos 30°]\dot{y} + [(F_g)_B \cos 30°]\dot{y}$$
$$= 2[k(0.866z) + c(0.866\dot{z})](0.866)\dot{y}$$

The standard form of the power input for this one-degree-of-freedom system is

$$\mathscr{P}_{\text{in}} = Q_1\dot{y}$$

so the generalized force is

$$Q_1 = 1.50(kz + c\dot{z})$$

Procedural Steps The developments required to implement the power balance method may be summarized in a series of general procedural steps:

1. Identify a set of unconstrained generalized coordinates q_j. The static equilibrium position of the system must be used as the reference location for defining these generalized coordinates, so that all $q_j = 0$ when the system is in static equilibrium. The number of degrees of freedom N is the number of unconstrained generalized coordinates.

2. Use the matching procedure to identify $[M]$. First use eq. (1.5.38) and/or eq. (1.5.39) to express T in for each body having significant mass. Then express as linear functions of the $\dot{q}_j$ variables the velocity $\bar{v}_G$ of each center of mass and the angular velocity ω of each body. Such linear relations are obtained by performing a kinematical analysis using the geometry of the system at its static equilibrium position. Substitute the descriptions of each $\bar{v}_G$ and ω into T, and collect like coefficients of the generalized velocities. When a $\frac{1}{2}$ factor is carried outside the expanded form, the coefficient of a square term $\dot{q}_j^2$ will be M_{jj}, while the coefficient of a mixed product $\dot{q}_j\dot{q}_n$ will be $2M_{jn}$. The symmetry property then gives $M_{nj} = M_{jn}$.

3. Use the matching procedure to identify $[C]$. First express $\mathscr{P}_{\text{in}}$ in terms of velocity components, as described by eqs. (1.5.45) and (1.5.47). Then express the velocity of the moving end of each dashpot as a linear function of the

generalized velocities. (It is likely that this step will follow the same type of kinematical analysis as that used for kinetic energy in step 2.) Follow eq. (1.5.44) by taking the component of the velocity parallel to the dashpot's orientation in the static equilibrium position. Substitute the linearized $\dot{u}$ descriptions to obtain $\mathcal{P}_{\text{dis}}$ as a quadratic sum, and collect like coefficients of generalized velocities. The coefficient of a square term $\dot{q}_j^2$ will be C_{jj}, while the coefficient of a mixed product $\dot{q}_j \dot{q}_n$ will be $2C_{jn}$. The symmetry property then gives $C_{nj} = C_{jn}$.

4. Construct the contribution of springs to $[K]$. Two alternatives are available:

   a. In most cases the stiff spring approximation will be valid, which permits usage of eq. (1.5.59). Express as a linear function of the generalized coordinate the displacement of each movable spring attachment point. An aid to this analysis is the similarity between the velocity and displacement of a point, as described by eq. (1.5.37). Collect like coefficients of the generalized coordinates in the resulting expression for V_{spring} to obtain a quadratic sum. Bring a common factor $\frac{1}{2}$ outside the quadratic terms in the expanded form. The coefficient of a square term q_j^2 will be K_{jj}, while the coefficient of a mixed product $q_j q_n$ will be $2K_{jn}$. The symmetry property then gives $K_{nj} = K_{jn}$.

   b. If the stiff spring approximation is not valid, it is necessary to derive a description of the potential energy valid for arbitrarily large values of the generalized coordinates. The stiffness coefficients associated with springs may then be obtained by differentiating V according to eq. (1.5.18).

5. Identify the contribution of gravity to $[K]$. There will be no such contribution when the motion occurs in the horizontal plane, when bodies translate along a straight path, or when the center of mass displaces vertically at the static equilibrium position. If gravity needs to be considered, describe as a function of the generalized coordinates the gravitational potential energy by adding eq. (1.5.60) for each body. Then apply eq. (1.5.18) to find the portion of the stiffness coefficients associated with gravity. Add each coefficient to the corresponding coefficient associated with springs.

6. Use the matching procedure to characterize $\{Q\}$. This begins by forming the power input to the system by external forces according to eq. (1.5.21). Express the force components for each dot product in terms of the orientation of the respective forces when the system is in its equilibrium position. Express the velocity of the point of application of each force by the same kinematical procedures as those for construction of the kinetic energy and power dissipation expressions. Substitute the force components and linearized velocities in the preceding step into the expression for power dissipation. Form each dot product and collect like coefficients of each generalized velocity. The coefficient of each linear term $\dot{q}_j$ will be the corresponding generalized force Q_j.

7. The differential equations of motion for the system are described by eq. (1.5.33).

EXAMPLE 1.9

The three coupled pendulums are in static equilibrium when they have the vertical orientation shown in the sketch. The mass of the longer bars is m, while the center bar's mass is $m/2$. The torque Γ causes the bars to move away from this equilibrium position. Determine the equations of motion for the system.

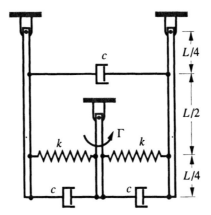

Solution This exercise provides practice in constructing the potential energy and power dissipation for spring and dashpot elements in which both ends move. It also illustrates the manner in which the coefficient matrices appearing in the equations of motion are identified from the energy and power dissipation expressions. We begin by drawing a free body diagram of each bar, in which the bars are placed at their equilibrium positions.

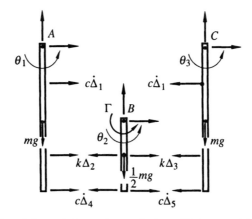

The rotation of each bar relative to its vertical equilibrium position forms a convenient set of generalized coordinates, so this system has three degrees of freedom: $q_1 = \theta_1$, $q_2 = \theta_2$, $q_3 = \theta_3$. Each bar executes a pure rotation about its pivot, and the moments of inertia about the respective pivots are

$$I_1 = I_3 = \frac{1}{3}mL^2, \qquad I_2 = \frac{1}{3}\left(\frac{1}{2}m\right)\left(\frac{L}{2}\right)^2 = \frac{1}{24}mL^2$$

The kinetic energy is the sum of the rotational energies of the bars about their respective pivots, so we have

$$T = \frac{1}{2}I_1\dot{\theta}_1^2 + \frac{1}{2}I_2\dot{\theta}_2^2 + I_3\dot{\theta}_3^2$$

The expanded quadratic form of kinetic energy for this three-degree-of-freedom system is

$$T = \frac{1}{2}(M_{11}\dot{\theta}^2 + M_{22}\dot{\theta}_2^2 + M_{33}\dot{\theta}_3^2 + 2M_{12}\dot{\theta}_1\dot{\theta}_2 + 2M_{13}\dot{\theta}_1\dot{\theta}_3 + 2M_{23}\dot{\theta}_2\dot{\theta}_3)$$

Matching this representation to the one derived for the system gives

$$M_{11} = M_{33} = \frac{1}{3}mL^2, \qquad M_{22} = \frac{1}{24}mL^2$$

$$M_{12} = M_{21} = M_{13} = M_{31} = M_{23} = M_{32} = 0$$

Contributions to the potential energy come from the springs and from gravity, which we treat separately. When either bar is at its static equilibrium position, the velocity of a point at distance r from that bar's pivot is $r\dot{\theta}$ in the horizontal direction. The corresponding displacement comes from replacing $\dot{\theta}$ with θ. Thus, the displacements at the ends of either spring are parallel to the orientation of that spring. We form the respective elongations by taking the difference of the horizontal displacements at each end:

$$\Delta_2 = \frac{L}{4}\theta_2 - \frac{3L}{4}\theta_1, \qquad \Delta_3 = \frac{3L}{4}\theta_3 - \frac{L}{4}\theta_2$$

The corresponding potential energy stored in the springs is

$$
\begin{aligned}
V_{\text{spring}} &= \frac{1}{2}k\Delta_2^2 + \frac{1}{2}k\Delta_3^2 = \frac{1}{2}k\left(\frac{L}{4}\theta_2 - \frac{3L}{4}\theta_1\right)^2 + \frac{1}{2}k\left(\frac{3L}{4}\theta_3 - \frac{L}{4}\theta_2\right)^2 \\
&= \frac{1}{2}kL^2\left[\frac{9}{16}\theta_1^2 + \frac{1}{8}\theta_2^2 + \frac{9}{16}\theta_3^2 + 2\left(-\frac{3}{16}\theta_1\theta_2 - \frac{3}{16}\theta_2\theta_3\right)\right]
\end{aligned}
$$

Equation (1.5.19) for a three-degree-of-freedom system is

$$V = \frac{1}{2}(K_{11}\theta_1^2 + K_{22}\theta_2^2 + K_{33}\theta_3^2 + 2K_{12}\theta_1\theta_2 + 2K_{13}\theta_1\theta_3 + 2K_{23}\theta_2\theta_3)$$

Comparing this with V_{spring} leads to

$$(K_{11})_{\text{sp}} = (K_{33})_{\text{sp}} = \frac{9}{16}kL^2, \qquad (K_{22})_{\text{sp}} = \frac{1}{8}kL^2$$

$$(K_{12})_{\text{sp}} = (K_{23})_{\text{sp}} = \frac{3}{16}kL^2, \qquad (K_{13})_{\text{sp}} = 0$$

For the gravitational potential energy we take the elevation of each pivot as the datum for the gravitational potential energy of the associated body, so we have

$$V_{\text{grav}} = -mg\frac{L}{2}\cos\theta_1 - \frac{mg}{2}\frac{L}{4}\cos\theta_2 - mg\frac{L}{2}\cos\theta_3$$

We determine the gravitational contribution to the stiffness coefficients by evaluating the partial derivatives in eqs. (1.5.18), which gives

$$(K_{11})_{\text{gr}} = \left(\frac{\partial^2 V}{\partial\theta_1^2}\right)_{\theta_j=0} = mg\frac{L}{2}\cos\theta_1\big|_{\theta_1=0} = \frac{1}{2}mgL$$

$$(K_{22})_{\text{gr}} = \left(\frac{\partial^2 V}{\partial\theta_2^2}\right)_{\theta_j=0} = \frac{1}{8}mgL, \qquad (K_{33})_{\text{gr}} = \left(\frac{\partial^2 V}{\partial\theta_3^2}\right)_{\theta_j=0} = \frac{1}{8}mgL$$

$$(K_{12})_{\text{gr}} = \left(\frac{\partial^2 V}{\partial\theta_1\partial\theta_2}\right)_{\theta_j=0} = 0, \qquad (K_{13})_{\text{gr}} = \left(\frac{\partial^2 V}{\partial\theta_1\partial\theta_3}\right)_{\theta_j=0} = 0$$

$$(K_{23})_{\text{gr}} = \left(\frac{\partial^2 V}{\partial\theta_2\partial\theta_3}\right)_{\theta_j=0} = 0$$

We combine these with the portion of the stiffness coefficients associated with springs to find that

$$K_{11} = K_{33} = \frac{9}{16}kL^2 + \frac{1}{2}mgL, \qquad K_{22} = \frac{1}{8}kL^2 + \frac{1}{8}mgL$$

$$K_{12} = K_{21} = K_{23} = K_{32} = -\frac{3}{16}kL^2, \qquad K_{13} = K_{31} = 0$$

Next, we form the power dissipation. At the equilibrium position, the velocity of each end of a dashpot is horizontal, and therefore parallel to the dashpot. The elongation rates are the differences between the velocities at the two ends:

$$\dot{\Delta}_1 = \frac{L}{4}\dot{\theta}_3 - \frac{L}{4}\dot{\theta}_1, \qquad \dot{\Delta}_4 = \frac{L}{2}\dot{\theta}_2 - L\dot{\theta}_1, \qquad \dot{\Delta}_5 = L\dot{\theta}_3 - \frac{L}{2}\dot{\theta}_2$$

We add the power dissipated in each dashpot to obtain

$$\mathscr{P}_{\text{dis}} = c\dot{\Delta}_3^2 + c\dot{\Delta}_4^2 + c\dot{\Delta}_5^2$$

$$= cL^2\left[\tfrac{17}{16}\dot{\theta}_1^2 + \tfrac{1}{2}\dot{\theta}_2^2 + \tfrac{17}{16}\dot{\theta}_3^2 - \tfrac{1}{8}\dot{\theta}_1\dot{\theta}_3 - \dot{\theta}_1\dot{\theta}_2 - \dot{\theta}_2\dot{\theta}_3\right]$$

The standard form of the power dissipation for this system is

$$\mathscr{P}_{\text{dis}} = C_{11}\dot{\theta}_1^2 + C_{22}\dot{\theta}_2^2 + C_{33}\dot{\theta}_3^2 + 2C_{12}\dot{\theta}_1\dot{\theta}_2 + 2C_{13}\dot{\theta}_1\dot{\theta}_3 + 2C_{23}\dot{\theta}_2\dot{\theta}_3$$

Matching the two forms leads to

$$C_{11} = C_{33} = \tfrac{17}{16}cL^2, \qquad C_{22} = \tfrac{1}{2}cL^2$$

$$C_{12} = C_{23} = -\tfrac{1}{2}cL^2, \qquad C_{13} = -\tfrac{1}{16}cL^2$$

It still remains to identify the generalized forces, which we do by forming the power input. The only external force to characterize is the disturbance torque Γ that acts on the center bar. The power input by a torque is $\Gamma\omega$, where ω is the rotation rate of the body to which the torque is applied, with positive ω corresponding to rotation in the sense of the torque. Thus, we have

$$\mathscr{P}_{\text{in}} = \Gamma(\dot{\theta}_2) = Q_1\dot{\theta}_1 + Q_2\dot{\theta}_2 + Q_3\dot{\theta}_3$$

from which we conclude that

$$Q_1 = Q_3 = 0, \qquad Q_2 = \Gamma$$

The equations of motion then are

$$[M]\left\{\begin{array}{c}\ddot{\theta}_1 \\ \ddot{\theta}_2 \\ \ddot{\theta}_3\end{array}\right\} + [C]\left\{\begin{array}{c}\dot{\theta}_1 \\ \dot{\theta}_2 \\ \dot{\theta}_3\end{array}\right\} + [K]\left\{\begin{array}{c}\theta_1 \\ \theta_2 \\ \theta_3\end{array}\right\} = \left\{\begin{array}{c}0 \\ \Gamma \\ 0\end{array}\right\}$$

EXAMPLE 1.10

The illustrated equilateral linkage moves in the vertical plane as the result of the application of two excitation forces: F_1 acting vertically and F_2 acting perpendicularly to bar BC. Each bar and the collar have mass m. The static equilibrium position of the system when F is not present is $\beta = 30°$, and the spring is compressed by 1% of its length at this position. Determine in terms of mg, k, and L the equation of motion governing small rotation $\theta = \beta - 30°$ relative to the static equilibrium position.

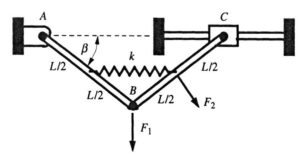

Solution This exercise is intended to exemplify implementation of the stiff spring approximation, as well as the evaluation of generalized forces. It also will give us practice in analyzing linkages. The system we consider is the whole linkage, including the collar. The connection forces exerted between the bars at pin B are internal to this system, so they do not appear in the free body diagram.

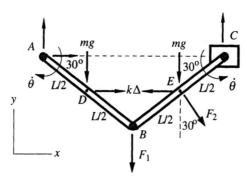

The position of the system is defined by the angle β, so the system has one degree of freedom. The specified generalized coordinate is the incremental rotation angle θ, which is zero when the system is at its static equilibrium position. Bar AB executes a pure rotation about pin A, so it has rotational kinetic energy about that point. Bar BC is in general rotation, so we represent its kinetic energy as the sum of a translational contribution relative to its center of mass E and a rotational part about its center of mass. In addition, collar C translates, so that

$$T = \tfrac{1}{2}I_A\omega_{AB}^2 + \tfrac{1}{2}m\,\bar{v}_E \cdot \bar{v}_E + \tfrac{1}{2}I_G\omega_{BC}^2 + \tfrac{1}{2}m\,\bar{v}_C \cdot \bar{v}_C$$

It is necessary to describe each kinematical quantity in terms of the generalized velocities. The angle of rotation of each bar is θ, in opposite senses, so $\bar{\omega}_{AB} = -\dot{\theta}\bar{k}$, $\bar{\omega}_{BC} = \dot{\theta}\bar{k}$. There are a number of ways in which we could describe $\bar{v}_E$ and $\bar{v}_C$. The simplest is to describe the corresponding position vectors of points as functions of β, differentiate those vectors with respect to time to obtain velocity expressions at arbitrary β, and then evaluate such velocity expressions at the static equilibrium position $\beta = 30°$. The weakness of this approach is that it can become cumbersome for more general linkages. The method we shall follow is a linkage analysis based on the rigid body velocity relations in eq. (1.5.34). We progress through the linkage, starting from the stationary end, so that

$$\bar{v}_B = \bar{\omega}_{AB} \times \bar{r}_{B/A}, \qquad \bar{v}_E = \bar{v}_B + \bar{\omega}_{BC} \times \bar{r}_{E/B}, \qquad \bar{v}_C = \bar{v}_B + \bar{\omega}_{BC} \times \bar{r}_{C/B}$$

We use the static equilibrium position $\beta = 30°$ to describe the position vectors:

$$\bar{r}_{B/A} = L\cos 30°\,\bar{i} - L\sin 30°\,\bar{j}, \qquad \bar{r}_{E/A} = \frac{L}{2}\cos 30°\,\bar{i} + \frac{L}{2}\sin 30°\,\bar{j}$$

$$\bar{r}_{C/A} = L\cos 30°\,\bar{i} + L\sin 30°\,\bar{j}$$

The corresponding velocities are

$$\bar{v}_B = -\dot{\theta}\bar{k} \times (L\cos 30°\,\bar{i} - L\sin 30°\,\bar{j}) = -0.5L\dot{\theta}\bar{i} - 0.866L\dot{\theta}\bar{j}$$

$$\bar{v}_E = -0.5L\dot{\theta}\bar{i} - 0.866L\dot{\theta}\bar{j} + \dot{\theta}\bar{k} \times \left(\frac{L}{2}\cos 30°\,\bar{i} + \frac{L}{2}\sin 30°\,\bar{j}\right) = -0.75L\dot{\theta}\bar{i} - 0.433L\dot{\theta}\bar{j}$$

$$\bar{v}_C = -0.5L\dot{\theta}\bar{i} - 0.866L\dot{\theta}\bar{j} + \dot{\theta}\bar{k} \times (L\cos 30°\,\bar{i} + L\sin 30°\,\bar{j}) = -L\dot{\theta}\bar{i}$$

The corresponding kinetic energy expression is

$$T = \tfrac{1}{2}(\tfrac{1}{3}mL^2)\dot{\theta}^2 + \tfrac{1}{2}m[(-0.75L\dot{\theta})^2 + (-0.433L\dot{\theta})^2] + \tfrac{1}{2}(\tfrac{1}{12}mL^2)\dot{\theta}^2 + \tfrac{1}{2}m(L\dot{\theta})^2$$

$$= \tfrac{1}{2}(3.167mL^2\dot{\theta}^2)$$

For a one-degree-of-freedom system, we have $T = \tfrac{1}{2}M_{11}\dot{\theta}^2$, which when matched to the preceding expression for T gives

$$M_{11} = 3.167mL^2$$

We obtain the stiffness coefficients by constructing the potential energy, which is the sum of the contributions of the spring and gravity. The fact that the spring is compressed by 1% of its unstretched length ℓ_0 means that its length in the equilibrium position is $L_0 = 0.99\ell_0$. This is sufficiently close to ℓ_0 to warrant using the stiff spring approximation. The first step is to describe the velocity at each end of the spring. We already have an expression for $\bar{v}_E$, and we find $\bar{v}_D$ at the static equilibrium position to be

$$\bar{v}_D = \bar{\omega}_{AB} \times \bar{r}_{D/A} = -0.25L\dot{\theta}\bar{i} - 0.433L\dot{\theta}\bar{j}$$

The displacements of these points result from replacing $\dot{\theta}$ by θ in the velocity expressions, so

$$\Delta\bar{r}_D = -0.25L\dot{\theta}\bar{i} - 0.433L\dot{\theta}\bar{j}$$

$$\Delta\bar{r}_E = -0.75L\dot{\theta}\bar{i} - 0.433L\dot{\theta}\bar{j}$$

The unit vector parallel to the spring is $\bar{e}_{E/D} = \bar{i}$, so the elongation is the difference between the $\bar{i}$ components of displacement at the two ends:

$$\Delta = \Delta\bar{r}_E \cdot \bar{i} - \Delta\bar{r}_D \cdot \bar{i} = -0.50L\theta$$

We identify the spring's contribution to K_{11} by matching $V_{sp} = \frac{1}{2}k\Delta^2$ to the standard $V = \frac{1}{2}K_{11}\theta^2$, which leads to

$$(K_{11})_{sp} = 0.25kL^2$$

The other set of conservative forces acting on the system is gravity. We define the datum as the elevation of pin A, and observe that the center of mass of each bar is at the same elevation, so that

$$V_{grav} = -mg\frac{L}{2}[\sin(30° + \theta)](2)$$

We use the first of eqs. (1.5.18) to find the corresponding contribution to the stiffness coefficient:

$$(K_{11})_{grav} = \left(\frac{d^2V}{d\theta^2}\right)_{\theta=0} = mgL\sin 30°$$

The total stiffness coefficient is the sum of the two contributions,

$$K_{11} = 0.25kL^2 + 0.5mgL$$

There are no dashpots in this system, so we set $\mathscr{P}_{dis} = 0$, which means that $C_{11} = 0$. The effect of the excitation forces is described by the power input, $\mathscr{P}_{in} = \bar{F}_1 \cdot \bar{v}_B + \bar{F}_2 \cdot \bar{v}_E$. We previously derived linearized expressions for both velocities. We use the equilibrium position to describe both forces, so we write

$$\mathscr{P}_{in} = -F_1\bar{j} \cdot (-0.5L\dot{\theta}\bar{i} - 0.866L\dot{\theta}\bar{j})$$
$$+ F_2(\sin 30°\,\bar{i} - \cos 30°\,\bar{j}) \cdot (-0.75L\dot{\theta}\bar{i} - 0.433L\dot{\theta}\bar{j})$$
$$= 0.866LF_1$$

Note that $\bar{F}_2$ does not input power to the system because the velocity of its point of application is perpendicular to the force. (Individuals who are familiar with the kinematics of linkages will observe that $\bar{F}_2$ intersects the instant center of bar BC.) For this system, the standard form is $\mathscr{P}_{in} = Q_1\dot{\theta}$, so we have

$$Q_1 = 0.866F_1L$$

The corresponding equation of motion is

$$2.167mL^2\ddot{\theta} + (0.25kL^2 + 0.5mgL)\theta = 0.866F_1L$$

EXAMPLE 1.11

The system in the illustration at the top of the next page is brought to static equilibrium in the position shown, after which vertical force F and torque Γ induce motion. The moments of inertia are I_1 and I_2, and the mass of the bar is m. Friction at the gears and the weight of the bar are negligible. Determine the equations of motion for vibratory response relative to the static equilibrium position.

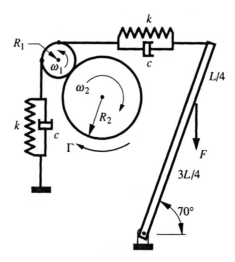

Solution By calling for most of the basic concepts, this exercise serves as a capstone for the derivation of equations by the power balance method. One of the new features here is the incorporation of gear ratios into the formulation. The first task is to identify a set of generalized coordinates. The presence of the parallel spring and dashpot on each cable segment causes the cable lengths connecting bodies to be variable. As a consequence, the gear rotations are not kinematically related to the motion of the bar. However, meshing of gear teeth imposes a kinematical relation on the gears. Thus, we select the rotation ψ of gear 1 and the rotation θ of the bar about its pivot as generalized coordinates, $\{q\} = [\psi\ \theta]^T$. Both are defined to be zero at the static equilibrium position. We need to relate the motion of gear 2 and the cable attachment points to the generalized velocities, so we draw a free body diagram of the system that also shows the appropriate motion parameters.

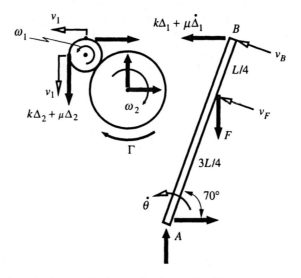

In this sketch, v_1 is the speed of a tooth of gear 1, which must match the gear 2 tooth speed. Thus, $v_1 = \omega_1 R_1 = \omega_2 R_2$. By definition, $\omega_1 = \dot\psi$, so $\omega_2 = \dot\psi(R_1/R_2)$. From this, we find that the kinetic energy in terms of the generalized velocities is

$$T = \frac{1}{2}\left[\left(\frac{1}{3}mL^2\right)\dot\theta^2 + I_1\omega_1^2 + I_2\omega_2^2\right] = \frac{1}{2}\left[\left(\frac{1}{3}mL^2\right)\dot\theta^2 + \left(I_1 + I_2\frac{R_1^2}{R_2^2}\right)\dot\psi^2\right]$$

Matching this to the explicit standard form of T, eq. (1.5.8), with $q_1 = \psi$, $q_2 = \theta$, leads to

$$M_{11} = I_1 + I_2 \frac{R_1^2}{R_2^2}, \qquad M_{22} = \tfrac{1}{3}mL^2, \qquad M_{12} = 0$$

To construct the power dissipation, we need to relate the extension rate of each dashpot to the generalized velocities. Because bar AB is in pure rotation, the velocity of end B is $L\dot\theta$ at 20° above the horizontal. The component of this velocity in the sense of the upper dashpot is $L\dot\theta \cos 20°$, which tends to shorten the dashpot. At the other end, the perimeter velocity v_1 tends to elongate the dashpot. Thus, $\dot\Delta_1$ is the difference between these velocity components. Also, v_1 is the rate at which dashpot 2 is shortening, so

$$\dot\Delta_1 = R_1 \dot\psi - L\dot\theta \cos 20°, \qquad \dot\Delta_2 = -R_1 \dot\psi$$

The corresponding power dissipation is

$$\mathscr{P}_{dis} = c\dot\Delta_1^2 + c\dot\Delta_2^2 = c(R_1 \dot\psi - 0.9397 L\dot\theta)^2 + c(R_1 \dot\psi)^2$$

$$= c[2R_1^2 \dot\psi^2 - 0.8830 L^2 \dot\theta^2 - 2(0.9397 L R_1)\dot\psi\dot\theta]$$

Comparison with the explicit standard form of power dissipation, eq. (1.5.12), shows that

$$C_{11} = 2cR_1^2, \qquad C_{22} = 0.8830cL^2, \qquad C_{12} = C_{21} = -0.9397cLR_1$$

It is stated that the weight of the bar is negligible, so the only conservative forces are those exerted by the springs. There is no need to carry out an analysis relating the elongations to the generalized coordinates, because each spring is parallel to a dashpot. We therefore integrate Δ_1 and Δ_2 with respect to time, which merely involves replacing each generalized velocity with the generalized coordinate, so that

$$\Delta_1 = R_1 \psi - L\theta \cos 20°, \qquad \Delta_2 = -R_1 \psi$$

The corresponding potential energy is

$$V = \tfrac{1}{2}k\Delta_1^2 + \tfrac{1}{2}k\Delta_2^2 = \tfrac{1}{2}k[(R_1 \psi - 0.9397 L\theta)^2 + (R_1 \psi)^2]$$

In view of the similarity of this form to the expression for $\mathscr{P}_{dis}$, we can directly identify the stiffness coefficients:

$$K_{11} = 2kR_1^2, \qquad K_{22} = 0.8830kL^2, \qquad K_{12} = K_{21} = -0.9397kLR_1$$

The last quantity to form is the power input associated with torque Γ and vertical force F. The power input by the torque Γ is $\Gamma\omega_2$, positive because both quantities are in the same rotational sense, so

$$\mathscr{P}_{in} = \Gamma\omega_2 + \bar{F} \cdot \bar{v}_F$$

We now proceed to express each motion variable in terms of the generalized velocities. The point at which $\bar{F}$ is applied has velocity $\bar{v}_F = \tfrac{3}{4}L\dot\theta$ perpendicular to the bar. The component of $\bar{F}$ parallel to $\bar{v}_F$ is $F\cos 70°$ opposite the sense of $\bar{v}_F$. We already have established that $\omega_2 = \omega_1(R_1/R_2)$ and $\omega_1 = \dot\psi$. These considerations lead to

$$\mathscr{P}_{in} = \Gamma\frac{R_1}{R_2}\dot\psi - (F\cos 70°)\left(\frac{3}{4}L\dot\theta\right)$$

The standard form of the power input for this two-degree-of-freedom system is $\mathscr{P}_{in} = Q_1\dot\psi + Q_2\dot\theta$. We identify the generalized forces by matching this to $\mathscr{P}_{in}$ for the system, which leads to

$$Q_1 = \frac{R_1}{R_2}\Gamma, \qquad Q_2 = -0.2565LF$$

The resulting equations of motion for this system are

$$[M]\left\{\begin{array}{c} \ddot{\psi} \\ \ddot{\theta} \end{array}\right\} + [C]\left\{\begin{array}{c} \psi \\ \theta \end{array}\right\} + [K]\left\{\begin{array}{c} \psi \\ \theta \end{array}\right\} = \left\{\begin{array}{c} (R_1/R_2)\Gamma \\ -0.2565LF \end{array}\right\}$$

where the elements of the inertia, damping, and stiffness matrices are as listed.

REFERENCES AND SELECTED READINGS

AISC. 1980. *Manual of Steel Construction*. American Institute of Steel Construction, New York.

COCHIN, I., & PLASS, H. J., JR. 1990. *Analysis and Design of Mechanical Systems*, 2nd ed. Harper & Row, New York.

GINSBERG, J. H. 1995. *Advanced Engineering Dynamics*, 2nd ed. Cambridge University Press, Cambridge, England.

HIBBELER, R. C. 1991. *Mechanics of Materials*, Macmillan, New York.

McGILL, D. J., & KING, W. W. 1995. *An Introduction to Dynamics*. 3rd ed. PWS, Boston.

SEIRIG, A. 1969. *Mechanical Systems Analysis*. International Textbook Company, Scranton, PA.

SHAMES, I. H., & DYM, C. L. 1985. *Energy and Finite Element Methods in Structural Mechanics*. Hemisphere, New York.

SMITH, D. L. 1994. *Introduction to Dynamic Systems Modeling for Design*. Prentice Hall, Englewood Cliffs, NJ.

THOMSON, W. T., & DAHLEH, M. D. 1993. *Theory of Vibration with Applications*, 5th ed. Prentice Hall, Englewood Cliffs, NJ.

EXERCISES

1.1 Determine the spring stiffness that is equivalent to the action of the four springs in the sketch.

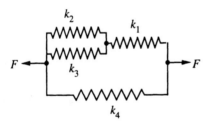

EXERCISE 1.1

1.2 The stiffnesses of the springs shown are $k_1 = 100$ N/m, $k_2 = 150$ N/m, $k_3 = 300$ N/m. The static force $F = 600$ N. Determine the corresponding static displacement of the block.

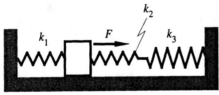

EXERCISE 1.2

1.3 Identical cantilevered elastic beams having flexural rigidity EI are combined with two identical springs k to support the block shown. Determine the static displacement of the block resulting from application of the static force F.

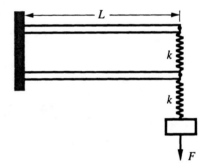

EXERCISE 1.3

1.4 The sketch depicts a model of a two-story building. Four columns (only two are visible), each having flexural rigidity EI, are used to support each floor, which is considered to be rigid. The lower columns are welded at both ends, so they act like beams whose upper end can displace horizontally, but not

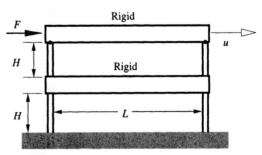

EXERCISE 1.4

rotate. The upper columns are welded at their lower end and pinned at their upper end, so they act like cantilever beams with regard to horizontal displacement of the floors. Determine the equivalent spring stiffness giving the ratio of the static force F to the displacement u.

1.5 Consider motion of the suspended block m in the vertical direction. Determine the equation of motion for this system.

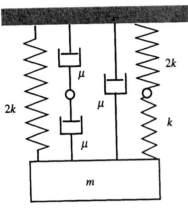

EXERCISE 1.5

1.6 (a) Determine the equations of vertical motion for the spring-mass-dashpot system.

(b) The external forces acting on each block are zero when the system is in static equilibrium. If the stiffness unit is $k = 20$ kN/m and the mass unit is $m = 20$ kg, what is the elongation of each spring in that state?

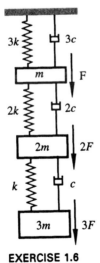

EXERCISE 1.6

1.7 The torsional spring k at the pivot of the bar is undeformed when the bar is in the upright position. Thus, $\theta = 0$ is a static equilibrium position.

The mass of the bar is m. Derive the linearized differential equation of motion governing θ.

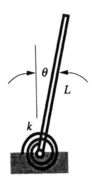

EXERCISE 1.7

1.8 The vertical orientation of the bar constitutes static equilibrium. The bar and the small sphere each have mass m. Derive the equation of motion for this system.

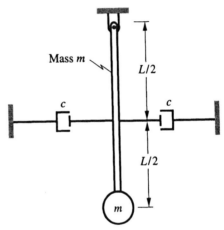

EXERCISE 1.8

1.9 The masses are m for the right bar and $\frac{3}{4}m$ for the left bar. The spring k is unstretched when the bars are oriented vertically, so the static equilibrium position is as shown in the sketch. Derive

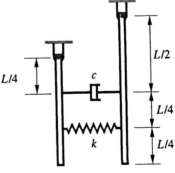

EXERCISE 1.9

linear differential equations governing the angle by which each bar rotates from this position.

1.10 When the system is in static equilibrium, the compressive force in spring 2 is 30 N. The spring stiffnesses are 2 kN/m, 1 kN/m, and 3 kN/m, and the unstretched lengths are 500 mm, 300 mm, and 400 mm for springs 1, 2, and 3, respectively. The masses are $m_1 = 3$ kg, $m_2 = 1.5$ kg.

(a) Determine the values of L_1, L_2, and L_3 at static equilibrium.

(b) Derive equations of motion for the horizontal displacement of each block measured from the static equilibrium position.

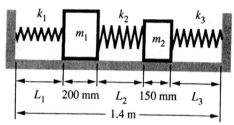

EXERCISE 1.10

1.11 When the system in the sketch is at its static equilibrium position, there is no axial force in each spring.

(a) Derive equations of motion for the horizontal displacements x_1 and x_2 measured from the equilibrium position.

(b) Prove that if $m_2 = 0$, the equation of motion for x_1 is the same as that obtained by replacing the four springs by a single equivalent spring. Identify the way in which the springs are connected—for example, series or parallel—according to the equivalent spring stiffness.

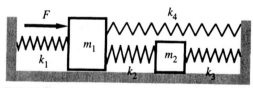

EXERCISE 1.11

1.12 The distances L_1, L_2, and L_3 appearing in the sketch are the unstretched lengths of the respective springs, so x, y, and z are the corresponding spring elongations. The sketch shows that the elongations may be used to locate each block, so they represent a set of generalized coordinates. Determine $[M]$ corresponding to these variables.

1.13 The torsional spring κ acts between gear 3 and the fixed arm to resist rotation. The rotation

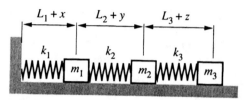

EXERCISE 1.12

rates ω_1, ω_2, and ω_3 are related by the kinematical condition that the velocities at contacting points on two gears must be equal, so this system has one degree of freedom.

(a) Determine the inertia coefficient corresponding to using θ_1 as the generalized coordinate.

(b) Determine the mass coefficient corresponding to using θ_3 as the generalized coordinate.

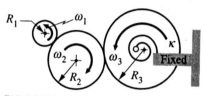

EXERCISE 1.13

1.14 The cylinder, whose mass is m_1, rolls without slipping under the restraint of a pair of springs. In addition, a small mass m_2 is attached to the cylinder at the top center position as shown. This position corresponds to static equilibrium. Determine the inertia coefficient corresponding to using as a generalized coordinate the horizontal displacement x of the cylinder's center.

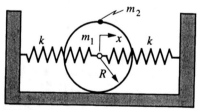

EXERCISE 1.14

1.15 A block of mass $3m$ is connected by a cable to a rigid bar whose mass is m. The bar executes small rotations relative to the horizontal orientation depicted in the sketch, which is the static equilibrium position when the disturbance force F is not present. Because the cable is inextensible, the angle of rotation θ of the bar defines the position of all points, so the system has one degree of freedom. Determine M_{11} corresponding to $q_1 = \theta$.

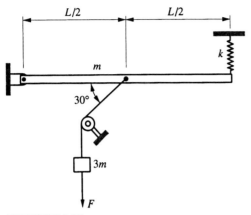

EXERCISE 1.15

1.16 The bar executes small rotations in the vertical plane relative to the static equilibrium position depicted in the sketch. Let the rotation of the bar be the generalized coordinate. Determine the damping coefficient C_{11}.

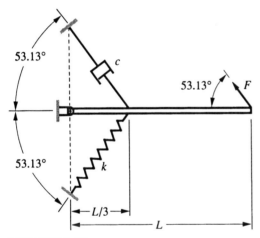

EXERCISE 1.16

1.17 The position depicted in the sketch corresponds to static equilibrium of the coupled pendu-

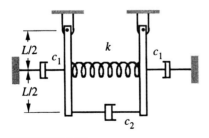

EXERCISE 1.17

lums. Generalized coordinates are selected to be the rotations of the respective bars. Determine the corresponding $[C]$.

1.18 The horizontal orientation of the bar depicted in the sketch corresponds to static equilibrium when the excitation force F is not present. Generalized coordinates are the displacement of the bar's center of mass and the rotation of the bar. Determine the corresponding $[C]$.

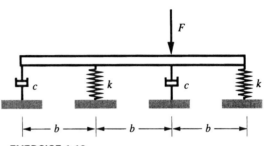

EXERCISE 1.18

1.19 Generalized coordinates for the system in Exercise 1.17 are the rotation of each bar from the vertical equilibrium position. The mass of each bar is m. Determine the corresponding $[K]$.

1.20 The diagram shows a piece of electronic equipment, whose mass is m, that is mounted on a lightweight flexible beam supported by springs k at each end. Only vertical translation of the equipment is to be considered, so the generalized coordinate for the system is the vertical displacement y of the equipment. A steel construction manual indicates that if k were infinite, the beam would displace downward by $mgL^3/(48EI)$ when the equipment is gently placed on the beam. In the case where $k = 12EI/L^3$, what is the stiffness coefficient associated with y?

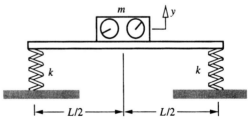

EXERCISE 1.20

1.21 When the block is in the position shown in the sketch, the system is in static equilibrium. Determine the stiffness coefficient corresponding to using the collar's displacement from this position as the generalized coordinate.

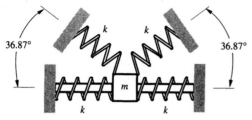

EXERCISE 1.21

1.22 The static equilibrium position of the rigid bar is the inclined orientation shown in the sketch. The mass of the bar is m_1, while m_2 is the suspended mass. Assume that m_2 only displaces vertically, so that a suitable set of generalized coordinates are the rotation θ of the bar and the vertical displacement y of block m_2. Determine the corresponding $[K]$.

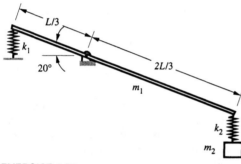

EXERCISE 1.22

1.23 The generalized coordinate for the system in Exercise 1.15 is the angle of rotation of the bar. Determine the corresponding generalized force.

1.24 A beam is suspended by two shock absorbers each consisting of a parallel spring and dashpot. The system was in a state of equilibrium with the beam horizontal when the point at which the right shock absorber is attached to the ceiling begins to displace upward in a known manner $z(t)$. Determine $\{Q\}$ corresponding to using as generalized

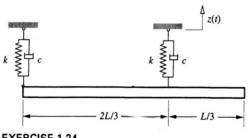

EXERCISE 1.24

coordinates the upward displacement of the center of mass of the beam and the rotation of the beam.

1.25 The sketch depicts two identical drums for rolling steel sheets, with the springs representing the elasticity of the sheets. A force F at distance r from the center of the right gear is applied at angle β from the radial line, as shown. Determine the generalized forces corresponding to using the rotation of each drum as a generalized coordinate.

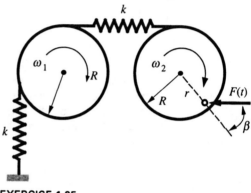

EXERCISE 1.25

1.26 Consider the system of two collars in Figure 1.10. Use the power balance method to determine the equations of motion corresponding to using as generalized coordinates:

(a) The displacements x_1 and x_2.

(b) The spring extensions y_1 and y_2.

(c) Suppose that spring k_2 is replaced by two springs k_2 and k_3 in series. How would this modify the analysis in part (a)?

1.27 A standard model for a wing that is often used for experiments has a translational spring k_Y and a torsional spring k_T representing the elastic rigidity. Point E represents the elastic center, because static application of a vertical force at that point results in upward displacement without an associated rotation. The design of the springs is such that horizontal movement of point E is negligible. The lift force L acts at point P, which is called the center of pressure. This force may be treated as a known excitation, even though it depends on the angular orientation of the wing relative to the direction of the airflow. Point G is the center of mass, and the radius of gyration about that point is r_G. When the wing is in its static equilibrium position, points G, E, and P form a horizontal line. Derive the equations of motion for the wing.

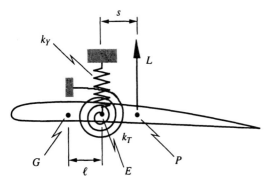

EXERCISE 1.27

1.28 Use the power balance method to derive the equation of motion for the system in Exercise 1.8.

1.29 Use the power balance formulation to determine the equations of vertical motion for the spring-mass-dashpot system in Exercise 1.6.

1.30 Both bars in the linkage are horizontal, as shown, when the system is in static equilibrium. Determine the linearized equations of motion for this system.

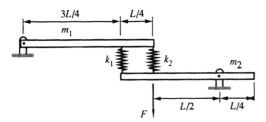

EXERCISE 1.30

1.31 Determine the equations of motion for the system in Exercise 1.18.

1.32 The system shown in the sketch moves in the horizontal plane. Both springs are undeformed when the system is in the position shown, so this position corresponds to static equilibrium when the excitation force $F(t)$ is not present. Identify generalized coordinates that locate the system's position relative to the equilibrium position, and derive the corresponding equations of motion.

1.33 Determine the equations of motion governing a pair of generalized coordinates that locate the position of the cart and the sliding block. Friction is negligible.

1.34 Use the power balance method to derive the equation of motion for the system in Example 1.3.

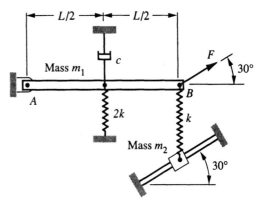

EXERCISE 1.32

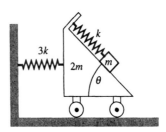

EXERCISE 1.33

1.35 Determine the equations of motion for the system in Exercise 1.25. Let I_0 denote the moment of inertia of a drum about its rotation axis.

1.36 The moment of inertia of the stepped pulley about its bearing is I_0. Cables wrapped around the inner and outer drums connect the pulley through a set of springs and despots to the wall and the suspended mass. Motion is caused by application of torque Γ, and horizontal displacement of the suspended mass is negligible. Determine the equations of motion for this system.

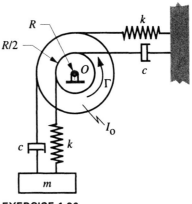

EXERCISE 1.36

1.37 Use the power balance method to derive the equation of motion for the system in Exercise 1.7.

1.38 The potential energy stored in each torsional spring is $V = \frac{1}{2}\kappa\Delta\theta^2$, where $\Delta\theta$ is relative rotation between the attached bodies. Neither spring is deformed when the bars are vertical, and $\kappa = \beta mgL$ for each spring, where β is a dimensionless parameter and m is the mass of each bar. Determine the equations of motion for small rotations of this system away from the vertical equilibrium position.

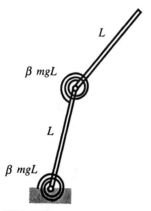

EXERCISE 1.38

1.39 The small sphere is suspended by identical springs. Static equilibrium corresponds to $\theta = 35°$. The spring stiffness is sufficiently large that the stiff spring approximation applies. Determine the linear equations of motion for the system.

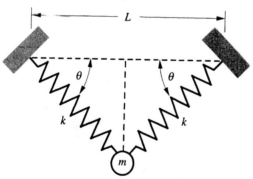

EXERCISE 1.39

1.40 When force F is not present, the static equilibrium position of the parallelogram linkage corresponds to $\theta = 60°$. The mass of each link is m. The system moves in the vertical plane, and the stiffness k is sufficiently large to permit application of the stiff spring approximation. Determine the equation of motion governing small rotation θ

of the angled links from their 60° orientation at equilibrium.

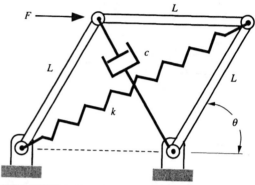

EXERCISES 1.40, 1.41

1.41 When force F is not present, the static equilibrium position of the parallelogram linkage corresponds to $\theta = 60°$. The mass of each link is m, and $k = mg/L$. This stiffness is sufficiently low that the stiff spring approximation is not valid.

(a) Determine the unstretched length of the spring.

(b) Determine the equation of motion corresponding to small movements relative to the static equilibrium position when F is present.

1.42 The total mass of the angle bar is m, and the cross-section of the bar is constant. The springs attached to the bar are undeformed in the position depicted in the sketch. Determine the equation of motion governing small rotation of the bar from this position. The system lies in the horizontal plane.

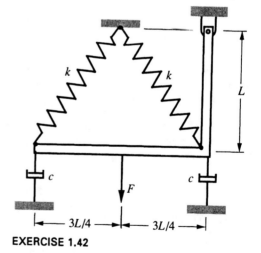

EXERCISE 1.42

1.43 A cylinder of radius r rolls without slipping in the interior of a cylindrical cavity whose radius is $5r$. The generalized coordinate for the system is selected to be the polar angle θ for the radial line from the center of the cavity to the center of the cylinder. Determine the equation of motion, assuming that θ is always a small value.

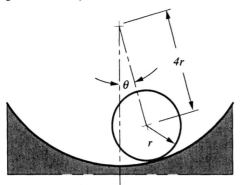

EXERCISE 1.43

1.44 The semi cylinder rolls without slipping over the horizontal surface. Determine the linearized equation of motion governing the angle of rotation relative to its static equilibrium position.

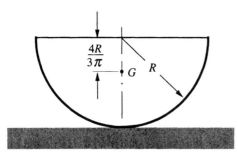

EXERCISE 1.44

1.45 Determine the equation of motion for the system in Exercise 1.15.

1.46 The sketch depicts a crude model of a truss structure, in which elastic bars are connected by pin joints. The elasticity of each bar is represented as a spring connected between the joints. A mechanics of materials analysis shows that the spring constant of a bar having length ℓ_j is $k_j = EA/\ell_j$, where E is Young's modulus and A is the cross-section. The value of EA is the same for all bars. Gravity is negligible compared with the forces generated within the bars. The inertia effects of the bars are represented by lumped masses m_1 and m_2 at the movable joints 1 and 2. Generalized coordinates for each mass is its horizontal and vertical displacement components x_n and y_n relative to the position shown. Determine the equations of motion.

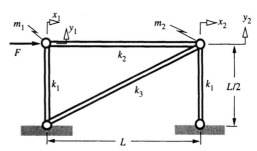

EXERCISE 1.46

1.47 The position shown in the sketch is the system's static equilibrium position when the force F is not present. The spring stuffiness k_1 and k_2 are much greater than $m_1 g/L$ and $m_2 g/L$. Determine the equations of motion for this system.

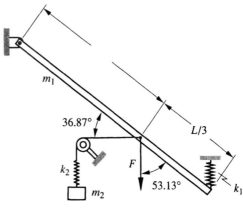

EXERCISE 1.47

1.48 The block is attached to bar AB by a cable that passes over an ideal pulley. The mass of bar AB equals the mass of the block. The system moves in the vertical plane, and the position

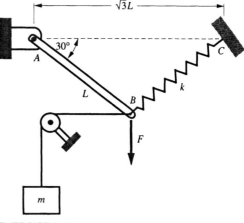

EXERCISE 1.48

depicted in the drawing is the static equilibrium position of the system when the excitation force F is not present. The unstretched length of the spring is 0.99L. Derive the linearized equation of motion governing rotation of bar AB relative to its static equilibrium position.

1.49 The sketch is a model of a suspension system for testing the launch vehicle of earth satellites. Generalized coordinates are the displacements of the center of mass x and y, and the rotation θ. The moment of inertia of the launcher about its center of mass is I_G. Derive the equations of motion.

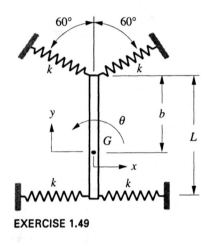

EXERCISE 1.49

1.50 When the vertical force F is not applied, the static equilibrium position corresponds to the orientation depicted in the sketch. The mass per unit length of each bar is m/L. Derive the equation of motion governing a small rotation θ of a bar relative to its equilibrium position.

1.51 The unstretched length of the springs in Exercise 1.39 is L, so the stiff spring approximation is not applicable. The static equilibrium position corresponds to $\theta = 36.87°$.

(a) Derive an expression for k in terms of mg and L corresponding to this equilibrium position.

(b) Derive the corresponding equations of motion.

1.52 Consider the bar in Exercise 1.16 in the case where the instructed length ℓ_0 of the spring is sufficiently different from the spring's length in the equilibrium position to invalidate the stiff spring approximation.

(a) Determine, in terms of the bar's length L, the unstretched length ℓ_0, and the other parameters, the linearized equation of motion governing the rotation θ of the bar.

(b) Consider the static equilibrium condition to derive an expression for ℓ_0.

1.53 Consider the equilateral linkage in Example 1.10 for the situation where the unstretched length ℓ_0 differs significantly from the length of the spring in the static equilibrium position.

(a) Determine, in terms of k, mg, and L, the unstretched length ℓ_0 for which $\beta = 30°$ is the static equilibrium position.

(b) If ℓ_0 has the value found in part (a), what is the stiffness coefficient K_{11} corresponding to the rotation angle θ being the generalized coordinate?

(c) For what range of the ratio kL/mg would the stiff spring approximation result in a stiffness coefficient K_{11} that is no more than 1% in error from the true value?

1.54 The linkage moves in the vertical plane under the action of force $\bar{F}$, which remains perpendicular to the bar. The static equilibrium of the system is $\theta = 36.87°$ when the force $\bar{F}$ is not

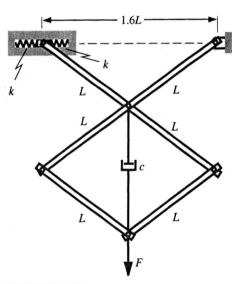

EXERCISE 1.50

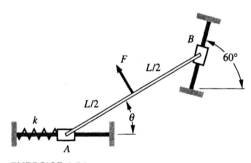

EXERCISE 1.54

applied to bar *AB*. The mass of the bar is *m,* and the mass of collars *A* and *B* is negligible. Derive a linear equation of motion describing rotation of bar *AB* relative to its equilibrium orientation. It is permissible to use the stiff spring approximation.

1.55 The linkage moves in the horizontal plane. The spring is undeformed when the system is in the illustrated position. Motion is initiated by application of the disturbance force *F,* which acts perpendicularly to bar *BD*. The masses are *m* and

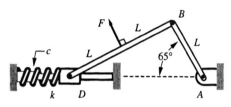

EXERCISE 1.55

2*m* for bars *AB* and *BD*, respectively, and the mass of collars *D* is negligible. Derive a linear equation of motion describing rotation of bar *AB* relative to its equilibrium orientation.

1.56 Cylinder *A*, whose mass is 2*m*, rolls without slipping over the horizontal surface. The mass of bar *AB* is *m*, and the mass of the piston *C* is negligible. The system is at its static equilibrium position, where $\theta = 0$, when the horizontal force *F* is

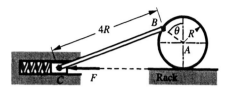

EXERCISE 1.56

applied to the piston. Determine the equation of motion governing θ in the resulting motion.

1.57 Solve Exercise 1.56 for the case where static equilibrium corresponds to $\theta = 90°$.

1.58 When the linkage occupies the illustrated position, neither spring is deformed. Gravity has negligible effect in comparison with the excitation resulting from application of the torque *M*. The masses are *m* for bar *CD* and 2*m* for bar *AB*. Derive the equations of motion governing small movements of the linkage relative to the illustrated position.

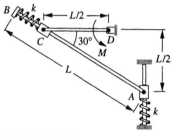

EXERCISE 1.58

CHAPTER 2

TRANSIENT RESPONSE OF ONE-DEGREE-OF-FREEDOM SYSTEMS

Given the simplicity of a one-degree-of-freedom model, it might seem surprising that such a representation is widely used to explore real-world systems. In part, the significance of this model lies in the fact that it captures many of the fundamental physical phenomena manifested in vibrations of all systems. One-degree-of-freedom models also are important because modal analysis methods, which are essential to the study of complicated systems, convert multiple-degree-of-freedom systems to an equivalent set of one-degree-of-freedom systems.

A vibratory response might be a free vibration stemming from some set of initial conditions, or a forced response resulting from dynamic excitation. In this chapter we consider a variety of forces whose basic time signature changes with time, in some cases disappearing entirely. We refer to the response to such forces, as well as free vibration response, as *transient*, because the response we observe evolves as time elapses.

We begin our study by developing an extremely useful mathematical tool. Sinusoidal-like fluctuation is a common feature of many vibratory phenomena. The application of complex variable concepts substantially simplifies analytical and computational tasks involving sinusoidal functions.

2.1 HARMONIC FUNCTIONS

Harmonic time dependence is a synonym for sinusoidal variation. The term *harmonic* arises from music, where pure tones vary sinusoidally. In engineering applications, harmonic variation is a hallmark of alternating current and electromagnetic waves. Mechanical and structural systems are excited harmonically by rotating machinery, as we will see. Numerous other excitations may be represented by either a single harmonic term or a sum of such terms. Correspondingly, harmonic features of vibratory response arise in a variety of situations.

2.1.1 Basic Properties

When we say that a function is harmonic, or sinusoidal, it need not vary as a sine function. Figure 2.1 displays a typical harmonic function $u(t)$. Its mathematical form is

$$\boxed{u = A\sin(\omega t - \phi)} \tag{2.1.1}$$

The coefficient A is the *amplitude*. The *frequency* of u is ω; if t is measured in units of seconds, then ω has units of radians per second. The argument of a sinusoidal function, that is, $\omega t - \phi$, indicates the *phase*—for example, whether the sinusoidal

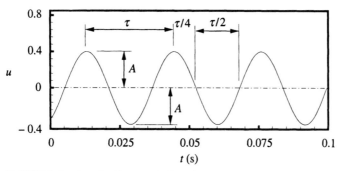

FIGURE 2.1 A typical harmonic function.

function is positive or negative, or whether it is close to a maximum or a zero. The *phase angle* is ϕ, which has units of radians.

The aforementioned parameters are manifested by the pattern in Figure 2.1. The amplitude is the peak excursion of u from the zero value, either positively or negatively, so that $A = \max|u|$. The frequency is directly related to the period τ, which is the time interval over which u repeats, such that $u(t + \tau) = u(t)$. The period appears in the figure as the time interval separating consecutive minima or maxima of u. If the pattern of u versus t is given, as it would be in an oscilloscope trace, then the most accurate value of τ would be obtained by measuring the time interval between adjacent zeroes, which would be $\tau/2$. This is so because it is difficult to identify precisely where a maximum occurs, but the zeroes are readily identified. To construct the relation between ω and τ, we observe that repetition of a sinusoidal function corresponds to an increase of the argument by 2π. The arguments at two instants separated by a period are $\omega t - \phi$ and $\omega(t + \tau) - \phi$. We take the difference and equate it to 2π, which leads to

$$\omega = \frac{2\pi}{\tau} \qquad (2.1.2)$$

It is standard practice to describe the frequency in units of hertz (Hz), which represents the number of cycles (that is, periods) that occur in a one-second interval. If τ seconds are required for one period, then $1/\tau$ periods occur in one second. We shall use the symbol f to denote *cyclical frequency* measured in hertz, so we have

$$f = \frac{1}{\tau} = \frac{\omega}{2\pi} \qquad (2.1.3)$$

Note that ω, rather than f, is the quantity to be used in any computations.

To understand the role of the phase angle consider the case where $\phi = 0$, so that the graph of u is a sine curve. The first zero would occur at $t = 0$, and the first maximum would occur at $t = \pi/2\omega$. When ϕ is nonzero, $u = 0$ when the phase $\omega t - \phi = 0$, which corresponds to $t = \phi/\omega$. Similarly, u has a maximum value when $\omega t - \phi = \pi/2$, which gives $t = \pi/2\omega + \phi/\omega$. Indeed, if $\phi > 0$, any feature of a sine function that occurs at instant t is displayed by u at a later time $t' = t + \phi/\omega$. The quantity ϕ/ω represents a time delay. We say that u *lags* relative to a sine by a time ϕ/ω, and ϕ is the *phase lag*. If ϕ were negative, we would say that u *leads* a sine, and $-\phi$ is the *phase lead*. It is common to describe a phase angle in degrees, but radians is the only acceptable unit for computations.

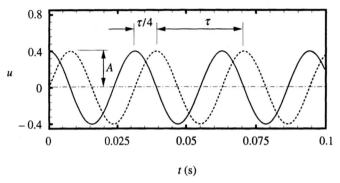

FIGURE 2.2 Phase delay of a sine relative to a cosine.

The phase angle is only meaningful if its reference function is specified. The reference is usually a sine or cosine function without a phase angle. We refer to such functions as a *pure sine* or *pure cosine*, respectively. For example, suppose that we wish to use a cosine function to describe $u(t)$ in Figure 2.1. The amplitude and period do not depend on whether we use a sine or cosine, so the plotted function would fit $u = A\cos(\omega t - \phi')$. The question is, What is ϕ' in terms of ϕ? A simple answer comes from matching the two forms at $t = 0$, which leads to $\sin(-\phi) = \cos(-\phi')$, that is, $-\sin(\phi) = \cos(\phi')$. This has multiple roots; we select $\phi' = \phi + \pi/2$. This choice is suggested by Figure 2.2, which shows that a sine function may be pictured as being delayed by $\tau/4$ relative to a cosine function. In other words, the phase lag of a sine relative to a cosine is 90°, or equivalently, the phase lead of a cosine relative to a sine is 90°. In a similar vein, we could say that the negative of a sine or cosine lags (or leads) the corresponding positive function by 180°.

EXAMPLE 2.1

Measurement of a harmonic function $F(t)$ leads to the observation that the maximum value of the function is 2 kN, that the elapsed time from a maximum to the first following zero value is 0.2 s, and that the earliest time $t > 0$ at which $F = 0$ and $\dot{F} > 0$ is 0.3 s. Determine the functional form of $F(t)$.

Solution This problem will enhance our familiarity with the fundamental properties of harmonic functions. We begin by noting that the elapsed time from maximum F to zero for a harmonic function is one quarter of the period, so $T = 4(0.2)$. Hence, the cyclical and circular frequencies are

$$f = \frac{1}{0.8} = 1.25 \text{ Hz} \quad \Rightarrow \quad \omega = 2.5\pi \text{ rad/s}$$

The standard form of a harmonic function is $F = A\sin(\omega t - \phi)$. The amplitude A is the maximum value, so $A = 2000$ N. To determine ϕ, we use the fact that the instant at which a harmonic function is zero and increasing corresponds to a zero value for the argument of the function. It is given that this instant is $t = 0.3$ s. Hence, it must be that

$$\omega(0.3) - \phi = 0 \quad \Rightarrow \quad \phi = 0.75\pi$$

The corresponding function is

$$F = 2000\sin(2.5\pi t - 0.75\pi) \text{ N}$$

2.1.2 Complex Variable Representation

The process of converting phase angles between sine and cosine functions is the first of many tasks that require the use of trigonometric identities. A more difficult one involves adding terms that have the same frequency, but different amplitudes and phase angles. We will use complex exponentials, rather than real functions, to describe harmonically varying quantities. Doing so will allow us to perform all operations with only a few identities. Furthermore, the use of complex exponentials will drastically simplify solving the differential equations of motion.

The foundation for the procedure is Euler's formula,

$$\exp(i\omega t) = \cos(\omega t) + i\sin(\omega t) \tag{2.1.4}$$

This follows from the definition of the cosine and sine functions in the complex plane,

$$\cos(\omega t) = \frac{1}{2}[\exp(i\omega t) + \exp(-i\omega t)]$$

$$\sin(\omega t) = \frac{1}{2i}[\exp(i\omega t) + \exp(-i\omega t)] \tag{2.1.5}$$

These definitions may always be used to replace the trigonometric functions. However, it is awkward and repetitive to carry around the second part of each definition, because it is merely the complex conjugate of the first part. For most operations, it is simpler to extract the desired function from eq. (2.1.4). When we wish to extract the cosine function from the complex exponential in eq. (2.1.4), we take the real part. To extract the sine function, we could extract the imaginary part, but doing so might lead to difficulties if we must combine terms, some of which are real parts and others are imaginary parts. We therefore will *make it standard practice to always use real parts.* Thus, to extract the sine function, we divide the complex exponential by i, and then take the real part. In other words

$$\cos(\omega t) = \text{Re}[\exp(i\omega t)]$$

$$\sin(\omega t) = \text{Re}\left[\frac{1}{i}\exp i\omega t\right] \equiv \text{Re}[-i\exp i\omega t] \tag{2.1.6}$$

Equivalent forms based on using $\exp(-i\omega t)$ are used by some practitioners. For this reason it is important to examine any treatment using complex functions to ascertain which convention (plus or minus sign) has been adopted.

Now consider the function u described by eq. (2.1.1). The argument of the sine must be the argument of the complex exponential, and the amplitude A is real, so it may be brought inside the bracket. We therefore have

$$u = A\sin(\omega t - \phi) = \text{Re}\left[\frac{A}{i}\exp(i\omega t - i\phi)\right] \tag{2.1.7}$$

We now use the property that the exponential of a sum is the product of the individual exponentials to rewrite the foregoing as

$$u = \text{Re}\left\{\frac{A}{i}\exp(-i\phi)\exp(i\omega t)\right\} \tag{2.1.8}$$

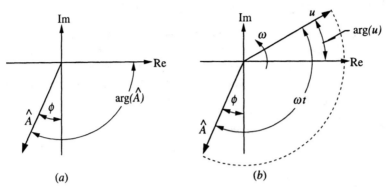

FIGURE 2.3 (a) Complex amplitude of a sine function with phase lag ϕ. (b) Complex plane representation of a sine function with phase lag ϕ.

The factor multiplying $\exp(i\omega t)$ is a complex constant, which we call the *complex amplitude*, $\hat{A}$. To plot this quantity in the complex plane, we use the polar representation. Because the polar form of i is $\exp(i\pi/2)$, we have

$$u = \mathrm{Re}[\hat{A}\exp(i\omega t)], \quad \hat{A} = \frac{A}{i}\exp(-i\phi) = A\exp\left[-i\left(\phi + \frac{\pi}{2}\right)\right] \quad (2.1.9)$$

Thus, the magnitude of $\hat{A}$ is A, and the polar angle of $\hat{A}$, which is called the *argument*, is $\phi + \pi/2$ below the positive real axis, as shown in Figure 2.3(a),

$$|\hat{A}| = A, \qquad \arg(\hat{A}) = -\phi - \frac{\pi}{2} \quad (2.1.10)$$

To construct the representation of u in the complex plane, we note that $|\exp(i\omega t)| = 1$, which means that we obtain u by merely adding the argument ωt to the argument of $\hat{A}$, as shown in Figure 2.3(b). This construction shows that using complex variables to represent a harmonic function at any instant is equivalent to depicting it as a rotating vector in the complex plane. Because the value of ωt increases linearly with elapsed time, the vector representing a complex function rotates counterclockwise at angular speed ω as t increases.

EXAMPLE 2.2

Write the following functions as complex exponentials. Express the corresponding complex amplitude in polar and rectangular form.

$$F = 5\cos(50t + 0.4), \quad G = 20\sin(10t - 0.5), \quad H = 30\sin\left(400t - \frac{2\pi}{3}\right)$$

Solution This exercise highlights the steps by which we convert harmonic functions to complex form. The basic idea is to apply the representations of sine and cosine functions in terms of complex functions, as given by eqs. (2.1.6), and then to match the converted form to the standard complex representation of a harmonic function,

$$F = \mathrm{Re}[A\exp(i\omega t)]$$

For the first function, we use the fact that a cosine is the real part of a complex exponential function, so that

$$F = 5\cos(50t + 0.4) = \text{Re}\{5\exp[i(50t + 0.4)]\}$$
$$= \text{Re}[5\exp(0.4i)\exp(i50t)]$$

We match this to the standard form, from which we conclude that

$$A = 5\exp(0.4i) = 4.605 + 1.9547i$$

A sine function is the imaginary part of a complex exponential, so the second function is

$$F = 20\sin(10t - 0.5) = \text{Re}\left\{\frac{20}{i}\exp[i(10t - 0.5)]\right\}$$

$$= \text{Re}\left[\frac{20}{i}\exp(-0.5i)\exp(i10t)\right]$$

This matches the standard representation of a harmonic function if the complex amplitude is

$$A = \frac{20}{i}\exp(-0.5i) = 20\exp(-0.5\pi i - 0.5i) = -9.589 - 17.552i$$

We follow the same procedure for the third of the given functions,

$$F = 30\sin\left(400t - \frac{2\pi}{3}\right) = \text{Re}\left\{\frac{30}{i}\exp\left[i\left(400t - \frac{2\pi}{3}\right)\right]\right\}$$

$$= \text{Re}\left[\frac{30}{i}\exp\left(-\frac{2\pi}{3}i\right)\exp(i400t)\right]$$

Matching this to the standard form gives

$$A = \frac{30}{i}\exp\left(-\frac{2\pi}{3}i\right) = 30\exp\left(-0.5\pi i - \frac{2\pi}{3}i\right) = -25.98 + 15i$$

As a sidebar to assist readers who are not comfortable with performing computations with complex numbers, let us review a few fundamental techniques. Many of the operations can be implemented directly on a calculator that recognizes complex numbers, but software like MATLAB and Mathcad offers an advantage, in that it retains a record of what one has done. In MATLAB, complex constants may be entered by writing them in the conventional manner, for example, $3 + 4i$ or $3 + 4/i$. If we wish to use variables in a similar manner, multiplication by i requires a multiplication sign, for example, $x + y*i$. Similarly, complex exponentials are obtained by following the written form, using the "exp" function. Complex conjugates come into play in some circumstances. The best strategy here is to use the "conj" function in MATLAB, rather than the prime operator ($'$), which also performs a transpose when applied to matrices. An important operation is conversion of complex numbers between polar and rectangular form. The latter is the internal format of a complex number. To find the magnitude of a complex number z, we write *mag* (z), while *angle* (z) gives the polar angle in radians relative to the positive real axis. Note that radians is also the angle measure that must be used for the exponential function.

Most of the preceding considerations also apply to Mathcad. One difference is that i must be accompanied by a numerical factor in all contexts if it is to be interpreted as $i = \sqrt{-1}$. Thus, we would write $x + y*1i$. Another difference is that the operation of finding the polar form of a complex number is achieved by writing |z| to determine the magnitude, and $arg(z)$ to obtain the polar angle.

2.1.3 Algebraic Operations

The complex exponential form simplifies many operations involving harmonic functions. For example, consider the earlier situation where we were given $u = A\sin(\omega t - \phi)$, and we wished to convert the expression to $u = A\cos(\omega t - \phi')$. Using a complex exponential to represent each form leads to

$$\frac{1}{i}\exp(i\omega t - i\phi) = \exp(i\omega t - i\phi') \tag{2.1.11}$$

Recall that $1/i = \exp(-i\pi/2)$. The equality must apply at all instants t, so we may cancel the common factor $\exp(i\omega t)$ on both sides. This yields

$$\exp(-i\pi/2)\exp(-i\phi) = \exp(-i\phi') \quad \Rightarrow \quad \phi' = \phi + \frac{\pi}{2} \tag{2.1.12}$$

which matches what we had deduced using real functions.

The notion that a complex function may be represented as a vector suggests that different functions at the same frequency may be added as vectors, based on a pictorial representation of the parallelogram law. Rather than doing so, we shall follow an algebraic procedure, which factors out the shared $\exp(i\omega t)$ dependence. For example, consider representing a harmonic function u that is known to be the sum of two other harmonics at the same frequency,

$$\begin{aligned} u &= \text{Re}[\hat{A}\exp(i\omega t)] \\ &= A_1\sin(\omega t - \phi_1) + A_2\cos(\omega t - \phi_2) \end{aligned} \tag{2.1.13}$$

To determine the value of the complex amplitude $\hat{A}$ given A_1, A_2, ϕ_1, and ϕ_2, we use eqs. (2.1.6) to represent the sine and cosine as complex exponentials,

$$\text{Re}[\hat{A}\exp(i\omega t)] = \text{Re}\left[\frac{A_1}{i}\exp(i\phi_1)\exp(i\omega t)\right] + \text{Re}[A_2\exp(i\phi_2)\exp(i\omega t)] \tag{2.1.14}$$

$$= \text{Re}\left\{\left[\frac{A_1}{i}\exp(i\phi_1) + A_2\exp(i\phi_2)\right]\exp(i\omega t)\right\}$$

In order for the real parts to match at all t, the complex coefficients of $\exp(i\omega t)$ must match, which leads to

$$\hat{A} = \frac{A_1}{i}\exp(-i\phi_1) + A_2\exp(i\phi_2) \tag{2.1.15}$$

If we have values for each A_j and ϕ_j, we may evaluate $\hat{A}$ numerically using a calculator or computer software. If the quantities are algebraic, we proceed by converting the terms in the preceding from polar to rectangular form,

$$\begin{aligned} \hat{A} &= \frac{A_1}{i}[\cos(\phi_1) - i\sin(\phi_1)] + A_2[\cos(\phi_2) + i\sin(\phi_2)] \\ &= [-A_1\sin(\phi_1) + A_2\cos(\phi_2)] + i[-A_1\cos(\phi_1) + A_2\sin(\phi_2)] \end{aligned} \tag{2.1.16}$$

If we wish, we may convert this rectangular representation of $\hat{A}$ to polar form using $\hat{A} = A\exp(-i\phi) = A\cos(\phi) - iA\sin(\phi)$. Matching real and imaginary parts of the two forms for $\hat{A}$ leads to

$$\begin{aligned} A\cos(\phi) &= -A_1\sin(\phi_1) + A_2\cos(\phi_2) \\ -A\sin(\phi) &= -A_1\cos(\phi_1) + A_2\sin(\phi_2) \end{aligned} \tag{2.1.17}$$

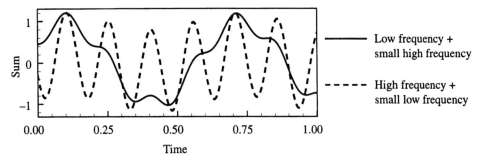

FIGURE 2.4 Summation of two harmonic functions at different frequencies.

We may determine the magnitude A of the complex amplitude by summing the squares of each of the above, while their ratio gives tan (ϕ). Note that in applying the arctangent function to the latter, we must be careful to place ϕ in the quadrant that is consistent with each of eqs. (2.1.17). Also note that the form of u corresponding to the polar representation of $\hat{A}$ is $u = \text{Re}[A \exp(i\omega t - i\phi)]$, which means that A is the (real) amplitude and ϕ is the phase lag of u relative to a cosine function.

Many situations lead to summations of harmonic terms having different frequencies. Let us consider the case where two harmonics at frequencies ω_1 and ω_2 are combined. Two typical situations are shown in Figure 2.4. If $\omega_2 \gg \omega_1$ and the amplitude of the higher frequency component is much smaller than that of the lower frequency component, the sum appears to be the lower frequency term with a superimposed high-frequency fluctuation. On the other hand if the low-frequency component has the smaller amplitude, then the alteration of the high-frequency signal appears to result in slow fluctuation of the amplitude of the high-frequency signal.

In general, not much can be done to simplify the mathematical representation of a combination of two harmonics. An exception, which we will encounter later in this chapter, arises when the amplitudes of the two terms are equal. Let us consider the sum of two harmonic functions at frequencies ω_1 and ω_2, whose phase lags relative to cosine functions are ϕ_1 and ϕ_2, respectively. We replace each harmonic function with its definition in terms of complex exponentials. Note that in this step, we do not use the real part notation because we wish to combine terms at different frequencies. Thus, we write

$$u = A\cos(\omega_1 t - \phi_1) + A\cos(\omega_2 t - \phi_2)$$

$$= \tfrac{1}{2}A\{\exp[i(\omega_1 t - \phi_1)] + \exp[-i(\omega_1 t - \phi_1)]\}$$

$$+ \tfrac{1}{2}A\{\exp[i(\omega_2 t - \phi_2)] + \exp[-i(\omega_2 t - \phi_2)]\} \qquad (2.1.18)$$

To combine these terms we define ω_{av} and ϕ_{av} to be the average frequency and phase lag, while Δ_ω and Δ_ϕ are the deviations from the average values,

$$\omega_{av} = \tfrac{1}{2}(\omega_1 + \omega_2), \qquad \Delta_\omega = \tfrac{1}{2}(\omega_2 - \omega_1)$$

$$\phi_{av} = \tfrac{1}{2}(\phi_1 + \phi_2), \qquad \Delta_\phi = \tfrac{1}{2}(\phi_2 - \phi_1) \qquad (2.1.19)$$

We use these definitions to replace the absolute frequencies and phase angles in eq. (2.1.18). For instance, we have $\omega_1 = \omega_{av} - \Delta\omega$ and $\omega_2 = \omega_{av} + \Delta_\omega$. After some manipulation, we find that

$$u = \tfrac{1}{2}A \exp[i(\omega_{av}t - \phi_{av})]\{\exp[-i(\Delta_\omega t - \Delta_\phi)] + \exp[+i(\Delta_\omega t - \Delta_\phi)]\}$$

$$+ \tfrac{1}{2}A \exp[-i(\omega_{av}t + \phi_{av})]\{\exp[+i(\Delta_\omega t - \Delta_\phi)] + \exp[-i(\Delta_\omega t - \Delta_\phi)]\}$$

$$(2.1.20)$$

The terms within each pair of braces are the same complex representation of a cosine function at frequency Δ_ω. Factoring them out leaves the complex representation of a harmonic function at frequency ω_{av}. Hence, we have

$$u = 2A \cos(\Delta_\omega t - \Delta_\phi) \cos(\omega_{av}t - \phi_{av}) \qquad (2.1.21)$$

Because $\omega_{av} > \Delta_\omega$, we interpret the above as varying harmonically at frequency ω_{av} with an amplitude $2A \cos(\Delta_\omega t - \Delta_\phi)$ that varies more slowly at frequency Δ_ω. We say that $u = \pm 2A \cos(\Delta_\omega t - \Delta_\phi)$ is the *envelope* function. When ω_1 and ω_2 are quite close, this combination is called a *beating signal*. It is readily produced musically by slightly mistuning one instrument relative to another, and then playing them with nearly equal intensities. A typical beating signal is displayed in Figure 2.5.

The interval π/Δ_ω over which the signal gets larger and then dies out is the *beat period*. Within each beat, the signal fluctuates at a frequency of ω_{av}, so the interval between zeroes is π/ω_{av}. The interval between successive minima or maxima of the signal is approximately $2\pi/\omega_{av}$.

In general, a beating signal is not periodic. The condition of periodicity requires that within the period T of the signal, each term contributing to the signal repeat an integer number of times. If any signal contains two harmonics, this requirement is $T = m(2\pi/\omega_1) = n(2\pi/\omega_2)$, where m and n are integers. This leads to $m/n = \omega_1/\omega_2$, which is possible only if ω_1/ω_2 is a rational fraction.

In the foregoing we found that a sum of two harmonic functions at different frequencies may be represented alternatively as a product. Occasionally, we need to go in the opposite direction by decomposing a product of harmonic functions into its individual components. It is imperative in such an operation to avoid a common mistake. The real part of a product of complex variables is not the product of the individual real parts, that is, if $u_1 = \text{Re}(z_1)$ and $u_2 = \text{Re}(z_2)$, then $u_1 u_2 \neq \text{Re}(z_1 z_2)$. Some treatments switch to real variables to handle the product, but that would require application of trigonometric identities. The complex definitions of the sine and cosine in eqs. (2.1.5) lead to the result directly. Equations

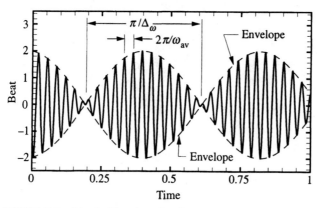

FIGURE 2.5 A typical beating signal.

(2.1.5) describe each real function as the sum of a complex exponential z and its complex conjugate z^*, $u_j = \frac{1}{2}(z_j + z_j^*)$). We then have

$$u_1 u_2 = \frac{1}{4}(z_1 + z_1^*)(z_2 + z_2^*) \equiv \frac{1}{4}(z_1 + z_1^*)z_2 + \frac{1}{4}[(z_1 + z_1^*)z_2]^*$$

$$\equiv \frac{1}{2}\mathrm{Re}[(z_1 + z_1^*)z_2] \equiv \frac{1}{2}\mathrm{Re}[z_1(z_2 + z_2^*)]$$

(2.1.22)

As a verification for this procedure let us consider $\sin(\omega_1 t)\cos(\omega_2 t)$. With the aid of eqs. (2.1.5), we have

$$\sin(\omega_1 t)\cos(\omega_2 t) = \frac{1}{2i}[\exp(i\omega_1 t) - \exp(-i\omega_1 t)]\frac{1}{2}[\exp(i\omega_2 t) + \exp(-i\omega_2 t)]$$

$$= \frac{1}{2}\mathrm{Re}\left\{\frac{1}{i}[\exp(i\omega_1 t) - \exp(-i\omega_1 t)]\exp(i\omega_2 t)\right\}$$

(2.1.23)

$$= \frac{1}{2}\mathrm{Re}\left\{\frac{1}{i}\exp[i(\omega_1 + \omega_2)t] - \frac{1}{i}\exp[i(-\omega_1 + \omega_2)t]\right\}$$

$$= \frac{1}{2}\{\sin[(\omega_1 + \omega_2)t] - \sin[(-\omega_1 + \omega_2)t]\}$$

It is not difficult to verify that the last form is equivalent to an identity associated with the sine of the sum of two angles.

An overview of the development thus far shows that Euler's formula and the fundamental rules of algebra are sufficient to manipulate the complex exponential representation of harmonic functions. Computations are readily implemented with scientific-type calculators and mathematical software. Most operations are done more directly than they would be if real functions were used.

EXAMPLE 2.3

A signal is measured to be $v = 12\sin(25t - 4.5)$. Decompose this signal into parts that are purely cosine and sine functions.

Solution   This exercise will enhance our proficiency in using complex representations of harmonic functions. We recall that a sine is the imaginary part of a complex exponential to write

$$v = 12\sin(25t - 4.5) = \mathrm{Re}\left\{\frac{12}{i}\exp[i(25t - 4.5)]\right\}$$

$$= \mathrm{Re}\left[\frac{12}{i}\exp(-4.5i)\exp(i25t)\right]$$

We now replace all polar forms of a complex quantity with their equivalent rectangular forms and combine real and imaginary parts, such that

$$v = \mathrm{Re}\left[\frac{12}{i}(-0.2108 + 0.97753i)\exp(i25t)\right]$$

$$= \mathrm{Re}\{(11.731 + 2.530i)[\cos(25t) + i\sin(25t)]\}$$

$$= 11.731\cos(25t) - 2.530\sin(25t)$$

EXAMPLE 2.4

Two harmonic functions are known to be $u_1 = 3\sin(40t)$ and $u_2 = 4\cos(40t + \pi/4)$. Express $u = u_1 - u_2$ as (a) a cosine function with a phase angle and (b) a sine function with a phase angle.

Solution   In addition to illustrating the basic operations, this exercise will improve our ability to perform computations with complex numbers. Regardless of the form we wish for the final result, we begin by converting the given real functions into complex form. Thus, we write

$$u_1 = \text{Re}\left[\frac{3}{i}\exp(i40t)\right]$$

$$u_2 = \text{Re}\left\{4\exp\left[i\left(40t + \frac{\pi}{4}\right)\right]\right\} = \text{Re}\left[4\exp\left(i\frac{\pi}{4}\right)\exp(i40t)\right]$$

We take the difference of the terms and collect the coefficients of $\exp(i40t)$,

$$u = u_1 - u_2 = \text{Re}\left\{\left[\frac{3}{i} - 4\exp\left(i\frac{\pi}{4}\right)\right]\exp(i40t)\right\}$$

For the sake of completeness, we shall explicitly display the arithmetic operations to simplify the coefficient. The reader is invited to perform the same operations solely with a calculator and with mathematical software. We convert each complex number from polar to rectangular form and combine like parts according to

$$u = \text{Re}\left\{\left[-3i - 4\cos\left(\frac{\pi}{4}\right) - 4\sin\left(\frac{\pi}{4}\right)i\right]\exp(i40t)\right\}$$

$$= \text{Re}[(-2.828 - 5.828i)\exp(i40t)]$$

How we proceed now depends on how u is to be represented. For a cosine function, the complex amplitude is the coefficient of $\exp(i40t)$. We convert this coefficient to polar form, which leads to

$$u = \text{Re}\{[-2.828 - 5.828i]\exp(i40t)\} = \text{Re}\{6.478\exp(-2.023i)\exp(i40t)\}$$

$$= 6.478\cos(40t - 2.023)$$

Note that the phase angle is in the third quadrant because both the real and imaginary parts of the complex amplitude are negative.

When we wish to represent u as a sine function, we use the identity $\sin(z) = \text{Re}\{(1/i)\exp(iz)\}$. To place the complex representation of u into real form, we place a factor i in the denominator, and introduce a compensating factor i into the complex amplitude. In addition, we use the fact that $i = \exp(i\pi/2)$. Hence,

$$u = \text{Re}\left[(-2.828 - 5.828i)i\frac{\exp(i40t)}{i}\right] = \text{Re}\left[(5.828 - 2.828i)\frac{\exp(i40t)}{i}\right]$$

$$= \text{Re}\left[6.478\exp(-0.4518i)\frac{\exp(i40t)}{i}\right] = 6.478\sin(40t - 0.4518)$$

2.2 FREE VIBRATION

A one-degree-of-freedom system may be represented by a standard mass-spring-dashpot system. The mechanical properties of such a system are the inertia, stiffness, and damping coefficients corresponding to the generalized coordinate we

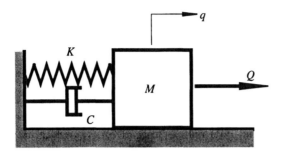

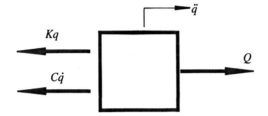

FIGURE 2.6 Standard one-degree-of-freedom system.

have selected. Because there is only one of each coefficient, we may dispense with subscripts, so the displacement q is affected by the mass M, spring K, and dashpot C. This leads to the generic one-degree-of-freedom system in Figure 2.6. Note that it is conventional to depict q as translational displacement of a small block that slides over a smooth horizontal surface, so the spring and dashpot are extensional types. In the event that q is an angular position quantity, M would be a moment of inertia, and the spring and dashpot would be rotational types. Depending on the type of quantity q represents, the generalized force Q exciting the system is either a force or a moment.

The equation of motion is the single differential equation,

$$M\ddot{q} + C\dot{q} + Kq = Q(t) \tag{2.2.1}$$

In general, the equation of motion governs the acceleration subsequent to the time when the motion is initiated. We usually set $t = 0$ as the initial time. Correspondingly, we must specify the initial displacement and velocity at $t = 0$, which we denote as q_0 and $\dot{q}_0$, respectively. Once the equation of motion and initial conditions are stated, the problem is well posed and ready to be solved.

The first type of response we consider is *free vibration*, which means that there are no external excitations, $Q \equiv 0$. The motion in this case results from the initial conditions, which state that the system was not initially at rest in its static equilibrium position. Because the right side is zero the standard equation of motion is a homogeneous linear differential equation. All solutions of this equation depend exponentially on time, so we seek a homogeneous solution in the form

$$q = B\exp(\lambda t) \tag{2.2.2}$$

where the constants B and λ must be determined. We find an equation for λ by substituting the trial solution into the homogeneous equation of motion. Because the equation must be satisfied at all t, we may factor out $B\exp(\lambda t)$ from the substituted form, which leads to a *characteristic equation* for λ, specifically

$$\boxed{M\lambda^2 + C\lambda + K = 0} \tag{2.2.3}$$

This is a quadratic equation, so there are two values of λ. Correspondingly, there are two homogeneous solutions. The nature of these solutions depends on the relative values of K, M, and C.

2.2.1 Undamped Systems

Dissipation is absent in an ideal system, which we model by setting $C = 0$. The roots of the characteristic equation in this case are purely imaginary, being given by

$$\lambda = \pm i\omega_{nat}, \qquad \omega_{nat} = \sqrt{\frac{K}{M}} \tag{2.2.4}$$

The parameter ω_{nat} is the *natural frequency*. (The reason for this term will soon become apparent.) It is a fundamental property of the system that will appear in most aspects of the response.

Because the characteristic roots are imaginary, there are two corresponding complex exponential solutions. The coefficient B associated with each may be complex, and they need not be the same. Hence, the general solution is

$$q = B_1 \exp(i\omega_{nat}t) + B_2 \exp(-i\omega_{nat}t) \tag{2.2.5}$$

This solution has a complex form, but q is a real quantity. To resolve this dilemma we observe that the two exponential functions are complex conjugates. If the coefficients are also complex conjugates, then the imaginary parts will cancel. We therefore set $B_1 = \frac{1}{2}\hat{A}$ and $B_2 = \frac{1}{2}\hat{A}^*$, where we introduce the half factor to simplify the final form of q. The corresponding solution is

$$q = \tfrac{1}{2}\hat{A}\exp(i\omega_{nat}t) + \tfrac{1}{2}\hat{A}^*\exp(-i\omega_{nat}t) = \text{Re}[\hat{A}\exp(i\omega_{nat}t)] \tag{2.2.6}$$

In other words the free response of an undamped one-degree-of-freedom system is a harmonic motion that occurs at the natural frequency of the system.

The complex amplitude is dictated by the initial conditions. To determine this quantity we set $\hat{A} = c_1 - ic_2$, where c_1 and c_2 are real and the minus sign is a matter of convenience. This leads to $q = \text{Re}[(c_1 - ic_2)\exp(i\omega_{nat}t)]$, which reduces to

$$q = c_1 \cos(\omega_{nat}t) + c_2 \sin(\omega_{nat}t) \tag{2.2.7}$$

In order to satisfy the initial conditions, we form $\dot{q}$ by differentiating the preceding with respect to t, and then evaluate q and $\dot{q}$ at $t = 0$, which leads to

$$c_1 = q_0, \quad c_2 = \frac{\dot{q}_0}{\omega_{nat}} \quad \Rightarrow \quad q = q_0\cos(\omega_{nat}t) + \frac{\dot{q}_0}{\omega_{nat}}\sin(\omega_{nat}t) \tag{2.2.8}$$

In a graph, the initial displacement q_0 is the intercept with the axis $t = 0$ of the curve showing q as a function of t. The slope of this curve is $\dot{q}$, so the initial velocity $\dot{q}_0$ is the slope at that intercept.

The preceding is the real form of the free vibration solution. The results may be converted to complex form. Using the polar form to represent $\hat{A} = c_1 - ic_2 = A\exp(-i\phi)$ leads to

$$A\cos(\phi) = q_0, \quad A\sin(\phi) = \frac{\dot{q}_0}{\omega_{nat}} \tag{2.2.9}$$

from which we find that the amplitude and phase angle are

$$q = A\cos(\omega_{nat}t - \phi)$$

$$A = \left[q_0^2 + \left(\frac{\dot{q}_0}{\omega_{nat}}\right)^2\right]^{1/2}, \quad \phi = \tan^{-1}\left(\frac{\dot{q}_0}{\omega_{nat}q_0}\right) \tag{2.2.10}$$

Note that evaluation of the arctangent requires that the quadrant be consistent with the values of $A\cos(\phi)$ and $A\sin(\phi)$.

The oscillatory nature of an undamped free vibration could have been predicted by using physical arguments. In the absence of dissipation, the system is conservative. The mechanical energy $T + V$ is therefore constant, equaling the value set by the initial conditions. When the system passes the static equilibrium position, $q = 0$, the potential energy is zero and the kinetic energy has its maximum value. The system's inertia causes it to continue past the equilibrium position. The spring, which always acts to return the mass to $q = 0$, slows the mass until it comes to rest at the maximum displacement, where $|q| = A$ and $\dot{q} = 0$. Thus, the kinetic energy is zero at this position and the potential energy is a maximum, corresponding to the maximum spring deformation. The spring then pulls the mass back to $q = 0$, after which the process is repeated on the other side of $q = 0$. This vibration continues periodically, with frequency $\omega_{nat} = \sqrt{K/M}$, because there is no damping to dissipate the mechanical energy.

The development thus far is based on the stiffness coefficient K being positive, but that is not always the case. Recall that the basic definition of K for a one-degree-of-freedom system is

$$K = \left(\frac{\partial^2 V}{\partial q^2}\right)_{q=0} \tag{2.2.11}$$

Consider the potential energy for a ball at the top of a rounded hill, such as the one depicted in Figure 2.7. Clearly, this is a case of unstable static equilibrium, for any disturbance will cause the ball to roll away from the top of the hill. If we use the elevation z as the generalized coordinate, it is evident that the potential energy is a maximum at the top position, so we have $\partial V/\partial z = 0$ and $\partial^2 V/\partial z^2 < 0$. Thus, the stiffness K is negative. In the next section we will examine the effects of damping. Because such forces oppose velocity, they cannot cause the system to return to its equilibrium position. Consequently, we conclude that, in general,

If K for a one-degree-of-freedom system is negative, then the equilibrium position is unstable.

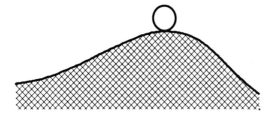

FIGURE 2.7 A ball at the top of a hill— a case of unstable static equilibrium.

It is instructive to observe that if $K < 0$, then the homogeneous solution to the equation of motion is

$$q = B_1 \exp\left(\sqrt{\frac{-K}{M}}t\right) + B_2 \exp\left(-\sqrt{\frac{-K}{M}}t\right) \tag{2.2.12}$$

Thus, one of the solutions grows with increasing time. We say that this is a *divergence instability*. The small displacement approximations leading to linearized equations of motion are valid only for a very short time when a system is unstable. One indication of this loss of validity is the fact that the response in eq. (2.2.12) does not conserve energy. We will encounter another type of instability in Chapter 11, where we study time-dependent systems.

EXAMPLE 2.5

The 200 gram sphere is not attached to the 50 gram piston. The stiffness of the spring is 1200 N/m. The spring is held 80 mm below the static equilibrium position, and then released. Determine the position of the piston above the equilibrium position, and the corresponding elapsed time, at which the sphere ceases to be in contact with the piston.

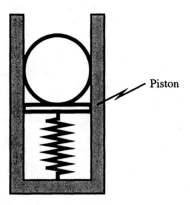

Solution   This exercise is intended to bring out the way in which static and dynamic forces might occur in a study, as well as to highlight interpretation of harmonic response. We begin by drawing two free body diagrams. The first, which we will use to derive the equation of motion, considers the sphere and the platform as a system, so that the contact force exerted between these bodies is an internal force that is not considered.

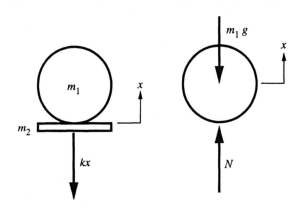

We consider the system to move away from its static equilibrium position a distance x upward, which serves as the generalized coordinate. Hence, the spring force acts downward. The gravity force has a static effect, which is irrelevant to the equation of motion because x is measured relative the static equilibrium position. Hence, the equation of motion resulting from the first free body diagram is

$$m\ddot{x} + kx = 0, \quad m = m_1 + m_2 = 0.25 \text{ kg}$$

The second free body diagram isolates the ball because we seek the conditions under which the contact force N applied by the platform becomes zero. We show the gravity force in this free body diagram because the weight affects the normal force N, and the condition we seek is $N = 0$. From the second free body diagram, we find that

$$N - m_1 g = m_1 \ddot{x}$$

Hence, $N = 0$ is marked by $\ddot{x} = -g$.

The idea now is to find the response corresponding to the given initial conditions, and then to determine when the acceleration condition $\ddot{x} = -g$ occurs. We write the equation of motion as

$$\ddot{x} + \omega_{\text{nat}}^2 x = 0$$

where the natural frequency is

$$\omega_{\text{nat}} = \left(\frac{k}{m}\right)^{1/2} = 69.28 \text{ rad/s}$$

The given initial conditions indicate that $x = -0.08$ m and $\dot{x} = 0$ at $t = 0$. The response matching these initial conditions is

$$x = -0.08 \cos(\omega_{\text{nat}} t)$$

We seek the value of t_1 for which $\ddot{x} = -g$, but the equation of motion states that $\ddot{x} = -\omega_{\text{nat}}^2 x$, so we seek the condition for which $\omega_{\text{nat}}^2 x = g$, or

$$-0.08 \, \omega_{\text{nat}}^2 \cos(\omega_{\text{nat}} t_1) = g$$

The solution of this relation is

$$t_1 = \frac{1}{\omega_{\text{nat}}} \cos^{-1}\left(-\frac{g}{0.08 \, \omega_{\text{nat}}^2}\right) = \frac{1}{69.28} \cos^{-1}\left(-\frac{9.807}{0.08 \times 69.28^2}\right)$$

$$= 0.014434 \cos^{-1}(-0.02554)$$

We select for the inverse cosine the smallest angle that gives $t_1 > 0$, $\cos^{-1}(-0.02554) = 1.5963$, which leads to

$$t_1 = 0.02304 \text{ s}$$

2.2.2 Underdamped Systems

If the amount of damping is small, the time scale over which an appreciable amount of energy is lost will be large relative to the natural period of free vibration, $\tau_{\text{nat}} = 2\pi/\omega_{\text{nat}}$. In that case it might be acceptable to ignore dissipation effects for a short time interval. However, any amount of damping will eventually quiet the system, as we will see here.

When damping is present, we use the natural frequency ω_{nat} to rewrite the equation of motion as

$$\ddot{q} + 2\zeta\omega_{\text{nat}}\dot{q} + \omega_{\text{nat}}^2 q = 0 \tag{2.2.13}$$

A comparison of this form with eq. (2.2.1) shows that the parameter ζ is

$$\zeta = \frac{C}{2\omega_{nat}M} \equiv \frac{C}{2\sqrt{KM}}$$

(2.2.14)

The corresponding characteristic equation is

$$\lambda^2 + 2\zeta\omega_{nat}\lambda + \omega_{nat}^2 = 0$$

(2.2.15)

The above characteristic equation is quadratic, so its roots are either both real, or complex conjugates, given by

$$\lambda = -\zeta\omega_{nat} \pm \omega_{nat}\sqrt{\zeta^2 - 1}$$

(2.2.16)

It is obvious that $\zeta = 1$ leads to a transition in the nature of the roots, so we refer to ζ as the *critical damping ratio*. We begin by considering the case $0 < \zeta < 1$. Because the damping is less than critical in this case, the system is said to be *underdamped*. (The undamped model we considered previously could be treated as a special case of an underdamped system, if we included $\zeta = 0$.)

The discriminant $\zeta^2 - 1$ of the characteristic equation is negative, so the roots are complex conjugates. We write them as

$$\lambda = -\zeta\omega_{nat} \pm i\omega_d$$

(2.2.17)

where ω_d is the *damped natural frequency*,

$$\omega_d = \omega_{nat}\sqrt{1 - \zeta^2}$$

(2.2.18)

As we did for undamped systems, we associate a different coefficient with the exponential function for each characteristic root. Because the exponential of a sum is the product of the individual exponentials, we factor out $\exp(-\zeta\omega_{nat}t)$ from each solution, which leads to

$$q = \exp(-\zeta\omega_{nat}t)[B_1\exp(i\omega_d t) + B_2\exp(i\omega_d t)]$$

(2.2.19)

Aside from the $\exp(-\zeta\omega_{nat}t)$ factor and the frequency being ω_d rather than ω_{nat}, this expression is just like eq. (2.2.5) for undamped free vibration. We follow that development to enforce the requirement that q is real. Depending on whether we write the bracketed term in the form of eq. (2.2.6), eq. (2.2.7), or eq. (2.2.10), the solution for q may be expressed as

$$q = \exp(-\zeta\omega_{nat}t)\text{Re}[\hat{A}\exp(i\omega_d t)]$$

$$= \exp(-\zeta\omega_{nat}t)[c_1\cos(\omega_d t) + c_2\sin(\omega_d t)], \quad \hat{A} = c_1 - ic_2$$

(2.2.20)

$$= A\exp(-\zeta\omega_{nat}t)\text{Re}\{\exp[i(\omega_d t - \phi)]\}, \quad \hat{A} = A\exp(-i\phi)$$

The unknown coefficients are set by the initial conditions, as they are for an undamped response. We use the second of the above representations to evaluate q and $\dot{q}$ at $t = 0$, which leads to

$$q_0 = c_1, \quad \dot{q}_0 = -\zeta\omega_{nat}c_1 + \omega_d c_2 \quad \Rightarrow \quad c_2 = \frac{\dot{q}_0 + \zeta\omega_{nat}q_0}{\omega_d}$$

(2.2.21)

The real form of the response is therefore

$$q = \exp(-\zeta\omega_{\text{nat}}t)\left[q_0\cos(\omega_{\text{d}}t) + \frac{\dot{q}_0 + \zeta\omega_{\text{nat}}q_0}{\omega_{\text{d}}}\sin(\omega_{\text{d}}t)\right] \tag{2.2.22}$$

Although the second form in eqs. (2.2.20) is the one we use to satisfy the initial conditions, the last form is most useful for discussions. One way of picturing the last of eqs. (2.2.20) is to consider the factor $A\exp(-\zeta\omega_{\text{nat}}t)$ as the decaying amplitude of a harmonic function at frequency ω_{d}. In Figure 2.3, we represented a harmonic function as a rotating vector in the complex plane. From this viewpoint the underdamped response appears as a vector that rotates counterclockwise at angular speed ω_{d}, with an amplitude that decays exponentially. Thus, the tip of the vector follows an inward spiral, as shown in Figure 2.8. Note that in the figure $\omega_{\text{d}}t$ measures the angle relative to the orientation at $t = 0$, whereas the angle $\omega_{\text{d}}t - \phi$ is the phase variable of the cosine term above.

The more conventional way of viewing the response is to plot q as a function of t. In advance of plotting a typical response, we may anticipate the qualitative aspects by recognizing that $\cos(\omega_{\text{d}}t - \phi)$ oscillates between -1 and 1. Hence, the largest possible positive value of q at any instant is $A\exp(-\zeta\omega_{\text{nat}}t)$, and the largest negative value is $-A\exp(-\zeta\omega_{\text{nat}}t)$. The curves $q = \pm A\exp(-\zeta\omega_{\text{nat}}t)$ are the *envelope* of the underdamped response. Within this envelope the signal oscillates at the damped natural frequency. The elapsed time between zeroes is one-half the *damped period* τ_{d},

$$\tau_{\text{d}} = 2\pi/\omega_{\text{d}} \tag{2.2.23}$$

Figure 2.9 indicates that τ_{d} also gives the interval over which the positive and negative peak values of q occur, which is an aspect that we shall prove because it has important implications for system identification. However, this repetition should not be taken to imply that the underdamped response is periodic, because a nonzero value $|q(t + \tau_{\text{d}})|$ is always smaller than $|q(t)|$.

We begin to investigate timing issues by observing that the $q(t)$ curve tangentially intersects the envelope whenever $|\cos(\omega_{\text{d}}t - \phi)| = 1$, so the interval between adjacent positive or negative intersections is τ_{d}. Because the zeroes occur when $|\cos(\omega_{\text{d}}t - \phi)| = 0$, intersections with the envelope occur at intervals that are separated from the zeroes by $\tau_{\text{d}}/4$. Next, we note that the slope of the envelope is negative,

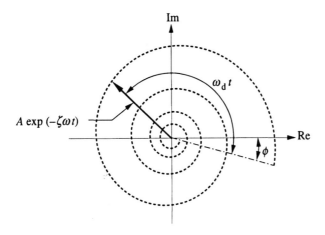

FIGURE 2.8 Underdamped response as a rotating vector in the complex plane.

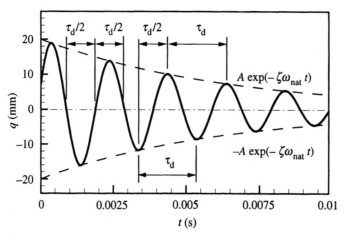

FIGURE 2.9 Typical underdamped response, $\omega_{nat} = 1000\pi$ rad/s, $\zeta = 0.05$.

so the intersections with the envelope occur later than the instant at which $|q|$ has a maximum value. In other words, *maxima or minima of $q(t)$ do not occur midway between the occurrence of zeroes.* (This aspect has been emphasized because failure to recognize it is a common mistake for novices.) To find the instants at which max $|q|$ occur, we must determine when $\dot{q} = 0$. We obtain an expression for $\dot{q}$ by differentiating eq. (2.2.22), which we simplify by using the definition of ω_d,

$$\dot{q} = \exp(-\zeta\omega_{nat}t)\left\{\dot{q}_0\cos(\omega_d t) - \left[\omega_d q_0 + \zeta\omega_{nat}\left(\frac{\dot{q}_0 + \zeta\omega_{nat}q_0}{\omega_d}\right)\right]\sin(\omega_d t)\right\}$$
(2.2.24)

$$\equiv \exp(-\zeta\omega_{nat}t)\left\{\dot{q}_0\cos(\omega_d t) - \frac{\omega_{nat}}{\omega_d}(\omega_{nat}q_0 + \zeta\dot{q}_0)\sin(\omega_d t)\right\}$$

Let us denote as $t_j, j = 1, 2, \ldots,$ the roots of $\dot{q} = 0$ for which q has a maximum positive value, while $t_{j+1/2}$ denotes instants at which $\dot{q} = 0$ and q has a maximum negative value. Because τ_d is the period of both trigonometric functions appearing in eq. (2.2.24), increasing t from t_j to $t_j + \tau_d$ leads to the same value of the sine and cosine functions, so we have $t_{j+1} = t_j + \tau_d$. Furthermore, if we increase t from t_j to $t_j + \tau_d/2$, the values of both functions are the negative of their values at $t = t_j$. This leads us to the conclusion that $t_{j+1/2} = t_j + \tau_d/2$, from which it follows that $t_{j+1/2} = t_{j-1/2} + \tau_d$. In summary, we find that

- Zeroes of the underdamped response occur at intervals of $\tau_d/2$.
- Adjacent maxima and minima are separated by intervals of $\tau_d/2$.
- Maxima and minima occur slightly earlier than $\tau_d/4$ following the previous zero.

These features are displayed in Figure 2.9.

The repetitive nature of the peaks leads us to a simple equation by which the critical damping ratio of a system may be determined from measurements of $q(t)$. Let us denote the maxima as $x_j \equiv q(t_j)$. We do not need to know the actual value of t_j, which would require solving eq. (2.2.24). Rather we only require a comparison of x_j and x_{j+1},

$$x_j = q(t_j) = A \exp(-\zeta\omega_{nat}t_j)\cos(\omega_d t_j - \phi)$$

$$x_{j+1} = q(t_j + \tau_d) = A \exp\left[-\zeta\omega_{nat}\left(t_j + \frac{2\pi}{\omega_d}\right)\right]\cos\left[\omega_d\left(t_j + \frac{2\pi}{\omega_d}\right) - \phi\right]$$

$$= \exp\left(-\frac{2\pi\zeta\omega_{nat}}{\omega_d}\right)A \exp(-\zeta\omega_{nat}t_j)\cos[\omega_d t_j - \phi] \tag{2.2.25}$$

$$= x_j\exp\left(-\frac{2\pi\zeta\omega_{nat}}{\omega_d}\right) \equiv x_j\exp\left[-\frac{2\pi\zeta}{(1-\zeta^2)^{1/2}}\right]$$

A similar analysis applied to a comparison of the minima, which we denote as $x_{j+1/2} \equiv |q(t_{j+1/2})|$, would show that

$$x_{j+1/2} = x_j\exp\left[-\frac{\pi\zeta}{(1-\zeta^2)^{1/2}}\right] \tag{2.2.26}$$

Suppose we have measured $q(t)$, which means that we know the values of several x_j. To find the value of ζ associated with the response, we form the ratio of the successive maxima, which leads to the *log decrement* δ, where

$$\boxed{\delta = \ln\left(\frac{x_j}{x_{j+1}}\right) = \frac{2\pi\zeta}{(1-\zeta^2)^{1/2}}} \tag{2.2.27}$$

We may determine the critical damping ratio from δ by solving this expression, which yields

$$\boxed{\zeta = \frac{\delta}{(4\pi^2 + \delta^2)^{1/2}}} \tag{2.2.28}$$

Many systems are said to be *lightly damped*, which means that their ratio of critical damping is small. In such systems the log decrement will also be a small value, so we may approximate the above relation as $\zeta \approx \delta/2\pi$, which is quite usable for $\zeta < 0.1$.

The smallness of δ in a lightly damped system brings up the issue of experimental error and its influence on the resulting value of ζ. A small δ leads to x_{j+1} being only slightly smaller than x_j. If this difference is of the order of magnitude of the measurement error, then the value of ζ derived from eqs. (2.2.27) and (2.2.28) will be quite inaccurate. We may improve the evaluation by comparing the maxima after a large number of cycles N. In view of the exponential decay of the successive values of q_j, we have

$$\boxed{\delta = \frac{1}{N}\ln\left(\frac{x_j}{x_{j+N}}\right)} \tag{2.2.29}$$

As a closure to this discussion, we should note that one could carry out the same evaluations equally well by using successive minimum values $x_{j+1/2}$.

Let us now consider a situation where $\zeta \ll 1$, in order to identify when it might be sufficient to use an undamped model to study free vibration of a system. In that case we may use power series in ζ to obtain alternative forms. Equation (2.2.18) indicates that the damped and undamped natural frequencies for small ζ are approximately related by $\omega_d \approx \omega_{nat}(1 - \frac{1}{2}\zeta^2)$. In other words light damping has a second-order effect on the observed oscillation rate. In contrast, when we expand the exponential in eq. (2.2.25), we find that $x_{j+1} \approx x_j(1 - 2\pi\zeta)$, which is a first-order effect. Thus, when we observe the response of a very lightly damped system, the fact that x_{j+1} is smaller than x_j is more observable than the fact that ω_d is smaller than ω_{nat}. It follows that we may use an undamped model to represent free vibration of a lightly damped system, provided that we limit the observation interval to be shorter than the number of damped periods required to obtain a reasonably precise value of δ from eq. (2.2.29).

For a different perspective on the significance of the log decrement, let us consider the mechanical energy that is dissipated in an underdamped free vibration. Because peak values x_j are defined to be maxima, at which $\dot{q} = 0$, it follows that the mechanical energy $E \equiv T + V$ corresponding to the peak values is solely stored as potential energy, so we have

$$E_j = \tfrac{1}{2}kx_j^2, \quad E_{j+1} = \tfrac{1}{2}kx_{j+1}^2 \tag{2.2.30}$$

From the definition of the log decrement, eq. (2.2.27), we have $x_{j+1} = x_j\exp(-\delta)$. The energy dissipated from one peak to the next is $E_j - E_{j+1}$, so we find that the fraction of the mechanical energy dissipated in a cycle is

$$\frac{E_j - E_{j+1}}{E_j} = 1 - \exp(-2\delta) \approx 2\delta \quad \text{if } \delta \ll 1 \tag{2.2.31}$$

In other words, in each damped cycle the fraction of mechanical energy lost to damping is approximately 2δ. Obviously, $1 - 2\delta$ is the fraction of energy that remains, so linear damping can never bring the system fully to rest in a finite amount of time.

Before we proceed to the case of an overdamped system, it is useful to examine the influence of increased damping in the underdamped case. Figure 2.10 shows responses for four large values of ζ; the natural frequency and initial conditions are the same as those for Figure 2.9. It is evident that the attenuation rate increases drastically as damping is increased, but the damped period is affected much less. (Recall that the effect of ζ on τ_d is second order.) The response for $\zeta = 0.8$ is interesting because it shows that there

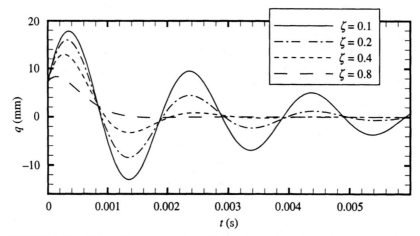

FIGURE 2.10 Effect of increasing damping for large damping ratios below critical, $\omega_{nat} = 1000\pi$ rad/s.

is almost no oscillation. Essentially what happens is that the *time constant* $1/(\zeta\omega_{nat})$ for the envelope, which is the time required for the envelope to decay by a factor of $1/e$, becomes comparable to the damped period, so that there is very little response at the end of a single cycle. Critical damping represents the limit of this trend. As we will see, overdamped and critically damped systems do not show oscillations in their free response.

EXAMPLE 2.6

A one-degree-of-freedom system has mass $M = 4$ kg and spring stiffness $K = 16(10^4)$ N/m. The system is known to be underdamped, but the damping constant C is unknown. The system is released from rest at $t = 0$ with an initial value of the generalized coordinate $q(t = 0) = -40$ mm. It is observed that the largest positive value of q in the free vibration occurs at $t = 0.02$ s. Determine (a) the value of q at $t = 0.02$ s, (b) the value of C, (c) the earliest instant at which $q = 0$ after the system is released, and the velocity $\dot{q}$ at that instant.

Solution   This exercise will highlight the relationship between the damped period, critical damping ratio, and features of the response as a function of time. We begin by placing the given aspects of the response in the context of the standard properties of underdamped system response. It is stated that at $t = 0$, $q_0 = -0.040$ m. We also know that at this instant $\dot{q}_0 = 0$, so a plot of q as a function of t has a horizontal slope at $t = 0$. Because q_0 is negative at this instant, it must be that $t = 0$ corresponds to a minimum value of q. Furthermore, $t = 0.02$ s is the time at which the first maximum occurs, because successive maxima decrease in magnitude and $t = 0.02$ s is stated to be the largest. We saw in Figure 2.9 that the instants when maxima occur are equally spaced between instants when minima occur. It follows that the interval from $t = 0$ to $t = 0.02$ s is one-half the damped period, so we have $\tau_d = (2)\,0.02$ s and

$$\omega_d = \frac{2\pi}{\tau_d} = \frac{2\pi}{0.04} = 157.08 \text{ rad/s}$$

We determine the undamped natural frequency from the given system parameters,

$$\omega_{nat} = \sqrt{\frac{K}{M}} = 200 \text{ rad/s}$$

The relation between the damped and undamped natural frequencies then yields

$$(1 - \zeta^2)^{1/2} = \frac{\omega_d}{\omega_{nat}} = 0.7854$$

From this, we find the critical damping ratio, which leads directly to the damping constant C,

$$\zeta = (1 - 0.7854^2)^{1/2} = 0.6190$$

$$\frac{C}{M} = 2\zeta\omega_{nat} \quad \Rightarrow \quad C = 4(2)(0.6190)(200) = 990.4 \text{ N-s/m}$$

Knowledge of the critical damping ratio also enables us to evaluate the log decrement, from which we may compute q at $t = 0.02$ s. First, we have

$$\delta = \frac{2\pi\zeta}{(1 - \zeta^2)^{1/2}} = 4.952$$

Because the interval from $t = 0$ to $t = 0.02$ s constitutes a half-cycle, we may employ eq. (2.2.29) with $N = 1/2$ and the minimum at the start of the interval set as $x_0 = q_0$,

$$\delta = \frac{1}{(1/2)} \ln\left(\left|\frac{q_0}{x_{1/2}}\right|\right) \quad \Rightarrow \quad |x_{1/2}| = |q_0|\exp\left(-\frac{\delta}{2}\right) = 3.363 \text{ mm}$$

The remaining properties to be evaluated pertain to the earliest instant at which $q = 0$. This condition does not occur at one-quarter of a damped period after the minimum. We identify the condition by examining the response as a function of time. The underdamped system response in eq. (2.2.22) corresponding to $\dot{q}_0 = 0$ is

$$q = q_0 \exp(-\zeta\omega_{nat}t)\left[\cos(\omega_d t) + \frac{\zeta\omega_{nat}}{\omega_d}\sin(\omega_a t)\right]$$

To determine the instant t' at which $q = 0$, we set

$$\cos(\omega_d t') + \frac{\zeta\omega_{nat}}{\omega_d}\sin(\omega_d t') = 0$$

from which we obtain

$$t' = \frac{1}{\omega_d}\tan^{-1}\left(-\frac{\omega_d}{\zeta\omega_{nat}}\right) \equiv \frac{1}{\omega_d}\tan^{-1}\left[-\frac{(1-\zeta^2)^{1/2}}{\zeta}\right] = 0.01425$$

Note that the argument of the arctangent is negative. A calculator or computer program is likely to return a negative angle for the function, so the computed value of the arctangent must be increased by π in order to obtain a positive value for t'. The velocity at t' is readily determined by differentiating the solution for q, which gives

$$\dot{q}' = q_0\exp(-\zeta\omega_{nat}t')\left\{-\zeta\omega_{nat}\left[\cos(\omega_d t') + \frac{\zeta\omega_{nat}}{\omega_d}\sin(\omega_d t')\right]\right.$$

$$\left.\left\{+\omega_d\left[-\sin(\omega_d t') + \frac{\zeta\omega_{nat}}{\omega_d}\cos(\omega_d t')\right]\right\}\right\} = 1445.5 \text{ mm/s}$$

EXAMPLE 2.7

A 0.8 kg platform is supported by a shock absorber. The stiffness of the spring is 600 N/m and the dashpot constant is 20 N-s/m. The platform is at rest in its static equilibrium position when a 4 kg package that is falling at 5 m/s impacts against it. The collision is perfectly plastic, so the platform and package move together in the subsequent motion. Determine the response after impact.

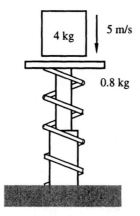

4 kg   5 m/s

0.8 kg

Solution A primary objective of this exercise is to illustrate the process of satisfying initial conditions. It will also emphasize the interplay between the static equilibrium position and the static effect of gravity. We begin by drawing a free body diagram of the platform and the package, which is the vibratory system following the impact.

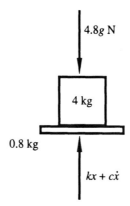

It is stated that the platform was in static equilibrium prior to the impact, but the addition of the package changes the static equilibrium position, because the weight supported by the spring is increased. In order to avoid any ambiguity as to which equilibrium position we are using, we shall deviate here from our usual practice by defining the generalized coordinate to be the downward displacement x measured from the location at which the spring is undeformed. Correspondingly, we must include the static gravitational force in the free body diagram. Thus, the equation of motion is

$$\sum F = mg - kx - c\dot{x} = m\ddot{x} \quad \Rightarrow \quad \ddot{x} + 2\zeta\omega_{nat}\dot{x} + \omega_{nat}^2 x = g$$

where $m = 4.8$ kg, $k = 600$ N/m, and $c = 20$ N-s/m. This corresponds to

$$\omega_{nat} = \left(\frac{k}{m}\right)^{1/2} = 11.1803 \text{ rad/s}$$

$$\zeta = \frac{c}{2(km)^{1/2}} = 0.1864$$

This last parameter is especially important because it indicates that the system is indeed underdamped.

Evaluation of the response requires initial conditions. We use the fact that the momentum of a system is conserved in a collision to determine the initial velocity. Prior to the impact, the package was falling with a speed of 5 m/s, and immediately after the impact the package and the platform both have speed v_0. Equating the momentum immediately before and after the collision leads to

$$4.8v_0 = 4(5) \quad \Rightarrow \quad v_0 = 4.1667 \text{ m/s}$$

The initial displacement is governed by the equations of static equilibrium. For $t < 0$, the system was in equilibrium under the weight of the platform, so the initial force in the spring equals the weight of the platform, $kx_0 = 0.8g$ N, from which we obtain

$$x_0 = \frac{0.8(9.807)}{600} = 0.013076 \text{ m}$$

The equation of motion has a constant term to the right of the equality sign, so we must add a constant particular solution to the complementary solution usually associated with free vibration. (We will review this aspect of solving equations of motion in a later section.) The general solution that results is

$$x = \frac{g}{\omega_{nat}^2} + \exp(-\zeta\omega_{nat}t)[c_1\cos(\omega_d t) + c_2\sin(\omega_d t)]$$

where

$$\omega_d = (1 - \zeta^2)^{1/2}\omega_{nat} = 10.984 \text{ rad/s}$$

We determine the coefficients c_1 and c_2 by satisfying the initial conditions. For displacement, we have

$$0.013076 = \frac{9.807}{11.1803^2} + c_1$$

while the initial velocity condition is

$$4.1667 = -\zeta\omega_{nat}c_1 + \omega_d c_2$$

Thus, we find

$$c_1 = -0.06537, \qquad c_2 = 0.3917$$

The response is

$$x = 0.7854 + \exp(-2.083t)[-0.06537\cos(10.984t) + 0.3917\sin(10.984t)] \text{ m}$$

2.2.3 Overdamped Systems

The nature of the solution of the equation of motion is dictated by the characteristic roots in eq. (2.2.16). We have already seen that these roots are complex conjugates when $\zeta < 1$. Let us consider a graph in which we plot the roots in the complex plane corresponding to different values of ζ with ω_{nat} held fixed. As shown in Figure 2.11, between $\zeta = 0$ and $\zeta = 1$, the roots are complex conjugates that lie on a circle whose radius is ω_{nat}. At $\zeta = 1$, the two roots merge. Increasing the critical damping ratio beyond $\zeta = 1$ causes the roots to split up, migrating in opposite directions along the real axis. As $\zeta \to \infty$, one root approaches zero, and the other becomes infinite.

When $\zeta > 1$, we say that the system is *overdamped*. Both characteristic roots are negative, which we emphasize by indicating the minus sign explicitly. Thus, the roots of the characteristic eq. (2.2.15) for $\zeta > 1$ are written as

$$\lambda = -\lambda_1, -\lambda_2$$
$$\lambda_1 = \zeta\omega_{nat} - \omega_{nat}\sqrt{\zeta^2 - 1}, \qquad \lambda_2 = \zeta\omega_{nat} + \omega_{nat}\sqrt{\zeta^2 - 1} \qquad (2.2.32)$$

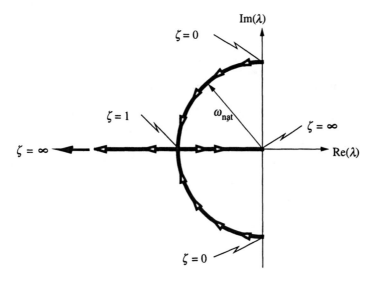

FIGURE 2.11 Characteristic roots in the complex plane as a function of ζ.

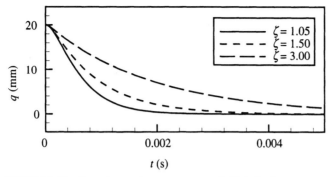

FIGURE 2.12 Overdamped response to an initial displacement.

Note that we have ordered the roots such that their magnitudes are $\lambda_1 < \lambda_2$. The corresponding general solution for the free response is a sum of decaying exponential functions with unspecified factors, specifically

$$q = c_1 \exp(-\lambda_1 t) + c_2 \exp(-\lambda_2 t) \qquad (2.2.33)$$

The coefficients c_1 and c_2 are defined by the initial conditions. Matching the general solution at $t = 0$ to the initial values of q and $\dot{q}$ leads to

$$q = \frac{\lambda_2 q_0 + \dot{q}_0}{\lambda_2 - \lambda_1} \exp(-\lambda_1 t) - \frac{\lambda_1 q_0 + \dot{q}_0}{\lambda_2 - \lambda_1} \exp(-\lambda_2 t) \qquad (2.2.34)$$

Overdamping is desirable in situations where it is necessary to avoid oscillation in response to a set of initial conditions. To examine the effects of damping, let us split up the effects of initial displacement and initial velocity. Figure 2.12 shows the response when the system is released from rest at a displaced position, $q_0 > 0$ and $\dot{q}_0 = 0$. Note that increasing the damping ratio leads to longer elapsed time until q diminishes to a specified value. To the novice this might seem counterintuitive, but it is readily explained. Increasing ζ with fixed ω_{nat} corresponds to making the dashpot progressively stronger. Thus, the dashpot exerts larger forces in opposition to the spring force that pulls the mass back to $q = 0$. From a mathematical perspective, as $\zeta \longrightarrow \infty$, $|\lambda_1| \longrightarrow 0$, so the first term in the general solution for q decays less rapidly with increasing damping.

Figure 2.13 displays overdamped responses to an initial velocity, $\dot{q}_0 > 0$ and $q_0 = 0$. There are two aspects of the effect of increased damping in this case. The maximum value of q is reduced and it occurs earlier when ζ is increased. However, increasing ζ leads to longer elapsed times for the response to become close to zero. These trends result from the fact that from $t = 0$ until maximum q is attained, the spring and dashpot both act to oppose the movement. After the maximum is attained, the dashpot opposes the action of the spring to return the mass to $q = 0$. Increasing ζ raises the dashpot force relative to the spring, hence the longer elapsed time.

In both Figures 2.12 and 2.13, the value of q never returns to zero, so there is no oscillation. It is possible to make q reach zero in a finite amount of time by letting $\dot{q}_0$ be a large negative value with $q_0 > 0$. (This corresponds to slamming an overdamped door.) In that case q reaches a negative maximum, and then asymptotically approaches $q = 0$ from below.

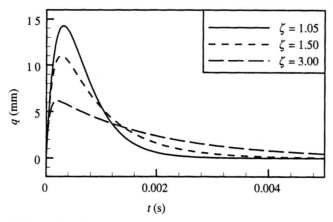

FIGURE 2.13 Overdamped response to an initial velocity.

2.2.4 Critically Damped Systems

Although the mathematical forms of the overdamped and underdamped responses are quite different, graphs of q as a function of t for ζ slightly larger or smaller than unity with all other parameters held constant do not differ greatly. The critical damping case, $\zeta = 1$, is the transition. Here, the characteristic equation gives a repeated root, $\lambda = -\omega_{nat}$. However, the differential equation, being second order, must have two independent solutions. The standard rule for such situations is to take the second solution as the first solution multiplied by t. There are several ways of demonstrating this. For us, the most meaningful view comes from taking the limit of either the overdamped or underdamped response as $\zeta \longrightarrow 1$. We will demonstrate the former here; the limiting form of the underdamped response is a topic in Exercise 2.38. The first step in finding the limiting form of the overdamped free response described by eq. (2.2.34) is to regroup terms,

$$q = q_0 \frac{\lambda_2 \exp(-\lambda_1 t) - \lambda_1 \exp(-\lambda_2 t)}{\lambda_2 - \lambda_1} + \dot{q}_0 \frac{\exp(-\lambda_1 t) - \exp(-\lambda_2 t)}{\lambda_2 - \lambda_1} \qquad (2.2.35)$$

When $\zeta > 1$, we have $\lambda_2 = \lambda_1 + \varepsilon$, where $\varepsilon = 2\omega_{nat}(\zeta^2 - 1)^{1/2}$. As $\zeta \longrightarrow 1$, $\varepsilon \longrightarrow 0$. We use the smallness of ε to write $\exp(-\lambda_2 t) = \exp[-(\lambda_1 + \varepsilon)t] = \exp(-\lambda_1 t)[1 - \varepsilon t + O(\varepsilon^2 t^2)]$, where we retain only the first two terms in the series expansion of $\exp(-\varepsilon t)$ because we are interested in the limit where ε vanishes. We use the foregoing to eliminate λ_2 and $\exp(-\lambda_2 t)$ from eq. (2.2.35), which leads to

$$q = q_0 \exp(-\lambda_1 t) \frac{(\lambda_1 + \varepsilon) - \lambda_1(1 - \varepsilon t)}{\varepsilon} + \dot{q}_0 \exp(-\lambda_1 t) \frac{1 - (1 - \varepsilon t)}{\varepsilon} + O(\varepsilon^2 t^2)$$

$$(2.2.36)$$

When we let $\varepsilon \longrightarrow 0$, so that $\lambda_1 \longrightarrow \omega_{nat}$, we find that

$$q = [q_0(1 + \omega_{nat} t) + \dot{q}_0 t] \exp(-\omega_{nat} t) \qquad (2.2.37)$$

The fact that $\exp(-\omega_{nat} t)$ and $t \exp(-\omega_{nat} t)$ are the fundamental solutions is evident in the above.

Our investigations of underdamped and overdamped responses suggested that critical damping will attenuate a free vibration most quickly. A corollary of the foregoing derivation is the observation that the response of a critically damped system is not drastically different from that of an overdamped system with $\zeta \approx 1$. This is illustrated by Figure 2.14.

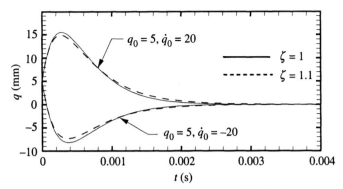

FIGURE 2.14 Comparison of the response of critically damped and slightly overdamped one-degree-of-freedom systems.

Because exponential decay only approaches the $q = 0$ axis asymptotically, no value of ζ ever brings the system to rest permanently. This is a fault of the linear damping model, but it is not a serious flaw. The nature of an exponential decay is such that eventually the value of q will be less than the error in any measurement we might make of an actual motion. The relative rapidity with which critical damping quiets a response makes it a particularly desirable case for controlling free vibration stemming from initial disturbances.

EXAMPLE 2.8

The system properties are $m = 400$ grams, $k = 1400$ N/m, and c is such that the system is critically damped. The block is given an initial velocity of 10 m/s to the left with the spring compressed by 30 mm. Determine the ensuing displacement $x(t)$ of the spring. Does there seem to be any instant at which $x = 0$?

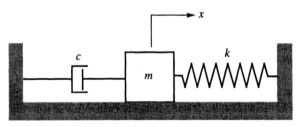

Solution This exercise highlights the evaluation and interpretation of critically damped response. The natural frequency is

$$\omega_{\text{nat}} = \left(\frac{k}{m}\right)^{1/2} = 59.16 \text{ rad/s}$$

The general solution for critical damping is

$$x = (c_1 + c_2 t)\exp(-\omega_{\text{nat}} t)$$

The initial displacement condition is that $x = 0.03$ m and $\dot{x} = -10$ m/s at $t = 0$, with the sign for both initial values selected according to the definition of positive x to the right. Matching the general solution to these initial values gives

$$c_1 = 0.03, \qquad -\omega_{\text{nat}}c_1 + c_2 = -10 \quad \Rightarrow \quad c_2 = -8.225$$

The corresponding response is

$$x = (0.03 - 8.225t)\exp(-59.16t) \text{ m}$$

To determine whether $x = 0$ is a possibility, we note that the exponential factor is always positive. Therefore, the only possibility for $x = 0$ is that the coefficient of that factor is zero. In most cases, a critically damped response will not attain a zero value in a finite amount of time, because the coefficients c_1 and c_2 in the general solution have the same sign. The present situation is an exception. Here we have

$$0.03 - 8.225t = 0 \quad \Rightarrow \quad t = 3.65 \text{ ms}$$

After this instant, the coefficient multiplying $\exp(-59.16t)$ in the expression for x remains negative, so we find that x remains negative in the subsequent motion. The plot of x as a function of t shows that the minimum value is comparable to the initial displacement.

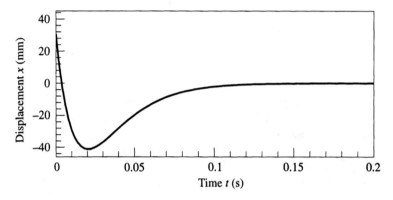

EXAMPLE 2.9

A one-degree-of-freedom system whose natural frequency is 2 Hz has an initial velocity of 5 m/s at $x = 300$ mm.

(a) Use mathematical software to determine the elapsed time required for the system to return to $x = 30$ mm as a function of critical damping ratios in the range $1 \le \zeta \le 20$. Show the result graphically. Which damping ratio seems to be optimal for bringing the system to rest?

(b) Repeat the analysis in part (a) to determine the minimum time in which the overdamped system will return to $x = 300$ mm.

Solution In addition to illustrating the procedure for satisfying initial conditions for overdamped systems, this exercise will provide an interesting perspective for the role of damping in calming a system. We begin with evaluation of the characteristic exponents. The natural frequency is stated to be $\omega_{\text{nat}} = 4\pi$ rad/s, but we treat the critical damping ratio as an algebraic parameter. Correspondingly, the characteristic roots in eq. (2.2.32) become functions of ζ,

$$\lambda_1(\zeta) = 4\pi[\zeta - \sqrt{\zeta^2 - 1}], \qquad \lambda_2(\zeta) = 4\pi[\zeta + \sqrt{\zeta^2 - 1}]$$

The initial conditions are $q_0 = 0.3$ m and $\dot{q}_0 = 5$ m/s. The response found from eq. (2.2.34) in this case is

$$q = \frac{1}{8\pi\sqrt{\zeta^2 - 1}}\{[0.3\lambda_2(\zeta) + 5]\exp[-\lambda_1(\zeta)t]$$
$$-[0.3\lambda_1(\zeta) + 5]\exp[-\lambda_2(\zeta)t]\}$$

When we equate this expression to a specified value $q = q'$, we obtain a complicated equation relating the value of ζ and the time t' required to reach that position. No closed form solution is possible, so we resort to numerical methods. MATLAB has the *fzero* function and Mathcad has the *root* function, either of which can find values of u that satisfy $F(u) = 0$. The function in this case is

$$F(t', \zeta) = \frac{1}{8\pi\sqrt{\zeta^2 - 1}}\{[0.3\lambda_2(\zeta) + 5]\exp[-\lambda_1(\zeta)t']$$
$$-[0.3\lambda_1(\zeta) + 5]\exp[-\lambda_2(\zeta)t']\} - q' = 0$$

We wish to solve this for the value of t' associated with a specified ζ.

Because we must find the root t' for a range of values of ζ, we shall follow an incremental procedure, starting at $\zeta = 1$, with increments $\Delta\zeta = 0.05$. When $\zeta = 1$, the response is described by eq. (2.2.37). For the stated initial conditions, the function to be solved is

$$F(t', 1) = [0.3(1 + 4\pi t') + 5t]\exp(-4\pi t') - q' = 0$$

The numerical algorithm requires a starting guess with which to initiate the procedure. We generate this value by plotting $F(t', 1)$ as a function of t', and visually identifying the location where the function crosses zero. We use the software equation solver to get an accurate value for the root t' at $\zeta = 1$. We then increment ζ by $\Delta\zeta$, so the new value is $\zeta = 1.05$. We find the value of t' satisfying $F(t', 1.05) = 0$ by invoking the equation solver, with a starting value set to the value of t' found for the previous ζ. Once this root has been determined, we increment ζ again and repeat the process. The starting value in this case is the value of t' found for the previous ζ. We continue this procedure in a program loop until we have found roots for each incremental value of ζ in the range $1 \le \zeta \le 20$. The results of this numerical solution process are graphed below.

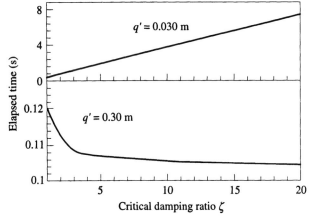

The case where $q' = 0.03$ m fits the notion that critical damping returns a system to its equilibrium position in the shortest amount of time. Increasing the value of ζ from $\zeta = 1$ results in an essentially linear increase in the time required to reach $q' = 0.003$. The situation for $q' = 0.3$ m, which is the same as the initial displacement, is quite different. In this case the time required to return to that value is approximately $t' = 0.1$ s, regardless of the value of ζ. To understand the reason for this, we plot q as a function of t for several values of ζ.

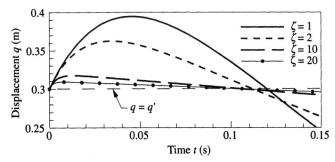

As displayed in the graphs, increasing damping reduces the time required for q to attain its maximum value, because $\dot{q} > 0$ in this interval. After the maximum is attained, $\dot{q}$ becomes negative, so the damping force resists the spring force, which acts to return q to zero. The value of q' in this case matches the initial value q_0, which leads us to conclude that whatever is gained in the initial phase of the motion by increasing damping is canceled by the retarding effect subsequent to maximum displacement.

2.2.5 Coulomb Friction

An important source of energy dissipation is friction generated by the relative motion of two surfaces that press against each other. The *Coulomb friction laws* offer a simple description of this complicated process. It is implicit to this discussion that the contacting surfaces are not lubricated. Indeed, the Coulomb friction model is sometimes referred to as *dry friction*. The friction laws have two aspects, depending on whether the surfaces are stationary or moving with respect to each other:

Static friction. The friction force $\overline{f}$ exerted between surfaces having no relative movement is known only to the extent that $|\overline{f}|$ cannot exceed a maximum value that is proportional to the magnitude of the normal force $\overline{N}$ pressing the surfaces together. The factor of proportionality is the *coefficient of static friction,* μ_s. Thus, the static friction law states that $|\overline{f}| \leq \mu_s N$.

Kinetic friction. In this case the contacting surfaces are sliding over each other. The magnitude of the friction force is then taken to be proportional to the magnitude of the normal force, $|\overline{f}| = \mu_k N$, where μ_k is the *coefficient of kinetic friction.* The sense of the friction force is such that it opposes the relative motion. For example, consider a case where the upper surface is sliding to the right as seen from the lower surface, which is equivalent to saying that the lower surface is sliding to the left relative to the upper surface. In that case the friction force on the upper surface will be oriented leftward, while the friction force on the lower surface will be oriented to the right.

These laws are crude first approximations of the true nature of friction, whose study is the specialty area of tribology. In keeping with its simplified nature, dynamics formulations often take the two coefficients to be equal, although reported values often indicate that μ_s can be as much as 30% larger than μ_k.

Let us use these laws to formulate the equation of motion for the system in Figure 2.15, where a block of mass m slides over a rough horizontal surface under the restraint of a spring. Because the surface is horizontal, the normal force is constant at $N = mg$. We select the position x, measured from the location at which the spring is unstretched, as the generalized coordinate. The Coulomb friction laws lead to three alternatives regarding the friction force $\overline{f}$:

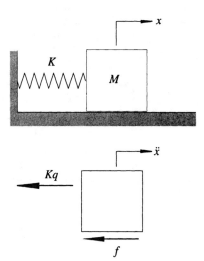

FIGURE 2.15 Spring-mass system with Coulomb friction.

- If the block is sliding to the right, $\dot{x} > 0$; then the friction force acting on the block is to the left, and its magnitude is $\mu_k N$.
- If the block is sliding to the left, $\dot{x} < 0$; then the friction force acting on the block is to the right, and its magnitude is $\mu_k N$.
- If the block comes to rest, $\dot{x} = 0$, at any instant, then there is static friction, $|\overline{f}| < \mu_s N$.

The sign of $\dot{x}$ defines the sense of the friction force. Because we have taken $\overline{f}$ to act to the left in the free body diagram, we have

$$f = \mu_k N \operatorname{sgn}(\dot{x}) \quad \text{if } \dot{x} \neq 0 \tag{2.2.38}$$

In this expression, $\operatorname{sgn}(\dot{x})$ denotes the signum function, which is implemented in many programming environments. If it is not, a simple way to evaluate it is $\operatorname{sgn}(\dot{x}) = \dot{x}/|\dot{x}|$, which obviously is not defined if $\dot{x} = 0$.

From Newton's Second Law, the equation of motion for the block is $m\ddot{x} = -kx - f$. In view of the above expression for f, this becomes

$$m\ddot{x} + kx = -\mu_k N \operatorname{sgn}(\dot{x}) \quad \text{if } \dot{x} \neq 0 \tag{2.2.39}$$

Instants when $\dot{x} = 0$ will be handled separately by considering the static friction law. We have left the normal force as N rather than replacing it with mg in order to accommodate situations where the system lies on an incline, or where forces other than gravity cause a body to press against a surface.

Although the term on the right side of eq. (2.2.39) depends on $\dot{x}$, its value is constant until the sign of $\dot{x}$ changes. This enables us to construct a solution in a piecewise fashion, based on subintervals in which $\dot{x}$ does not change sign. In each subinterval, the solution is the sum of a constant particular solution plus a complementary solution in the form of an undamped free vibration, specifically

$$x = -\Delta \operatorname{sgn}(\dot{x}) + A\cos(\omega_{\text{nat}}t) + B\sin(\omega_{\text{nat}}t), \qquad \Delta = \frac{\mu_k N}{k} \tag{2.2.40}$$

We could construct the solution for arbitrary initial conditions, but it is easier to consider a specific case where the system is initially at rest, so we have $x = x_0$ and $\dot{x} = 0$ at $t = 0$. Also, let us assume that x_0 is very large, so that we may defer considering the possibility that the system will freeze due to static friction. Because the spring force is kx_0, it pulls the mass back to $x = 0$, so the initial motion corresponds to $\dot{x} < 0$. The initial condition that $\dot{x} = 0$ requires that $B = 0$ in eq. (2.2.40), and we obtain A by setting $x = x_0$ at $t = 0$. The result is

$$x = \Delta + (x_0 - \Delta)\cos(\omega_{\text{nat}}t) \tag{2.2.41}$$

As depicted in the initial interval of Figure 2.16, the quantity Δ represents the mean value of the response, and the amplitude $x_0 - \Delta$ is the difference between the value of x at the start of the interval and the mean value. This solution is valid as long as x continues to decrease, which is true when $\omega_{\text{nat}}t < \pi$. Hence at $t = \pi/\omega_{\text{nat}}$, the value of x attains a minimum. Consistent with our analysis of underdamped systems, let us define minimum values of x with a subscript that is an odd multiple of $\frac{1}{2}$, so we have

$$x_{1/2} = -x_0 + 2\Delta, \qquad t_{1/2} = \frac{\pi}{\omega_{\text{nat}}} \tag{2.2.42}$$

If x_0 is very large, $x_{1/2}$ will also be large negatively, which means that the spring is compressed greatly. As a result it pushes the block back to $x = 0$, so the

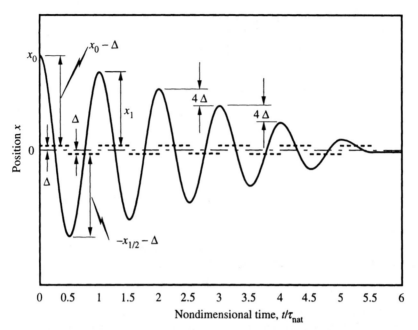

FIGURE 2.16 Response of a spring-mass system with Coulomb friction, $\Delta = 0.109$, $x_0 = 22.5\Delta$.

next interval of motion commences, with $\dot{x} > 0$. The position and velocity must be continuous at $t = t_{1/2}$, so we construct a solution according to eq. (2.2.40), with $\text{sgn}(\dot{x}) = 1$ and initial conditions set as $x = x_{1/2}$ and $\dot{x} = 0$ at $t = \pi/\omega_{\text{nat}}$. Once again, the coefficient $B = 0$. Thus, we obtain

$$x = -\Delta - (x_{1/2} + \Delta)\cos(\omega_{\text{nat}}t) \equiv -\Delta + (x_0 - 3\Delta)\cos(\omega_{\text{nat}}t) \qquad (2.2.43)$$

As depicted in Figure 2.16 for the second interval, the mean value of x is now $-\Delta$ and the amplitude $A = -x_{1/2} - \Delta$ is the difference between this mean value and $x_{1/2}$. This solution is valid as long as $\dot{x} > 0$. When $\cos(\omega_{\text{nat}}t) = 1$, x attains a maximum value. The displacement and time at the last instant for the preceding solution are

$$x_1 = x_0 - 4\Delta, \qquad t_1 = \frac{2\pi}{\omega_{\text{nat}}} \qquad (2.2.44)$$

We could go on in this manner constructing alternating intervals in which $\dot{x} < 0$ is followed by $\dot{x} > 0$, but a simpler approach is available. We observe that at $t = t_1$, we have the same situation as at $t = 0$, except that the position is x_1 rather than x_0. We therefore introduce the retarded time $t' = t - t_1$. Because the interval from $t = 0$ to $t = t_1$ is a natural period, we have $\cos[\omega_{\text{nat}}(t - t_1)] \equiv \cos(\omega_{\text{nat}}t)$. Thus, we may apply eqs. (2.2.41) and (2.2.43) alternately, with subscript $n - 1$ used to denote the starting maximum, subscript $n - \frac{1}{2}$ for the minimum, and subscript n for the ending maximum,

$$x = \begin{cases} \Delta + (x_{n-1} - \Delta)\cos(\omega_{\text{nat}}t), & t_{n-1} < t < t_{n-1/2} \\[2mm] -\Delta + (x_{n-1} - 3\Delta)\cos(\omega_{\text{nat}}t), & t_{n-1/2} < t < t_n \end{cases} \qquad (2.2.45)$$

The extreme values of x and the times at which they occur are given by

$$x_n = x_{n-1} - 4\Delta, \qquad t_n = n\frac{2\pi}{\omega_{\text{nat}}}$$

$$x_{n-1/2} = -x_{n-1} + 2\Delta \equiv x_{n-3/2} + 4\Delta, \qquad t_{n-1/2} = \left(n - \frac{1}{2}\right)\frac{2\pi}{\omega_{\text{nat}}}$$

(2.2.46)

From these considerations, we are able to fill in the remainder of Figure 2.16. Note that each maximum decreases relative to the previous one by a constant difference 4Δ, which also is the increment by which a minimum increases relative to the previous one. The intervals separating adjacent maxima or minima are the natural period of the undamped system, $\tau_{\text{nat}} = 2\pi/\omega_{\text{nat}}$. In other words, *the response decays linearly with time.*

One consequence of a linear decay is that the system will come to rest in a finite amount of time, unlike the response due to viscous damping. To ascertain when motion ceases, we must consider the static friction phenomenon. The instants at which $\dot{x} = 0$ are $t = t_{n-1/2}$ and $t = t_n$. At these instants, the magnitude of the spring force is $k|x_{n-1/2}|$ or kx_n, respectively. The maximum friction force that can be developed is $\mu_s N$. If the spring force, which tends to return the block to $x = 0$, does not exceed the maximum friction force, then the block will freeze at that position, and the motion ceases. Thus, the conditions for which motion ceases are

$$x_n < \frac{\mu_s N}{k} \quad \text{or} \quad x_{n-1/2} > -\frac{\mu_s N}{k}$$

(2.2.47)

If we take $\mu_s = \mu_k$, these limits will be collinear with the mean values $x = \pm\Delta$ for each half-cycle. Otherwise, because $\mu_s > \mu_k$, the limits will be slightly farther from the t axis than the mean values. For the case depicted in Figure 2.16, the motion ends after $5\frac{1}{2}$ cycles. For a given system (that is, fixed N, μ_s, μ_k, m, and k), the number of half-cycles required to bring the system to rest is set by the x_0 value.

For many situations, the Coulomb friction model is more realistic than the viscous damping model. Although we have considered a simple situation in which a block rubs over a plane, it is possible to generalize it to other systems, such as a shaft in an unlubricated bearing. Despite the virtue of realism, Coulomb friction is used much less often than viscous damping as a model for energy dissipation, primarily because Coulomb friction is much more difficult to analyze. If we wish to, we could define an equivalent critical damping ratio by using peak values x_n that are separated by several cycles to form the logarithmic decrement according to eq. (2.2.29).

EXAMPLE 2.10

Parameters for the spring-mass system in Figure 2.15 are $M = 10$ kg, $\omega_{\text{nat}} = 200$ rad/s, and $\mu_s = \mu_k = 0.2$. The system is released from rest with an initial displacement of 20 mm at $t = 0$. If energy dissipation is solely due to dry friction, how long will it be before the system comes to rest, and what will be the final displacement?

Solution This exercise is a straightforward demonstration of the application of the decay results for Coulomb friction. From the given parameters, we find $K = M\omega_{\text{nat}}^2 = 4(10^5)$ N/m. With $N = Mg = 98.1$ N, this gives $\Delta = \mu_k N/K = 4.905(10^{-5})$ m. Because the initial velocity is zero, we have $x_0 = 0.02$ m. From eqs. (2.2.46) we find $x_{1/2} = -x_0 + 2\Delta = -1.9902(10^{-2})$ m. For linear decay, the successive maxima and minima are found from eqs. (2.2.46) to be

$$x_n = x_0 - n(4\Delta), \qquad x_{n+1/2} = x_{1/2} + n(4\Delta)$$

It is given that $\mu_s = \mu_k$, so eq. (2.2.47) indicates that the conditions for static equilibrium are

$$x_n < 4.905(10^{-5})\,\text{m} \quad \text{or} \quad x_{n+1/2} > -4.905(10^{-5})\,\text{m}$$

For cessation of motion at a maximum, the first of the preceding is satisfied if

$$x_0 - n(4\Delta) < 4.905(10^{-5}) \quad \Rightarrow \quad n < \frac{x_0 - 4.905(10^{-5})}{4\Delta} \quad \Rightarrow \quad n = 102$$

A similar analysis for cessation of motion at a minimum leads to

$$x_{1/2} + n(4\Delta) > -4.905(10^{-5}) \quad \Rightarrow \quad n > \frac{-x_{1/2} - 4.905(10^{-5})}{4\Delta} \quad \Rightarrow \quad n = 102$$

Both calculations give the same number of intervals, but x_{102} occurs before $x_{102+1/2}$. Hence, the system comes to rest at the 102nd maximum. The natural period is the time interval between successive maxima, so the elapsed time is

$$t = 102\left(\frac{2\pi}{\omega_{\text{nat}}}\right) = 3.204\,\text{s}$$

2.3 TRANSIENT RESPONSE TO BASIC EXCITATIONS

Our concern here is with the determination of the response to excitations that are present for a limited time interval. Such excitation is produced by an external force, which appears in the equation of motion as an inhomogeneous term $Q(t)$ on the right-hand side, $M\ddot{q} + C\dot{q} + Kq = Q$. Because this differential equation is linear and its coefficients are constant, a variety of techniques are available for solving it. Here, we shall invoke a fundamental technique, the method of undetermined coefficients. It will lead us to a fundamental concept for designing systems to withstand shock loadings, specifically, shock spectra. However, the method of undetermined coefficients is limited in the types of excitations it can address, so we will return to the evaluation of transient response later in this chapter.

2.3.1 Method of Undetermined Coefficients

The equation of motion for the system is eq. (2.2.1). Its solution must satisfy an appropriate set of initial conditions, which we shall consider to apply at $t = 0$. The general procedure entails finding two types of solutions:

1. The complementary (or homogeneous) solution q_c gives zero when it is substituted into the left side of the equation of motion. Thus, the form of the homogeneous solution will be the same as that of the free response—undamped, underdamped, critically damped, or overdamped.

2. The particular solution q_p, which when substituted into the left side of the differential equation, gives the same terms as $Q(t)$.

We may determine q_p by the method of undetermined coefficients only in situations where the inhomogeneous term $Q(t)$ is composed of terms that are products of integer powers of t, exponentials in t, and/or sinusoidals in t, that is

$$Q(t) = t^n \exp(\mu t)[a_1 \sin(\omega t) + a_2 \cos(\omega t)] \qquad (2.3.1)$$

where n is a nonnegative integer and μ, ω, a_1, and a_2 are real. The basic method may be found in any standard text on ordinary differential equations, so we shall only summarize it here. The first step is to identify the general form of q_p. Let us suppose that $Q(t)$ consists of a single term. We identify this term as a *variable part* after we set any constant factor in $Q(t)$ to unity. We differentiate this variable part with respect to t, and identify any new type of function with coefficient equal to one as another variable part. We continue to differentiate all variable parts until no new types of functions appear. If *any* of the variable parts identified in this manner match *any* of the terms forming the homogeneous solution, then *all* variable parts should be multiplied by the lowest power of t that makes all of them differ from the fundamental homogeneous solutions. We then form the trial particular solution by multiplying each variable part by an arbitrary constant. We substitute this trial solution into the differential equation. To determine the values of these constants we collect the coefficients of each variable part in the left side of the equation of motion, and match each to the coefficient of the corresponding variable part of $Q(t)$. In the event that the excitation $Q(t)$ consists of several terms in the form of eq. (2.3.1), we may superpose the particular solution for each.

The foregoing fully specifies the particular solution. It still remains to determine the coefficients appearing in the complementary solution. As was true for free vibration, these are determined by satisfying the initial conditions. For that step it is important to recognize that the particular solution might contribute to the initial conditions, that is, q_p or $\dot{q}_p$ might be nonzero at $t = 0$. Because the total solution is

$$q = q_p + q_c \qquad (2.3.2)$$

we determine the coefficients of the homogeneous solution by requiring that

$$\left. q_p \right|_{t=0} + \left. q_c \right|_{t=0} = q_0$$
$$\left. \dot{q}_p \right|_{t=0} + \left. \dot{q}_c \right|_{t=0} = \dot{q}_0 \qquad (2.3.3)$$

EXAMPLE 2.11

A waveform often used to represent an excitation that rises rapidly, then decays slowly, is $Q(t) = F_0 \omega_{\text{nat}} t \exp(-\lambda t)$. A force such as this is applied at $t = 0$ to an undamped system that is at rest in its static equilibrium position. Determine the response $q(t)$.

Solution This exercise will illustrate most of the basic operations entailed in using the method of undetermined coefficients. The system is undamped, so the complementary solution is

$$q_c = c_1 \cos(\omega_{\text{nat}} t) + c_2 \sin(\omega_{\text{nat}} t)$$

The first variable part for the particular solution is $Q(t)$ with the constant factors set to one, $q_1 = t \exp(-\lambda t)$. The derivative of this term is $\dot{q}_1 = \exp(\lambda t) - \lambda t \exp(\lambda t)$. The first term is unlike the first variable part, so the second variable part is $q_2 = \exp(-\lambda t)$. Further differentiations of q_1 and q_2 reproduce those functions, so there are no other variable parts to consider. Both variable parts are different from the terms in the complementary solution; thus we multiply each variable part by a coefficient to form the particular solution,

$$q_p = At \exp(-\lambda t) + B \exp(-\lambda t)$$

We substitute this trial solution into the differential equation for the undamped system, which leads to

$$\ddot{q}_p + \omega_{nat}^2 q_p = [(-2\lambda + \lambda^2 t + \omega_{nat}^2 t)A + (\lambda^2 + \omega_{nat}^2)B]\exp(-\lambda t)$$

$$= \frac{Q(t)}{M} = \frac{F_0}{M}\omega_{nat}t \, \exp(-\lambda t)$$

Matching like terms on either side of the last equality requires that

$$-2\lambda A + (\lambda^2 + \omega_{nat}^2)B = 0, \qquad (\lambda^2 + \omega_{nat}^2)A = \frac{F_0}{M}\omega_{nat}$$

from which we find that

$$A = \frac{F_0}{M}\frac{\omega_{nat}}{(\lambda^2 + \omega_{nat}^2)}, \qquad B = \frac{F_0}{M}\frac{2\lambda\omega_{nat}}{(\lambda^2 + \omega_{nat}^2)^2}$$

The last step is to determine the constants in the complementary solution by satisfying the initial conditions. Here, we require that $q = q_c + q_p$ satisfy $q = \dot{q} = 0$ at $t = 0$,

$$q(t = 0) = B + c_1 = 0, \qquad \dot{q}(t = 0) = A - \lambda B + \omega_{nat}c_2 = 0$$

We solve these relations for c_1 and c_2. The response is the sum of the complementary and particular solutions,

$$q = At\exp(-\lambda t) + B\exp(-\lambda t) + c_1\cos(\omega_{nat}t) + c_2\sin(\omega_{nat}t)$$

$$= \frac{F_0}{M(\lambda^2 + \omega_{nat}^2)}\left\{\left[\omega_{nat}t + \frac{2\lambda\omega_{nat}}{(\lambda^2 + \omega_{nat}^2)}\right]\exp(-\lambda t)\right.$$

$$\left. - \frac{2\lambda\omega_{nat}}{(\lambda^2 + \omega_{nat}^2)}\cos(\omega_{nat}t) + \left(\frac{\lambda^2 - \omega_{nat}^2}{\lambda^2 + \omega_{nat}^2}\right)\sin\omega_{nat}t\right\}$$

An aspect of the result worth noting is that, although both initial conditions are zero, the complementary solution is nonzero. Omission of the complementary solution is a common error.

2.3.2 Superposition and Time Delay

The method of undetermined coefficients treats each excitation as a different problem to be done anew. Many individuals learn Laplace transforms as a way of solving ordinary differential equations. The elementary application of the concept uses tabulated transforms for basic functions, in combination with fundamental transform properties, to obtain a solution. We will develop in this section a technique for determining response that has many of the virtues of Laplace transforms, without requiring consideration of the transform procedure.

The approach relies on the linearity property, which permits decomposition of the response into parts other than complementary and particular solutions. The first decomposition considers q to be the sum of a term q_{IC} that is the homogeneous solution satisfying the initial conditions for q, and a term q_F that represents the forced response corresponding to zero initial conditions,

$$q = q_{IC} + q_F \tag{2.3.4}$$

where

$$M\ddot{q}_{IC} + C\dot{q}_{IC} + Kq_{IC} = 0, \quad q_{IC} = q_0 \text{ and } \dot{q}_{IC} = \dot{q}_0 \text{ at } t = 0$$

$$M\ddot{q}_F + C\dot{q}_F + Kq_F = Q, \quad q_F = \dot{q}_F = 0 \text{ at } t = 0 \tag{2.3.5}$$

Clearly, q_{IC} is the response that would be obtained without excitation; it is given by either of eqs. (2.2.22) for $\zeta < 1$, (2.2.37) for $\zeta = 1$, or (2.2.34) for $\zeta > 1$. In most cases of transient excitation, the system will initially be at rest in its equilibrium position, in which case q_{IC} will be identically zero.

The method of undetermined coefficients may be used to derive and tabulate solutions for q_F corresponding to a set of standard excitations $Q(t)$. To indicate that these excitations start at $t = 0$, we introduce the *Heaviside step function*. The definition of this function is that it is zero if its argument is negative, and one if its argument is positive. It is conventional to use the symbol h to denote the step function. It soon will be necessary to consider functions that start at some instant $T > 0$, so a step function starting at $t = T$ is $h(t - T)$, which is defined as

$$h(t - T) = \begin{cases} 0 \text{ if } t < T \\ 1 \text{ if } t > T \end{cases} \tag{2.3.6}$$

(The function is not defined at $t = T$, where it is discontinuous.)

Appendix B gives the response for zero initial conditions resulting from several types of excitation, all of which start at $t = 0$. The transient excitations described there include a *step*, $Q(t) = h(t)$; a *ramp*, $Q(t) = th(t)$; a *quadratic*, $t^2h(t)$; an *exponential*, $Q(t) = \exp(-\beta t)h(t)$; a *sinusoidal*, $Q(t) = \sin(\omega t)h(t)$; and a *cosinusoidal*, $Q(t) = \cos(\omega t)h(t)$. In each case multiplying the analytical function by $h(t)$ serves to emphasize that the excitation starts at $t = 0$. Examination of the tabulated responses makes it obvious that they contain both the particular solution and the complementary solution required to give zero initial conditions.

Let us use the symbol $w_a(t)$ to denote a solution taken from Appendix B for a specified excitation $F_a(t)h(t)$. The coefficient multiplying each excitation function in Appendix B is unity. Linearity leads to the conclusion that if the excitation is $\alpha F_a(t)h(t)$, then $q_F = \alpha w_a(t)$. In other words, the coefficient multiplying the tabulated excitation function scales the response. Now suppose that the excitation of interest is like one of the functions described by Appendix B, except that it does not commence until some later instant $T > 0$. We say that such a function is *delayed by an interval T*. The instant that we consider to be the time origin is arbitrary—what is important is the time elapsed from the instant at which the excitation began. The elapsed time is $t - T$. If $Q(t) = \alpha F_a(t)h(t)$ describes the force starting at $t = 0$, then starting the same force at $t = T$ corresponds to $Q(t) = \alpha F_a(t - T)h(t - T)$. In other words, wherever t occurs in the original force, we replace it by $t - T$ to describe the delayed force. Because delaying the force is equivalent to shifting the time origin, it follows that if $\alpha w_a(t)$ is the response to $Q(t) = \alpha F_a(t)h(t)$, then the response to the delayed force will be $q_F(t) = \alpha w_a(t - T)$,

$$Q(t) = \alpha F_a(t - T)h(t - T) \quad \Rightarrow \quad q_F = \alpha w_a(t - T) \tag{2.3.7}$$

This property may be implemented by replacing t with $t - T$ in the response entry of Appendix B.

The next situation to consider is one in which the excitation of interest can be represented as the sum of two types of functions $F_a(t)$ and $F_b(t)$, with both functions being excitations described by Appendix B. Linearity of the equations of motion enables us to represent the response as a superposition of responses associated with each excitation function. In other words, we have

$$Q(t) = \alpha F_a(t)h(t) + \beta F_b(t)h(t) \quad \Rightarrow \quad q_F = \alpha q_a(t) + \beta q_b(t) \tag{2.3.8}$$

Because $q_a(t)$ and $q_b(t)$ represent the response for zero initial conditions, it is evident that eq. (2.3.8) produces the response associated with the full excitation $Q(t)$ and zero initial conditions. The delay property in eq. (2.3.7) and the superposition property in eq. (2.3.8) may be used jointly. For example, if the second excitation in eq. (2.3.8) commences at $t = T$, rather than at $t = 0$, we would have

$$Q(t) = \alpha F_a(t)h(t) + \beta F_b(t - T)h(t - T) \quad \Rightarrow \quad q_F = \alpha q_a(t) + \beta q_b(t - T)$$

$$(2.3.9)$$

EXAMPLE 2.12

An undamped system is subject to a negative constant force for an interval T, after which the force rises linearly, as shown in the graph. Derive an expression for the response. Graph the resulting nondimensional response $q(M\omega_{nat}^2 / F_0)$ as a function of the nondimensional time parameter $\omega_{nat}t$ for the case where $T = 4\pi / \omega_{nat}$.

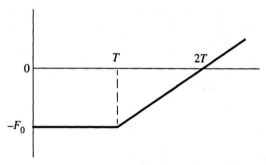

Solution The use of the superposition method in conjunction with Appendix B is the primary feature of this example. We also will discuss using mathematical software to graph responses containing step functions. The first operation is to identify the given excitation as a superposition of simpler functions. Until $t = T$, the excitation is constant at $-F_0$, which fits a step function starting at $t = 0$ and having $-F_0$ as its amplitude. At $t = T$, $Q(t)$ is modified by adding to the step an amount that increases proportionally to $t - T$. The proportionality factor is the slope F_0/T. This part of the force is a ramp starting at $t = T$, so the total excitation is

$$Q = -F_0 h(t) + \frac{F_0}{T}(t - T)h(t - T)$$

A useful check is to verify that $Q = 0$ at $t = 2T$.

Let us denote the responses in Appendix B for unit step and unit ramp excitations as $s(t)$ and $r(t)$. Because the system of interest is undamped, we set $\zeta = 0$ and $\omega_d = \omega_{nat}$ in those entries, which become

$$s(t) = \frac{1}{M\omega_{nat}^2}[1 - \cos(\omega_{nat}t)]h(t)$$

$$r(t) = \frac{1}{M\omega_{nat}^3}[\omega_{nat}t - \sin(\omega_{nat}t)]h(t)$$

The ramp is delayed in $Q(t)$, so the response is given by

$$q = -F_0 s(t) + \frac{F_0}{T}r(t - T)$$

$$= -\frac{F_0}{M\omega_{nat}^2}[1 - \cos(\omega_{nat}t)]h(t) + \frac{F_0}{TM\omega_{nat}^3}\{\omega_{nat}(t - T)$$

$$- \sin[\omega_{nat}(t - T)]\}h(t - T)$$

We wish to plot $q(M\omega_{nat}^2/F_0)$ as a function of a nondimensional time $\omega_{nat}t$, which we denote as τ. Because $t > T$ corresponds to $\tau > \omega_{nat}T$, we rewrite the second step function as $h(\tau - \omega_{nat}t)$, so that the solution becomes

$$q\left(\frac{M\omega_{nat}^2}{F_0}\right) = -[1 - \cos(\tau)]h(\tau) + \frac{1}{\omega_{nat}T}[(\tau - \omega_{nat}T) - \sin(\tau - \omega_{nat}T)]h(\tau - \omega_{nat}T)$$

The value of $\omega_{nat}T$ is 4π for the present case, so we are ready to plot this function.

Regardless of whether we use Mathcad, MATLAB, or other software, we must select the number of instants for evaluating q and the time between those instants. In terms of the τ variable, the sinusoidal terms have period 2π. Breaking a period into 20 parts should enable us to obtain a smooth curve, so the separation between instants will be $\Delta = \pi/10$. The excitation crosses zero at $\tau = 2\omega_{nat}T = 8\pi$, so stopping the evaluation at $\tau = 12\pi$ should be sufficient to see any interesting features.

In Mathcad, one defines a function anywhere in the worksheet prior to the place where it is needed. We define functions $s(\tau)$ and $r(\tau)$ as listed above. It is important to include the step function in this definition. This may be done with a logical term, $(\tau > 0)$, which is zero when false, or one if true. Thus, we would write $s(t) := (1 - \cos(\tau))*(\tau > 0)$ and $r(\tau) = (\tau - \sin(\tau))*(\tau > 0)$. We then set $T = 4*\pi$, $\Delta = 0.1*\pi$, and $N = 12*\pi/\Delta$. The last parameter is the number of time instants for the computation, corresponding to index $j := 1; N$. The time instants are $\tau_j = j*\Delta$. We may obtain a column of q values corresponding to the τ values by using the vectorize operator as follows:

$$\text{Mathcad: } q := \overrightarrow{\left(-s(\tau) + \frac{1}{T}*r(\tau - T)\right)}$$

The arrow above the right-side term designates that the operation is vectorized, corresponding to t being a column.

To carry out comparable operations in MATLAB, we define the unit step and ramp responses in function M-files. The first lines in these files are *function[q] = step(t)* and *function[q] = ramp(t)*, where q is defined according to the respective formula. In the MATLAB command window, we obtain the response by writing

$$\text{MATLAB: } t = 0 : 0.1*pi : 12*\pi; \quad q = -step(t) + \frac{1}{T}*ramp(t - T)$$

Note that because t is an array, the evaluation of *step(t)* and *ramp(t)* must be carried out in the M-files on the column of t values. In general, such operations might require using MATLAB's dot operator to indicate an element by element evaluation. For example, if it were necessary to evaluate t^2 at each instant in some other response, we would write the operation as t.^2. Note that the dot operator is not required to evaluate terms like $\cos(\tau)$.

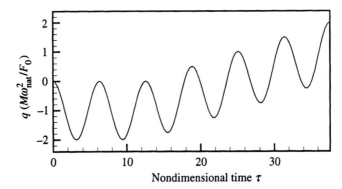

The graph displays the result of these computations. We see that the mean value of the response seems to track the excitation, with a superposed oscillation associated with the homogeneous solution. Note that the solution satisfies the continuity conditions at $t = 4\pi$, where the displacement and the slope are continuous.

2.3.3 Pulse Excitation

An important type of excitation is a *pulse*, which refers to any excitation that lasts a finite duration τ and then is terminated. Let $F(t)$ denote the time dependence of the generalized force Q between $t = 0$ and $t = \tau$. One way of analyzing pulse response is to consider two time intervals. In the forced phase, $0 \le t \le \tau$, the pulse excitation is active. The fact that the excitation will cease later is irrelevant to this interval, so the response is governed by the same equations as those for any other transient response,

$$\ddot{q} + 2\zeta\omega_{\text{nat}}\dot{q} + \omega_{\text{nat}}^2 q = \frac{F(t)}{M}; \quad 0 \le t \le \tau$$

$$q|_{t=0} = q_0, \qquad \dot{q}|_{t=0} = \dot{q}_0$$

(2.3.10)

If the forcing function $F(t)$ fits one or more of the entries in Appendix B, we may use the methods of the preceding section to analyze the response in the forced interval. Otherwise, we may obtain $q(t)$ by the method of undetermined coefficients, if $F(t)$ is suitable to that technique.

The second interval is $t > \tau$. Because there is no excitation, the system vibrates freely in this interval. The position and velocity cannot change discontinuously. Therefore, the initial conditions for the free vibration interval must be the same as the position and velocity at the end of the forced interval. Let us denote these values as q_τ and $\dot{q}_\tau$, which we obtain by evaluating the forced response at $t = \tau$. It follows that the free vibration response is governed by

$$\ddot{q} + 2\zeta\omega_{\text{nat}}\dot{q} + \omega_{\text{nat}}^2 q = 0, \quad t \ge \tau$$

$$q|_{t=\tau} = q_\tau, \qquad \dot{q}|_{t=\tau} = \dot{q}_\tau$$

(2.3.11)

The nonzero value of the initial time can complicate algebraic evaluation of the homogeneous solution's coefficients because the exponential and harmonic functions are not merely unity or zero at the initial time. One way of circumventing this minor difficulty is to shift the time origin, such that the new time t' is zero when $t = \tau$, so $t' = t - \tau$. Derivatives with respect to t' and t are the same, so the free vibration solution appropriate to the value of ζ may be applied directly to the present situation.

An alternative line of analysis for pulse excitation uses the superposition concept of the previous section to treat termination of the force as a delayed excitation. This procedure will yield a single expression describing the response at any time instant. The fundamental step is to identify how terminating the pulse may be decomposed into separate terms starting at various instants in order to invoke the superposition relations, eq. (2.3.9). In some cases the decomposition may be achieved graphically. A simple case illustrating this is the uniform pulse in Figure 2.17. We may picture the excitation as being a Heaviside step starting at $t = 0$ with amplitude three, which is cancelled by subtracting an amplitude three Heaviside step that begins at $t = 2$—that is, $F(t) = 3h(t) - 3h(t - 2)$. If $s(t)$ is the response to a unit amplitude step function starting at $t = 0$, then the response of the system, assuming the initial conditions are zero, would be $q = 3s(t) - 3s(t - 2)$.

In many situations it is difficult to perform the decomposition graphically, in which case a mathematical approach is available. Consider a pulse whose

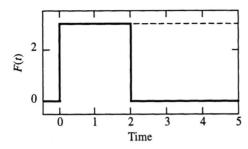

FIGURE 2.17 Typical uniform pulse.

functional form is $F(t)$ until it is removed at $t = \tau$, after which it is zero. The combination of Heaviside steps given by $h(t) - h(t - \tau)$ has the property of being unity for $0 < t < \tau$, and zero thereafter. Thus, the excitation may be written as

$$\text{Pulse:} \quad Q = F(t)[h(t) - h(t - \tau)] \tag{2.3.12}$$

We split this up into two terms. In the second term, we introduce the elapsed time $t' = t - \tau$, which is the argument of its step function. This is achieved by substituting $t = t' + \tau$, so that

$$Q = F(t)h(t) - F(t' + \tau)h(t'), \quad t' = t - \tau \tag{2.3.13}$$

The next step is to break up $F(t' + \tau)$ into simpler terms that depend only on t'. To illustrate this process consider $F(t) = \beta t^2$. Then $F(t' + \tau) = \beta(t' + \tau)^2 = \beta(t')^2 + 2\beta\tau t' + \beta\tau^2$. Let us denote as w_0, w_1, and w_2 basic solutions of the equation of motion when the excitation is $Q(t) = 1$, t, and t^2, respectively. (The initial conditions for each basic solution are zero.) When we write the quadratic pulse in the form of eq. (2.3.13), we obtain four basic excitations,

$$Q = \beta t^2 h(t) - \beta[(t')^2 + 2\tau t' + \tau^2]h(t') \tag{2.3.14}$$

The first term is a quadratic starting at $t = 0$, while the three terms in the brackets consist of quadratic, linear, and constant terms starting at $t' = 0$. Correspondingly, we superpose the basic responses, according to

$$q = \beta w_2(t)h(t) - \beta[w_2(t') + 2\tau w_1(t') + \tau^2 w_0(t')]h(t')$$

$$\equiv \beta w_2(t)h(t) - \beta[w_2(t - \tau) + 2\tau w_1(t - \tau) + \tau^2 w_0(t - \tau)]h(t - \tau) \tag{2.3.15}$$

The essential aspect of decomposing $F(t' + \tau)$ is that the individual terms should correspond to excitations for which the solution is known. Such a decomposition is always possible for the excitations covered in Appendix B. For example, if $F(t) = \exp(-\lambda t)$, then $F(t' + \tau) = \exp(-\lambda\tau)\exp(-\lambda t')$. Note that $\exp(-\lambda\tau)$ is a constant factor. If the pulse varies sinusoidally, so that $F(t) = \sin(\omega t)$, we would have $F(t' + \tau) = \sin[\omega(t' + \tau)] = \cos(\omega\tau)\sin(\omega t') + \sin(\omega\tau)\cos(\omega t')$.

EXAMPLE 2.13

The sketch is a simplified model of an automobile and its suspension. When the vertical displacement of the sprung mass is $y = 0$, the compressive force in the spring equals the weight. The system is underdamped, and the model is constrained to remain vertical. The horizontal speed v is constant. At $t = 0$, the wheel hits a bump in the shape of one lobe of a sine curve, which is defined as $z = H\sin(\pi x/L)$ for $0 < x < L$. Prior to contacting the bump the vehicle was not moving in the vertical direction, so $y = \dot{y} = 0$ at $t = 0$. The natural frequency is 1 Hz and the critical damping ratio is 0.2. Determine and graph the response y as a function of t when $H = 100$ mm, $L = 500$ mm, and $v = 5$ m/s.

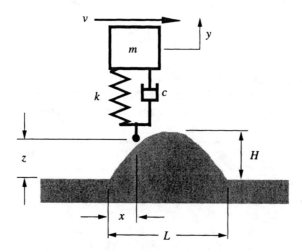

Solution The primary purpose of this exercise is to illustrate the procedure for decomposing a pulse excitation into simpler parts. It also will demonstrate application of the technique developed in Section 1.5.4 for modeling situations where the excitation of a system stems from inducing motion at the end of a spring or dashpot. According to that technique, we decompose the forces in these devices into a part that would result if the end was stationary, and a part associated with the imposed motion. Both contributions are shown in the free body diagram. An upward displacement z at the ground shortens the spring and dashpot by that amount, thereby inducing compressive forces kz and $c\dot{z}$ that push the mass upward. The forces that would exist even if the ground were stationary are the tensile forces $k\Delta$ and $c\dot{\Delta}$, corresponding to $\Delta > 0$ and $\dot{\Delta} > 0$. Gravity is omitted in the free body diagram, because it represents a static effect.

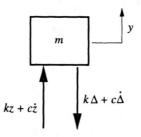

We select as the generalized coordinate the vertical displacement y, positive upward and measured from the static equilibrium elevation. Newton's Second Law for motion is readily applied in the present situation. Thus, we have $kz + c\dot{z} - k\Delta - c\dot{\Delta} = m\ddot{y}$. Upward displacement elongates the spring, so $\Delta = y$. We bring all terms containing y to the left side, which leads to

$$\ddot{y} + 2\zeta\omega_{nat}\dot{y} + \omega_{nat}^2 y = \omega_{nat}^2 z + 2\zeta\omega_{nat}\dot{z}$$

The ground elevation for $0 < x < L$ is given as $z = H\sin(\pi x/L)$, and $z = 0$ for $x > L$. We use step functions starting at $x = 0$ and at $x = L$ to turn on and cut off the sine function, specifically

$$z = H\sin\left(\frac{\pi x}{L}\right)[h(x) - h(x - L)]$$

The horizontal distance $x = vt$, and $h(vt - L) \equiv h(t - L/v)$. Hence, the time dependence of the ground elevation is

$$z(t) = H \sin\left(\frac{\pi v}{L} t\right)\left[h(t) - h\left(t - \frac{L}{v}\right)\right]$$

In anticipation of forming the excitation from this term, we split it up into two terms,

$$z = H \sin\left(\frac{\pi v}{L} t\right)h(t) - H \sin\left(\frac{\pi v}{L} t\right)h\left(t - \frac{L}{v}\right)$$

If we are to follow the superposition method for solving this differential equation, we need to convert the second sine term to a delayed function that depends on $t' = t - L/v$. Toward that end we substitute $t = t' + L/v$, which gives

$$\sin\left(\frac{\pi v}{L} t\right) = \sin\left[\frac{\pi v}{L}\left(t' + \frac{L}{v}\right)\right] = \sin\left(\frac{\pi v}{L} t' + \pi\right) \equiv -\sin\left(\frac{\pi v}{L} t'\right)$$

Thus the ground elevation is defined by

$$z(t) = H \sin\left(\frac{\pi v}{L} t\right)h(t) + H \sin\left[\frac{\pi v}{L}\left(t - \frac{L}{v}\right)\right]h\left(t - \frac{L}{v}\right)$$

We substitute this expression into the equation of motion, and collect the terms according to whether they start at $t = 0$ or $t = L/v$,

$$\ddot{y} + 2\zeta\omega_{nat}\dot{y} + \omega_{nat}^2 y = H\left[\omega_{nat}^2 \sin\left(\frac{\pi vt}{L}\right) + 2\zeta\omega_{nat}\frac{\pi v}{L}\cos\left(\frac{\pi vt}{L}\right)\right]h(t)$$

$$+ H\left\{\omega_{nat}^2\sin\left[\frac{\pi v}{L}\left(t - \frac{L}{v}\right)\right]\right.$$

$$\left. + 2\zeta\omega_{nat}\frac{\pi v}{L}\cos\left[\frac{\pi v}{L}\left(t - \frac{L}{v}\right)\right]\right\}h\left(t - \frac{L}{v}\right)$$

(In this situation, z is a continuous function of t, so $\dot{z}$ merely requires differentiation of the sine functions. If z had a discontinuity, it would be necessary to employ an impulse function, developed in the next section, to represent the infinite slope of z at the discontinuity.)

The right side is the sum of sine and cosine excitations starting at $t = 0$ and at $t = L/v$. Let us denote the responses for unit amplitude excitations of each type as $s(t)$ and $c(t)$, which are given in Appendix B. We multiply each unit response by the coefficient of the corresponding force term in the equation of motion, with the mass factor $M = 1$ because it has canceled out. The unit responses for the second set of terms are delayed by L/v, so the response is given by

$$\frac{q}{H} = \omega_{nat}^2 s(t) + 2\zeta\omega_{nat}\frac{\pi v}{L}c(t) + \omega_{nat}^2 s\left(t - \frac{L}{v}\right)$$

$$+ 2\zeta\omega_{nat}\frac{\pi v}{L}c\left(t - \frac{L}{v}\right)$$

This expression is adequate to evaluate and graph the response following the procedure in Exercise 2.12. We define $s(t)$ and $c(t)$ as functions directly in the Mathcad worksheet, or as function M-files for MATLAB. In the workspace we enter the values $\omega_{nat} = 2\pi$ rad/s, $\zeta = 0.2$, $H = 0.1$ m, $L = 0.5$ m, and $v = 5$ m/s. The time increment for evaluating the response is $0.1\,\pi/\omega_{nat}$, based on 20 points per undamped period. We set the maximum time for the evaluation to a value sufficiently large to permit the homogeneous solution to decay, $\zeta\omega_{nat}t_{max} = 5$, which leads to $t_{max} = 4$.

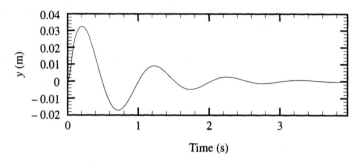

The duration during which the wheel contacts the bump is $L/v = 0.10$ s, which is much smaller than the natural period of 1 s. Thus, most of the response appearing in the graph corresponds to free vibration. In fact, the response looks like that of a heavily underdamped system to an initial velocity. We see that hitting the bump represents an excitation that acts for a short interval of time, yet induces a substantial motion. We will encounter such a force again in the next section.

2.3.4 Step and Impulse Responses

Two types of pulse excitation are very useful for interpreting the response to any pulse. An impulse excitation is one that is very large, but has an extremely short duration relative to the natural period, as exemplified by a hammer blow. A step excitation rises discontinuously from zero to a constant value that is sustained indefinitely. Such an excitation corresponds to suddenly applying a constant force to a system. In discussing both cases, we consider the system to be initially at rest, $q_0 = \dot{q}_0 = 0$ at $t = 0$.

The *unit step response* is the result of applying an excitation in the form of a Heaviside step function, $Q = h(t - \tau)$, where τ is the instant when the unit force is applied. Because the system was at rest at $q = 0$ when $t = 0$, it will remain at $q = 0$ until the force is applied. To determine the response after the application of the force, we use the elapsed time variable $t' = t - \tau$, so we solve

$$M\ddot{q} + C\dot{q} + Kq = 1, \quad q = \dot{q} = 0 \text{ at } t' = 0 \tag{2.3.16}$$

To see how the Appendix B entry for a unit step excitation was obtained, we observe that the particular solution is a constant, and the complementary has the form of the free vibration solution appropriate to the specific value of ζ. In the case of an underdamped system, the solution is

$$q = \frac{1}{K}\left\{1 - \exp(-\zeta\omega_{\text{nat}}t')\left[\cos(\omega_d t') + \frac{\zeta}{(1 - \zeta^2)^{1/2}}\sin(\omega_d t')\right]\right\} \tag{2.3.17}$$

The preceding may be converted to a function of t by substituting $t' = t - \tau$. In order to have a compact solution that is applicable for all t we multiply the solution by a Heaviside step function $h(t - \tau)$ that nullifies the solution for $t < \tau$. Standard notation to denote the unit step response is $u(t - \tau)$, so we have

$$u(t - \tau) = \frac{1}{K}\left\{1 - \exp(-\zeta\omega_{\text{nat}}(t - \tau))\left[\cos(\omega_d(t - \tau))\right.\right.$$

$$\left.\left. + \frac{\zeta}{(1 - \zeta^2)^{1/2}}\sin(\omega_d(t - \tau))\right]\right\} \tag{2.3.18}$$

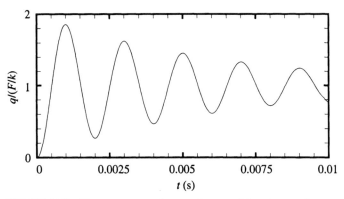

FIGURE 2.18 Step response of a one-degree-of-freedom system, $\omega_{nat} = 1000\pi$ rad/s, $\zeta = 0.05$.

Figure 2.18 displays a typical step response. We see that the response rises quickly and then oscillates about the mean value F/K. Eventually, the system comes to rest at this value, which is the steady-state static displacement.

Now let us consider a uniform pulse starting at $t = \tau$ and having unit amplitude. We let the duration of this pulse be T. As described by eq. (2.3.12), an excitation whose magnitude is one between $t = \tau$ and $t = \tau + T$ may be written as $Q = h(t - \tau) - h(t - \tau - T)$. The excitation is the difference between two step functions, so linear superposition leads to a response that is the difference between step responses associated with each excitation term, specifically

$$q = [u(t - \tau) - u(t - \tau - T)] \tag{2.3.19}$$

An impulse excitation is the limit of a uniform pulse in which the duration T of the pulse approaches zero, but the impulse of the pulse remains constant. The strength of an impulse function is its impulse, $I = \int Q dt$. We express such a function as $Q(t) = I\delta(t - \tau)$, where $\delta(t - \tau)$ is the *Dirac delta function* and τ is the instant at which the impulse acts.

We may obtain the response to a unit impulse excitation, $I = 1$, by taking the limit as $T \longrightarrow 0$ of the uniform pulse response. In order to have an impulse of unity as the duration of the pulse decreases, we let the amplitude of the pulse be $F = 1/T$. The pulse starts at $t = \tau$, and cuts off at $t = \tau + T$. We denote as $g(t - \tau)$ the response to a unit impulse applied at $t = \tau$, so superposition indicates that

$$g(t - \tau) \lim_{T \to 0} \left\{ \frac{1}{T}[u(t - \tau) - u(t - \tau - T)] \right\} \tag{2.3.20}$$

The right side is a somewhat unconventional way of writing a time derivative, specifically,

$$g(t - \tau) = \frac{d}{dt}u(t - \tau) \tag{2.3.21}$$

Rather than evaluating this derivative, we shall determine $g(t - \tau)$ from the impulse-momentum principle for the mass M, because doing so emphasizes the physical aspects of the impulse response. In the model of an impulse excitation, the force is infinite, but the duration is zero. One consequence of the zero duration is that any finite force, such as those generated by springs and dashpots, may be ignored while the impulsive force acts, because the impulse of such finite forces will be zero. Let us denote as $t = \tau^-$ and $t = \tau^+$ the instants immediately before and after the application of the impulse. It therefore must be that the momentum change equals the unit impulse,

$M\dot{q}(\tau^+) - M\dot{q}(\tau^-) = 1$. The velocity changes instantaneously in this model, corresponding to an infinite acceleration. However, the position cannot change instantaneously, because the velocity is always finite. It follows that $q(\tau^+) = q(\tau^-) = 0$. Furthermore, the system was at rest at $t = 0$, so it will not start to move until $t = \tau$. These considerations lead to the following equations governing the impulse response,

$$\ddot{g} + 2\zeta\omega_{\text{nat}}\dot{g} + \omega_{\text{nat}}^2 g = 0 \text{ for } t > \tau, \quad g = 0 \text{ and } \dot{g} = \frac{1}{M} \text{ at } t = \tau, \quad g = 0 \text{ if } t < \tau$$

(2.3.22)

In other words, $g(t - \tau)$ is the free response to an initial velocity condition. Clearly, the form of this response will depend on the value of ζ. In the case of an underdamped system, the impulse response is

$$g(t - \tau) = \frac{1}{M\omega_{\text{d}}} \exp\left[-\zeta\omega_{\text{nat}}(t - \tau)\right] \sin\left[\omega_{\text{d}}(t - \tau)\right] h(t - \tau), \quad \zeta < 1 \quad (2.3.23)$$

The step and impulse responses are basic building blocks for understanding transient response. For example, in Example 2.13 we saw that the interval during which the wheel is in contact with the bump in the road is short compared to the natural period. The response was found to resemble the damped free vibration resulting from an initial velocity. We now recognize that the bump represents a nearly impulsive excitation when the vehicle speed is high.

2.3.5 Shock Spectra

Evaluation of pulse response has an important application to the design of equipment to resist shock. A shock loading is one that has a large magnitude, but a short duration, as exemplified by a hammer blow or an impact when a body falls. In the design stage, it is assumed that we have some idea of the manner in which the excitation varies. The question is, How should we design the system, especially its natural frequency, so as to best withstand the excitation? Before we can answer this question, we must identify what response parameter or set of parameters indicate the severity of the shock. One measure often used is the maximum displacement, on the basis that this affects the clearance between vibrating internal parts and an exterior housing. Another shock severity measure is the peak acceleration, which indicates the stresses likely to be encountered. Let us denote as χ the shock parameter that must meet some design criteria. The only significant aspect of the response for this consideration is the magnitude of χ; the instant at which it occurs, and whether it is positive or negative, usually are not relevant. Let τ denote the pulse duration. Thus, it is necessary to examine the response for $0 \le t \le \tau$ and for $t > \tau$ to identify the peak value of χ, which we denote as χ_{max}.

A graph of χ_{max} as a function of ω_{nat} is called a *shock spectrum*. Such spectra are usually displayed in nondimensional form, with the horizontal axis being the ratio of the pulse duration τ to the natural period τ_{nat}, that is, $\tau/\tau_{\text{nat}} \equiv \omega_{\text{nat}}\tau/2\pi$. It is standard practice to nondimensionalize the vertical scale by plotting χ_{max} relative to a comparable measure of the excitation's strength. For displacement we usually divide χ_{max} by the static displacement associated with the force, $F/k \equiv |F|/\omega_{\text{nat}}^2 M$, while a shock spectrum of acceleration is nondimensionalized by dividing the maximum acceleration by $|F|/M$. The critical cases for design purposes would be the parameter ranges where these spectra are maxima.

The process of constructing a shock spectrum may be carried out by setting the basic system parameters, M, ω_{nat}, and ζ. The actual values of the first two will not matter because of the way in which the plotting scales are nondimensionalized. However, changing ζ will alter the curves, so the damping ratio should be set to the actual value of the system of interest. (If ζ is also a design parameter, several shock spectra for a range of possible ζ will need to be constructed.) With the system properties held constant, we consider pulse durations τ ranging from zero to several multiples of the natural period. For each value of τ, we solve for q as a function of t over a time interval sufficiently long that we are sure that we have observed the maximum response $|q|$ or $|\ddot{q}|$. We scan the solution for q or $\ddot{q}$ to identify χ_{max}. Dividing that maximum by F/K or F/M, as appropriate, gives one data point to plot versus the value of τ/τ_{nat} for which that response was evaluated.

Because of the manner in which the plots are nondimensionalized, the shock spectra for a one-degree-of-freedom system depend only on the shape of the pulse excitation and the critical damping ratio. Displacement and acceleration spectra for a uniform pulse with light and heavy damping are shown in Figure 2.19. We conclude from the shock spectra for $\zeta = 0.05$ that the peak displacement will be minimized if τ_{nat} is as large as possible, which means that the spring should be relatively soft. Large τ_{nat} also lessens the severity of the acceleration. However, the acceleration can never be reduced below F/m, which is the amount that would result if there were no spring and dashpot. In other words, suspending a mass with a spring and dashpot cannot insulate the mass from accelerating due to the pulse. The curves for $\zeta = 0.5$ are consistent with the criteria established for the lightly damped case, but we see that the worst case displacement and acceleration are both lessened by increasing damping.

For small $\tau < \tau_{nat}/4$, the pulse acts over a relatively short interval. In that case the force appears to the system to act impulsively. The amplitude of the step is fixed at unity, so decreasing τ decreases the impulse. Correspondingly, the amplitude of the response decreases almost linearly as τ decreases below $\tau = \tau_{nat}/4$.

The superposition method offers a different perspective on the features of the shock spectra in Figure 2.19. We saw in Figure 2.18 that the largest overall displacement for a step force at $t = 0$ occurs at $t \approx \tau_d/2$. Because the second step forming a uniform pulse is subtracted from the first, cutting off the pulse at $t > \tau_d/2$ cannot affect this maximum displacement. In contrast, cutting off the pulse at $t < \tau_d/2$ will result in the second step force acting against the velocity, thereby lessening the maximum displacement. To understand the acceleration effect, we observe that the maximum acceleration magnitude due to a step occurs at $t' = 0$, before the spring and dashpot forces build up to oppose the applied force. Application of a negative second

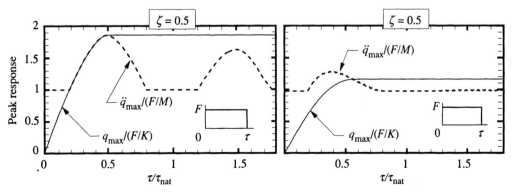

FIGURE 2.19 Shock spectra for a uniform pulse applied to a damped one-degree-of-freedom system.

step in order to form a uniform pulse cannot affect this initial acceleration, so the smallest possible peak acceleration is $\ddot{q}_{max} = F/m$. The range of values of τ for which the maximum acceleration is enhanced beyond this value corresponds to a time interval in which the response to the initial step excitation has a negative acceleration. Application of the negative second step in this interval adds negatively to the acceleration that would be obtained if the first step were not cut off, thereby increasing the magnitude of the acceleration.

REFERENCES AND SELECTED READINGS

CRAIG, R. R., Jr. 1981, *Structural Dynamics*. John Wiley & Sons, New York.

HARRIS, C. M., & CREDE, C. E. 1976. *Shock and Vibration Handbook*. McGraw-Hill, New York.

KAPLAN, W. 1964. *Elements of Ordinary Differential Equations*. Addison-Wesley, Reading, MA.

MEIROVITCH, L. 1986. *Elements of Vibration Analysis*, 2nd ed. McGraw-Hill, New York.

PIPES, L. A., & HARVILL, L. R. 1970. *Applied Mathematics for Engineers and Physicists*, 3rd ed. McGraw-Hill, New York.

THOMSON, W. T., & DAHLEH, M. D. 1993. *Theory of Vibration with Applications*, 5th ed. Prentice Hall, Upper Saddle River, N.J.

TONGUE, B. H. 1996. *Principles of Vibrations*. Oxford University Press, New York.

EXERCISES

2.1 Measurement of a harmonically varying force $F(t)$ reveals that its maximum value is 500 N, that it is zero at intervals of 200 ms, and that $F = 150$ N when $t = 0$. Describe F as a cosine function. What are the possible values of the phase angle that fit the measured properties?

2.2 A harmonically varying displacement y is such that the earliest maximum is $y_{max} = 40$ mm at $t = 0.002$ s. The earliest minimum value occurs at $t = 0.010$ s.

(a) Express this displacement as a cosine function.

(b) Express this displacement as a sine function.

2.3 The initial conditions of a variable are known only to the extent that $x(0) > 0$ and $\dot{x}(0) < 0$, which results in the time trace depicted in the figure.

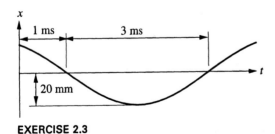

x

1 ms   3 ms

20 mm

t

EXERCISE 2.3

Use the properties in this trace to determine

(a) The initial values of x and $\dot{x}$.

(b) The value of t at which the minimum value of x first occurs.

(c) The maximum positive value of $\dot{x}$ and the earliest value of t at which this maximum occurs.

(d) The maximum positive value of $\ddot{x}$ and the earliest value of t at which this maximum occurs.

2.4 The following are complex amplitudes of harmonic functions q. Determine the amplitude A and phase angle ϕ. Also determine the earliest positive t at which $q = 0$.

(a) $\hat{A} = 0.002 + 0.004i$, $q = A\cos(50t + \phi)$

(b) $\hat{A} = -5 + 3i$, $q = A\cos(1200t + \phi)$

(c) $\hat{A} = -25 - 10i$, $q = A\sin(2t + \phi)$

(d) $\hat{A} = 0.4 - 0.3i$, $q = A\sin(\Omega t - \phi)$

2.5 An oscilloscope trace indicates that the voltage output v from a sensor varies harmonically, with zeroes occurring every 8 ms. The first zero of v occurs at $t = 5.5$ ms, the amplitude of the signal is 1.2 V, and $v > 0$ at $t = 0$.

(a) Express this signal as a complex exponential. Write the complex amplitude in polar and rectangular forms.

(b) Express the time rate of change of the voltage as a complex exponential. Write this quantity in polar and rectangular forms.

(c) What is the maximum rate of change of the signal, and at what instants does that maximum occur?

2.6 The following lists complex amplitudes of harmonic functions q whose phase angle is measured relative to a function that is neither a pure sine nor a pure cosine, as indicated. Determine the amplitude A and phase angle ϕ for the corresponding real function. Also determine the earliest positive t at which q attains its maximum value.

(a) $\hat{A} = 0.003 + 0.001i$,
  $q = A\cos(50t + \pi/3 + \phi)$
(b) $\hat{A} = -0.5 + 2i$,
  $q = A\cos(1200t - \pi/4 + \phi)$
(c) $\hat{A} = -800 - 1200i$, $q = A\sin(2t - 5 + \phi)$
(d) $\hat{A} = 0.4 - 0.3i$, $q = A\sin(\Omega t + 4.5 - \phi)$

2.7 It is given that $x = 18\sin(\Omega t + \pi/3) - 21\cos(\Omega t - 0.4\pi)$.

(a) Determine the amplitude and phase angle of x relative to a pure sine response.
(b) Determine the earliest time at which $x = 0$.
(c) Determine the earliest time at which $\dot{x} = 0$.

2.8 Suppose $q = 0.01\sin(50t) - 0.02\cos(50t - 0.3\pi)$.

(a) Write q in complex exponential form. What is the complex amplitude?
(b) What is the time interval separating instants at which $q = 0$?
(c) What is the earliest positive t at which $q = 0$?
(d) What is the largest value of q that will occur, and what is the earliest positive t at which this maximum occurs?

2.9 A force is observed to vary as $F = 400\cos(250t - \pi/3)$.

(a) What is the earliest instant $t > 0$ at which $F = 0$?
(b) Express F as the sum of pure cosine and sine functions.

2.10 Two parts of the harmonic motion of a system are $x_1 = 8\sin(10t - 5\pi/6)$, $x_2 = 12\cos(10t + \phi)$. Find the phase angle ϕ for which $x = x_1 + x_2$ is a pure sine function. What is the amplitude of x in that case?

2.11 A response is given by $q = 5\cos(50t) + B\sin(52t)$ mm. Consider $B = 10$, 6, and 5. Use mathematical software to plot q as a function of t for a sufficiently long interval to see a repeating pattern. Interpret the response for each value of B from the perspective of beating phenomena.

2.12 A beating response is known to be $q = 0.03\cos(0.5t - 0.4)\sin(200t)$. Represent this product as the sum of two harmonic functions. What are the amplitude, frequency, and phase angle for each?

2.13 A microphone records a beating sound that seems to be at 440 Hz, with zeroes occurring every 0.5 s. The peak instantaneous pressure in this sound is 0.01 Pa.

(a) Determine the amplitude and frequency of the two harmonic tones equivalent to this beating sound.
(b) Suppose the first instant at which the envelope of the beating signal vanishes is $t = 0.3$ s. What can one deduce from such information regarding the phase lead of each harmonic tone?

2.14 Consider the sum of three harmonics whose frequencies are spaced equally, $u = c_1 \sin[(\omega - \dot{\Delta})t] + c_2 \sin(\omega t) + c_1 \sin[(\omega + \dot{\Delta})t]$. For what combinations of nonzero amplitudes c_1, c_2, and c_3 does this sum exhibit a beating type of behavior? What is the envelope of the beat in that case?

2.15 Use complex exponentials to evaluate

$$u = \int_0^T \sin(30t + 0.6)\cos(20t)dt$$

where T is a real constant.

2.16 The stiffness and unstretched length of each spring are $k_1 = 2$ kN/m, $\ell_1 = 90$ mm, and $k_2 = 1$ kN/m, $\ell_2 = 140$ mm. The mass of the block is 0.5 kg.

(a) Determine the value of s at the static equilibrium position.
(b) Determine the cyclical natural frequency.

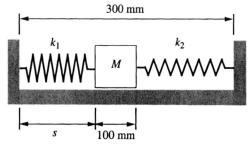

EXERCISE 2.16

2.17 An undamped one-degree-of-freedom system has a mass coefficient of 50 kg and a natural frequency of 80 Hz. At $t = 0$ it is released from $q = 20$ mm with $\dot{q} = -50$ m/s.

(a) Determine the maximum positive value of q that occurs in the ensuing vibration, and the earliest instant at which it occurs.

(b) Determine the maximum positive value of $\dot{q}$ that occurs in the ensuing vibration, and the earliest instant at which it occurs.

2.18 Block m is suspended from the elastic beam by spring k. The mass of the beam is negligible compared to m. Suppose the natural frequency observed in a free vibration is ω_{nat}. Derive an expression in terms of m, k, and ω_{nat} for the stiffness of the beam at the location where the spring is attached. What is the smallest possible value of k consistent with a specified value of ω_{nat}?

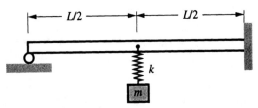

EXERCISE 2.18

2.19 A block of mass m is mounted on a spring having stiffness k. The block moves in the vertical direction. When the system is at rest, a 2 kg block is placed gently on the original block. It is observed that the static length of the spring after insertion of the additional block is 50 mm less than it was prior to the addition. It also is observed that the natural frequency with the additional mass is 5 Hz less than it was originally. Determine k and m.

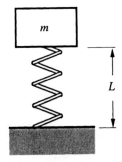

EXERCISE 2.19

2.20 The torsional spring at the pivot of the bar is undeformed when the bar is in the upright position. Thus, $\theta = 0$ is a static equilibrium position. Derive an expression for the natural frequency of rotation relative to this position. Show that there is a minimum value of k below which the natural frequency is not a real number. Explain the significance of this minimum stiffness.

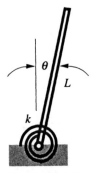

EXERCISE 2.20

2.21 A cylinder, whose mass is m_1, rolls without slipping under the restraint of a pair of springs. In addition, a small mass m_2 is attached to the cylinder at the top center position shown. This position corresponds to static equilibrium. The horizontal displacement x of the cylinder's center is selected as the generalized coordinate. What is the system's natural frequency? Is there a range of values of m_2 for which the upright equilibrium position is unstable?

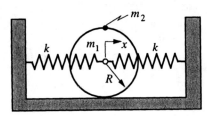

EXERCISE 2.21

2.22 A one-degree-of-freedom system has an undamped natural frequency of 20 Hz and a critical damping ratio of 0.005. It is released with no initial velocity at 60 mm from the static equilibrium position. Approximately how much time will elapse before the amplitude of the free vibration does not exceed 10 mm?

2.23 The graph at the top of page 115 shows the measured displacement of a one-degree-of-freedom system vibrating freely. The equivalent mass of the system is 400 kg. Determine the corresponding equivalent dashpot and spring constants.

2.24 A one-degree-of-freedom underdamped system is observed to have the following properties when it is released from rest at $t = 0$ with a positive displacement q_0:

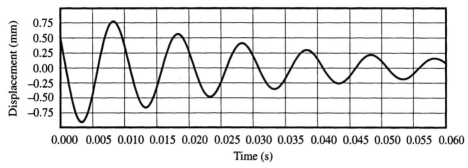

EXERCISE 2.23

1. The first instant at which $q = 0$ is $t = 8$ ms.
2. The maximum value of q in the tenth cycle of vibration is $q = 0.5q_0$.

Determine the natural frequency and ratio of critical damping.

2.25 A one-degree-of-freedom system has mass $M = 4$ kg, spring stiffness $K = 160$ kN/m, and damping constant $C = 320$ N-s/m. The system is released from rest at $q = 12$ mm.

(a) Determine the natural frequency and critical damping ratio.

(b) Determine the response q as a function of t.

(c) Determine the most negative value of q in the free vibration, and the value of t at which it occurs.

(d) Determine the earliest instant after release at which $q = 0$.

2.26 Solve Exercise 2.25 in the case where the initial conditions are $q = -3$ mm and $\dot{q} = 0.4$ m/s at $q = 0$.

2.27 A one-degree-of-freedom system has mass $M = 0.08$ kg and spring stiffness $K = 3.2$ kN/m. The system is known to be underdamped, but the damping constant C is unknown. The system is released from rest at $t = 0$, with an initial value $q = -2$ mm. It is observed that $t = 0.01$ s is the earliest instant at which $q = 0$ after the system is released. Determine the value of C. Also determine the largest positive value of q in the free vibration, and the value of t at which it occurs.

2.28 The following measurements of the response of an underdamped one-degree-of-freedom system have been made:

1. $x = 0$ when $t = 0$.
2. $\dot{x} < 0$ when $t = 0$.
3. Maximum positive displacement in the first cycle of oscillation occurs when $t = 40$ ms

4. $x = 15$ mm when $t = 40$ ms.
5. Maximum positive displacement in the third cycle of oscillation is $x = 2$ mm.

Use this information to deduce the cyclic natural frequency (Hz), the ratio of critical damping, and the initial velocity.

2.29 The measured free vibration response of a one-degree-of-freedom system is as shown in the graph.

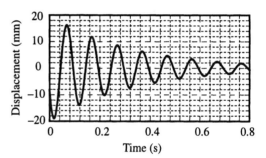

EXERCISE 2.29

(a) Deduce from this measurement the log decrement, the natural frequency, and the critical damping ratio of the system.

(b) Estimate the value of t beyond which the displacement magnitude $|q|$ will not exceed 0.01 mm.

(c) If the damping constant C is held fixed, while the system is modified by doubling the stiffness K and halving the generalized mass M, how would that alter the answer to part (b)?

(d) The initial displacement, at $t = 0$, is $q_0 = -10$ mm. What is the initial velocity?

2.30 The properties m, k, and c of a mass-spring-dashpot system are unknown. Experiments have determined that the free vibration response has the following properties:

1. The motion is oscillatory.
2. The interval between instants when the mass returns to its static equilibrium position is 0.022 s.
3. Removing the dashpot from the system reduces the time between instants when the mass returns to its static equilibrium position to 0.015 s.
4. Adding 2 kg to the mass, with the dashpot connected, increases the interval between instants when the mass returns to its static equilibrium position to 0.033 s.

Determine m, k, and c.

2.31 A one-degree-of-freedom system (generalized coordinate q) is given an initial velocity v_0 when the system is at its static equilibrium position. Observation of the free vibration shows that it is oscillatory, with the following properties:

1. The earliest maximum value of q occurs at $t = t_a$.
2. The value of q when $t = t_a$ is q_{max}.
3. The earliest time after release at which the system returns to the static equilibrium position is $t_b > 2t_a$.

In terms of t_a, t_b, and q_{max}, derive expressions for the natural frequency ω, ratio of critical damping ζ, the value of v_0, and the time at which the third maximum (positive) value of q occurs.

2.32 The cushioning for a package of mass m may be represented as a spring k and dashpot c. After falling some distance, the package hits the ground with a known initial velocity v. The system is underdamped.

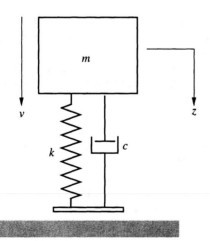

EXERCISE 2.32

(a) Determine the downward displacement $z(t)$, based on $z = 0$ being the center of mass position at the instant when the package first contacts the ground. *Hint:* Gravity cannot be ignored because $z = 0$ is not the static equilibrium position.

(b) Use the solution in part (a) to derive an expression for the force exerted by the cushioning on the package mass m. How can this expression be used to determine the instant t' at which the package will rebound from the ground?

(c) Consider the case where $m = 1$ kg, $\omega_{nat} = 5$ Hz, and $v = 4$ m/s. Use mathematical software to evaluate the maximum cushioning force in part (b) for all t at a fixed critical damping ratio. Consider $0 < \zeta < 1$ for this evaluation. Which case leads to the most protection for the package?

2.33 A testing device consists of a piston that compresses a spring inside a cylinder filled with a viscous fluid. The fluid flows from one side of the piston to the other through orifices in the piston. The mass of the piston is 5 kg, the spring stiffness is 500 N/m, and the system is critically damped. At $t = 0$, the spring is unstretched, and the piston is moving to the right at 60 m/s. Determine the maximum compression the spring undergoes, and the time at which that maximum occurs.

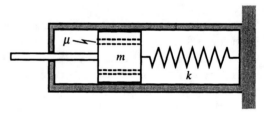

EXERCISE 2.33

2.34 Solve Exercise 2.33 for the case where $c = 200$ N-s/m.

2.35 The model of a one-story building considers a rigid floor that is displaced by u horizontally. The floor, whose mass is $M = 20,000$ kg, is supported by four columns having flexural rigidity EI. Two diagonal dashpots c, only one of which is visible in the sketch (page 117), provide damping. What values of EI and c will lead to a critically damped system whose natural frequency is 5 Hz?

2.36 Parameters for a one-degree-of-freedom system are $M = 10$ kg, $K = 16$ kN/m, and $C = 320$ N-s/m.

(a) By what factor should C be multiplied, with M and K held constant, to achieve critical damp-

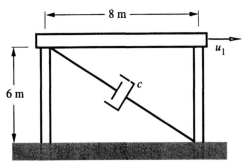

EXERCISE 2.35

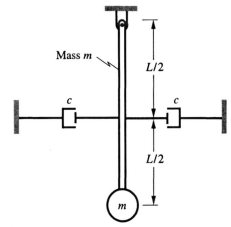

EXERCISE 2.39

ing? What is the natural frequency of the new design?

(b) By what factor should M be multiplied, with C and K held constant, to achieve critical damping? What is the natural frequency of the new design?

(c) By what factor should K be multiplied, with M and C held constant, to achieve critical damping? What is the natural frequency of the new design?

2.37 The sketch shows the top view of a 2.5 m high door. The mass of the door is 30 kg. The hinge is fitted with a torsional shock absorber, which consists of a spring whose stiffness is 20 N-m/rad acting in parallel with a torsional dashpot c_T.

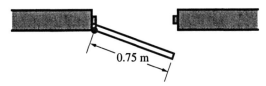

EXERCISE 2.37

(a) Determine the value of c_T for which the door is critically damped as it returns to the closed position.

(b) If the critically damped door is opened to 20° and released, how much time will be required to return it to within 1° of closure.

(c) Repeat the analysis in part (b) for the case where the door is released from 40°.

2.38 Consider the response of an underdamped system for arbitrary initial conditions q_0 and $\dot{q}_0$ at $t = 0$. Derive eq. (2.2.37) from that solution by evaluating its limit as $\zeta \longrightarrow 1$.

2.39 The system hangs in the vertical plane. The bar and the sphere have equal masses $m = 2$ kg, and the moment of inertia of the attached sphere about its own center of mass is negligible. The natural frequency for this system is 2 Hz and the ratio of critical damping is $\zeta = 0.6$. The bar is in the

vertical position when it is given a counterclockwise angular velocity of 3 rad/s at $t = 0$.

(a) Determine the values of the length L and damping constant c.

(b) At what instant $t = t_1$ does the rotation first attain a maximum value? What is the corresponding value of θ?

(c) Suppose an electromagnet fastening the sphere to the bar is turned off at $t = t_1$, causing the sphere to fall away. Derive an expression for θ as a function of time when $t > t_1$.

2.40 A one-degree-of-freedom system slides over a horizontal surface in opposition to dry friction. The system mass is 3 kg. Measurements of the vibratory response of this system indicates that the initial maximum value of q is 90 mm and the maximum value of q after 10 cycles is 10 mm. The time required for 10 cycles to occur is 10 s. Determine the system's natural frequency, the stiffness coefficient, and the coefficient of sliding friction.

2.41 A one-degree-of-freedom system slides over a horizontal surface in opposition to dry friction. The mass is 200 kg, the natural frequency is 60 Hz, $\mu_s = 0.4$, and $\mu_k = 0.35$. The system is released from rest at $q = 50$ mm. Determine the elapsed time until the system comes to rest.

2.42 Parameters for the spring-mass system in Figure 2.15 are $M = 10$ kg, $\omega_{nat} = 200$ rad/s, $\mu_s = 0.3$, and $\mu_k = 0.2$. The system is released from its static equilibrium position with an initial velocity of 20 m/s at $t = 0$. If energy dissipation is solely due to dry friction, how long will it be before the system comes to rest, and what will be the final displacement?

2.43 Measurement of the motion of a one-degree-of-freedom system reveals that a maximum displacement occurs at $t = 105$ ms, after which 200 ms is required for 10 more maxima to occur. The maximum displacements at $t = 105$ ms and 305 ms are 8.0 mm and 7.9 mm, respectively.

(a) If the sole source of dissipation is dry friction, what is the coefficient of sliding friction for this system?

(b) Determine the maximum displacement that will occur after $t = 500$ ms if the sole source of dissipation is dry friction. Compare this value to the result of viscous damping as the sole source of dissipation.

2.44 An undamped system is initially at rest at its static equilibrium position when it is subjected to an excitation given by $Q(t) = F_0(1 - \omega_{nat}^2 t^2)$. Use the method of undetermined coefficients to derive an expression for the response. Graph the response $q/(F_0/k)$ as a function of $\omega_{nat} t$. Based on those results, identify at what instant the displacement is a maximum.

2.45 Solve Exercise 2.44 for the case where the system is underdamped, with $\zeta = 0.4$.

2.46 A critically damped one-degree-of-freedom system is subjected to a harmonically varying force $F_0 \sin(\omega t)$. The system initially had velocity v_0 at the static equilibrium position. Use the method of undetermined coefficients to derive an expression for the response.

2.47 Solve Exercise 2.46 for the case of an overdamped system with $\zeta = 1.10$.

2.48 The response of a one-degree-of-freedom system to a periodic force can be obtained by Fourier series techniques, as is done in Section 3.7.5. An interesting alternative method for finding a periodic solution considers the initial conditions at the start of a period to be unstated algebraic parameters. The value of these parameters should be such that the response at the end of the period matches the response at the beginning. Consider using such a procedure to determine the response of an undamped system to a periodic force $Q(t) = \alpha t/T$ for $0 \le t < T$, with $Q(t + T) = Q(t)$. Let $q = q_p + q_c$, where q_p is the particular solution for $0 \le t < T$. Determine q_p, then determine the coefficients of q_c by requiring that $q(t = T) = q(t = 0)$ and $\dot{q}(t = T) = \dot{q}(t = 0)$. Are there any values of T for which this solution is not valid?

2.49 A force that varies with time as a beat, $Q(t) = F_0 \sin(\omega_1 t) \sin(\omega_2 t)$, $\omega_2 \gg \omega_1$, is applied to an undamped one-degree-of-freedom system. The initial conditions are $q = \dot{q} = 0$ at $t = 0$. Derive an expression for the value of q at the end of one beat period, $t = \pi/\omega_1$. Are there any values of ω_1 and ω_2 for which the solution is not valid?

2.50 A critically damped one-degree-of-freedom system having natural frequency ω_{nat} is subjected to an exponentially decaying force given by $Q = F\tau^2 \exp(-\tau)$, where $\tau = \omega_{nat} t$ is a nondimensional time variable and F is constant. The system was initially at rest in the static equilibrium position. Determine and graph the response.

2.51 An underdamped system is subjected to an exponentially decaying excitation $Q = F_0 \exp(-\lambda t)$. The system was initially at rest in the static equilibrium position. Determine the response. Graph it for a case where $\omega_{nat} = 1$ rad/s, $\zeta = 0.05$, and $\lambda = 0.1$.

2.52 An undamped system is subjected to a sequence of step excitations, $Q(t) = 100$ N if $0 < t < 3\pi/\omega_{nat}$, $Q(t) = 200$ N if $t > 3\pi/\omega_{nat}$. The system mass is 5 kg and $\omega_{nat} = 50$ rad/s. Determine and graph the response.

2.53 An underdamped system is initially at rest in the static equilibrium position when it is subjected to a force that increases linearly until $t = \tau$, at which time it begins to decrease. Specifically, $Q(t) = \beta t$ if $0 \le t \le \tau$, $Q(t) = \beta(2\tau - t)$ if $t > \tau$. Use superposition to determine the response.

2.54 An undamped system is subjected to a constant force for an interval T, after which it begins to roll off exponentially. Specifically, $Q(t) = F_0$ if $0 < t < T$, $Q(t) = F_0 \exp[-\beta(t - T)]$ if $t > T$. Derive an expression for the response.

2.55 An underdamped system is initially at rest in its static equilibrium position when it is subjected to an excitation that decreases from F_0 to zero in a time interval τ, that is, $Q(t) = F_0(1 - t/\tau)$ if $0 \le t \le \tau$, $Q(t) = 0$ if $t > \tau$. The mass is 2 kg, the (undamped) natural frequency is 50 Hz, and the ratio of critical damping is $\zeta = 0.30$. Determine the response.

2.56 An undamped one-degree-of-freedom system is initially at rest at its static equilibrium position when it is subjected to a cosinusoidal force $Q = F_0 \cos(\pi t/2\tau)$ during the interval $0 < t < \tau$, after which $Q = 0$. The frequency factor $\pi/2\tau$ is different from the system's natural frequency.

(a) Use superposition and Appendix B to derive an expression for the response as a function of t.

(b) Use the method of undetermined coefficients and the continuity requirements at $t = \tau$ to obtain an alternative description of the response.

(c) Assess the ease of each analysis.

2.57 An undamped system is subjected to a triangular pulse defined by $Q(t) = 10t$ kN if $0 \leq t \leq 0.02$ s, $Q(t) = 0$ if $t > 0.02$ s. The mass is 2 kg and the natural frequency is 50 Hz. The system is initially at rest in its static equilibrium position. Use superposition of ramp and step responses to determine this system's response. Graph the result.

2.58 An undamped system is subjected to a parabolic pulse defined by $Q(t) = F(Tt - t^2)/T^2$ for the interval $0 < t < T$, after which there is no excitation. The system was initially at rest in its static equilibrium position. Use superposition and Appendix B to derive an expression for the response as a function of t.

2.59 An underdamped system was initially at rest in its static equilibrium position. At $t = 0$, it is subjected to a cosinusoidal force at its damped natural frequency. The duration for this force is one-quarter of the damped natural period, so that $Q(t) = F\cos(\omega_d t)[h(t) - h(t - \pi/2\omega_d)]$. Use superposition and Appendix B to derive an expression for the response as a function of t.

2.60 An underdamped one-degree-of-freedom system is initially at rest at its static equilibrium position when it is subjected to a force $Q = F_0 \exp(-\zeta\omega_{nat}t)$, where ω_{nat} is the system's natural frequency and ζ is the critical damping ratio.

(a) Derive an expression for the response q as a function of t.

(b) Suppose the force is removed at some time $\tau > 0$. Use the principle of superposition to derive an expression for the response $q(t)$ in the interval $t > \tau$.

(c) Draw a graph of the response in parts (a) and (b) for the case where $\omega_{nat} = 1$ rad/s, $\zeta = 0.10$, $F_0 = 1$ N, $\tau = 10$ s, and the system mass is $M = 1$ kg.

(d) What is the smallest value of $\tau > 0$, if any, for which removal of the force at time τ will result in the cessation of motion, such that $q \equiv 0$ for $t > \tau$?

2.61 The sketch shows a test stand for automobile bumpers. Each of the two shock absorbers consists of a spring having stiffness $k = 3000$ N/m that acts concentrically with a dashpot c. The bumper's mass is 40 kg. The system is at rest in the static equilibrium position when a force F having an impulse of 2000 N-s acting over a very short interval is applied at the centerline of the bumper.

(a) Determine the value of c that will bring the bumper to rest in the shortest possible time without rebound.

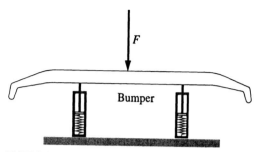

EXERCISE 2.61

(b) If $c = 1500$ N-s/m, determine the displacement $x(t)$ of the bumper.

2.62 The escapement mechanism in an undamped one-degree-of-freedom linkage applies a repeated sequence of identical taps. Each tap acts over a very short interval, and P is the impulse imparted to the system by each tap. Consider a case where the system was at rest initially in the static equilibrium position when the first tap is applied. The time interval separating the taps is τ. Derive an expression for the displacement immediately after the third tap, that is, at $t = 2\tau^+$. Then examine this expression to identify the value of τ that will maximize this displacement.

2.63 Consider the system in Exercise 2.62. Derive an expression for the displacement q as a function of t. Graph the nondimensional displacement $q(M\omega_{nat}/P)$ as a function of the nondimensional time $\omega_{nat}\tau$ over the interval $0 < t < 3\tau$ corresponding to $\omega_{nat}\tau = \pi/2, \pi, 2\pi$, and 3π. Which case leads to the largest displacement?

2.64 An undamped one-degree-of-freedom system is initially at rest when it is subjected to a sequence of two step excitations. The first step, which acts in the interval $0 < t < \pi/2\omega_{nat}$, has magnitude F_0. The second step, of magnitude $2F_0$, acts in the interval $\pi/2\omega_{nat} < t < \pi/\omega_{nat}$. Derive expressions for q and $\dot{q}$ at $t = \pi/\omega_{nat}$. Then compare this value of $\dot{q}$ to the one that would be obtained from the impulse-momentum principle if the spring were not present.

2.65 The equation for the output voltage from an accelerometer is $\ddot{u} + \omega_{nat}^2 u = -\alpha\ddot{z}$, where α is a constant and z is the displacement of the body to which the accelerometer is attached. Consider a situation where $z = 0$ prior to $t = 0$, after which the base displacement is $z = c\omega_{nat}(t - t^2/\tau)$. Prior to $t = 0$, the accelerometer was not moving, that is, $u \equiv 0$ for $t < 0$. Determine the output of the seismometer. (*Hint:* It is necessary to account for the discontinuity of $\ddot{z}$ at $t = 0$.)

2.66 Consider a pulse in the form of a ramp $Q(t) = Ft/\tau$ for $0 < t < \tau$. Construct the shock spectrum for maximum displacement as a function of natural frequency when $\zeta = 0.2$. What range of natural frequencies is likely to have the greatest severity?

2.67 Consider a pulse in the form of a sine lobe $Q(t) = F\sin(\pi t/\tau)$ for $0 < t < \tau$. The system is underdamped, with $\zeta = 0.05$. Construct the shock spectrum for maximum acceleration as a function of natural frequency. What range of natural frequencies is likely to have the greatest severity?

CHAPTER 3

STEADY-STATE RESPONSE TO HARMONIC EXCITATION

Excitations that vary harmonically arise from a number of sources, notably imbalance in rotating parts. Unlike a transient excitation, harmonic excitation persists until it is cut off. The response to this type of excitation will consist of complementary and particular solutions. However, because it is acceptable to ignore dissipation only if the observation period is not too long, the complementary solution will eventually disappear. For this reason the particular solution of equations of motion driven harmonically is generally referred to as the *steady-state response*, which we denote as q_{ss}.

In addition to being important for its own sake, our study of the response to harmonic excitation will lead us to the ability to analyze far more general situations. The first extension will be to Fourier series for periodic excitation, which consist of many frequencies. Computational implementation of Fourier series will lead us to the Fast Fourier Transform (FFT). We then will see how FFT techniques enable us to efficiently determine the response to arbitrary transient excitation. The end of this chapter will address experimental techniques using FFT technology, which is the foundation for much of the electronic diagnostic instrumentation used to identify and correct vibration problems.

3.1 COMPLEX FREQUENCY RESPONSE

In principle, the determination of the steady-state response of a one-degree-of-freedom system could be performed according to the method of undetermined coefficients using real functions. However, a complex variable representation has numerous advantages. Thus, we shall express the excitation as

$$Q(t) = \text{Re}[F \exp(i\omega t)]$$ (3.1.1)

where F is a complex coefficient whose argument defines the phase of the excitation relative to a cosine function. According to the method of undetermined coefficients, the particular solution corresponding to a harmonic excitation is also harmonic at the excitation frequency. (The only exception to this is the case where the system is undamped and the excitation frequency ω is identical to the natural frequency. We will consider that situation separately.) The complex representation of a harmonic steady-state response is

$$q_{ss} = \text{Re}[X \exp(i\omega t)]$$ (3.1.2)

We determine the complex amplitude X by requiring that the above form be the particular solution of the equation of motion. Thus, we have

$$\boxed{\text{Re}[(-M\omega^2 + i\omega C + K)X \exp(i\omega t)] = \text{Re}[F \exp(i\omega t)]} \qquad (3.1.3)$$

This equation must be satisfied for all values of t, so we match the terms contained within the brackets. This leads to

$$\boxed{X = \frac{F}{K + i\omega C - M\omega^2}} \qquad (3.1.4)$$

The denominator is the *dynamic stiffness*, whose behavior we will consider when we interpret the underlying phenomena. Some vibrations texts refer to $K + i\omega C - M\omega^2$ as the mechanical impedance. However, the term impedance in other subject areas, such as AC circuits and acoustics, denotes the ratio of the complex forcelike amplitude to the complex velocity-type amplitude. In view of the fact that the latter is $V = i\omega X$, the *impedance* of a one-degree-of-freedom system is

$$Z = \frac{F}{i\omega X} = \frac{K}{i\omega} + C + i\omega M \qquad (3.1.5)$$

We may obtain a more useful relation for the steady-state amplitude by recalling the definitions of the natural frequency and the critical damping ratio to eliminate M and C. When we substitute $M = K/\omega_{\text{nat}}^2$ and $C = 2\zeta\omega_{\text{nat}}M \equiv 2\zeta K/\omega_{\text{nat}}$, we obtain

$$\boxed{X = D(r, \zeta)\frac{F}{K}} \qquad (3.1.6)$$

where $D(r, \zeta)$ is a nondimensional quantity called the *complex frequency response*. Its value depends only on the ratio r of the excitation and natural frequencies and on the critical damping ratio, according to

$$\boxed{D(r, \zeta) = \frac{1}{1 + 2i\zeta r - r^2}, \quad r = \frac{\omega}{\omega_{\text{nat}}}} \qquad (3.1.7)$$

To interpret this relation let us express the denominator, $1/D$, in polar form using the complex plane representation in Figure 3.1. Note that the real part $1 - r^2$ may be either positive or negative, but the imaginary part $2\zeta r$ is always positive. Correspondingly, the argument of $1/D$ always lies in the range $0 \leq \phi \leq \pi$. Hence, we have

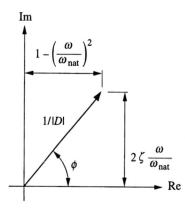

FIGURE 3.1 Real and imaginary parts of the complex frequency response.

$$D(r, \zeta) = |D(r, \zeta)|\exp[-i\phi(r, \zeta)]$$

$$|D(r, \zeta)| = \frac{1}{[(1 - r^2)^2 + 4\zeta^2 r^2]^{1/2}}$$

$$\phi(r, \zeta) = \tan^{-1}\left(\frac{2\zeta r}{1 - r^2}\right), \quad 0 \le \phi \le \pi$$

(3.1.8)

These relations are the *amplitude-frequency-phase lag relations*. The corresponding steady-state response is

$$q = \text{Re}\left\{\frac{F}{K}|D(r, \zeta)|\exp[i(\omega t - \phi(r, \zeta))]\right\}$$

(3.1.9)

Our first observation is that $|D(r, \zeta)| \longrightarrow 1$ and $\phi(r, \zeta) \longrightarrow 0$ as $r \longrightarrow 0$, so that the low-frequency limit of the steady-state response is $\text{Re}[(F/K)\exp(i\omega t)]$. In other words, at very low frequencies the response is in phase with the excitation, and the amplitude is $|F|/K$. This amplitude is the same as what would result if a force $|F|$ were applied statically, so we say that $r \longrightarrow 0$ is the *quasi-static limit*. Thus the magnitude $|D|$ of the complex frequency response represents the amount by which the quasi-static displacement is amplified by the dynamical aspects of the system. From this we deduce that $|D|$ is a *magnification factor* indicating the amount by which the steady-state amplitude is amplified by the system's dynamics. In the same vein, $\phi(r)$ indicates the phase angle of the response relative to the excitation. Because ϕ is positive, it follows from eq. (3.1.9) that ϕ is a *phase lag*, with a particular phase of the response occurring at ϕ/ω after the force has the same phase. Typical plots of $|D|$ versus $\omega/\omega_{\text{nat}}$ are displayed in Figure 3.2, while plots of ϕ as a function of $\omega/\omega_{\text{nat}}$ are depicted in Figure 3.3. Because the peak value of $|D|$ can be quite large, it often is the practice to use a log scale for the ordinate axis.

Several features of the magnification factor $|D|$ and phase lag ϕ are important. As we noted, the amplification factor is unity and the phase lag is zero when $\omega = 0$, regardless of the damping ratio. Let us investigate the undamped case first. When $\zeta = 0$, we see in Figure 3.2 that $|D|$ rises monotonically from unity to an infinite value at $\omega = \omega_{\text{nat}}$. This singularity is not the actual response. Rather, the case where $\omega = \omega_{\text{nat}}$ and $\zeta = 0$, which we refer to as *resonance*, requires a separate analysis; see Section 3.3. Beyond this singularity, the magnification factor decreases monotonically to zero as the frequency is increased. The phase lag is zero when $\omega < \omega_{\text{nat}}$, and then switches to 180° for $\omega > \omega_{\text{nat}}$. Both behaviors may be understood by returning to the

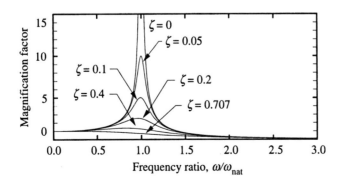

FIGURE 3.2 Magnification factor of the steady-state response as a function of excitation frequency.

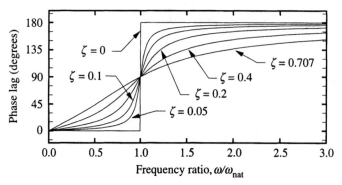

FIGURE 3.3 Phase lag of the steady-state response as a function of excitation frequency.

dynamic stiffness concept associated with eq. (3.1.4). As the frequency increases from zero to the natural frequency, the dynamic stiffness $K - \omega^2 M$ decreases but remains positive, so the amplitude increases and the response remains in phase with the force. Beyond the natural frequency, the inertial effect exceeds the stiffness, so the dynamic stiffness becomes increasingly large negatively. Correspondingly, the response is 180° out of phase from the force. At large r, the inertial contribution to the dynamic stiffness is dominant, which causes the amplitude to decrease as the inverse square of the excitation frequency.

When $\zeta \neq 0$, the dynamic stiffness is $K + i\omega C - \omega^2 M$. Because the effect of damping is 90° out-of-phase from the response, the response is no longer in phase with the force. As the frequency is increased from zero, the inertial effect, which depends on ω^2, grows more rapidly than the damping effect, which depends linearly on ω. In the vicinity of the natural frequency, where the inertial effect is comparable to the elastic effect, damping is most important. In fact, at $\omega = \omega_{nat}$ the inertial and stiffness effects cancel, so only damping can carry the force. Correspondingly, the dynamic stiffness is purely imaginary and the response lags 90° behind the force, as shown in Figure 3.3. The fact that the steady-state response lags 90° behind the force when $\omega = \omega_{nat}$, regardless of the value of ζ, is useful for experimentally identifying the natural frequency. The amplitude in this situation is $|X| = |F|/(\omega C)$, and the corresponding magnification factor is $|D| = 1/(2\zeta)$. Beyond the natural frequency, the inertial effect grows most rapidly, and the amplitude decreases. At very large frequencies the inertial resistance becomes dominant, and the response tends to be 180° out of phase from the force.

To identify the frequency at which the amplitude is largest we note that $|D|$ actually depends on r^2. Hence, we seek values of r for which $d|D|/d(r^2) = 0$, which gives

$$\frac{2[1 - r^2] + 4\zeta^2}{[(1 - r^2)^2 + 4\zeta^2 r^2]^{3/2}} = 0 \tag{3.1.10}$$

Setting the numerator to zero leads to

$$r = (1 - 2\zeta^2)^{1/2} \text{ for max } (|D|) \tag{3.1.11}$$

Thus, the peak amplitude always occurs at a frequency that is below the natural frequency. However, the difference between ω_{nat} and the frequency at which the peak occurs is negligible if $\zeta < 0.02$. In lightly damped systems, where $\zeta \ll 1$, we refer to the response close to $\omega = \omega_{nat}$ as the *resonant response*. We use this terminology even though the phenomenon of resonance occurs only when $\omega = \omega_{nat}$ and $\zeta = 0$. As displayed by Figure 3.2, the peak value of $|D|$ decreases, and the frequency at which it occurs decreases, with increasing damping. When $\zeta > 1/\sqrt{2}$, the peak ceases to exist, and $|D|$ decreases monotonically with increasing frequency.

EXAMPLE 3.1

The steady-state response of a system to a harmonic force $20\cos(\omega t) - 10\sin(\omega t)$ is observed to lag the excitation by 90° when the frequency is 50 Hz. At 48 Hz, the response is observed to be $x = 2.654\cos(\omega t) + 2.993\sin(\omega t)$ mm. Determine the stiffness, mass, and critical damping ratio of this system.

Solution A number of aspects of the analysis of steady-state response, including the use of the relations for amplitude and phase lag as a function of frequency, as well as the use of complex variables to combine terms having different phase, are highlighted by this example. The key to solving the problem is the stated fact that the phase lag of the response relative to the excitation is 90° when the frequency is 50 Hz. A 90° phase lag only occurs when $\omega = \omega_{nat}$ (regardless of the damping ratio), so we conclude that $\omega_{nat} = 100\pi$ rad/s. To proceed further we combine the two excitation terms into a complex exponential,

$$F = 20\cos(\omega t) - 10\sin(\omega t) = \text{Re}\left\{20e^{i\omega t} - \frac{10}{i}e^{i\omega t}\right\} = \text{Re}\{Fe^{i\omega t}\}$$

where

$$F = 20 - \frac{10}{i} = 22.36e^{0.4637i} \text{ N}$$

We express the given response at 48 Hz in a similar form,

$$q = 2.654\cos(\omega t) + 2.993\sin(\omega t)$$

$$= \text{Re}\left\{\left(2.654 + \frac{2.993}{i}\right)\exp(i\omega t)\right\} = \text{Re}\{X\exp(i\omega t)\}$$

Matching the factors of the complex exponential shows that the complex amplitude is

$$X = 2.654 + \frac{2.993}{i} = 4.000e^{-0.8454i} \text{ mm}$$

We know that in general,

$$X = \frac{F}{K}D(r, \zeta)$$

For the given response $\omega = 48(2\pi)$ rad/s, so $r = \omega/\omega_{nat} = 48/50 = 0.96$. The above relation for X then requires that

$$4.000e^{-0.8454i} \text{ mm} = \frac{22.36e^{0.4637i}}{K}D$$

Thus, we know that

$$\frac{D}{K} = 0.17889e^{-1.3091i} \text{ mm/N}$$

We now use the fact that the stiffness K is real, from which it follows that

$$\arg D \equiv -\phi = -1.3091, \qquad \frac{|D|}{K} = 0.17889 \text{ mm/N}$$

The next step is to invoke the expressions relating $|D|$ and ϕ to the frequency ratio. We consider the phase angle first, because the only unknown parameter in that relation is the damping ratio. The last of eqs. (3.18) gives

$$\tan(\phi) = \tan(1.3091) = \frac{2\zeta(0.96)}{1 - 0.96^2} \quad \Rightarrow \quad \zeta = 0.15246$$

The corresponding magnification factor obtained from the second of eqs. (3.1.8) is

$$|D| = \frac{1}{[(1 - r^2)^2 + 4\zeta^2 r^2]^{1/2}} = 3.300$$

We determined the value of $|D|/K$ earlier, from which it follows that

$$K \equiv \frac{|D|}{(|D|/K)} = \frac{3.300}{0.17889} = 18.615 \text{ N/mm} = 18.615(10^3) \text{ N/m}$$

We then find the mass by using the relation for the natural frequency,

$$M = \frac{K}{\omega_{\text{nat}}^2} = 0.18861 \text{ kg}$$

EXAMPLE 3.2

The sketch shows a one-degree-of-freedom model for the vertical displacement y of an automobile as it travels along a rough road. The road elevation varies sinusoidally according to $z = H\cos(2\pi x/L)$, where L is the distance over which the pattern repeats. The vehicle speed is v, which is constant. System parameters are $m = 1600$ kg, $k = 80$ kN/m, $H = 100$ mm, and $L = 2$ m.

(a) For the case where there is very little damping, $\zeta = 0.001$, determine the speed v_{cr} at which the amplitude of the vibration will be largest, and the amplitude of y at that speed.

(b) Suppose $v = v_{\text{cr}}$, but $\zeta = 0.50$. Determine the amplitude of y in this case.

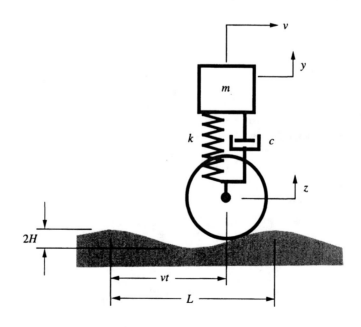

Solution In addition to demonstrating calculation of harmonic response, this exercise shows why the concept of excitation from ground motion is particularly significant for moving vehicles. As shown in the given diagram, the distance x measured from a peak elevation is vt, so the time dependence of the ground elevation under the wheel is $z = H\cos(2\pi vt/L)$. This represents a harmonic excitation at frequency $\omega = 2\pi v/L$. To account for its effect on the vehicle, we follow the development in Section 1.5.4, where the effect is added to what would be present if the grounded ends of springs and dashpots were stationary. The elongations of the spring and dashpot are both $\Delta = y - z$. Thus, positive z represents compressive effects, while positive y is tensile. Accordingly, the free body diagram shows the spring and dashpot forces $kz + c\dot{z}$ acting upward, while the corresponding forces $ky + c\dot{y}$ induced by upward displacement of the vehicle act downward.

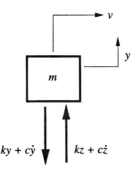

Direct application of Newton's Second Law in the vertical direction yields

$$m\ddot{y} + c\dot{y} + ky = kz + c\dot{z}$$

The value of k is given, and we could determine c from the given ζ. However, it is easier to divide the equation of motion by m, and then use $k/m = \omega_{nat}^2$ and $c/m = 2\zeta\omega_{nat}$. This converts the equation of motion to

$$\ddot{y} + 2\zeta\omega_{nat}\dot{y} + \omega_{nat}^2 y = \omega_{nat}^2 z + 2\zeta\omega_{nat}\dot{z}$$

$$= \omega_{nat}^2 H\cos(\omega t) - 2\zeta\omega_{nat}\omega H\sin(\omega t), \quad \omega = \frac{2\pi v}{L}$$

We convert the trigonometric functions to their complex exponential equivalent,

$$\ddot{y} + 2\zeta\omega_{nat}\dot{y} + \omega_{nat}^2 y = \omega_{nat}^2 H\,\mathrm{Re}[\exp(i\omega t)] - 2\zeta\omega_{nat}\omega H\,\mathrm{Re}\left[\frac{1}{i}\exp(i\omega t)\right]$$

$$= \mathrm{Re}[H(\omega_{nat}^2 + i2\zeta\omega_{nat}\omega)\exp(i\omega t)]$$

We represent the steady-state response as a complex exponential,

$$y = \mathrm{Re}[Y\exp(i\omega t)]$$

Substitution of this expression into the last form of the equation of motion leads to

$$(\omega_{nat}^2 + i2\zeta\omega_{nat}\omega - \omega^2)Y = H(\omega_{nat}^2 + i2\zeta\omega_{nat}\omega)$$

The critical damping ratio is very small in the first case. This corresponds to a situation where the peak magnification factor is a very large value occurring essentially at $r = 1$. Thus, the frequency leading to the largest $|Y|$ when ζ is very small is $\omega = \omega_{nat}$. The natural frequency is $\omega_{nat} = \sqrt{k/m} = 7.071$ rad/s, which leads to

$$v_{cr} = \frac{L\omega_{nat}}{2\pi} = 2.25 \text{ m/s}$$

Setting $\omega = \omega_{nat}$ in the equation for Y yields

$$Y = H\frac{1 + 2i\zeta}{2i\zeta}$$

When $\zeta = 0.001$, we find that

$$Y = 1 - 50i \quad \Rightarrow \quad |Y| = 50 \text{ m}$$

This large value comes from the fact that only the dashpot can resist an excitation at the natural frequency. If we carry out the same computation for $\zeta = 0.50$, we find

$$Y = H\frac{1 + 1.0i}{1.0i} = 0.1 - 0.1i \quad \Rightarrow \quad |Y| = 0.141 \text{ m}$$

Clearly, a reasonably strong dashpot is essential to prevent the automobile from vibrating excessively. Further increase of ζ would reduce $|Y|$, with $|Y| \rightarrow H$ as $\zeta \rightarrow \infty$. In practice, ζ cannot be made very large because of other dynamic considerations, notably passenger comfort.

EXAMPLE 3.3

The grooved actuator A is made to oscillate horizontally by pin B on the flywheel, which is rotating clockwise at constant speed Ω. The actuator is connected by a spring to sliding collar D; the masses of the actuator and of the collar are each m. Derive expressions for the steady-state response $x(t)$ and the corresponding torque that must be applied to the flywheel. Friction is negligible. What is the period of the response and of the torque?

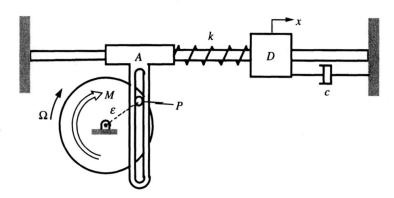

Solution   The objective of this example is to emphasize that a vibration analysis might entail more than merely determining the response, especially because many aspects of design require knowledge of forces generated by vibratory motion. In order to examine the forces exerted between parts of the system, we draw free body diagrams of the block D, actuator A, and the flywheel.

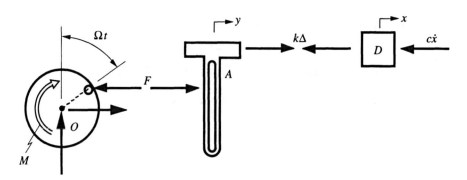

The block and the actuator translate horizontally, so we let x and y denote their respective displacements from the static equilibrium position. The force F is the reaction force exerted between pin P in the flywheel and the groove in the actuator, whose effect is to make y match the horizontal position of the pin. If $x > y$, the spring is elongated, so $\Delta = x - y$, and we show $k\Delta$ as a tensile force acting on the block and the actuator. The equations of motion for the translating bodies are

$$\text{Block } D: \Sigma F = -k\Delta - c\dot{x} = m\ddot{x} \quad \Rightarrow \quad m\ddot{x} + c\dot{x} + kx = ky$$

$$\text{Actuator } A: \Sigma F = F + k\Delta = m\ddot{y} \quad \Rightarrow \quad m\ddot{y} + ky = F + kx$$

The flywheel executes a pure rotation about center O. The rate of this rotation is specified to be constant, so there must be moment equilibrium. We define $t = 0$ to coincide with an instant when the pin moving the actuator is at the uppermost position, so Ωt is the vertical angle of the radial line to the pin. The lever arm of F about the center of rotation is $\varepsilon \cos(\Omega t)$, so the moment equilibrium equation is

$$\Sigma M_O = F[\varepsilon \cos(\Omega t)] - M = 0$$

We noted that the displacement y must match the horizontal movement of the pin, so we set

$$y = \varepsilon \sin(\Omega t) = \text{Re}\left\{\frac{\varepsilon}{i}\exp(i\Omega t)\right\}$$

We express the response as a complex exponential, by writing $x = \text{Re}\{X \exp(i\Omega t)\}$. Upon substitution of the descriptions of x and y, the equations of motion for each body become

$$\text{Block } D: \text{Re}\{(-m\Omega^2 + ic\Omega + k)X \exp(i\Omega t)\} = \text{Re}\left\{k\frac{\varepsilon}{i}\exp(i\Omega t)\right\}$$

$$\text{Actuator } A: F = \text{Re}\left\{\left[(k - m\Omega^2)\frac{\varepsilon}{i} - kX\right]\exp(i\Omega t)\right\}$$

Because these relations must be valid at all t, we may match like coefficients of the complex exponentials, which leads to an expression for X,

$$X = \frac{k\varepsilon}{i(k + ic\Omega - m\Omega^2)}$$

In turn, the actuator equation yields an expression for the pin force. Substitution of the expression for X, followed by some algebraic manipulations, yields

$$F = \text{Re}\left\{\frac{\varepsilon}{i}\frac{(k - m\Omega^2)(k + ic\Omega - m\Omega^2) - k^2}{(k + ic\Omega - m\Omega^2)}\exp(i\Omega t)\right\}$$

The equation of motion for the flywheel indicates that $M = F\varepsilon\cos(\Omega t)$. Because F varies harmonically, it is not correct to merely use real parts to form M. Instead, we replace the cosine term with its definition in terms of complex exponentials, and explicitly introduce the complex conjugate part into the expression for F. As a convenience in performing these steps, we define $\hat{F}$ to be the complex amplitude of F,

$$\hat{F} = \frac{\varepsilon}{i}\frac{(k - m\Omega^2)(k + ic\Omega - m\Omega^2) - k^2}{(k + ic\Omega - m\Omega^2)}$$

$$F = \text{Re}(\hat{F}e^{i\Omega t}) \equiv \frac{1}{2}[\hat{F}\exp(i\Omega t) + \hat{F}^*\exp + (-i\Omega t)]$$

Then the expression for the couple M becomes

$$M = \frac{1}{2}[\hat{F}\exp(i\Omega t) + \hat{F}^*\exp(-i\Omega t)]\frac{\varepsilon}{2}[\exp(i\Omega t) + \exp(-i\Omega t)]$$

$$= \frac{1}{4}\hat{F}\varepsilon\exp(2i\Omega t) + \frac{1}{4}\hat{F}\varepsilon + \frac{1}{4}\hat{F}^*\exp(-2i\Omega t) + \frac{1}{4}\hat{F}^*\varepsilon$$

$$= \frac{1}{2}\text{Re}\{\hat{F}\varepsilon\exp(2i\Omega t)\} + \frac{1}{2}\text{Re}\{\hat{F}\varepsilon\}$$

This expression provides several useful insights into the system. First, we note that M fluctuates at twice the rate of rotation. This is a result of the fact that the flywheel is presented with the same situation when the pin is above and below the center point. The mean value of the moment is $\frac{1}{2}\text{Re}\{\hat{F}\varepsilon\}$. This mean value is zero only if $\hat{F}$ is purely imaginary, but the expression for $\hat{F}$ indicates that it is purely imaginary only if there is no dashpot, $c = 0$. The mean value of M represents the average torque that is required to sustain the rotation, so the average power input required for the motion is $\frac{1}{2}\text{Re}\{\hat{F}\varepsilon\}\Omega$. In other words, when damping is present, power must be input to the system by the motor that causes the flywheel to rotate. This power balances the power that is dissipated in the dashpot. Further consideration of the expression for $\hat{F}$ shows that, if c is relatively small, it is a maximum when $\Omega \approx (k/m)^{1/2}$, which is the natural frequency of the block and spring system when $\Omega = 0$. Thus, the largest torque and the maximum power are required when the system is close to resonance. We will look at power dissipation more carefully in the next section.

3.2 POWER DISSIPATION

We may obtain a different perspective for the role of a dashpot by considering the energy it dissipates. The power balance law states that

$$\dot{T} + \dot{V} = \mathscr{P}_{\text{in}} - \mathscr{P}_{\text{dis}} \tag{3.2.1}$$

The integral of this relation over time interval Δt leads to the work energy law,

$$\Delta T + \Delta V = \Delta W_{\text{in}} - \Delta E_{\text{dis}} \tag{3.2.2}$$

where ΔW_{in} is the work done by the excitation forces and ΔE_{dis} denotes the energy dissipated in the dashpots over that time interval. Recall that the power input to the system is the product of the generalized force and generalized velocity, while the instantaneous power dissipation is $C\dot{q}^2$. Hence, we have

$$\Delta W_{\text{in}} = \int_{t_1}^{t_1+\Delta t} Q\dot{q}\,dt, \quad \Delta E_{\text{dis}} = \int_{t_1}^{t_1+\Delta t} C\dot{q}^2\,dt \tag{3.2.3}$$

Let us consider the situation when the time interval Δt equals the steady-state period $2\pi/\omega$. The response repeats in this interval, so $\Delta T = \Delta V = 0$. From eq. (3.2.2), it follows that the work done by the excitation in this interval equals the energy dissipated by the dashpot. Correspondingly, we find from eqs. (3.2.3) that there are two ways by which we may compute the energy dissipated in a cycle,

$$E_{\text{dis}} = \int_{t_1}^{t_1+2\pi/\omega} C\dot{q}^2\,dt = \int_{t_1}^{t_1+2\pi/\omega} Q\dot{q}\,dt \tag{3.2.4}$$

To evaluate the integrals we use the complex representation of the harmonically varying force and the response. Because each term contains a product, we must explicitly include the complex conjugate part of each term, rather than merely indicating that the real part should be taken. The cyclical nature of the response allows us to set $t_1 = 0$ without loss of generality. The direct evaluation of the energy dissipation gives

$$E_{dis} = \int_0^{2\pi/\omega} \frac{1}{4}C[i\omega X\exp(i\omega t) - i\omega X^*\exp(-i\omega t)]^2 \, dt$$

$$= \frac{1}{4}\omega^2 C \int_0^{2\pi/\omega} [-X^2\exp(2i\omega t) + 2XX^* - (X^*)^2\exp(-2i\omega t)] \, dt \tag{3.2.5}$$

$$= \pi\omega CXX^*$$

The product of a complex number and its complex conjugate gives the square of the magnitude, so

$$\boxed{E_{dis} = \pi\omega C|X|^2} \tag{3.2.6}$$

When we follow a similar procedure to analyze the work done by the external force, we obtain

$$E_{dis} = \int_0^{2\pi/\omega} \frac{1}{4}[F\exp(i\omega t) + F^*\exp(-i\omega t)][i\omega X\exp(i\omega t) - i\omega t X^*\exp(-i\omega t)] \, dt$$

$$= \frac{i}{4}\omega \int_0^{2\pi/\omega} [FX\exp(2i\omega t) - FX^* + F^*X - F^*X^*\exp(-2i\omega t)] \, dt \tag{3.2.7}$$

$$= \frac{i\pi}{2}(-FX^* + F^*X)$$

Because the definition of $t = 0$ is arbitrary, we may consider F to be real, so that $F = F^* = |F|$. The phase lag of the response relative to the force is ϕ, so $X = |X|\exp(-i\phi)$. With this substitution, the above expression becomes

$$E_{dis} = \pi|F||X|\sin(\phi) \tag{3.2.8}$$

Equations (3.2.6) and (3.2.8) are alternative descriptions of the energy dissipated in a cycle of steady-state response. The two representations should both be correct; matching them leads to

$$\omega C|X| = |F|\sin(\phi) \tag{3.2.9}$$

Although this relation appears to be new, it could have been derived directly from the differential equation of motion. To do so, we would substitute the complex representations in which the force leads the response, $Q = \text{Re}[|F|\exp(i\omega t)]$ and $q = \text{Re}[|X|\exp(i\omega t - \phi)]$, into $M\ddot{q} + C\dot{q} + Kq = Q$, then match the imaginary parts on either side of the equality. Thus, we may use both representations of E_{dis} interchangeably.

3.2.1 Quality Factor of Resonance

The manner in which E_{dis} depends on frequency leads to an important concept for measurements. To identify the frequency dependence we substitute $|X| = (|F|/K)$ $|D(r, \zeta)|$, $C = 2\zeta\omega_{nat}M$, and $\omega = r\omega_{nat}$ into eq. (3.2.6). The result is

$$E_{dis} = \pi\frac{|F|^2}{K}\left[\frac{2\zeta r}{(1 - r^2)^2 + 4\zeta^2 r^2}\right] \tag{3.2.10}$$

The *bandwidth* of a resonance is the frequency interval over which E_{dis} is no less than half the maximum value. Identification of the bandwidth requires that we determine the value of r that maximizes E_{dis}. Toward that end we set $dE_{dis}/dr = 0$, which leads to

$$\max(E_{dis}) \quad \Leftrightarrow \quad 1 + 2(1 - 2\zeta^2)r^2 - 3r^4 = 0 \tag{3.2.11}$$

We could find the roots algebraically, but a much simpler solution results if we focus on lightly damped systems, $\zeta \ll 1$. In that case, we have

$$\max(E_{dis}) \approx \pi \frac{|F|^2}{K} \frac{1}{2\zeta} \quad \text{at} \quad r \approx 1 \tag{3.2.12}$$

It follows from eq. (3.2.10) that the bandwidth of a lightly damped system is such that

$$\left[\frac{2\zeta r}{(1 - r^2)^2 + 4\zeta^2 r^2} \right] \geq \frac{1}{2}\left(\frac{1}{2\zeta} \right) \tag{3.2.13}$$

The equation for the limiting values of r at which this becomes an equality is a quartic polynomial. Rather than attempting to find the roots exactly, we reason that because $\zeta \ll 1$, a plot of $|D(r, \zeta)|$ will feature a very narrow peak near the natural frequency. Consequently, the values of r satisfying the inequality should be close to unity. This suggests that we substitute $r = 1 + \delta$, with $\delta \ll 1$. We keep only the leading-order terms in δ, so the denominator becomes $[1 - (1 + \delta)^2]^2 + 4\zeta^2(1 + \delta)^2 \approx 4\delta^2 + 4\zeta^2$. The bandwidth inequality then leads to

$$\frac{2\zeta}{4\delta^2 + 4\zeta^2} \geq \frac{1}{4\zeta} \quad \Rightarrow \quad \delta \leq \pm\zeta \quad \Rightarrow \quad \omega_{nat}(1 - \zeta) \leq \omega \leq \omega_{nat}(1 + \zeta) \tag{3.2.14}$$

The limiting frequencies are called the *half-power points*, and the corresponding bandwidth is

$$\Delta\omega \approx 2\zeta\omega_{nat} \tag{3.2.15}$$

The *quality factor* QF indicates the narrowness of a resonant peak. Its definition and approximate value for a lightly damped one-degree-of-freedom system are

$$\boxed{\text{QF} \equiv \frac{\omega_{nat}}{\Delta\omega} \approx \frac{1}{2\zeta} = |D(r = 1, \zeta)| \approx |D|_{max}} \tag{3.2.16}$$

Furthermore, substituting $r = 1 \pm \zeta$ into the expression for $|D(r, \zeta)|$ shows the magnification factor at the half-power points to be

$$\boxed{|D|_{1/2} \equiv |D(1 \pm \zeta, \zeta)| \approx \frac{1}{\sqrt{2}}|D|_{max}} \tag{3.2.17}$$

In other words, the amplitude at the half-power points is approximately 70.7% of the peak amplitude.

Equations (3.2.16) and (3.2.17) lead to a concept for experimental identification of system properties. Suppose we have measured and plotted the steady-state amplitude $|X|$ as a function of the excitation's frequency for a one-degree-of-freedom system. If the peak is reasonably narrow, we may identify the natural frequency as the value of ω at which $|X|$ attains its maximum. We measure $|X|_{max}$ at that frequency, and then draw a horizontal line at 70.7% of the maximum, as typified by Figure 3.4. The

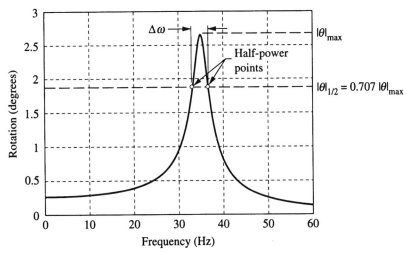

FIGURE 3.4 Evaluation of the bandwidth.

points of intersection of this line with the plot of $|X|$ versus ω mark the half-power points. We measure the bandwidth $\Delta\omega$ as the difference between the frequencies of these points. Substitution of $\Delta\omega$ and ω_{nat} into Eq. (3.2.16) then enables us to ascertain ζ for the system.

It might seem as though determining ζ from the bandwidth is more cumbersome than using the equation for $|D(r, \zeta)|$, which has the properties that $|D| = 1$ at $\omega = 0$, and $|D| = 1/(2\zeta)$ at $\omega = \omega_{nat}$. This leads to the steady-state amplitudes being related to ζ by

$$\zeta = \frac{X|_{\omega=0}}{2X|_{\omega=\omega_{nat}}} \tag{3.2.18}$$

Both amplitudes in the above are available from a graph like Figure 3.4. In practice, eq. (3.2.16) is preferable to eq. (3.2.18) because the latter requires knowing the amplitude at a very low frequency. This can be a difficult measurement for many motion sensors.

EXAMPLE 3.4

A different way in which the complex frequency response may be displayed is with a *Nyquist plot*. In it, the real and imaginary parts of $D(r, \zeta)$ are plotted against each other in the complex plane, with r (or ω) treated as a variable parameter.

(a) Construct the Nyquist plots for a one-degree-of-freedom system for the cases where $\zeta = 0.01$ and $\zeta = 0.6$.

(b) Prove that when $\zeta \ll 1$, the Nyquist plot is essentially circular.

(c) Identify the points on the circle corresponding to the natural frequency and the half-power points.

Solution This exercise will show a manifestation of the half-power points that leads to a widely employed technique for experimentally identifying ζ. Obtaining the Nyquist plot from the standard form of $D(r, \zeta)$ is straightforward. We define a set of evenly spaced values of r in the range $0 \leq r \leq 4$. (A large number of such points is necessary to obtain a smooth curve.)

We then compute $D(r, \zeta)$ at each r, and use the values of Re $[D(r, \zeta)]$ and Im $[D(r, \zeta)]$ at each r as the abscissa and ordinate, respectively. The results are displayed in the graphs.

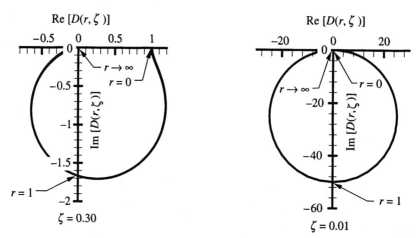

The curve for $\zeta = 0.30$ resembles a circle, whereas the one for $\zeta = 0.01$ is indistinguishable from a circle. The two curve intersects the imaginary axis at $D = -1/2i\zeta$ when $r = 1$. Other points of interest are $r = 0$ where $D = 1$ and $r \rightarrow \infty$ where $D \rightarrow 0$. The latter two points seem to have merged in the lightly damped case because the overall scale is greatly enlarged due to the large magnitude of D encountered in the vicinity of $r = 1$.

To identify the properties of the limiting circle we define variables u and v to be the real and imaginary parts of D. Clearing the complex quantity from the denominator of $D(r, \zeta)$ gives

$$D(r, \zeta) = u + iv, \quad u = \frac{(1 - r^2)}{(1 - r^2)^2 + 4\zeta^2 r^2}, \quad v = \frac{-2\zeta r}{(1 - r^2)^2 + 4\zeta^2 r^2}$$

We wish to eliminate r in order to see the relationship between v and u. The dependence on r is quite complicated. To simplify it we observe that large values of u and v correspond to the vicinity of $r = 1$. In that region it is acceptable to set $1 - r^2 = (1 + r)(1 - r) \approx 2(1 - r)$ and $2\zeta r \approx 2\zeta$, so that

$$u \approx \frac{(1 - r)}{2[(1 - r)^2 + \zeta^2]}, \quad v \approx \frac{-\zeta}{2[(1 - r)^2 + \zeta^2]}$$

The expression for v may be solved for the value of $1 - r$, which when substituted into the equation for u gives

$$u^2 + \left(v + \frac{1}{4\zeta}\right)^2 = \left(\frac{1}{4\zeta}\right)^2$$

This is the equation for a circle having radius $1/4\zeta$ centered at $u = 0$ and $v = -1/4\zeta$, that is, at $D(r, \zeta) = -i/4\zeta$.

An interesting aspect of this circle is the half-power points, which occur at $r = 1 \pm \zeta$. At that frequency, $u = \pm 1/4\zeta$ and $v = 1/4\zeta$. These are the points on the horizontal diameter of the circle, at which $u = $ Re $[D(r, \zeta)]$ has its largest magnitude. These observations lead to an alternative method for identifying ω_{nat} and ζ. Suppose we have measured the complex amplitude Y as a function of frequency with the force amplitude fixed. Rather than searching for a peak in a plot of $|Y(\omega)|$ as a function of ω, we can construct the Nyquist plot of Im(Y) against Re(Y). Then, we may fit a circle to the data. The intersection of the fitted circle with the negative imaginary axis will correspond to $Y(\omega_{nat})$. The difference of the frequencies at the horizontal diametral points will be the bandwidth, from which ζ can be extracted according to eq. (3.2.16). One of the beneficial aspects of such a procedure is that it tends to average out measurement errors, especially if one uses a least squares procedure to fit the circle.

3.2.2 Structural Damping

Equation (3.2.6) indicates that the energy dissipated by a dashpot in a harmonic motion is proportional to $\omega |X|^2$. However, experiments often indicate that E_{dis} is independent of ω when the amplitude is held fixed. (Performing such an experiment requires that the force amplitude be adjusted in order to maintain a constant displacement amplitude as ω is altered.) To see how it might happen that E_{dis} remains constant, let us consider the spring and mass in a one-degree-of-freedom system as a metaphor for the relationship between stress σ and strain ε in a material. Figure 3.5 is a Kelvin model of the constitutive equation in a viscoelastic material. In it, the spring represents the static Young's modulus, while the dashpot accounts for the dependence of the stress σ on the strain rate $\dot{\varepsilon}$. The constitutive equation represented by this model is $\sigma = E\varepsilon + \mu\dot{\varepsilon}$. Suppose a testing machine varies the strain harmonically, such that $\varepsilon = \varepsilon_0 \sin(\omega t)$. In this circumstance, we may express $\dot{\varepsilon}$ as a function of ε by eliminating t according to

$$\dot{\varepsilon} = \varepsilon_0 \cos(\omega t) = \pm\varepsilon_0 \omega [1 - \sin(\omega t)^2]^{1/2} = \pm\omega(\varepsilon_0^2 - \varepsilon^2)^{1/2} \qquad (3.2.19)$$

This leads to $\sigma = E\varepsilon \pm \mu\omega(\varepsilon_0^2 - \varepsilon^2)^{1/2}$. This dependence, which is depicted in Figure 3.6, is elliptical, with $\sigma = E\varepsilon$ as the major axis. The upper part of the ellipse corresponds to $\dot{\varepsilon} > 0$, because $\sigma > E\varepsilon$ in that case.

The stress-strain curve in Figure 3.6 depicts a *hysteresis phenomenon*. The energy dissipated by the viscoelastic effect is the area of the ellipse, because the work going from $-\varepsilon_0$ to $+\varepsilon_0$ along the upper curve is positive ($d\varepsilon > 0$), while the work on

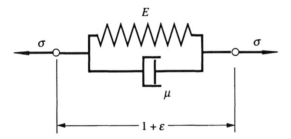

FIGURE 3.5 Kelvin model of a viscoelastic stress-strain relation.

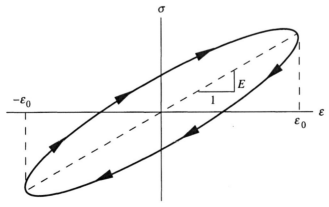

FIGURE 3.6 Stress-strain relation for a Kelvin model of a viscoelastic material.

the lower curve is negative ($d\varepsilon < 0$). From eq. (3.2.6) we have $E_{\mathrm{dis}} = \pi\omega\mu\varepsilon_0^2$. We now see that this dependence on ω is a consequence of the fact that the width of the deviation of the ellipse from the major axis is $\mu\dot{\varepsilon}$, which is proportional to ω.

If the area enclosed by the hysteresis curve is independent of ω for a specified amplitude ε, the hysteresis curve must be independent of the magnitude of ε. One situation where this occurs is in an elastic-plastic constitutive equation, as depicted in Figure 3.7. When the harmonic variation of strain begins at $t = 0$ with $\varepsilon = 0$, the stress follows $\sigma = E\varepsilon$ until it reaches the yield stress σ_Y. At that point, the material undergoes additional strain without changing the stress, until $\varepsilon = \varepsilon_0$. At this juncture, the stress decreases linearly with the reduction in strain in the portion of the cycle where $\dot{\varepsilon} < 0$, until $\sigma = -\sigma_Y$. At this point, the strain continues to decrease until $\varepsilon = -\varepsilon_0$, with no reduction in stress. The portion of the cycle in which $\dot{\varepsilon} > 0$ now begins, and the stress increases linearly with increasing strain until $\sigma = \sigma_Y$, after which the remainder of the cycle brings ε to ε_0 with no change in stress. The cycle then repeats. Clearly, this stress-strain curve depends only on the sense of $\dot{\varepsilon}$, so the area enclosed by the hysteresis curve is independent of ω.

In mechanical systems a similar effect can be produced with mechanical friction, which is generated at joints connecting bodies. Figure 3.8 depicts a simple model reproducing the hysteresis behavior. The spring is connected in series with a Coulomb friction element that is loaded by a normal force N. The total deformation Δ is the sum of the spring deformation F/K and the amount by which the friction element slips. As the magnitude of the axial force rises from zero to $\mu_s N$, where μ_s is the coefficient of static friction, the friction element is locked and the spring deforms. Further deformation is associated with slipping of the friction element, with no change in the force F.

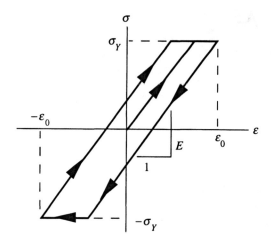

FIGURE 3.7 Stress-strain relation for an elastic-plastic material.

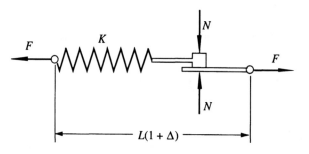

FIGURE 3.8 Spring-friction element that exhibits a constant-area hysteresis loop.

When the motion ceases prior to reversing, the friction element locks up, and stays locked until $|F|$ again attains the value $\mu_s N$ and slipping occurs. Hence, the spring stiffness K is analogous to E in Figure 3.7, and $\mu_s N$ is analogous to σ_Y. (If the static and kinetic coefficients are not the same, the axial force F drops to $\mu_k N$ when slipping begins. Nevertheless, the nature of the hysteresis loop in that case would still depend only on the sense of Δ, but not its magnitude. Consequently, the area contained within the hysteresis loop would still be independent of frequency.)

Friction processes are quite difficult to model, but we can perform experiments to determine E_{dis}. Suppose that through such an experiment, we determine that

$$E_{dis} = \alpha |X|^2 \tag{3.2.20}$$

where α is a coefficient that is independent of ω. We refer to this as *structural damping*. We may determine an equivalent dashpot constant C_{eq} by equating the measured E_{dis} to eq. (3.2.6), which gives

$$C_{eq} = \frac{\alpha}{\pi \omega} \tag{3.2.21}$$

This equivalence applies only in situations where the displacement varies harmonically. The dynamic stiffness for harmonic motion is given by eq. (3.1.4). When we substitute C_{eq} from above, we have

$$X = \frac{F}{K + i\dfrac{\alpha}{\pi} - M\omega^2} \tag{3.2.22}$$

The terms in the denominator that are independent of frequency represent the *complex stiffness*, K',

$$\boxed{K' = K(1 + i\gamma)} \tag{3.2.23}$$

where γ is the *structural damping loss factor*,

$$\boxed{\gamma = \frac{\alpha}{\pi K}} \tag{3.2.24}$$

The corresponding solution for the complex amplitude of the response is

$$\boxed{X = \frac{F}{K}\left[\frac{1}{1 - (\omega/\omega_{nat})^2 + i\gamma}\right]} \tag{3.2.25}$$

The quality factor of a structural damping resonance is $1/\gamma$. Note that the peak value of $|X|$ in a system with structural damping occurs at $\omega = \omega_{nat}$, regardless of the value of γ.

EXAMPLE 3.5

A spring-mass system slides over a rough horizontal surface, for which the coefficient of sliding friction is μ. A harmonic force $F\cos(\omega t)$ is applied.

(a) Derive an expression for the energy dissipated by friction during one cycle of steady-state vibration by considering the work done by the friction force. Use that expression to identify the equivalent linear dashpot constant and structural damping loss factor.

(*b*) Use the equivalent dashpot constant to derive an expression for X as a function of ω.

(*c*) Derive an expression for the work done by the excitation force in one cycle and use that result to derive a relation for the phase lag ϕ.

Solution This exercise illustrates a process for balancing alternative descriptions of energy dissipation in order to fit unusual dissipation mechanisms into the linear dashpot model. As we saw in Section 2.2.5, the friction force applied to a block sliding over a rough surface may be written as

$$f = -\mu_k N \operatorname{sgn}(\dot{x}) \text{ if } \dot{x} \ne 0$$

where sgn () is the signum function, and the negative sign indicates that the force is opposite the sense of $\dot{q}$. If the sliding friction force μN is small compared to the excitation amplitude F, the steady-state response will be essentially harmonic. We set

$$q = X \cos(\omega t - \phi) = \operatorname{Re}[X \exp(-i\phi) \exp(i\omega t)]$$

If we consider a cycle to begin at an instant when x is the largest *negatively*, which corresponds to $\cos(\omega t - \phi) = -1$, then $\dot{q}$ will be positive in the first half of the cycle, and negative in the second half. Correspondingly, $f = -\mu N$ in the first half-cycle, and $f = \mu N$ in the second half. Thus, we have

$$f = \begin{cases} -uN \text{ when } (\phi - \pi)/\omega < t < \phi/\omega \\ uN \text{ when } \phi/\omega < t < (\phi + \pi)/\omega \end{cases}$$

The power dissipation is the negative of $f\dot{x}$. The energy dissipated over a cycle is the integral of this quantity over one period. In view of the piecewise description of f, we find that

$$E_{\text{dis}} = -\int_{(\phi-\pi)/\omega}^{\phi/\omega} f\dot{x}\,dt - \int_{\phi/\omega}^{(\phi+\pi)/\omega} f\dot{x}\,dt = \mu N \int_{(\phi-\pi)/\omega}^{\phi/\omega} \dot{x}\,dt - \mu N \int_{\phi/\omega}^{(\phi+\pi)/\omega} \dot{x}\,dt$$

$$= \mu N[x|_{\phi/\omega} - x|_{(\phi-\pi)/\omega}] + \mu N[x|_{\phi/\omega} - x|_{(\phi+\pi)/\omega}] = 4\mu NX$$

According to eq. (3.2.6), the energy dissipated by a linear dashpot is $E_{\text{dis}} = \pi \omega C_{\text{eq}} X^2$. Matching this to the expression we derived yields

$$C_{\text{eq}} = \frac{4\mu N}{\pi \omega X}$$

This expression shows that the damping effect of Coulomb friction for steady-state response decreases inversely to both the amplitude and the frequency of that response. According to the structural damping model, the energy dissipated per cycle would be $E_{\text{dis}} = \alpha X^2$, so $\alpha = 4\mu N/X$ for Coulomb friction. We find from eq. (3.2.24) that the corresponding loss factor is

$$\gamma_{\text{eq}} = \frac{\alpha}{\pi K} = \frac{4\mu N}{\pi KX}$$

From this, we see that the effect of Coulomb friction resembles structural damping, in the sense that both models lead to a loss factor that is independent of frequency when the displacement amplitude is specified. However, unlike structural damping, the loss factor for Coulomb friction decreases inversely with the amplitude. The observation that the loss factor is proportional to the ratio of the normal force pressing together the surfaces to the maximum spring force KX led Caughey and Vijayaraghavan (1970) to suggest that structural damping could be simulated with a device in which the normal force increases linearly with displacement, for example, by using wedges.

To derive an expression for the amplitude at a specified frequency, we use C_{eq} to form the dynamic stiffness. Setting $x = \operatorname{Re}[X \exp(-i\phi) \exp(i\omega t)]$ leads to

$$X \exp(i\phi) = \frac{F}{K - \omega^2 M + i\omega C_{\text{eq}}} = \frac{F}{K - \omega^2 M + i\dfrac{4\mu N}{\pi X}}$$

Taking the magnitude of this relation and solving for X gives

$$X = \frac{\left[F^2 - \left(\frac{4\mu N}{\pi}\right)^2\right]^{1/2}}{K - \omega^2 M} = \frac{F}{K}\left[1 - \left(\frac{4\mu N}{\pi F}\right)^2\right]^{1/2} \frac{1}{1 - \left(\frac{\omega}{\omega_{\text{nat}}}\right)^2}$$

The analysis is valid only if $\mu N/F$ is small, so X is always real. Interestingly, the dry friction model of dissipation leads to a true resonance when $\omega = \omega_{\text{nat}}$.

The analysis thus far has examined the friction force to describe the energy dissipation. We may also describe E_{dis} by equating it to the negative of the work done by the excitation in a cycle. The analysis leading to eq. (3.2.8) remains valid here, because both the force and the response are harmonic. Thus, we have

$$E_{\text{dis}} = 4\mu NX = \pi FX \sin(\phi) \quad \Rightarrow \quad \phi = \arcsin\left(\frac{4\mu N}{\pi F}\right)$$

In other words, the phase lag resulting from a specified excitation amplitude is independent of the frequency of the response.

3.3 RESONANT RESPONSE

Conventional terminology refers to the steady-state response of a system excited at its natural frequency as the *resonant response*, regardless of whether there is damping. However, in the strictest sense, *resonance* should be reserved for the case where *the system has no damping and the frequency matches the natural frequency*. As noted earlier, the response in this case is not harmonic. The equations governing resonance of a one-degree-of-freedom system are

$$\ddot{q} + \omega_{\text{nat}}^2 q = \frac{F_0}{M}\cos(\omega_{\text{nat}}t + \psi) = \text{Re}\left[\frac{F}{M}\exp(i\omega_{\text{nat}}t)\right], \quad F = F_0\exp(i\psi)$$

$$q(0) = q_0, \qquad \dot{q}(0) = \dot{q}_0 \tag{3.3.1}$$

The forcing function matches the complementary solution, so the method of undetermined coefficients dictates that the particular solution be multiplied by a factor t,

$$q_p = \text{Re}\{tX\exp(i\omega_{\text{nat}}t)\} \tag{3.3.2}$$

Substitution of this trial form into eq. (3.3.1) yields

$$X = \frac{1}{2i\omega_{\text{nat}}}\frac{F}{M} \tag{3.3.3}$$

The factor of i in the denominator indicates that the harmonic term in the response will lag by 90° relative to the excitation. In terms of real functions, the resonant response is

$$q = \frac{|F|}{2M\omega_{\text{nat}}}t\sin(\omega_{\text{nat}}t + \psi) + c_1\cos(\omega_{\text{nat}}t) + c_2\sin(\omega_{\text{nat}}t) \tag{3.3.4}$$

The coefficients c_1 and c_2 for the homogeneous solution are obtained by satisfying the initial conditions. A typical response for the case where $q_0 = \dot{q}_0 = \psi = 0$, which leads to $c_1 = c_2 = 0$, is depicted in Figure 3.9.

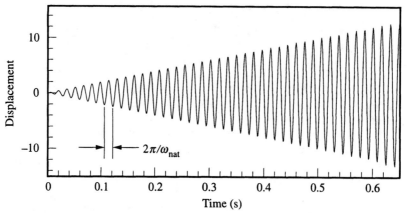

FIGURE 3.9 Resonant response of an undamped system, $\omega_{nat} = 400$ rad/s.

Equation (3.3.2) indicates that the steady-state response consists of an oscillation at the natural frequency whose amplitude is defined by an envelope function that increases linearly with elapsed time. This is sensible from a physical viewpoint. Consider the work-energy principle. If the response actually was harmonic at the natural frequency, the spring force, $-Kq$, would match the inertial effect, $M\ddot{q}$, so the system would be unable to resist the external harmonic force. In order to balance the work done by the force, the mechanical energy of the system must increase, which is manifested by growth of the response's amplitude.

From a different perspective, the above particular solution is troubling, because it suggests that the basic nature of the system's response changes suddenly if ω is gradually changed to match ω_{nat}. This actually is not the case. To see why, let us consider the response when ω is close to, but not equal to, the natural frequency. The analysis requires that we evaluate the full solution, including the homogeneous terms that are used to satisfy the initial conditions. When the excitation frequency is different from ω_{nat}, the solution of eqs. (3.3.1) may be written as

$$q = \frac{1}{M(\omega_{nat}^2 - \omega^2)}\text{Re}\{F[\exp(i\omega t) - \exp(i\omega_{nat}t)]\} \\ + c_1\cos(\omega_{nat}t) + c_2\sin(\omega_{nat}t)$$
(3.3.5)

The constants c_1 and c_2 depend on the initial conditions, as well as the amplitude and phase of F. The reason for writing the solution in the above form is that the bracketed term isolates the portion of the harmonic complementary solution that has the same amplitude as the particular solution. When ω is close to ω_{nat}, these two terms combine to form a beating oscillation, which was discussed in Section 2.1.3. We say that the portion of the homogeneous solution within the brackets becomes *entrained* with the particular solution. To employ the aforementioned description of a beating response, we express the bracketed term as real functions,

$$u = \text{Re}\{F[\exp(i\omega t) - \exp(i\omega_{nat}t)]\} \\ = F_0[\cos(\omega t + \psi) - \cos(\omega_{nat}t + \psi)]$$
(3.3.6)

If we set $\omega_1 = \omega$, $\phi_1 = -\psi$, $\omega_2 = \omega_{nat}$, and $\phi_2 = -\psi - \pi$, then this matches eq. (2.1.18). (Note that the additional 180° phase lag in ϕ_2 results from taking the difference of two harmonics.) The beat frequency then is $\Delta_\omega = \frac{1}{2}(\omega_{nat} - \omega)$, and the center frequency is $\omega_{av} = \frac{1}{2}(\omega_{nat} + \omega)$. We also have $\phi_{av} = -\psi - \pi/2$, $\Delta_\phi = -\pi/2$, and

$\omega_{nat}^2 - \omega^2 \equiv (\omega_{nat} - \omega)(\omega_{nat} + \omega) \equiv 4\omega_{av}\Delta_\omega$. It follows from eq. (2.1.21) that eq. (3.3.5) may be rewritten as

$$q = \frac{F_0}{2M\omega_{av}\Delta_\omega} \sin(\Delta_\omega t) \sin(\omega_{av} t + \psi) + q_c(t) \qquad (3.3.7)$$

As shown in Figure 3.10, the first two factors constitute an envelope function describing the slowly varying amplitude of a high-frequency oscillation at the natural frequency.

Equation (3.3.7) enables us to examine how the undamped response changes when ω approaches ω_{nat}. As Δ_ω decreases, the beat interval π/Δ_ω increases, and the primary oscillation rate ω_{av} approaches ω_{nat}. Let us observe the above response over an interval that is short compared to the beat interval, but large compared to the period of the primary oscillation, $2\pi/\omega_{av}$. In other words, we consider the response for $0 < t < \varepsilon\pi/\Delta_\omega$, where $2\Delta_\omega/\omega_{av} \ll \varepsilon \ll 1$; this is the region depicted by the insert in Figure 3.10. Within this interval, it is reasonable to set $\sin(\Delta_\omega t) \approx \Delta_\omega t$, but we must retain the sinusoidal appearance of the average frequency term. Hence, the response in this interval may be approximated as

$$q \approx \frac{F_0}{2M\omega_{av}} t \sin(\omega_{av} t + \psi) + q_c(t), \quad 0 < t < \frac{\varepsilon\pi}{\Delta_\omega} \qquad (3.3.8)$$

The above shows that the envelope function is approximately proportional to t when the response begins to grow, which is a feature that is exhibited by the insert in Figure 3.10. It is evident that in the limit as $\Delta_\omega \longrightarrow 0$, this expression matches the resonant response in eq. (3.3.4). We conclude that the resonant response of an undamped system is the limit of a beating response formed when a portion of the homogeneous solution becomes entrained with the particular solution. As the excitation frequency approaches the natural frequency, the beat period becomes increasingly large. In the limit $\omega \longrightarrow \omega_{nat}$, only the initial part of the beat, in which the envelope function is proportional to t, would be seen.

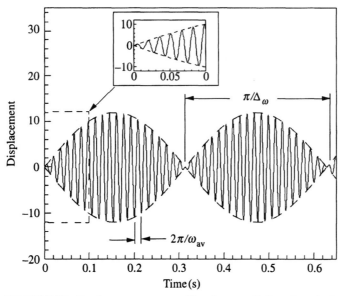

FIGURE 3.10 Beating response resulting from excitation close to the natural frequency, $\omega_{nat} = 400$ rad/s, $\omega = 390$ rad/s.

From a practical standpoint, the different appearance of the resonant response does not occasion concern. Obviously, all systems have some sort of energy dissipation. In fact, we will see in the next example that the resonant response may also be interpreted as the limiting behavior as damping decreases with ω fixed at the natural frequency. Even if there were no damping, resonance requires an exact equality of ω and ω_{nat}, which is not likely to occur. In addition, we should note that the magnitude of the resonant response cannot grow in proportion to t without limit. As the displacement increases, nonlinear effects might come into play, thereby invalidating the entire analysis. It also is likely that the system will fracture due to the cyclical displacement at increasingly large amplitudes.

EXAMPLE 3.6

Consider a lightly damped system that is subjected to harmonic excitation at its natural frequency, $Q = F\cos(\omega_{nat}t)$. The system is initially at rest at its equilibrium position, so $q = \dot{q} = 0$ at $t = 0$. To nondimensionalize the analysis, let $F = K = M = 1$. Plot the response q as a function of t for ratios of critical damping $\zeta = 0.001$ and $\zeta = 0.01$. What does a comparison of the responses indicate regarding the role of damping in the occurrence of a resonant response?

Solution   This example will provide insight into the true significance of resonant response. It is a straightforward matter to determine the response of the system in the stated conditions, because the system is damped. We construct the response as the sum of a complementary solution, in the form of the free response of an underdamped system, and a harmonic particular solution. Hence, we have

$$q = \exp(-\zeta\omega_{nat}t)[c_1\cos(\omega_d t) + c_2\sin(\omega_d t)] + \mathrm{Re}\left\{\frac{F}{K}|D|\exp[i(\omega t - \phi)]\right\}$$

It is given that $\omega = \omega_{nat}$, so the excitation frequency ratio is $r = 1$. It follows that $|D| = 1/(2\zeta)$ and $\phi = \pi/2$. The preceding expression then becomes

$$q = \exp(-\zeta\omega_{nat}t)[c_1\cos(\omega_d t) + c_2\sin(\omega_d t)] + \mathrm{Re}\left\{\frac{F}{K}\frac{1}{2\zeta i}\exp(i\omega t)\right\}$$

$$= \exp(-\zeta\omega_{nat}t)[c_1\cos(\omega_d t) + c_2\sin(\omega_d t)] + \frac{F}{K}\frac{1}{2\zeta}\sin(\omega_{nat}t)$$

We determine the coefficients c_1 and c_2 by satisfying the initial conditions that $q = \dot{q} = 0$ at $t = 0$. The first condition gives $c_1 = 0$, while satisfying the second requires that $-\zeta\omega_{nat}c_1 + \omega_d c_2 + (F/K)(1/2\zeta)\omega_{nat} = 0$. Thus, the response is

$$q = \frac{F}{K}\frac{1}{2\zeta}\left\{\sin(\omega_{nat}t) - \frac{1}{(1-\zeta^2)^{1/2}}\exp(-\zeta\omega_{nat}t)\sin[(1-\zeta^2)^{1/2}\omega_{nat}t]\right\}$$

The given parameters state that $F = K = M = 1$; thus $\omega_{nat} = 1$. To plot a graph of q as a function of t, we observe that the harmonic functions vary much more rapidly than the exponential when ζ is small. Consequently, we select Δt for plotting by dividing the natural period into 20 intervals,

$$\Delta t = \frac{2\pi}{20}$$

We plot the graphs up to a time for which the exponential factor is negligible, so that all that remains is the steady-state response. For example $\exp(-\zeta\omega_{nat}t) \approx 0.02$ when $\zeta\omega_{nat}t = 4$. Thus, we use $t_{max} = 4/\zeta$ to construct the graphs. The first result shows the very lightly damped case, $\zeta = 0.001$, for the interval up to $t_{max}/20$. It is clear that the response has the form of a sinusoidally varying term whose amplitude grows proportionally to t, which is the nature of the undamped resonance response. In the next graph, we see the same response up to t_{max}. The blur results from the large number of cycles that occur within this interval. We see that the response levels off to a constant amplitude, corresponding to the steady-state condition. The magnification factor $1/(2\zeta)$ is quite large. The third graph shows that increasing damping reduces the duration of the initial interval in which the amplitude seems to grow linearly. In addition, the steady-state amplitude is much smaller and occurs much sooner.

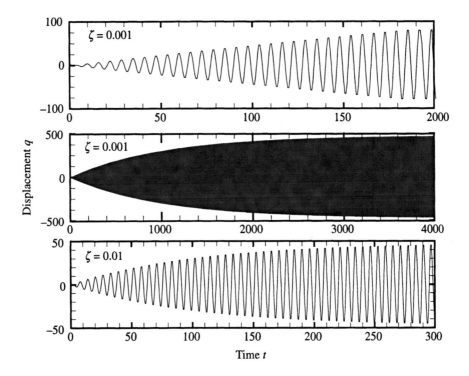

This result further illustrates our previous observation that a model in which dissipation has been ignored should be used only if it is applied to a restricted early interval of the motion. For example, in the first case ($\zeta = 0.001$), the first graph is essentially identical to the graph that would be obtained if ζ were zero. Hence, it might be adequate to use the undamped model if we knew that the system would fail when the amplitude exceeded the levels described by the first graph.

3.4 ROTATING IMBALANCE

3.4.1 Basic Model

A common source of harmonic excitation is a rotating machine in which the rotating mass is not situated exactly on the rotation axis. A prototypical situation is depicted in Figure 3.11. A motor turns a flywheel of mass m at constant angular speed ω. The angle θ of the radial line from the axis of rotation to the flywheel's center of mass is measured clockwise relative to the highest possible position, called top-dead-center. Hence, $\dot{\theta} = \omega$. The distance ε from the rotation axis to the flywheel's center of mass is the *eccentricity*. The motor is attached to a platform, which is supported by a spring and a dashpot. The combined mass of the platform and the nonrotating

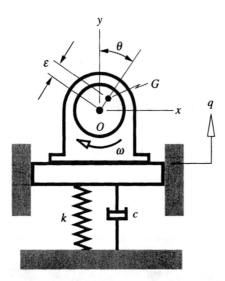

FIGURE 3.11 A motor mounted on a platform with an unbalanced flywheel.

parts of the motor is M. The vertical guides prevent sidesway, so the system has one degree of freedom. We denote the vertical position of the platform by the generalized coordinate q.

This system is not time invariant because we must know the value of t in order to determine the angle θ locating the eccentric center of mass. We could obtain the equations of motion for this system by applying Lagrange's equations. However, the Newton-Euler laws will lead to a physical interpretation that can readily be incorporated into the power formulation. Figure 3.12(a) is a free body diagram of the flywheel. (Gravity will generally have negligible effect if the rotation rate is sufficiently large that the dynamic effect of imbalance is important.) The forces F_x and F_y are the bearing reactions. The kinematical relation between the acceleration of two points in a rigid body that is rotating at a constant rate is $\bar{a}_G = \bar{a}_O + \bar{\omega} \times (\bar{\omega} \times \bar{r}_{G/O})$. The first term is the translational acceleration of the flywheel's shaft, $\bar{a}_O = \ddot{q}\bar{j}$, while the second term is the centripetal acceleration, which may be written alternatively as $-\omega^2 \bar{r}_{G/O}$. When we apply $\sum \bar{F} = m\bar{a}_G$ to the flywheel, we find that the bearing force is given by

$$\boxed{\bar{F} = m\bar{a}_O - m\omega^2 \bar{r}_{G/O}} \qquad (3.4.1)$$

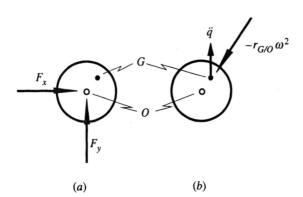

(a)   (b)

FIGURE 3.12 An imbalanced flywheel. (a) Free body diagram. (b) Acceleration of the center of mass.

This force contains two effects. Regardless of the rotation rate, the flywheel's mass must be made to translate following the motion of the shaft. The force required is $m\bar{a}_O$. In addition, the rotation causes the flywheel's center of mass to follow a circular path of radius $|\bar{r}_{G/O}| = \varepsilon$ relative to the center of rotation. This requires a centripetal force $-m\omega^2\bar{r}_{G/O}$.

The bearing forces act on the motor in the opposite sense. The motor and its base, isolated from the flywheel, represent a rigid body that translates in the vertical direction with an acceleration $\ddot{q}$, which is the same as the above acceleration $\bar{a}_O$. The corresponding equation of motion is

$$\Sigma F_y = -\bar{F}\cdot\bar{j} - ky - c\dot{y} = M\ddot{q} \tag{3.4.2}$$

We substitute eq. (3.4.1), with $\bar{a}_O = \ddot{q}\bar{j}$ and $\bar{r}_{G/O} = \varepsilon\sin(\theta)\bar{i} + \varepsilon\cos(\theta)\bar{j}$, into this basic equation of motion, which leads to

$$(M + m)\ddot{q} + C\dot{q} + Kq = m\varepsilon\omega^2\cos\theta \tag{3.4.3}$$

This equation of motion shows that the unbalanced mass enters into the system's dynamics in two ways. The mass m combines with the rest of the system's mass to form the inertial coefficient. In addition, the flywheel's rotation creates a centripetal acceleration exciting the system. These observations lead to an alternative free body diagram from which the equations of motion could have been derived. In Figure 3.13, the flywheel is mounted on the motor. In addition to the actual external forces, which are exerted by the spring and dashpot, we show the "centrifugal force" $m\varepsilon\omega^2$ representing the reaction to the centripetal force that is exerted on the flywheel by the shaft. The component of this forcing effect in the vertical direction is $m\varepsilon\omega^2\cos(\theta)$, which corresponds to a power input of $[m\varepsilon\omega^2\cos(\theta)]\dot{q}$. The mass of this system is $M + m$, which corresponds to a system kinetic energy of $\frac{1}{2}(M + m)\dot{q}^2$ if we neglect the rotational kinetic energy of the flywheel.

A generalization of the preceding development enables us to treat any system that has an unbalanced rotating mass. It involves a minor modification of the power balance method. We construct a free body diagram of the system with the flywheel mounted on its shaft. Forces that appear in that diagram are due to springs, dashpots, and the "centripetal force" $m\omega^2\bar{r}_{G/O}$ associated with the rotation of the eccentric center of mass G about point O on the shaft. To account for these forces we describe V, $\mathscr{P}_{dis}$, and $\mathscr{P}_{in}$, respectively, in terms of the generalized coordinates. In

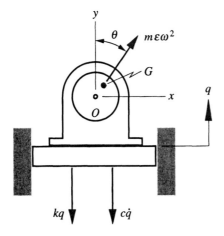

FIGURE 3.13 Unified free body diagram for a system containing an imbalanced rotating flywheel.

particular, because the centripetal force is applied at the shaft, whose velocity is $\bar{v}_O$, the associated power is

$$\boxed{\mathcal{P}_{\text{in}} = m\omega^2 \bar{r}_{G/O} \cdot \bar{v}_O} \tag{3.4.4}$$

To obtain the generalized force (or forces in the case of a multi-degree-of-freedom system), we linearize $\bar{v}_O$ in terms of the generalized velocity, as we would for any other force. Matching the resulting expression for $\mathcal{P}_{\text{in}}$ to the standard form $\mathcal{P}_{\text{in}} = \sum Q_j \dot{q}_j$ yields the required expression for the generalized forces.

The inertia coefficient(s) are obtained by adding the kinetic energy of the flywheel when it does not rotate to the kinetic energy of the other parts of the system. In other words, the kinetic energy T to be used in the power balance formulation is

$$T = T_{\omega=0} \tag{3.4.5}$$

where

$T_{\omega=0}$ *represents the kinetic energy of the whole system, including the flywheel, when the motor does not rotate.*

We determine the inertia coefficients by expressing this T in terms of the generalized coordinates, as we would for a time-invariant system. One of the attractive features of this modified procedure is that it is readily extended to systems having several motors and flywheels. To do so, we add to T the kinetic energy contribution of each flywheel in the absence of rotation, and we add to $\mathcal{P}_{\text{in}}$ a power term associated with each centripetal acceleration. These operations will be featured in Example 3.8.

3.4.2 Frequency Response

Let us now turn to the evaluation of the steady-state response. In Figure 3.11, θ is measured from the top-dead-center position. We take this position to correspond to $t = 0$, so $\theta = \omega t$. Consequently, the excitation resulting from the unbalanced flywheel varies harmonically as $m\varepsilon\omega^2 \cos(\omega t)$. (This would be a sine function if θ were measured from the horizontal.) The system mass in eq. (3.4.3) is $M + m$, so the natural frequency is

$$\boxed{\omega_{\text{nat}} = \left(\frac{K}{M + m}\right)^{1/2}} \tag{3.4.6}$$

The amplitude of the excitation is $F = m\varepsilon\omega^2$, so the steady-state response obtained from eqs. (3.1.7) and (3.1.9) is

$$\boxed{\begin{aligned} q &= \text{Re}\left\{\frac{m\varepsilon\omega^2}{K}|D(r, \zeta)| \exp[i(\omega t - \phi(r, \zeta))]\right\} \\ &= \varepsilon\left(\frac{m}{M + m}\right)\frac{r^2}{[(1 - r^2)^2 + 4\zeta^2 r^2]^{1/2}} \cos[\omega t - \phi(r, \zeta)] \end{aligned}} \tag{3.4.7}$$

where $r = \omega/\omega_{\text{nat}}$ as before. The last form highlights the fact that the ratio $|q|/\varepsilon$ is proportional to the ratio of the unbalanced mass to the total system mass, $m/(M + m)$. To reduce $|q|$ one would like to minimize both ε and m; the product εm is sometimes referred to as the *imbalance*.

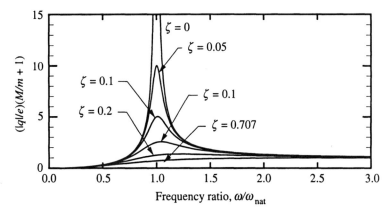

FIGURE 3.14 Steady-state amplitude of the response of a motor with an unbalanced flywheel.

The phase angle $\phi(r, \zeta)$ is the same as for steady-state response to a constant amplitude force; see eqs. (3.1.8). Because we defined $t = 0$ to be the instant at which the center of mass G is at the top-dead-center position, we see that ϕ gives the angle of the radial line to point G relative to vertical, at the instant when the displacement q is largest.

The amplitude factor in eq. (3.4.7) is plotted as a function of rotation rate in Figure 3.14. These curves resemble their counterparts in Figure 3.2, but there are basic differences. At low rotation rates, the response to rotating imbalance is very small, because the centripetal acceleration of the unbalanced mass is small. At rotation rates that are well above the natural frequency, the steady-state amplitude approaches the constant amplitude $\varepsilon m/(M + m)$. This is readily explained by recalling that the primary load-carrying capability of the system at high frequencies is its inertial effect. Here, the mass is $(M + m)$, so $(M + m)\ddot{q}$ must balance the vertical component of the centripetal force $m\varepsilon\omega^2$ required to make the center of mass G rotate in a circle relative to the center of rotation O. For harmonic motion $\ddot{q} = -\omega^2 q$, so the force balance requires $-(M + m)\omega^2 q = m\varepsilon\omega^2$; the minus sign shows that ϕ tends to 180° in the high frequency limit. A less obvious difference between Figures 3.14 and 3.2 is the fact that in the present case the peak amplitude occurs at a rotation rate that is higher than the natural frequency. Differentiation of the ω-dependent factor in eq. (3.4.7) would show that the peak occurs at $\omega/\omega_{nat} = 1/(1 - 2\zeta^2)$. Thus, the peak occurs at a frequency that exceeds ω_{nat}, but the difference is negligible if $\zeta \ll 1$. Any damping ratio above $\zeta = \sqrt{2}/2$ will lead to a steady-state displacement amplitude that increases monotonically to the high-speed limit $\varepsilon m/(M + m)$. In contrast, exceeding that level of damping for a *constant-amplitude* force leads to a displacement amplitude that decreases monotonically to zero.

EXAMPLE 3.7

The motor shown on the next page has a flywheel whose center of mass is 5 mm from the rotation axis. The mass of the stationary parts of the motor is 10 kg and the rotating parts have a mass of 1 kg; the mass of the beam is negligible. It is known that the downward static deflection of the beam due to the weight of the motor is 2 mm. The beam is coated by a damping material that gives a ratio of critical damping, $\zeta = 0.10$. The flywheel rotates at a constant speed of 675 rpm.

(a) Determine the natural frequency of the system when the motor is turned off.

(b) Determine the amplitude of the steady-state vibration of the motor at the given rotation rate.

(c) Determine the phase lag of the response relative to the excitation. Use the phase lag to determine the elapsed time after top-dead-center at which the beam reaches its highest position in the vibration.

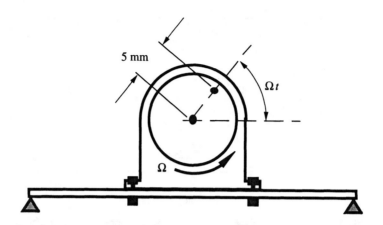

Solution This exercise provides familiarization with the frequency dependence of amplitude and phase lag associated with rotating imbalance, as well as insight into the significance of the phase lag. We create a one-degree-of-freedom model based on considering the beam to act as a simple spring that resists vertical displacement of the motor. We let y, positive upward, be the generalized coordinate for translational motion of the motor. The free body diagram is essentially as shown in Figure 3.13. The only difference is that here the angle of rotation has been defined relative to a horizontal line, so $\sin(\Omega t)$ gives the component of the centripetal force in the vertical direction. Thus, the equation of motion for the system is

$$(M + m)\ddot{y} + \mu\dot{y} + ky = m\varepsilon\Omega^2\sin(\Omega t)$$

where M is the mass of the nonrotating parts, m is the mass of the flywheel, and K is the stiffness of the beam. The natural frequency and ratio of critical damping are

$$\omega_{\text{nat}} = \sqrt{\frac{k}{M + m}}, \qquad \zeta = \frac{\mu}{2(M + m)\omega_{\text{nat}}}$$

We could solve this problem by merely applying the formulas, but following the derivation steps in the specific case of this system will enhance our proficiency. Dividing the equation of motion by $M + m$ converts it to

$$\ddot{y} + 2\zeta\omega_{\text{nat}}\dot{y} + \omega_{\text{nat}}^2 y = \frac{m}{M + m}\varepsilon\Omega^2\,\text{Re}\left[\frac{1}{i}\exp(i\Omega t)\right]$$

We use a complex representation of y that matches the form of the term on the right side,

$$y = \text{Re}\left\{\frac{Y}{i}\exp(i\Omega t)\right\}$$

Our reason for taking the solution to have this form is that the argument of the complex amplitude Y will then represent the lag of the response. We substitute the trial form into the equation of motion and match like coefficients of the complex exponentials, which leads to

$$Y = \frac{m}{M + m}\frac{\varepsilon\Omega^2}{\omega_{\text{nat}} - \Omega^2 + 2i\zeta\omega_{\text{nat}}\Omega}$$

$$\equiv \varepsilon\frac{m}{M + m}\frac{r^2}{1 - r^2 + 2i\zeta r}$$

Not surprisingly, this is the same as the relation for the complex amplitude found earlier.

We are now ready to use the stated system properties. The static deflection under the total weight of the system is

$$\frac{(M + m)g}{k} = 0.002 \text{ m}$$

From this we compute the natural frequency to be

$$\omega_{nat} = \left(\frac{k}{M + m}\right)^{1/2} = \left(\frac{g}{0.002}\right)^{1/2} = 70.025 \text{ rad/s}$$

The rotation rate of interest is 675 rev/min, which when converted to rad/s gives an excitation frequency of

$$\Omega = 675\frac{2\pi}{60} = 70.686 \text{ rad/s}$$

Correspondingly, we have $r = \Omega/\omega_{nat} = 1.0094$. The next step is direct substitution into the expression for Y using the given values of ε and ζ,

$$Y = \frac{1}{11}(0.005)\left(\frac{1.01896}{-0.01896 + 0.2091i}\right) = 2.284(10^{-3})e^{-1.664i} \text{ m}$$

This resulting expression for the steady-state response is

$$y = \text{Re}\left\{\frac{1}{i}2.284(10^{-3})e^{-1.664i}e^{I\Omega t}\right\} \text{ m}$$

$$= 2.284\sin(\Omega t - 1.664) \text{ mm}$$

The angle of the radial line to the center of mass is Ωt, measured from the horizontal. Thus, the top-dead-center position of the center of mass corresponds to $\Omega t_1 = \pi/2$. For maximum elevation of the beam, we seek the instant t_2 at which y has its maximum value. We accordingly set its sine function to unity,

$$\sin(\Omega t_2 - 1.664) = 1 \quad \Rightarrow \quad \Omega t_2 - 1.664 = \frac{\pi}{2}$$

Thus, the elapsed time is

$$t_2 - t_1 = \frac{1.664}{\Omega} = 0.0235$$

EXAMPLE 3.8

In the system shown on the next page, two disks, each of whose mass is m, rotate at constant angular speed about bearings in the rigid angled bar. Each disk has a small lead insert whose masses m_1 and m_2 are much less than m; they are mounted at distance ε from the bearing. Both inserts are positioned such that the radial line from the bearing to the insert forms an angle ωt from the vertical, as shown. The system lies in the horizontal plane, so gravity has no effect, and the drawing shows the system in its static equilibrium position. Also, the moment of inertia of the angled bar about the pivot is $5ma^2$, and the moment of inertia of each disk about its bearing is negligible.

(a) The equation of motion governing the rotation θ of the angled bar has the general form

$$M\ddot{\theta} + C\dot{\theta} + K\theta = Q(t)$$

Derive expressions for M, C, K, and $Q(t)$.

(b) Is there any combination of the imbalances εm_1 and εm_2 for which their individual effects cancel, such that $Q(t) \equiv 0$?

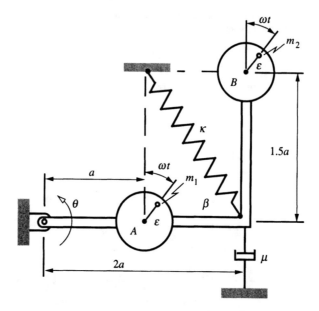

Solution This exercise will illustrate the general steps required to incorporate the effects of rotating imbalance into equations of motion. It also will improve understanding of the physical effect. A free body diagram of the angled bar with attached disks shows the bearing forces, the spring and dashpot forces, and centrifugal forces associated with the centripetal acceleration of each unbalanced disk.

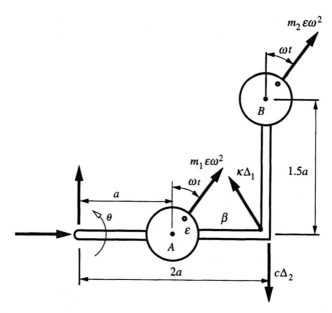

We begin by constructing the kinetic energy in order to identify the generalized mass M. Recall that for this step, we ignore the rotation of the eccentric bodies relative to the rest of the system. It is stated that the moment of inertia of each disk about its bearing is negligible, and the inserts have small mass compared to the mass of the bar. Hence, for the purpose of constructing T, we may consider each disk to be a point mass situated at the respective centers of rotation. We add the translational kinetic energy of these bodies to the rotational kinetic energy of the angled bar,

$$T = \tfrac{1}{2}I_O\dot{\theta}^2 + \tfrac{1}{2}mv_A^2 + \tfrac{1}{2}mv_B^2$$

Because points A and B follow circular paths centered on the pivot, we have

$$v_A = R_A\dot{\theta} = a\dot{\theta}$$

$$v_B = R_B\dot{\theta} = [(2a)^2 + (1.5a)^2]^{1/2}\dot{\theta} = 2.5a\dot{\theta}$$

We substitute these descriptions into T. Matching the result to the standard form yields the inertia coefficient M,

$$T = \tfrac{1}{2}(5ma^2)\dot{\theta}^2 + \tfrac{1}{2}(m)(a\dot{\theta})^2 + \tfrac{1}{2}(m)(2.5a\dot{\theta})^2 = \tfrac{1}{2}M\dot{\theta}^2 \quad \Rightarrow \quad M = 12.25ma^2$$

Potential energy, which we construct in order to identify the stiffness coefficient K, is stored in the spring. For circular motion, the displacement of the point at which the spring is attached to the bar is $2a\theta$. The elongation of the spring is the component along the spring of the corresponding displacement,

$$\Delta_k = 2a\theta\sin(\beta)$$

The value of $\sin(\beta)$ is related to the length parameters by the Pythagorean theorem,

$$\sin(\beta) = \frac{1.5a}{[(1.5a)^2 + a^2]^{1/2}}$$

so we find that

$$V = \frac{1}{2}\kappa\left[2a\theta\frac{1.5a}{(3.25a^2)^{1/2}}\right]^2 = \frac{1}{2}K\dot{\theta}^2 \quad \Rightarrow \quad K = 2.769\kappa a^2$$

We find the damping constant C in a similar manner. The velocity of the end of the dashpot is $2a\theta$, which is aligned with the dashpot. Thus, the power dissipation gives

$$\mathcal{P}_{dis} = \mu(2a\dot{\theta})^2 = C\dot{\theta}^2 \quad \Rightarrow \quad C = 4\mu a^2$$

The generalized force contains the effect of the centripetal acceleration of the mass centers. A variety of ways are available to describe the power input. We shall form the total moment exerted by the forces about the angle bar's pivot. The power input will then be $\mathcal{P}_{in} = (\Sigma M)\dot{\theta}$. We compute that moment by decomposing each centripetal force into its x and y components, from which we find that

$$\Sigma M = [m_1\varepsilon\omega^2\cos(\omega t)]a + [m_2\varepsilon\omega^2\cos(\omega t)]2a - [m_2\varepsilon\omega^2\sin(\omega t)]1.5a$$

$$= \varepsilon a\omega^2[(m_1 + 2m_2)\cos(\omega t) - 1.5m_2\sin(\omega t)]$$

Note that a positive moment corresponds to counterclockwise rotation $\dot{\theta}$. We use this expression to form $\mathcal{P}_{in}$, and equate the result to the standard form,

$$\mathcal{P}_{in} = \varepsilon a\omega^2[(m_1 + 2m_2)\cos(\omega t) - 1.5m_2\sin(\omega t)]\dot{\theta} = Q\dot{\theta}$$

Cancellation of the common factor $\dot{\theta}$ yields Q, which clearly is the moment about the pivot of the centripetal forces. The resulting equation of motion is

$$12.25ma^2\ddot{\theta} + 4a^2\mu\dot{\theta} + 2.769\kappa a^2\theta = \varepsilon a\omega^2[(m_1 + 2m_2)\cos(\omega t) - 1.5m_2\sin(\omega t)]$$

We may now examine whether it is possible to cancel the imbalance effect. For this to be true, it must be that the generalized force is identically zero regardless of ω. It is evident that this is not possible. If we were to reverse the radial orientation of insert m_1, that is, place it 180° relative to current radial position, the sign of the m_1 term in the generalized force would be negative. In that case, selecting $m_1 = 2m_2$ would cancel the cosine term, but the sine term would still remain. In essence, this results from the fact that the x component of the second centripetal force is 90° out of phase from the y component of both forces, so there is no opportunity to cancel its moment.

3.5 WHIRLING OF SHAFTS

Rotating machinery, which ranges in size from tiny gyroscopes used to guide aircraft, to water turbines for power generation, features a number of vibratory phenomena that can lead to catastrophic failure. One case that bears some resemblance to the effect of imbalance is described by the *Jeffcott model*, depicted in Figure 3.15. The model's name recognizes the contribution of the English scientist Jeffcott (1919), although Dimarogonas (1996) suggests that it would be more appropriate to recognize the earlier contributions of Laval and Föppel (1895), who previously successfully analyzed the model. It consists of a large rotor mounted at its geometric center C on a low-mass, flexible shaft that rotates at constant angular speed ω about bearings at either end. The center of mass is off center by distance ε. Because of imbalance, the shaft at any instant will be deflected outward. To locate the rotor's position at any instant we define a fixed reference frame XYZ whose X axis aligns with the axis of the bearings. As shown in the figure, the position coordinates Y_C and Z_C locate point C at any instant.

Torsional deformation of the shaft is assumed to be negligible, so any line in the rotor rotates at angular speed ω. Thus, the line CG from the shaft to the rotor's center of mass rotates at angular speed ω. We define $t = 0$ such that the angle between this line and the Y axis is ωt. Figure 3.16 is a free body diagram of the rotor. This diagram assumes that the shaft has a circular cross-section, so the stiffness coefficients for each direction are the same, and the elastic forces are kY_C and kZ_C. (The stiffness coefficients for a noncircular shaft are different values defined relative to the shaft. Such situations are analyzed in Chapter 12.)

The acceleration of the center is $\bar{a}_C = \ddot{Y}_C \bar{j} + \ddot{Z}_C \bar{k}$. At an arbitrary instant we have $\bar{r}_{G/C} = \varepsilon \cos(\omega t)\bar{j} + \varepsilon \sin(\omega t)\bar{k}$, where ε is the offset of the mass center of the flywheel. Because ω is constant, the angular acceleration of the rotor is zero. Therefore, the acceleration of the center of mass is given by

$$\bar{a}_G = \bar{a}_C + \bar{\omega} \times (\bar{\omega} \times \bar{r}_{G/C}) = (\ddot{Y}_C - \omega^2 \varepsilon \cos(\omega t))\bar{j} + (\ddot{Z}_C - \omega^2 \varepsilon \sin(\omega t))\bar{k}$$

$$(3.5.1)$$

We then find from Newton's Second Law that

$$m\ddot{Y}_C - m\omega^2 \varepsilon \cos(\omega t) = -kY_C, \qquad m\ddot{Z}_C - m\omega^2 \varepsilon \sin(\omega t) = -kZc \qquad (3.5.2)$$

Although the motion is described by two generalized coordinates, the equations are uncoupled and identical, except that the excitations are 90° out of phase. Let us as-

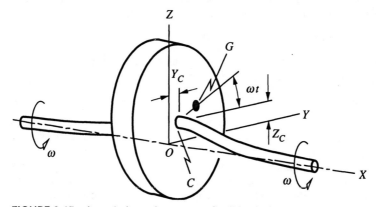

FIGURE 3.15 An unbalanced rotor on a flexible shaft undergoing whirling.

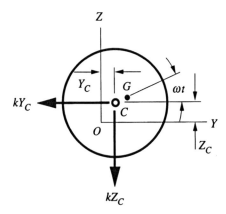

FIGURE 3.16 Free body diagram of an unbalanced rotor on a flexible shaft.

sume that there is damping from some external source, such as air resistance. The equations of motion in that case become

$$\ddot{Y}_C + 2\zeta\omega_{\text{nat}}\dot{Y}_C + \omega_{\text{nat}}^2 Y_C = \varepsilon\omega^2\cos(\omega t)$$
$$\ddot{Z}_C + 2\zeta\omega_{\text{nat}}\dot{Z}_C + \omega_{\text{nat}}^2 Z_C = \varepsilon\omega^2\sin(\omega t)$$

(3.5.3)

After a startup interval, the homogeneous solution for Y_C and Z_C will die out due to damping. The steady-state response is essentially as given by eq. (3.4.7) with $M = 0$,

$$Y_C = \varepsilon r^2 D(r, \zeta)\cos[\omega t - \phi(r, \zeta)]$$
$$Z_C = \varepsilon r^2 D(r, \zeta)\sin[\omega t - \phi(r, \zeta)]$$

(3.5.4)

where $r = \omega/\omega_{\text{nat}}$. Because $Y_C^2 + Z_C^2$ is a constant, the steady-state response is such that the attachment point C follows a circle whose radius is

$$R_C = \sqrt{Y_C^2 + Z_C^2} = \varepsilon r^2 |D(r, \zeta)| = \varepsilon\frac{r^2}{\sqrt{(1 - r^2)^2 + 4\zeta^2 r^2}}$$

(3.5.5)

The instantaneous position of the center of the disk as it follows its circular path is defined by the relative values of Y_C and Z_C. Let us define θ as the angle from the Y axis to line OC. We find that

$$\theta = \tan^{-1}\left(\frac{Z_C}{Y_C}\right) = \omega t - \phi(r, \zeta)$$

(3.5.6)

From this, we deduce that line OC rotates at angular speed ω, with θ lagging by ϕ behind the rotation of line CG. This motion is depicted in Figure 3.17 for cases where $r \ll 1$ ($R_0 \ll \varepsilon$ and $\phi \approx 0$); $r = 1$ ($R_0 \gg \varepsilon$ and $\phi = \pi/2$); and $r \gg 1$ ($R_0 \approx \varepsilon$ and $\phi \approx \pi$). Note in the figure that the origin of XYZ is fixed on the bearing axis. These observations indicate that at any instant the bent axis of the shaft lies in a plane that rotates at the same rate as the disk, although the line between the attachment point and the mass center does not coincide with this plane. Such a motion is called *synchronous whirl*.

The factor multiplying ε in eq. (3.5.5) is the same as the quantity plotted in Figure 3.14. As we found for unbalanced motors, the resonant angular speed for lightly damped systems occurs at $r = 1$, or

$$\omega_{\text{cr}} = \left(\frac{k}{m}\right)^{1/2}$$

(3.5.7)

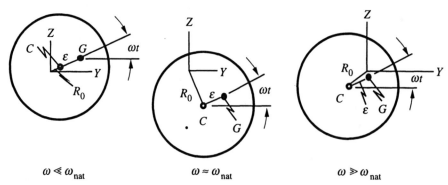

FIGURE 3.17 Synchronous whirl according to the Jeffcott model.

Clearly, $(k/m)^{1/2}$ is the natural frequency of the system when the system is not rotating. The resonant condition $\omega = \omega_{nat}$ is commonly called a *critical speed*. This condition is one of several dynamic conditions that can damage a rotor system. In many situations, it is desirable to operate rotating machinery above the critical speed. This is done by increasing ω at a sufficiently high rate that the resonance is passed before the vibration amplitude grows to excessive levels.

3.6 FORCE TRANSMISSION

An important parameter for the design of suspension systems is the force transmitted between the vibrating mass and the ground. The magnitude of the force has obvious significance for the suspended mass. In addition, although we consider the ground to be rigid and infinitely stiff, in reality vibrations at the suspension location can generate stress waves in the ground that might lead to troublesome vibration at distant locations. The force transmitted between the vibrating mass and the ground is indicated by the free body diagram, Figure 3.6, to be the sum of the forces in the spring and in the dashpot,

$$\boxed{F_{tr} = Kq + C\dot{q}} \tag{3.6.1}$$

The steady-state response to a harmonic excitation $\text{Re}[F \exp(i\omega t)]$ is given by eq. (3.1.9). We substitute that expression into F_{tr}, and use the fact that $C/K \equiv 2\zeta/\omega_{nat}$ to obtain

$$F_{tr} = \text{Re}[FD(r, \zeta)(1 + 2i\zeta r)\exp(i\omega t)] \tag{3.6.2}$$

The *transmissibility*, tr, is the ratio of the transmitted and exciting force amplitudes,

$$\boxed{\text{tr}(r, \zeta) \equiv \frac{|F_{tr}|}{|F|} = |D(r, \zeta)||(1 + 2i\zeta r)| = \left[\frac{1 + 4\zeta^2 r^2}{(1 - r^2)^2 + 4\zeta^2 r^2}\right]^{1/2}} \tag{3.6.3}$$

Figure 3.18 displays transmissibility as a function of the frequency ratio for several values of ζ. The value of r at which tr $= 1$ for any ζ may be found from eq. (3.6.3). This condition occurs when $r^2 - 1 = \pm 1$; that is, $r = \sqrt{2}$. When $0 < r < \sqrt{2}$, we find that tr > 1, regardless of the value of ζ. In other words, the force applied by the system to its base is greater than the exciting force causing the vibration. One aspect of vibration control requires reduction of the force transmitted to the ground. We see that making r exceed $\sqrt{2}$ brings tr below unity. Hence, if a certain frequency leads to excessive force transmission, we would redesign the system

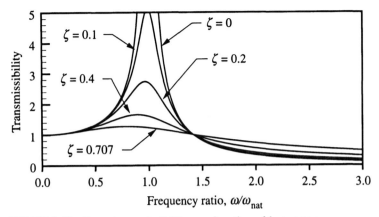

FIGURE 3.18 Force transmissibility as a function of frequency.

to decrease the natural frequency by making the suspension system as soft as possible. Notice also that raising the damping ratio when $r > \sqrt{2}$ increases the force transmission, because the dashpot force amplitude goes up in proportion to the frequency. Thus, the ideal suspension system for isolating the ground from a vibrating system is soft and lightly damped.

3.7 PERIODIC EXCITATIONS

The ability to determine the response to harmonic excitation provides a foundation for addressing other types of excitations in an efficient manner. The first generalization is the case of periodic excitation, which means that the excitation repeats in a constant time interval T, such that $Q(t + T) = Q(t)$ for any t. A harmonic force is a special case, but we retain the term *period* to refer to T for any periodic function.

3.7.1 Complex Fourier Series

Fourier series techniques are the logical approach for dealing with periodic functions. Most individuals come to be familiar with Fourier series as sums of sines and cosines. Our approach here will exploit our capabilities with complex exponentials, but the underlying concept is the same. We form a general periodic function by adding sinusoidally varying terms that repeat an integer number of times within the period. The term that repeats only once in an interval T is the *fundamental harmonic*, whose frequency is

$$\boxed{\omega_1 = \frac{2\pi}{T}} \tag{3.7.1}$$

We refer to ω_1 as the *fundamental frequency*. The variation of the fundamental harmonic is described by $\exp(i\omega_1 t)$. A term that repeats n times in an interval of one period has a frequency $\omega_n = 2\pi/(T/n) \equiv n\omega_1$. This term is the nth *harmonic*, whose variation is described by $\exp(in\omega_1 t)$. Each harmonic may have a complex amplitude F_n, whose argument defines the phase of the variation relative to a pure cosine function. In addition to adding every harmonic, we use a constant term $\frac{1}{2}F_0$ to allow for the fact that the function might have a mean value. Thus, we have

$$Q(t) = \frac{1}{2}F_0 + \mathrm{Re}\left[\sum_{n=1}^{\infty} F_n \exp(in\omega_1 t)\right] \tag{3.7.2}$$

Mathematical derivations proceed more easily if we avoid taking the real part, which we can do if we add each term and its complex conjugate. Thus, the terms representing the nth harmonic are $\frac{1}{2}[F_n \exp(in\omega_1 t) + F_n^* \exp + (-in\omega_1 t)]$, and the series becomes

$$Q(t) = \frac{1}{2}F_0 + \frac{1}{2}\sum_{n=1}^{\infty} [F_n \exp(in\omega_1 t) + F_n^* \exp + (-in\omega_1 t)] \tag{3.7.3}$$

A simpler way in which we may write this expression is to assign the complex conjugate terms to negative values of the index n, with $F_{-n} = F_n^*$. This leads to a *complex Fourier series*,

$$Q(t) = \frac{1}{2}\sum_{n=-\infty}^{\infty} F_n \exp(in\omega_1 t), \quad F_{-n} = F_n^* \tag{3.7.4}$$

This expression is equivalent to eq. (3.7.2). We will find it convenient to use eq. (3.7.4) in derivations. Equation (3.7.2) is the one we use to evaluate a series, because there is no need then to evaluate the contributions of the negative indexed terms.

Each of the terms for $n \neq 0$ has a zero mean value when averaged over a period T. Thus, the real value F_0 is twice the mean value of $Q(t)$. The task now is to ascertain the other coefficients corresponding to specified $Q(t)$. When real Fourier series are employed, this is achieved by exploiting the orthogonality property of cosine and sine functions. Complex exponentials also are orthogonal. Let us select a specific term $\exp(-ik\omega_1 t)$ associated with harmonic number k. We multiply each side of eq. (3.7.4) by this term, and integrate over a time interval matching the period. The starting time for this integration does not matter in a formal sense, so we will take the integration interval to be $-T/2 \leq t \leq T/2$. (We will see that if $Q(t)$ is defined in a piecewise manner over a period, then it may be advantageous to shift this interval.) These operations lead to

$$\int_{-T/2}^{T/2} Q(t) \exp(-ik\omega_1 t)\, dt = \frac{1}{2}\sum_{n=-\infty}^{\infty} F_n \int_{-T/2}^{T/2} \exp(in\omega_1 t) \exp(-ik\omega_1 t)\, dt$$

$$\tag{3.7.5}$$

$$= \frac{1}{2}\sum_{n=-\infty}^{\infty} F_n \int_{-T/2}^{T/2} \exp[i(n-k)\omega_1 t]\, dt$$

If $n = k$, the integrand on the right side is unity, so the integral equals T. To evaluate the integral for $n \neq k$ we recall that $\omega_1 T = 2\pi$, so that

$$\int_{-T/2}^{T/2} \exp[i(n-k)\omega_1 t]\, dt = \frac{\exp[i(n-k)\omega_1 t]\big|_{-T/2}^{T/2}}{i(n-k)\omega_1}$$

$$= \frac{\{\exp[i(n-k)\pi] - \exp[-i(n-k)\pi]\}}{i(n-k)\omega_1} = 0, \quad n \neq k$$

$$\tag{3.7.6}$$

That this integral should be zero when $n \neq k$ is obvious when we consider Euler's equation, which gives the integrand as the sum of a cosine and a sine that extend over $n - k$ periods. It follows from the above that the integration prescribed in the right side of eq. (3.7.5) may be expressed in terms of the *Kronecker delta* as $T\delta_{nk}$, where

$$\delta_{nk} = 1 \text{ if } n = k, \qquad \delta_{nk} = 0 \text{ if } n \neq k \tag{3.7.7}$$

This serves to filter out all terms in the sum except the one that matches the value of k we selected. Specifically, we find that

$$\int_{-T/2}^{T/2} Q(t)\exp(-ik\omega_1 t)\, dt = \frac{1}{2}\sum_{n=-\infty}^{\infty} F_n T\delta_{nk} = \frac{1}{2}F_k T$$

$$\boxed{F_k = \frac{2}{T}\int_{-T/2}^{T/2} Q(t)\exp(-ik\omega_1 t)\, dt} \tag{3.7.8}$$

This relation confirms our heuristic reasoning that F_0 is twice the mean value of $Q(t)$, as well as the fact that $F_{-k} = F_k^*$. The latter property enables us to determine the coefficients F_k corresponding to negative indices without actually evaluating them.

The easiest Fourier series to determine describes a periodic sequence of impulse functions separated by an interval T. This ease stems from the integration properties of a Dirac delta function. We previously characterized $\delta(t - \tau)$ as the limit of a uniform pulse of height $1/\varepsilon$ that acts between $t = \tau$ and $t = \tau + \varepsilon$. If $F(t)$ is an analytical function and $t = \tau$ falls in an interval between two instants t_1 and t_2, then we have

$$\int_{t_1}^{t_2} F(t)\delta(t - \tau)\, dt = \lim_{\varepsilon \to 0}\int_{\tau}^{\tau+\varepsilon} F(t)\frac{1}{\varepsilon}\, dt \tag{3.7.9}$$

By the central limit theorem $F(t)$ is essentially $F(\tau)$ during the very short time interval, so we find that

$$\int_{t_1}^{t_2} F(t)\delta(t - \tau)\, dt = F(\tau) \text{ if } t_1 < \tau < t_2 \tag{3.7.10}$$

When applied to eq. (3.7.8), this property indicates that a periodic sequence of impulse excitations corresponds to

$$Q(t) = \sum_{n=-1}^{\infty} \delta(t - nT) \Leftrightarrow F_k = \frac{2}{T} \tag{3.7.11}$$

In other words all of the Fourier coefficients are equal for a periodic train of impulses. Such a Fourier series will converge extremely slowly to zero away from the instants when the impulses act, and it will not converge at those instants. For other excitations, the coefficients F_k generally will decrease in magnitude with increasing k. Depending on the rate of decrease, it may be possible to construct a reasonable approximation to the function with only a few harmonics.

The integral for the Fourier series coefficients may be evaluated over any interval of duration T. If $Q(t)$ is an analytical function for all t, we most likely would use either $0 \leq t \leq 2\pi/\omega$ or $-\pi/\omega \leq t \leq \pi/\omega$. However, the more usual situation is one where a function defined over an interval of duration T is pieced together repetitively to construct the full time function. For example, consider a square wave, which is a periodic sequence of rectangular pulses and pauses. Suppose we define one cycle of the square wave to be $Q(t) = Q_0$ if $-T/4 \leq t \leq T/4$ and $Q(t) = 0$ if $T/4 \leq t \leq 3T/4$. Because this function's definition starts at $t = -T/4$, we would perform the integrals in eq. (3.7.8) over $-T/4 \leq t \leq 3T/4$. (For this function, the integrand is zero for the second half of a period.)

The concept that a periodic function may be defined in a piecewise manner gives rise to some interesting issues regarding the behavior of Fourier series at discontinuities. To explore this, let us consider the square wave described above. The series coefficients obtained from eq. (3.7.8) are

$$F_k = \frac{2}{T}Q_0 \int_{-T/4}^{T/4} \exp(-ik\omega_1 t)\, dt = \begin{cases} \dfrac{2}{k\pi}Q_0 \sin\left(\dfrac{1}{2}k\pi\right), & k \neq 0 \\[2mm] Q_0, & k = 0 \end{cases} \tag{3.7.12}$$

Let us denote as N the value of k at which we truncate the Fourier series, and let $S(t, N)$ denote the truncated series,

$$\boxed{\, S(t, N) = \frac{1}{2}F_0 + \mathrm{Re} \sum_{n}^{N} F_n \exp(in\omega_1 t) \,} \tag{3.7.13}$$

Increasing N brings $S(t, N)$ into closer agreement with $Q(t)$, but the improvement does not occur in a uniform manner at all values of t. As shown in Figure 3.19, the discrepancy between $S(t, N)$ and $Q(t)$ is greater near the discontinuities. We also see that the peak error relative to Q_0 does not depend on N. Increasing N decreases the extent of this region of discrepancy, and thereby causes the peak error to occur closer to the discontinuity. This is known as the *Gibbs phenomenon*. Another aspect worth noting is that at a discontinuity, $S(t, N)$ evaluates to a value that is the average of the function values on either side of the discontinuity.

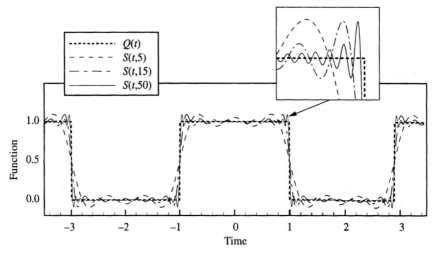

FIGURE 3.19 Fourier series for a square wave, $T = 4$.

The Fourier series in eq. (3.7.12) describes a train of pulses that is an even function of t. In some cases, it is desirable to consider a pulse to begin at $t = 0$. Rather than evaluating the Fourier coefficients anew, we may shift the time base of the function by an interval $T/4$. Suppose we know the Fourier coefficients F_k corresponding to function $f(t)$, and the delayed function is $f(t) = f(t - \tau)$, where τ is the time by which $f(t)$ is delayed. The Fourier coefficients for the delayed function are given by

$$\hat{F}_k = \frac{2}{T}\int_{-T/2}^{T/2} f(t - \tau)\exp(-ik\omega_1 t)dt = \frac{2}{T}\int_{-T/2}^{T/2} f(t')\exp[-ik\omega_1(t' - \tau)]dt'$$

$$\boxed{\hat{F}_k = F_k \exp(-ik\omega_1\tau)} \tag{3.7.14}$$

In other words, delaying a periodic function shifts the argument of the complex harmonic amplitudes by an amount that is proportional to the number of the harmonic. Note that $\tau = T$, which reproduces the function, shifts the arguments by multiples of 2π so that the coefficients are unchanged.

3.7.2 The Discrete Fourier Transform

The periodic excitations we have considered thus far were known in functional form, which enabled us to evaluate the harmonic amplitudes by analytically evaluating the associated integrals. Suppose the function were sufficiently complicated that we decided to use numerical integration instead. In that case we would only need to know the excitation values at a discrete set of instants. A similar situation arises in experimental applications, because modern data acquisition systems store digitized data taken at discrete, uniformly spaced instants. We will see here that powerful techniques are available for applying Fourier analysis to such data.

Figure 3.20 depicts a periodic excitation $g(t)$ whose variation is represented by a set of values at uniformly spaced time instants. The period T is divided into N intervals, so the time increment for one interval is

$$\Delta = \frac{T}{N} \tag{3.7.15}$$

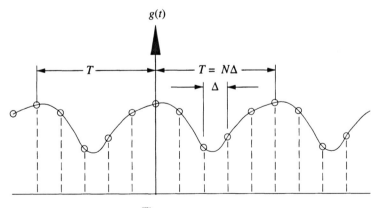

FIGURE 3.20 Sampling of a periodic function at uniform intervals.

It is convenient to use an interval covering $0 \leq t \leq T$ to form the Fourier coefficients, because doing so avoids negative subscripts. We define the instants in this interval as $t_n = n\Delta$, and define $g_n = g(t_n)$. Because $g(T) = g(0)$ due to periodicity of the data, the data set required to identify the periodic pattern consists of the N values $g_0, \ldots, g_{N-1}$. A wide variety of algorithms are available for approximating integrals in terms of discrete function values. We will use the simplest, specifically a strip rule based on approximating each interval as a rectangle whose height is the function value at the instant when the strip begins. With $f(t)$ representing an arbitrary function, the strip rule states that

$$\int_0^T f(t)\,dt \approx f(t_0)\Delta + f(t_1)\Delta + \cdots = \Delta \sum_{n=0}^{N-1} f_n \tag{3.7.16}$$

We use the foregoing to approximate the integral for the Fourier series coefficients. It is standard in most discretized implementations to evaluate Fourier coefficients that are multiplied by the factor $N/2$. Correspondingly, we replace the actual Fourier coefficient $\hat{G}_k$ with $(2/N)G_k$ in the left side of eq. (3.7.8), which yields

$$\hat{G}_k = \frac{2}{N}G_k = \frac{2}{T}\int_0^T g(t)\exp(-ik\omega_1 t)\,dt \approx \frac{2}{T}\Delta \sum_{n=0}^{N-1} g_n\exp(-ik\omega_1 t_n) \tag{3.7.17}$$

By definition, $\Delta/T = 1/N$. Also, the fundamental frequency is $\omega_1 = 2\pi/T$ and $t_n = n(T/N)$, so we have

$$\boxed{G_k = \sum_{n=0}^{N-1} g_n\exp\left(-2\pi i\frac{kn}{N}\right)} \tag{3.7.18}$$

The frequency associated with G_k is $k\omega_1$, or

$$\boxed{\omega_k = \frac{2\pi k}{T} = \frac{2\pi k}{\Delta N}} \tag{3.7.19}$$

Note that by replacing the approximation sign with an equality in eq. (3.7.18), we are defining the G_k to be the result of computing the indicated sum, rather than an approximation of the integral. Although this distinction is subtle, it is of primary importance. For example, we now consider eq. (3.7.18) to be a linear transformation from a set of N time values to a set of harmonic amplitudes. The g_n constitute a set of N real values, but the G_k are complex. This leads to the question, How many coefficients G_k can be computed? To answer this, we observe that two real numbers are required to form a complex number. According to eq. (3.7.18), $G_{-k} = G_k^*$, so each pair of positive and negative indices entails two independent real numbers. However, $k = 0$ is a special case because G_0 is real. The value of $G_{N/2}$ also is real, and $G_{N/2} = G_{-N/2}$, because the exponential factor in eq. (3.7.18) when $k = \pm N/2$ is $\exp[-2\pi i(\pm n/2)] = (-1)^n$. (It is implicit to the discussion that N is even.) Hence, $G_{(-N/2+1)} \cdots G_{N/2}$ are constituted from N real numbers. We say that eq. (3.7.18) is the *discrete Fourier transform* (DFT) of the time data, and $G_{(-N/2+1)}$ to $G_{n/2}$ are the DFT coefficients.

Matrix notation offers a convenient way to symbolically represent the DFT definition. We use g_0 to g_{N-1} to form a vector array $\{g\}$ of time data,

$$\{g\} = [g_0 \ g_1 \cdots g_{N-1}]^\mathrm{T} \tag{3.7.20}$$

while $\{G\}$ contains the DFT coefficients

$$\{G\} = [G_{(-N/2+1)} \ \cdots \ G_{-1} \ G_0 \ G_1 \ \cdots \ G_{N/2}]^\mathrm{T} \tag{3.7.21}$$

The DFT computations in eq. (3.7.18) may then be written as

$$\boxed{\{G\} = [E_\mathrm{DFT}]\{g\}} \tag{3.7.22}$$

where $[E_\mathrm{DFT}]$ is an $N \times N$ array. In view of the manner in which the elements of $\{g\}$ and $\{G\}$ are defined, the column index for $[E_\mathrm{DFT}]$ ranges from 0 to $N - 1$, and the row index ranges from $-N/2 + 1$ to $N/2$, so that the elements of $[E_\mathrm{DFT}]$ are defined to be

$$(E_\mathrm{DFT})_{kn} = \exp\left(-2\pi i \frac{kn}{N}\right); \quad k = -\frac{N}{2} + 1, -\frac{N}{2} + 2, \ldots, -\frac{N}{2};$$

$$n = 0, 1, \ldots, N - 1 \tag{3.7.23}$$

Equation (3.7.22) defines a unique, one-to-one relation between time domain data g_n and frequency domain data G_k. Inverting the relationship yields the time data $h_n \equiv h(t_n)$ corresponding to a set of DFT coefficients H_k. The remarkable fact, proven in the next section, is that the inverse of $[E_\mathrm{DFT}]$ may be found without computing a matrix inverse. The process of converting from DFT coefficients to time data is called the *inverse discrete Fourier transform* (IDFT),

$$\boxed{\{h\} = \frac{1}{N}[E_\mathrm{IDFT}]\{H\}} \tag{3.7.24}$$

The introduction of the $1/N$ factor anticipates later developments, where it will be proven that the elements of $[E_\mathrm{IDFT}]$ are

$$(E_\mathrm{IDFT})_{nk} = \exp\left(2\pi i \frac{kn}{N}\right); \quad k = -\frac{N}{2} + 1, -\frac{N}{2} + 2, \ldots, -\frac{N}{2};$$

$$n = 0, 1, \ldots, N - 1 \tag{3.7.25}$$

The preceding discussion leads us to an important observation. If the time data constitutes a set of N values sampled at interval Δ, then $G_{N/2}$ corresponds to the highest frequency we may compute, which is $(N/2)\omega_1$. This frequency is called the *Nyquist critical frequency*. Because the period is $T = N\Delta$ and $\omega_1 = 2\pi/T$, we see that this critical frequency is π/Δ. It is common to describe the Nyquist frequency in units of hertz, so we have

$$\boxed{f_\mathrm{cr} = \frac{1}{2\Delta} \ (\mathrm{Hz})} \tag{3.7.26}$$

Suppose now that we know that a certain number of harmonics are required to adequately represent the time function as a Fourier series. Let f_max denote the highest cyclical frequency (Hz) in such a series. The foregoing states that the Nyquist critical frequency must be equal to, or exceed, f_max. This leads to an upper limit on the sample interval,

$$\boxed{\Delta \leq \frac{1}{2f_\mathrm{max}}} \tag{3.7.27}$$

This is the *Nyquist sampling criterion*. Note that $1/f_{max}$ is the period of the highest harmonic. Thus, the Nyquist sampling criterion states that the sampling interval must be sufficiently small to capture at least two data values within the period of the highest harmonic. A common source of experimental error is undersampling of the time data, which can readily occur if one has no idea of the underlying functional form of the process that is being measured. We will investigate this type of error, which is called *aliasing,* later in this chapter when we apply Fourier theory to evaluate transient response. It is possible to identify aliasing after the fact by examining the Fourier coefficients; the coefficients of the several highest harmonics will be small relative to the largest coefficients if the sample interval Δ was adequately short.

It is important to recognize that the time interval Δ defines the maximum frequency at which we perform evaluations, whereas the period T defines the fundamental frequency, which is also the increment $\Delta\omega$ between adjacent harmonics. Also, we should note that evaluating the DFT according to eq. (3.7.18) requires only knowledge of the value of N, and of the N values f_n. We obtain the same coefficients, regardless of the period T over which the data is defined.

Equations (3.7.22) and (3.7.24) are useful as a way of conceptually visualizing the manner in which data is processed to evaluate a DFT or IDFT. If N is a very large value (it is common to deal with $N > 1000$), evaluating all of the coefficients of either $[E_{DFT}]$ or $[E_{IDFT}]$ and then performing the matrix product would be computationally intensive. For this reason, highly efficient numerical algorithms have been developed. These carry out the calculations without evaluating matrix elements. Several algorithms have been developed; they are referred to as *Fast Fourier Transforms* (FFTs). We will explore one of these algorithms in a later section.

3.7.3 The Inverse Discrete Fourier Transform

Our objective here is to prove eqs. (3.7.24) and (3.7.25). We suppose that we know a set of DFT coefficients H_k, and we wish to compute values of the corresponding periodic function $h(t)$. We could evaluate $h(t)$ at any instant we choose by computing the complex Fourier series definition in eq. (3.7.4). Instead, we shall consider only function values at the discrete instants t_n. We recall that the Fourier coefficients are scaled by $2/N$ in eq. (3.7.17), that $\omega_1 = 2\pi/T$, and that $t_n = n\Delta = nT/N$. The complex Fourier series, eq. (3.7.4), which does not explicitly account for the complex conjugate properties of the Fourier coefficients, then gives

$$h(t_n) = \frac{1}{N}\sum_{k=-\infty}^{\infty} H_k \exp\left(2\pi i \frac{kn}{N}\right) \qquad (3.7.28)$$

The summation index k must be truncated. The DFT operation gives coefficients $H_{(-N/2+1)}$ to $H_{N/2}$, so we consider all H_k outside this range to be zero. Thus, we have

$$\boxed{h(t_n) = \frac{1}{N}\sum_{k=-N/2+1}^{N/2} H_k \exp\left(2\pi i \frac{kn}{N}\right)} \qquad (3.7.29)$$

In view of the fact that we assumed that some of the H_k data is unimportant to arrive at the preceding, it might seem as though eq. (3.7.29) is merely an approximation. In fact, *it is exact.* To see why, let us use eq. (3.7.18) to represent the DFT coefficients H_k in the preceding. The result is

$$h(t_n) = \frac{1}{N} \sum_{k=-N/2+1}^{N/2} \left[\sum_{m=0}^{N-1} h_m \exp\left(-2\pi i \frac{km}{N}\right) \right] \exp\left(2\pi i \frac{kn}{N}\right)$$

(3.7.30)

$$= \sum_{m=0}^{N-1} h_m \left[\frac{1}{N} \sum_{k=-N/2+1}^{N/2-1} \exp\left(2\pi i k \frac{n-m}{N}\right) \right]$$

The term within the brackets is the discrete analog of the orthogonality property for complex exponential functions. Specifically, it is

$$\frac{1}{N} \sum_{k=-N/2+1}^{N/2} \exp\left(2\pi i k \frac{n-m}{N}\right) = \begin{cases} 0 \text{ if } m \neq n \\ 1 \text{ if } m = n \end{cases}$$

(3.7.31)

The validity of this expression when $m = n$ is obvious. To prove it for $m \neq n$, we rewrite the sum in the left side as

$$\frac{1}{N} \sum_{k=-N/2+1}^{N/2} \exp\left(2\pi i k \frac{n-m}{N}\right) = \frac{1}{N} \sum_{k=-N/2+1}^{N/2} \exp(ik\theta), \quad \theta = 2\pi \frac{n-m}{N}$$

(3.7.32)

Both n and m range from zero to $N - 1$, so we have $-N + 1 \leq n - m \leq N - 1$. It follows that if $m \neq n$, then $\exp(i\theta)$ is a unit vector in the complex plane at some angle θ relative to the real axis, with $|\theta| < 2\pi$. Similarly, $\exp(ik\theta)$ is a unit vector whose angle from the real axis is $k\theta$. These unit vectors are illustrated in Figure 3.21 for the case where $n - m = 4$ and $N = 10$. We form the summation of the unit vectors by placing them tip to tail successively, which creates an N-sided closed polygon (a pentagon in the case of Figure 3.21). Closure of the polygon yields a zero sum. In view of eq. (3.7.31), the right side of eq. (3.7.29) reduces identically to h_n. In other words, DFT processing of a set of time data, followed by IDFT processing of the DFT data, identically reproduces the original time data. Furthermore, it is evident that eq.

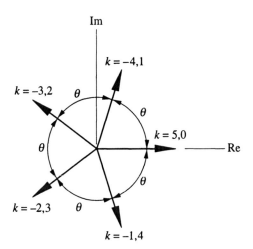

FIGURE 3.21 Unit vectors $\exp(ik\theta)$, where $\theta = 2\pi(n - m)/N$ with $n - m = 4$ and $N = 10$.

(3.7.29) is the summation form of eqs. (3.7.24) and (3.7.25), which proves the validity of the matrix representation.

An interesting sidelight of the fact that an IDFT reproduces the original data arises when we consider the Gibbs phenomenon exhibited by the square wave in Figure 3.19. The sampled data values will be either one or zero. Applying the DFT-IDFT sequence to this data will reproduce these values exactly, so the processed time data will not exhibit the Gibbs phenomenon. We may deduce from this that the Gibbs phenomenon arises from using a Fourier series to construct the function in a time increment that is shorter than that permitted by the Nyquist sampling criterion, eq. (3.7.27), for the number of terms in that series.

To summarize the above development, we begin by sampling a function having period T to construct a discrete data set of N time values. The DFT, eq. (3.7.18), transforms the time data into a set of N complex numbers in the frequency domain. These will be reasonably good approximations of the complex Fourier coefficients, multiplied by $N/2$, at frequencies $k(2\pi/T)$ if the sampling interval Δ satisfies the Nyquist sampling criterion. We return to the time domain from frequency domain data by applying the IDFT, eq. (3.7.29). This yields, without approximation, the original time data corresponding to the frequency domain data.

3.7.4 Explanation and Implementation of the FFT Algorithm

The numerical operations required for a DFT or IDFT are quite straightforward, but they are numerous, particularly if N is large. Several algorithmic shortcuts have been developed to accelerate the computations. These improvements constitute a class of numerical procedures called *Fast Fourier Transforms* (FFTs). Several algorithms have been developed; Press et al. (1992) provides a good entry into the area. They also offer programs that implement the most common algorithms. FFT algorithms are contained in standard mathematical software (Maple, Mathematica, MATLAB, Mathcad, etc.), and subroutines for performing the operation in a variety of computer languages are contained in mathematical libraries, such as IMSL and NAG. Our purpose here is to give only a flavor of what the algorithms entail and, in the process, see how the data should be set up to use these algorithms. We will refer to the data and its transform as DFTs and IDFTs, while FFT and IFFT will refer to the processes applied to the data.

The first step originates from the observation that the DFT, eq. (3.7.18), and the IDFT, eq. (3.7.29), have a similar appearance, except that the latter uses indices that are both positive and negative. We therefore translate the DFT data G_k for negative k to the right of the positive-indexed data by incrementing the former's index by N. Because the negative-indexed coefficients are the complex conjugate of the corresponding positive-indexed coefficients, this translation is performed by defining

$$G_k = G_{k-N} = G_{N-k}^*, \quad k = \frac{N}{2} + 1, \frac{N}{2} + 2, ..., N - 1 \tag{3.7.33}$$

When we need to refer to the frequency of this extended data, as we will when we evaluate the complex frequency response for each harmonic, we use the original values,

$$\omega_k = \omega_{k-N} = (k - N)\left(\frac{2\pi}{T}\right), \quad k = \frac{N}{2} + 1, \frac{N}{2}, ..., N - 1 \tag{3.7.34}$$

The DFT-IDFT pair resulting from the above is

$$G_m = \sum_{n=0}^{N-1} g_n \exp\left(-2\pi i \frac{mn}{N}\right), \quad g_n = \frac{1}{N}\sum_{m=0}^{N-1} G_m \exp\left(2\pi i \frac{mn}{N}\right) \qquad (3.7.35)$$

As we noted earlier, the preceding represent basic operations that are performed on sets of numbers, without approximation. The approximate aspect of the DFT is associated with its use to determine the Fourier series coefficients $\hat{G}_m$ corresponding to a function $g(t)$. Specifically, if g_n is the data resulting from sampling $g(t)$, then eq. (3.7.17) indicates that

$$\hat{G}_n \approx \frac{2}{N} G_n \qquad (3.7.36)$$

Aside from the leading factor and the difference in sign within the exponential, the DFT and IDFT defined in eq. (3.7.35) are alike. Consequently, the same subroutine can be used to perform both sets of operations. In its raw form, either transformation entails N^2 sums, products, and evaluations of complex exponentials. A very simple reduction in the number of such operations comes from recognizing the periodicity of the complex exponential, because we have

$$\exp\left(2\pi i \frac{mn}{N}\right) = W^{mn} = W^{\mathrm{mod}(mn,N)}, \quad W = \exp(2\pi i/N) \qquad (3.7.37)$$

where the mod function subtracts from mn an integer number of multiples of N. Thus, rather than evaluating N^2 values of the exponential, one can merely evaluate and reuse N powers of the complex constant W.

Although the foregoing does reduce some operations, more significant improvements come about by splitting up the data set based on the parity (even-odd) of its index. It is not our purpose here to go into the details of these steps, in that detailed explanations may be found in several texts, including those by Brigham (1974) and Press et al. (1992). However, it is useful to have some appreciation of what they entail. We shall consider the DFT here, since the IDFT is essentially the same.

Rather than performing the operations on the entire data set g_n, we can split the sum over N into two parts consisting of contributions from even n and odd n. Applying this grouping to the first of eqs. (3.7.35) leads to

$$G_m = \sum_{n=0}^{N/2-1} g_{2n} \exp\left[-2\pi i \frac{m(2n)}{N}\right] + \sum_{n=0}^{N/2-1} g_{2n+1} \exp\left[-2\pi i \frac{m(2n+1)}{N}\right]$$

$$= \sum_{n=0}^{N/2-1} g_{2n} \exp\left(-2\pi i \frac{mn}{N/2}\right) + \exp\left(-2\pi i \frac{m}{N}\right) \sum_{n=0}^{N/2-1} g_{2n+1} \exp\left(-2\pi i \frac{mn}{N/2}\right)$$

$$(3.7.38)$$

A comparison of the preceding with eqs. (3.7.35) shows that each summation has the appearance of a DFT applied to a data set composed of $N/2$ values. Let us use a superscript 0 or 1 according to whether a quantity is exclusively formed from even- or odd-indexed data, respectively. We thus have two sets of DFTs, G_m^0 and G_m^1,

$$G_m^0 = \sum_{n=0}^{N/2-1} g_{2n} \exp\left(-2\pi i \frac{mn}{N/2}\right), \quad G_m^1 = \sum_{n=0}^{N/2-1} g_{2n+1} \exp\left(-2\pi i \frac{mn}{N/2}\right) \quad (3.7.39)$$

Note that the even and odd DFTs are formed from data sets whose length is half as long as the original data. If we know the even and odd DFTs, we may find the original DFT according to

$$G_m = G_m^0 + \exp\left(-2\pi i \frac{m}{N}\right) G_m^1 \quad (3.7.40)$$

The even and odd DFTs may be subdivided further, based on the parity of the data from which they are formed. This leads to four transforms, G_m^{00}, G_m^{01}, G_m^{10}, and G_m^{11}, each of which is formed from a data set whose length is $N/4$. If N is a multiple of 2—that is, if $N = 2^p$, where p is an integer—the subdivided data set eventually reduces to N DFTs, each of which is one of the elements of the time data g_n. The original G_m are computed from these single length values by going in the opposite direction, in a process that successively combines even- and odd-indexed transforms at any length with the W factor defined above to construct transforms whose length is larger by 1.

The superscripts for these reduced transforms, $G^{00000\cdots}$, are a binary representation of the transform's index. It turns out that the index of the g_n value corresponding to a specific one of these single element transforms may be obtained by bit reversal of the binary number of the transform. For example, G^{10010} consists of element g_m where the binary value of m is 01101. This leads to the *Cooley-Tukey FFT algorithm,* which is valid when N is a multiple of 2. Other algorithms also are available to handle cases where $N = 2^p$. They all reduce the number of operations from order N^2 to order $N \log_2 N = pN$, which is a substantial reduction when N is very large. Cases where N is not a multiple of 2 may be handled by similar procedures, except that they stop whenever the subdivided transform's length is no longer divisible by 2. The resulting algorithms are less efficient, but still better than a direct DFT computation.

An individual working with FFT routines does not need to know how the algorithm is programmed, but must know how to prepare the data and how to interpret the results. Several subroutines are usually available in any mathematics package. A crucial question here is whether N is a multiple of 2. If we have experimental data taken by equipment using a fixed Δ to subdivide a periodic process, it is likely that $N = T/\Delta$ will not be a multiple of 2. In most other applications, it is possible to use $N = 2^p$. Routines that are specialized to $N = 2^p$ will not be valid for arbitrary N. Also, some routines scale the data, which is equivalent to multiplying the G_k data by a factor for the FFT, and then dividing the inverse FFT by that factor to recover the time data. Some routines allow the g_n data, as well as the G_k data, to be complex, in which case the full set of G_k will be computed. In contrast, the routines for real g_n data usually compute only the G_k for $k = 0, \ldots, N/2$ because of the complex conjugate property in eq. (3.7.33). Another possible difference stems from our use of $\exp(+i\omega t)$ to represent harmonic functions. In other areas, such as acoustics, individuals often prefer to use $\exp(-i\omega t)$, which reverses the DFT-IDFT pair. Prior to using any implementation of the FFT, one should become familiar with these features. Once we have such information, we may treat computer routines as "black boxes" that calculate DFTs and IDFTs in a very efficient manner. Thus, although we might represent the DFT and IDFT operations in matrix form, as described by eqs. (3.7.22) and (3.7.24), in actuality we use the FFT algorithm and associated computer routines to carry out the operations.

EXAMPLE 3.9

Consider the square wave excitation described in Figure 3.19. Use FFTs to determine the Fourier series coefficients of this excitation, and compare the results to the analytical values.

Solution   This exercise will illustrate the basic FFT operations and how to account for a scaling factor. For the signal in Figure 3.19, $T = 2$ s. The primary decision is selection of the number of samples N, which is equivalent to selecting the sampling interval T/N. A reasonable estimate of the lowest N that would be acceptable is to consider the Nyquist sampling criterion qualitatively. If we divide the positive and negative regions of the signal into four intervals each, we would see the basic nature of the waveform. This would suggest that $N = 8$, which is a power of 2, is the minimum acceptable value. To be safe let us use $N = 16$.

We shall first implement the solution with MATLAB, and then Mathcad. The sampling instants are multiples of T/N, and the function is one for the first and last quarter cycles, so we write

$$\text{MATLAB:} \quad \begin{aligned} &for\ j = 0 : N - 1;\ t = j\ *\ T/N; \\ &f\ (j + 1) = (t\ <=\ T/4) + (t\ >\ 3\ *\ T/4);\ end \\ &F\_trans = fft(f) \end{aligned}$$

Unlike the derivation of the FFT algorithm, MATLAB requires that array indices begin at 1. This is handled in the preceding program fragment by running index j over the range in the derivation, after which the actual subscript for a quantity is incremented by 1 in order to satisfy MATLAB's requirement. The F_trans data consists of N complex numbers. The MATLAB fft function follows eq. (3.7.18) without any multiplicative factors, and it satisfies the complex conjugate mirroring operation in eq. (3.7.33). Thus, the Fourier series coefficients for zero and positive harmonics should be comparable to the first $N/2 + 1$ of the F_trans (j) data, multiplied by $2/N$, which is the scaling factor in eq. (3.7.17). (Instead of considering only a portion of the F_trans data, the MATLAB user can invoke the *fftshift* function to rearrange the data from $-N/2 + 1$ to $N/2$.)

Mathcad has several FFT functions. The $FFT(f)$ routine implements the DFT definition in eq. (3.7.18), except that it multiplies the summation by $1/N$. This means that the DFT values it returns will be $1/N$ times the DFT values that would result from application of the definition in the text. According to eq. (3.7.36), the actual Fourier series coefficients are $2/N$ times the DFT values associated with the text definition. This leads to the conclusion that the true Fourier coefficients should be twice the Mathcad DFT values. Mathcad allows the user to define the base index for arrays at any value. We could set $ORIGIN = 0$ in the setup menu, which would allow us to implement the equations defining FFT operation without adjusting subscripts. However, in all discussions of FFT operation using Mathcad, we shall keep the default $ORIGIN = 1$, both in order to be consistent with MATLAB, and also to avoid adjustments when we perform matrix operations in later chapters. Thus, the Mathcad steps are

$$\text{Mathcad:} \quad \begin{aligned} &j: = 0; N - 1 \quad f_{j+1} := (j\ *\ T/N \leq T/4) + (j\ *\ T/N > 3\ *\ T/4) \\ &F_{trans} := FFT(f) \end{aligned}$$

The F_{trans} data obtained by these operations consists of only $N/2 + 1$ values. These correspond to the DFT coefficients for zero and positive subscripts sequentially. In essence, Mathcad's FFT function is restricted to transforming real data. It recognizes that there is no need to compute the DFT data for negative subscripts. As a result, eq. (3.7.33) can be ignored when we use this Mathcad function. (We will see in the next section that the companion $IFFT$ function in Mathcad takes $N/2 + 1$ complex DFT values and returns N real inverse values.)

The results of these computations appear in the first tabulation:

n	0	1	2	3	4
DFT (scaled)	1	0.6284 + 0.125i	0	−0.1871 − 0.125i	0
Analytical	1	0.6366	0	−0.2122	0

n	5	6	7	8
DFT (scaled)	0.0835 + 0.125i	0	−0.0249 − 0.125i	0
Analytical	0.1273	0	−0.0909	0

Although the DFT and analytical values are close, the disagreement in the real parts is noticeable. Another error is the imaginary part of the DFT, which becomes the dominant term at the higher harmonics. These discrepancies arise because N actually is not sufficiently high, as evidenced by the fact that the Fourier coefficient F_7 is 20% of F_1. This suggests that the Nyquist sampling criterion has not actually been met.

The first nine DFT values obtained from $N = 128$ appear in the second tabulation:

n	0	1	2	3	4
DFT (scaled)	1	0.6365 + 0.016i	0	−0.2118 − 0.016i	0
Analytical	1	0.6366	0	−0.2122	0

n	5	6	7	8
DFT (scaled)	0.1267 + 0.016i	0	−0.0249 − 0.016i	0
Analytical	0.1273	0	−0.0909	0

The DFT and analytical values now are much closer. Furthermore, the DFT value at $N/2$ is $−4(10^{-4}) − 0.0156i$, and the analytical value is $−0.0101$. The smallness of both values suggests that $N = 128$ satisfies the Nyquist criterion. However, raising N has not eliminated the imaginary part of the DFT values. This portion of the coefficients arises from the fact that the discontinuities occur at sampling intervals, and it is unclear what function value should be assigned at those instants.

One of the aspects revealed by the preceding computations is that the Nyquist sampling criterion requires more than a visual examination when the function to be sampled possesses discontinuities. The only truly reliable way of ensuring that the criterion has been met is to check that the DFT data for the highest harmonics is substantially smaller in magnitude than the largest coefficients.

3.7.5 Periodic Response Using FFTs

We consider here a one-degree-of-freedom system that is subjected to an arbitrary periodic excitation. We only need to consider the case where the system is initially at rest in the static equilibrium position, because nonzero initial conditions lead to adjustments in the complementary solution. It is logical to employ a Fourier series to describe a periodic excitation. Thus the system's response is governed by

$$M\ddot{q} + C\dot{q} + Kq = \frac{1}{2}\sum_{n=-\infty}^{\infty} F_n \exp(in\omega_1 t), \quad q(0) = \dot{q}(0) = 0 \qquad (3.7.41)$$

Linear superposition indicates that the general solution is a sum of particular solutions that match the respective terms in the excitation's Fourier series, and a complementary solution. Eventually, the complementary solution decays away. This leaves the particular solution, which is a steady-state response having the period of the excitation. We write this solution as

$$q_{ss} = \frac{1}{2} \sum_{n=-\infty}^{\infty} X_n \exp(in\omega_1 t) \qquad (3.7.42)$$

The corresponding velocities and accelerations are also periodic, being given by

$$\dot{q}_{ss} = \frac{1}{2} \sum_{n=-\infty}^{\infty} V_n \exp(in\omega_1 t), \quad V_n = in\omega_1 X_n$$

$$\qquad (3.7.43)$$

$$\ddot{q}_{ss} = \frac{1}{2} \sum_{n=-\infty}^{\infty} A_n \exp(in\omega_1 t), \quad A_n = -(n\omega_1)^2 X$$

To determine the complex amplitudes X_n we substitute the above into the equation of motion. Because the result of this substitution must apply at all values of t, it must be satisfied term by term. Hence, we have

$$X_n = \frac{F_n}{K + in\omega_1 C - (n\omega_1)^2 M} \qquad (3.7.44)$$

In other words, each displacement harmonic is the steady-state response to a force having complex amplitude F_n at frequency $n\omega_1$. We may write the steady-state response in an alternative form by using the complex frequency response. The frequency ratio for harmonic number n is $\omega_n/\omega_{\text{nat}} = nr_1$, where $r_1 = \omega_1/\omega_{\text{nat}} = 2\pi/(\omega_{\text{nat}}T)$ is the fundamental frequency ratio. It follows from eqs. (3.1.6) and (3.1.7) that the steady-state response is

$$q_{ss} = \frac{1}{2} \sum_{n=-\infty}^{\infty} \frac{F_n}{K} D(nr_1, \zeta) \exp(in\omega_1 t) \qquad (3.7.45)$$

The terms in q_{ss} corresponding to negative n are the complex conjugate of those at positive n, so the result is real. To emphasize this, let us rewrite the solution using real variables,

$$q_{ss} = \frac{1}{2} \frac{F_0}{K} + \text{Re}\left[\sum_{n=1}^{\infty} \frac{F_n}{K} D(nr_1, \zeta) \exp(in\omega_1 t) \right]$$

$$= \frac{1}{2} \frac{F_0}{K} + \sum_{n=1}^{\infty} \frac{|F_n|}{K} |D(nr_1, \zeta)| \cos[n\omega_1 t - \phi(nr_1, \zeta) + \arg(F_n)] \qquad (3.7.46)$$

where $|D(nr_1, \zeta)|$ and $\phi(nr_1, \zeta)$ are the magnification factor and phase lag in eqs. (3.1.8).

The steady-state response in eq. (3.7.46) leads to several important observations. Consider a lightly damped system. The largest value of $|D(nr_1)|$ will correspond to the frequency $n\omega_1$ that is closest to ω_{nat}, that is, $n' = \text{int}(\omega_{nat}/\omega_1)$. If $n'\omega_1$ is very close to ω_{nat}, harmonic n' will be resonant, and therefore dominate the steady-state response, unless the force amplitude $|F_{n'}|$ is very small in comparison to the other force amplitudes. Even if all of the harmonic frequencies are significantly different from the natural frequency, the steady-state response will usually vary with time in a manner that is vastly different from that of the excitation, because the magnification factors $|D(nr_1)|$ change significantly from harmonic to harmonic. (There are situations for which q_{ss} is nearly proportional to $Q(t)$. We will explore such systems in the next section on transducers for vibration measurement.)

The DFT-IDFT pair is eminently suitable for the task of evaluating the response of a one-degree-of-freedom system to a periodic excitation $Q(t)$. We begin by sampling the excitation using an interval Δ that satisfies the Nyquist sampling criterion for the excitation, thereby creating a set of time data Q_j. We next apply the FFT to obtain data set F_n,

$$\{F\} = [E_{DFT}]\{Q\} \tag{3.7.47}$$

Coefficient F_n approximates the harmonic amplitude of the excitation at the frequency $\omega_n = n(2\pi/T)$. The nth harmonic amplitude X_n of the steady-state response is obtained by applying the complex frequency response $D(r_n, \zeta)$,

$$X_n = \frac{F_n}{K}D(nr_1, \zeta), \quad r_1 = \frac{2\pi}{\omega_{nat}T} \tag{3.7.48}$$

If our software requires the full set of N DFT values in order to perform the inverse DFT, we must explicitly implement the complex conjugate shift in eq. (3.7.33). Once we have formed the X_k values, we apply the IFFT to obtain the time response,

$$\{q\} = [E_{IDFT}]\{X\} \tag{3.7.49}$$

It is interesting to observe that for these operations, the value of the period T is significant only to the extent that it defines the fundamental frequency ratio r_1.

EXAMPLE 3.10

The Fourier series for the periodic force applied to a one-degree-of-freedom system is given by

$$F(t) = \sum_{n=1}^{\infty} \frac{P_0}{n^2 - 8.9} \sin(n\omega t)$$

where ω is the fundamental frequency and P_0 is a constant. The natural frequency of the system is ω_{nat} and the ratio of critical damping is ζ.

(a) Derive a Fourier series representation of the response.

(b) The nth response harmonic in the Fourier series may be written as $Y_n \sin(n\omega t - \psi_n)$. Consider the case where the fundamental frequency is set at $\omega = 20$ rad/s and $\zeta = 0.25$. Determine the value of ω_{nat} that maximizes the amplitude Y_2 of the second harmonic. Also determine the corresponding values of Y_2 and ϕ_2.

(c) Suppose that ω_{nat} has the value in part (b). Which harmonic of the response has the largest amplitude?

Solution   The intent of this exercise is to demonstrate the basic theory governing periodic response evaluations, as well as to illustrate some of the factors influencing that response. We begin by converting the given series for the excitation to complex form. We use the complex exponential and its complex conjugate in order to form a series whose index values are positive and negative. Thus, we write

$$F(t) = \sum_{n=1}^{\infty} \frac{P_0}{n^2 - 8.9}\left[\frac{\exp(in\omega t) - \exp(-in\omega t)}{2i}\right]$$

Inspection of the coefficient of $\exp(in\omega t)$ shows that this fits the standard complex Fourier series,

$$F(t) = \frac{1}{2}\sum_{n=-\infty}^{\infty} F_n \exp(in\omega t)$$

$$F_n = \frac{P_0}{i(n^2 - 8.9)} \text{ if } n > 0, \quad F_0 = 0, \quad F_n = F_{-n}^* \text{ if } n < 0$$

We use a complex Fourier series to represent the steady-state response,

$$x = \frac{1}{2}\sum_{n=-\infty}^{\infty} X_n \exp(in\omega t), \quad X_n = X_{-n}^* \text{ if } n < 0$$

We substitute the series for F and x into the standard equation of motion, $m\ddot{x} + c\dot{x} + kx = F$, and match the coefficients of like complex exponentials. This leads to

$$X_n = \frac{F_n}{k - m(n\omega)^2 + c(in\omega)}$$

In order to convert this to a form that depends on the ratio of the frequency of the harmonic to the natural frequency, we use the definition of the natural frequency and critical damping ratio, that is, $k = m\omega_{nat}^2$ and $c = 2\zeta\omega_{nat}m$, to rewrite the preceding expression as

$$X_n = \left(\frac{1}{k}\right)\left[\frac{P_0}{i(n^2 - 8.9)}\right]\left[\frac{1}{1 - \left(\dfrac{n\omega}{\omega_{nat}}\right)^2 + 2i\zeta\left(\dfrac{n\omega}{\omega_{nat}}\right)}\right]$$

$$= \frac{1}{i}\left(\frac{P_0}{k}\right)\left(\frac{1}{n^2 - 8.9}\right)\left|D\left(\frac{n\omega}{\omega_{nat}}, \zeta\right)\right|\exp\left[-i\phi\left(\frac{n\omega}{\omega_{nat}}, \zeta\right)\right]$$

To explicitly describe the nth harmonic, we isolate the terms corresponding to $+n$ and $-n$ in the Fourier series for the response. These are

$$x_n = \tfrac{1}{2}[X_n \exp(in\omega t) + X_n^* \exp(-in\omega t)] = \text{Re}[X_n \exp(in\omega t)]$$

The requested form is

$$x_n = Y_n \sin(n\omega t - \psi_n) = \text{Re}\left[\frac{Y_n}{i}\exp(-i\psi_n)\exp(in\omega t)\right]$$

Comparison of this form with the one obtained from the Fourier series shows that

$$Y_n = \left(\frac{P_0}{k}\right)\frac{1}{|n^2 - 8.9|}\left|D\left(\frac{n\omega}{\omega_{nat}}, \zeta\right)\right|$$

When $n < 3$, $n^2 - 8.9$ is positive, in which case ψ_n matches $\phi(n\omega/\omega_{nat}, \zeta)$. In contrast, for $n \geq 3$, the denominator is negative, which corresponds to a 180° phase shift. Thus, the phase angle is

$$\psi_n = \begin{cases} \phi(n\omega/\omega_{nat}, \zeta) \text{ if } n = 0, 1, 2 \\ \phi(n\omega/\omega_{nat}, \zeta) + \pi \text{ if } n \geq 3 \end{cases}$$

It is requested that we identify the value of ω_{nat} that maximizes Y_2 when $\omega = 20$ rad/s and $\zeta = 0.2$. The only part of any Y_n that depends on the value of ω_{nat} is the complex frequency response. Because ζ is reasonably large, we use the precise location of the peak magnification factor, eq. (3.1.11), which gives $2\omega/\omega_{nat} = (1 - \zeta^2)^{1/2}$. Thus, the maximum value of X_2 corresponds to

$$\omega_{nat} = \frac{2(20)}{(1 - \zeta^2)^{1/2}} = 40.825 \text{ rad/s}$$

Hence, for $n = 2$ we have $n\omega/\omega_{nat} = 0.9798$, which gives

$$Y_2 = 0.6346$$

To identify the largest amplitude for the specified values of ω and ω_{nat}, we observe that the expression for Y_n depends on n in two factors. The first, $1/(n^2 - 8.9)$, comes from the force amplitude; the second is the magnification factor. The largest force amplitude is obtained at $n = 3$, while the largest magnification factor is at $n = 2$. The earlier expression gives

$$Y_3 = 0.7690$$

This exercise illustrates a fundamental aspect of the response of one-degree-of-freedom systems to periodic excitation. Which harmonic has the largest amplitude depends on the amplitudes of the excitation's harmonics and on the proximity of the harmonic's frequency to the natural frequency. In the case of the exercise, each factor was largest at a different n.

EXAMPLE 3.11

An excitation in the form of a sawtooth waveform, as plotted below, is applied to a one-degree-of-freedom system having natural frequency ω_{nat} and ratio of critical damping ζ.

(a) Use an FFT to evaluate the complex Fourier series for this excitation, and compare those values to the analytical results.

(b) Cases of interest are $\omega_{nat}T = 0.2\pi, 2\pi$, and 20π. The ratio of critical damping in each case is $\zeta = 0.05$. For each set of system parameters, use FFT techniques to determine the steady-state response to the sawtooth excitation.

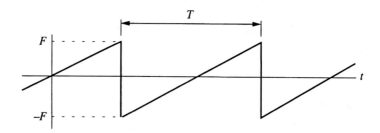

Solution The primary purpose of this exercise is to illustrate the ease with which Fourier series analysis of periodic response may be carried out with the aid of FFT software. It is specified that the discretization should use $N = 256$; we will be able to verify that this value of N is sufficient by examining the resulting DFT coefficients. It is convenient to set $T = 1$, which is equivalent to defining t nondimensionally relative to the period. Setting $F = 1$ will yield the response per unit force.

We begin by developing the equations for the quantities to be evaluated with the software. We take the lowest subscript to be one for consistency between Mathcad and MATLAB. The time increment and data points for the discretization are therefore

$$\Delta = \frac{1}{N}, \quad t_{j+1} = j\Delta, \quad j = 0, \ldots, N - 1$$

To sample the excitation at the discrete instants we break the discontinuous sawtooth curve into two intervals, so that

$$f_{j+1} = f(t_{j+1}) = \begin{cases} 2j\Delta & \text{if } j < N/2 \\ 0 & \text{if } j = N/2 \\ 2j\Delta - 2 & \text{if } j > N/2 \end{cases}$$

At the discontinuity, the preceding assigned a zero value to the function. In part, this is based on the property that a Fourier series at a discontinuity evaluates to the mean of the values at either side of the discontinuity. The effect of not taking this extra step will be discussed at the conclusion of the exercise.

The matrix notation for the FFT operation is

$$\{F\} = [E_{\mathrm{DFT}}]\{f\}$$

The specifics of carrying out this operation were addressed in Exercise 3.9. Regardless of whether we use Mathcad or MATLAB, the first $N/2 + 1$ elements of $\{F\}$ are the mean value and positive-indexed DFT coefficients of the excitation. The fundamental frequency is $2\pi /T$. Hence, the nonnegative frequencies are

$$\omega_{n+1} = \frac{2\pi}{T}n, \quad n = 0, 1, \ldots, \frac{N}{2}$$

It was requested that we compare the Fourier coefficients obtained from the FFT to the analytical values. Because $Q(t)$ is a straight line in the interval $-T/2 < t < T/2$, it is convenient to integrate over that interval in applying eq. (3.7.8). Thus,

$$\begin{aligned} (F_n)_{\text{analytical}} &= \frac{2}{T}\int_{-T/2}^{T/2} \left(F\frac{2t}{T}\right)\exp(-in\omega_1 t)\, dt \\ &= \frac{4F}{T^2}\left[\frac{in\omega_1 t + 1}{n^2\omega_1^2}\exp(-in\omega_1 t)\right]_{t=-T/2}^{t=T/2} \quad \text{if } n \neq 0 \\ &= F\frac{2i}{\pi n}(-1)^n \quad \text{if } n \neq 0 \end{aligned}$$

Note that the final form results from the fact that $\exp(\pm in\pi) \equiv [\exp(\pm i\pi)]^n \equiv (-1)^n$. When $n = 0$, so that $\exp(-in\omega_1 t) \equiv 1$, the integral gives $F_0 = 0$, which is consistent with $Q(t)$ having a zero mean value over a period.

To compare the computed DFT coefficients F_n to the analytical values, we multiply the DFT values by the scaling factor identified in Exercise 3.9. In view of the offset subscript for the programs, we have

Mathcad: $\quad 2F_{n+1} \approx (F_n)_{\text{analytical}}, \quad n = 0, 1, \ldots, \frac{N}{2}$

MATLAB: $\quad \frac{2}{N}F_{n+1} \approx (F_n)_{\text{analytical}}, \quad n = 0, 1, \ldots, \frac{N}{2}$

The graph at the top of the next page shows no visible difference between the imaginary parts of these analytical values and the scaled DFT values over a range of n. In fact, the largest difference between the imaginary parts is 0.005, which occurs at $n = N/2 = 128$, and the real part of the DFT values never exceeds $3(10^{-13})$.

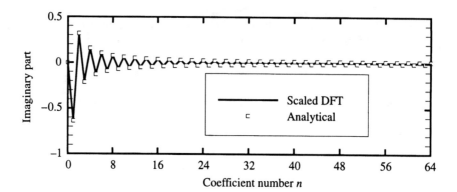

Even if we did not evaluate the analytical Fourier coefficients, we may verify directly from the DFT values that $N = 256$ is adequate. The Nyquist sampling criterion is such that if the sampling interval $\Delta = T/N$ satisfies the criterion, then the highest harmonics—that is, the F_n values for $n \approx N/2$—should have relatively small magnitudes, which is evident in the graph. In fact, the graph indicates that the F_n values are small for $n > 64$, which suggests that $N = 2(64) = 128$ would have been adequate.

We now proceed to evaluate the steady-state response. The stiffness is not specified, so we set $K = 1$. (In combination with setting $F = 1$ and $T = 1$, the computation will be equivalent to evaluating the nondimensional displacement qK/F as a function of nondimensional time t/T.) Thus we compute $N/2 + 1$ complex frequency response values according to

$$D_{n+1} \equiv D\left(\frac{\omega_{n+1}}{\omega_{\text{nat}}}, \zeta\right) = D\left(\frac{\omega_{n+1}T}{\omega_{\text{nat}}T}, \zeta\right) = D\left(\frac{2\pi n}{\omega_{\text{nat}}T}, \zeta\right), \quad n = 0, 1, \ldots, \frac{N}{2}$$

This form is useful because the values of $\omega_{\text{nat}}T$ are specified. The DFT coefficients of the response are the product of the force coefficient at a discrete frequency and the complex frequency response at that frequency, so we have

$$X_{n+1} = D_{n+1}F_{n+1}$$

where the range for n matches the preceding definition of the D_n values.

Mathcad users would use these values as the input to the inverse FFT routine *IFFT*. The program fragment for evaluating the time response in Mathcad may be written as

$$\text{Mathcad:} \quad n := 0; N/2 \quad r_{n+1} = \frac{2\pi n}{T * \omega_{\text{nat}}} \quad X_{n+1} = \frac{F_{n+1}}{1 - r_{n+1}^2 + 2i * r_{n+1} * \zeta}$$

$$q := IFFT(X)$$

A program fragment for evaluating the DFT response coefficients in MATLAB is

> MATLAB: *for n = 0 : N/2; r = n * 2 * pi/(omega_T)*
> *X_dft(n + 1) = F(n + 1)/(1 − r^2 + 2 * i * r * ζ); end*

To take the inverse FFT of the X data, MATLAB users must account for the mirrored nature of the second half of the data set, as described by eq. (3.7.33). To carry out this operation here, as well as in later exercises, let us define an M-file "IFFT_real.m" as follows:

> MATLAB: *function x_time = IFFT_real(X_dft)*
> *N = 2 * (length(X_dft) − 1); X = X_dft*
> *for k = N/2 + 1 : N − 1; X(k + 1) = conj(X_dft(N − k + 1)); end*
> *x_time = ifft(X)*

We then find the response data by calling this M-file: $q = IFFT\_real(X\_dft)$. Note that the scaling factors in either program are relevant only to the comparison with the analytical Fourier coefficients.

The output of the inverse FFT is the desired response q_n corresponding to the t_n. To produce the graphs, the output of IFFT was repeated periodically in order to see two periods of the steady-state response.

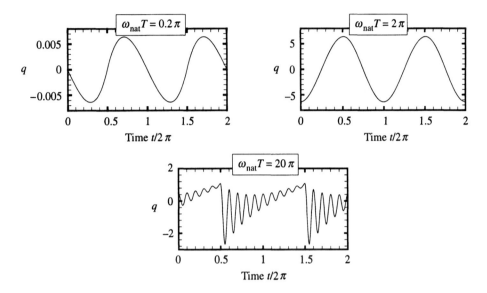

We see that in the first case, where the fundamental frequency $2\pi/T$ is a factor of 10 greater than the natural frequency, the response is very small, and dominated by the fundamental frequency. This is a consequence of the fact that all harmonics are well above the natural frequency, so the complex frequency response $D(\omega_n/\omega_{nat}, \zeta)$ decreases rapidly with increasing n. As a result, the fundamental harmonic gives the largest value, but it nevertheless is much less than unity. In the second case the fundamental frequency matches the natural frequency. Consequently, the fundamental harmonic once again dominates the response. However, in this case the amplitude is large, and it is 90° out of phase from the excitation's fundamental harmonic, which is a sine. The third case corresponds to a fundamental frequency that is one-tenth of the natural frequency. As a result, the complex frequency response for the lower harmonics is close to unity, because $n\omega_1 \ll \omega_{nat}$. Thus, the main details of the sawtooth waveform are reproduced in the response. However, as n approaches 10, the complex frequency response increases. Note that the ringing part of the response has 10 peaks in the interval of a period, which confirms the significance of the tenth harmonic.

As a closure, it is instructive to consider the effect of not setting the excitation value to zero at the discontinuity, as we did when we sampled the sawtooth excitation. In that case, the computed mean value F_1 would not be zero. This would show up as a noticeable mean value in the response in the first case, where the actual response is small. (Raising N would lessen this error.) In the other cases, where the response is much larger, there would be little indication of a mean value.

3.7.6 Accelerometers and Seismometers

Accelerometers and seismometers are sensors that measure the vibration of bodies. Their responses are governed by the same equations of motion, but their design parameters are vastly different. The general model consists of a one-degree-of-freedom system mounted on a base that is vibrating due to some other cause, such as vibrational excitation of a machine part, or seismic waves in an earthquake. The model for these systems appears in Figure 3.22, where the base displacement $z(t)$ is specified and the absolute displacement of the suspended mass is u. Rather than using the absolute position as the

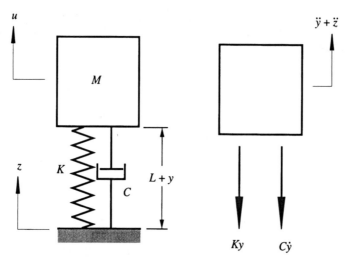

FIGURE 3.22 One-degree-of-freedom system mounted on a moving base.

generalized coordinate for the system, we use the extension y of the spring relative to the spring length L in the static equilibrium position. The extension rate of the dashpot is $\dot{y}$. Because the displacement of the mass is the sum of the base displacement and the amount by which the spring is extended, the acceleration of the mass is $\ddot{y} + \ddot{z}$. No external forces act on the suspended mass, so the equation of motion may be written as

$$M(\ddot{y} + \ddot{z}) + C\dot{y} + Ky = 0 \tag{3.7.50}$$

A more useful form results from introducing the natural frequency and the ratio of critical damping,

$$\boxed{\ddot{y} + 2\zeta\omega_{\text{nat}}\dot{y} + \omega_{\text{nat}}^2 y = -\ddot{z}} \tag{3.7.51}$$

Thus the excitation is the base acceleration $\ddot{z}$. The concept is that measurement of y, which is the amount that the suspended mass moves relative to the system housing, may be used to infer information regarding z. The qualitative aspects of both an accelerometer and a seismometer become apparent when we consider limiting cases. If the natural frequency is very high, meaning that the system is very stiff, then a reasonable first-order approximation of eq. (3.7.51) would ignore the inertial and damping terms. This leads to a particular solution $y \approx -\ddot{z}/\omega_{\text{nat}}^2$, so that the response is proportional to the foundation acceleration. This is the case of an accelerometer. The opposite limit is that of a system whose natural frequency is very low, meaning that the suspension is very soft. In this case, we would neglect the damping and stiffness terms in eq. (3.7.51), which leads to $\ddot{y} \approx -\ddot{z}$. Correspondingly, the particular solution is $y \approx z$, so that the response is proportional to the foundation displacement. Such a system is a seismometer.

The above qualitative discussion does not address the question of what conditions must be met to consider ω_{nat} to be very high or very low. To examine the relevant issues, let us consider a base motion that is a periodic vibration, which we represent as a Fourier series,

$$z(t) = \frac{1}{2}Z_0 + \text{Re} \sum_{n=1}^{\infty} Z_n \exp(in\omega_1 t) \tag{3.7.52}$$

The corresponding foundation acceleration is

$$\ddot{z}(t) = \mathrm{Re}\left[\sum_{n=1}^{\infty} A_n \exp(in\omega_1 t)\right], \quad A_n = -(n\omega_1)^2 Z_n \tag{3.7.53}$$

The steady-state response has the same period, and the mean value of $\ddot{z}$ is zero. Hence, the Fourier series for the response must have the form

$$y = \mathrm{Re}\left[\sum_{n=1}^{\infty} Y_n \exp(in\omega_1 t)\right] \tag{3.7.54}$$

The harmonic amplitudes Y_n are described by eq. (3.7.45), with $-MA_n$ replacing the force amplitudes F_n. Because $K/M = \omega_{\mathrm{nat}}^2$, we find that

$$Y_n = -\frac{MA_n}{K} D(nr_1, \zeta) = (nr_1)^2 D(nr_1, \zeta) Z_n, \quad r_1 = \frac{\omega_1}{\omega_{\mathrm{nat}}} \tag{3.7.55}$$

The design parameters for which the system acts like an accelerometer depend on the nature of the base motion. Let N denote the number of harmonics required to form $\ddot{z}$ in eq. (3.7.53) to the desired level of accuracy. We want y to be proportional to $\ddot{z}$, which is equivalent to requiring that each Y_n in eq. (3.7.55) be in a constant ratio to the corresponding A_n. This condition is met if $D(nr_1, \zeta)$ is very close to unity for $n \leq N$. Examination of Figure 3.2 shows that $|D(r, \zeta)|$ is essentially unity and the phase lag $\phi(r, \zeta)$ is essentially zero if $r \ll 1$. For example, if $r = 0.2$ and $\zeta = 0.05$, then $|D| = 1.04$ and $\phi = 1.2°$. The highest frequency of interest in the base acceleration is $N\omega_1$, so each acceleration harmonic will be reproduced in y with an accuracy of 4% in amplitude and 1° in phase angle if $Nr_1 < 0.2$, that is, if $N\omega_1 < 0.2\omega_{\mathrm{nat}}$. In general, the highest frequency ω_{max} ($= N\omega_1$) for which the accelerometer will have an error less than a specified amount will be found to be $\omega_{\mathrm{max}} < \beta\omega_{\mathrm{nat}}$, where $\beta < 1$ is a parameter whose value depends on the desired accuracy.

We will see in Example 3.12 that increasing ζ substantially with ω_{nat} held constant allows larger values of β relative to the undamped case. Although this would seem to imply that accelerometers should be highly damped, there is a trade-off, because the desire to measure high frequencies with very small accelerometers requires stiff materials for the suspension, and such materials tend to have little dissipation. Most modern accelerometers are small devices in which elastic suspension of a tiny block is provided by a piezoelectric ceramic material. Such a material generates an electric voltage when it is deformed. (It also exhibits the reciprocal behavior of deforming when a voltage is applied to it.) This voltage is then processed electronically to obtain the acceleration measurement. Piezoelectric ceramics have very little internal dissipation.

The case of a seismometer may also be investigated by examining eq. (3.7.55). Here, the natural frequency is supposed to be low compared to the fundamental frequency of the excitation, so that all of the frequencies contained in the excitation will be substantially larger than ω_{nat}. The complex frequency response in this case is approximately $D(nr_1, \zeta) \approx -1/(nr_1)^2$, in which case eq. (3.7.55) leads to $Y_n \approx -Z_n$. In other words, the response amplitudes are approximately the negative of the base displacement amplitudes, so that $y \approx -z$. (This near equality excludes the mean value Z_0, which does not excite the suspended mass.) It follows that the accuracy of a seismometer is set by the degree to which $D(r_1, \zeta) \approx -1/r_1^2$. This condition is met for frequencies that are moderately larger than the fundamental frequency. For example, if $\omega_1 > 3.3\omega_{nat}$, then $1 < r_1^2|D| < 1.1$, and the phase lag differs from $180°$ by less than $2°$.

As implied by its name, a common application for a seismometer is to measure ground motion in earthquakes. The lowest frequency in such cases can be 2 Hz or less. A device whose fundamental frequency is lower than this will be quite massive, with a relatively soft spring suspension. The motion of the suspended mass is typically measured by fabricating the system as a ferrous block that translates inside an electromagnetic coil. The motion of the block induces a voltage change across the coil that is processed electronically to measure the base displacement.

In summary, an accelerometer is used to measure base accelerations whose highest frequency is well below the seismometer's natural frequency. In contrast, a seismometer measures base displacement whose fundamental frequency is above the system's resonance. An accelerometer suitable for a specified range of frequencies will be smaller than a seismometer. It therefore can be used to measure vibration of other systems without significantly adding to the inertia of the system. Seismometers are massive and tend to be used only in large-scale structures, such as buildings. In its favor, a seismometer's electrical sensitivity tends to be higher, and the overall system, including electronics, tends to be somewhat simpler.

Thus far, we have considered the relative displacement because we were interested in the possibility of using a suspended mass to measure base displacement. A different issue is the effect of base displacement on the motion of a suspended body. We already touched on this effect in Example 3.2, which analyzed a one-degree-of-freedom model of an automobile as it moved along a rough road. Let us reformulate the equation of motion for the measurement devices in terms of the absolute vertical displacement u of the suspended mass, which is the quantity of concern for an automobile. The spring and dashpot forces appearing in the free body diagram, Figure 3.22, would then be described as $k(u - z)$ and $c(\dot{u} - \dot{z})$, and the equation of motion would be

$$m\ddot{u} + C\dot{u} + Ku = Kz + C\dot{z} \tag{3.7.56}$$

We consider a base displacement that is periodic in time, and seek the resulting displacement u. The Fourier series for z is given by eq. (3.7.52), and we use a similar Fourier series to describe u. Matching like harmonics leads to

$$u(t) = \frac{1}{2}U_0 + \text{Re}\left[\sum_{n=1}^{\infty} U_n \exp(in\omega_1 t)\right] \tag{3.7.57}$$

$$U_n = \frac{K + in\omega_1 C}{K - in\omega_1 C - n^2\omega_1^2 M}Z_n \equiv \frac{1 + inr_1\zeta}{1 + inr_1\zeta - n^2r_1^2}Z_n$$

The significant aspect of this result is the fact that the ratio of amplitudes $|U_n|/|Z_n|$ is the same as the transmissibility we encountered earlier in eq. (3.6.3),

$$\frac{|U_n|}{|Z_n|} = \text{tr}(nr_1) \tag{3.7.58}$$

Thus, whatever measures we take to isolate the ground from a vibrating mass that is subjected to an external force also serve to isolate a suspended mass from the vibration of the ground. Specifically, if we wish to isolate the mass, we should make the suspension system soft and lightly damped; this is the parameter criterion for a seismometer.

EXAMPLE 3.12

A common test for an accelerometer is its ability to reproduce a base acceleration that is a periodic square wave given by $\ddot{z} = a$ if $0 < t < T/2$, or $\ddot{z} = -a$ if $T/2 < t < T$. In order to explore the role of natural frequency and damping ratio, determine the square-wave response of an accelerometer for the following alternative cases: $\omega_{nat}T = 80\pi$ and $\zeta = 0.01$, $\omega_{nat}T = 40\pi$ and $\zeta = 0.1$, $\omega_{nat}T = 10\pi$ and $\zeta = 0.7071$. Which of these cases gives the most faithful reproduction?

Solution In addition to further familiarizing us with FFT evaluation of response to periodic excitation, this exercise will illustrate the issues involved in designing an accurate device. We sample $\ddot{z}$ at N instants over one period of the square wave. The largest value of T among the cases presented is $T = 40(2\pi/\omega_{nat})$, which means that we may expect the $n = 40$ harmonic to resonate the system. For that reason we should obtain an adequate representation of the response if the highest harmonic is at least twice the resonant one, which leads to a criterion that $N/2 > 80$. The lowest power of 2 satisfying this condition is $N = 256$, but we shall use $N = 512$ to be safe. The sampling interval is $\Delta = T/512$, and the discretized times are $t_{j+1} = j\Delta$, $j = 0, 1, ..., N - 1$. The discontinuities in $\ddot{z}$ occur at $t = 0$ and $t = T/2$, at which instants we set the $\ddot{z}$ value to zero, which is the average of the values at either side of the discontinuity. The amplitude factor a merely scales the acceleration, so we define $C(t) = \ddot{z}/a$ to construct the sampled set of acceleration values,

$$C_1 = C_{N/2+1} = 0; \quad C_{j+1} = 1, j = 1, ..., N/2 - 1; \quad C_{j+1} = -1, j = N/2 + 1, ..., N - 1$$

The operation of using our software's FFT routine to find DFT values A_n corresponding to the C_j acceleration data is indicated as

$$\{A\} = [E_{DFT}]\{C\}$$

We use the first form of eqs. (3.7.55), which describes the DFT of the displacement when the DFT of the base acceleration is known. In view of the fact that a scales the acceleration and we are using $n = 1$ as the lowest subscript, that expression becomes

$$Y_{n+1} = -\frac{aA_{n+1}}{\omega_{nat}^2}D(nr_1, \zeta), \quad n = 0, 1, ..., N/2$$

The fundamental frequency is $2\pi/T$, so

$$r_1 = \frac{2\pi}{\omega_{nat}T}$$

We do not have specific values of a and ω_{nat}, so we use them to create a nondimensional DFT. Thus we compute

$$\frac{Y_n\omega_{nat}^2}{a} = U_{n+1} = -\frac{A_{n+1}}{1 - (nr_i)^2 + 2i\zeta(nr_1)}, \quad n = 0, 1, ..., N/2$$

To determine the response in MATLAB, we use the M-file "FFT_real.m" described in Example 3.11, $u = FFT\_real(U)$, which takes care of the complex conjugate mirror image property, whereas in Mathcad we write $u = FFT(U)$. The result is an array $\{u\}$ of N values at the successive instants t_j. The factors used to define U_j also apply to the inverted data, so the u_j data are related to the displacement by $u_j = \omega_{nat}^2 y(t_j)/a$.

The response for the case where $\omega_{nat}T = 80\pi$, so that the fundamental frequency is $\omega_{nat}/40$, and $\zeta = 0.01$ shows that the seismometer's displacement y exhibits ringing. In essence, the sudden change of the base acceleration at a discontinuity results in a damped step response. Before the oscillation decays away, the value of $\ddot{z}$ changes again, and the process repeats.

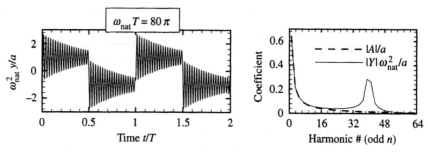

An alternative perspective for the difference between $\ddot{z}$ and y is to examine the respective Fourier coefficients A_{n+1} and Y_{n+1} as functions of harmonic number n. The graph only describes the magnitudes of the odd-numbered harmonics, which are A_{2n} and Y_{2n}, because the even-numbered harmonics of $\ddot{z}$ are zero. (The analytical Fourier series for the base acceleration is $A_n = 4a/(i\pi n)$ if n is odd.) The second graph shows that the displacement harmonics close to the resonance condition $nr_1 = 1$, that is, $n = 40$, are large, unlike the corresponding coefficients of the acceleration. Note also that because $n = 40$ is an even harmonic, and only the odd harmonics are excited, none of the DFT frequencies for the excitation exactly matches the system's natural frequency. If $\omega_{nat}T/2\pi$ was odd, the ringing would be much worse.

The second case, $\omega_{nat}T = 40\pi$ and $\zeta = 0.1$, corresponds to a design that is less stiff (lower ω_{nat} for a specified T) but has more damping than in the first case. This design is better, in that there is less ringing. The plots of the series coefficient magnitudes show that although the resonant harmonics occur at a smaller n, where the A_n coefficients are somewhat larger, this is compensated by the increased damping, which lowers the Y_n values near resonance.

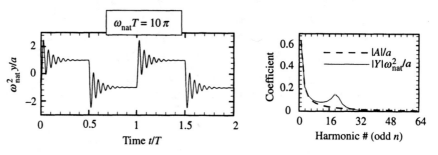

The third design further raises the damping to $\zeta = 0.7071$. The plot of y shows that even though the design is less stiff, $\omega_{nat}T = 10\pi$, there is little ringing. However, the penalty is that the output cannot rise as rapidly. The plot of the displacement coefficients shows no evidence of Y_n peaking at resonance, which corresponds to $n = 5$. There is a noticeable difference between the A_n and Y_n values in a broad range of n, but the plot of the time response suggests that of the three designs, this seems to be the best.

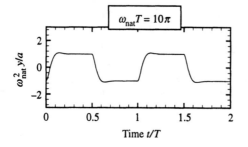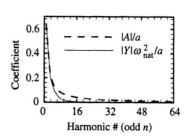

3.8 TRANSIENT RESPONSE REVISITED

We return here to the analysis of transient response. Our purpose is to address situations where the methods of undetermined coefficients and superposition are not suitable. By deferring consideration of this task until now, we will be able to synergistically make use of previous developments. Specifically, we will see that the impulse response leads us to a convolution integral, which describes the response to an arbitrary excitation. We then will see that FFT techniques enable us to numerically evaluate convolution integrals in a manner that is both efficient and accurate. The FFT development will then lead us to an experimental procedure by which the complex frequency response may be extracted from measured data.

It also is possible to employ finite difference techniques to solve differential equations of motion. Such solutions are fairly simple to implement and may be used for nonlinear as well as linear equations of motion. The advantage of frequency domain analysis is that, in addition to providing response data, proper examination of the results will provide physical insight into the factors influencing the response.

3.8.1 Convolution Integral

In this section we shall see how the principle of superposition and knowledge of the impulse response lead us to a general expression for the response of a system to any excitation. This expression has the form of an integral that combines the impulse response at one instant with the force at another instant in a process of *convolution*. Convolution integrals usually arise in mathematics texts in the context of integral transforms. However, our heuristic derivation offers a degree of physical understanding that is not conveyed by a formal mathematical derivation.

We begin by recalling that the impulse response $g(t - \tau)$ is the response of a system when an impulsive force $Q(t) = \delta(t - \tau)$ is applied at instant τ. Two of the basic properties of the Dirac delta function were discussed in Section 2.3.4. We saw that its value is zero, except for the instant $t = \tau$ at which the function's argument is zero. We also established that the integral of a Dirac delta function is unity if the interval over which it is integrated spans the instant τ. We established the third fundamental property of a Dirac delta function in eq. (3.7.10). It states that the integral of $f(t)\delta(t - \tau)$, where $f(t)$ is an analytical function, is the analytical function's value at instant τ. These properties are summarized as

$$\delta(t - \tau) = 0 \text{ if } t \neq \tau$$

$$\int_{t_1}^{t_2} \delta(t - \tau) \, dt = 1 \text{ if } t_1 < \tau < t_2 \qquad (3.8.1)$$

$$\int_{t_1}^{t_2} F(t)\delta(t - \tau) \, dt = F(\tau) \text{ if } t_1 < \tau < t_2$$

We will derive the convolution integral for transient response in two ways. The first makes use of the above definitions for the Dirac function and the properties of linear differential equations. This derivation is concise, but it does not provide much fundamental insight, which is the reason for the second derivation. By definition, the impulse response g is the solution of the following problem:

$$M\ddot{g} + C\dot{g} + Kg = \delta(t - \tau), \quad g(0) = \dot{g}(0) = 0, \quad \tau > 0 \qquad (3.8.2)$$

On the other hand, the response we seek satisfies the differential equation

$$M\ddot{q} + C\dot{q} + Kq = Q(t), \quad q(0) = \dot{q}(0) = 0 \tag{3.8.3}$$

Note that we are restricting our attention to the case of zero initial conditions, because nonzero initial conditions merely modify the homogeneous solution. From the last of eqs. (3.8.1), we may express the excitation as

$$Q(t) = \int_0^\infty Q(\tau)\delta(t - \tau) \, d\tau \tag{3.8.4}$$

where we select the integration region to be $0 \le \tau < \infty$ in order to assure that the expression is valid for any finite t. We now note that integration is essentially a summation process. If we multiply or add terms on the right side of a linear differential equation, the variable on the left side will be modified in the same way. Thus, if we multiply the right side of eq. (3.8.2) by $Q(\tau)$ and integrate that term over $0 \le \tau < \infty$, the solution of the differential equation will be modified in the same way. Because these operations produce $Q(t)$ on the right side, it must be that the variable in the left side is the response q we seek:

$$q(t) = \int_0^\infty Q(\tau)g(t - \tau) \, d\tau \tag{3.8.5}$$

This is not quite the final form, because we know that $g(t - \tau) = 0$ if $t < \tau$. This is an expression of the basic principle of causality obeyed by all dynamical systems. Specifically, time is irreversible, so a system's response at any instant cannot depend on the nature of the excitation after that instant. (In other words, the future cannot affect the past.) As a consequence of causality, the value of g in the above integral is zero for $\tau > t$, so we have

$$\boxed{q(t) = \int_0^t Q(\tau)g(t - \tau)d\tau} \tag{3.8.6}$$

This is the *impulse response convolution integral*. We may write it in an alternate form by introducing the change of variables $\tau' = t - \tau$, which leads to

$$\boxed{q(t) = \int_0^t Q(t - \tau)g(\tau)d\tau} \tag{3.8.7}$$

Depending on the nature of the functions, it might be easier to evaluate the second form.

Although the derivation might seem to be mathematical trickery, the resulting convolution integral has a simple explanation. Suppose we discretize a function into a set of impulses spaced uniformly at interval $d\tau$, as we have done in Figure 3.23. In order that the discretized force history have the same effect as the continuous one, we require that the impulse at each interval match that of the actual excitation. By the central limit theorem, the value of Q during the infinitesimal interval $d\tau$ may be taken to be its value at the instant of the impulse because the interval is infinitesimal. Thus, the excitation at any instant $t = \tau$ is $[Q(\tau)d\tau]\delta(t - \tau)$. Because the system is linear, the response to this specific excitation is $[Q(\tau)d\tau]g(t - \tau)$. We then obtain the response at instant t resulting from all

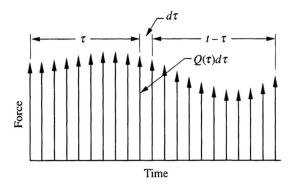

FIGURE 3.23 Representation of the excitation affecting the response at time t as a sequence of impulses.

of the excitations by adding the $[Q(\tau)d\tau]g(t - \tau)$ values for every instant τ up to the instant of interest. The process of adding differential quantities is integration, which leads us directly to eq. (3.8.6).

The convolution integral may be used in a variety of ways. The impulse response for an underdamped system is given in eq. (2.3.23). Its mathematical form is not too forbidding, especially in the case where $\zeta = 0$. If the functional form of the excitation $Q(t)$ also is not too complicated, we might have a reasonable chance of evaluating the convolution integral analytically, especially if we make use of symbolic mathematics programs. If we take this approach, we need to be careful when we treat a pulse excitation, which cuts off at $t = T$. We use a step function to describe this excitation, $Q(t) = F(t)[h(t) - h(t - T)]$, where $F(t)$ is an analytical function. Substitution of this excitation into the convolution integral leads to

$$q(t) = \begin{cases} \displaystyle\int_0^t F(\tau)g(t - \tau)\,d\tau, & t < T \\[2em] \displaystyle\int_0^T F(\tau)g(t - \tau)\,d\tau, & t > T \end{cases} \qquad (3.8.8)$$

This expression emphasizes an implicit aspect of either derivation. In the general convolution integral of eq. (3.8.6), the variable t in the integrand is the instant at which the response is being evaluated, while the parameter t in the upper limit of the integral denotes the last instant at which the excitation affects the response.

Equation (3.8.8) leads us to a basic insight regarding the effect of pulse excitation on an underdamped system. Suppose the pulse duration T is small compared to $\tau_d/2$, where τ_d is the damped period. Consider the response for $t > T$. In the integration over $0 \leq \tau < T$, the value of $g(t - \tau)$ will not change much from $g(t)$, because a cycle of free vibration requires a duration much greater than T. This leads us to

$$q(t) \approx g(t)\int_0^T F(\tau)d\tau, \quad T \ll \tau_d, \quad t > T \qquad (3.8.9)$$

This approximation shows quite clearly that *the primary effect of a short pulse is its impulse.* Another observation is that after the pulse expires, the response is a free vibration, whose peak occurs at $t > T$. If ζ is not too large, we therefore can anticipate that the peak acceleration will be approximately $\omega_{nat}^2 |q_{max}|$. This can be very useful for designing suspension systems to withstand shock loadings.

If the excitation is too complicated to evaluate the convolution integral analytically, it might seem appropriate to evaluate the integral numerically. This could be an acceptable procedure, especially if it is only necessary to determine the response at a few specific instants. However, it can be a difficult task to get a convergent answer using numerical integration. The primary reason for this is the fact that $g(t - \tau)$ for a lightly damped system is highly oscillatory. If we endeavor to use numerical methods to evaluate the convolution integral when t is many times greater than the damped period, it will be necessary to integrate over many plus and minus lobes of the impulse response. The contributions from the positive lobes often approximately cancel those of negative lobes, thereby necessitating many small integration steps. We will see in the next section that FFTs allow us to evaluate the convolution integral in a rapid, yet accurate, manner.

3.8.2 FFTs for Transient Response

It is possible to evaluate transient response with the aid of FFTs. We will first develop the approach heuristically, based on our treatment in Section 3.7.5 of the steady-state response to a periodic excitation. Then, in order to understand the possible sources of error in the procedure, we will introduce the mathematical tool of Fourier transforms. In turn, this will enable us to tie the FFT approach to the convolution integral in the previous section. It also will lead to some useful experimental procedures.

Basic Concept Consider a transient excitation and associated response that commence at $t = 0$ and are negligibly small if $t > T'$, where the latter value depends on the nature of the excitation and the system properties. In order to employ Fourier series concepts we consider the response to be periodic. Avoidance of a type of error known as *wraparound*, which we will explore later, requires that the period T be at least twice the value of T'. The excitation $Q(t)$ and response $q(t)$ may be represented as Fourier series based on period T. Such series will periodically reproduce $Q(t)$ and $q(t)$ if t is outside the time interval $0 < t < T$. In contrast, the actual functions are zero outside that interval. We refer to the interval $0 < t < T'$ as the *time window*, which is the interval within which the data are defined. Because the actual and windowed functions can differ for $t < 0$ and $t > T'$, *all results obtained from the FFT algorithms are valid only within the time window*. The basic concept is that by considering the transient process to be periodic, we may use the procedure in Section 3.7.5 to calculate the response within the time window, as though the process actually were periodic.

The first step in using FFT techniques is to sample the generalized force at discrete instants $t_n = n\Delta$, where Δ is the sampling interval. All quantities are negligible for $t > T'$, which means that we may replace the actual values by zeroes. This is a process of *zero padding*.

$$
\begin{array}{ll}
Q_n = Q(n\Delta) & \text{if } 0 \leq n\Delta < T' \\
Q_n = 0 & \text{if } T' \leq n\Delta < T
\end{array}
\tag{3.8.10}
$$

Criteria for selecting T', T, and Δ are addressed later in the development. The number of time samples will be $N = T/\Delta$.

Once the time data set has been established, we may evaluate its DFT. We symbolically represent this operation in terms of the DFT weighting function matrix as

$$\{F\} = [E_{\text{DFT}}]\{Q\}$$
(3.8.11)

The operation described by the preceding relation would actually be carried out by setting $\{Q\}$ as the input to an FFT computational routine. In effect, the DFT coefficients F_j account for the excitation's time dependence within the FFT's time window, scaled by whatever factor is used in the specific FFT routine. The fundamental frequency is $2\pi/T$, so the discrete frequencies are

$$\omega_j = j\left(\frac{2\pi}{T}\right)$$
(3.8.12)

We have already seen how to evaluate the response corresponding to a DFT representation of an excitation. For each frequency ω_j, we use the complex frequency response to determine the Fourier coefficient X_j of q,

$$X_j = \frac{F_j}{K}D\left(\frac{\omega_j}{\omega_{\text{nat}}}, \zeta\right), \quad j = 0, 1, ..., \frac{N}{2}$$

$$X_j = X_{N-j}^*, \quad j = \frac{N}{2} + 1, ..., N - 1$$
(3.8.13)

The response at the discretized time instants, that is, $q(n\Delta)$, may be determined by using the X_j values as the input to the inverse FFT routine associated with the routine we employed to determine the F_j coefficients. Symbolically, this yields

$$\{q\} = [E_{\text{IDFT}}]\{X\}$$
(3.8.14)

Analytical Justification—Fourier Transforms The basic concept laid out in the preceding section is easy to implement. However, the operations of windowing, sampling, and using the complex frequency response have potential errors. Exploring these issues requires that we consider the theoretical foundation for employing FFTs to analyze transient response, specifically the *Fourier transform*. This transform is derived from the Fourier integral theorem (Bracewell, 1986), which also is the basis for other integral transforms, such as Laplace and Hankel. The Fourier transform converts a function $y(t)$ that is defined over $-\infty < t < \infty$ into a function $Y(\omega)$, where ω is a frequency parameter. The definition is

$$Y(\omega) = \int_{-\infty}^{\infty} y(t)\exp(-i\omega t)dt$$
(3.8.15)

(Most mathematics books define the transform with a positive sign in the exponential. We have adopted the above definition in order to be consistent with our use of $\exp(i\omega t)$ to describe a harmonic function of t.) If the transform $Y(\omega)$ is known, the corresponding time function may be found by applying the *inverse Fourier transform*,

$$y(t) = \frac{1}{2\pi}\int_{-\infty}^{\infty} Y(\omega)\exp(i\omega t)d\omega$$
(3.8.16)

We say that $y(t)$ and $Y(\omega)$ form a Fourier transform pair, which we indicate with a correspondence arrow as $y(t) \Leftrightarrow Y(\omega)$.

The fact that $y(t)$ is an analytic function does not guarantee that its transform is analytic in ω. Most important in this category is a harmonic function, $\mathrm{Re}[A\exp(i\Omega t)]$. Let us evaluate its Fourier integral, eq. (3.8.15), based on the function being approximated as being zero for $|t| > b$,

$$\int_{-b}^{b} \mathrm{Re}[A\exp(i\Omega t)]\exp(-i\omega t)\,dt$$

$$= \frac{1}{2}\int_{-b}^{b}[A\exp(i\Omega t) + A^*\exp(-i\Omega t)]\exp(-i\omega t)\,dt$$

$$= \begin{cases} \dfrac{A}{(\Omega - \omega)}\sin[(\Omega - \omega)b] + \dfrac{A^*}{(\Omega + \omega)}\sin[(\Omega + \omega)b], & |\omega| \neq \Omega \quad (3.8.17) \\[2mm] Ab + \dfrac{A^*}{2\Omega}\sin(2\Omega b), & \omega = \Omega \\[2mm] \dfrac{A}{2\Omega}\sin(2\Omega b) + A^*b, & \omega = -\Omega \end{cases}$$

If $|\omega| \neq \Omega$, this integral is certainly finite as b increases, but its value oscillates rapidly, so the limit $b \longrightarrow \infty$ is not defined. Furthermore, if $|\omega| = \Omega$, the integral's value increases without limit as b increases. In a formal sense, the Fourier transform of a harmonic function does not exist. However, the basic properties of a Dirac delta function allow us to circumvent this difficulty. Given that the integral blows up as we try to extend the limits of integration, let us consider the possibility that the transform we seek is proportional to a Dirac delta function. We take the argument of this function to be $\omega \mp \Omega$ because the infinite value occurs at $\omega = \pm\Omega$. In view of the third of eqs. (3.8.1), we find from eq. (3.8.16) that the inverse transform of $\delta(\omega \mp \Omega)$ is

$$\frac{1}{2\pi}\int_{-\infty}^{\infty}\delta(\omega \mp \Omega)\exp(i\omega t)\,d\omega = \frac{1}{2\pi}\exp(\pm i\Omega t) \tag{3.8.18}$$

This establishes the transform pair

$$\boxed{\exp(\pm i\Omega t) \Leftrightarrow 2\pi\delta(\omega \mp \Omega)} \tag{3.8.19}$$

When we write the harmonic function as the sum of a complex exponential and its complex conjugate, we are led to the following Fourier transform pair,

$$\boxed{\mathrm{Re}[A\exp(i\Omega t)] \Leftrightarrow \pi[A\delta(\omega - \Omega) + A^*\delta(\omega + \Omega)]} \tag{3.8.20}$$

In other words, the Fourier transform of a harmonic function will show up as spikes equidistant from the origin when the transform is plotted versus frequency.

Another important transform pair is the one for an impulse in the time domain. The integration property for a Dirac delta function, in conjunction with eq. (3.8.15), yields

$$\boxed{\delta(t - \tau) \Leftrightarrow \exp(i\omega\tau)} \tag{3.8.21}$$

From this, we conclude that an impulse excites all frequencies with equal magnitude, with all transformed values being real if the impulse occurs at $\tau = 0$. The above is a special case of the time delay transform pair. Suppose that $y(t) \Leftrightarrow Y(\omega)$. If we substitute $y(t - \tau)$ into the Fourier transform's definition, then substitute $t' = t - \tau$, we are led to the realization that

$$y(t - \tau) \Leftrightarrow Y(\omega) \exp(i\omega\tau) \text{ if } y(t) \Leftrightarrow Y(\omega) \tag{3.8.22}$$

Of particular relevance to our efforts will be the fact that there is a transform pair associated with a convolution integral. Specifically, it can be shown that if $y(t) \Leftrightarrow Y(\omega)$ and $z(t) \Leftrightarrow Z(\omega)$, then

$$\int_{-\infty}^{\infty} y(\tau)z(t - \tau)\,d\tau \equiv \int_{-\infty}^{\infty} y(t - \tau)z(\tau)\,d\tau \equiv y * z \Leftrightarrow Y(\omega)Z(\omega) \tag{3.8.23}$$

(The use of an asterisk to denote a convolution integral is standard abbreviated notation.)

This convolution integral is strongly reminiscent of eqs. (3.8.6) and (3.8.7), by which we may evaluate transient response, except that the integration domain here is $-\infty < \tau < \infty$. This difference stems from the generality of the present functions $y(t)$ and $z(t)$. Suppose we let $y(t)$ be the generalized force and let $z(t)$ be the impulse response function, in which case we have $y(\tau) = Q(\tau)$ and $z(t - \tau) = g(t - \tau)$. For transient response, we consider the excitation to begin at $t = 0$, so we take $Q(\tau) = 0$ if $\tau < 0$. Furthermore, the impulse response is zero prior to the instant at which the impulse is applied, so $g(t - \tau) = 0$ if $\tau > t$. Hence, in this case the only interval over which both functions in the Fourier convolution integral, eq. (3.8.23), are nonzero is $0 < \tau < t$. Because the Fourier convolution integral in this case is the same as eqs. (3.8.6) and (3.8.7), it follows that the Fourier transform of any response $q(t)$ is the product of transforms of the excitation and of the impulse response,

$$q(t) \Leftrightarrow F(\omega)G(\omega) \text{ if } Q(t) \Leftrightarrow F(\omega) \text{ and } g(t) \Leftrightarrow G(\omega) \tag{3.8.24}$$

A useful application of this transform pair uses $Q(t) = \delta(t)$, so that $q(t)$ in the preceding is the impulse response. According to eq. (3.8.21), the Fourier transform of this impulse function is $F(\omega) = 1$. It follows from eq. (3.8.24) that $G(\omega)$, which is the Fourier transform of the impulse response, may also be viewed as the complex amplitude of the response to a harmonic excitation having unit amplitude at all frequencies, $F(\omega) = 1$. The latter is the complex frequency response, so we have the following transform pair,

$$g(t) \Leftrightarrow G(\omega) = \frac{1}{K}D(\omega, \zeta) \tag{3.8.25}$$

Using this relation to replace $G(\omega)$ in eq. (3.8.24) leads to

$$q(t) = Q * g \Leftrightarrow \frac{1}{K}F(\omega)D(\omega, \zeta) \tag{3.8.26}$$

The similarity of this relation to eq. (3.8.13) is not coincidental. It is a manifestation of the general relationship between the Fourier transform of a function and the function's DFT. To identify this relationship, let us evaluate the Fourier transform of a windowed function, meaning that we consider $y(t) = 0$ for $t < 0$ and $t > T$. When ω equals the jth discrete frequency, the Fourier transform is

$$Y(j\omega_1) \approx \int_0^T y(t) \exp(-ij\omega_1 t)\, dt \qquad (3.8.27)$$

where the approximate sign indicates that the transform is based on replacing the actual time function by its windowed representation. We encountered the same integral, with a shifted time window, in eq. (3.7.8). Thus, the right side is the Fourier series coefficients $\hat{Y}_j$ of the windowed function, multiplied by $T/2$. According to eq. (3.7.36), Fourier series coefficients are scaled by a factor $2/N$ to obtain DFT values. Combining both factors leads to

$$Y(j\omega_1) \approx \Delta Y_j \qquad (3.8.28)$$

where Y_j are the DFT values of the windowed time function.

Sampling and Windowing   Identification of the DFT as an approximation of a Fourier transform has several implications. First, we observe that the scaling factor applies equally to the transform of the excitation $Q(t)$ and the transform of the response $q(t)$. Thus, we verify that eq. (3.8.13) is the correct approximation of the Fourier convolution pair in eq. (3.8.26). As a corollary of this observation, we conclude that the time window and sampling interval must be adequate to obtain accurate FFTs of the excitation and of the impulse response. This is true, even though eq. (3.8.25) makes it unnecessary to actually evaluate the FFT of the impulse response.

At the beginning of this section, T' was defined as an instant that is sufficiently large that the excitation can be considered to be negligible for $t > T'$. The same must be true for the impulse response. Consider an underdamped system. The decay envelope for the impulse response is $\exp(-\zeta\omega_{nat}t)$; see eq. (2.3.23). If we set $\zeta\omega_{nat}T' = 4$, then the impulse response for $t > T'$ will be less than a factor $e^{-4} \approx 0.018$ smaller than its maximum value. Because the DFT period T should be at least twice T', we shall require that

$$\boxed{T \ge 2T' \ge \frac{8}{\zeta\omega_{nat}}} \qquad (3.8.29)$$

Of course, this criterion must be weighed against the duration of the excitation. Whichever is larger is the one that applies.

The criterion above has an interesting interpretation in terms of the complex frequency response. When we examined dissipation in harmonic motion, we saw in eq. (3.2.16) that the bandwidth of a resonance is $\Delta\omega = 2\zeta\omega_{nat}$, so eq. (3.8.29) may equivalently be written as

$$\boxed{T \ge 2T' \ge \frac{16}{\Delta\omega}} \qquad (3.8.30)$$

If this condition is satisfied, the fundamental frequency, which is also the interval between the DFT frequencies, will satisfy $\omega_1 = 2\pi/T < (\pi/8)\Delta\omega$. This means that the discrete frequency increment will be less than half the bandwidth, which ensures that the complex frequency response in the vicinity of the natural frequency will be represented by at least two discrete values.

Note that eq. (3.8.30) is a minimal requirement. Failure to satisfy it will introduce an error into the DFT spectrum called *leakage*. Leakage error, and techniques to lessen such error, are explored in the conclusion of this chapter.

Recognition that the DFT approximates the Fourier transform also provides a guideline for selecting the sampling interval Δ. The highest (circular) frequency that may be evaluated for a specified value of Δ is the Nyquist critical frequency $2\pi f_{cr}$, which is given by eq. (3.7.27) as π/Δ. Let us consider using Δ to sample a tone burst excitation at frequency Ω, that is, a sinusoidally varying force at frequency Ω that exists for several cycles. Suppose the frequency of the tone burst exceeds the Nyquist critical frequency, $\Omega > \pi/\Delta$. Evaluating the DFT value F_j for frequency ω_j entails adding terms of the form $Q_n\exp(-i\omega_j n\Delta)$, where $n\Delta$ are the sampling instants, $Q_n = Q(n\Delta)$, and $-\pi/\Delta \leq \omega_j \leq \pi/\Delta$. During the duration of the tone burst $Q_n = A\exp(i\Omega n\Delta)$, so the same values of Q_n result if $\Omega\Delta$ is decreased by any multiple of 2π. Consequently, rather than being attributed to frequency Ω, which is outside the DFT frequency band, the DFT will be attributed to frequency $\Omega' = \Omega - p(2\pi/\Delta)$, where p is the integer for which $-\pi/\Delta \leq \Omega' \leq \pi/\Delta$. In other words, a signal that has a frequency component at a frequency that exceeds the Nyquist critical frequency is repeated periodically until it shows up in the frequency band whose limits are $\pm 2\pi f_{cr}$. It then adds to the amplitude that actually is associated with frequency Ω'. Figure 3.24 depicts the general situation for a function $y(t)$ whose Fourier transform is significant over a range of frequencies exceeding $2\pi f_{cr}$. We say that the Fourier transform in the spectrum out of the DFT's frequency band is *aliased* into the DFT spectrum. (The transform values in Figure 3.24 are for a complex function $y(t)$ because they do not satisfy $Y(-\omega) = Y^*(\omega)$. Such a function emphasizes the manner in which aliasing manifests itself. If $y(t)$ is real, the real part of $Y(\omega)$ would be an even function of ω, while the imaginary part would be odd in ω.)

It is obvious that aliasing must be avoided. We often can examine the excitation before we sample it, as would be the case when we use FFTs to solve an analytical problem. Then we can estimate the required sampling interval based on the fluctuations in the function that occur most rapidly, and therefore indicate the highest frequency to be captured. If we already have taken the FFT of the excitation, as might be the case if we are using an automated data acquisition system, we can identify aliasing by examining the transformed data as a function of k. If the values in the vicinity of $k = N/2$ (that is, $\omega_k \approx 2\pi f_{cr}$) are not small, we should suspect that aliasing has occurred. If possible, we then should resample the excitation using a smaller Δ. If that is not possible, we should use a low-pass filter that removes the high-frequency components *before the signal is processed*. (The usage of DFT's to filter data is dis-

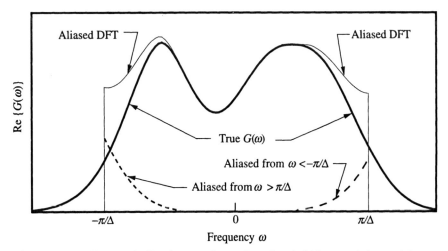

FIGURE 3.24 Aliasing of a Fourier transform whose bandwidth extends beyond the Nyquist critical frequency, $2\pi f_{cr} = \pi/\Delta$.

cussed by Press et al. (1992).) Doing so would lead to DFT values that are reasonably correct, even though they do not fully describe the function's frequency spectrum. If circumstances dictate that we proceed with the faulty data, we must recognize that high-frequency errors might corrupt any subsequent operations.

The original excitation data was zero padded in eq. (3.8.10) in order to avoid *wraparound error*. This numerical phenomenon arises from the fact that eqs. (3.8.13) and (3.8.14) are approximations of a convolution integral. The remainder of this section is an optional sidebar that examines the cause of wraparound error.

Consider two arbitrary functions $y(t)$ and $z(t)$ whose DFT coefficients ΔY_j and ΔZ_j after scaling are good approximations of the Fourier transforms of these functions. Let $(y * z)_n$ denote discrete values of the convolution integral of these functions at instants $t_n = n\Delta$, $n = 0, \ldots, N - 1$. According to eq. (3.8.23), the inverse DFT of the product of Y_j and Z_j at each ω_j approximately gives the convolution values at the various instants. Let us examine this evaluation using the definition of the inverse DFT, eq. (3.7.29). In view of the Δ scaling factor relating DFT and Fourier transforms, we have

$$(y * z)_n = \frac{\Delta}{N} \sum_{j=0}^{N-1} Y_j Z_j \exp\left(2\pi i \frac{jn}{N}\right) \tag{3.8.31}$$

Each Y_j and Z_j may be expressed in terms of the sampled time data with the aid of the DFT definition, eq. (3.7.18). A different index must be used for each sum in the DFT definition. Hence, the substitution leads to

$$(y * z)_n = \frac{\Delta}{N} \sum_{j=0}^{N-1} \left[\sum_{m=0}^{N-1} y_m \exp\left(-2\pi i \frac{jm}{N}\right) \right] \left[\sum_{p=0}^{N-1} z_p \exp\left(-2\pi i \frac{jp}{N}\right) \right] \exp\left(2\pi i \frac{jn}{N}\right)$$

$$\tag{3.8.32}$$

We rearrange the order of the summations, and collect the exponential factors, such that

$$(y * z)_n = \frac{\Delta}{N} \sum_{m=0}^{N-1} \sum_{p=0}^{N-1} y_m z_p \sum_{j=0}^{N-1} \exp\left[2\pi i \frac{j(n - m - p)}{N}\right] \tag{3.8.33}$$

The last sum is the discrete orthogonality property, eq. (3.7.31), which leads to

$$\frac{1}{N} \sum_{k=0}^{N-1} \exp\left[2\pi i \frac{j(n - m - p)}{N}\right] = \begin{cases} 0 \text{ if } m + p \neq n + \ell N \\ 1 \text{ if } m + p = n + \ell N \end{cases} \tag{3.8.34}$$

where ℓ is an integer.

Now let us consider how we use the above to evaluate eq. (3.8.33) for a specified n. For each value of m in the first sum, we let p range from 0 to $N - 1$. If $p \neq n - m + \ell N$, we obtain no contribution according to eq. (3.8.34). Because $0 \leq n, p$, $m \leq N - 1$, the nonzero terms for which we obtain a contribution when $m \leq n$ correspond to $p = n - m$, that is, $\ell = 0$, while $p = n - m + N (\ell = +1)$ gives a contribution when $m > n$. With these considerations we are able to reduce eq. (3.8.33) to the *discrete convolution theorem*,

$$(y * z)_n = \Delta \sum_{m=0}^{n} y_m z_{n-m} + \Delta \sum_{m=n+1}^{N-1} y_m z_{n-m+N} \qquad (3.8.35)$$

Note that the second sum should not be computed if $n = N - 1$.

Comparison of the discrete convolution theorem with the Fourier convolution integral, eq. (3.8.23), reveals an anomaly. In the integral, one selects an instant t, and then forms the integrand from values of $y(t)$ at progressively larger instants multiplied by values of $z(t)$ at progressively smaller instants. In contrast, only the first sum in the above expression has this property. For example, for $n = 0$ and $n = 1$, the discrete convolution theorem gives

$$(y * z)_0 = \Delta(y_0 z_0) + \Delta(y_1 z_{N-1} + y_2 z_{N-2} + \cdots + y_{N-2} z_2 + y_{N-1} z_1)$$
$$(y * z)_1 = \Delta(y_0 z_1 + y_1 z_0) + \Delta(y_2 z_{N-1} + y_3 z_{N-2} + \cdots + y_{N-2} z_3 + y_{N-1} z_2) \qquad (3.8.36)$$

where the terms in the second set of parentheses correspond to the second sum in eq. (3.8.35). The $z(t)$ values in each occur at later times than the $z(t)$ values associated with the first sum. This is called wraparound error because it results from the periodic manner in which the DFT formulation treats time data. Specifically, when any DFT operation requires a value for an instant outside of $0 < t < T$, it shifts the data periodically into the window.

Zero padding extends the original data sets for both variables with zeroes at the end. The number of zeroes required depends on the nature of the data. If the time window is very long, such that either $y(t) \approx 0$ or $z(t) \approx 0$ for a substantial interval at the beginning or end of the time window, then many of the data values in the wrapped portions of eqs. (3.8.36) will be zero anyway. In that case we only need to pad with a few zeroes. However, if the time window is the minimum allowable, such that most of the wrapped terms are nonzero, we will need to pad with many zero values. A fail-safe procedure pads the time data with N zeroes, thereby making the data set length $2N$. To recognize why this is so, consider the $(y * z)_0$ case in eqs. (3.8.36), which has the most wrapped terms. We replace N in that expression with $2N$. The first $N - 1$ terms in the wrapped data, that is, $y_1 z_{2N-1}$ to $y_N z_N$, will be zero because they contain the zero pads of the z_n data. The remaining wrapped terms, $y_{n+1} z_{n-1}$ to $y_{2N-1} z_1$, will be zero because they contain the zero pads of the y_N data. It follows that each wrapped term contributing to the first N values of $(y * z)_j$ will have a zero value.

Physical insight into zero padding may be obtained by recalling that whenever a DFT operation requires a time value outside the time window, the data is periodically reproduced from inside the time window. Suppose the function in Figure 3.25 represents an impulse response. If the time window is $0 < t < 4$ s, the DFT in effect treats the impulse response for $t < 0$ as though it were not identically zero. Doing so obviously ignores the principle of causality. In contrast, padding the data with zeroes extends the time window to $0 < t < 8$ s. The zeroes that are inserted for $t > 4$ s are equivalent for a DFT to placing zeroes in the interval $-4 < t < 0$. Thus, the impulse response with zero padding is recognized by the DFT as being a function that is zero prior to application of the impulse.

If we have identified an elapsed time T' sufficient to see the response decay and a sampling interval Δ, the sampled time data will consist of $N' = T'/\Delta$ values. Padding the data with an equal number of zeroes gives $2N'$ values corresponding to a DFT window of $2T'$. However, the most efficient FFT algorithms operate on data sets whose length is a power of 2. Let p be the smallest integer for which $N = 2^p \geq 2T'/\Delta$. We

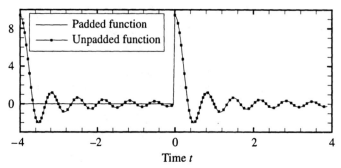

FIGURE 3.25 Periodic replication and zero padding of a windowed function.

have two options. We can retain the original Δ, and set the DFT period such that $T = N\Delta$. In that case, $T > 2T'$ and the number of zero pads will exceed the original data length. Alternatively, we can set $T = 2T'$, so that the number of zero pads equals the length of the data. This redefines the sampling interval to be $\Delta = T/N$.

Procedural Steps   The developments in the preceding sections may be synthesized into a procedure by which FFT techniques may be used to evaluate the convolution integral for a one-degree-of-freedom system, eq. (3.8.6). The steps required to implement the procedure are:

1. Identify the elapsed time T' as the *larger* of the times for the excitation to die off and the time $4/(\zeta\omega_{nat})$ for the impulse response to decay to a negligible value.

2. Let Δ' denote the *maximum* allowable sampling interval. The value of Δ' must satisfy the Nyquist sampling criterion relative to the most rapid variation evident in the excitation, and in the impulse response. Evaluate an integer p such that $2T'/\Delta' \leq 2^p$. Set $N = 2^p$. Set $\Delta = \Delta'$ and $T = N\Delta$, or alternatively, set $T = 2T'$ and $\Delta = T/N$.

3. Sample the excitation such that $Q_n = Q(n\Delta)$ for $0 \leq n\Delta \leq T'$.

4. Zero pad the excitation data by setting $Q_n = 0$ for $T' < n\Delta \leq T - \Delta$.

5. Use an FFT routine to evaluate the DFT coefficients F_j corresponding to the Q_n data.

6. Form the DFT coefficients X_j of the response by evaluating eq. (3.8.13) for $j = 0, \ldots, N/2$. If the inverse FFT routine requires the full set of N DFT coefficients (e.g., MATLAB), use eq. (3.7.33) to fill in the X_j for $j = N/2 + 1, \ldots, N - 1$.

7. Evaluate the inverse FFT of the X_j data to compute the responses $q_n = q(n\Delta)$ for $0 \leq n \leq N - 1$. Then discard the values of q_n for $n > N/2 - 1$, because they will be contaminated by wraparound error. Note that no scaling factors are required for any of the operations if the *FFT* and *IFFT* routines are consistent, as they would be if one software package is used for all operations.

Although this procedure might seem to be lengthy, it actually is extremely efficient. It will require much less computational time to evaluate the response at each discrete instant than would be required to use conventional numerical procedures, such as Simpson's rule, to evaluate a convolution integral for each instant. In fact, as soon as one has become familiar with the operations, the procedure will be found to be an easy way in which to evaluate transient response.

EXAMPLE 3.13

Consider a one-degree-of-freedom system whose mass is 1000 kg and whose natural frequency is 2 Hz. The ratio of critical damping for the system is $\zeta = 0.10$. The system is at rest at its static equilibrium position when, at $t = 0$, it is subjected to an excitation force $Q = 2(t/\tau)\exp(-t/\tau)$ $\sin[\lambda(t/\tau)^2]$ kN, which is called a *chirp*. The coefficients are $\tau = 1$ s and $\lambda = 4\pi$. This type of excitation is very broadband, which means that its Fourier transform is significant over a wide range of frequencies well beyond the natural frequency. Use FFT techniques to determine the system response. Plot the DFT coefficients of the excitation as a function of frequency, and then plot the response as a function of t. In addition, explain what considerations influenced selection of the window T and sampling interval Δ with which the solution was implemented.

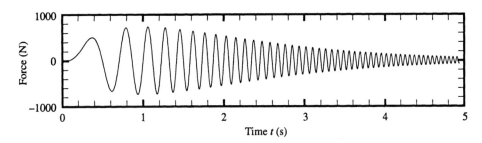

Solution   In addition to illustrating the basic operation required to implement an FFT solution for transient response, this exercise will provide an interesting perspective on the factors affecting transient response. We begin by evaluating the stiffness and natural frequency from the given properties, specifically

$$\omega_{nat} = 4\pi \text{ rad/s}, \quad K = M\omega_{nat}^2 = 157.91 \text{ kN/m}$$

The first step in the FFT implementation is to select the time window and the sampling interval. Inspection of a graph of the excitation indicates that it is very small for $t > 8$ s. Doubling this leads to $T = 16$ s. The other criterion for selecting T is to consider the time for the impulse response to decay to a small value, as given by eq. (3.8.29), which gives $T > 8/(\zeta\omega_{nat}) = 6.37$ s. Thus, we select $T = 16$ s. For the preliminary guess Δ', we estimate from the graph of the excitation that 40 cycles occur in four seconds of the initial interval, where the excitation is largest. In accord with the Nyquist sampling criterion, we select a sampling interval equal to one-half the average cyclical period, $\Delta' = 1/20$ s. This corresponds to $N' = T/\Delta' = 320$. The length N of the DFT data set should be the smallest power of 2 greater than N'. This gives $N = 512$, which leads to $\Delta = T/N = 0.03125$ s. The first $N/2$ values correspond to the sampling instants

$$t_{n+1} = n\Delta, \quad n = 0, 1, \ldots, 255$$

where we have incremented the subscript by 1 to fit MATLAB's requirement. The corresponding sampled values of the excitation are

$$Q_{n+1} = 2(t_{n+1}/\tau)\exp(-t_{n+1}/\tau)\sin[\lambda(t_{n+1}/\tau)^2] \text{ kN}$$

We pad this data set with a sufficient number of zeroes to extend the length of the data set to $N = 512$,

$$Q_{n+1} = 0, \quad n = 256, \ldots, 511$$

We take the FFT of this data, using $F = fft(Q)$. The first $N/2 + 1$ DFT values as a function of the index number appear in the first graph (top of page 194). It is evident that the DFT values near $N/2$ are not very small compared to the largest values, which suggests that our initial choice for Δ did not meet the Nyquist sampling criterion. We therefore double N to 1024 with the time window held at $T = 16$ s. This halves Δ, so we re-sample the excitation at

$$t_{n+1} = n\Delta, \quad n = 0, 1, \ldots, 511, \quad \Delta = 0.01563 \text{ s}$$

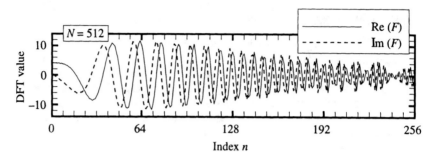

We pad this data with 512 zeroes. The new DFT values for this data appear in the second graph. The values now taper off nicely with increasing n, so we may proceed to the convolution step.

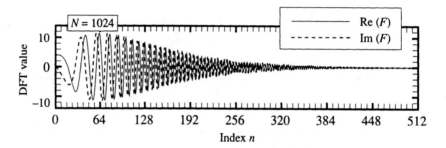

The fundamental frequency is $2\pi/T$. We evaluate the complex frequency response at the DFT frequencies according to

$$D_{n+1} = D(nr_1, \zeta), \quad r_1 = \frac{2\pi}{\omega_{\text{nat}}T}, \quad n = 0, 1, \ldots, N/2$$

Convolution in the frequency domain multiplies the DFT excitation values by the respective D_{n+1},

$$X_{n+1} = F_{n+1}D_{n+1}, \quad n = 0, 1, \ldots, N/2$$

Note that we only computed the products for the first half of the DFT values. In Mathcad, this is the data that the *IFFT* function needs, so we evaluate $q := IFFT(X)$. To use MATLAB's *ifft* function the second half of the DFT data must satisfy the complex conjugate mirror image requirement in eq. (3.7.33). This operation and the inverse were incorporated into the M-file "IFFT_real.m" described in Example 3.11, so we write $q = IFFT\_real(X)$.

Because of wraparound error, only the first $N/2$ values of the time responses q_n are significant. The resulting plot of these values versus t_n is as shown.

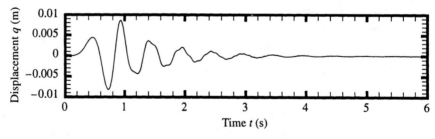

An interesting aspect of the response is obtained by estimating its primary frequency based on elapsed time between peaks. Starting from the maximum at $t \approx 0.4$ s, the fifth succeeding maximum occurs at $t \approx 3$ s, which corresponds to a frequency of $[2\pi/(3 - 0.4)]/5 \approx 12.1$ rad/s. For comparison, the damped natural frequency is $\omega_d = 4\pi(1 - \zeta^2)^{1/2} \approx 12.5$ s. In essence, the displacement consists of the impulse response of the system multiplied by the envelope of the chirp.

Experimental Techniques   The focus thus far has been on using FFTs to perform analytical evaluation of response. The developments also lead to a useful procedure for experimentally identifying the complex frequency response $D(r, \zeta)$. Such information may be used to predict the manner in which the system will respond to a variety of excitations.

A standard experimental method for determining complex frequency response, both for one- and multiple-degree-of-freedom systems, is to measure it directly using a shaker to create a harmonic excitation. Common designs for a shaker use a reciprocating electric motor or a rotatory motor that has an imbalance εm. We shall focus on the rotational type, but the development is easily modified to consider a reciprocating shaker. When the motor is mounted on the test system and rotated at frequency ω, the effective force applied to the system has an amplitude $\varepsilon m \omega^2$, with orthogonal components that vary harmonically at frequency ω. The corresponding generalized force will be Re $[\beta \varepsilon m \omega^2 \exp (i \omega t)]$, where β is a constant that depends on how the generalized coordinate is defined. If the motion sensor measures displacement, its output will indicate the system's steady-state response X at that frequency. We then obtain the complex frequency response associated with that ω by evaluating $D/K = X/(\beta \varepsilon m \omega^2)$. In the case where accelerometers measure the motion, the sensor output is the acceleration $A = -\omega^2 X$, which leads to $D/K = -A/(\beta \varepsilon m)$. Note in this regard that the phase lag for D may be determined by monitoring the time delay between the imbalance being at top-dead-center and the time at which the displacement or acceleration reaches its maximum.

The measurement gives the value of D/K at the motor's rotational frequency. Determining the corresponding frequency ratio, $r \equiv \omega/\omega_{nat}$, requires that we identify ω_{nat}. This is done by performing a frequency sweep, in which we apply a small increment to ω, wait until the transient part of the response decays, and then measure $|D|/K$ and $\phi = \arg(D)$. In this manner we can plot $|D|/K$ and ϕ as functions of ω within the swept frequency band. We may then determine ω_{nat} and ζ by fitting the measured plots to the standard behavior given by eq. (3.1.8). For example, we know that $\phi = 90°$ when $\omega = \omega_{nat}$.

Another method for lightly damped systems is to identify ω_{nat} as the frequency at which $|D|/K$ has its maximum value. From a plot of $|D|/K$ versus ω in the vicinity of ω_{nat}, we may measure the bandwidth $\Delta\omega$. (According to eq. (3.2.17), $\Delta\omega$ is the frequency increment between the half-power points on a plot of measured D versus ω.) From $\Delta\omega$, we may determine the damping factor for the system by using eq. (3.2.16) for the quality factor, which gives $\zeta = \Delta\omega/(2\omega_{nat})$.

Either of the preceding methods for identifying ω_{nat} and ζ requires a reasonably accurate resolution of the behavior around $\omega = \omega_{nat}$. If the damping is very light, the resonance will have a high quality factor, which would require reducing the sweep's frequency increment in the vicinity of the natural frequency. Even then, it might be difficult to determine the peak precisely. Such situations may be handled by using the properties of the Nyquist plot identified in Example 3.4. The process of fitting a circle through the data tends to compensate for sparsity of data. This procedure is described in texts on experimental modal analysis, such as Ewins (1984).

Regardless of which scheme is used, the objective is to estimate ω_{nat} and ζ from the measurements. Such knowledge allows us to compute $D(\omega/\omega_{nat}, \zeta)/K$ at any ω, even if it is outside the frequency band in which the sweep was performed. We then may predict the response of the system to any excitation. For instance, we can predict a transient response by employing convolution, as described by eqs. (3.8.13) and (3.7.33).

One potential difficulty with the foregoing technique is that it is invasive: By mounting a shaker on a system, we modify the system's inertial properties. If the shaker's mass, including rotating parts, is a large fraction of the system's mass, the measured natural frequency will be different from the system's true natural frequency. In turn, this will lead to an erroneous estimate of ζ.

An alternative technique is founded on eq. (3.8.13), which relates the DFT coefficients of an excitation and the corresponding response. It requires generation of an impulsive excitation, which can be done with a light blow from a hammer. The hammer used for this purpose is instrumented with a load cell (for example, a set of strain gauges), which is calibrated to measure the instantaneous contact force exerted by the hammer. This contact force $f(t)$ and the system motion $s(t)$ are measured by sensors and stored at a discrete set of instants. Depending on the type of sensor, s will be either displacement or acceleration.

The sampling interval Δ in this process must fit the Nyquist sampling criterion for the interval over which the hammer force is in contact with the test specimen; this value will be much smaller than the minimum Δ value required to sample the impulse response of the system. The number of samples should be selected to be a power of two for greatest efficiency, $N = 2^p$. Under ideal circumstances, the corresponding time window $T = N\Delta$ will satisfy eq. (3.8.29) in order to avoid leakage. (The exponential windowing technique discussed in the next section can be used if this value of T is deemed to be excessive.) We thereby assemble two data sets, $f_n = f(n\Delta)$ and $s_n = s(n\Delta)$.

We zero-pad this time data in order to be consistent with previous procedures. The force and sensor measurements are input to an FFT routine, from which we obtain the corresponding DFT coefficients F_j and S_j. Recall that β is a proportionality factor converting a physically applied force to the generalized force. Thus, βF_j is the DFT of the generalized force. In the case of a displacement sensor, S_j is the DFT of the displacement, $X_j = S_j$. If an accelerometer is used, then S_j is the DFT of the acceleration, so that $X_j = -S_j/\omega_j^2$. According to eq. (3.8.24), the DFT values of the system's impulse response are given by

$$G_j = \frac{X_j}{\beta F_j} \tag{3.8.37}$$

Let us assume temporarily that the response and force were measured over a time window that is sufficiently large to see the response decay to a very small magnitude. In that case, we may consider the DFT of the response to yield a good approximation of the Fourier transform. If such is the case, then the complex frequency response at the discrete DFT frequencies may be found from Fourier transform theory, as given by eq. (3.8.25),

$$\frac{D(jr_1, \zeta)}{K} \approx \Delta G_j = \Delta \frac{X_j}{\beta F_j}, \quad r_1 = \frac{2\pi}{\omega_{nat} T} \tag{3.8.38}$$

where Δ is the scaling factor relating DFTs and Fourier transforms, as described by eq. (3.8.28). At this stage, we have a sampled set of complex frequency response values. We therefore may identify the natural frequency and damping ratio by the same procedures as those applied to comparable data from a shaker test.

Now suppose that the time window in which the impulse response was measured is too short. For example, for an underdamped system, the envelope of the impulse response is $\exp(-\zeta\omega_{nat}t) = 0.018$ when the elapsed time is four time constants, that is, $t = 4/(\zeta\omega_{nat})$. If T is substantially smaller than $4/(\zeta\omega_{nat})$, we can anticipate that the DFT of the response will not be representative of the response's Fourier transform, so eq. (3.8.38) will not be accurate.

Consequently, we cannot use the short-window measurements to identify the system's natural frequency and damping ratio. Nevertheless, we can use such measurements to predict the manner in which the system will respond to excitations other than a hammer blow. The method for performing such a computation is based on ignoring the fact that eq. (3.8.38) is not accurate! If $f'(t)$ is the alternative excitation, we sample it using the same values of Δ and N that led to the G_j values. We then pad the $f'(t_n)$ data with N zeroes. We denote as F'_n the DFT coefficients corresponding to the excitation of interest. To determine the DFT of the corresponding response, we invoke eq. (3.8.13). Even though eq. (3.8.38) is not accurate, we use it to describe $D(jr_1, \zeta)/K$. Hence, the DFT coefficients of the response resulting from this new excitation are

$$X'_j = \Delta G_j F'_j \tag{3.8.39}$$

We then obtain the response in the time domain by employing the inverse FFT routine to evaluate the time data q'_n corresponding to the X'_j values,

$$\{q'\} = [E_{\text{IDFT}}]\{X'\} \tag{3.8.40}$$

As always when working with DFT convolution, the time data is only valid within the original time window. Hence, computed q_n values for $n > N$, corresponding to zero padding of the original data taken from the hammer test, should be discarded.

An important aspect of eqs. (3.8.39) and (3.8.40) is that the results are accurate, regardless of the value of T. Specifically, it is not necessary that the value of T be sufficiently large to see either the impulse response or the excitation decay to negligible values. The explanation for this unexpected outcome lies in eq. (3.8.24). When T is too small, the G_j values obtained from that equation are the DFT of the impulse response taken over the short time window. Correspondingly, eqs. (3.8.39) and (3.8.40) constitute a discretized approximation of the time domain convolution integral, eq. (3.8.6). Evaluation of that integral for any time $t \leq T$ only requires values of the impulse response and of the excitation in the interval $0 \leq t \leq T$, which matches the data used to implement eqs. (3.8.39) and (3.8.40). This feature will be demonstrated in the next example.

Some individuals are surprised by the success of the impulse response method when the time window fails to meet the usual criteria for evaluating a DFT. The ultimate reason why it works is the causality principle, which states the obvious fact that the response at a current instant cannot be affected by a subsequent excitation. It follows that measurement of the impulse response in any time window provides all the information required to determine the response to other forces during the same time window.

EXAMPLE 3.14

Consider the response of a one-degree-of freedom system to a uniform pulse $Q(t) = F_0$ $[h(t) - h(t - \tau)]$ starting at $t = 0$ in a case where the pulse duration τ is one-half the undamped natural period, that is, $\tau = \pi/\omega_{\text{nat}}$.

(a) Determine the DFT of the excitation. For this analysis use $N = 32$ and 256, with T = 20τ in both cases. Assess the degree to which each satisfies the Nyquist sampling criterion.

(b) Use $T = 20\tau$ and $N = 256$ to evaluate the DFT of the impulse response given by eq. (2.3.23). Perform this evaluation for $\zeta = 0.2$ and $\zeta = 0.001$. Assess the degree to which the first part of eq. (3.8.38) correctly gives the complex frequency response associated with each value of ζ.

(c) Determine the response at the discretized instants according to the following procedures. For each analysis, double T and N from the values in part (b) in order to implement zero padding.

1. Use eqs. (3.8.39) and (3.8.40), with the DFT of the impulse response obtained by sampling eq. (2.3.23).

2. Use the true complex frequency response, instead of the FFT of the impulse response. Specifically, use eq. (3.8.13) to generate the DFT values X_k of the response. Assess the correctness of each result in comparison to the analytical solution.

(d) Repeat the analysis in part (c) for the case where $\zeta = 0.001$.

Solution This exercise will illustrate the procedure for using impulse response measurements to predict system response. It also will demonstrate the accuracy of the method, and the need to distinguish between the complex frequency response and a short time window FFT of the impulse response. The formulation uses the impulse response of a one-degree-of-freedom system, which we know, to generate artificial experimental data. We then proceed as though this data were actually obtained by measuring the response of a system. The value of ω_{nat} is not specified, so we set $\omega_{nat} = 1$, which is equivalent to defining a nondimensional time given by $\omega_{nat}t$. For a similar reason, we take the pulse amplitude to be $F_0 = 1$.

The specified window for the FFT is $T = 20\pi$. The sampling intervals for the first part of the analysis are $\Delta = 20\pi/32 = 1.9635$ and $\Delta = 20\pi/256 = 0.24544$, corresponding to $N = 32$ and 256, respectively. For this exercise we shall focus on the application of the relations and definitions in the text, supplemented where appropriate with discussions of MATLAB or Mathcad implementations. We define discrete times and sampled data for the excitation according to

$$t_{n+1} = n\Delta, \quad Q_{n+1} = \begin{cases} 1 \text{ if } t_{n+1} < \tau \\ 0 \text{ if } t_{n+1} > \tau \end{cases}, \quad n = 0, 1 \ldots, N-1$$

where $\tau = \pi$ because we set $\omega_{nat} = 1$. In the case where we use $N = 32$, the above leads to only the first two instants falling in the interval of the pulse, while the first 13 instants give nonzero values when $N = 256$.

We next take the FFT of the excitation. Because we wish to compare DFT values obtained from different N, we multiply each set by Δ, which is the factor of proportionality in eq. (3.8.28). (Mathcad users need to multiply values obtained from $FFT(Q)$ by an additional factor of N to compensate for the $1/N$ factor built into its algorithm.) The time window $T = 20\pi$ for both values of N, so the frequencies corresponding to the DFT values are defined in either case by

$$\omega_{j+1} = j\frac{2\pi}{T} = 0.1j, \quad j = 0, 1, \ldots, \frac{N}{2}$$

A plot of the scaled DFT values against the corresponding ω_j shows that values in the highest range of j are substantial for $N = 32$, which means that this number of samples fails to satisfy the Nyquist sampling criterion. In contrast, the DFT values become small at the highest n when $N = 256$, which means that the excitation has been adequately sampled in that case.

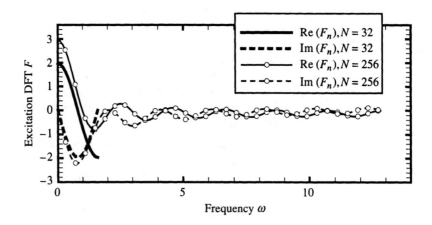

To examine the relation between the DFT of the impulse response and the complex frequency response, we recall eq. (2.3.23) for the impulse response of an underdamped system. When a unit impulse is applied at $\tau = 0$, the response is

$$g(t) = \frac{1}{\omega_d} \exp(-\zeta\omega_{nat}t)\sin(\omega_d t)h(t)$$

Note that we have set the mass coefficient $M = 1$, consistent with setting $\omega_{nat} = 1$. We need to compare the DFT of this function to $D(\omega/\omega_{nat}, \zeta)$. We therefore create a set of time data

$$g_{n+1} = g(t_{n+1}), \quad n = 0, 1, \ldots, N - 1$$

with $N = 256$ and $\zeta = 0.2$ or $\zeta = 0.001$. Application of the FFT process leads to the DFT values $G_{j+1}, j = 0, 1, \ldots, N$. (Users of the Mathcad $FFT(g)$ function will obtain only the first $N/2 + 1$ values of G_n, because complex conjugate mirroring is built into the routine.) According to eq. (3.8.38), the DFT values should be related to the complex frequency response at the discrete frequencies by

$$D\left(\frac{\omega_{j+1}}{\omega_{nat}}, \zeta\right) = D(0.1j, \zeta) = \frac{1}{1 + 2i\zeta(0.1j) - 0.01j^2}$$

$$= \Delta G_{j+1} = \frac{20\pi}{256}G_{j+1}, \quad j = 0, 1, \ldots, N/2$$

(Mathcad users must multiply by N the DFT values obtained by invoking $FFT(g)$ in order to compensate for the difference in definitions of the DFT.)

In the case where $\zeta = 0.20$, there is very close agreement between ΔG_{j+1} and $D(0.1j, \zeta)$, as shown in the first of the following plots. In contrast, there is very little agreement in the case of very light damping, $\zeta = 0.001$. The latter resonance has a very high quality factor, and the DFT vastly underpredicts the peak.

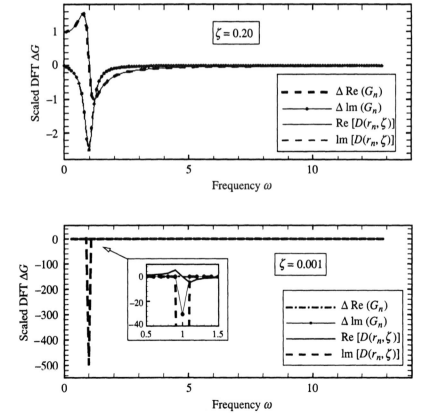

The success of the highly damped case and the failure of the lightly damped one are both readily explained by considering the decay of the impulse response. Recall that in order for

ΔG_{j+1} to match the complex frequency response, it is necessary that the measured $g(t)$ values decay to a very small value. Specifically, eq. (3.8.29) calls for $T = 8/(\zeta \omega_{nat})$ as the minimum window length after zero padding. This requires $T > 40$ in the case where $\zeta = 0.2$, and $T > 4000$ in the case where $\zeta = 0.001$. The time window used for the computations, $T = 20\pi$, exceeds the first required value, but it is much too short for the second.

The next task is to implement convolution operations leading to the response. In the first method, we need DFT values for the impulse response, as well as the excitation. The specified values of T and Δ correspond to $N = 256$. Thus, we create time data that consists of $N/2$ samples of $Q(t)$ and $g(t)$ followed by $N/2$ zeroes:

$$Q_{n+1} = Q(t_{n+1}), \quad g_{n+1} = g(t_{n+1}), \quad n = 0, 1, \ldots, 127$$

$$Q_{n+1} = g_{n+1} = 0, \quad n = 128, \ldots, 255$$

Let us denote the DFT values associated with this data as F'_n and G'_n. In accord with eq. (3.8.39), the DFT values of the response are given by

$$X_{j+1} = \Delta G'_{j+1} F'_{j+1}, \quad j = 0, 1, \ldots, 128$$

$$X_{j+1} = X^*_{255-j}, \quad j = 129, \ldots, 255$$

(Mathcad users may omit the complex conjugate mirror image operation. However, they need to multiply the product by N, essentially because $\Delta G'_n$ approximates a Fourier transform, and Mathcad's *FFT* algorithm divides the text definition of a DFT by N.) The last step in the evaluation is to apply the inverse FFT to determine the response. In order to avoid confusion with the response values resulting from the second method, let us denote these values as $(x_G)_n$, with $n = 0, 1, \ldots, 255$. Because the last half of the inverse transform data might be polluted by wraparound error, we discard the values of $(x_G)_n$ corresponding to $n \geq 128$.

The second specified method is intended to examine the error introduced when ΔG_n values obtained with a DFT are replaced by the complex frequency response $D(\omega_n/\omega_{nat}, \zeta)/K$. For the present situation, we set $\omega_n = 0.1n$ and $K = \omega_{nat} = 1$, as we did previously. We replace ΔG_n by $D(0.1n, \zeta)$ only for the first half of the data set, because the second half is the complex conjugate mirror image of the lower half, as described by eq. (3.7.33). Hence, the DFT values of the response are given by

$$X_n = \frac{1}{1 + 2i\zeta(0.1n) - (0.1n)^2} F_n, \quad n = 0, 1, \ldots, 128$$

$$X_n = X^*_{N'-n}, \quad n = 129, \ldots, 255$$

(The inverse routine *IFFT*(X) accompanying the implementation of *FFT*(Q) in Mathcad returns N real values corresponding to an input of $N/2 + 1$ complex values, so the last part of the specification may be skipped in that implementation.) Let us denote as $(x_D)_n$ the N inverse DFT values resulting from an IFFT routine. As we did in the evaluation of $(x_G)_n$, we discard the last half of this data set.

An analytical solution for the response to a uniform pulse excitation is given by the combination of eqs. (2.3.18) and (2.3.19). In view of the closeness of ΔG_n and the complex frequency response when $\zeta = 0.20$, it is reasonable to expect that either DFT formulation will reproduce the analytical solution. This expectation is confirmed by the first of the following graphs. In contrast, in the lightly damped case, the short time window DFT of the impulse response gives an erroneous prediction of the complex frequency response. Clearly, this will lead to different response data. The last graph (top of page 201) shows the closeness between the x_G data and the analytical solution.

As we noted earlier, the short-window impulse response method succeeds because of the causality principle. Nevertheless, obtaining a correct result requires care in the execution of the procedure, especially in being sure to use the same time window and sampling interval for the excitation as those used to obtain the DFT of the impulse response. Zero padding is another key element of the procedure. In fact, the reader is encouraged to repeat the evalua-

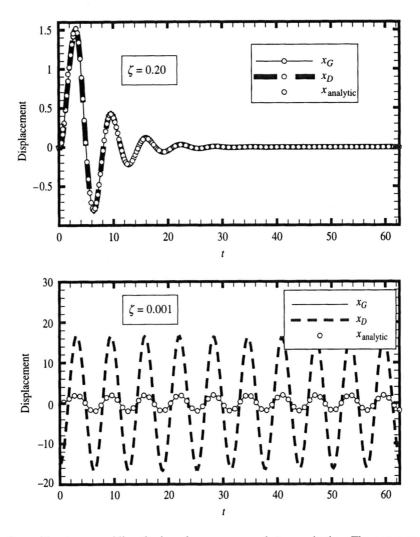

tion of x_G without zero padding the impulse response and step excitation. The consequences of this omission are the subject of Exercise 3.62.

Leakage and Windowing Functions Some data acquisition systems are limited in the amount of data N that can be stored. Setting the sampling interval Δ to a value that satisfies the Nyquist sampling criterion might lead to a situation where the time window $T = N\Delta$ is not sufficient to see the signal decay toward zero. In the same vein, the maximum observation T might be limited by environmental conditions, such as other disturbances. If T is not sufficiently large, processing the data will result in leakage.

To understand why leakage is used to describe this error, consider sinusoidal signals, $x = \sin(\omega t)$, corresponding to several ω. The window T and sampling interval Δ used to obtain each signal are fixed. Figure 3.26 depicts the DFT coefficients X_n obtained for three such signals, $\omega_a = 1.00$ rad/s, $\omega_b = 0.95$ rad/s, $\omega_c = 0.90$ rad/s. The DFT values are plotted against the corresponding discrete frequency, $2\pi n/T$. (The data is scaled to be actual Fourier series coefficients; see eq. (3.7.36). Also, for the sake of clarity, only the lower portion of the range of DFT frequencies is plotted.) The DFT for each signal is computed with a window $T = 20\pi$. In cases a and c, the value of T is an integer multiple of the period of the signal, while in the second case T falls at a

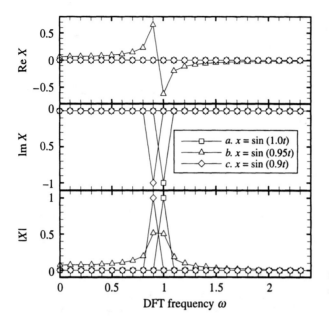

FIGURE 3.26 DFT of sinusoidal functions, $T = 20\pi$, $N = 128$.

half-period of the signal. The DFT in cases a and c correctly identifies that the signal consists of a single sinusoidal signal (Re $X = 0$, Im $X = -1$) at the appropriate frequency. This is to be expected, because replicating the signal periodically outside the time window, as is implied by the DFT algorithm, produces a continuous sinusoidal signal. In case b, replicating the signal periodically outside the window does not produce a sine function. As a result the FFT of this signal is incorrect. The DFT of this data is distributed into a set of discrete frequencies surrounding ω_b. In other words, it has "leaked" into frequencies adjacent to the correct one. In addition, the phase angles of the DFT values are wrong.

Fourier transform theory provides the explanation for the error. The data for the DFT may be considered to be the product of the true sine function and a square windowing function:

$$\text{Square window: } w_{sq}(t) = H(t) - H(t - T) = \begin{cases} 1, & 0 < t < T \\ 0, & t < 0 \text{ or } t > T \end{cases} \quad (3.8.41)$$

The DFT approximates the Fourier transform of the product $x(t)w_{sq}(t)$. The Fourier transform of this product in the time domain is a convolution in the frequency domain of the transforms of $x(t)$ and $w_{sq}(t)$. Such convolution in the case of the $x_b(t)$ data has the effect of shifting the DFT of x_b to the wrong frequencies.

Leakage error can be reduced by using a different windowing function. One that is often used is the Hanning window,

$$\text{Hanning window: } w_{\text{Hann}}(t) = \left[\sin\left(\frac{\pi t}{T}\right) \right]^2 [H(t) - H(t - T)] \quad (3.8.42)$$

This and other windows are discussed by Press et al. (1992). Multiplication by $w_{\text{Hann}}(t)$ tapers $x(t)$ near $t = 0$ and $t = T$. Figure 3.27 displays the results of applying the Hanning window to the DFT of each $x(t)$. Processing $x_b(t)$ results in far less leakage than in Figure 3.27, as indicated by the fact that $|X_b|$ is much more tightly concentrated around

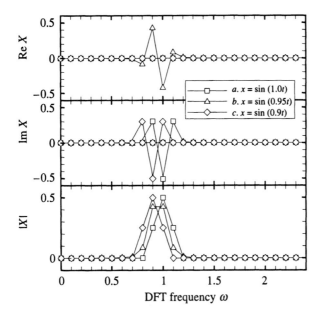

FIGURE 3.27 DFT of sinusoidal functions with a Hanning window, $T = 20\pi$, $N = 128$.

$\omega = 0.95$ rad/s. However, the windowing function causes leakage in $X_a(\omega)$ and $X_c(\omega)$, when there is none without windowing. In addition, none of the amplitudes and phase angles for DFT data are correct now.

An application where windowing can be useful concerns efforts to identify the spectral power distribution in a measured vibratory signal. Recall that for a harmonic function $x(t) = \text{Re } [X \exp (i\omega t)]$, the energy dissipation per unit cycle is $E_{\text{dis}} = \pi C \omega |X|^2$. By reducing leakage, windowing enables one to more accurately identify the frequency ranges in which power is being dissipated. The merits of various window functions are discussed in a report by Gade and Herlufsen (1987).

A primary use of the DFT in vibrations is to identify natural frequencies and critical damping ratios from measured impulse response data. If a system is very lightly damped, a long time interval will be required to see the response decay to small values. If this interval exceeds the largest possible measurement window T, leakage will occur. As we saw in Example 3.14, the consequence will be an erroneous transfer function $G(\omega)$. A remedy for this is to use an exponential window,

$$\text{Exponential window: } w_{\text{exp}}(t) = \exp\left(-\beta\frac{t}{T}\right)h(t) \qquad (3.8.43)$$

The coefficient β is an adjustable nondimensional parameter whose selection we will soon discuss.

The reason the exponential window can be successful becomes obvious when we consider the impulse response of a system, windowed with $w_{\text{exp}}(t)$,

$$g't \equiv g(t)w_{\text{exp}}(t) = \frac{1}{M\omega_{\text{d}}}\exp\left[-\left(\zeta + \frac{\beta}{\omega_{\text{nat}}T}\right)\omega_{\text{nat}}t\right]\sin(\omega_{\text{d}}t), \quad t > 0 \qquad (3.8.44)$$

This shows that $\beta/\omega_{\text{nat}}T$ plays the role of artificial damping. If we select β such that the exponential factor is small in the vicinity of $t = T$, we will have (altered) response data that meets the criterion for avoiding leakage. DFT processing of the windowed time data leads to a transfer function $G'(\omega)$ corresponding to $g'(t)$. We may use $G'(\omega)$ to identify ω_{nat} and the effective damping ratio, $\zeta' = (\zeta + \beta/\omega_{nat}T)$. For example, because the system is lightly damped (otherwise, we would not be concerned with leakage), we may identify ω_{nat} as the frequency at which $|G'(\omega)|$ peaks. We then may use the half-power points to determine ζ'. The value of β was selected in advance, so we may recover the true damping ratio from

$$\zeta = \zeta' - \frac{\beta}{\omega_{nat}T} \tag{3.8.45}$$

This procedure might seem to be a general way to avoid having leakage affect parameter identification. However, the discussion avoids the issue of measurement error and noise, which will introduce an error in the value of ζ'. Suppose the system is very lightly damped, $\zeta \ll 1$. In that case both ζ' and $\beta/\omega_{nat}T$ will be relatively large values having nearly equal magnitudes. Taking their difference, as required by eq. (3.8.45), leads to a loss of precision that greatly magnifies the error contained in ζ'.

EXAMPLE 3.15

Artificial experimental data is generated by sampling the impulse response for a system having unit natural frequency, $\omega_{nat} = 1$ rad/s, and critical damping ratio $\zeta = 0.01$. The time window is $0 < t < 20\pi/\omega_{nat}$, and $N = 128$ for the data. Compare the natural frequency and critical damping ratio obtained from the half-power method using an exponential window to the parameters obtained when the time data is not windowed.

Solution This step will demonstrate the usefulness of the exponential window function, provided that the decay factor β is properly selected. The analysis is fairly straightforward. We set $T = 20\pi$, and the time instants for sampling are $t_{n+1} = nT/N$. The amplitude of the impulse response does not affect the process of identifying ω_{nat} and ζ, so the time data are computed as

$$x_{n+1} = \exp(-0.01t_{n+1})\sin(\sqrt{0.9999}\,t_{n+1})$$

At $t = T$, the exponential factor is $\exp(-0.2\pi) = 0.533$, so the x_n values will not have decayed to small values in the time window. We take the DFT of this data using the FFT function of our software,

$$\{G\} = [E_{DFT}]\{x\}$$

The associated positive DFT frequencies are $\omega_{j+1} = j(2\pi/T) = 0.1j, j = 0,1, \ldots, N/2$.

A plot of $|G_j|$ versus ω_j exhibits a peak in the vicinity of $\omega = 1$ (top of page 205). We scan the data to evaluate the maximum $|G_j|$ value. In a zoomed graph of the vicinity of the peak, we draw a dashed horizontal line at max $(|G_j|)/\sqrt{2}$, which is the half-power amplitude. This construction is shown in the first row of the graph set. The peak amplitude occurs at $\omega_{11} = 1$ rad/s, so we identify that value as ω_{nat}. This matches the value used to generate the data.

The intersections of the half-power line with straight lines connecting the DFT points are approximately the half-power points. The intersections occur at $\omega = 0.968$ rad/s and $\omega = 1.033$ rad/s, so the bandwidth is $\Delta\omega = 0.065$. The corresponding critical damping ratio is

$$\zeta = \frac{\Delta\omega}{2\omega_{nat}} = 0.033$$

Clearly, this value is far from the value $\zeta = 0.01$ with which the x_n data was computed.

To correct the leakage error resulting from having an inadequately large T, we apply an exponential window. Let us begin the analysis using $\beta = 2$. This gives an exponential function

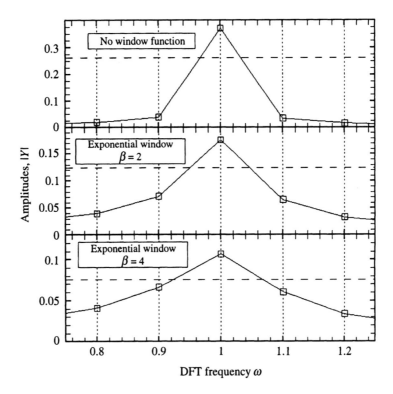

at $t = T$ of $\omega_{exp} = \exp(-2) = 0.13534$, which is not too small. We multiply the x_n data by $\omega_{exp}(t_n)$,

$$y_n = \exp\left(-2\frac{t_n}{T}\right)x_n$$

The DFT of this data is G_n', which also shows a single peak. The graph in the second row zooms in on this peak, and also shows the corresponding half-power line. The peak still appears at $\omega_{11} = 1$ rad/s, so the exponential window has not altered the correct identification of the natural frequency. The intersections of the half-power line with straight lines connecting the DFT values occur at $\omega = 0.950$ rad/s and $\omega = 1.048$ rad/s, which corresponds to a 0.098 rad/s bandwidth. This yields the effective damping constant,

$$\zeta' = \frac{0.098}{2(1)} = 0.049$$

We then find from eq. (3.8.45) that, for $\beta = 2$,

$$\zeta = \zeta' - \frac{2}{(1)(20\pi)} = 0.017$$

This value is much better than the one without the exponential window function.

The identification can be improved by selecting a larger value of β. The graph in the third row shows the DFT data when $\beta = 4$, which corresponds to $\omega_{exp} = 0.0183$ at $t = T$. The peak $|X_n|$ value still occurs at $\omega_{11} = 1$ rad/s, so the identification of the natural frequency is still correct. We follow the preceding analysis to measure the bandwidth, which is $\Delta\omega = 0.144$. From this, we find

$$\zeta = \frac{0.144}{2(1)} - \frac{4}{(1)(20\pi)} = 0.0083$$

This value is quite good compared to the original $\zeta = 0.01$.

It is natural to conjecture whether further increase of β will give a better result. In fact, repetition of the analysis for $\beta = 6$ identifies $\zeta = 0.014$. This is less close to the true ζ value, although it is better than the result obtained from $\beta = 2$. The lesser quality of the answer when $\beta = 6$ is a consequence of overly attenuating the signal. The exponential window is so small when $t \approx T$ that the products $x_n \omega_{\exp}(t_n)$ at the end of the time window cease to be significant. In other words, using an exponential window function that decays too rapidly obliterates significant response data.

Before we leave this exercise, it is appropriate to note that $|Y_n|$ maximized at the natural frequency because T was selected fortuitously. In a realistic situation, it is unlikely that T will be an integer multiple of $2\pi/\omega_{\mathrm{nat}}$. In that case, ω_{nat} will not coincide with one of the DFT frequencies. By virtue of having a comparatively small T, the interval between DFT frequencies will be relatively large. As a result, only a few DFT frequencies will fall in the range of the peak. In such circumstances, it is preferable to identify the bandwidth by following the Nyquist plot construction in Example 3.4.

REFERENCES AND SELECTED READINGS

BRACEWELL, R. N. 1986. *The Fourier Transform and Its Applications.* McGraw-Hill, New York.

BRIGHAM, E. 1974. *The Fast Fourier Transform.* Prentice-Hall, Englewood Cliffs, NJ.

CAUGHEY, T. K., & VIJAYARAGHAVAN, A. 1970. "Free and Forced Oscillations of a Dynamic System with 'Linear Hysteretic Damping' (Non-linear Theory)." *International Journal of Non-Linear Mechanics*, 5, 533–535.

CRAIG, R. R. JR. 1981. *Structural Dynamics.* John Wiley & Sons, New York.

DIMAROGONAS, A. 1996. *Vibration for Engineers*, 2nd ed. Prentice Hall, Upper Saddle River, NJ.

EWINS, D. J. 1984. *Modal Testing: Theory and Practice.* Research Studies Press, Letchworth, England.

FÖPPEL, A. 1895. "Das Problem der Laval'schen Turbinewelle." *Civilingenier*, 41, 332–342.

FORMENTI, D., & MACMILLAN, B. 1999. "The Exponential Window." *Sound and Vibration*, 33, 10–15.

GADE, S., & HERLUFSEN, H. 1987. "Use of Weighting Functions in FFT/DFT Analysis." *Technical Review*, 3, Bruel and Kjaer (pamphlet).

JEFFCOTT, N. 1919. "Lateral Vibration of Loaded Shafts in the Neighborhood of a Whirling Speed—The Effect of Want of Balance." *Philosophical Magazine*, 37, 304–314.

MEIROVITCH, L. 1986. *Elements of Vibration Analysis*, 2nd ed. McGraw-Hill, New York.

NASHIF, A. D., JONES, D. I. G., & HENDERSON, J. P. 1985. *Vibration Damping.* John Wiley & Sons, New York.

NEWLAND, D. E. 1985. *An Introduction to Random Vibrations and Spectral Analysis.* John Wiley & Sons, New York.

NEWLAND, D. E. 1989. *Mechanical Vibration Analysis and Computation.* Longman Scientific and Technical, Harlow, England.

OPPENHEIM, A. V., & SHAFER, R. W. 1975. *Digital Signal Processing.* Prentice Hall, Englewood Cliffs, NJ.

PAPOULIS, A. 1962. *The Fourier Integral and its Applications.* McGraw-Hill, New York.

PRESS W. H. TEUKOLSKY, S. A., VETTERLING, W. T., & FLANNERY, B. P. 1992. *Numerical Recipes*, 2nd ed. Cambridge University Press, Cambridge, England.

STEIDEL, R. F., JR., 1989. *An Introduction to Mechanical Vibrations*, 3rd ed. John Wiley & Sons, New York.

TONGUE, B. H. 1996. *Principles of Vibrations.* Oxford University Press, New York.

VERNON, J. B. 1967. *Linear Vibration Theory*, John Wiley & Sons, New York.

WEAVER, W., TIMOSHENKO, S., & YOUNG, D. H. 1990. *Vibration Problems in Engineering*, 5th ed. John Wiley & Sons, New York.

EXERCISES

3.1 A system has mass $M = 20$ kg and natural frequency $\omega_{\mathrm{nat}} = 100$ rad/s. It is observed that the steady-state response is $q = 20 \cos (110t - 1.5)$ mm, where t has units of seconds. Determine the harmonic excitation causing this response for $\zeta = 0$ and $\zeta = 0.4$.

3.2 A radar display is to be tested by mounting it on a spring-dashpot suspension and subjecting it to a harmonic force, $Q = F\cos (\omega t)$. The mounted mass is 8 kg and the critical damping ratio is known to be $\zeta = 0.25$. A free vibration measurement shows that the damped natural frequency of the spring-mass-dashpot system is 5 Hz. It also is observed that when the force is applied at a very low frequency, the displacement amplitude is 2 mm. The test is to be performed at 5.2 Hz. What will be the steady-state response?

3.3 The signal lights for a rail line may be modeled as an 80 kg mass mounted 3 m above the ground on an elastic post. The natural frequency of the system is observed to be 12.2 Hz. Wind buffet generates a horizontal harmonic force F at 12 Hz. The light filaments will break if their peak acceleration exceeds 15g. Determine the maximum acceptable force amplitude $|F|$ when $\zeta = 0$ and $\zeta = 0.05$.

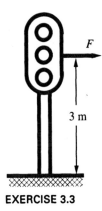

EXERCISE 3.3

3.4 A system has a stiffness of 5 kN/m and a mass of 2 kg. When a force of 20 $\sin(\omega t)$ N is applied to the system, the response is $q = -10 \cos(\omega t)$ mm. Determine ω and ζ.

3.5 A one-degree-of-freedom system whose mass is 10 kg and whose natural frequency is 1 kHz is subjected to a harmonic excitation 1.20 $\sin(\omega t)$ kN. The steady-state amplitude when $\omega = 1$ kHz is observed to be 2.4 mm. Determine the steady-state response at $\omega = 0.95$ kHz and $\omega = 1.05$ kHz.

3.6 An excitation 60 $\sin(\omega t) + 80 \cos(\omega t)$ N produces a steady-state response whose quasi-static amplitude is 50 mm. The natural frequency of the system is 20 Hz, and $\omega = 18$ Hz produces a

steady-state response that is in phase with a sine response. Determine the amplitude of the displacement when $\omega = 18$ Hz.

3.7 A one-degree-of-freedom system is subjected to a harmonic excitation $Q(t) = 400\cos(\omega t)$ N. Measurement of the response at several frequencies has led to the following observations regarding the steady-state response:

- When the excitation frequency is 40 Hz, the velocity $\dot{q}$ is always in phase with the force, so that $\dot{q} = \beta Q$, where β is an unknown constant.
- The amplitude of the vibration is 16 mm when the frequency is 25 Hz.
- At a frequency of 42 Hz, an oscilloscope trace of the manner in which q varies with time is as shown below. In this graph, the vertical scale is not known.

Based on these observations, determine the following:

(a) The natural frequency of the system.

(b) The ratio of critical damping.

(c) The inertia coefficient M, stiffness coefficient K, and damping constant C.

(d) The maximum value of q at 42 Hz.

(e) The steady-state velocity $\dot{q}$ as a function of t when the frequency is 40 Hz.

3.8 The right end of the spring (top of page 208) executes a steady-state displacement $y = Y \sin(\omega t)$, where Y and ω are known values.

(a) For the steady-state condition, derive an expression for the displacement $x(t)$ of block m and the force $F(t)$ causing the motion.

(b) For the case where ω is 10% greater than the natural frequency of the system and the damping is 20% of critical, derive expressions for the amplitude of x, and for the phase lag of $x(t)$ relative to $y(t)$.

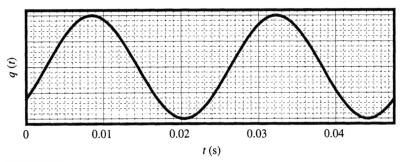

EXERCISE 3.7

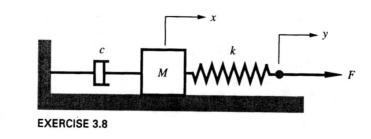

EXERCISE 3.8

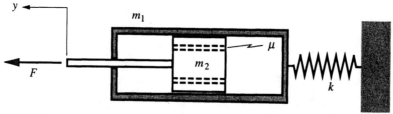

EXERCISE 3.9

3.9 The shock absorber consists of an outer tube of mass m_1 that is restrained by spring k and an inner piston of mass m_2. Orifices in the piston permit passage of a viscous fluid that fills the tube; the coefficient of damping between the piston and the tube is μ. An unknown force $F(t)$ is applied to the piston, with the result that the absolute displacement of the piston is $y = A\sin(\omega t)$. The parameters of the system are: $m_1 = 0.5$ kg, $m_2 = 1.0$ kg, $k = 3.2$ kN/m, $\mu = 40$ N-s/m, and $A = 20$ mm. Determine the amplitude and phase of the force $F(t)$ relative to the displacement $y(t)$ when $\omega = 75$ rad/s and $\omega = 85$ rad/s.

3.10 Consider the amplitude-frequency diagram in Figure 3.4. The moment of inertia for the system is 4 kg-m². Estimate the natural frequency and ratio of critical damping of the system. Then use those estimates to determine the torsional spring stiffness K, torsional damping constant C, and torsional moment amplitude $|\Gamma|$ associated with that diagram.

3.11 Measurement of the steady-state response of a one-degree-of-freedom system to a harmonic excitation $F\cos(\omega t)$ indicates that at a frequency of 100 Hz, the response is $x = 4\sin(\omega t)$. It also is observed that 105 Hz is a half-power point.

(a) Determine the phase lag of the response relative to the excitation at 105 Hz.

(b) Determine the amplitude and phase lag of the response at 110 Hz.

3.12 A 50 kg block is supported by a lightweight elastic beam. The static downward displacement of the beam when the block is placed on the beam is 4 mm. A force $F_0\sin(\omega t)$ applied at $\omega \approx 0$

results in a steady-state displacement whose amplitude is 20 mm, while applying this force at the natural frequency leads to an amplitude of 400 mm.

(a) Determine as a function of ω the maximum power dissipation, averaged over a cycle, that will be observed if the value of F_0 is held constant as the frequency is changed.

(b) Determine the bandwidth of the resonance.

(c) Determine the power dissipation, averaged over a cycle, when $\omega = 0.9\,\omega_{nat}$.

3.13 A 40 kg mass is attached to a lightweight elastic beam, which is observed to cause the beam to sag by 5 mm at the attachment point. A harmonically varying force whose amplitude is 240N is applied to the mass, and the steady-state response is measured as a function of frequency. The maximum amplitude is observed to be 60 mm.

(a) Based on the assumption that dissipation is caused by structural damping, determine the loss factor and the amplitude of the vibration at 9 Hz.

(b) What would be the amplitude of the vibration at 9 Hz if dissipation were due to viscous damping?

3.14 A 1 kW power source is used to excite a 0.5 kg system over a frequency range from 100 Hz to 1 kHz. It is observed that the steady-state amplitude across this range is constant at 8 mm. It also is noted that at 200 Hz the response lags the excitation by 90°.

(a) How does the equivalent loss factor for this source depend on frequency?

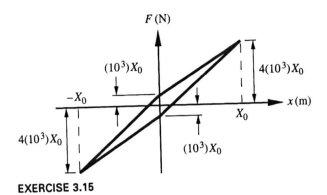

EXERCISE 3.15

(b) Determine the amplitude of the force generated by the power source at 100 Hz, 200 Hz, and 1 kHz.

(c) Determine the frequency at which the force amplitude is smallest.

3.15 A shock-absorbing suspension supports a 10 kg mass. A testing machine induces a harmonic displacement of the mass at frequency ω and amplitude X_0. Measurement of the force exerted by the support as a function of the displacement provides the force deflection curve shown above, regardless of the value of ω.

(a) What are the values of stiffness k and damping constant c for an equivalent linear spring-dashpot suspension system?

(b) Consider a situation where a force $400 \sin(50t) + 300 \cos(50t)$ N is applied to the mass. Use the equivalent linear suspension properties resulting from the analysis in part (a) to determine the corresponding steady state response $x(t)$.

3.16 A block, whose mass is m, is suspended from the wall by a lightweight cantilevered beam that is fabricated of a polymeric composite. When the strain field in the beam varies harmonically, the Young's modulus E is a function of frequency because of the viscoelastic behavior of the material. This dependence fits the standard solid model, for which

$$E(\omega) = E_0 + (E_\infty - E_0)\frac{\omega^2\tau^2}{\omega^2\tau^2 + 1}\left(1 + \frac{i}{\omega\tau}\right)$$

In this expression E_0 is the static Young's modulus, corresponding to very low values of the frequency ω; E_∞ is the Young's modulus for very large ω; and $\tau > 0$ is the relaxation time. It is observed that when the block was first attached to the beam and released gently, so that E_0 was the effective modulus, the downward displacement of the end of the beam was δ. Thus, the stiffness at nonzero frequencies is the static stiffness multiplied by $E(\omega)/E_0$,

$$K(\omega) = \frac{mg}{\delta}\frac{E(\omega)}{E_0}$$

It is possible by adjusting the polymer mix to fabricate materials having different values of E_0, E_∞, and τ. Consider a situation where E_0 and E_∞ are fixed, with $E_\infty = 1.5E_0$. It is desired to identify the value of τ that will best reduce the amplitude of the block's displacement when the force acts at the elastic natural frequency,

$$[K(0)/m]^{1/2} \equiv (g/\delta)^{1/2}.$$

(a) Show that the complex amplitude of the block's steady-state response to an excitation $\mathrm{Re}[F \exp(i\omega t)]$ may be written as

$$X = \frac{F\delta}{mg}\frac{1}{\alpha(r, \omega_{nat}\tau) - r^2}$$

where $r = \omega/\omega_{nat}$, $\omega_{nat}^2 = g/\delta$, and $\alpha(r, \omega_{nat}\tau) = E(\omega)/E_0$.

(b) Consider frequencies ranging over $0.75\,\omega_{nat} < \omega < 1.25\omega_{nat}$. Evaluate and graph $|X|$ as a function of r for cases where $\omega_{nat}\tau = 0.01$, 0.5, 1.0, 2.0, and 10.0. Identify the maximum in each case. Which case seems to be best for reducing the vibration? Also estimate the quality factor of the resonance in each case.

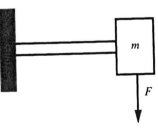

EXERCISE 3.16

3.17 The air resistance in turbulent flow is $f = \beta|\dot{q}|^2$. Consider a spring-mass system that slides over a smooth horizontal surface driven by a harmonic force $F\cos(\omega t)$ in the presence of air resistance. If β is sufficiently small the response will essentially be harmonic, $q = X\cos(\omega t - \phi)$.

(a) Derive an expression for the energy dissipated by friction during one cycle of steady-state vibration by considering the work done by the friction force. Use that expression to identify the equivalent linear dashpot constant and structural damping loss factor.

(b) Use the equivalent dashpot constant to derive an equation whose solution would give X at a specified ω.

3.18 A one-degree-of-freedom system, in which damping is negligible, is initially at rest in its static equilibrium position. At $t = 0$ a resonant excitation $Q = F\cos(\omega_{nat}t)$ is applied. The mass of the system is 100 kg, and the static displacement due to the system's own weight is 5 mm. Also, static application of force F would result in a static displacement of 0.4 mm. The springs supporting the system will fail if the displacement from the equilibrium position attains the value of 20 mm.

(a) Determine the response as a function of time. From that solution, determine the maximum time duration over which the resonant excitation may be allowed to continue without causing the springs to break.

(b) Would the result of part (a) change if the excitation were $Q = F\sin(\omega_{nat}t)$? Explain why.

3.19 The total mass of a motor is 80 kg. It is observed that gently placing the motor on a beam produces a static downward displacement of 40 mm. When the motor rotates at an angular speed of 145 rev/min, the steady-state amplitude of the beam is 10 mm. It also is observed that at a rotation rate of 145 rev/min, the radial line to the center of mass of each rotor is 75° above horizontal when the beam is at its static reference position (where $q = 0$).

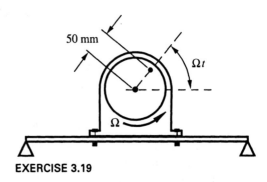

EXERCISE 3.19

(a) Determine the damping ratio ζ for the system.

(b) Determine the imbalance εm.

(c) Determine the smallest possible amplitude of vibration of the beam if the motor turns at a rate that is much larger than the natural frequency of the system.

3.20 The sketch depicts a motor mounted on a cantilevered beam that is coated with a polymeric damping material. The mass of the nonrotating parts of the motor is 75 kg, and the mass of the rotating parts is 5 kg. Point G is the center of mass of the rotor. The mass of the beam is sufficiently small to consider its effect to be equivalent to a spring and dashpot supporting the motor. It is observed that when the motor is off, applying a static 1000 N force vertically to the shaft O causes the beam to deflect by 20 mm. This force is removed, after which the motor is turned on and the system is brought to the

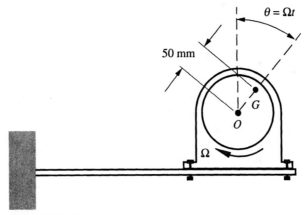

EXERCISE 3.20

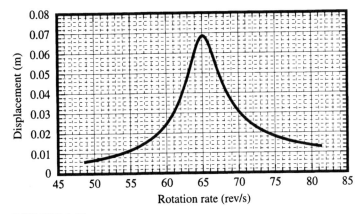

EXERCISE 3.21

steady-state condition corresponding to a rotation rate of 260 rev/min. It is observed that the steady-state response at this frequency is such that at the instant when the beam reaches its highest upward position, the angle between vertical and the radial line OG is $\theta = 120°$. From this information, determine the damping ratio ζ, the amplitude of the displacement of point O when the motor rotates at 260 rev/min, and the largest possible displacement amplitude of point O, and the frequency at which it would occur. (The last answer may be approximate.)

3.21 The graph shows the rotation rate dependence of the steady-state amplitude of a one-degree-of-freedom system supporting a motor. The generalized mass of the system is $M = 500$ kg. From this data, deduce the natural frequency, the bandwidth of the resonance, the critical damping ratio, the imbalance εm, and the smallest possible amplitude when the motor rotates above 80 rev/s.

3.22 The diagram is a schematic top view of a washing machine, whose motion in the x direction is to be studied. In the spin-dry cycle, the round tub of radius R and mass m_1 rotates at a constant angular speed Ω, and the clothes may be considered to be folded into a ball of mass m_2 whose center of mass is situated at distance r from the rotation axis of the drum. The mass of the framework supporting the drum bearings is m_3, and snubbers permit movement of the framework in the x direction only. The undamped natural frequency is 5 Hz, and the critical damping ratio is 0.1.

(a) Find the equation of motion governing the displacement x.

(b) Determine the smallest steady-state amplitude that can be obtained if the system is operated above resonance.

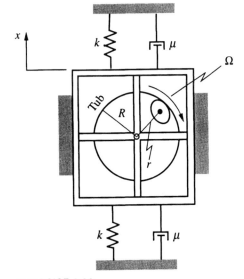

EXERCISE 3.22

(c) Determine the largest amplitude of x that can occur, and the rotation rate Ω at which it will occur.

3.23 A counter-rotating eccentric mass exciter is attached to a block, which is supported by a light-weight beam. Stroboscopic measurement at an angular speed $\omega = 900$ rev/min indicates that the block passes its static equilibrium position with an upward velocity at the instant when the eccentric masses are at their highest position. The amplitude of vertical displacement at this speed is 8.5 mm. The total mass of the system is $m = 200$ kg and the rotating imbalance of each rotor is 0.5 kg-m. Determine

(a) The natural frequency of the system.

(b) The damping constant c.

(c) For the case where $\Omega = 1000$ rev/min, the amplitude of the vertical displacement of the mass and the angular position of the eccentric masses when the block passes its equilibrium position.

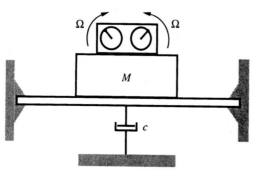

EXERCISES 3.23, 3.24

3.24 A counter-rotating eccentric mass exciter is attached to a block, which is supported by a lightweight beam. The total mass of the block and the motor is 80 kg. It is observed that gently placing the block and motor on the beam produces a static downward displacement of 40 mm. When the motor rotates at an angular speed of 145 rev/min, the steady-state amplitude of the platform is 10 mm. It also is observed that at a rotation rate of

145 rev/min, the radial line to the center of mass of each rotor is 75° above horizontal when the beam is at its static reference position (where $q = 0$).

(a) Determine the damping ratio ζ for the system.

(b) Determine the imbalance εm.

(c) Determine the smallest possible amplitude of vibration of the beam if the motor turns at a rate that is much larger than the natural frequency of the system.

3.25 A pair of unbalanced disks are mounted on a rigid bar such that, at an arbitrary instant t, their centers of mass have the radial orientations shown in the sketch. Each disk has mass m, and the eccentricity of its center of mass is ε. The moment of inertia about the pivot point O of the supporting bar and the electric motors driving the disks is I_1, and each disk is sufficiently small to consider it to have negligible moment of inertia about its shaft. Derive the equation of motion for this system. Then express the steady-state rotation amplitude of the bar as a function of the frequency ratio $r = \omega/\omega_{\text{nat}}$.

3.26 Two counter-rotating motors are mounted on a lightweight, rigid framework. The nonrotating mass of each motor is 50 kg, and the flywheel

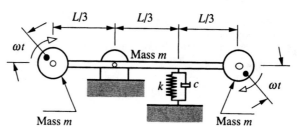

EXERCISE 3.25

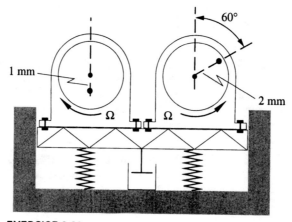

EXERCISE 3.26

masses are 2 kg for the left motor, and 3 kg for the right motor. The centers of mass are 1 mm (left flywheel) and 2 mm (right flywheel) from the respective shafts. Also, the radial line to the center of mass of the right flywheel is 60° right of vertical when the left flywheel is at its lowest position. The natural frequency of the system is 20 Hz, and the ratio of critical damping is 0.2. For the case where the flywheels rotate at 25 Hz, determine

(a) The amplitude of the vibration of the framework.

(b) The angular position of the center of mass of the left flywheel at the instant when the framework is moving upward at its static equilibrium position.

3.27 Consider the system in Example 3.8, when $a = 250$ mm, $m = 5$ kg, $\kappa = 3000$ N/m, and $\varepsilon m_1 = \varepsilon m_2 = 0.02$ kg-m.

(a) Suppose the disks rotate at 12 rad/s and $\zeta = 0.01$. What is the steady-state response of θ? Give its amplitude and phase lag relative to a sine function.

(b) Determine the minimum value of μ that will assure that the rotational amplitude of the bar in a steady-state motion does not exceed 0.10 rad, regardless of the value of ω.

3.28 Observation of the radial distance to the point where a 50 kg rotor is mounted on an elastic shaft shows that at very high rotation rates, the distance is 1 mm, while the distance is 25 mm when $\omega = 800$ rev/min.

(a) Determine the critical damping ratio.

(b) Determine the stiffness coefficient.

(c) Determine the radial distance when $\omega = 600$ rev/min.

3.29 Consider the Jeffcott model when the eccentricity is $\varepsilon = 2$ mm and the critical damping ratio is $\zeta = 0.05$. Angular speeds of interest are $\omega = 0.5\omega_{cr}$, $\omega = \omega_{cr}$, and $\omega = 2\omega_{cr}$. For each case determine the distance from the attachment point to point O on the bearing axis. Also draw a sketch showing the plane in which the shaft is bent and the position of the mass center relative to that plane.

3.30 A 100 kg rotor must rotate at a speed of 6000 rev/min. The distance between bearings is $L = 800$ mm, and the shaft is composed of steel. The stiffness of a circular shaft for the particular type of bearings is $k = 192EI/L^3$, where E is Young's modulus and $I = (\pi/4)R^4$ is the area moment of inertia. What is the minimum radius R of the shaft for which ω_{cr} will exceed the operational speed by at least 10%?

3.31 It is desired to design a circular steel shaft to support a 50 kg rotor. The span is $L = 400$ mm, and the ends are supported by ball bearings. As a result the shaft is effectively a simply supported beam, for which the stiffness coefficient relating force and displacement at the rotor is $k = 48EI/L^3$, where $I = \pi R^4/4$. Design criteria are that the critical speed should be 6000 rev/min, the radial distance from the bearing axis to the rotor attachment point at the critical speed should not exceed 5 mm, and the radial distance at 5900 rev/min should be 70% of the distance at the critical speed. What is the cross-sectional radius R and the maximum value of ε that will meet these criteria?

3.32 Application of a harmonic force at a very low frequency is observed to produce a steady-state amplitude δ_0. Application of the same force amplitude at a frequency that matches the system's natural frequency produces a displacement amplitude that is $10\delta_0$. Determine the range of frequency ratios for which the transmissibility will satisfy tr < 0.8.

3.33 Measurement of the force transmitted to the ground by a damped one-degree-of-freedom system indicates that at a certain frequency ratio ω/ω_{nat}, tr $= 1.75$. It also is observed that at this frequency the steady-state amplitude is 50% larger than the amplitude when the frequency is very low. Determine the value of ω/ω_{nat} and ζ.

3.34 Suppose $|D(r, \zeta)|$ and tr are known values. Derive expressions for the corresponding values of r and ζ. From those expressions determine the range of values of tr that can be obtained from a specified value of $|D(r, \zeta)|$.

3.35 Consider the model in Figure 3.11, which depicts a platform and suspension system supporting a motor rotating at angular speed ω. Derive an expression for the magnitude of the force transmitted to the ground as a function of the ratio ω/ω_{nat}. Graph the result for critical damping ratios of $\zeta = 0, 0.05$, and 0.50. On the basis of these graphs, what guidelines should be followed for selecting ω_{nat} and ζ so as to minimize the force transmitted to the ground when the rotation rate is a specified constant value?

3.36 Consider the automobile model in Example 3.2. The system parameters are as stated there, with $\zeta = 0.50$. Derive an expression for the time dependence of the total force (static plus dynamic) exerted between the wheels and the ground at a specified speed v. Based on this result determine the maximum value of v for which a wheel will remain in contact with the ground.

3.37 A harmonic force whose amplitude is 1000 N must be applied to a block whose mass is 12 kg. It is desired to design a suspension system for the block consisting of a parallel spring k and dashpot c. The design must be such that the transmissibility is tr = 0.50 when the excitation frequency is 60 Hz. In addition, the steady-state amplitude of the response should be as small as possible when the excitation frequency matches the system's natural frequency. Determine the values of k and c that meet these objectives. (*Hint*: Let σ denote the frequency ratio corresponding to a 60 Hz excitation. It is possible to solve algebraically for the value of ζ as a function of σ for which tr = 0.50. Furthermore, if the excitation frequency and the system mass are fixed, then k will depend only on σ. Considering k and ζ to be functions of σ enables one to construct a graph of the steady-state amplitude at resonance as a function of σ. Such a graph will lead to recognition of the optimal conditions.)

3.38 The Fourier series for a periodic array of impulses is described by eqs. (3.7.11). Evaluate and graph as a function of nondimensional time t/T the truncated series $S(t, N)$ over the interval $-T \le t < T$. Perform these operations for $N = 5$, 10, and 20. What does a comparison of the results suggest regarding convergence of the series?

3.39 The forcing function depicted below is a periodic sequence of reversed ramps. Derive the complex Fourier series representing this force.

3.40 An undamped system is subjected to a periodic excitation consisting of the positive lobes of a sine function, that is, $Q(t) = F\sin(\Omega t)$ if $0 \le t < \pi/\Omega$, $Q(t) = 0$ if $\pi/\Omega \le t < 2\pi/\Omega$, $Q(t \pm 2\pi/\Omega) = Q(t)$. Determine the complex Fourier series representing this force. What is the mean value of this excitation? At what harmonic is the amplitude less than 1% of this mean value?

3.41 A periodic disturbance consists of a sequence of exponentially decaying pulses repeated at interval T, such that $Q(t) = F\exp(-\lambda t/T)$ for $0 < t < T$, $Q(t \pm T) = Q(t)$. The parameter λ is nondimensional. Determine the complex Fourier series representing this force. Evaluate the first five series coefficients when $\lambda = 0.1$, 1, and 10. What does

this reveal regarding the influence of λ on the frequency spectrum?

3.42 A sequence of small bumps in a road may be modeled as a periodic set of parabolic bumps having height h and width w, spaced at intervals L, such that

$$z(x) = \begin{cases} h(1 - x^2/w^2) \text{ if } -w \le x \le w \\ 0 \text{ if } -L/2 \le x < -w \text{ or } w \le x < L \\ z(x \pm L) \text{ if } |x| > w \end{cases}$$

Derive the coefficients of the complex Fourier series representing $z(x)$. Evaluate these coefficients for the case where $w = L/50$. For what range of harmonics, if any, do these coefficients resemble those for a periodic sequence of impulses, as described by eqs. (3.7.11)?

3.43 An undamped system is subjected to the periodic excitation in Exercise 3.40. The value of Ω is not an integer multiple of the natural frequency. Derive an analytical expression for the steady-state response of the system.

3.44 The system shown below is a simple model for the vertical motion of an automobile as it travels along a road. The excitation when the vehicle travels over a periodic strip of bumps

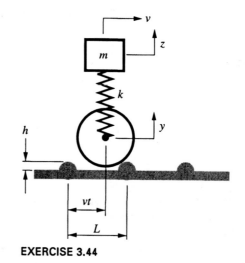

EXERCISE 3.44

EXERCISE 3.39

spaced at distance L may be represented as a periodic sequence of Dirac delta functions $y = A\delta(t)$, $A\delta(t - L/v)$, $A\delta(t - 2L/v)$, where A is the vertical area of a bump. Damping is negligible. Derive an analytical expression for the corresponding steady-state vertical displacement z of the sprung mass as a function of t. Use this result to determine critical values of v. Then graph $z(t)$ for the case where v is 90% of the lowest critical value of v.

3.45 A one-degree-of-freedom, underdamped system having mass m, natural frequency ω_{nat}, and critical damping ratio $\zeta = 0.04$ is subjected to cyclical triangular pulse excitation, as shown below. What is the largest harmonic in the response when $\tau = \pi/3\omega_{nat}$?

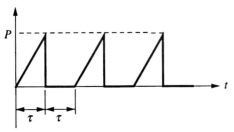

EXERCISES 3.45, 3.46

3.46 Use FFT techniques to determine and graph the steady-state displacement and acceleration of the system in Exercise 3.45 for the parameters stated there.

3.47 A flywheel is welded to an elastic, low-mass shaft. Due to voltage fluctuations, the torque applied to the flywheel by a motor is $\Gamma = 4 + 2 \sin (10 \cos (60t))$ kN-m, where t is measured in seconds. This function has a period $T = 2\pi/60$ s. The system properties are $I_0 = 50$ kg-m^2, $\kappa = 750$ kN-m/rad. The critical damping ratio is observed to be $\zeta = 0.08$. Use FFT techniques to determine the harmonic amplitudes of Γ and the steady-state rotation θ of the flywheel. Then plot Γ and θ as functions of time over an interval covering two periods.

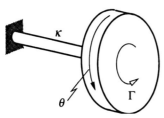

EXERCISE 3.47

3.48 An excitation in the form of a periodic sequence of uniform pulses—defined by $Q(t) = F$ if $0 \le t < T/3$, $Q(t) = 0$ if $T/3 \le t < T$, and $Q(t + T) = Q(t)$—is applied to a one-degree-of-freedom system having natural frequency ω_{nat} and ratio of critical damping ζ. Determine the harmonic amplitudes of the steady-state response to this force for cases where $\omega_{nat}T = 0.2\pi$, 2π, and 20π. The ratio of critical damping in each case is $\zeta = 0.2$.

3.49 Graph the response of the system in Exercise 3.48 for each set of parameters stated there.

3.50 The sketch depicts a one-degree-of-freedom model of an automobile traveling to the right at constant speed v when the road is not smooth. The mass is 1200 kg, the natural frequency of the system is 5 Hz, and the critical damping ratio is 0.4. The elevation of a certain road is a sequence of periodic 50 mm high bumps spaced at a distance of 4 m, specifically, $z = (x - 5x^2)$ if $0 < x < 0.2$ m, $z = 0$ if $0.2 < x < 4$ m, $z(x + 4) = z(x)$.

(a) What speeds v would cause the vertical displacement y to be resonant if the dashpot were not present?

(b) Determine the steady-state displacement y when $v = 5$ m/s.

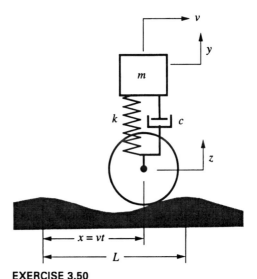

EXERCISE 3.50

3.51 Alternative designs for a seismometer use $\omega_{nat} = 100$ Hz and $\zeta = 0.05$, $\omega_{nat} = 150$ Hz and $\zeta = 0.05$, and $\omega_{nat} = 150$ Hz and $\zeta = 0.5$. Assess the accuracy of each design for a base displacement given by $z = 20\cos(\omega t) + 10\sin(10\omega t)$ mm when $\omega = 200$ Hz.

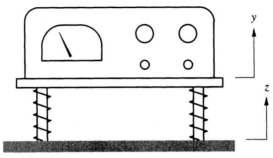

EXERCISE 3.52

3.52 An electronic instrument is isolated from the vibratory motion of the floor by a set of four springs that are collinearly mounted on dashpots. The mass of the instrument and platform is 20 kg, and the stiffness of each spring is 3 kN/m. The critical damping ratio is $\zeta = 0.05$. The upward displacement of the floor is $z = 5\sin(\omega t)$, where z is measured in millimeters. Determine the maximum steady-state acceleration of the instrument and the time delay of that maximum relative to the maximum floor acceleration when $\omega = 3.5$ Hz, 4.0 Hz, and 4.5 Hz.

3.53 A periodic ground displacement consists of a sequence of exponentially decaying pulses repeated at interval T, such that $\ddot{z}(t) = a\exp(-\lambda(t/T))$ for $0 < t < T$, $\ddot{z}(t \pm T) = \ddot{z}(t)$. The parameter λ is nondimensional. Consider an accelerometer for which $\omega_{nat} = 20\pi/T$ and $\zeta = 0.2$. Three cases of interest are $\lambda = 1$, $\lambda = 5$, and $\lambda = 25$. For each case compare the DFT of the accelerometer output to the DFT of $\ddot{z}$. Also compare the time dependence of the accelerometer output to $\ddot{z}$. What do these comparisons suggest regarding the influence of λ on the accuracy of the accelerometer?

3.54 Consider measuring the ground motion in Exercise 3.53 with a seismometer, for which $\omega_{nat} = \pi/T$ and $\zeta = 0.05$. Three cases of interest are $\lambda = 1$, $\lambda = 5$, and $\lambda = 25$. For each case compare the DFT of the seismometer output to the DFT of z. Also compare the time dependence of the seismometer output to z. What do these comparisons suggest regarding the influence of λ on the accuracy of the seismometer?

3.55 A MEMS (micro-electro-mechanical system) accelerometer is mounted on a head of a hard disk drive whose platter rotates at 10,000 rev/min. It is observed that the elevation of the platter at a certain radial distance deviates from planarity by $z = 2\sin(3\cos(\theta))\ \mu m$, where θ is the polar angle. This function is periodic such that $z(\theta \pm 2\pi) = z(\theta)$. Because $\theta = \omega t$ at the drive's

head, the platter elevation at that location is a periodic function of time. The accelerometer has a natural frequency of 150 kHz and its critical damping ratio is $\zeta = 0.05$. It is reasonable to assume that the head maintains a constant distance above the platter. Determine the output of the accelerometer in this situation. Is this a good design for the accelerometer?

3.56 Consider an undamped system that is subjected to a sine pulse starting at $t = 0$, such that $Q(t) = F_0\sin(2\pi(t/T))[h(t) - h(t - T)]$. Use the impulse response and the convolution integral to derive expressions for the response in the intervals $0 \le t < T$ and $t > T$. Does the case $T = 2\pi/\omega_{nat}$ require special attention? Identify in each expression the complementary and particular solutions that would have been obtained if the method of undetermined coefficients were employed to determine the response. Also, examine the response for $T \ll 2\pi/\omega_{nat}$ to demonstrate the validity of eq. (3.8.9).

3.57 An undamped one-degree-of-freedom system having mass m and natural frequency ω_{nat} is subjected to the force shown on page 217. At $t = 0$ the system is at rest at its static equilibrium position.

(a) Derive expressions for $x(t)$ valid for the time intervals $0 < t < \tau$ and $t > \tau$ by evaluating a convolution integral.

(b) Compare the results of part (a) to those obtained by the superposition method in conjunction with Appendix B.

3.58 A one-degree-of-freedom system is subjected to a transient excitation. Derive a convolution integral for the acceleration $\ddot{q}$ of the mass element. The result may be left in symbolic form with an integrand that depends on the impulse response function $g(t)$, the force function $Q(t)$, and their derivatives.

3.59 A one-degree-of-freedom system whose natural frequency is 50 Hz and whose ratio of

$$P(1 - t^2/\tau^2)$$

Force

Time t

EXERCISE 3.57

critical damping is 0.005 is subjected to resonant tone burst, in which a harmonic force at the natural frequency is applied at $t = 0$ and cut off after 10 cycles; thus $Q(t) = F_0 \sin(\omega_{nat}t)$ $[h(t) - h(t - 20\pi/\omega_{nat})]$. Use FFT techniques to determine the nondimensional displacement qF_0/K as a function of time.

3.60 A one-degree-of-freedom system is subjected to a pulse excitation in the form of a parabola,

$$Q = 2000\frac{t(T - t)}{T^2}[h(t) - h(t - T)] \quad N$$

The system mass is 0.5 kg, and the natural frequency is 100 Hz. The pulse duration T equals the undamped period of free vibration period. The system is at rest in the equilibrium position at $t = 0$.

(a) Use FFT techniques to evaluate the response when $\zeta = 0.20$.

(b) Use FFT techniques to evaluate the response when $\zeta = 0.002$.

(c) Use superposition and Appendix B to derive the analytical solution for this pulse. Compare the analytical and FFT results for $\zeta = 0.2$ and $\zeta = 0.002$.

3.61 A one-degree-of-freedom system for which $\omega_{nat} = 15$ rad/s and $\zeta = 0.2$ is subjected to a chirp defined by $(Q(t) = F_0\sin[\pi t(1 - t)]) \times \sin(\pi t/5)[h(t) - h(t - 5)]$, where t is measured in seconds. Use FFT techniques to determine the nondimensional displacement qF_0/K as a function of time.

3.62 Consider the analysis for part (d) of Example 3.14. Repeat that work using the discretized impulse response, but do not zero pad the time data. Show that the result is a solution that is incorrect in an initial interval, after which it matches the analytical solution. Also show that the interval over which the error occurs matches the duration of the pulse.

3.63 Consider a one-degree-of-freedom system whose natural frequency is 500 Hz. Critical damping ratios of interest are $\zeta = 0.2$ and $\zeta = 0.002$. For each value of ζ use the impulse response function $g(t)$ to create a set of artificial experimental data. The time window should be 10 damped periods in duration, and the sampling rate should be chosen to be consistent with the Nyquist sampling criterion. Construct an approximate frequency domain transfer function from the DFT of the sampled impulse response data. Then use the bandwidth method to identify the natural frequency and critical damping ratio. Compare the results to the actual ω_{nat} and ζ used to construct the sampled data.

3.64 Consider a one-degree-of-freedom system whose natural frequency is 24 rad/s and whose critical damping ratio is $\zeta = 0.25$. Use the superposition method and Appendix B to construct the response to a ramp pulse lasting two undamped periods,

$$Q = F_0\frac{\omega_{nat}t}{4\pi}\left[h(t) - h\left(t - \frac{4\pi}{\omega_{nat}}\right)\right]$$

Then evaluate the DFT of the excitation and of the response, using a value for the time window T that is adequate to process both sets of data. Perform deconvolution in the frequency domain by using both sets of DFT data to determine the complex frequency response. Then use the bandwidth method to identify the natural frequency and critical damping ratio. Compare the results to the actual ω_{nat} and ζ of the system.

3.65 Perform the analysis in Exercise 3.64 for the case where $\zeta = 0.025$.

3.66 Consider a two-frequency response, $x = A_1\cos(\omega_1 t) + A_2\sin(\omega_2 t)$, $\omega_2 > \omega_1$. Sample this data in the time interval $0 < t < 20\pi/\omega_1$. Compare the magnitude of the DFT of this data obtained when a Hanning window is used to the DFT when a Hanning window is not used. Cases to consider are

$A_1 = 1$, $\omega_1 = 100$ rad/s and $A_2 = 0.5$, $\omega_2 = 200$ rad/s; $A_1 = 1$, $\omega_1 = 100$ rad/s and $A_2 = 0.5$, $\omega_2 = 120$ rad/s; and $A_1 = 1$, $\omega_1 = 100$ rad/s and $A_2 = 0.1$, $\omega_2 = 120$ rad/s; Use these results to assess limitations on using the Hanning window to isolate frequencies where power is concentrated.

3.67 Consider the impulse response of a one-degree-of-freedom system for which $\omega_{nat} = 1$ rad/s and $\zeta = 0.005$. Construct artificial experimental data by sampling the response function over the time interval $0 < t < 100$. Compare the DFT of this data obtained using a Hanning window to the DFT when a Hanning window is not used. From this, assess the degree to which a Hanning window might be useful for identifying ω_{nat} and ζ.

3.68 In order to understand the role of the time window and sample size, repeat the analysis in Example 3.15 using $T = 20\pi$ and $N = 512$, $T = 80$ and $N = 128$.

3.69 Consider Example 3.15 when $T = 80$ and $N = 128$. Because T is not a multiple of $2\pi/\omega_{nat}$, the natural frequency will not correspond to any of the discrete FFT frequencies.

(a) Apply exponential windows with $\beta = 2, 4$, and 6, and construct the Nyquist plot of the DFT data in each case.

(b) Identify which β leads to a Nyquist plot that most closely resembles a circle, as described by Example 3.4. Determine the bandwidth of the windowed data by fitting a circle to the Nyquist plot.

(c) Identify the natural frequency and critical damping ratio ζ from this data set.

3.70 Consider the impulse response in Example 3.15 when there are experimental errors. To model this situation, let the impulse response data be $x_n = g(t_n)[1 + \varepsilon R_n]$, where $-1 < R_n < 1$ is a random number with uniform probability distribution, and $1/\varepsilon$ is the signal-to-noise ratio. Values to consider are $\varepsilon = 0.01$, $\varepsilon = 0.10$, and $\varepsilon = 1.0$. What are the identified values of ω_{nat} and ζ in each case?

3.71 Consider a displacement that combines the impulse responses of two systems having one degree of freedom. Specifically, consider $x = g(t, 80, 0.005) + 2g(t, 150, 0.01)$, where $g(t, \omega_{nat}, \zeta)$ denotes the impulse response for the respective systems. Generate a sampled set of data x_n covering the interval $0 < t < 1$ s. The DFT magnitudes of that data will exhibit two peaks. Use an exponential window function and the half-power method to identify the two natural frequencies and damping ratios indicated by the DFT data. (*Hint*: Consider each peak as though it corresponded to an isolated one-degree-of-freedom system.)

CHAPTER 4

MODAL ANALYSIS OF MULTI-DEGREE-OF-FREEDOM SYSTEMS

A one-degree-of-freedom system displays many, but not all, of the fundamental vibratory phenomena. Most engineering systems contain a multitude of parts that move independently, subject to kinematical restrictions imposed by their interconnections. It is necessary in such situations to know the instantaneous values of several generalized coordinates q_j and of the corresponding generalized velocities $\dot{q}_j$ if we wish to fully describe the state of motion. A primary new feature of multi-degree-of-freedom systems is the existence of preferred free vibration patterns, called *vibration modes*. Modal analysis is a two-step process in which the basic features of free vibration are identified, and then used to transform the equations of motion. The development ultimately will enable us to represent the response of any system in terms of a set of one-degree-of-freedom systems. Thus, the developments in the previous chapter are crucial to modal analysis of multi-degree-of-freedom systems.

The notion that a reasonably complicated model can be converted to a less complicated one will also be manifested in our later study of continuous systems. Many of the vibration analysis methods for continuous systems are based on deriving approximately equivalent discrete models. Thus, the techniques and procedures we develop here are central to our ability to analyze and understand the manner in which all systems vibrate.

4.1 BACKGROUND

We require that the generalized coordinates be defined such that all $q_j = 0$ when the system is at its equilibrium position. If the displacement of the system from this position is sufficiently small, we may linearize the equations of motion based on the smallness of the values of q_j and $\dot{q}_j$. The power balance formulation developed in Chapter 1 is quite useful for this task. Recall that it requires derivation of expressions in the form of quadratic sums representing the system's kinetic and potential energy, and instantaneous power dissipation in terms of the N generalized coordinates q_j and associated generalized velocities $\dot{q}_j$. Matching these expressions to the respective standard forms for any time-invariant system,

$$T = \frac{1}{2}\sum_{j=1}^{N}\sum_{n=1}^{N} M_{jn}\dot{q}_j\dot{q}_n$$

$$V = \frac{1}{2}\sum_{j=1}^{N}\sum_{n=1}^{N} K_{jn}q_j q_n + \sum_{j=1}^{N} F_0 q_j + V_0$$

$$\mathscr{P}_{\text{dis}} = \sum_{j=1}^{N}\sum_{n=1}^{N} C_{jn}\dot{q}_j\dot{q}_n \tag{4.1.1}$$

leads to identification of the elements of the inertia matrix $[M]$, stiffness matrix $[K]$, and damping matrix $[C]$, each of which is symmetric. The vector $\{F_0\}$ represents the conservative forces at static equilibrium. The corresponding general equations of motion for a multi-degree-of-freedom system were shown in Chapter 1 to be expressible in matrix form as

$$[M]\{\ddot{q}\} + [C]\{\dot{q}\} + [K]\{q\} = \{Q\} - \{F_0\} \tag{4.1.2}$$

The elements of $\{Q\}$ are the generalized forces. The definition of these quantities is associated with the standard description of the power input to a system by the excitation forces:

$$\mathscr{P}_{\text{in}} = \sum_{j=1}^{N} Q_j\dot{q}_j \tag{4.1.3}$$

We identify the generalized forces Q_j by matching this representation to the expression describing the actual power input in terms of the generalized velocities. The term $\{Q\} - \{F_0\}$ appearing in the right side of eq. (4.1.2) represents the disturbance forces causing the system to move away from its static equilibrium position. If no portion of the nonconservative force system is responsible for holding the system in equilibrium, then we will find that $\{F_0\} = \{0\}$.

Before we address the analytical task, we shall consider free vibration results for a simple two-degree-of-freedom system. Doing so will demonstrate that there is a relatively simple set of properties underlying responses that appear to be quite complicated. We consider the two-degree-of-freedom system depicted in Figure 4.1, in which two masses are connected to each other and the ground with springs and dashpots. We select as generalized coordinates the displacements x_1 and x_2 of each mass. We could build this system and measure its response. However, we consider here solutions for $x_1(t)$ and $x_2(t)$ that are obtained by solving nu-

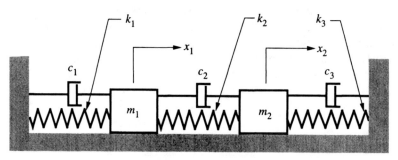

FIGURE 4.1 Typical two-degree-of-freedom system.

merically the differential equations of motion. (The Runge-Kutta method, which is described in most texts on numerical methods—see Press et al. (1992)—and is implemented in MATLAB and Mathcad, is well suited to such an evaluation.) Data generated in this manner is often referred to as *artificial experimental data*, because it enables one to test the procedures for analyzing experimental data prior to actually performing the experiments. The two blocks are coupled by spring k_2 and dashpot c_2, because motion of either block generates forces within these elements that are transmitted to the other block, thereby inducing motion of both bodies.

The system parameters for the specific situation we shall consider are $m_1 = 1$ kg, $m_2 = 2$ kg, $k_1 = 100$ N/m, $k_2 = 50$ N/m, $k_3 = 250$ N/m, $c_1 = 0.8$ N-s/m, $c_2 = 0.4$ N-s/m, and $c_3 = 0.6$ N-s/m. We consider two cases of free vibration, which means that no external excitation of the system is present. In both cases the system is released from rest. In the first case the block on the left is initially displaced and the block on the right is not disturbed initially, such that $x_1 = 10$ mm and $x_2 = 0$. The second case we consider switches these initial conditions, such that $x_1 = 0$ and $x_2 = 10$ mm.

The response for the first set of initial conditions covering the initial 10 s interval is depicted in Figure 4.2. The first feature we note is that both blocks do displace, even though only the left block was disturbed at $t = 0$, which illustrates the coupling effect. We also see that the response is quite irregular, with no indication of a repeated pattern. Also, the envelope containing either displacement function does not have the exponential form we encountered in underdamped one-degree-of-freedom systems although it does seem as though both responses are decaying with time. Such decay is due to the dissipative action of the dashpots. A careful examination of the response curves suggests that there is a degree of beating, in that the amplitude of each variable seems to grow and then reduce in a nearly periodic manner.

FFT techniques, which break a function of time into its frequency components, offer a useful perspective for investigating the processes underlying the function. The techniques for using FFTs to analyze functions of time, which were developed in the previous chapter, require that the time window in which the function is viewed be sufficiently large to see the function decay to a relatively small value. This is not the case for Figure 4.2. We therefore consider the results of numerical solution of the equations of motion over an interval of 20π s. As shown in Figure 4.3, which depicts

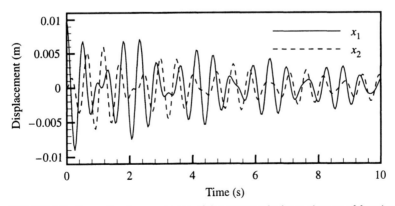

FIGURE 4.2 Free vibration response of the prototypical two-degree-of-freedom system corresponding to $x_1 = 10$ mm, $x_2 = \dot{x}_1 = \dot{x}_2 = 0$ at $t = 0$.

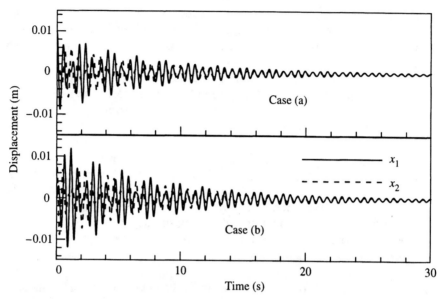

FIGURE 4.3 Free response of the prototypical two-degree-of-freedom system for two cases of initial conditions at $t = 0$: case (a) $x_1 = 10$ mm, $x_2 = \dot{x}_1 = \dot{x}_2 = 0$; case (b) $x_2 = 10$ mm, $x_1 = \dot{x}_1 = \dot{x}_2 = 0$.

the first 30 s, this interval is adequate to see the response for either set of initial conditions decay to very small values relative to the maxima. Although the finite line width of the graphs blurs details, it is evident that the responses for each set of initial conditions are different, although they seem to have similar patterns.

We wish to perform FFT evaluations of the output, which are most efficiently performed when the time data length is a power of two. To accommodate this wish, the numerical computations leading to Figure 4.3 were adjusted to break the time window $T = 20\pi$ s into 1023 steps, which leads to a data set including the initial displacement whose length is $N = 1024$. The corresponding time step is $\Delta = T/N = 0.0307$ s, and the discrete time instants for the data are $t_n = n\Delta, n = 0,1,\ldots, N-1$. A visual examination of Figure 4.2 suggests that this value of Δ is sufficiently small to resolve the wiggles in each response, which means that we may anticipate that it will satisfy the Nyquist sampling criterion for the data set.

Figure 4.4 displays the results of applying FFT analysis to the x_1 and x_2 time data for initial condition cases (a) and (b). The variables $(X_1)_j$ and $(X_2)_j$ denote the DFT of $x_1(t_n)$ and $x_2(t_n)$, respectively. Because the time data is real, it is only necessary to consider the values for $j = 0,\ldots, N/2$. The fundamental frequency for a DFT is $\omega_1 = 2\pi/T$, so the discrete frequency corresponding to each pair of $(X_1)_j$ and $(X_2)_j$ values is $\omega_j = j\omega_1 = j(2\pi/T)$.

In Figure 4.4 the magnitude of the transformed data is plotted against the corresponding frequency. To improve visibility only the band $0 < \omega < 50$ rad/s is depicted in the main frame. The DFT values for $\omega > 50$ rad/s are essentially zero, which confirms that the choice for Δ met the Nyquist sampling criterion. The most notable feature of the DFT data is that both cases of initial conditions lead to two peaks, at indexes 107 and 136, which correspond to $\omega = 10.7$ and 13.6 rad/s, respectively. A noteworthy feature of these peaks is the ratio of the DFT values, which are $|(X_2)_{107}|/|(X_1)_{107}| = 0.695$ and $|(X_2)_{136}|/|(X_1)_{136}| = 0.707$ in case (a), and $|(X_2)_{107}|/|(X_1)_{107}| = 0.720$ and $|(X_2)_{136}|/|(X_1)_{136}| = 0.710$ in case (b). These

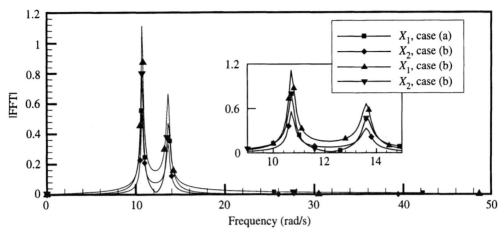

FIGURE 4.4 FFTs of the responses in Figure 4.3.

values are quite close. Perhaps most surprising is the observation that $|(X_2)_j|$ in case (b) seems to be the same as $|(X_1)_j|$ in case (a).

At this juncture, one might wonder whether the similarities of the DFT data noted in the preceding paragraph are the result of a fortuitous selection of the system parameters. The answer is an emphatic no! They are manifestations of a fundamental aspect of the vibratory properties of linearized systems. Such systems have preferred types of free vibration, called *modes*. These modes correspond to oscillations at the system's natural frequencies, which are the locations of the peaks in Figure 4.4 (to within the resolution of the DFT). We will see in the following developments that viewing system response from a modal perspective leads to physical insight, as well as analytical and computational efficiencies.

4.2 EIGENVALUE PROBLEM—UNDAMPED SYSTEMS

Coupling of the generalized coordinates arises if either $[M]$, $[K]$, or $[C]$ is not diagonal. It then is necessary that the individual equations of motion be satisfied simultaneously. We will build up our capability by considering progressively more complicated situations. The first consideration is the case of free vibration in the absence of damping, which parallels the way we approached the analysis of one-degree-of-freedom systems.

4.2.1 Natural Frequencies and Modes

We set $\{Q\} = \{0\}$ and $[C] = [0]$, so the equation of motion reduces to

$$[M]\{\ddot{q}\} + [K]\{q\} = \{0\} \tag{4.2.1}$$

This is a set of linear differential equations with constant coefficients. Because differentiation of an exponential function reproduces that function, we may generate a solution by assuming that each generalized coordinate varies exponentially as $\exp(\lambda t)$. In principle, the coefficient λ could be any constant, but we also know that in the absence of damping the system is conservative. If $\text{Re}(\lambda) > 0$, the total mechanical energy $T + V$ will grow, while $\text{Re}(\lambda) < 0$ leads to a decaying response; both cases violate conservation of energy. Accordingly, we anticipate that

$\lambda = i\omega$, corresponding to harmonic motion, so we construct a solution based on the trial form

$$\{q\} = \text{Re}[B\{\phi\}\exp(i\omega t)] \tag{4.2.2}$$

where B and $\{\phi\}$ are constants. Every term produced by the substitution of the trial solution exhibits the same time dependence, so the equation of motion will be satisfied at all instants only if the coefficients of the exponential terms match. Furthermore, the constant factor B is common to every term, so it cancels. The remaining terms are

$$\boxed{[[K] - \omega^2[M]]\{\phi\} = \{0\}} \tag{4.2.3}$$

We now have two alternatives to consider. Suppose we select the value of ω arbitrarily. In that case the square array $[[K] - \omega^2[M]]$ multiplying $\{\phi\}$ will be invertible, so that the solution of the above is $\{\phi\} = \{0\}$. This leads to a trivial solution, $\{q\} = \{0\}$. A nontrivial solution can exist only if the value of ω is such that $[[K] - \omega^2[M]]$ is not invertible; in other words, we must find the value of ω for which the determinant of this set of coefficients vanishes,

$$\boxed{[K] - \omega^2[M]| = 0} \tag{4.2.4}$$

Equation (4.2.3) constitutes a (matrix) *eigenvalue problem*, and eq. (4.2.4) is its *characteristic equation*. The prefix "eigen" is German for "own"—it refers to the idea that the values of ω are intrinsic properties that the system possesses. We say that they are the system's *natural frequencies*. It is clear that for a one-degree-of-freedom system, for which $M_{11} = M$ and $K_{11} = K$, $\omega \equiv \omega_{\text{nat}}$. A system with N degrees of freedom has N natural frequencies.

Equation (4.2.3) is known as a *general eigenvalue problem*. The simpler, and more common, *standard eigenvalue problem* is $[[A] - \lambda[I]]\{x\} = \{0\}$. The developments in this section discuss the solution of the eigenvalue problem from a theoretical perspective. The differences between the two types of equations will not be an issue. Recognition of the differences does become important when we employ computer-aided methods, which we will discuss in a later section.

Suppose we have numerical values for $[M]$ and $[K]$. Evaluation of the determinant in eq. (4.2.4) by any standard method, such as cofactors, is quite tedious when N is large because the parameter ω^2 occurs algebraically. Nevertheless, contemplation of how such an evaluation would proceed makes it clear that the characteristic equation will eventually reduce to an Nth-order *characteristic polynomial* in ω^2,

$$a_N(\omega^2)^N + a_{N-1}(\omega^2)^{N-1} + \cdots + a_1(\omega^2) + a_0 = 0 \tag{4.2.5}$$

This equation has N roots for ω^2. In general the roots of a real polynomial must either be real or complex conjugate pairs. We could pursue a mathematical proof that the roots have certain properties. Indeed, we will see such a proof later. However, for now it is simpler to rely on our physical knowledge that because the system is conservative, the values of ω must be real. Because the values of ω only occur in the characteristic equation as squares, the roots ω^2 must be positive or zero. (This would be recognizable in an actual characteristic polynomial by the fact that its coefficients have alternating signs.) We use the positive sign for the square root, and order the roots such that $0 \leq \omega_1 \leq \omega_2 \leq \cdots \leq \omega_N$. In this sequencing we are allowing for the possibility of repeated and zero natural frequencies, but we shall defer consideration of such possibilities until we have developed the basic concept for natural frequencies that are positive and distinct.

A special term, *fundamental frequency*, is reserved for the lowest nonzero natural frequency, which is usually ω_1. The reason for selecting a specific name for this frequency lies in a result we will see in later sections: specifically, that a harmonic excitation at any natural frequency will result in resonant response. Thus, if we know the fundamental frequency of a system, and make sure that the excitation frequency never exceeds that value, we know that we will avoid resonating the system.

Each natural frequency has a different solution for the coefficient vector $\{\phi\}$ in eq. (4.2.3). Substitution of a specific value ω_j converts this equation to

$$[[K] - \omega_j^2[M]]\{\phi_j\} = \{0\} \qquad (4.2.6)$$

Mathematically, $\{\phi_j\}$ is referred to as the *eigenvector* associated with the jth natural frequency. Vibration engineers refer to it as the jth *mode of free vibration*, or more briefly as the jth *mode*. We shall adopt the convention of referring to the nth element of the jth eigenvector as ϕ_{nj}. (An equivalent interpretation of this notation is to consider it to be the element in the nth row and jth column of a square array $[\phi]$. This view will come to the fore in later developments.)

We cannot solve eq. (4.2.6) directly, because the determinant of the coefficient matrix has been made to vanish by setting $\omega = \omega_j$. Recall that when the determinant of a square array is zero, one or more rows of the array are a linear combination of the other rows. The number of rows that are combinations of the others is the reduction in rank of the matrix. When the natural frequency ω_j is not a repeated root of the characteristic equation, the rank is reduced by one. This means that eq. (4.2.6) represents $N - 1$ independent equations for the elements of $\{\phi_j\}$. Because $\{\phi_j\}$ consists of N elements, we deduce that one of the elements of the eigenvector must be arbitrary. We must select one of the elements of $\{\phi_j\}$ as the arbitrary one, and then solve for the other $N - 1$ elements in terms of that one. Note that all of the quantities in eq. (4.2.6) are real, so all of the elements of $\{\phi_j\}$ are real numbers.

We now are faced with two questions: Which element should we designate as arbitrary? Which row of the coefficient matrix should we consider to be a combination of the others, and therefore discardable? The answer to both questions is that usually it does not matter! Barring special circumstances, it is convenient to consider the first element, ϕ_{1j}, to be arbitrary and to discard the last row of the coefficient matrix. Let us list the $N - 1$ remaining rows explicitly,

$$\begin{bmatrix} (K_{11} - \omega_j^2 M_{11}) & \cdots & (K_{1N} - \omega_j^2 M_{1N}) \\ \vdots & & \vdots \\ (K_{(N-1)1} - \omega_j^2 M_{(N-1)1}) & \cdots & (K_{(N-1)N} - \omega_j^2 M_{(N-1)N}) \end{bmatrix} \begin{Bmatrix} \phi_{1j} \\ \phi_{2j} \\ \vdots \\ \phi_{Nj} \end{Bmatrix} = \{0\} \qquad (4.2.7)$$

If ϕ_{1j} is arbitrary, we may assign any value to it. Let us set $\phi_{1j} = 1$. We move the terms associated with the arbitrary element to the right side, which leads to

$$
\begin{bmatrix}
(K_{12} - \omega_j^2 M_{12}) & \cdots & (K_{1N} - \omega_j^2 M_{1N}) \\
\vdots & & \vdots \\
(K_{(N-1)2} - \omega_j^2 M_{(N-1)2}) & \cdots & (K_{(N-1)N} - \omega_j^2 M_{(N-1)N})
\end{bmatrix}
\begin{Bmatrix}
\phi_{2j} \\
\vdots \\
\phi_{2Nj}
\end{Bmatrix}
$$

$$
= - \begin{Bmatrix}
(K_{11} - \omega_j^2 M_{11}) \\
\vdots \\
(K_{(N-1)1} - \omega_j^2 M_{(N-1)1})
\end{Bmatrix}
\tag{4.2.8}
$$

This represents a set of $N - 1$ simultaneous algebraic equations for the unknown elements ϕ_{nj}, $n = 2, \ldots, N$.

In a later section we will address the application of computer programs to solve the eigenvalue problem. The eigenvectors returned by a numerical routine will generally be a set of numbers, with no indication of which element was considered to be arbitrary. The arbitrariness of the selection and the value of the arbitrary element will ultimately be seen not to affect any response we actually evaluate.

The above procedure can fail in two special circumstances. It might happen that the element ϕ_{1j} we considered to be arbitrary actually is zero. Another possibility is that we retained a trivial equation, corresponding to an entire row of zeroes, and by doing so discarded an independent equation. Neither possibility is very common, but their occurrence is readily identified by the fact that the coefficient matrix in the left side of eq. (4.2.8) has a zero determinant. A simple remedy for both cases is to discard a different equation, and to alter the element of $\{\phi_j\}$ that is considered to be arbitrary. These alterations merely require resequencing the coefficient matrices on both sides of eq. (4.2.8). The more troublesome possibility is ill-conditioning, in which case the aforementioned determinant is very small, but not zero. Such a situation usually significantly increases numerical round-off error. As we will soon see, the properties of the mode functions lead to a self-contained check that will identify such situations.

At this juncture, we have determined the N natural frequencies ω_j and the elements of the corresponding eigenvectors $\{\phi_j\}$. According to eq. (4.2.2), the solution for $\{q\}$ associated with a specific natural frequency ω_j is

$$
\boxed{\{q\} = \text{Re}\,[B_j\{\phi_j\}\exp\,(i\omega_j t)]}
\tag{4.2.9}
$$

All of the generalized coordinates exhibit the same time signature in this response. We say that the response is *synchronous*. The elements of $\{\phi_j\}$ give the proportions between the various generalized coordinates in the synchronous motion. For this reason, many individuals refer to $\{\phi_j\}$ as a *mode shape*. The elements of $\{\phi_j\}$ are real, so a negative element ϕ_{nj} indicates that a generalized coordinate is 180° out of phase from those generalized coordinates associated with positive elements. The coefficient B_j may be complex. Its magnitude represents a scaling factor that depends on how the arbitrary element of $\{\phi_j\}$ was defined, while arg (B_j) is the phase angle of the displacement relative to a cosine variation. We will see in a later section how to determine these coefficients.

The procedure we have developed here is the manner in which we might proceed when the number of degrees of freedom is not large, specifically $N = 2$ or 3. Evaluation of the characteristic equation becomes significantly more cumbersome as N increases. For larger N, we usually rely on numerical algorithms available in mathematical software.

EXAMPLE 4.1

Determine the natural frequencies and mode shapes of the system in Figure 4.1. How do these results compare to the FFT data in Figure 4.4?

Solution In addition to illustrating the setup and solution of the eigenvalue problem, this exercise will demonstrate the relation between a free vibration and the modal properties. The first step is to determine the inertia and stiffness matrices for the system, which we obtain from the kinetic and potential energy functions. The displacements of the blocks are the generalized coordinates $q_1 = x_1$ and $q_2 = x_2$. The elongations of the springs are $\Delta_1 = x_1$, $\Delta_2 = x_2 - x_1$, and $\Delta_3 = -x_2$. Thus, we have

$$T = \tfrac{1}{2}(m_1 \dot{x}_1^2 + m_2 \dot{x}_2^2)$$

$$V = \tfrac{1}{2}[k_1 x_1^2 + k_2(x_2 - x_1)^2 + k_3 x_2^2] = \tfrac{1}{2}[(k_1 + k_2)x_1^2 - 2k_2 x_1 x_2 + (k_2 + k_3)x_2^2]$$

The corresponding system matrices are

$$[M] = \begin{bmatrix} m_1 & 0 \\ 0 & m_2 \end{bmatrix}, \qquad [K] = \begin{bmatrix} k_1 + k_2 & -k_2 \\ -k_2 & k_2 + k_3 \end{bmatrix}$$

The given values of the parameters are $m_1 = 1$ kg, $m_2 = 2$ kg, $k_1 = 100$ N/m, $k_2 = 50$ N/m, and $k_3 = 250$ N/m, so the eigenvalue equation is

$$[[K] - \omega^2[M]]\{\phi\} = \begin{bmatrix} (150 - \omega^2) & -50 \\ -50 & (300 - 2\omega^2) \end{bmatrix} \begin{Bmatrix} \phi_1 \\ \phi_2 \end{Bmatrix} = \begin{Bmatrix} 0 \\ 0 \end{Bmatrix}$$

We obtain the characteristic equation by setting the determinant $|[K] - \omega^2[M]| = 0$. For a 2×2 matrix, a convenient rule for the determinant is to take the product of the terms on the diagonal, then subtract the product of terms along the skew diagonal (from the 1,2 to the 2,1 element). This leads to

$$(150 - \omega^2)(300 - 2\omega^2) - 50^2 \equiv 2\omega^4 - 600\omega^2 + 42{,}500 = 0$$

The roots of this quadratic equation are

$$\omega^2 = 114.64,\ 185.36$$

We take the positive square root of these values and assign the smaller root to the fundamental frequency,

$$\omega_1 = 10.707 \text{ rad/s}, \qquad \omega_2 = 13.615 \text{ rad/s}$$

Now that we have determined the natural frequencies, we proceed to evaluate the mode shapes. When $\omega = \omega_1$ or ω_2, at least one of the scalar equations described by $[[K] - \omega_j^2[M]]\{\phi\} = \{0\}$ is not independent of the other. Let us retain the first equation, so we have

$$\begin{bmatrix} (150 - \omega_j^2) & -50 \\ \times & \times \end{bmatrix} \begin{Bmatrix} \phi_{1j} \\ \phi_{2j} \end{Bmatrix} = \begin{Bmatrix} 0 \\ \times \end{Bmatrix}$$

where $\times$ denotes a term that is discarded. If we assign a unit value to ϕ_{1j}, the remaining equation is

$$(150 - \omega_j^2)(1) - 50\phi_{2j} = 0 \quad \Rightarrow \quad \phi_{2j} = \frac{150 - \omega_j^2}{50}$$

We obtain the eigenvector element associated with each natural frequency by substituting the appropriate value of ω^2,

$$\phi_{21} = 0.7071, \quad \phi_{22} = -0.7071$$

The corresponding eigenvectors are

$$\{\phi_1\} = \begin{Bmatrix} 1 \\ 0.7071 \end{Bmatrix}, \quad \{\phi_2\} = \begin{Bmatrix} 1 \\ -0.7071 \end{Bmatrix}$$

According to our analysis, a vibration in the first mode occurs at 10.707 rad/s, with the amplitude of x_2 0.7071 times that of x_1. Because ϕ_{21}/ϕ_{11} is positive, both blocks move in phase in this modal vibration. The second mode occurs at 13.615 rad/s, with the amplitude of x_2 also being 0.7071 times that of x_1. In the second mode ϕ_{22}/ϕ_{12} is negative, which means that at each instant the blocks are moving in opposite directions.

The peak amplitudes in Figure 4.4 were found to occur at 10.7 rad/s and 13.6 rad/s, which are very close to the computed natural frequencies. The FFT amplitudes at the lower frequency were found to be $|X_2|/|X_1| = 0.695$ for the first set of initial conditions and 0.720 for the second. The corresponding amplitude ratios for the second frequency were found to be 0.720 and 0.710. All values are very close to the ratio $|\phi_{2j}|/|\phi_{1j}| = 0.7071$ found in the analysis. Our study of the eigenvalue problem enables us now to recognize that the FFT data processing of the synthetic responses in Figure 4.3 revealed the modal contributions to each case of free vibration.

EXAMPLE 4.2

A crude model of an automobile and its suspension system appears below. The chassis has a mass of 1200 kg, and the radius of gyration about the center of mass G is 1.1 m. The spring stiffnesses are 15 kN/m for each of the two front springs and 10 kN/m for the two rear springs. Consider vertical displacement only, which will be described by letting the vertical displacement y_G of the center of mass and the counterclockwise rotation θ about the center of mass be generalized coordinates. Determine the natural frequencies and mode shapes of the vehicle. What is the displacement ratio of the front end relative to the rear end in each mode?

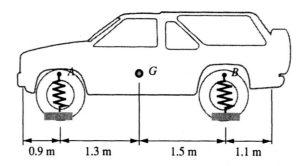

| 0.9 m | 1.3 m | 1.5 m | 1.1 m |

Solution   This exercise will illustrate the relationship between a modal solution for a system's generalized coordinates and the corresponding physical displacement of points. We do not need to derive the equations of motion here, because their derivation was illustrated in Example 1.4. The resulting inertia and stiffness coefficients corresponding to selecting $q_1 = y_G$ and $q_2 = \theta$ were found to be

$$M_{11} = m, \qquad M_{22} = 1.21m, \qquad M_{12} = 0$$
$$K_{11} = 2(k_A + k_B), \qquad K_{12} = 2(-1.3k_A + 1.5k_B), \qquad K_{22} = 2(1.69k_A + 2.25k_B)$$

For the given values of m, k_A, and k_B, the eigenvalue problem for free vibration is

$$[[K] - \omega^2[M]]\{\phi\} = \begin{bmatrix} 50{,}000 - 1200\omega^2 & -9000 \\ -9000 & 95{,}700 - 1452\omega^2 \end{bmatrix} \{\phi\} = \begin{Bmatrix} 0 \\ 0 \end{Bmatrix}$$

Before we form the characteristic equation, it is useful to divide each row of $[K] - \omega^2[M]$ by 100 in order to avoid the occurrence of extremely large numbers in products. Thus, we form

$$\left\| \begin{bmatrix} 500 - 12\omega^2 & -90 \\ -90 & 957 - 14.52\omega^2 \end{bmatrix} \right\| = (500 - 12\omega^2)(957 - 14.52\omega^2) - 90^2 = 0$$

The roots of this quadratic equation are

$$\omega^2 = 39.88, 67.70$$

from which we find the natural frequencies to be

$$\omega_1 = 6.315 \text{ rad/s} = 1.0146 \text{ Hz}$$
$$\omega_2 = 8.228 \text{ rad/s} = 1.3095 \text{ Hz}$$

To find the eigenvectors we substitute each ω_j value into the eigenequation, and set the first element $\phi_{1j} = 1$, which gives

$$\begin{bmatrix} 500 - 12\omega_j^2 & -90 \\ -90 & 957 - 14.52\omega_j^2 \end{bmatrix} \begin{Bmatrix} 1 \\ \phi_{2j} \end{Bmatrix} = \begin{Bmatrix} 0 \\ 0 \end{Bmatrix}$$

We discard the second row above, because it only differs by a factor from the first when ω_j^2 is an eigenvalue, so we find that

$$\phi_{2j} = \frac{500 - 12\omega_j^2}{90}$$

Substituting the two eigenvalues ω^2 found above gives $\phi_{21} = 0.2381$, $\phi_{22} = -3.470$. Note that because q_1 is a displacement, whereas q_2 is a rotation, the physical units are m for ϕ_{1j}, and rad for ϕ_{2j}. Thus, the mode shapes are

$$\{\phi_1\} = \begin{Bmatrix} 1 \text{ m} \\ 0.2381 \text{ rad} \end{Bmatrix}, \quad \{\phi_2\} = \begin{Bmatrix} 1 \text{ m} \\ -3.470 \text{ rad} \end{Bmatrix}$$

The generalized coordinates were defined such that positive y_G corresponds to upward displacement and positive θ is counterclockwise. Both elements of the first mode shape have the same sign, which means that y_G and θ are in phase in the first mode. Hence, we interpret the fundamental mode to consist of a vibration at 1.0146 Hz with the generalized coordinates amplitudes being such that 0.2381 rad of counterclockwise rotation occurs for every meter of upward displacement of the center of mass. Because the elements of the second mode shape have opposite signs, the rotation in the second mode is 180° out of phase from the vertical displacement of the center of mass. Thus, a vibration following the second mode occurs at 1.3095 Hz, with 3.470 rad of clockwise rotation for each meter of upward displacement.

In order to better understand the implication of these mode shapes, we examine the corresponding motions of the front and rear of the automobile. For small displacements and rotations, these points (at the same elevation as G) displace vertically. A counterclockwise rotation causes the front to displace downward and the rear to displace upward relative to the center of mass, so we have at any instant

$$y_F = y_G - r_{F/G}\theta = y_G - 2.2\theta, \quad y_R = y_G + r_{R/G}\theta = y_G + 2.6\theta$$

In the first mode, the proportions are $+0.2381$ rad rotation for 1 m of y_G, while the corresponding proportions for the second mode are -3.470 rad rotation for 1 m of y_G. Thus, the ratio of displacements is

$$\left(\frac{y_F}{y_R}\right)_1 = \frac{1 - 2.2(0.2381)}{1 + 2.6(0.2381)} = 0.294$$

$$\left(\frac{y_F}{y_R}\right)_2 = \frac{1 - 2.2(-3.470)}{1 + 2.6(-3.470)} = -1.077$$

These results may be shown pictorially by assigning an arbitrary value to y_R, evaluating the corresponding value of y_F, and plotting the two values over the total 4.8 m length from the rear to the front. The sketch corresponds to modal vibrations in which the amplitude of y_R is 1 m. (Although 1 m would be an unrealistically large displacement inconsistent with small displacement approximations, we use it for the sake of pictorial clarity.) The dark lines represent the extremes of each modal vibration.

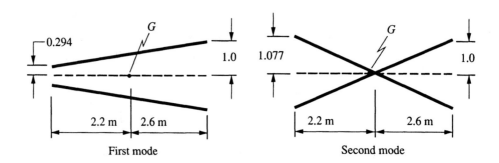

First mode   Second mode

We see in the sketch that the first mode consists of a translation with a relatively small rotation. This is often referred to as the *bounce mode*. The second mode has a point between the front and rear at which there is no displacement called a *node*. The location of the node is close to, but not coincident with, the center of mass. As shown in the sketch, we may visualize the displacement in the second mode as being a rotation about the node. Because u is the pitch angle, the second mode is often referred to as the *pitching mode*. If the center of mass were midway between the wheels and the front and rear springs had equal stiffness, we would find that no rotation occurs in the bounce mode, while the node of the pitching mode would coincide with the center of mass.

4.2.2 Orthogonality and Normalization

The fact that a system has natural frequencies and vibration modes has obvious relevance to free vibration. What makes these features important to the analysis of forced vibration is their orthogonality properties. From a linear algebra viewpoint, this property is comparable to the orthogonality of unit vectors in a three-dimensional space. By referring an arbitrary vector to these favored directions, and thereby defining components, we are able to perform many operations that would otherwise be cumbersome. In the same manner, we will find that decomposing an arbitrary response into its modal components leads to substantial simplifications.

We begin by considering two eigensolutions, associated with natural frequencies ω_j and ω_n, where j and n are any indices in the range from 1 to N. The respective equations may be written as

$$\omega_j^2[M]\{\phi_j\} = [K]\{\phi_j\}, \qquad \omega_n^2[M]\{\phi_n\} = [K]\{\phi_n\} \qquad (4.2.10)$$

In order to make these two relations look more alike, let us premultiply the first equation by $\{\phi_n\}^T$, while $\{\phi_j\}^T$ is the premultiplication factor for the second equation.

$$\omega_j^2\{\phi_n\}^T[M]\{\phi_j\} = \{\phi_n\}^T[K]\{\phi_j\},$$

$$\omega_n^2\{\phi_j\}^T[M]\{\phi_n\} = \{\phi_j\}^T[K]\{\phi_n\} \qquad (4.2.11)$$

Each product leads to a scalar number. We now employ an identity from matrix algebra that states that the transpose of a product may be obtained by multiplying the transpose of the factors in reverse order,

$$[[a][b]]^T = [b]^T[a]^T \qquad (4.2.12)$$

Because both $[K]$ and $[M]$ are symmetric, so that $[K]^T = [K]$ and $[M]^T = [M]$, application of this identity to the second of eqs. (4.2.11) leads to

$$\omega_n^2\{\phi_n\}^T[M]\{\phi_j\} = \{\phi_n\}^T[K]\{\phi_j\} \qquad (4.2.13)$$

The last step is to subtract this relation from the first of eqs. (4.2.11), which results in

$$(\omega_j^2 - \omega_n^2)\{\phi_n\}^T[M]\{\phi_j\} = 0 \qquad (4.2.14)$$

This is an identity that must be satisfied for any pair of j and n indices. There are two possibilities: Either $\omega_n = \omega_j$, or else the matrix product is zero. We start with the usual case where the system's natural frequencies are distinct. Then if $n \neq j$, so that $\omega_n \neq \omega_j$, eq. (4.2.14) requires that

$$\{\phi_n\}^T[M]\{\phi_j\} = 0, \quad n, j = 1, \dots, N, \qquad n \neq j \qquad (4.2.15)$$

An immediate consequence of this relation resulting from eq. (4.2.13) is

$$\{\phi_n\}^T[K]\{\phi_j\} = 0, \quad n, j = 1, \dots, N, \qquad n \neq j \qquad (4.2.16)$$

We describe eqs. (4.2.15) and (4.2.16) by saying that the *vibration modes are orthogonal with respect to the inertia and stiffness matrices*. The term "orthogonal" stems from the properties of physical vectors. If we let $\{x\}$ and $\{y\}$ denote two vectors, their dot product is $\{x\}^T\{y\}$. If this product is zero, the vectors are orthogonal.

If we consider the same mode, corresponding to $n = j$, eq. (4.2.14) is satisfied identically, because $\omega_n \equiv \omega_j$. The mass orthogonality product in this case gives a set of numbers

$$\{\phi_j\}^T[M]\{\phi_j\} \equiv \mu_j, \quad j = 1, \dots, N \qquad (4.2.17)$$

Then, as a consequence of eq. (4.2.13), it must be that

$$\{\phi_j\}^T[K]\{\phi_j\} \equiv \mu_j\omega_j^2, \quad j = 1, \dots, N \qquad (4.2.18)$$

The parameters μ_j are called the *modal masses*. Developments in the following sections will explain the reason for this terminology. It also will be proven that the modal masses are always positive, $\mu_j > 0$.

Each element of an eigenvector is scaled by its arbitrary element, so it follows that the modal mass values depend on the choice of that element. The modal masses occur throughout the evaluation of any response, thereby compensating for the arbitrariness contained in the evaluation of the eigenvector. However, rather than having to account for the modal mass in each instance where it occurs, we shall scale the modes such that the modal masses become unity. To do so, we compute μ_j for the original modes, and then divide each mode by $1/\sqrt{\mu_j}$. The result is a set of *normal modes*. We shall denote them with an upper-case "phi," that is, as $\{\Phi_j\}$, in order to distinguish modes that have been normalized from those that have not. Thus, we have

$$\{\Phi_j\} = \frac{\{\phi_j\}}{\sqrt{\{\phi_j\}^T[M]\{\phi_j\}}} \qquad (4.2.19)$$

Correspondingly we may combine the orthogonality properties and modal mass definition for the normal modes, such that

$$\{\Phi_n\}^T[M]\{\Phi_j\} = \delta_{nj}$$
$$\{\Phi_n\}^T[K]\{\Phi_j\} = \omega_n^2 \delta_{nj} \qquad (4.2.20)$$

where δ_{nj} is the Kronecker delta: $\delta_{nj} = 0$ if $n \neq j$, $\delta_{nj} = 1$ if $n = j$.

There is a shorthand way of checking all of the orthogonality properties based on the technique of partitioning a matrix. The idea of partitioning is to break a column or rectangular matrix into submatrices that are *partitions* of the full array. These partitions may be visualized by drawing vertical lines through the full matrix to indicate its decomposition into groups of columns, while horizontal lines through the full matrix would indicate decomposition into groups of rows. If the submatrices forming partitions are dimensioned in a conformable manner, the partitions behave like single elements for sums, differences, and products. For example, multiplication of two partitioned matrices proceeds as follows:

$$\begin{bmatrix} [a][b] \\ [c][d] \end{bmatrix} \begin{bmatrix} [e] & [f] \\ [g] & [h] \end{bmatrix} \equiv \begin{bmatrix} [a][e] + [b][g] & [a][f] + [b][h] \\ [c][e] + [d][g] & [c][f] + [d][h] \end{bmatrix} \qquad (4.2.21)$$

Clearly, the above product is defined only if the number of columns of the first matrix equals the number of rows of the second matrix. Conformability of the partitions is satisfied if the number of columns of $[a]$ and $[c]$ equals the number of rows of $[e]$ and $[f]$.

We use partitions to define an array $[\Phi]$ such that column n is the normal mode $\{\Phi_n\}$. In partitioned notation the successive normal modes form a single row, specifically

$$[\Phi] = [\{\Phi_1\} \cdots \{\Phi_N\}] \qquad (4.2.22)$$

We call the $N \times N$ matrix $[\Phi]$ the *normal mode matrix*. Note that the element of $[\Phi]$ in the jth row and nth column is Φ_{jn}, which is consistent with the notation we adopted earlier to indicate the jth element of $\{\phi_n\}$.

Let us use partitioning to evaluate the product $[\Phi]^T[M][\Phi]$. Because the partitioned form of $[\Phi]$ is a row whose elements are $\{\Phi_n\}$, the corresponding partitioned form of $[\Phi]^T$ is a column whose elements are the sequence of $\{\Phi_n\}^T$. That is, the partitions of $[\Phi]^T$ are rows containing the transpose of the mode vectors. The individual partitions behave as single elements in a product, so we have

$$[\Phi]^T[M][\Phi] = \begin{bmatrix} \{\Phi_1\}^T \\ \vdots \\ \{\Phi_N\}^T \end{bmatrix} [M][\{\Phi_1\} \cdots \{\Phi_N\}]$$

$$= \begin{bmatrix} \{\Phi_1\}^T \\ \vdots \\ \{\Phi_N\}^T \end{bmatrix} [[M]\{\Phi_1\} \cdots [M]\{\Phi_N\}] \qquad (4.2.23)$$

$$= \begin{bmatrix} \{\Phi_1\}^T[M]\{\Phi_1\} \cdots \{\Phi_1\}^T[M]\{\Phi_N\} \\ \vdots \qquad\qquad \vdots \\ \{\Phi_N\}^T[M]\{\Phi_1\} \cdots \{\Phi_N\}^T[M]\{\Phi_N\} \end{bmatrix}$$

Each element in the above result is a scalar whose value is described by the first of eqs. (4.2.20). Specifically, the off-diagonal values are zero, while the diagonal terms are the unit modal masses associated with modes that have been normalized. Hence, we have

$$\boxed{[\Phi]^T[M][\Phi] = [I]} \qquad (4.2.24)$$

A diagonal form also results from following the same operations with $[M]$ replaced by $[K]$. Specifically, application of the second of eqs. (4.2.20) yields

$$\boxed{[\Phi]^T[K][\Phi] = [\omega_{\text{nat}}^2] \equiv \begin{bmatrix} \omega_1^2 & 0 & \cdots & 0 \\ 0 & \omega_2^2 & \cdots & 0 \\ \vdots & \vdots & \vdots & \vdots \\ 0 & 0 & \cdots & \omega_N^2 \end{bmatrix}} \qquad (4.2.25)$$

The mathematical terminology for the products in the left side of eqs. (4.2.24) and (4.2.25) is that they are *similarity transformations*.

After we have evaluated the normal modes, it is wise to compute $[\Phi]^T[M][\Phi]$ and $[\Phi]^T[K][\Phi]$. Because of truncation errors stemming from finite precision calculations, the off-diagonal terms will probably not be zero, but they should be much smaller than the diagonal terms if the modes were correctly determined. (In fact, the square root of the ratio of the magnitude of the former to the latter is a measure of the precision of the eigensolution.) Assuming that the off-diagonal terms are sufficiently small, the diagonal terms of the first product should be extremely close to unity, while those of the second product should be very close to the squares of the natural frequencies. It is possible to prove that satisfaction of eqs. (4.2.24) and (4.2.25) is a necessary and sufficient condition for the natural frequencies and normal modes to be solutions of the generalized eigenvalue problem, $[[K] - \omega_j^2[M]]\{\Phi_j\} = \{0\}$. Thus, verifying that the computed modes actually satisfy both orthogonality properties provides a robust check that the eigensolution was correctly computed.

EXAMPLE 4.3

The mass matrix and first two (nonnormalized) modes of a three-degree-of-freedom system are

$$[M] = \begin{bmatrix} 4 & 0 & 0 \\ 0 & 6 & 0 \\ 0 & 0 & 5 \end{bmatrix}, \quad \{\phi_1\} = \begin{Bmatrix} 1 \\ -1.9543 \\ -1.6779 \end{Bmatrix}, \quad \{\phi_2\} = \begin{Bmatrix} 1 \\ 0.46157 \\ -0.15519 \end{Bmatrix}$$

(a) Determine the third normal mode of free vibration.

(b) If the natural frequencies are $\omega_1 = 20$, $\omega_2 = 30$, and $\omega_3 = 40$ rad/s, what is the stiffness matrix $[K]$?

Solution This exercise will illustrate the operations by which modes are normalized. It also will demonstrate that knowledge of the modal properties can be used to infer the system properties, which is an important concept underlying some experimental techniques. We begin by using the given $[M]$ to normalize the two known modes. Thus, we compute the modal mass factors

$$\{\phi_1\}^T[M]\{\phi_1\} = 40.992, \qquad \{\phi_2\}^T[M]\{\phi_2\} = 5.399$$

Dividing the given modes by the square root of these modal mass factors gives

$$\{\Phi_1\} = \frac{\{\phi_1\}}{\sqrt{40.992}} = \begin{Bmatrix} 0.1562 \\ -0.3052 \\ -0.2621 \end{Bmatrix}, \quad \{\Phi_2\} = \frac{\{\phi_2\}}{\sqrt{5.399}} = \begin{Bmatrix} 0.4304 \\ 0.1986 \\ -0.0668 \end{Bmatrix}$$

We assign algebraic values to the three unknown elements of the third normal mode, $\{\Phi_3\} = [a\ b\ c]^T$, and require that this mode be orthogonal to the other two modes,

$$\{\Phi_1\}^T[M]\{\Phi_3\} = 0 \quad \text{and} \quad \{\Phi_2\}^T[M]\{\Phi_3\} = 0$$

We use the normal mode values above to compute

$$\{\Phi_1\}^T[M] = [0.6248 \quad -1.8314 \quad -1.3103]$$
$$\{\Phi_2\}^T[M] = [1.7215 \quad -1.1919 \quad -0.3340]$$

from which it follows that

$$\{\Phi_1\}^T[M]\{\Phi_3\} = 0.6248a - 1.8314b - 1.3103c = 0$$
$$\{\Phi_2\}^T[M]\{\Phi_3\} = 1.7215a - 1.1919b - 0.3340c = 0$$

Because these are two equations for three unknown values, we solve for b and c in terms of a:

$$-1.8314b - 1.3103c = -0.6248a$$
$$-1.1919b - 0.3340c = -1.7215a$$
$$b = -0.9418a, \qquad c = 1.7932a$$

Enforcing the requirement that $\{\Phi_3\}$ is normal removes the arbitrariness of a. We follow the earlier procedure to compute

$$\{\Phi_3\} = \frac{\{\phi_3\}}{\sqrt{\{\phi_3\}^T[M]\{\phi_3\}}} = \frac{1}{\sqrt{25.40}} \begin{Bmatrix} 1 \\ -0.9418 \\ 1.7932 \end{Bmatrix} = \begin{Bmatrix} 0.1984 \\ -0.1869 \\ 0.3558 \end{Bmatrix}$$

Now that we have established the third mode, we check orthogonality using the normal mode matrix, which we form by arranging the eigenmodes sequentially in columns,

$$[\Phi] = \begin{bmatrix} 0.1562 & 0.4304 & 0.1984 \\ -0.3052 & 0.1986 & -0.1869 \\ -0.2621 & -0.0668 & 0.3558 \end{bmatrix}$$

The first orthogonality product is

$$[\Phi]^T[M][\Phi] = \begin{bmatrix} 1.000 & -0.007 & -1.8(10^{-5}) \\ -0.007 & 1.000 & 2.0(10^{-5}) \\ 1.8(10^{-5}) & 2.0(10^{-5}) & 1.000 \end{bmatrix}$$

This is close to being the identity matrix, which confirms that we have performed the operations correctly.

We now proceed to the determination of the stiffness matrix. The natural frequencies are given, so we know the result that would be obtained from the second orthogonality product if we knew $[K]$,

$$[\Phi]^T[K][\Phi] = [\omega^2] = \begin{bmatrix} \omega_1^2 & 0 & 0 \\ 0 & \omega_2^2 & 0 \\ 0 & 0 & \omega_3^2 \end{bmatrix}$$

We may solve this relation for $[K]$ by postmultiplying it by $[\Phi]^{-1}$ and premultiplying it by the inverse of $[\Phi]^T$. An identity from matrix algebra states that $[[\Phi]^T]^{-1} = [[\Phi]^{-1}]^T$, so we write either term as $[\Phi]^{-T}$. These operations yield

$$[K] = [\Phi]^{-T}[\omega^2][\Phi]^{-1}$$

We compute the inverse of the normal mode matrix,

$$[\Phi]^{-1} = \begin{bmatrix} 0.6376 & -1.8227 & -1.3129 \\ 1.7262 & 1.1784 & -0.3437 \\ 0.7937 & -1.1213 & 1.7790 \end{bmatrix}$$

Substitution of this quantity into the above identity for $[K]$ yields

$$[K] = \begin{bmatrix} 3831.3 & -34.8 & 1414.3 \\ -34.8 & 4631.8 & -2589.9 \\ 1414.3 & -2589.9 & 5851.1 \end{bmatrix}$$

An interesting aspect of the determination of $\{\Phi_3\}$ is that the same procedure may be extended to any situation where we know $[M]$ and all except one of the modes of an N-degree-of-freedom system. Also, the fact that we were able to determine $[K]$ from knowledge of $[M]$ and $\{\Phi\}$ emphasizes the close relationship between the modal properties and the inertia and stiffness matrices.

4.2.3 Computerized Implementations

The formal procedure for solving the eigenvalue problem governing free vibration relies on determination of the characteristic equation. As we saw in the context of eq. (4.2.5), this determination involves a symbolic expansion of the determinant $|[K] - \omega^2[M]|$, with ω^2 represented algebraically. If the system has only two or three degrees of freedom, this task is not too difficult, because there are convenient shortcuts for evaluating determinants of 2×2 and 3×3 matrices.

The task becomes much more intricate as the number of degrees of freedom is increased beyond $N = 3$. Symbolic programming languages such as Mathematica, Maple, and Macsyma could be used to determine the characteristic equation for somewhat larger N, but the result would quickly become unwieldy. Our interest in the characteristic equation is primarily as a path leading to the eigenvalues. If we are willing to give up the notion of solving the eigenvalue problem algebraically in terms of system parameters, we may employ numerical methods for solving matrix eigenvalue problems. Rather than approaching the problem from the viewpoint of solving the characteristic equation, these methods are founded on fundamental mathematical properties of eigenvalue problems. In fact, the characteristic equation does not occur at all in the procedures. These methods generally fall into two categories: sweeping methods and orthogonal transformations. It is beyond the scope of this text to describe the algorithms, which are described in a reasonably accessible manner in a number of texts, notably Press et al. (1992) and Meirovitch (1997).

The important feature for us is that matrix eigenvalue routines have been implemented as part of most mathematics software, as well as in libraries of numerical subroutines for higher-level computer languages (Fortran, C++, etc.). Our purpose here is to examine how those routines may be employed. The first factor to note is that eq. (4.2.3) constitutes a *general (matrix) eigenvalue problem*. Much of the popular mathematical software has routines that yield the eigenvalues and eigenvectors for the general eigenvalue problem. In MATLAB, the routine is called "eig," which should be called using the syntax $[phi, lambda] = eig(K, M)$, where K and M are numerical stiffness and inertia matrices. After execution of this statement, *lambda* will be the diagonal matrix $[\omega^2]$, and *phi* will hold the eigenvectors in successive columns matching the sequence in which $[\omega^2]$ holds the eigenvalues. Mathcad performs the computation using two routines: $\lambda := genvals\ (K, M)$ and $\phi := genvecs\ (K, M)$. (Usage of the software is addressed in more detail in the next example.) It is important to recognize that the modes contained in the columns of the mode matrix returned by either program are not necessarily normalized according to eq. (4.2.19).

There will be many occasions where it is useful to embed algebraic scaling factors, which we denote as ν and κ, in $[M]$ and $[K]$. Algebraic quantities are not compatible with numerical (as opposed to symbolic) software. When we account explicitly for these factors, the eigenvalue problem becomes

$$[\kappa[K'] - \omega^2 \nu[M']]\{\phi\} = \{0\} \tag{4.2.26}$$

where $[M']$ and $[K']$ are purely numerical values. Division by κ leads to

$$[[K'] - \lambda[M']]\{\phi\} = \{0\}, \qquad \omega = \left(\lambda \frac{\kappa}{\nu}\right)^{1/2} \tag{4.2.27}$$

Note that this is the eigenvalue problem we would have if ν and κ were set to unity. We use the software to determine numerical values of λ, after which we compute the corresponding natural frequencies by applying the definition of λ.

It is possible that a circumstance will arise in which one does not have access to a general eigenvalue solution routine. Two strategies are possible if the available routines can only solve the standard problem, $[[A] - \lambda[I]]\{x\} = \{0\}$. The simplest entails inverting either $[M]$ or $[K]$. Premultiplying eq. (4.2.3) by either inverse yields

$$[[M]^{-1}[K] - \omega^2[I]]\{\phi\} = \{0\}$$
$$[[I] - \omega^2[K]^{-1}[M]]\{\phi\} = \{0\} \tag{4.2.28}$$

In the first case we would use the standard eigenvalue solver by setting $[A] = [M]^{-1}[K]$ and then find the natural frequencies from the eigenvalues returned by the solver as $\omega_j = \sqrt{\lambda_j}$. In the second case above, we would input $[A] = [K]^{-1}[M]$ to the solver and then find the natural frequencies corresponding to the computed eigenvalues by evaluating $\omega_j = 1/\sqrt{\lambda_j}$.

A note of caution applies if we choose to recast the problem into the second of eqs. (4.2.28). It obviously assumes that $[K]$ is invertible. As we will see in a later section that treats systems capable of rigid body motion, this is not always the case. In contrast, $[M]$ will always be invertible. Thus, the first of eqs. (4.2.28) represents the more robust of the approaches.

Although converting the problem to either of eqs. (4.2.28) is quite straightforward, its suitability depends on the specific requirements of the standard eigenvalue solver to be used. Some numerical algorithms require that the coefficient matrix $[A]$ in the standard eigenvalue problem be symmetric. Even though $[M]$ and $[K]$ are both symmetric, the products $[M]^{-1}[K]$ and $[K]^{-1}[M]$ will usually not be symmetric. Also, the lack of symmetry can be a source of difficulty, even if the eigensolver is capable of handling nonsymmetric $[A]$. Press et al. (1992) note that the methods used for solving nonsymmetric standard eigenvalue problems might be highly sensitive to round-off error, depending on the nature of $[A]$. For this reason the reader is advised to avoid using either of eqs. (4.2.28), although the method seems to work well for $N < 10$.

When a general eigenvalue solver is not available, $[[K] - \omega^2[M]]\{\phi\} = \{0\}$ can be transformed to an equivalent symmetric standard eigenvalue problem. The procedure applies *Cholesky decomposition,* which is sometimes referred to as the *square root of a matrix.* Only nonsingular matrices may be decomposed in this manner, so we select $[M]$ for this treatment. Cholesky decomposition factorizes $[M]$ such that

$$[M] = [\ell][\ell]^{\mathrm{T}} \qquad (4.2.29)$$

The factor $[\ell]$ is a square lower triangular matrix, which means that it is populated with zeroes above and to the right of the diagonal, $\ell_{jk} = 0$ if $k > j$. Subroutines to perform Cholesky decomposition are widely available.

When we substitute the factorized form of $[M]$ into the modal eigenvalue problem, we obtain

$$[K]\{\phi\} - \omega^2[\ell][\ell]^{\mathrm{T}}\{\phi\} = \{0\} \qquad (4.2.30)$$

The next step is to introduce a change of variables,

$$\{y\} = [\ell]^{\mathrm{T}}\{\phi\} \quad \Rightarrow \quad \{\phi\} = [\ell]^{-\mathrm{T}}\{y\} \qquad (4.2.31)$$

where $[\ell]^{-\mathrm{T}}$ denotes the inverse of the transpose, or equivalently, the transpose of the inverse. The lower triangular nature of $[\ell]$ makes finding $[\ell]^{-\mathrm{T}}$ a simple task. We substitute eq. (4.2.31) into the eigenvalue problem. In order to obtain $[I]$ as the coefficient matrix multiplying ω^2, we also premultiply the equation by $[\ell]^{-1}$, which leads to

$$[[\ell]^{-1}[K][\ell]^{-\mathrm{T}} - \omega^2[I]]\{y\} = \{0\} \qquad (4.2.32)$$

Because $[[\ell]^{-\mathrm{T}}]^{\mathrm{T}} \equiv [\ell]^{-1}$, the first coefficient matrix is symmetric. Thus, we would input $[\ell]^{-1}[K][\ell]^{-\mathrm{T}}$ as the coefficient matrix for the symmetric standard eigenvalue solver. The routine would return the values of ω_j^2 as the eigenvalues, and the eigenvectors would be the associated values of $\{y_j\}$. We would then need to evaluate the vibration modes by applying eq. (4.2.31), which gives $\{\phi_j\} = [\ell]^{-T}\{y_j\}$. An interesting aspect of Cholesky decomposition is that it apparently is implemented internally when the user calls the general eigenvalue solvers in MATLAB and Mathcad.

The point of the foregoing discussion is that numerous computational aids are available to assist us with the evaluation of a system's natural frequencies and vibration modes. For systems having a truly large number of degrees of freedom, for example, $N > 1000$, the computational effort will be quite large, but the evaluation would not be possible without the availability of computers.

Different computer implementations might not scale the eigenvectors by the same scaling factors, even if they use the same algorithms. In any event, we wish to employ normal modes, which are defined in eq. (4.2.19). Thus, one should make it standard practice to apply eq. (4.2.19) to the computed modes immediately after those modes have been determined.

EXAMPLE 4.4

The system illustrated below is a lumped mass model used to explain extensional vibrations of bars. It is based on dividing a bar of length L into N equal length segments. The mass of each segment m/N is represented as a point mass situated at the midpoint of the segment, where m is the total mass of the bar. Static application of an axial force F to a bar having length ℓ elongates that bar by $F\ell/EA$, so the corresponding stiffness is EA/ℓ. In the lumped mass model the masses are separated by distance L/N, so the stiffness of springs connected between masses is Nk, where $k = EA/L$. The springs at the left and right ends have length $L/2N$, so the stiffness of those springs is $2Nk$. It is desired to explore the modal properties of this system for various N using the horizontal displacement x_n of each mass from its static equilibrium position. Compare the lowest five natural frequencies resulting from setting $N = 5, 10, 15,$ and 20. Then examine the first four mode shapes obtained from each value of N.

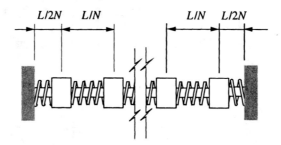

Solution   This exercise will illustrate the considerations necessary to employ mathematical software to solve the eigenvalue problem and to normalize the modes. The results will be a precursor to later chapters, where we will examine the vibratory properties of bars from a more precise formulation. The arrangement of springs is sufficiently simple to use Newton's Second Law. We let n denote the number of the mass under consideration, sequenced left to right. We consider the displacements to be positive if they are rightward. The spring to the left of mass m_n is k_n, whose elongation is $\Delta_n = x_n - x_{n-1}$. The spring to the right of m_n is k_{n+1}; its elongation is $\Delta_{n+1} = x_{n+1} - x_n$. The equation of motion for this mass is $\Sigma F = -k_n\Delta_n + k_{n+1}\Delta_{n+1} = m_n\ddot{x}_n$. The m_n values are all m/N, and the k_n values are all Nk, except for the end springs, for which $k_1 = k_{N+1} = 2Nk$. For the interior masses, the equations of motion that result are

$$\frac{m}{N}\ddot{x}_n - Nkx_{n-1} + 2Nx_n - Nkx_{n+1} = 0, \quad n = 2, 3, ..., N - 1$$

The masses at the ends represent special cases because the spring constants are different and also because the ends are fixed, $\Delta_1 = x_1$, $\Delta_{N+1} = -x_N$. Thus, the equations of motion for these bodies are

$$\frac{m}{N}\ddot{x}_1 + 3Nkx_1 - Nkx_2 = 0$$

$$\frac{m}{N}\ddot{x}_N - Nkx_{N-1} + 3Nkx_1 = 0$$

We see that the corresponding inertia matrix is m/N times an identity matrix. The stiffness matrix is said to be tridiagonal, which means that its only nonzero elements occur on the diagonal and in the elements on either side of the diagonal. Specifically,

$$[M] = \frac{m}{N}[I], \qquad [K] = Nk \begin{bmatrix} 3 & -1 & 0 & \cdots & 0 & 0 & 0 \\ -1 & 2 & -1 & \cdots & 0 & 0 & 0 \\ 0 & -1 & 2 & \cdots & 0 & 0 & 0 \\ \vdots & \vdots & \vdots & & \vdots & \vdots & \vdots \\ 0 & 0 & 0 & \cdots & 2 & -1 & 0 \\ 0 & 0 & 0 & \cdots & -1 & 2 & -1 \\ 0 & 0 & 0 & \cdots & 0 & -1 & 3 \end{bmatrix}$$

We wish to employ numerical software to determine the natural frequencies and modes, but k and m are algebraic parameters. This situation is addressed by eq. (4.2.27). We define nondimensional inertia and stiffness matrices and a nondimensional eigenvalue parameter, such that

$$[M'] = \frac{1}{m}[M], \qquad [K'] = \frac{1}{k}[K]$$

$$\lambda = \frac{m\omega^2}{k} \equiv \frac{mL\omega^2}{EA}$$

The resulting eigenvalue problem is

$$[[K'] - \lambda[M']]\{\phi\} = \{0\}$$

Let us first consider the step required to use MATLAB. The syntax for the function call that returns both the eigenvalues and eigenvectors is *[phi,lambda] = eig(K_prime,M_prime)*. The result of this operation is a diagonal array *lambda* whose elements are the values λ_n, and an $N \times N$ array *phi*, whose nth column is the mode $\{\phi_n\}$ corresponding to λ_n. We may obtain nondimensional natural frequencies by using the *sqrt* function, which operates on the individual elements of an array, such that *freq = sqrt(diag(lambda))*. Dimensional values could then be obtained by multiplying the *freq* array by $\sqrt{k/m}$ if the system properties were stated.

In this problem, we need to compare the lowest values of the natural frequencies for different values of N, and then compare the associated modes. This brings up a basic aspect of the *eig* function, which is that the natural frequencies it obtains are not arranged in any specific order. If we wish to run our MATLAB program for different N and save the lowest modal properties for each N case in an automated manner, we need to rearrange the data. The MATLAB *sort* function provides the tool for this purpose. If we use *[freq_sort,n_sort] = sort(freq)*, then *freq_sort* will contain the natural frequency values arranged in ascending order. Equally important is *n_sort*, whose nth element holds the position in *freq* of the elements of *freq_sort*. The programmatic loop *for j = 1 : N; phi_sort(:, j) = phi(:, n_sort(j)); end* will rearrange the columns of *phi* to be consistent with the sorted sequence of natural frequencies.

In order to compare the mode shapes for different N, we normalize the modes by evaluating the first orthogonality product according to *mu = phi_sort' *M_prime*phi_sort; for*

$j = 1 : N; phi\_norm = phi\_prime(:, j)/sqrt(mu(j)); end.$ The syntax of these statements may be converted directly into programs for other software, such as C++ and Fortran.

Nondimensional natural frequencies may be obtained in Mathcad by using the *genvals* function,

$$\Omega = \overrightarrow{\sqrt{genvals(K', M')}}$$

where an arrow above a quantity is Mathcad's way of denoting a vectorized operation that is performed individually on each element. The corresponding modes are computed with the *genvecs* function, $\phi = genvecs(K', M')$. Using $\omega = sort(\Omega)$ would rearrange the natural frequencies in ascending order, but Mathcad does not provide a built-in function that tracks the indicial correspondence between sorted and unsorted values. A simple workaround for this is to form an array in which the transpose of the column array Ω is placed as the first row in a matrix whose lower partition is the eigenvector array ϕ. The *stack* function does this operation. The function $rsort(A, n)$ returns an array formed by rearranging columns of A until row n is in ascending order. We then may extract the eigenvectors in the sequence matching the sorted ω values by using the submatrix function. All of these operations may be executed as a single step using the following syntax:

$$\phi := submatrix(rsort(stack(\Omega^{\mathrm{T}}, \phi), 1), 2, rows(\Omega) + 1, 1, rows(\Omega))$$

Note that the foregoing assumes that Mathcad has been set up to number the lowest subscript for arrays as 1. The steps to normalize the modes in Mathcad are $\mu = \phi^{\mathrm{T}} * M' * \phi$ $\Phi^{<j>} = \phi^{<j>} / \sqrt{\mu_{j,j}}$.

The results for the nondimensional frequencies are displayed in the first graph. We see that the lowest two natural frequencies obtained from the smallest value of N agree well with those obtained from larger N, but the discrepancies grow as the mode number increases. This is a general property of discrete models of continuous systems, as we will see in later chapters. A more exact analysis of the axial vibration of this bar according to the methods in Chapter 7 would show that the natural frequencies are $\omega'_n = n\pi$. For the first five modes, the maximum error from these values obtained for $N = 20$ is 2.5% for the fifth frequency.

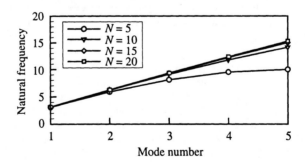

The mode shapes corresponding to the first four natural frequencies may be displayed graphically. To do so, we use as the abscissa the position of mass number j from the right end, which is $(j - 0.5)/N$, and plot that value against each element $\phi_{j,n}$ of mode n. The modes are described in the next set of graphs. The mode shapes display an increasing number of wiggles as the mode number increases. The values for each mode at a specific axial position agree, but a higher N provides a much smoother appearing curve. The analysis using the methods in Chapter 7 indicates that these curves should depend on the axial position y according to $\phi_n = \sin(n\pi y/L)$, which closely fits the graphs.

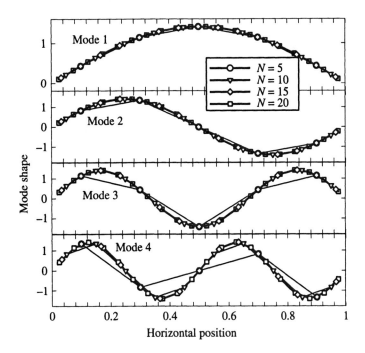

4.2.4 Flexibility Coefficients

One of the approaches noted in the previous section for converting the general eigenvalue problem $[[K] - \omega^2[M]]\{\phi\} = \{0\}$ to a standard eigenvalue problem entailed inverting the stiffness matrix. This inverse, which we shall denote as $[L]$, is called the *flexibility matrix*, or, in some references, the *influence coefficient matrix*,

$$\boxed{[L] \equiv [K]^{-1}} \qquad (4.2.33)$$

The physical significance of the flexibility coefficients may be recognized by considering a system carrying a set of constant forces $\{Q\}$. Eventually the system will return to rest under such loads. The equations of motion for this static position are

$$[K]\{q\} = \{Q\} \qquad (4.2.34)$$

To determine the displacements corresponding to the known set of forces, we solve the above, which leads to

$$\{q\} = [L]\{Q\} \qquad (4.2.35)$$

It is clear from the preceding that if two designs for a system are compared, the one whose elements of $[L]$ are larger will be displaced more in response to a specified static force system. Hence, the overall magnitude of $[L]$ is a measure of the system's flexibility. This is in contrast to $[K]$, which indicates the forces required to produce a given displacement field, and therefore is a measure of the system's stiffness. For a one-degree-of-freedom system, we have $L_{11} = 1/K_{11}$, which makes the inverse relationship obvious.

To understand the meaning of a flexibility coefficient, consider the lightweight elastic beam in Figure 4.5, which supports four blocks. If we ignore the mass of the beam and define the vertical displacement x_j of each supported mass to be a generalized coordinate, then the generalized forces are the vertical force F_j acting on each

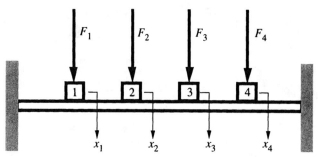

FIGURE 4.5 An elastic beam supporting several masses.

mass. Suppose a force F_n is applied to mass number n, and none of the other masses carry a force. In that case $\{Q\}$ contains F_n in row n and zeroes everywhere else. The corresponding displacement field given by eq. (4.2.35) is $x_j = L_{jn}F_n$, $j = 1, \ldots, N$. Thus, a coefficient L_{jn} represents the value of the jth generalized coordinate when the nth generalized force having unit magnitude is applied statically to a system, with all of the other generalized forces set to zero.

This interpretation demonstrates that it is a comparatively straightforward matter to measure $[L]$ experimentally. All that one need do is sequentially apply a single generalized force Q_n and measure all the generalized coordinates q_j. Dividing each q_j by Q_n yields the corresponding L_{jn}. Compare this procedure to what would be required to apply eq. (4.2.34) to measure stiffness coefficients experimentally. According to that relation, a column of $[K]$ constitutes the set of forces required to impose a displacement field in which one generalized coordinate has a unit value and all other generalized coordinates are zero. Implementing such a procedure for the beam in Figure 4.5 would require using trial and error or servo-control to identify the set of static forces required to produce such a displacement field.

Once we have determined $[L]$, we may use it to determine the natural frequencies and modes. Premultiplying the general eigenvalue equation by $[L]$ leads to

$$[\lambda[I] - [L][M]]\{\phi\} = \{0\}, \quad \omega^2 = \frac{1}{\lambda} \tag{4.2.36}$$

The product $[L][M]$ will be nonsymmetric. If solution of a nonsymmetric standard eigenvalue does not present a difficulty for the available eigenvalue solver, then the eigenvalues λ_n and modes $\{\phi_n\}$ may be obtained directly from that solver. To do so, one would compute $[A] = [L][M]$ and solve $[[A] - \lambda[I]]\{\phi\} = \{0\}$. The natural frequencies in this procedure would be obtained by evaluating $(1/\lambda_n)^{1/2}$. On the other hand, if the lack of symmetry is a difficulty, one could evaluate the stiffness by computing $[K] = [L]^{-1}$, and then proceed to solve the general eigenvalue problem, as we have up to now. It is important to recognize that, regardless of which form of the eigenvalue problem is solved, orthogonality and normalization of the modes is governed by $\{\Phi_j\}^T[M]\{\Phi_n\} = \delta_{jn}$.

EXAMPLE 4.5

A steel construction manual gives the static displacement w at distance x from the left end of a simply supported beam loaded by a transverse force P as

$$w = \frac{Pbx}{6EIL}(L^2 - b^2 - x^2) \quad \text{if } x \le a$$

In this relation, L is the length of the beam, a is the distance of the force from the left end, $b = L - a$, and EI is the bending rigidity. Use this information to obtain the flexibility coefficients corresponding to a set of generalized coordinates that are the vertical displacement of each of the three blocks supported by the beam. Then evaluate the corresponding natural frequencies and normal modes. The masses are m, $3m$, and $2m$ for blocks A, B, and C, respectively, where m is a basic mass unit, and the mass of the beam is negligible.

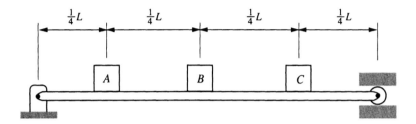

Solution In this example we will combine the formulation in terms of flexibility coefficients with our ability to solve eigenvalue problems using mathematical software. The generalized coordinates are $q_1 = w(x_A = L/4)$, $q_2 = w(x_B = L/2)$, and $q_3 = w(x_C = 3L/4)$. The given function for w is the static displacement due to a specified force. The flexibility is this displacement divided by the force. We need the flexibility values corresponding to forces and displacements at the locations of the blocks. The function describes situations where the force is at, or to the right of, the displacement point. In the context of the present problem, this enables us to evaluate the coefficients L_{jn} for $j \leq n$ as follows:

$$j = 1, \quad n = 1 \quad \Rightarrow \quad x = L/4, \quad a = L/4, \quad b = 3L/4 \quad \Rightarrow \quad L_{11} = \frac{w}{P} = 0.01172D$$

$$j = 1, \quad n = 2 \quad \Rightarrow \quad x = L/4, \quad a = L/2, \quad b = L/2 \quad \Rightarrow \quad L_{12} = \frac{w}{P} = 0.01432D$$

$$j = 1, \quad n = 3 \quad \Rightarrow \quad x = L/4, \quad a = 3L/4, \quad b = L/4 \quad \Rightarrow \quad L_{13} = \frac{w}{P} = 0.00911D$$

$$j = 2, \quad n = 2 \quad \Rightarrow \quad x = L/2, \quad a = L/2, \quad b = L/2 \quad \Rightarrow \quad L_{22} = \frac{w}{P} = 0.02083D$$

$$j = 2, \quad n = 3 \quad \Rightarrow \quad x = L/2, \quad a = 3L/4, \quad b = L/4 \quad \Rightarrow \quad L_{23} = \frac{w}{P} = 0.01432D$$

$$j = 3, \quad n = 3 \quad \Rightarrow \quad x = 3L/4, \quad a = 3L/4, \quad b = L/4 \quad \Rightarrow \quad L_{33} = \frac{w}{P} = 0.01172D$$

where $D = L^3/EI$. The preceding fills in the elements of $[L]$ on and to the right of the diagonal. We fill in the remaining elements by using the symmetry property, according to which $L_{jn} = L_{nj}$ for $j > n$. Thus, we find that

$$[L] = D \begin{bmatrix} 0.01172 & 0.01432 & 0.00911 \\ 0.01432 & 0.02083 & 0.01432 \\ 0.00911 & 0.01432 & 0.01172 \end{bmatrix}$$

The inertia matrix is substantially easier to obtain. The generalized coordinates are the displacement of each block, so the kinetic energy is

$$T = \tfrac{1}{2}(m_A \dot{x}_A^2 + m_B \dot{x}_B^2 + m_C \dot{x}_C^2)$$

Thus, we have

$$[M] = \begin{bmatrix} m_A & 0 & 0 \\ 0 & m_B & 0 \\ 0 & 0 & m_C \end{bmatrix} = m \begin{bmatrix} 1 & 0 & 0 \\ 0 & 3 & 0 \\ 0 & 0 & 2 \end{bmatrix}$$

Rather than inverting the flexibility matrix to obtain a symmetric general eigenvalue problem, we directly solve the nonsymmetric problem,

$$[[I] - \omega^2[L][M]]\{\phi\} = \{0\}$$

As a preliminary, we remove the D and m parameters and place the equation into standard form by defining a nondimensional eigenvalue and corresponding nondimensional inertia and stiffness matrices, such that

$$\lambda = \frac{1}{mD\omega^2}, \qquad [M'] = \frac{1}{m}[M], \qquad [L'] = \frac{1}{D}[L]$$

The eigenvalue equation that results is

$$[\lambda[I] - [A]]\{\phi\} = \{0\}$$

where

$$[A] = [L'][M'] = \begin{bmatrix} 0.01172 & 0.04296 & 0.01822 \\ 0.01432 & 0.06249 & 0.02864 \\ 0.00911 & 0.04296 & 0.02344 \end{bmatrix}$$

The next operations entail determining the eigenvalues λ_n and eigenvectors $\{\phi_n\}$. Because the preceding is a standard eigenvalue problem, Mathcad users may obtain the eigenvalues by writing $\lambda := eigenvals(A)$; the eigenvectors arranged in column sequence matching that of the eigenvalues is computed by writing $\phi = eigenvecs(A)$. MATLAB users may solve the problem in a single step by writing $[phi, lambda] = eig(A)$. The values returned by MATLAB are

$$diag(lambda) = \begin{Bmatrix} 0.092773 \\ 0.001032 \\ 0.003889 \end{Bmatrix}, \qquad phi = \begin{bmatrix} 0.4903 & -0.8741 & 0.7118 \\ 0.7104 & 0.3571 & 0.1603 \\ 0.5409 & -0.3292 & -0.6838 \end{bmatrix}$$

We compute the nondimensional natural frequencies in unsorted order by evaluating $\omega_n = 1/\sqrt{\lambda_n}$, which gives

$$\omega_1 = 3.284, \qquad \omega_2 = 31.151, \qquad \omega_3 = 16.054$$

From this juncture, the procedure for normalizing the modes follows Example 4.4. First, the natural frequencies are sorted in ascending order, and the column numbers of $[\phi]$ are rearranged to match the new sequence. Then the mass orthogonality condition is checked by evaluating

$$[\mu] = [\Phi]^T[M'][\Phi] = \begin{bmatrix} 2.2643 & 0 & 0 \\ 0 & 1.5192 & -2.72(10^{-15}) \\ 0 & -2.72(10^{-15}) & 1.3635 \end{bmatrix}$$

We see that the off-diagonal terms are very small, which confirms the correctness of the eigenvalue solution. We obtain normalized modes by dividing each column of $[\phi]$ by the corresponding square root of the diagonal element of $[\mu]$. This leads to the following modal properties, in which the dimensional natural frequencies are obtained by recalling that the nondimensionalizing factor is $\sqrt{mD}$, and $D = L^3/EI$,

$$\text{mode 1:} \quad \omega_1 = 3.284 \sqrt{\frac{EI}{mL^3}}, \qquad \{\Phi_1\} = \begin{Bmatrix} 0.3285 \\ 0.4721 \\ 0.3355 \end{Bmatrix}$$

$$\text{mode 2:} \quad \omega_2 = 16.054 \sqrt{\frac{EI}{mL^3}}, \qquad \{\Phi_2\} = \begin{Bmatrix} 0.5774 \\ 0.1301 \\ -0.5549 \end{Bmatrix}$$

$$\text{mode 3:} \quad \omega_3 = 31.151 \sqrt{\frac{EI}{mL^3}}, \qquad \{\Phi_3\} = \begin{Bmatrix} 0.7486 \\ -0.3058 \\ 0.2820 \end{Bmatrix}$$

4.2.5 Repeated Natural Frequencies

Our derivation of the mass and stiffness orthogonality relations began with the assumption that the natural frequencies are distinct. To ascertain the modifications required to handle situations where such is not the case, let us consider a specific system. Figure 4.6 depicts three springs equally arranged to support a small mass m. As generalized coordinates we use the horizontal and vertical displacements x and y of the mass. These displacements are measured relative to the static equilibrium position, so gravity may be ignored in formulating the potential energy. We obtain the deformation of each spring by taking the component of each displacement parallel to the spring, which leads to

$$\Delta_1 = -y, \qquad \Delta_2 = -x \cos 30° + y \cos 60°,$$

$$\Delta_3 = x \cos 30° + y \cos 60° \tag{4.2.37}$$

The energy expressions for this system are

$$T = \tfrac{1}{2}m(\dot{x}^2 + \dot{y}^2) \tag{4.2.38}$$

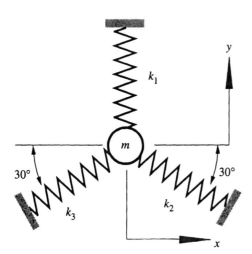

FIGURE 4.6 Two-degree-of-freedom system.

$$V = \frac{1}{2}k_1\Delta_1^2 + \frac{1}{2}k_2\Delta_2^2 + \frac{1}{2}k_3\Delta_3^2$$

$$= \frac{1}{2}\left[\frac{3}{4}(k_2 + k_3)x^2 + \frac{\sqrt{3}}{2}(k_3 - k_2)xy + \left(k_1 + \frac{1}{4}k_2 + \frac{1}{4}k_3\right)y^2\right] \qquad (4.2.39)$$

from which we find that the mass and stiffness matrices are

$$[M] = \begin{bmatrix} m & 0 \\ 0 & m \end{bmatrix}, \qquad [K] = \begin{bmatrix} \dfrac{3}{4}(k_2 + k_3) & \dfrac{\sqrt{3}}{4}(k_3 - k_2) \\ \dfrac{\sqrt{3}}{4}(k_3 - k_2) & k_1 + \dfrac{1}{4}(k_2 + k_3) \end{bmatrix} \qquad (4.2.40)$$

For arbitrary values of the spring stiffnesses the x and y displacements are coupled by $[K]$. However, if $k_3 = k_2$, then we find that $[K]$ is diagonal, so that the displacement components are uncoupled. The eigenvalue problem in that case is

$$\begin{bmatrix} (\frac{3}{2}k_2 - \omega^2 m) & 0 \\ 0 & (k_1 + \frac{1}{2}k_2 - \omega^2 m) \end{bmatrix}\{\phi\} = \{0\} \qquad (4.2.41)$$

The characteristic equation consists of the upper left and lower right terms as two factors. Correspondingly, we find that the first mode is governed by

$$\omega_1^2 = \frac{3k_2}{2m} \quad \Rightarrow \quad \begin{bmatrix} 0 & 0 \\ 0 & (k_1 - k_2) \end{bmatrix}\{\phi_1\} = \{0\} \qquad (4.2.42)$$

We discard the first row, and recognize that the second row requires that $\phi_{21} = 0$. Similarly, the second mode is governed by

$$\omega_2^2 = \frac{k_1}{m} + \frac{1}{2}\frac{k_2}{m} \quad \Rightarrow \quad \begin{bmatrix} (k_2 - k_1) & 0 \\ 0 & 0 \end{bmatrix}\{\phi_2\} = \{0\} \qquad (4.2.43)$$

Here we discard the second row, and find from the first row that $\phi_{12} = 0$. Thus, the modal properties when $k_3 = k_2 \neq k_1$ are

$$\omega_1^2 = \frac{3k_2}{2m}, \qquad \{\phi_1\} = \begin{Bmatrix} 1 \\ 0 \end{Bmatrix}$$

$$\omega_2^2 = \frac{k_1}{m} + \frac{1}{2}\frac{k_2}{m}, \qquad \{\phi_2\} = \begin{Bmatrix} 0 \\ 1 \end{Bmatrix} \qquad (4.2.44)$$

These modes reflect the fact that when $k_3 = k_2$, the x and y displacements are fully uncoupled, so the modes consist of displacement that is solely in either of these orthogonal directions.

Now suppose that we let $k_1 \rightarrow k_2$. In the limit, the natural frequencies become equal. Until the limit is actually attained, the modes in eqs. (4.2.44) are the only possible solutions of the eigenvalue problem, regardless of how small is the difference between k_2 and k_1. However, if $k_2 = k_1$, the coefficient matrices in eqs. (4.2.42) and (4.2.43) both vanish. In that case, the elements of $\{\phi_1\}$ and $\{\phi_2\}$ may be assigned any set of values. In other words, both elements become arbitrary. In such a situation, any linearly independent set of values may be used to construct the modes. It is reasonable, but not mandatory, to consider the modes for this case to be like those when k_2 is not quite equal to k_1. If we follow this dictum, then we select the two modes as indicated in eq. (4.2.44). However, we could just as well select any other linearly independent combination, such as $\{\phi_1\} = [1 \quad 0]^T$ and $\{\phi_2\} = [1 \quad -1]^T$. Thus, it seems as though the modes are ambiguously defined when the natural frequencies are equal.

We may, however, impose other requirements. Because $\omega_2 = \omega_1$, eq. (4.2.14), which is the basic identity leading to the orthogonality properties, will be satisfied regardless of how $\{\phi_1\}$ and $\{\phi_2\}$ are defined. Nevertheless, let us require that $\{\phi_1\}$ and $\{\phi_2\}$ be mutually orthogonal, as though the natural frequencies were not repeated. This gives an additional equation to compensate for our loss of an eigenequation. For example, suppose we decide that rather than employing the modes listed in eqs. (4.2.42) and (4.2.43), we wish to define $\{\phi_1\} = [1 \quad 2]^T$. We require that the corresponding second mode $\{\phi_2\} = [\phi_{12} \quad \phi_{22}]^T$ be orthogonal to $\{\phi_1\}$,

$$\{\phi_1\}^T[M]\{\phi_2\} = \begin{bmatrix} 1 & 2 \end{bmatrix} \begin{bmatrix} m & 0 \\ 0 & m \end{bmatrix} \begin{Bmatrix} \phi_{12} \\ \phi_{22} \end{Bmatrix} = 0 \quad \Rightarrow \quad \phi_{22} = -\frac{1}{2}\phi_{12} \qquad (4.2.45)$$

This defines ϕ_{22} in terms of ϕ_{12}, which remains arbitrary.

The preceding specific situation reflects the general case. The rank of the eigenequation is reduced by the number of times the natural frequency is encountered. Thus, if ω_j is not repeated, the rank of $[[K] - \omega_j^2[M]]$ will be $N - 1$ and $\{\phi_j\}$ contains one arbitrary element. If the natural frequency is found to occur p times as a root of the characteristic equation, such that $\omega_j = \omega_{j+1} = \cdots = \omega_{j+p-1}$, then the rank of $[[K] - \omega_j^2[M]]$ will be $N - p$, and ϕ_j will contain p arbitrary elements. We then impose the further restriction that the modes associated with the repeated eigenvalue be mutually orthogonal, as though the natural frequencies were not repeated. Thus, we require that

$$\{\phi_j\}^T[M]\{\phi_{j+1}\} = \{\phi_j\}^T[M]\{\phi_{j+2}\} = \cdots = \{\phi_j\}^T[M]\{\phi_{j+2}\} = 0 \qquad (4.2.46)$$

A technique for implementing these conditions in an orderly fashion is the *Gram-Schmidt orthogonalization procedure,* which constructs the mutually orthogonal modes in a recursive manner. The sequence of operations entailed in this procedure may be recognized by following an example in which an eigenvalue occurs three times, $p = 3$. We define a new set of modes $\{\phi_j'\}$, $\{\phi_{j+1}'\}$, $\{\phi_{j+2}'\}$, as a sum of the original mode at the new mode's number plus a linear combination of the new modes at lower numbers, that is,

$$\boxed{\begin{aligned} \{\phi_j'\} &= \{\phi_j\} \\ \{\phi_{j+1}'\} &= \{\phi_{j+1}\} + \alpha_{11}\{\phi_j'\} \\ \{\phi_{j+2}'\} &= \{\phi_{j+2}\} + \alpha_{21}\{\phi_j'\} + \alpha_{22}\{\phi_{j+1}'\} \end{aligned}} \qquad (4.2.47)$$

We obtain the α_{jn} coefficients by successively enforcing the orthogonality of each mode relative to the modes at lower numbers. For $\{\phi_{j+1}'\}$ we have

$$\{\phi_j'\}^T[M]\{\phi_{j+1}'\} \equiv \{\phi_j'\}^T[M]\{\phi_{j+1}\} + \alpha_{11}\{\phi_j'\}^T[M]\{\phi_j'\} = 0$$

(4.2.48)

$$\alpha_{11} = -\frac{\{\phi_j'\}^T[M]\{\phi_{j+1}\}}{\{\phi_j'\}^T[M]\{\phi_j'\}}$$

Now that modes $\{\phi_j'\}$ and $\{\phi_{j+1}'\}$ are set, we require that $\{\phi_{j+2}'\}$ be orthogonal to those modes, which leads to

$$\{\phi_j'\}^T[M]\{\phi_{j+2}'\} \equiv \{\phi_j'\}^T[M]\{\phi_{j+2}\} + \alpha_{21}\{\phi_j'\}^T[M]\{\phi_j'\} = 0$$

(4.2.49)

$$\{\phi_{j+1}'\}^T[M]\{\phi_{j+2}'\} \equiv \{\phi_{j+1}'\}^T[M]\{\phi_{j+2}\} + \alpha_{22}\{\phi_{j+1}'\}^T[M]\{\phi_j'\} = 0$$

This yields

$$\alpha_{21} = -\frac{\{\phi_j'\}^T[M]\{\phi_{j+2}\}}{\{\phi_j'\}^T[M]\{\phi_j'\}}$$

(4.2.50)

$$\alpha_2 = -\frac{\{\phi_{j+1}'\}^T[M]\{\phi_{j+2}\}}{\{\phi_{j+1}'\}^T[M]\{\phi_j'\}}$$

Note that the numerator and denominator in the preceding are scalars. The procedure is readily extended to any number of repeated eigenvalues. Once the new modes have been set, they replace the respective original modes and may be normalized in the same manner as any other mode.

Because repeated natural frequencies usually correspond to special values of system parameters, the most common circumstance in which one might observe the phenomenon is execution of a parameter study. A troublesome aspect of Gram-Schmidt orthogonalization is that the resulting modes depend on which mode was selected to be $\{\phi_j'\}$, with which the procedure is initiated. Consequently, some of the modes for the special parameter case might be drastically different from those for a parameter set that is only slightly different. One way to meet this objection is to find the modal solution for a very slightly different set of parameters. We then would select the arbitrary elements of $\{\phi_j'\}$, which correspond to the first of the repeated natural frequencies, such that the mode matches the corresponding mode for the slightly different system.

In closing, it should be noted that the Gram-Schmidt orthogonalization procedure is often incorporated into a software's eigensolver. If that is so, all computed modes will be mutually orthogonal, even when the natural frequencies are repeated. In any event, the best course is to verify that computed modes are indeed mutually orthogonal, and then to invoke the Gram-Schmidt procedure if they are not.

EXAMPLE 4.6

Mode localization refers to a phenomenon in which mode shapes become confined to one part of a system, rather than being distributed over the system. One source of such behavior is the introduction of small defects into a structure whose features nominally are spatially periodic. This has been, and continues to be, a significant source of concern for turbomachinery, in which turbine blades are supposed to be arranged periodically on a rotating hub. Wei and Pierre (1988) considered a set of 10 lightweight flexible bars arranged on a disk with point masses at the free tips. Elastic coupling between the blades was modeled by weak springs attached to adjacent masses. The sketch shows 3 such blade models in a set of 10.

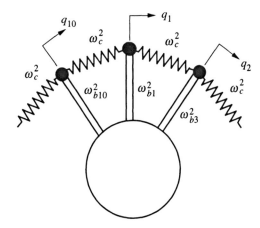

The equations of motion for generalized coordinates q_j that are the transverse displacement at the tip of each blade were provided by Pierre and Wei,

$$\ddot{q}_j + \omega_{bj}^2 q_j + \omega_c^2(2q_j - q_{j-1} - q_{j+1}) = 0, \quad j = 1, \ldots, N$$

where the radial arrangement of blades leads to $q_0 \equiv q_N$ and $q_{N+1} \equiv q_1$. The parameter ω_c^2 measures the strength of the interblade coupling, and ω_{bj} is the natural frequency of a single blade in the absence of interblade coupling. The deviation from periodicity of the structure was described by allowing the blade stiffness to vary about a mean value, such that

$$\omega_{bj}^2 = \omega_b^2(1 + \Delta f_j)$$

where Δf_j is a set of small values having a zero mean. A parameter defining the interblade coupling strength was defined as

$$R^2 = \frac{\omega_c^2}{\omega_b^2}$$

The actual value of ω_b is unimportant if time t is defined nondimensionally such that $t = \omega_b t'$, where t' is the dimensional quantity. Wei and Pierre presented results for a 10-bladed system ($N = 10$) that is very weakly coupled, $R = 0.05$. The imperfections in the blade stiffness were $\Delta f_j = 0.06581$, 0.03339, -0.03685, -0.00011, -0.05306, -0.00875, 0.04960, -0.04441, 0.00854, and -0.01416. This Δf_j distribution has a zero mean value and a 4% standard deviation. The idealized version of this system corresponds to $R = 0.05$ and all $\Delta f_j = 0$, so that the properties are constant from one blade to the next.

(a) Determine the natural frequencies of the ideal and imperfect systems. Plot each set of ω_j values against the mode number j on a common graph. What can one conclude from this graph regarding the effect of small imperfections on the natural frequencies?

(b) Determine the mode shapes of the ideal and imperfect systems. Plot each mode of the perfect system and the corresponding imperfect system modes versus the blade number on a set of 10 graphs.

Solution This system is an extreme example of a case of repeated natural frequencies. The phenomenon of mode localization is a fascinating one that is a subject of ongoing research. As we will see, small changes in a system parameter result in large alterations of the mode shapes. Consequently, it is extremely difficult to design such systems to meet vibration criteria, because the parameters in actual systems are seldom known to the precision required to predict the actual mode shapes. We begin the analysis by converting the equations of motion to nondimensional form, which we do by converting the dimensional t' derivatives to derivatives with respect to t, according to

$$\dot{q}_j = \frac{dq_j}{dt'} = \frac{dq_j}{dt}\frac{dt}{dt'} = \omega_b \frac{dq_j}{dt}, \qquad \ddot{q}_j = \omega_b^2 \frac{d^2q_j}{dt^2}$$

The corresponding equations of motion for $N = 10$ are

$$\frac{d^2q_j}{dt^2} + (1 + \Delta f_j)q_j + R^2(2q_j - q_{j-1} - q_{j+1}) = 0, \quad j = 1, ..., N$$

$$q_0 = q_{10}, \qquad q_{11} = q_1$$

We identify the inertia and stiffness matrix from these equations. The mass matrix for this system is the identity matrix. The stiffness matrix is essentially tridiagonal, except for elements in the upper right corner, corresponding to the influence of blade 10 on blade 1, and the lower left corner, which describes the influence of blade 1 on blade 10,

$$M_{j,n} = \begin{cases} 1, & j = n \\ 0, & j \ne n \end{cases}, \quad j, n = 1, ..., 10$$

$$K_{j,j} = 1 + \Delta f_j + 2R^2$$

$$K_{j,j+1} = K_{j+1,j} = -R^2, \quad j = 1, ..., 9$$

$$K_{1,10} = K_{10,1} = -R^2$$

We now solve $[[K] - \lambda[M]]\{\phi\} = \{0\}$ by employing our generalized eigenvalue solver. We sort the eigenvalues λ_n in ascending order, and apply that sort order to the columns of the matrix *phi* returned by the solver in the same order. The procedure for performing this task was outlined in Exercise 4.4. The natural frequencies are the square root of the sorted eigenvalues. The eigenvalues obtained from MATLAB for the perfect system (all $\Delta f_j = 0$) are the same as those obtained from Mathcad. Both evaluations show that there are four pairs of repeated eigenvalues,

$$\lambda_1 = 1, \qquad \lambda_2 = \lambda_3 = 1.00095, \qquad \lambda_4 = \lambda_5 = 1.00345$$
$$\lambda_6 = \lambda_7 = 1.00655, \qquad \lambda_8 = \lambda_9 = 1.00905, \qquad \lambda_{10} = 1.01$$

We next check the orthogonality of the (sorted) modes by computing $\mu = \phi^T M \phi$. The result obtained from MATLAB is a diagonal 10×10 matrix. In contrast, the off-diagonal terms $\mu_{n,n+1} = \mu_{n+1,n}$ for $n = 2, 4, 6,$ and 8 obtained in Mathcad are comparable in size to the diagonal terms. The other off-diagonal terms are essentially zero, as they should be. This informs us that the pair of modes associated with the repeated eigenvalues are not orthogonal to each other, although they are orthogonal to all other modes. Hence, Mathcad requires that we apply the Gram-Schmidt orthogonalization procedure. Because the eigenvalues are repeated once, we only need to apply the first orthogonalization, eq. (4.2.48), to modes 3, 5, 7, and 9. We perform this operation by defining an index k whose values are 3, 5, 7, and 9 and then writing, in the syntax of Mathcad,

$$\phi^{<k>} := \phi^{<k>} - \frac{(\phi^{<k>T} * M * \phi^{<k-1>})_{1,1}}{(\phi^{<k-1>T} * M * \phi^{<k-1>})_{1,1}} * \phi^{<k-1>}$$

where $\phi^{<k>}$ is Mathcad's way of specifying column k of ϕ. After this operation is performed, evaluation of the orthogonality product $\mu = \phi^T M \phi$ using the corrected modes yields a diagonal result. As a final confirmation of the computation for either Mathcad or MATLAB we compute the second orthogonality product, $\phi^T K \phi$. The result also is very close to the diagonal $[\omega^2]$ matrix, so we know that the eigenvectors have been correctly determined.

The first graph displays the nondimensional natural frequencies, which are the square root of the λ_j values, as a function of the mode number. The parameter $R = 0.05$ for both curves. All $\Delta f_j = 0$ for the curve denoted as the perfect system, while Δf_j are the specified values for the curve denoted as the imperfect system. We see that the imperfection results in a 3.5% maximum change of the natural frequency commensurate with the 4% standard deviation of the Δf_j values.

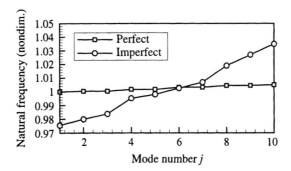

Unlike the natural frequencies, the imperfection has a drastic effect on the mode shapes. To display these quantities graphically we divide each $\{\phi_j\}$ by its maximum element (in the absolute value sense). This yields a mode shape whose maximum positive value is unity. The results for the jth mode are graphed by plotting the ϕ_{nj} values for fixed j as a function of n, which represents the blade number.

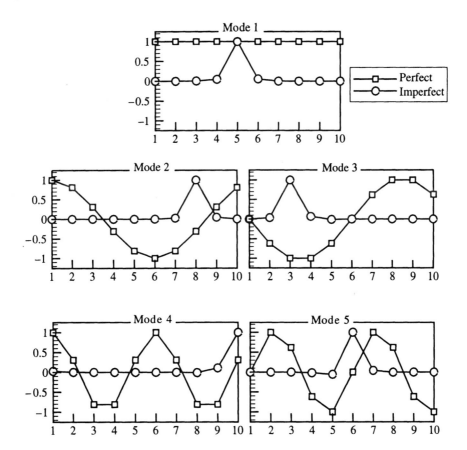

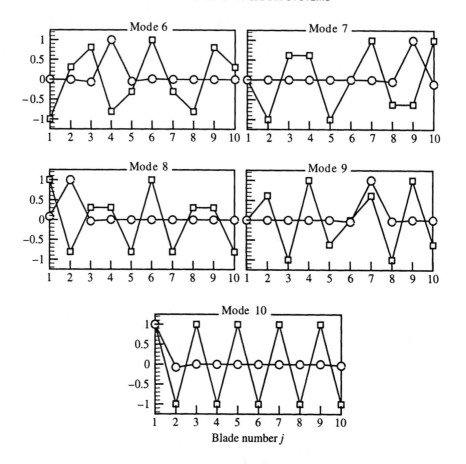

Blade number j

The modes of the perfect system show an increasing number of nodes as the mode number increases. A careful examination of these modes reveals that those associated with equal natural frequencies are somewhat similar, except that the features of the second in each pair are shifted to the left or right by approximately three blades. The latter feature is associated with the rotational symmetry of the system, which means that a 90° rotation of a mode shape produces an orthogonal pattern.

The mode shapes of the imperfect system are entirely different. In each mode essentially only one blade responds, which is the phenomenon of mode localization mentioned in the problem statement. It is obvious from this analysis that the mode shapes are extremely sensitive to the imperfection. These are important issues for the design of turbomachinery, as well as other systems featuring a periodic arrangement of mechanical elements. Chen and Ginsberg (1992) showed that the extreme parameter sensitivity, as well as localization, may be regarded as the result of mixing the mode shapes of the perfect system to obtain imperfect system modes, with the coefficients of the mix depending strongly on the amount of imperfection. In Chapter 9, we will see another example of mode localization involving a multispan beam.

4.2.6 Rigid Body Modes

The orthogonality properties are valid for any set of natural frequencies, even zero values. However, the modes associated with zero natural frequency have a special significance that makes it worthwhile to consider that case explicitly. As we did in the previous section, we begin with a specific system. Figure 4.7 features a pair of blocks that are interconnected by a spring, but otherwise are free to slide over the ground.

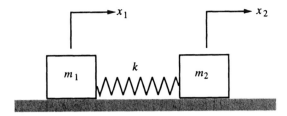

FIGURE 4.7 A two-degree-of-freedom system having a zero natural frequency.

When the displacements x_1 and x_2 are used as generalized coordinates, the eigenvalue problem is

$$\begin{bmatrix} (k - \omega^2 m_1) & -k \\ -k & (k - \omega^2 m_2) \end{bmatrix} \{\phi\} = \{0\} \tag{4.2.51}$$

The roots of the characteristic equation are $\omega^2 = 0$, $k(m_1 + m_2)/m_1 m_2$, which leads to the following equations for the vibration modes:

$$\omega_1 = 0 \quad \Rightarrow \quad \begin{bmatrix} k & -k \\ -k & k \end{bmatrix} \begin{Bmatrix} \phi_{11} \\ \phi_{21} \end{Bmatrix} = \begin{Bmatrix} 0 \\ 0 \end{Bmatrix}$$

$$\omega_2 = \sqrt{\frac{k(m_1 + m_2)}{m_1 m_2}} \quad \Rightarrow \quad \begin{bmatrix} -k\dfrac{m_1}{m_2} & -k \\ -k & -k\dfrac{m_2}{m_1} \end{bmatrix} \begin{Bmatrix} \phi_{12} \\ \phi_{22} \end{Bmatrix} = \begin{Bmatrix} 0 \\ 0 \end{Bmatrix} \tag{4.2.52}$$

If we consider the first element of each mode to be arbitrary, we find that

$$\{\phi_1\} = \phi_{11} \begin{Bmatrix} 1 \\ 1 \end{Bmatrix}, \qquad \{\phi_2\} = \phi_{12} \begin{Bmatrix} 1 \\ -\dfrac{m_1}{m_2} \end{Bmatrix} \tag{4.2.53}$$

Let us interpret these properties. In the zero frequency mode, both masses are displaced by equal amounts. We say that this is a *rigid body mode*, because the elastic element undergoes no deformation. The potential energy stored in the spring is identically zero in the rigid body mode, so there is no exchange between kinetic and potential energy to give rise to an oscillation.

The second mode is a *deformational mode,* in which the spring length oscillates. An interesting aspect of this motion comes from writing the relation satisfied by $\{\phi_2\}$ as $m_1 \phi_{12} + m_2 \phi_{22} = 0$. Because the displacement of each generalized coordinate in a free vibration at the second natural frequency is given by $x_j = \mathrm{Re}[B_2 \phi_{j2} \exp(i\omega_2 t)]$, it follows that at every instant, a displacement in the second mode is characterized by $m_1 x_1 + m_2 x_2 = 0$. Now consider the location of the center of mass of the system at an arbitrary instant. Let s_1 and s_2 be the horizontal positions of masses at the static equilibrium position. The position s_G of the center of mass in equilibrium is found by taking the first moment of mass, which leads to $(m_1 + m_2)s_G = m_1 s_1 + m_2 s_2$. If we denote the displacement of the center of mass from its equilibrium position as x_G, we find that the first moment of mass in this case is $(m_1 + m_2)(s_G + x_G) = m_1(s_1 + x_1) + m_2(s_2 + x_2)$. Subtracting the first moment of mass at equilibrium leads to $(m_1 + m_2)x_G = m_1 x_1 + m_2 x_2$. Consequently, we find that $x_G \equiv 0$ when the system executes a vibration in the second mode, which means that the center of mass remains in its static equilibrium position. This behavior is most apparent in the special case where the masses are equal, $m_2 = m_1$, in which case the modal displacements are equal in magnitude but opposite in direction. For unequal masses, the displacements are still opposite in direction, but the amplitudes are inversely proportional to the magnitude of the corresponding mass (that is, the bigger the mass is, the less it displaces).

These properties are manifestations of the general behavior of systems. The occurrence of a zero natural frequency always corresponds to a rigid body mode. To prove this we note that if $[[K] - \omega^2[M]]\{\phi\} = \{0\}$ has a nontrivial solution for $\omega = 0$, it must be that $|[K]| = 0$, so K is not invertible. However, the inverse of $[K]$ is the flexibility matrix $[L]$. When a set of forces are applied statically to the system, the equations of motion reduce to $\{Q\} = [K]\{q\}$. From this relation, we may determine the set of forces required to produce a specified static displacement. On the other hand, if $[K]$ is not invertible, we cannot select $\{Q\}$ arbitrarily, and expect that it will produce a static displacement field $\{q\}$. The eigenvector equation for a rigid body mode is $[K]\{\phi\} = \{0\}$; this means that $\{\phi\}$ is in the *null space* of $[K]$. We may interpret the resulting $\{\phi\}$ to be the displacement produced when no force is applied; in other words, the rigid motion that is possible if a system is not fully restrained.

A rigid body mode corresponds to a motion in which the relative distance between all parts remains constant, so that *none* of the elastic elements are deformed. If the system is restricted to planar motion, then there can be as many as three rigid body modes, corresponding to translation in either of two orthogonal directions and rotation. Thus, the eigenvalue $\omega^2 = 0$ can occur as many as three times in planar motion— once for each possible type of rigid body motion. The number of arbitrary elements in a rigid body mode that satisfies $[K]\{\phi\} = \{0\}$ will equal the number of times $\omega^2 = 0$ occurs as an eigenvalue. We may select these elements such that the rigid body modes are mutually orthogonal, $\{\phi_j\}^T[M]\{\phi_k\} = 0$ if $k \neq j$. The three mutually orthogonal rigid body modes we would obtain in this manner will correspond to translation in one of two orthogonal directions, and rotation about the center of mass. (Note that regardless of the choice of the arbitrary elements, the rigid body modes will always be orthogonal to the deformational modes.) The same properties apply to systems that may execute rigid body motion in three dimensions. Then there can be as many as six rigid body degrees of freedom: translation in any of three orthogonal directions and rotation about any of three orthogonal axes.

Many systems feature rigid body modes, notably, automobiles, airplanes, and other transportation systems. They also are encountered in various machine components. It is important to identify rigid body modes and to include them in the full set of eigensolutions, because in many cases the rigid body motion is the largest contributor to the displacement field.

EXAMPLE 4.7

A block of mass m slides along the inclined face of the cart, whose mass is $2m$. Frictional resistance is negligible. Determine the natural frequencies and mode shapes of this system.

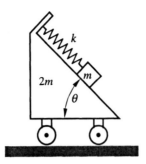

Solution This system is obviously capable of rigid body motion, but the special features of deformational modes will be disguised by our choice of generalized coordinates. We select as one generalized coordinate the horizontal displacement q_1 of the cart. The second generalized coordinate is the displacement q_2 of the block relative to the cart, with q_2 taken to be zero at the static equilibrium position, where the spring force balances the component of the weight parallel to the cart's surface. These variables are depicted in the free body diagram.

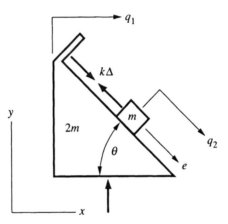

We shall use the power balance method to develop the equations of motion. The velocities of the two bodies are

$$\bar{v}_{2m} = \dot{q}\,\bar{i}, \quad \bar{v}_m = \bar{v}_{2m} + \dot{q}_2\bar{e} = (\dot{q}_1 + \dot{q}_2\cos\theta)\bar{i} - (\dot{q}_2\sin\theta)\bar{j}$$

The corresponding kinetic energy is

$$T = \tfrac{1}{2}(2m)\bar{v}_{2m}\cdot\bar{v}_{2m} + \tfrac{1}{2}m\bar{v}_m\cdot\bar{v}_m = \tfrac{1}{2}m[2\dot{q}_1^2 + (\dot{q}_1^2 + 2\dot{q}_1\dot{q}_2\cos\theta + \dot{q}_2^2)]$$

Matching this to the standard quadratic form leads to

$$M_{11} = 3m, \quad M_{22} = m, \quad M_{12} = m\cos\theta$$

We defined q_2 to be zero at the static equilibrium position of the spring, so we set $\Delta = q_2$, which gives

$$V = \tfrac{1}{2}kq_2^2 \quad \Rightarrow \quad K_{11} = K_{12} = 0, \quad K_{22} = k$$

The eigenvalue problem corresponding to the above inertia and stiffness is

$$[[K] - \omega^2[M]]\{\phi\} = \begin{bmatrix} -3m\omega^2 & -m(\cos\theta)\omega^2 \\ -m(\cos\theta)\omega^2 & (k - m\omega^2) \end{bmatrix}\begin{Bmatrix} 1 \\ \phi_2 \end{Bmatrix} = \begin{Bmatrix} 0 \\ 0 \end{Bmatrix}$$

The characteristic equation is

$$|[K] - \omega^2[M]| = -(3m\omega^2)(k - m\omega^2) - [m(\cos\theta)\omega^2]^2$$
$$= (3m^2 - m^2\cos^2\theta)\omega^4 - 3km\omega^2 = 0$$

The roots of this equation are

$$\omega_1^2 = 0, \qquad \omega_2^2 = \frac{3k}{m(3 - \cos^2\theta)}$$

The first equation of the eigenvalue problem corresponding to each natural frequency is

$$-3m\omega_n^2 - m(\cos\theta)\omega_n^2\phi_{2n} = 0$$

When $\omega_n = 0$, this equation is trivial, so we turn to the second equation, which gives

$$-m(\cos\theta)\omega_n^2 + (k - m\omega^2)\phi_{2n} = 0 \quad\Rightarrow\quad \phi_{2n} = \frac{m(\cos\theta)\omega_n^2}{(k - m\omega_n^2)}$$

Substitution of the two natural frequencies yields

$$\phi_{21} = 0, \qquad \phi_{22} = \frac{\cos\theta\left(\dfrac{3k}{3 - \cos^2\theta}\right)}{k - \dfrac{3k}{3 - \cos^2\theta}} = -\frac{3}{\cos\theta}$$

Thus, the modal properties are

$$\omega_1 = 0, \qquad \{\phi_1\} = \left\{\begin{array}{c} 1 \\ 0 \end{array}\right\}$$

$$\omega_2 = \left[\frac{3k}{m(3 - \cos^2\theta)}\right]^{1/2}, \qquad \{\phi_2\} = \left\{\begin{array}{c} 1 \\ -\dfrac{3}{\cos\theta} \end{array}\right\}$$

We see from this solution that in the first mode the block does not displace relative to the cart. In other words, in the rigid body mode the cart and block move in unison. The second mode involves motion of both the cart and the block. The expression for the velocity $\bar{v}_m$ of the block shows that the horizontal displacement of the block is $\Delta x_m = q_1 + q_2\cos\theta = \phi_{1n} + \phi_{2n}\cos\theta$. For the second mode this reduces to $\Delta x_m = -2$. The horizontal displacement of the cart in the second mode is the first element of the mode, so $\Delta x_{2m} = \phi_{21} = 1$. The first moment of mass for this displacement is $m\Delta x_m + 2m\Delta x_{2m} = 0$, which shows that the horizontal displacement of the system's center of mass is zero in the second mode.

4.3 MODAL EQUATIONS

4.3.1 Coordinate Transformations

On several occasions we selected generalized coordinates that were the displacement of a body's center of mass and/or the rotation of a body. Doing so led to a diagonal $[M]$ matrix. Diagonality means that the generalized coordinates are *uncoupled inertially*, so that the generalized acceleration $\ddot{q}_n$ associated with one variable does not occur in the differential equations of motion for a different q_j. In a few situations we selected generalized coordinates that measured the elongation of springs, with the result that the stiffness matrix was diagonal. Such generalized coordinates are *uncoupled elastically*, which means that the elasticity effects associated with a specific generalized coordinate only occur in the equation of motion for that variable.

Our approach has been to formulate the equations of motion using a single set of generalized coordinates. However, let us contemplate a situation where we have identified two alternative sets of generalized coordinates: $\{q\}$ is fully coupled, and $\{\hat{q}\}$ is uncoupled either inertially or elastically. These generalized coordinates must be related geometrically. Because we have linearized the analysis, these geometrical relations must have the form of linear algebraic equations, which we write in matrix form as

$$\{q\} = [R]\{\hat{q}\} \tag{4.3.1}$$

We say that $[R]$ is a *generalized coordinate transformation matrix*.

To relate the inertia matrices for these alternative descriptions we observe that the kinetic energy must be the same regardless of which set is used, so that

$$T = \tfrac{1}{2}\{\dot{\hat{q}}\}^{\mathrm{T}}[\hat{M}]\{\dot{\hat{q}}\} = \tfrac{1}{2}\{\dot{q}\}^{\mathrm{T}}[M]\{\dot{q}\} \tag{4.3.2}$$

Substitution of eq. (4.3.1) into this relation leads to

$$T = \tfrac{1}{2}\{\dot{\hat{q}}\}^{\mathrm{T}}[\hat{M}]\{\dot{\hat{q}}\} = \tfrac{1}{2}\{\dot{\hat{q}}\}^{\mathrm{T}}[R]^{\mathrm{T}}[M][R]\{\dot{\hat{q}}\} \tag{4.3.3}$$

These alternate descriptions must be true for any values of the generalized coordinates, which means that

$$\boxed{[\hat{M}] = [R]^{\mathrm{T}}[M][R]} \tag{4.3.4}$$

A similar analysis of potential energy leads to

$$\boxed{[\hat{K}] = [R]^{\mathrm{T}}[K][R]} \tag{4.3.5}$$

We will have frequent use for transformations such as these when we study finite element and substructuring concepts in later chapters. When $\{\hat{q}\}$ is inertially uncoupled, we say that $[R]$ *diagonalizes* $[M]$. Similarly, if $\{\hat{q}\}$ is elastically uncoupled, then $[R]$ diagonalizes $[K]$.

4.3.2 Modal Coordinates

In view of the orthogonality properties of the mode matrix $[\Phi]$, it is apparent from eqs. (4.3.4) and (4.3.5) that setting $[R] = [\Phi]$ will lead to equations of motion that are neither inertially nor elastically coupled. Such equations should be much easier to solve. For a different perspective of the nature of such a coordinate transformation, let us consider the general solution for free vibration. Each of the N eigensolutions features a natural frequency ω_j and normal vibration mode $\{\Phi_j\}$. A vibration following one of these modes is described by eq. (4.2.9), which we may write using the normal modes. Because the equations of motion are linear, the most general free vibration will consist of a superposition of these N possible solutions. In other words a free vibration is given by

$$\{q\} = \mathrm{Re}[B_1\{\Phi_1\}\exp(i\omega_1 t)] + \cdots + \mathrm{Re}[B_N\{\Phi_N\}\exp(i\omega_N t)]$$
$$= \{\Phi_1\}\mathrm{Re}[B_1\exp(i\omega_1 t)] + \cdots + \{\Phi_N\}\mathrm{Re}[B_N\exp(i\omega_N t)] \tag{4.3.6}$$

Note that bringing $\{\Phi_j\}$ outside the real parts is permissible because the modes are real quantities. In this perspective the complex exponential factors describe the amount each normal mode contributes to a free vibration.

Partitioning makes eq. (4.3.6) look like a generalized coordinate transformation because it allows the relation to be written as

$$
\{q\} = \left[\{\phi_1\} \cdots \{\phi_N\} \right] \left\{ \begin{array}{c} \text{Re}[B_1 \exp(i\omega_1 t)] \\ \vdots \\ \text{Re}[B_N \exp(i\omega_N t)] \end{array} \right\}
$$

$$
= [\Phi] \left\{ \begin{array}{c} \text{Re}[B_1 \exp(i\omega_1 t)] \\ \vdots \\ \text{Re}[B_N \exp(i\omega_N t)] \end{array} \right\} \tag{4.3.7}
$$

From the perspective of a generalized coordinate transformation, the preceding corresponds to $[R] = [\Phi]$ and the complex exponentials constitute a set of one-degree-of-freedom oscillators.

Equation (4.3.7) suggests that we consider *any response*, not just free vibration, to consist of a superposition of the contributions from each normal mode. We express this assumption by replacing the complex exponentials by a new set of time-dependent position variables η_j. This leads to the *modal transformation*,

$$
\{q\} = [\Phi]\{\eta\} \tag{4.3.8}
$$

Because the columns of $[\Phi]$ are the normal modes, which are linearly independent, $[\Phi]$ is invertible. This means that we could determine $\{\eta\}$ if we knew $\{q\}$.

To understand the significance of this transformation let us expand it,

$$
\{q\} = \left[\{\Phi_1\} \cdots \{\Phi_N\} \right] \left\{ \begin{array}{c} \eta_1 \\ \vdots \\ \eta_N \end{array} \right\}
$$

$$
= \{\Phi_1\}\eta_1 + \cdots + \{\Phi_N\}\eta_N = \sum_{j=1}^{N} \{\Phi_j\}\eta_j \tag{4.3.9}
$$

Thus, the η_j variables are generalized coordinates that give the contribution of each normal mode to the response. For this reason, the N variables $\eta_j(t)$ are called *modal coordinates*. They also are sometimes called *principal coordinates*, because they are uncoupled, just as principal stresses are uncoupled from shear.

We may picture the modal decomposition in an alternative way by considering the second form of eq. (4.3.9) from the perspective of the component representation of vectors. Suppose $\{q\}$ was a three-dimensional vector and we used $\{\Phi_j\}$ to denote a unit vector parallel to either the x, y, or z axis. In that case, we would say that η_j is the component of $\{q\}$ parallel to $\{\Phi_j\}$. Thus, in this sense we may consider the normal modes to be favored directions that facilitate the description of $\{q\}$.

We obtain the equations of motion governing the modal coordinates by substituting eq. (4.3.8) into the standard equations of motion. We will discuss the effect of damping in a later section, so we set $[C] = [0]$ for the development. Because $[\Phi]$ is a constant matrix, we have $\{\ddot{q}\} = [\Phi]\{\ddot{\eta}\}$, so that

$$[M][\Phi]\{\ddot{\eta}\} + [K][\Phi]\{\eta\} = \{Q\} \tag{4.3.10}$$

In order to make use of the orthogonality conditions, we premultiply this relation by $[\Phi]^T$. Equations (4.2.24) and (4.2.25) lead to

$$\{\ddot{\eta}\} + [\omega^2_{nat}]\{\eta\} = [\Phi]^T\{Q\} \tag{4.3.11}$$

In view of the diagonal nature of $[\omega^2_{nat}]$ and the partitioned representation of $[\Phi]^T$, the scalar equations described by the above are

$$\boxed{\ddot{\eta}_j + \omega^2_j \eta_j = \{\Phi_j\}^T\{Q\}, \quad j = 1, ..., N} \tag{4.3.12}$$

Decoupling of a system's equations of motion as a result of the modal transformation is an enormous advantage. The differential equation for each modal coordinate is exactly like that for an undamped one-degree-of-freedom oscillator. The oscillator has unit mass, and ω_j is its natural frequency. In essence, we have converted an N-degree-of-freedom system into N one-degree-of-freedom systems. As we saw in the previous chapters, numerous techniques are available for solving for the response of such systems.

The right side of the differential equations indicates that the excitation received by each modal coordinate depends on the degree to which the generalized force vector looks like a mode. Suppose that $[Q]$ is exactly proportional to $[M]\{\Phi_n\}$, where n is a specific index. In that case, the orthogonality conditions tell us that only modal coordinate n will be excited, from which it follows from eq. (4.3.9) that the proportions of the response $\{q\}$ will resemble mode n.

Unlike the situation for one-degree-of-freedom systems, the response analysis does not end with solution of the equations of motion. The last step is to use the modal transformation, eq. (4.3.8), to determine the generalized coordinates, which are the physically meaningful variables. Suppose we wish to draw graphs showing the time dependence of some or all generalized coordinates. There are two strategies one may follow. If N is not too large, the product $[\Phi]\{\eta\}$ may be expanded to obtain explicit expressions for each q_j as a function of t. This is the approach that is employed in the next exercise. For large N, it is more efficient to employ the numerical capabilities of software. Let us define a rectangular array of generalized coordinates $[q]$, such that the jth row holds the values of q_j at a succession of equally spaced time instants. Hence, the nth column of $[q]$ will hold all q_j values at time t_n. We use partitioning to describe the latter viewpoint, and use the modal transformation to relate the q_j values to the modal coordinates, specifically

$$[q] = \left[\{q_1(t_1)\} \{q_2(t_2)\} \cdots\right] = \left[[\Phi]\{\eta(t_1)\} [\Phi]\{\eta(t_1)\} \cdots\right] \tag{4.3.13}$$

The normal mode matrix is a common factor, so

$$\boxed{[q] = [\Phi]\left[\{\eta(t_1)\} \{\eta(t_2)\} \cdots\right]} \tag{4.3.14}$$

In other words, the procedure creates a rectangular array $[\eta]$ in which the values of all modal coordinates at each of a succession of time instants is stored in column-wise order. Then the rectangular array $[q] = [\Phi][\eta]$ holds in row-wise order each generalized coordinate at the succession of time instants.

4.3.3 Modal Coordinate Description of Mechanical Energy and Power

We may gain further insight into the significance of modal coordinates by regarding eq. (4.3.8) to define a new set of generalized coordinates η_j. According to eqs. (4.3.4) and (4.3.5), the inertia and stiffness matrices corresponding to the modal transformation $\{q\} = [\Phi]\{\eta\}$ are

$$[\hat{M}] = [\Phi]^T[M][\Phi] = [I], \qquad [\hat{K}] = [\Phi]^T[K][\Phi] = [\omega^2]$$

Correspondingly, the kinetic energy becomes

$$T = \frac{1}{2}\{\dot{\eta}\}^T[I]\{\dot{\eta}\} = \frac{1}{2}\sum_{j=1}^{N}\dot{\eta}_j^2 \tag{4.3.15}$$

while the potential energy is transformed to

$$V = \frac{1}{2}\{\eta\}^T[\omega_{nat}^2]\{\eta\} = \frac{1}{2}\sum_{j=1}^{N}\omega_j^2\eta_j^2 \tag{4.3.16}$$

Substitution of the modal coordinate definition into the power input leads to

$$\mathscr{P}_{in} = \{\dot{\eta}\}^T[\Phi]^T\{Q\} \tag{4.3.17}$$

The coefficient of $\{\dot{\eta}\}^T$ constitutes the generalized forces $\{\hat{Q}\}$ associated with the modal coordinates, such that

$$\mathscr{P}_{in} = \{\dot{\eta}\}^T\{\hat{Q}\} \quad \Rightarrow \quad \hat{Q}_j = \{\Phi_j\}^T\{Q\} \tag{4.3.18}$$

In other words, the modal transformation converts the physical system to a set of one-degree-of-freedom systems, whose mass coefficient is unity, whose stiffness coefficient is ω_j^2, and whose excitation is the *modal generalized force* $\{\Phi_j\}^T\{Q\}$. These properties will be seen to be a central theme of modal analysis for continuous as well as discrete systems.

The expressions for kinetic and potential energy in terms of modal coordinates, eqs. (4.3.15) and (4.3.16), also provide a useful insight into the possible existence of rigid body modes. Because the kinetic energy T is a sum of $\frac{1}{2}mv^2$ terms for every particle in the system, it is clear that T is *positive definite*. This means that $T > 0$, unless there is no motion (all $\dot{\eta}_j = 0$), in which case $T = 0$. Positive definiteness of T enables us to normalize all modes such that their modal mass is unity.

The situation for potential energy is different, because V consists of a sum of $\omega_j^2\eta_j^2$. This means that V is positive definite if, and only if, all $\omega_j > 0$. Then any free motion of the system will cause the elastic elements to deform, thereby storing potential energy. In contrast, if any $\omega_j = 0$, then nonzero values of the modal coordinate(s) associated with the rigid body modes do not result in a contribution to V. In that case V is *positive semidefinite*, which means that we can only be sure that it is not negative. Systems whose potential energy is positive semidefinite may displace without deforming the elastic elements.

Because the question of whether T or V is positive definite depends solely on the properties of $[M]$ and $[K]$, we use the same terms to discuss the coefficient matrices. For example, we say that $[M]$ is always positive definite. Definite matrices are invertible. A semidefinite matrix, like $[K]$ for a system that can move as a rigid body, is not invertible.

4.3.4 Free Vibration Response

The differential equations governing the principal coordinates, eq. (4.3.12), are like those for an undamped one-degree-of-freedom system. These equations are much easier to solve than the original equations of motion. The first response we shall address is the free vibration resulting when a system is released with a specified set of initial conditions. Assuming that $t = 0$ is the instant at which the system is released, the required conditions are the initial values $\{q(0)\}$ and $\{\dot{q}(0)\}$. Solving the differential equations for the modal coordinates requires that we determine the corresponding initial conditions for each η_j. To do so we observe that the modal transformation gives $\{q(0)\} = [\Phi]\{\eta(0)\}$ and $\{\dot{q}(0)\} = [\Phi]\{\dot{\eta}(0)\}$.

It is a straightforward task to solve for $\{\eta(0)\}$ and $\{\dot{\eta}(0)\}$. The modal orthogonality property simplifies this operation. Postmultiplying eq. (4.2.24) by $[\Phi]^{-1}$ leads to

$$\boxed{[\Phi]^{-1} = [\Phi]^T[M]} \qquad (4.3.19)$$

Thus, we determine the initial conditions for the modal coordinates according to

$$\boxed{\begin{aligned} \{\eta(0)\} &= [\Phi]^T[M]\{q(0)\} \\ \{\dot{\eta}(0)\} &= [\Phi]^T[M]\{\dot{q}(0)\} \end{aligned}} \qquad (4.3.20)$$

For free vibration we set $\{Q\} = \{0\}$ in eq. (4.3.12). The corresponding homogeneous solution is

$$\boxed{\eta_j = \text{Re}[B_j \exp(i\omega_j t)]} \qquad (4.3.21)$$

We match this to the initial conditions, which leads to

$$\text{Re}(B_j) = \eta_j(0), \qquad \text{Re}(i\omega_j B_j) = \dot{\eta}_j(0) \qquad (4.3.22)$$

so that

$$B_j = \eta_j(0) - \frac{i}{\omega_j}\dot{\eta}_j(0) \qquad (4.3.23)$$

Knowledge of the B_j coefficients enables us to evaluate the modal coordinates at any instant. We then use eq. (4.3.8) to determine the corresponding generalized coordinates $\{q(t)\}$.

It is interesting to consider the circumstances under which we may induce a free vibration that matches a specific vibration mode. This requires that the initial values $\eta_j(0)$ and $\dot{\eta}_j(0)$ be zero for all j except the mode of interest. It is evident from eq. (4.3.20) that this condition will be attained if the initial displacement $\{q(0)\}$ and initial velocity $\{\dot{q}(0)\}$ are exactly proportional to the mode.

EXAMPLE 4.8

The system depicted on the next page is a double pendulum, in which identical suspended bars having mass m are coupled by a spring. The system is constructed such that the spring is unstretched when the bars are vertical. At $t = 0$, the system is released from rest with counterclockwise rotations of 0.2 rad for the left bar and 0.1 rad for the right bar. Determine the ensuing motion for two cases $k/m = 2g/L$ and $k/m = 0.02g/L$.

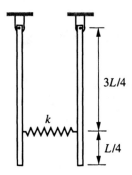

Solution This exercise will illustrate the procedure for obtaining a free response that satisfies specified initial conditions. In addition, by examining the response for the two values of the spring constants, we will observe an interesting phenomenon pertaining to the exchange of energy between parts of a system. We select as generalized coordinates the rotation angle θ_1 of the left bar and θ_2 of the right bar, both of which are defined to be zero at the (vertical) static equilibrium position. The inertia and stiffness coefficients for this two-degree-of-freedom system are

$$M_{11} = M_{22} = \tfrac{1}{3}mL^2, \qquad M_{12} = 0$$

$$K_{11} = K_{22} = mg\frac{L}{2} + \frac{9}{16}kL^2, \qquad K_{12} = -\frac{9}{16}kL^2$$

As an aid to solving the eigenvalue problem algebraically, we define a parameter α that scales the spring stiffness relative to the weight of a bar,

$$\alpha = \frac{kL}{mg}$$

so that $\alpha = 2.0$ in the first case, and $\alpha = 0.02$ in the second.
The resulting eigenvalue problem is

$$\left[mgL \begin{bmatrix} \left(\dfrac{1}{2} + \dfrac{9}{16}\alpha\right) & -\dfrac{9}{16}\alpha \\[2mm] -\dfrac{9}{16}\alpha & \left(\dfrac{1}{2} + \dfrac{9}{16}\alpha\right) \end{bmatrix} - \omega^2 mL^2 \begin{bmatrix} \dfrac{1}{3} & 0 \\[2mm] 0 & \dfrac{1}{3} \end{bmatrix} \right] \{\phi\} = \{0\}$$

As we have done previously, we convert the problem to nondimensional form by defining the eigenvalue parameter to be

$$\lambda = \frac{\omega^2 L}{g}$$

The characteristic equation is

$$\left(\frac{1}{2} + \frac{9}{16}\alpha - \frac{1}{3}\lambda\right)^2 - \left(\frac{9}{16}\alpha\right)^2 = 0$$

whose roots are

$$\lambda_1 = \frac{3}{2}, \qquad \lambda_2 = \frac{3}{2} + \frac{27}{8}\alpha$$

When λ is one of the eigenvalues, only one of the eigenvalue equations is independent. We select the first to evaluate the modes, so we have

$$\left[\begin{matrix} \left(\dfrac{1}{2} + \dfrac{9}{16}\alpha - \dfrac{1}{3}\lambda_j\right) & -\dfrac{9}{16}\alpha \\ \times & \times \end{matrix}\right]\left\{\begin{matrix} 1 \\ \phi_{2j} \end{matrix}\right\} = \left\{\begin{matrix} 0 \\ 0 \end{matrix}\right\}$$

When we substitute the two eigenvalues, we obtain

$$\lambda_1 = \frac{3}{2} \quad \Rightarrow \quad \phi_{21} = 1, \qquad \lambda_2 = \frac{3}{2} + \frac{27}{8}\alpha \quad \Rightarrow \quad \phi_{22} = -1$$

These modal properties stem from the fact that the left and right bars are identical. Consequently, the modes must be either symmetric or antisymmetric relative to a vertical plane through the middle. In the first mode, both bars swing in the same direction. This represents an antisymmetric motion, in the sense that the mirror image of the motion on one side is opposite to the motion on the other side. Because both bars move in phase in the first mode, the spring remains undeformed in that mode, and the natural frequency is independent of the spring constant. The second mode is a symmetric motion. The gravitational effect on each bar is the same as it is in the first mode, so the gravitational term in λ_2, that is, the one that does not depend on α, is the same as λ_1. The spring affects the second natural frequency because the bars move in opposite directions in the second mode, which means that the spring deforms.

To normalize the modes we compute the first orthogonality product,

$$[\mu] = [\phi]^{\mathrm{T}}[M][\phi] = \begin{bmatrix} 1 & 1 \\ 1 & -1 \end{bmatrix}\begin{bmatrix} \dfrac{ML^2}{3} & 0 \\ 0 & \dfrac{ML^2}{3} \end{bmatrix}\begin{bmatrix} 1 & 1 \\ 1 & -1 \end{bmatrix} = \frac{ML^2}{3}\begin{bmatrix} 2 & 0 \\ 0 & 2 \end{bmatrix}$$

Dividing the jth column of $[\phi]$ by the square root of the jth diagonal element of $[\mu]$ yields the normal mode matrix,

$$[\Phi] = \sqrt{\frac{3}{2mL^2}}\begin{bmatrix} 1 & 1 \\ 1 & -1 \end{bmatrix}$$

We begin the solution of the equations of motion by introducing the modal transformation, which leads to uncoupled modal equations for free vibration,

$$\{q\} = [\Phi]\{\eta\} \quad \Rightarrow \quad \ddot{\eta}_j + \omega_j^2 \eta_j = 0$$

The homogeneous solution is

$$\eta_j = C_{1j}\cos(\omega_j t) + C_{2j}\sin(\omega_j t)$$

The coefficients C_{1j} and C_{2j} are set by satisfying initial conditions. The initial generalized coordinates and velocities are specified to be

$$t = 0: \quad \{q\} = \left\{\begin{matrix} 0.2 \\ 0.1 \end{matrix}\right\} \quad \text{and} \quad \{\dot{q}\} = \left\{\begin{matrix} 0 \\ 0 \end{matrix}\right\} \text{ rad/s}$$

We compute the inverse of the mode matrix according to eq. (4.3.19),

$$[\Phi]^{-1} = [\Phi]^{\mathrm{T}}[M] = \sqrt{\frac{mL^2}{6}}\begin{bmatrix} 1 & 1 \\ 1 & -1 \end{bmatrix}$$

From this, we determine the initial conditions for the modal coordinates,

$$t = 0: \quad \{\eta\} = [\Phi]^{-1}\{q\} = \sqrt{\frac{mL^2}{6}}\left\{\begin{matrix} 0.3 \\ 0.1 \end{matrix}\right\}, \quad \{\dot{\eta}\} = [\Phi]^{-1}\{\dot{q}\} = \left\{\begin{matrix} 0 \\ 0 \end{matrix}\right\}$$

The requirement that the homogeneous solution for the modal coordinates satisfy these initial conditions leads to

$$C_{11} = 0.12247\sqrt{mL^2}, \qquad C_{12} = 0.04082\sqrt{mL^2}, \qquad C_{21} = C_{22} = 0$$

Thus, the modal coordinate responses are

$$\eta_1 = 0.12247\sqrt{mL^2}\cos(\omega_1 t), \qquad \eta_2 = 0.04082\sqrt{mL^2}\cos(\omega_2 t)$$

The last step is to form the generalized coordinate responses by synthesizing the modal transformation, $\{q\} = [\Phi]\{\eta\}$, at a sequence of time instants in order to graph each rotation. There is no need to implement the matrix computation described by eq. (4.3.14) because there are only two modes. Thus, we have

$$\{q\} = [\Phi]\{\eta\} = \sqrt{\frac{3}{2mL^2}}\begin{bmatrix} 1 & 1 \\ 1 & -1 \end{bmatrix}\begin{Bmatrix} 0.12247\sqrt{mL^2}\cos(\omega_1 t) \\ 0.04082\sqrt{mL^2}\cos(\omega_2 t) \end{Bmatrix}$$

$$= \begin{Bmatrix} 0.150\cos(\omega_1 t) + 0.050\cos(\omega_2 t) \\ 0.150\cos(\omega_1 t) - 0.050\cos(\omega_2 t) \end{Bmatrix} \text{rad}$$

Note that the spring constant, as represented by α, affects only the natural frequencies, which are related to the eigenvalues by $\omega_j = \sqrt{\lambda_j g/L}$, so that

$$\alpha = 2 \quad \Rightarrow \quad \omega_1 = 1.2247\sqrt{\frac{g}{L}}, \qquad \omega_2 = 2.8723\sqrt{\frac{g}{L}}$$

$$\alpha = 0.02 \quad \Rightarrow \quad \omega_1 = 1.2247\sqrt{\frac{g}{L}}, \qquad \omega_2 = 1.2520\sqrt{\frac{g}{L}}$$

The responses θ_1 and θ_2 are plotted as functions of the nondimensional time parameter $t\sqrt{g/L}$, which we identify by examining the dimensionality of $\omega_j t$. The time increment for such plots should be such that the shortest natural period, which is $2\pi/\omega_2$, is adequately sampled, so we set $\Delta t = 0.1(2\pi/\omega_2)$. The results for the larger value of α are unremarkable, except that the responses almost seem to be periodic over an interval larger than either natural period $2\pi/\omega_1$ or $2\pi/\omega_2$.

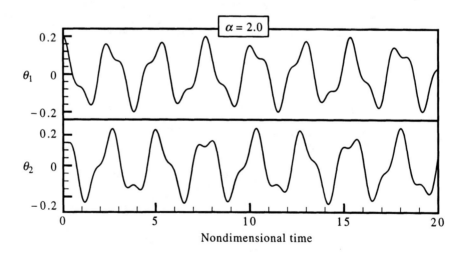

The responses for $\alpha = 0.02$ indicate that the two bars alternately show larger and smaller vibration amplitudes. This suggests the presence of a beating phenomenon, which should not be surprising because the two natural frequencies are close, and our analysis has shown that the response of either bar consists of a combination of responses at the two natural frequencies.

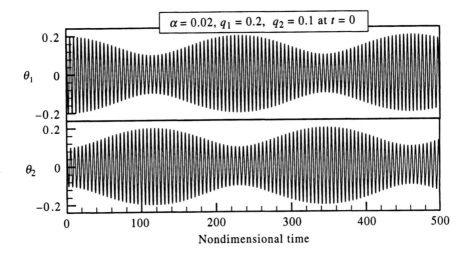

A true beating phenomenon occurs when the two frequencies have equal amplitude. We can obtain such a condition by altering the initial conditions so that only one bar is displaced initially. For example, if $q_1 = 0.2$ rad, $q_2 = \dot{q}_1 = \dot{q}_2 = 0$, we would find that the response is

$$\{q\} = \begin{Bmatrix} 0.100 \cos(\omega_1 t) + 0.100 \cos(\omega_2 t) \\ 0.100 \cos(\omega_1 t) - 0.100 \cos(\omega_2 t) \end{Bmatrix} \text{ rad}$$

The responses in this case are seen to feature beats that are 90° out of phase between the two bars, so that when one bar is experiencing its maximum excursion, the other is almost quiescent.

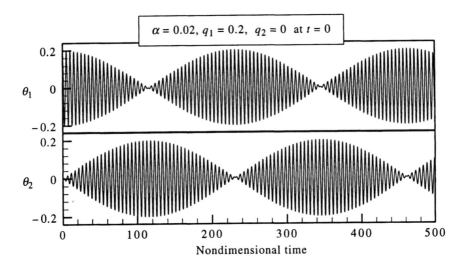

This phenomenon is characteristic of systems that feature a weak coupling between individual parts. In the coupled pendulum, the individual bars would vibrate at their pendulum frequency if the spring were not present. Attaching a very weak spring joins them, such that energy slowly (relative to the pendulum period) transfers from the one that is in motion to the one that is not. Once most of the energy has transferred out of one bar, the process reverses.

4.3.5 Forced Response

When a system is subjected to a specified excitation $\{Q(t)\}$, we determine the forced response by considering eq. (4.3.12) to represent a set of N one-degree-of-freedom systems. Any of the methods we used in Chapters 2 and 3 are equally suitable for determining η_j. One approach decomposes each η_j into complementary and particular solutions. The former has the form of eq. (4.3.21), so the general solution for the modal coordinates is

$$\boxed{\eta_j = \text{Re}[B_j \exp(i\omega_j t)] + (\eta_j)_p} \tag{4.3.24}$$

If all $Q_j(t)$ consist of combinations of powers of t and/or real or complex exponentials, we may use the method of undetermined coefficients to determine $(\eta_j)_p$. The superposition method in conjunction with Appendix B also is useful. Alternatively, we may use the complex frequency response if all $Q_j(t)$ are harmonic functions. For periodic excitations, Fourier series analysis is suitable, while the transient response for zero initial conditions may be obtained by applying the convolution integral in conjunction with FFT concepts.

In situations where η_j is decomposed into complementary and particular parts, the total solution must be used to satisfy the initial conditions for the modal coordinates. To do so, we determine the values of $\eta_j(0)$ and $\dot{\eta}_j(0)$ by applying eqs. (4.3.20). The corresponding initial values of the complementary solution are obtained from eq. (4.3.24), which allows us to apply eq. (4.3.23), to determine the constant factors,

$$B_j = [\eta_j - (\eta_j)_p]\big|_{t=0} - \frac{i}{\omega_j}[\dot{\eta}_j - (\dot{\eta}_j)_p]\big|_{t=0} \tag{4.3.25}$$

In contrast, if we are using the superposition method, Appendix B gives responses corresponding to zero initial conditions. In that case the additional complementary solution for nonzero initial conditions is defined directly by eq. (4.3.23).

EXAMPLE 4.9

The sketch models an automobile as a box of mass m that is translating to the left along level ground at constant speed v. At $t = 0$ the front wheel encounters a ramp, for which the ground elevation is $w = \beta x h(x)$, where $h(x)$ denotes a step function. The slope β is small, so it is permissible to consider the horizontal position of the wheels to be $x_1 = vt$, $x_2 = vt - L$.

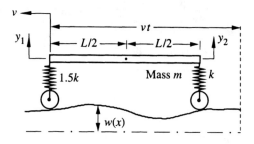

(a) Show that the equations of motion for this system are

$$[M]\begin{Bmatrix} \ddot{y}_1 \\ \ddot{y}_2 \end{Bmatrix} + [K]\begin{Bmatrix} y_1 \\ y_2 \end{Bmatrix} = [K]\begin{Bmatrix} w(vt) \\ w(vt - L) \end{Bmatrix}$$

(b) Suppose the system matrices are

$$[M] = m \begin{bmatrix} 3/8 & 1/8 \\ 1/8 & 3/8 \end{bmatrix}, \qquad [K] = k \begin{bmatrix} 1.5 & 0 \\ 0 & 1.0 \end{bmatrix}$$

(The centroidal radius of gyration corresponding to $[M]$ is $\kappa_G = L/\sqrt{2}$.) Derive algebraic expressions for y_1 and y_2 as functions of the nondimensional time $t' = (k/m)^{1/2}t$ and the nondimensional speed $\alpha = (v/L)(m/k)^{1/2}$.

(c) Graph the dependence of y_1 and y_2 on t' when $\alpha = 0.1$ and $\alpha = 1$.

Solution In this example, we will encounter all of the tasks that must be performed to determine the transient response of systems. It also will bring to the fore some basic phenomena encountered in the dynamic response of moving vehicles. It is specified that the generalized coordinates should be $q_1 = y_1$, $q_2 = y_2$. Because these variables are defined independently of the ground elevation, we may employ the technique described in Section 1.5.4, wherein the force in each spring is decomposed into a part that is induced by displacement of the vehicle, and a part that is induced by the changing elevation of the ground. The first effect is described by $[M]$ and $[K]$, which are the same as they would be for motion over level ground. If the ground elevation $w(x)$ at a wheel is positive, the spring at that location shortens, resulting in additional compressive forces $k_1 w(x_1)$ at the front wheel and $k_2 w(x_2)$ at the rear wheel. These forces are transmitted to the vehicle body. The upward velocity at the respective locations is $\dot{y}_1$ and $\dot{y}_2$, so the power input is

$$\mathcal{P}_e = [k_1 w(x_1)]\dot{y}_1 + [k_2 w(x_2)]\dot{y}_2$$

Matching this to the standard form for the power input leads to

$$Q_1 = k_1 w(x_1), \qquad Q_2 = k_2 w(x_2)$$

In view of the form of $[K]$, and the fact that $x_1 = vt$ and $x_2 = vt - L$, it follows that the generalized force matrix may be written as

$$\{Q\} = [K] \begin{Bmatrix} w(vt) \\ w(vt - L) \end{Bmatrix}$$

This completes the derivation of the equations of motion.

It is requested that the responses be obtained in nondimensional form. Solving nondimensionalized equations of motion will enable us to derive such forms directly. We scale the generalized coordinates by L, so that $q_j' = q_j/L$. Then, because $t' = (k/m)^{1/2}t$, we have $d^2 q_j/dt^2 = (kL/m)^2 \ddot{q}_j'$, where an overdot will now denote differentiation with respect to t'. Also, because the speed is nondimensionalized as $\alpha L(k/m)^{1/2}$, we have $vt = \alpha L t'$, which leads to $w(vt) = \beta(\alpha L t')h(t')$ and $w(vt - L) = \beta(\alpha L t' - L)h(t' - 1/\alpha)$. Thus, the nondimensional equations to be solved are

$$[M']\{\ddot{q}'\} + [K']\{q'\} = \left(\frac{1}{L}\right)[K'] \begin{Bmatrix} w(vt) \\ w(vt - L) \end{Bmatrix}$$

$$= \beta\alpha[K'] \begin{Bmatrix} t'h(t') \\ \left(t' - \dfrac{1}{\alpha}\right)h\left(t' - \dfrac{1}{\alpha}\right) \end{Bmatrix}$$

where $[K']$ and $[M']$ are the inertia and stiffness matrices corresponding to $k = m = 1$.

Solution of the equations of motion begins with the eigenvalue problem

$$[[K'] - (\omega')^2[M']]\{\phi\} = \{0\}$$

from which we find that the natural frequencies and mode shapes are

$$\omega_1' = 1.521, \qquad \omega_2' = 2.277$$

$$\{\phi_1\} = \begin{Bmatrix} 1 \\ 2.186 \end{Bmatrix}, \qquad \{\phi_2\} = \begin{Bmatrix} 1 \\ -0.686 \end{Bmatrix}$$

We use the first orthogonality product, $\{\phi\}^T[M']\{\phi\}$, to normalize each mode, from which we find the normal mode matrix,

$$[\Phi] = \begin{bmatrix} 0.6072 & 1.6222 \\ 1.3272 & -1.1128 \end{bmatrix}$$

We may now form the modal coordinate equations. We define the modal transformation

$$\{q'\} = [\Phi]\{\eta\}$$

In view of the form of the generalized forces for this system, the standard modal equations become

$$\ddot{\eta}_j + (\omega_j')^2 \eta_j = \{\Phi_j\}^T\{Q\} = R_j(t')$$

where the modal excitations $R_j(t')$ are

$$R_j(t') = \{\Phi_j\}^T \beta\alpha[K'] \begin{Bmatrix} t'h(t') \\ \left(t' - \dfrac{1}{\alpha}\right)h\left(t' - \dfrac{1}{\alpha}\right) \end{Bmatrix}$$

The fact that $[K']$ is diagonal reduces these terms to

$$R_j(t') = \beta\alpha\left[\Phi_{1j}(1.5)t'h(t') + \Phi_{2j}\left(t' - \dfrac{1}{\alpha}\right)h\left(t' - \dfrac{1}{\alpha}\right)\right]$$

This shows that each modal coordinate is subjected to a ramp excitation at $t' = 0$, followed by a second ramp excitation that begins to act at $t' = 1/\alpha$ (that is, $t = L/v$), which is the time when the rear wheel arrives at the ramp. The strength of the excitation imparted to each mode in the initial ramp is proportional to the stiffness of the front spring and the modal value at the front wheel. Similarly, the second ramp excitation is scaled by the rear spring stiffness and the modal value at the rear wheel.

We shall use the superposition method to describe the modal response. Appendix B gives the unit ramp response for a one-degree-of-freedom system that is initially at rest in its static equilibrium position. We set the mass in that formula to unity, because the coefficient of $\ddot{\eta}_j$ in the modal differential equation is one. We also replace t with t', and ω_{nat} with ω_j'. The damping ratio is zero, so the tabulated ramp response becomes

$$r(t', \omega_j') = \frac{1}{(\omega_j')^3}[\omega_j't' - \sin(\omega_j't')]h(t')$$

To evaluate the solution for each η_j, we combine unit ramp responses starting at $t' = 0$ and $t' = 1/\alpha$ in the manner indicated by the respective $R_j(t')$ terms,

$$\eta_j = \beta\alpha[1.5\Phi_{1j}r(t', \omega_j) + \Phi_{2j}r(t' - 1/\alpha, \omega_j)]$$

This expression and the preceding description of the ramp response prescribe η_1 and η_2 at any specified value of t'. From those values, we may determine the generalized coordinates at that instant by recalling the modal transformation, $\{q\} = [\Phi][\eta]$, which gives

$$y_1 = \Phi_{11}\eta_1 + \Phi_{12}\eta_2, \qquad y_2 = \Phi_{21}\eta_1 + \Phi_{22}\eta_2$$

Rather than evaluating this sum at each t', let us evaluate the responses at a sequence of t' values by following the matrix-oriented computation in eq. (4.3.14). The procedure is somewhat different in Mathcad and MATLAB. In Mathcad, we define the ramp response as a function $r(t,\omega_j)$, being sure that it evaluates to zero if $t < 0$. An appropriate time increment for sampling the response divides the shortest natural period into 10 or more subintervals; we shall use $\Delta = 0.1 \pi/\omega_2'$. We define an index n and corresponding time values $t_n = (n-1)\Delta$. We also define an index $j := 1; 2$. These definitions allow us to evaluate a two-row rectangular array of modal coordinates $\eta_{j,n} = 1.5 * \Phi_{1,j} * r(t_n, \omega_j) + \Phi_{2,j} * r(t_n - 1/\alpha, \omega_j)$. Computation of $q = \Phi * \eta$ then gives two rows holding y_1 and y_2 at each instant. In MATLAB, we use a function M-file *ramp.m*, whose first line is $r = ramp(t, omega)$, to compute the ramp function. The computation of the modal coordinates defines an array of time instants, $t = 0:delta:t\_end$. Within a loop $j = 1:2$, we evaluate a row of η_j values according to $eta(j,:) = 1.5 * phi(1,j) * ramp(t, omeg(j)) + phi(2,j) * ramp(t - 1/\alpha, omeg(j))$. Upon completion of this loop, the rectangular array of generalized coordinates is found to be $y = phi * eta$.

The results for the two values of α are shown in the graph. Each displacement is scaled by the reciprocal of α so the data have comparable magnitudes in both cases. When $\alpha = 0.1$, which corresponds to a low-speed v, the displacements are quite small. (The plotted values must be multiplied by α.) The front end begins to move upward when the front wheel contacts the ramp. The rear wheels contact the ramp at $t' = 10$, at which time the rear wheel also moves upward. The case where $\alpha = 1$ corresponds to a high speed. Both ends almost move in unison in this case. In essence, when $\alpha = 1$, the time between the front and rear wheels contacting the ramps is so small relative to the period of each mode that the two events merge. An interesting aspect of the response for both values of α is the fact that the initial tendency of y_2 is to move downward. This is a consequence of the role of $\{\Phi_2\}$, which is the pitching mode. Pushing the front upward in this mode produces a downward motion at the rear.

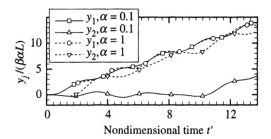

somewhat different in Mathcad

The results for the two values

4.3.6 Modal Damping

The modal approach is a core concept for vibration analysis. At the same time, we must recognize that it is of limited usefulness if we continue to ignore dissipation. This section develops some approaches by which the modal coordinate description may be modified to include the effects of damping, $[C] \neq [0]$. The important thing to consider in weighing the alternative approaches is that our knowledge of dissipation effects can vary strongly between systems. If we are dealing with a system that has shock absorbers in the form of dashpots, we are likely to be confident in our knowledge of the damping characteristics. On the other hand, if we are studying a linkage with several joints (Coulomb friction and viscous friction), or a material whose elasticity is not perfect (plasticity and viscoelasticity), or a system with significant air resistance (fluid drag and acoustic radiation), we might not have high-quality information regarding damping. In such cases, it would not make sense to expend great analytical effort to drastically modify the simple procedures we already have developed.

Proportional Damping The term *proportional damping* refers to situations where the damping matrix $[C]$ is exactly proportional to the inertia and/or stiffness matrix. We let α and β be the coefficients of proportionality, such that

$$\boxed{[C] = \alpha[M] + \beta[K]} \tag{4.3.26}$$

The physical interpretation of this relation is enlightening. If $[C]$ is proportional to $[K]$, it must be that each spring is accompanied by a parallel dashpot. Furthermore, the ratio of dashpot constant to spring stiffness must be the same for all dashpot-spring pairs. Identifying that $[C]$ is proportional to $[M]$ is more subtle. The damping and inertia constants are both obtained from quadratic sums of generalized velocities, $\mathscr{P}_{\text{dis}}$ and T, respectively. However, the velocities from which the kinetic energy is obtained are absolute quantities relative to an inertial reference frame. It follows that if $[C]$ is proportional to $[M]$, then dashpots must be connected between each mass element and the ground. As an illustration, consider the two-degree-of-freedom system in Figure 4.1. If the dashpots parallel to each spring are such that $c_j = \beta k_j$, then $[C] = \beta[K]$. A damping matrix proportional to $[M]$ results if the dashpots attached between each block and the ground are $c_1 = \alpha m_1$, $c_3 = \alpha m_2$, and $c_2 = 0$.

The usual situation is one where we have constructed $[M]$, $[K]$, and $[C]$, and we wish to ascertain if $[C]$ fits the proportional damping definition. Equation (4.3.26) relates $N \times N$ matrices, so satisfying the equality requires that the N^2 elements on either side of the equality match. We may solve any two nonzero elements for tentative values of α and β. If damping is truly proportional, then we will find that all of the other elemental equations are satisfied by these values. In practice, this will seldom be found to be true. It might seem reasonable in that situation to find a pair of α and β values by solving the N^2 elemental equations in a least squares approach that minimizes the overall error. However, there are better ways of treating nonproportional damping, as we will see.

Let us proceed by assuming that we have identified values of α and β in eq. (4.3.26). The forced vibration analysis in this case closely follows the treatment of undamped systems. It requires that we first evaluate the undamped vibration modes, in effect ignoring damping for the eigenvalue problem. We presume that at this stage we have determined the natural frequencies and normal modes, so $[\omega_{\text{nat}}^2]$ and $[\Phi]$ are known.

We begin by substituting the modal coordinate transformation, eq. (4.3.8), into the equations of motion. In order to exploit the modal orthogonality properties, we then premultiply the equation by $[\Phi]^T$, which leads to

$$[\Phi]^T[M][\Phi]\{\ddot{\eta}\} + [\Phi]^T[\alpha[M] + \beta[K]][\Phi]\{\dot{\eta}\} + [\Phi]^T[K][\Phi]\{\eta\}$$
$$= [\Phi]^T\{Q\} \tag{4.3.27}$$

The orthogonality relations for the normal mode matrix, eqs. (4.2.24) and (4.2.25), then lead to

$$\{\ddot{\eta}\} + [\alpha + \beta[\omega_{\text{nat}}^2]]\{\dot{\eta}\} + [\omega_{\text{nat}}^2]\{\eta\} = [\Phi]^T\{Q\} \tag{4.3.28}$$

The corresponding equations for each element of $\{\eta\}$ are uncoupled, because $[\omega_{\text{nat}}^2]$ is diagonal. Specifically, we have

$$\ddot{\eta}_j + (\alpha + \beta\omega_j^2)\dot{\eta}_j + \omega_j^2\eta_j = \{\Phi_j\}^T\{Q\} \tag{4.3.29}$$

It is evident that proportional damping leads to a set of uncoupled differential equations for the modal coordinates. The form of these equations is like that of a

This value of $[\Phi]$ is accurate to four significant figures. A check of the orthogonality relations would show that $[\Phi]^T[M][\Phi]$ is very close to the identity matrix, and $[\Phi]^T[K][\Phi]$ is very close to the diagonal $[\omega^2]$ matrix.

We next use the normal mode to transform the damping matrix to modal coordinates, according to

$$[\Phi]^T[C][\Phi] = \begin{bmatrix} 0.7071 & 0.7071 \\ 0.5 & -0.5 \end{bmatrix}^T \begin{bmatrix} 1.2 & -0.4 \\ -0.4 & 1.0 \end{bmatrix} \begin{bmatrix} 0.7071 & 0.7071 \\ 0.5 & -0.5 \end{bmatrix}$$

$$= \begin{bmatrix} 0.5671 & 0.34998 \\ 0.34998 & 1.13282 \end{bmatrix}$$

In accord with the light damping approximation, we evaluate approximate modal damping ratios ζ_j by dropping the off-diagonal terms, then equating the diagonal terms to the respective values of $2\zeta_j\omega_j$,

$$2\zeta_1\omega_1 = 0.5671 \quad \Rightarrow \quad \zeta_1 = 0.02648$$
$$2\zeta_2\omega_2 = 1.1328 \quad \Rightarrow \quad \zeta_2 = 0.04161$$

Both damping ratios are small compared to unity, so we should expect the response predicted by this light damping approximation to be close to the actual one.

The differential equations for the modal coordinates are like those for free vibration of one-degree-of-freedom oscillators,

$$\ddot{\eta}_j + 2\zeta_j\omega_j\dot{\eta}_j + \omega_j^2\eta_j = 0$$

We find the initial conditions for these variables by inverting the modal transformation, for which we can use the identity

$$[\Phi]^{-1} = [\Phi]^T[M] = \begin{bmatrix} 0.7071 & 1.0 \\ 0.7071 & -1.0 \end{bmatrix}$$

Thus, the initial values of the modal coordinates are

$$\begin{Bmatrix} \eta_1(0) \\ \eta_2(0) \end{Bmatrix} = [\Phi]^{-1}\begin{Bmatrix} x_1(0) \\ x_2(0) \end{Bmatrix} = \begin{Bmatrix} 0.007071 \\ 0.007071 \end{Bmatrix}$$

The initial values $\dot{\eta}_1(0)$ and $\dot{\eta}_2(0)$ are zero because the system was initially at rest. The solution of the differential equations matching these initial conditions is

$$\eta_j(t) = \eta_j(0)\exp(-\zeta_j\omega_j t)\left\{\cos[(\omega_d)_j t] + \frac{\zeta_j\omega_j}{(\omega_d)_j}\sin[(\omega_d)_j t]\right\}$$

where the parameters $(\omega_d)_j$ are damped natural frequencies for each mode,

$$(\omega_d)_1 = \omega_1(1 - \zeta_1^2)^{1/2} = 10.703$$
$$(\omega_d)_2 = \omega_2(1 - \zeta_2^2)^{1/2} = 13.603$$

The values of x_1 and x_2 at any instant may be evaluated from the modal transformation, $\{x\} = [\Phi]\{\eta\}$. To select an appropriate time increment for sampling the solution in preparation for graphing it, we observe that the highest frequency for the modal coordinates is $(\omega_d)_2$. Division of one damped period into ten subintervals would result if we set

$$\Delta t = \frac{2\pi}{(\omega_d)_2} = 0.046 \text{ s}$$

The graphs appearing here were obtained with $\Delta t = 0.025$ s. The numerical solutions depicted in these graphs were obtained by the Runge-Kutta method, which accounts for the full $[C]$ array.

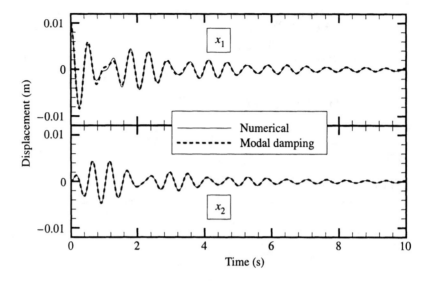

It is clear that the two types of solutions are in extraordinarily good agreement. The light damping approximation is particularly good for this system, even for larger damping levels. To illustrate this, we have graphed the approximate and numerical responses for damping constants that are a factor of 10 larger, so that the maximum modal damping ratio is approximately 42% of critical. The discrepancies between the approximate and numerical solutions are quite acceptable for many applications. One reason for this is that, although there is greater instantaneous error entailed in applying the light damping approximation to the differential equations, the error does not have much opportunity to accumulate because the higher level of damping results in more rapid decay of the response.

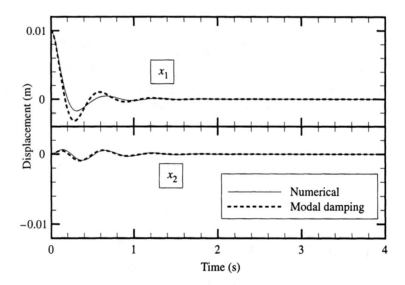

4.3.7 Harmonic Steady-State Response

We may obtain specific solutions to the differential equations for the modal coordinates in the case of harmonic excitation. In addition to being important for their own sake, the results will lead to insight into the factors affecting the response of a multiple-degree-of-freedom system. We consider a situation where all of the generalized forces vary at the same frequency ω. It is not necessary that all of the forces be in phase, so we let F_j denote the complex amplitude of each force. Correspondingly, the excitation is

$$\boxed{\{Q\} = \text{Re}[\{F\}\exp(i\omega t)]} \tag{4.3.35}$$

Substitution of this form into eq. (4.3.34) yields

$$\ddot{\eta}_j + 2\zeta_j\omega_j\dot{\eta}_j + \omega_j^2\eta_j = \{\Phi_j\}^{\text{T}}\text{Re}[\{F\}\exp(i\omega t)], \qquad j = 1, \ldots, N \tag{4.3.36}$$

The particular solution of this equation is the steady-state response. The differential equation is exactly like that of a damped one-degree-of-freedom system that is subjected to harmonic excitation, with ω_{nat} and ζ replaced by the natural frequency ω_j and damping ratio ζ_j of the modal coordinate under consideration. The complex amplitude of the excitation is $\{\Phi_j\}^{\text{T}}\{F\}$. Expressing the harmonic steady-state solution as

$$(\eta_j)_{\text{ss}} = \text{Re}[X_j\exp(i\omega t)] \tag{4.3.37}$$

leads to

$$X_j = \frac{\{\Phi_j\}^{\text{T}}\{F\}}{\omega_j^2 + 2i\zeta_j\omega_j\omega - \omega^2} \tag{4.3.38}$$

The corresponding steady-state response of the generalized coordinates is

$$\{q\} = [\Phi]\{\eta\} = \text{Re}[[\Phi]\{X\}\exp(i\omega t)] \tag{4.3.39}$$

where we are permitted to multiply by $[\Phi]$ prior to taking the real part because $[\Phi]$ is real.

A useful way of viewing the preceding is that it describes the complex amplitude of each q_j variable. Specifically, we may write the solution as

$$\{q\} = \text{Re}[\{Y\}\exp(i\omega t)], \quad \{Y\} = [\Phi]\{X\} \tag{4.3.40}$$

If we write $[\Phi]$ in its partitioned form, we will be able to see the individual modal contributions to the complex amplitudes. The result is

$$\boxed{\{Y\} = \sum_{j=1}^{N}\{\Phi_j\}X_j = \sum_{j=1}^{N}\{\Phi_j\}\frac{\{\Phi_j\}^{\text{T}}\{F\}}{\omega_j^2 + 2i\zeta_j\omega_j\omega - \omega^2}} \tag{4.3.41}$$

The equivalent scalar form of this expression is

$$\boxed{Y_n = \sum_{j=1}^{N}\Phi_{nj}\frac{\{\Phi_j\}^{\text{T}}\{F\}}{\omega_j^2 + 2i\zeta_j\omega_j\omega - \omega^2}} \tag{4.3.42}$$

Both descriptions disclose that the contribution of any one mode to a specific generalized coordinate's complex amplitude is affected by three factors. The first is the complex modal force amplitude $\{\Phi_j\}^{\text{T}}\{F\}$, which drives the modal coordinates. This quantity is the steady-state analog of the modal generalized forces in eq. (4.3.18). Our discussion of those quantities suggested that they represent a mapping of the actual forces into the modal variables. The degree to which $\{F\}$ is like a mode dictates the magnitude of the excitation that mode receives. For example, if $\{F\}$ is exactly proportional to $[M]\{\Phi_p\}$, then only mode p will participate in the response, because $\{\Phi_j\}^{\text{T}}[M]\{\Phi_p\} \equiv 0$ if $j \neq p$.

The second factor affecting the modal contribution is the closeness of the excitation frequency to the natural frequency associated with that mode. When $\omega \longrightarrow \omega_j$, the denominator decreases, so the contribution of mode j increases. If the system is lightly damped, so that all $\zeta_j \ll 1$, the contribution of a specific mode p to the response will be dominant when ω is very close to that mode's natural frequency. If ω exactly equals ω_p, the imaginary term in the denominator is all that remains. In that case, damping, which represents a 90° out-of-phase addition to the complex stiffness of each mode, is the sole resistance the mode can present to the excitation.

The two factors described thus far affect the overall contribution of a mode to each generalized coordinate. The third factor Φ_{nj} affects the relative contribution of a mode to a specific generalized coordinate. Recall that $\{\Phi_j\}$ gives the relative proportions between the various generalized coordinates in a modal vibration. Hence, Φ_{nj} in eq. (4.3.42) describes the proportional contribution of mode j to each generalized coordinate. If Φ_{nj} is small relative to other elements, then generalized coordinate q_n will show little evidence of the mode's contribution, even though it might be quite evident in the other generalized coordinates.

It is the interplay between the three factors above that makes it difficult to predict intuitively the response of multiple-degree-of-freedom systems. Let us consider a situation in which ω is very close to a specific ω_p, so that mode p is resonant. If the generalized force is dissimilar to the mode, such that $\{\Phi_p\}^T\{F\}$ is close to zero, then the contribution of mode p to the response will be small, even though the associated complex frequency response is large. Even if $\{\Phi_p\}^T\{F\}$ is substantial, the nth generalized coordinate will show little evidence of this mode's contribution if Φ_{np} is small. In that case the nth generalized coordinate will display the contributions of the other modes, whereas the other generalized coordinates will be dominated by the resonant mode.

A corollary of these considerations is that system response often is highly sensitive to redesign. Suppose we change a system to shift a natural frequency away from resonance. This would lead to a large change in the complex frequency response. Redesigning a system to change its vibration modes can also strongly alter the influence of a set of generalized forces, especially if the modal force $\{\Phi_p\}^T\{F\}$ in the original design is small. Thus, a redesign that alters the modes is doubly significant. The high degree to which a system's response may be altered by small modifications is not necessarily bad. For example, we will see in the next chapter that it is crucial to the concept of a vibration absorber.

We say that a system is lightly damped if $\zeta_j \ll 1$ for all modes. Recall from our study of one-degree-of-freedom systems that a low value of ζ leads to a high quality factor for the resonant peak, that is, $|D(\omega, \zeta)|$ displays a narrow tall peak in the vicinity of $\omega = \omega_{nat}$. Consider a situation in which we gradually change ω to sweep through the range of natural frequencies. According to eq. (4.3.42), the corresponding plot of some or all steady-state amplitudes $|Y_k|$ as a function of ω will display a narrow peak at each natural frequency ω_j if ζ_j is small. We refer to these peaks as *modal resonances*.

EXAMPLE 4.11

The force applied to the left block in the system depicted on the next page is $F(t) = F_0 \sin(\omega t)$. Derive expressions for the steady-state amplitude of each block as a function of ω. Graph the results for $0 < \omega < 1.5\omega_3$, where ω_3 is the highest natural frequency of the system.

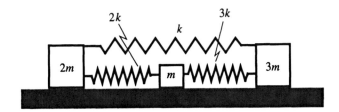

Solution This system has a rigid body mode. Thus, in addition to demonstrating the general procedure for analyzing steady-state response, this exercise will illustrate the manner in which deformational effects are superposed onto rigid body motion. We select as generalized coordinates the horizontal displacements x_1, x_2, and x_3 of the blocks, each measured to the right from a reference position in which the springs are undeformed. The inertia and stiffness matrices are

$$[M] = m \begin{bmatrix} 2 & 0 & 0 \\ 0 & 1 & 0 \\ 0 & 0 & 3 \end{bmatrix}, \qquad [K] = k \begin{bmatrix} 3 & -2 & -1 \\ -2 & 5 & -3 \\ -1 & -3 & 4 \end{bmatrix}$$

The natural frequencies are

$$\omega_1 = 0, \qquad \omega_2 = 1.3540 \left(\frac{k}{m}\right)^{1/2}, \qquad \omega_3 = 2.4495 \left(\frac{k}{m}\right)^{1/2}$$

The corresponding normal mode matrix is

$$\Phi = \frac{1}{m^{1/2}} \begin{bmatrix} 0.40825 & 0.54772 & 0.18257 \\ 0.40825 & 0 & -0.91287 \\ 0.40825 & -0.36515 & 0.18257 \end{bmatrix}$$

The first natural frequency is zero and the elements of the first mode are identical, which makes it apparent that the first mode is a rigid body mode.

The generalized forces corresponding to our selection of generalized coordinates are $Q_1 = F_0 \sin(\omega t) \equiv \text{Re}[(F_0/i) \exp(i\omega t)]$, $Q_2 = Q_3 = 0$. To determine the steady-state response to this excitation, we introduce the modal transformation $\{Q\} = [\Phi]\{\eta\}$. The resulting differential equations for the modal coordinates are

$$\ddot{\eta}_j + \omega_j^2 \eta_j = \{\Phi_j\}^T \{Q\} = \Phi_{1j} Q_1 = \Phi_{1j} \text{Re}\left[\frac{F_0}{i} \exp(i\omega t)\right]$$

The steady-state solution for η_j will have the frequency of the excitation. In order to match the phase of the response to the excitation, we include a factor i in the response, so that

$$\eta_j = \text{Re}\left[\frac{X_j}{i} \exp(i\omega t)\right]$$

Substitution of this representation into the modal differential equations yields

$$X_j = \frac{\Phi_{1j} F_0}{\omega_j^2 - \omega^2}$$

Note that this solution is valid for the rigid body mode, for which $\omega_1 = 0$. We obtain an expression for the generalized coordinates by recalling the modal transformation, which yields

$$\{q\} = [\Phi]\{\eta\} = [\Phi]\text{Re}\left(\frac{1}{i}\begin{Bmatrix} X_1 \\ X_2 \\ X_3 \end{Bmatrix} \exp(i\omega t)\right)$$

We wish to evaluate this result for a range of ω, with all other parameters fixed. The highest value of ω according to the problem statement should be $1.5\omega_3 \approx 3.7(k/m)^{1/2}$. For the purpose of numerical evaluation we need to nondimensionalize the expression for $\{q\}$. Toward that end we define nondimensional frequency parameters and a nondimensional normal mode matrix,

$$\Omega_j = \frac{\omega_j}{(k/m)^{1/2}}, \qquad \Omega = \frac{\omega}{(k/m)^{1/2}}$$

$$[\hat{\Phi}] = m^{1/2}[\Phi]$$

When we use these definitions to eliminate dimensional quantities, the solution for the generalized coordinates becomes

$$\{q\} = \frac{F_0}{k}\mathrm{Re}\left(\frac{1}{i}\{Y\}\exp(i\omega t)\right)$$

where the elements of $\{Y\}$ are nondimensional amplitudes of the respective generalized coordinates,

$$\{Y\} = [\hat{\Phi}]\begin{Bmatrix} \dfrac{\hat{\Phi}_{11}}{\Omega_1^2 - \Omega^2} \\[2mm] \dfrac{\hat{\Phi}_{12}}{\Omega_2^2 - \Omega^2} \\[2mm] \dfrac{\hat{\Phi}_{13}}{\Omega_3^2 - \Omega^2} \end{Bmatrix} = \begin{bmatrix} 0.40825 & 0.54772 & 0.18257 \\ 0.40825 & 0 & -0.91287 \\ 0.40825 & -0.36515 & 0.18257 \end{bmatrix}\begin{Bmatrix} \dfrac{0.40825}{-\Omega^2} \\[2mm] \dfrac{0.54772}{1.8333 - \Omega^2} \\[2mm] \dfrac{0.18257}{6 - \Omega^2} \end{Bmatrix}$$

The nondimensional amplitudes $\{Y\}$ are the same as the dimensional values that would result if k and m were equal to unity. The result for the specified range of frequencies is as plotted.

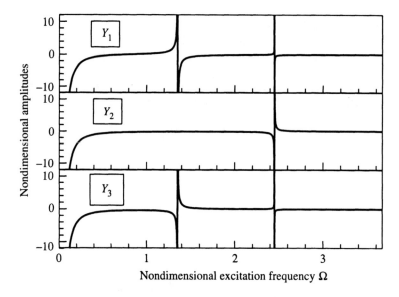

An examination of the preceding expression for $\{Y\}$ reveals that if Ω has an arbitrary value not close to one of the nondimensional natural frequencies, the Y_j values have comparable orders of magnitude. Near $\Omega = 0$ the rigid body modal amplitude factor is almost singular. This may be explained by noting that if an oscillatory force is applied to a body that may move freely, the force is balanced solely by the body's inertial resistance, $m\bar{a}$, and $\bar{a}$, being proportional to Ω^2, is negligible at low frequencies. Up to $\Omega \approx 0.5$, the three displacements are approximately equal, which indicates that the system essentially moves as a rigid body up to that frequency. The first deformational mode, that is, $\{\Phi_2\}$, becomes important as the frequency approaches ω_2, but the middle block does not respond at that frequency because it is a node for the second mode. Near the third natural frequency, all blocks are close to resonance. Also observe that as the frequency is increased across a natural frequency, amplitudes involved in the resonance change sign, a consequence of crossing a zero in the denominator of the amplitude factor X_j for each modal coordinate. As is true for one-degree-of-freedom systems, this zero corresponds to the transition of the dynamic stiffness from being spring-like to being mass-like.

Another interesting feature is that Y_1 has a zero value at $\Omega \approx 0.8$ and $\Omega \approx 2.4$, which surprises some individuals because block 1 is the one to which the force is applied. At these frequencies the forces in the springs connecting block 1 to blocks 2 and 3 balance the applied force, without any motion of block 1. We will study this phenomenon in greater detail later, when we study vibration absorbers.

Away from the natural frequencies, all amplitudes are comparatively small. The narrowness of the frequency intervals in which the amplitudes are large presents a difficulty for experimental measurements, because it indicates that we must be sure to change the frequency in very small increments if we wish to observe resonances associated with natural frequencies.

EXAMPLE 4.12

Harmonic forces that are 90° out of phase are applied transversely to blades 1 and 6 of the 10-bladed disk assembly in Example 4.6. The nondimensional representation of these forces is

$$F_1 = Fm\omega_b^2 \cos(\omega t), \qquad F_6 = Fm\omega_b^2 \sin(\omega t)$$

where m is the blade mass, ω_b is the natural frequency of an ideal single blade when there is no connective spring, and F is a nondimensional constant. The case to be considered is the imperfect one, in which the isolated blade natural frequencies are related to the ideal values by

$$\omega_{bj}^2 = \omega_b^2(1 + \Delta f_j)$$

with Δf_j being small statistical parameters whose values are specified in Example 4.6. In order to account for dissipation, each blade is considered to act on the blade's mass as a spring whose stiffness is ω_{bj}^2 and a parallel dashpot whose constant is $2\zeta\omega_{bj}$. The equations of motion in this model are

$$\ddot{q}_j + 2\zeta\omega_{bj}\dot{q}_j + \omega_{bj}^2 q_j + \omega_c^2(2q_j - q_{j-1} - q_{j+1}) = \begin{cases} F\omega_b^2\cos(\omega t), & j = 1 \\ F\omega_b^2\sin(\omega t), & j = 6 \\ 0 \text{ otherwise} \end{cases}$$

where $j = 1, \ldots, N \ (=10)$. As described in Example 4.6, the radial arrangement of the blades requires that $q_0 \equiv q_N$ and $q_{N+1} \equiv q_1$. Determine the steady-state amplitudes of each blade's vibration when the blades are weakly coupled, $R \equiv \omega_c/\omega_b = 0.05$, and damping is light, $\zeta = 0.005$. The frequency range for this evaluation is $0.8\omega_1 < \omega < 1.2\omega_N$, where ω_1 and ω_N are the lowest and highest natural frequencies.

Solution   Although this exercise will demonstrate the procedures for using software to evaluate steady-state response, its primary objective is to shed light on the variety of phenomena that influence such response. The importance of these phenomena is magnified by the close spacing of the natural frequencies, as well as the localized nature of the mode shapes. We begin by nondimensionalizing time t using $t = \omega_b t'$, which introduces a factor ω_b to every time derivative. The given equations of motion then become

$$\frac{d^2 q_j}{dt^2} + 2\zeta\sqrt{1 + \Delta f_j}\,\frac{dq_j}{dt} + (1 + \Delta f_j)q_j + R^2(2q_j - q_{j-1} - q_{j+1}) = \text{Re}[F_j\exp(i\omega t)]$$

$$F_1 = F, \qquad F_6 = \frac{F}{i}, \qquad F_j = 0 \text{ otherwise}$$

where $j = 1, \ldots, 10$, $q_0 = q_{10}$, and $q_{11} = q_1$. The inertia and stiffness matrices corresponding to this nondimensional form were identified in Example 4.6 as being

$$M_{j,n} = \begin{cases} 1, & j = n \\ 0, & j \neq n \end{cases}, \quad j, n = 1, \ldots, 10$$

$$M_{j,j} = 1 + \Delta f_j + 2R^2$$

$$K_{j,j+1} = K_{j+1,j} = -R^2, \quad j = 1, \ldots, 9$$

$$K_{1,10} = K_{10,1} = -R^2$$

The damping matrix is diagonal, with elements that are the coefficients of the first derivative term in each of the equations of motion,

$$C_{j,j} = 2\zeta\sqrt{1 + \Delta f_j}$$

The values of R and each Δf_j are identical to those used previously to evaluate the natural frequencies ω_j and mode shapes $\{\phi_j\}$, so we may directly employ the earlier eigensolution. It will be useful for our later discussions to have quick reference to the natural frequency values. In addition, the localized nature of the mode shapes, in which a different blade responds at each frequency, will also prove to be important. Hence, we repeat in tabular form the natural frequencies and the blade at which each mode is localized.

Mode #j	ω_j	Localized at blade
1	0.976	5
2	0.980	8
3	0.984	3
4	0.995	10
5	0.998	6
6	1.003	4
7	1.007	9
8	1.019	2
9	1.027	7
10	1.035	1

The modes calculated previously were not normalized, but we need the normalized values to analyze forced response. Thus, we evaluate normal modes according to

$$\{\Phi_j\} = \frac{\{\phi_j\}}{\sqrt{\{\phi_j\}^T[M]\{\phi_j\}}}, \quad j = 1, \ldots, 10$$

As always, we use these to fill in the columns of the normal mode matrix $[\Phi]$. We then check the correctness of our work at this stage by verifying that both orthogonality conditions are satisfied by $[\Phi]$ and ω_j.

The next step is to evaluate the modal damping ratios ζ_j. Toward that end we compute $[\Phi]^T[C][\Phi]$. The result, being a 10×10 matrix, is too lengthy to list here, but it is very close to being diagonal. In fact, if the system had ideal properties, such that all Δf_j values are zero, we would find that the damping is proportional, being given by $[C] = 2\zeta[M]$. In that case, $[\Phi]^T[C][\Phi]$ would actually be diagonal. Because the Δf_j values here are very small, the off-diagonal terms in the modal damping transformation are a factor of 100 or more smaller than the diagonal terms. We obtain the modal damping ratios by equating the jth diagonal term of $[\Phi]^T[C][\Phi]$ to $2\zeta_j\omega_j$. All of the factors are the same, to three significant figures, being $\zeta_j = 0.0490$, $j = 1, \ldots, 10$.

The modal coordinate differential equations corresponding to the modal transformation $\{q\} = [\Phi]\{\eta\}$ are

$$\ddot{\eta}_j + 2\zeta_j\omega_j\dot{\eta}_j + \omega_j^2\eta_j = \{\Phi_j\}^T\text{Re}[\{F\}\exp(i\omega t)]$$

where the nonzero elements of $\{F\}$ are $F_1 = F$ and $F_6 = F/i$. The steady-state blade displacement is indicated by eq. (4.3.42) to be

$$\{q\} = \sum_{j=1}^{10} \{\Phi_j\}(\eta_j)_{ss} = \mathrm{Re}\left[\sum_{j=1}^{10} \{\Phi_j\}\frac{\{\Phi_j\}^{\mathrm{T}}\{F\}}{\omega_j^2 + 2i\zeta_j\omega_j\omega - \omega^2}\exp(i\omega t)\right]$$

The problem statement requests evaluation of the amplitude of each q_n. This requires that we form the coefficient of $\exp(i\omega t)$, then take the magnitude of each coefficient, that is,

$$|q_n| = |Y_n|, \quad \text{where } \{Y\} = \sum_{j=1}^{10}\{\Phi_j\}\frac{\{\Phi_j\}^{\mathrm{T}}\{F\}}{\omega_j^2 + 2i\zeta_j\omega_j\omega - \omega^2}$$

Note that the operation of taking the magnitude of each element of a complex vector may be carried out in MATLAB by preceding the *abs* function with a period, for example, using the program step $amp = .abs(Y)$. In Mathcad, operations are performed element by element by vectorizing the operation, so the corresponding program step would be $amp = \overrightarrow{|Y|}$. It will be necessary to perform these computations at many frequencies ω. The efficiency of such computations may be increased greatly by recognizing that the only part of the preceding expression for $\{Y\}$ that depends on ω is the denominator. We therefore define a square array $[P]$ that holds the frequency-independent terms, such that the *j*th column of $[P]$ is

$$\{P_j\} = \{\Phi_j\}\{\Phi_j\}^{\mathrm{T}}\{F\}$$

The syntax for this operation in MATLAB would be $P(:,j) = PHI(:,j) * PHI(:,j)' * F$, while Mathcad users would write $P^{<j>} = \Phi^{<j>} * \Phi^{<j>\mathrm{T}} * F$. One then only needs to define a function giving the complex frequency response for each mode, which is combined with the columns of $[P]$ to form the complex amplitudes $[Y]$ as follows,

$$D(\omega, j) \equiv \frac{1}{\omega_j^2 + 2i\zeta_j\omega_j\omega - \omega^2}$$

$$\{Y(\omega)\} = \sum_{j=1}^{10}\{P_j\}D(\omega, j)$$

An important consideration when we wish to construct a graph of amplitude as a function of frequency is the frequency increment to use for the computation. The dependence on ω is described by the complex frequency response $D(\omega, j)$, which is comparable to the quantity for a one-degree-of-freedom system having natural frequency ω_j and critical damping ratio ζ_j. The region over which this quantity changes most rapidly is the vicinity of $\omega = \omega_j$. Recall that the bandwidth (the frequency interval separating the half-power points) of a resonant peak is $\Delta\omega = 2\zeta_j\omega_j$. If we select the frequency increment to be no larger than one-quarter of the bandwidth, then we will have at least three frequency evaluations around the peak value of D. The smallest $\Delta\omega$ corresponds to the smallest combination $2\zeta_j\omega_j$, all of which are approximately 0.010 for this system, so we require that the frequency increment be smaller than $(0.01)/4$.

There are several features of the graphs (see the next page) to discuss. It is a common practice to use a logarithmic scale to display the amplitudes because they cover a large range when a system is lightly damped. In most systems peaks associated with resonant excitation are exhibited by all generalized coordinates, and all of the generalized coordinates have comparable magnitude at a resonance. The bladed disk model does not exhibit this behavior. We see that the amplitudes of blades 1 and 6 are much larger than the others, and that each only displays one peak. What we have here is a combination of effects. We anticipated that the bandwidth of each of the resonances would be $\Delta\omega = 0.01$, but that is comparable to the average interval separating natural frequencies. As a consequence, the width of each peak is bigger than the interval that would separate the peaks if there were less damping. This causes the peaks to merge, which is a phenomenon known as *modal coupling*. Despite its name, the term does not imply that the modal coordinates are mutually

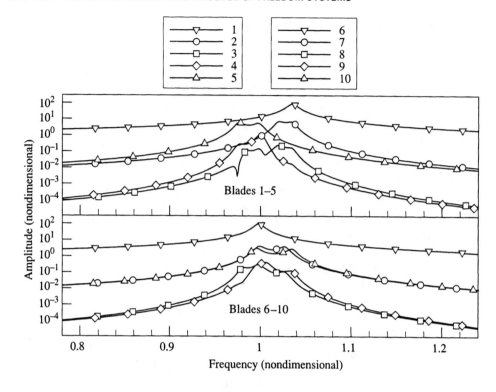

dependent. Rather, it connotes a situation in which the individual contributions of the modes cannot be distinguished. The occurrence of this phenomenon causes difficulty when one performs experiments to measure the modal properties of a system.

Modal coupling will occur in any system that has closely spaced natural frequencies if the damping is sufficiently large. What is special about the bladed disk assembly is that its modes are localized. Recall that in any case the amount by which generalized forces excite the jth mode is described by $\{\Phi_j\}^T\{F\}$. Hence, the force at blade 1 most strongly excites the mode that is localized at blade 1, which is the 10th mode. Similarly, the force at blade 6 excites the mode that is localized at blade 6, which coincidentally is mode 6. The modal values at blades 1 and 6 for the other modes are very small, so the excitation those modes receive is also very small. The excitations received by modes 6 and 10 are comparable in magnitude, but the other factor affecting a modal coordinate's amplitude is its complex frequency response $D(\omega, j)$. When ω is close to ω_j, this quantity is close to its maximum value. Thus, modal coordinate η_6 shows a maximum near $\omega_6 = 0.998$, while η_{10} maximizes at $\omega_{10} = 1.035$. The mode shape then reenters consideration when we convert from the modal coordinates to the generalized coordinates by forming $[\Phi]\{\eta\}$, which we have seen is equivalent to summing the various products $\{\Phi_j\}\eta_j$. For a given value of a modal coordinate, the largest element of $\{\Phi_j\}$ indicates the generalized coordinate that responds the most. Thus, localization of the mode further magnifies the response of the blade at which localization occurs.

We may see the effect of localization without mode coupling by decreasing the damping parameter. The next set of graphs describes the response when $\zeta = 5(10^{-5})$. The range of frequencies for this evaluation is much smaller in order to highlight the resonances, whose bandwidth is very narrow. Unlike the previous case, there now are amplitude peaks near each of the natural frequencies. The vast differences in the amplitudes at a particular natural frequency, as well as the fact that some blades do not manifest all of the resonances, are attributable to the localized nature of the mode shapes through the factors discussed earlier.

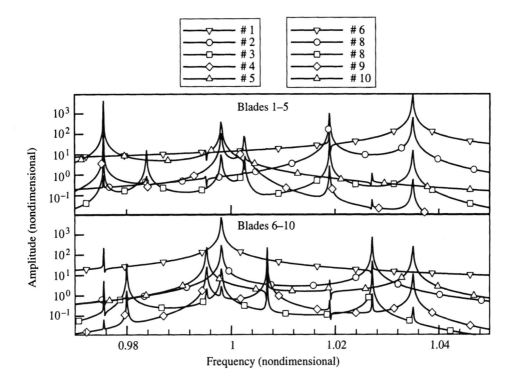

REFERENCES AND SELECTED READINGS

CHEN, P. T., & GINSBERG, J. H. 1992. "On the Relationship between Veering of Eigenvalue Loci and Parameter Sensitivity of Eigenfunctions." *ASME Journal of Vibration and Acoustics*, 114, 141-148.

CRAIG, R. R., JR. 1981. *Structural Dynamics*. John Wiley and Sons, Inc., New York.

DIMARGONAS, A. 1996. *Vibration for Engineers*, 2nd ed. Prentice Hall, Englewood Cliffs, NJ.

INMAN, D. J. 1989. *Vibration*. Prentice Hall, Englewood Cliffs, NJ.

JAMES, M. L., SMITH., G. M., WOLFORD, J. C., & WHALEY, P. W. 1989. *Vibration of Mechanical and Structural Systems*. Harper & Row, New York.

KRAUS, A. D. 1987. *Matrices for Engineers*. Hemisphere, Washington, DC.

MEIROVITCH, L. 1967. *Analytical Methods in Vibrations*. Macmillan, New York.

MEIROVITCH, L. 1986. *Elements of Vibration Analysis*, 2nd ed. McGraw-Hill, New York.

MEIROVITCH, L. 1997. *Principles and Techniques of Vibrations*. Prentice Hall, Englewood Cliffs, NJ.

NEWLAND, D. E. 1989. *Mechanical Vibration Analysis and Computation*. Longman Scientific and Technical, White Plains, NY.

PRESS, W. H., TEUKOLSKY, S. A., VETTERLING, W. T., & FLANNERY, B. P. 1992. *Numerical Recipes*, 2nd ed. Cambridge University Press, Cambridge, England.

TONGUE, B. H. 1996. *Principles of Vibrations*. Oxford University Press, New York.

WEI, S.-T., & PIERRE, C. 1988. "Localization Phenomena in Mistuned Assemblies with Cyclic Symmetry." *ASME Journal of Vibration, Acoustics, Stress, and Reliability in Design*, 110, 429–438.

EXERCISES

4.1 The inertia and stiffness matrices for a system are

$$[M] = \begin{bmatrix} 4 & 0 \\ 0 & 2 \end{bmatrix} \text{ kg}, \quad [K] = \begin{bmatrix} 200 & 200 \\ 200 & 800 \end{bmatrix} \text{ N/m}$$

Determine the corresponding natural frequencies and modes of free vibration.

4.2 Determine the natural frequencies and corresponding mode shapes of the system shown on the next page for the case where $k_1 = k_2 = k$ and

$m_1 = m_2 = m$, where k and m are scaling factors for stiffness and mass.

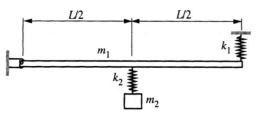

EXERCISE 4.2

4.3 Determine the natural frequencies and modes of free vibration for this system of bars and springs for the case where $m_1 = m$, $m_2 = 2m$, $k_1 = k$, and $k_2 = k/2$, where m and k are basic units of mass and stiffness.

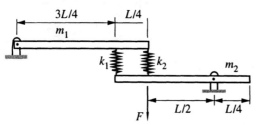

EXERCISE 4.3

4.4 Solve Example 4.2 for the case where the vertical displacements of the spring attachment points A and B are used as generalized coordinates.

4.5 The bar is horizontal when the system is in static equilibrium. Derive the characteristic equation and an equation for the mode vector corresponding to a specified natural frequency. Then compute the natural frequencies and mode shapes for two cases (a) $m_1 = 0.5$ kg, $m_2 = 2$ kg, $k = 200$ N/m, $L = 400$ mm, and (b) $m_1 = 0.5$ kg, $m_2 = 0.02$ kg, $k = 200$ N/m, $L = 400$ mm. Determine the natural frequencies and mode shapes of the system. What simpler system does the fundamental frequency suggest in the second case?

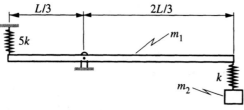

EXERCISE 4.5

4.6 In the mechanism shown, the stiffnesses are $k_1 = 4$ and $k_2 = 6$ kN/m. The masses are $m_1 = 5$ and $m_2 = 10$ kg, and $L = 600$ mm. Determine the natural frequencies and mode shapes.

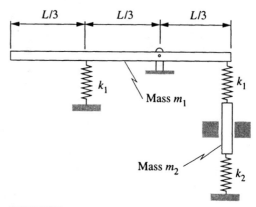

EXERCISE 4.6

4.7 The linkage shown consists of two rigid bars having length L and mass m. The torsional springs, whose stiffness is βmgL (β is a nondimensional parameter), are undeformed when the bars are vertical. For the case where $\beta = 4$, determine the natural frequencies and modes of free vibration. Then determine the natural frequencies when $\beta = 2$. Explain the significance of the result of the second case.

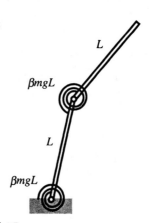

EXERCISE 4.7

4.8 The system shown in the sketch represents a scale model used to study the vibration of a three-story building. The masses for the model are $m_1 = 100$, $m_2 = 200$, and $m_3 = 300$ kg. The springs are selected such that $k_j = \omega_j^2 m_j$, where $\omega_j = 40$, 50, and 60 rad/s. Determine the characteristic equation, natural frequencies, and mode shapes of this system.

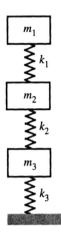

EXERCISE 4.8

4.9 Three identical bars having mass m are suspended from the ceiling and interconnected by a pair of identical springs. The unstretched length of each spring equals the spacing between the bars when they are vertical. For the purpose of performing algebraic manipulations the spring constant will be represented as $k = \alpha mg/L$, where α is a nondimensional constant. Generalized coordinates are the counterclockwise rotation of each bar.

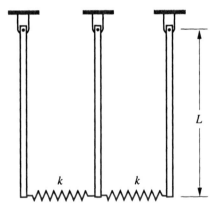

EXERCISE 4.9

(a) Derive the characteristic equation whose solution would yield the natural frequencies.

(b) Derive a set of algebraic equations whose solution would yield the free vibration mode corresponding to a specific natural frequency.

(c) Evaluate the system's natural frequencies and modes corresponding to $\alpha = 2$ and $\alpha = 0.02$.

4.10 The diagram depicts a model of an automobile that accounts for vertical displacement of the center of mass y_G, pitching rotation θ_y, and rolling rotation θ_x. The top view shows the arrangement of the spring attachment points relative to the center of mass. The inertial properties of the chassis, which is modeled as a rigid body, are mass, $m = 1400$ kg, and centroidal radii of gyration, $\kappa_x = 0.4$ m, $\kappa_y = 1.2$ m. The stiffness of each of the two front springs is $k_A = k_B = 30$ kN/m, while the stiffness of each of the rear springs is $k_C = k_D = 20$ kN/m. A linearized approximation of the vertical displacement of the point at which each spring-dashpot pair is attached to the chassis is

$$y_A = y_G - 1.3\theta_y - 0.9\theta_x,$$

$$y_B = y_G - 1.3\theta_y + 0.9\theta_x,$$

$$y_C = y_G + 1.5\theta_y - 0.9\theta_x,$$

$$y_D = y_G + 1.5\theta_y + 0.9\theta_x$$

Determine the natural frequencies and free vibration modes of this vehicle model.

4.11 The mass and stiffness matrices of a system are

$$[M] = \begin{bmatrix} 4 & 1 \\ 1 & 3 \end{bmatrix} \text{kg}, \quad [K] = \begin{bmatrix} 300 & 0 \\ 0 & 200 \end{bmatrix} \text{kN/m}$$

Determine the system's natural frequencies and normal vibration modes.

4.12 The following questions pertain to the orthogonality properties of vibration modes:

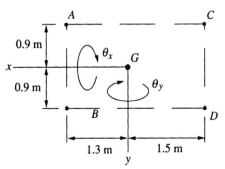

EXERCISE 4.10   Top view

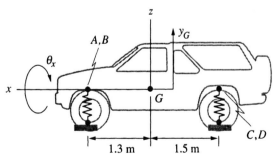

Side view

(a) The inertial matrix and first normal mode of a system are

$$[M] = \begin{bmatrix} 2 & 1 \\ 1 & 3 \end{bmatrix} \text{kg}, \qquad \{\Phi_1\} = \begin{Bmatrix} 0.4 \\ a \end{Bmatrix}$$

Determine the possible values of a.

(b) The inertia matrix and unnormalized modes of a system are known to be

$$[M] = \begin{bmatrix} 90 & 30 \\ 30 & 60 \end{bmatrix} \text{kg}, \qquad \{\phi_1\} = \begin{Bmatrix} a \\ b \end{Bmatrix},$$

$$\{\phi_2\} = \begin{Bmatrix} 1 \\ -3 \end{Bmatrix}$$

What relation(s) must exist between the values of a and b?

(c) The stiffness matrix and unnormalized modes of a system are known to be

$$[K] = \begin{bmatrix} 2000 & 0 \\ 0 & 4000 \end{bmatrix} \text{N/m}, \qquad \{\phi_1\} = \begin{Bmatrix} 2 \\ 1 \end{Bmatrix},$$

$$\{\phi_2\} = \begin{Bmatrix} a \\ b \end{Bmatrix}$$

What relation(s) must exist between the values of a and b?

(d) The stiffness matrix, second natural frequency, and corresponding unnormalized mode of a system are known to be

$$[K] = \begin{bmatrix} 4000 & 2000 \\ 2000 & 3000 \end{bmatrix} \text{kg}, \qquad \omega_2 = 80 \text{ rad/s}$$

$$\{\phi_2\} = \begin{Bmatrix} 1 \\ -0.4 \end{Bmatrix}$$

Determine the *normal* mode $\{\phi_2\}$.

4.13 The following properties of a two-degree-of-freedom-system are known from measurements taken in a set of experiments:

- The mass matrix for these generalized coordinates is $m_{11} = 2$, $m_{12} = 0$, and $m_{22} = 3$ kg.
- The only stiffness coefficient that has been measured is $k_{11} = 25$ kN/m.
- The amplitude of q_1 is three times that of q_2 in one of the modes of free vibration, and q_2 is 180° out of phase from q_1 in that mode.
- The natural frequency for the mode described in part (c) is 100 rad/s.

Use this information to determine the natural frequency of the other mode, the normal modes for

both natural frequencies, and the other stiffness coefficients.

4.14 The diagram below models an automobile and its suspension as a rigid block on springs. The mass is m, and the radius of gyration relative to the center of mass G is r_G. Generalized coordinates are the vertical displacements y_1 and y_2 of the ends relative to the static equilibrium position. The natural frequencies and unnormalized modes of this system are

$$\omega_1 = 1.5(k/m)^{1/2}, \qquad \omega_2 = 2(k/m)^{1/2}$$

$$\{\phi_1\} = \begin{Bmatrix} 1 \\ 3 \end{Bmatrix}, \qquad \{\phi_2\} = \begin{Bmatrix} 1 \\ -0.5 \end{Bmatrix}$$

where k and m are basic stiffness and mass units. For these properties, determine the normal modes corresponding to the given properties and the radius of gyration relative to the center of mass G.

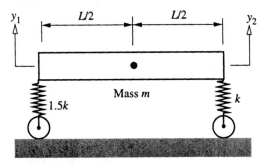

EXERCISE 4.14

4.15 Consider the system in Exercise 4.1. Use mathematical software to determine the natural frequencies and normal modes of the system.

4.16 Consider the system in Exercise 4.8. Use mathematical software to determine the natural frequencies and normal vibration modes of the system.

4.17 Consider the situation in which the left end of the bar in Example 4.4 is free. Such a situation may be represented by setting the leftmost spring constant to zero. Repeat the analysis of the natural frequencies and mode shapes for this system.

4.18 A simple model for a truss structure was constructed in Exercise 1.46. Consider a case where $m_1 = 50,000$ kg, $m_2 = 75,000$ kg, $E = 210(10^9)$ Pa, $A = 0.02$ m^2, and $L = 20$ m. Determine the natural frequencies and free vibration modes of the truss. Draw a sketch showing amplified pictures of the displacement patterns.

4.19 Because the center of mass G of the vehicle in Exercise 4.10 is situated laterally midway between the wheels, the rolling rotation θ_X is uncoupled from the mass center's vertical displacement y_G and the pitching rotation θ_y. If the center of mass is shifted 100 mm to the right (looking forward), the displacement of the springs will be related to these generalized coordinates by

$$y_A = y_G - 1.3\theta_y - 0.8\theta_x,$$
$$y_B = y_G - 1.3\theta_y + 1.0\theta_x$$
$$y_C = y_G + 1.5\theta_y - 0.8\theta_x,$$
$$y_D = y_G + 1.5\theta_y + 1.0\theta_x$$

All other parameters are as stated in Exercise 4.10. Use mathematical software to determine the natural frequencies and normal modes of the system.

4.20 The static displacement of a cantilever beam at distance x from the fixed end resulting from application of a unit transverse force at distance y from the fixed end is

$$\omega = \frac{y^2}{6EI}(3x - y) \quad \text{if } x \ge y$$

A lumped mass model for the vibration of such a beam, whose total mass is m, may be constructed by dividing the beam into N segments of length L/N. The mass distribution is replaced by N particles having mass $m_j = m/M$ situated at the midpoint of each segment, that is, at distance $x_j = (2j-1)L/(2N)$, as depicted below. Use the above relation for static displacement to determine the lowest four natural frequencies and corresponding mode shapes resulting from setting $N = 4$ and $N = 8$. What does a comparison of the results indicate regarding the validity of the lumped mass model?

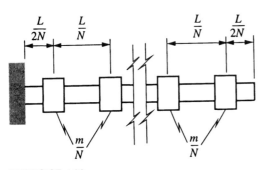

EXERCISE 4.20

4.21 The sketch depicts a model of a four-floor building, in which the columns are treated as massless beams and the floors are treated as rigid slabs whose masses in kilograms are $m_1 = 50{,}000$, $m_2 = 50{,}000$, $m_3 = 40{,}000$, and $m_4 = 35{,}000$ kg. A static horizontal force of 5 kN is applied succes-

sively to each floor. The following are measurements of the static horizontal displacements produced by each application of the force:

	Displacement (mm) at floor no.:			
Force location	1	2	3	4
1	3	3	3	3
2	×	8	8	8
3	×	×	20	20
4	×	×	×	35

Entries marked as × denote measurements that were not made. Determine the natural frequencies and modes of this system.

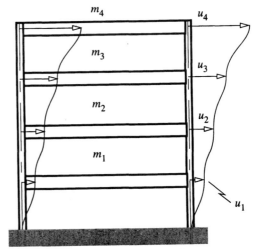

EXERCISE 4.21

4.22 Consider the system in Figure 4.6. Use the Gram-Schmidt orthogonalization procedure to determine a set of modes for $k_1 = k_2 = k_3 = k$ that match the modes for two cases where the stiffnesses are not identical:

(a) $k_1 = 0.99k$, $k_2 = k_3 = k$
(b) $k_1 = k_3 = k$, $k_2 = 0.99k$

4.23 Consider the automobile modeled in Example 4.2. All parameters except the rear spring constant k_B and the distance from the center of mass to the rear wheels L_2 are as specified there. Is there any combination of values of k_B and L_2 for which both modes have the natural frequency? If so, what are the natural frequencies and mode shapes in that case?

4.24 The system in the sketch consists of blocks having mass m that slide over smooth guidebars and are connected to the rigid bar by four identical springs. The mass of the bar is σm, and the system lies in the horizontal plane.

(a) Prove that in order for the system to have three equal natural frequencies, it is necessary that $\sigma = 6$ and $a = L/2$.

(b) Determine the single natural frequency and three normal modes in the case where $\sigma = 6$ and $a = L/2$.

(c) Compare the results of part (b) to those obtained from mathematical software when $\sigma = 6$ and $a = 0.499L$.

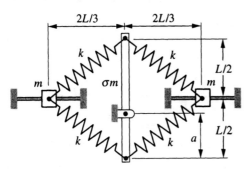

EXERCISE 4.24

4.25 Three blocks are interconnected by springs and slide over the ground with negligible resistance from friction. One of the modes is a rigid body motion. It is possible to determine the deformational modes solely by using the facts that the system is physically symmetric, and that the center of mass of the system remains stationary in the deformational modes. Determine the modes in this manner, along with a brief explanation of your reasoning. Use those results to determine the natural frequencies of the system. Then, as a separate solution, determine the natural frequencies of this system by solving the eigenvalue problem.

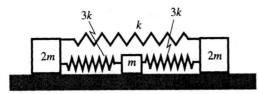

EXERCISE 4.25

4.26 Determine the natural frequencies and mode shapes of the pair of carts and attached springs. Friction at all contacting surfaces is negligible.

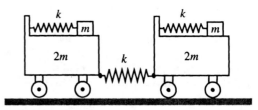

EXERCISE 4.26

4.27 Three identical disks, whose mass is m, are interconnected by identical springs k. The disks slide without friction over a horizontal table. This system has six degrees of freedom. The displacement components x_j, y_j of each mass relative to the respective initial positions constitute a suitable set of generalized coordinates. Determine the natural frequencies and mode shapes.

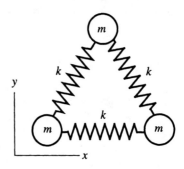

EXERCISE 4.27

4.28 The bars in Exercise 4.3 were at rest in their static equilibrium position, when at $t = 0$ the upper bar is given an initial clockwise rotation of $4°$. Determine the rotation of each bar as a function of the nondimensional time parameter $\tau = (k/m)^{1/2} t$.

4.29 The diagram models an automobile and its suspension as a rigid block on springs. The mass of the bar is m, and the radius of gyration relative to the center of mass G is $\kappa_G = 0.4L$. Generalized coordinates are the vertical displacements y_1 and y_2 of the ends relative to the static equilibrium

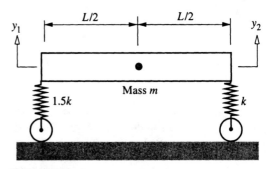

EXERCISE 4.29

position. Consider a situation where the vehicle is released from rest with $y_1 = mg/k$, $y_2 = 0$. Determine the ensuing free vibration as a function of the nondimensional time $t = (k/m)^{1/2}t$.

4.30 Three identical bars having mass m are suspended from the ceiling and interconnected by a pair of identical springs. The unstretched length of each spring equals the spacing between the bars when they are vertical. The spring stiffness is $k = 0.05mg/L$. At $t = 0$, all bars are vertical. The left and right bars are not moving at this instant, while the middle bar is rotating at 2 rad/s. Determine and graph the vibration of each bar.

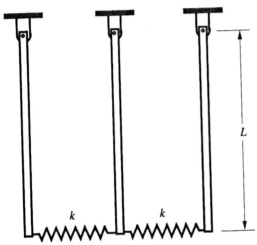

EXERCISE 4.30

4.31 At $t = 0$ the building in Exercise 4.21 is at its static equilibrium position. At this instant, a gust of wind imparts to each floor a horizontal velocity of 5 m/s to the right. Determine and graph the response in an initial interval of 2 s.

4.32 Consider the three blocks in Exercise 4.25. At $t = 0$, all bodies are initially at rest, except that the right block has velocity v to the right. At that instant, the springs are unstretched. Determine the displacement of each block relative to its initial position in the ensuing free vibration.

4.33 The inertia matrix, natural frequencies, and modes of an undamped two-degree-of-freedom system are

$$[M] = \begin{bmatrix} 3 & 0 \\ 0 & 4.5 \end{bmatrix} \text{kg}, \quad \omega_1 = 2 \text{ rad/s}, \quad \omega_2 = 3 \text{ rad/s}$$

$$\{\phi_1\} = \begin{Bmatrix} 1 \\ 2 \end{Bmatrix}, \quad \{\phi_2\} = \begin{Bmatrix} 3 \\ -1 \end{Bmatrix}$$

The system is at rest at the reference position $q_1 = q_2 = 0$ when $t = 0$, at which instant the following generalized forces are applied: $Q_1 = 0$,

$Q_2 = 10\exp(-t/2)$ N, where t is measured in seconds. Determine q_1 and q_2 when $t = 2$ s.

4.34 The mass matrix, excitation, and modal properties of a two-degree-of-freedom system are known to be

$$[m] = \begin{bmatrix} 3 & 1 \\ 1 & 1 \end{bmatrix} \text{kg}, \quad \{Q\} = \begin{Bmatrix} 0 \\ 20t \end{Bmatrix} N$$

$$\omega_1 = 5.42 \text{ rad/s}, \quad \omega_2 = 13.00 \text{ rad/s},$$

$$\{\phi_1\} = \begin{Bmatrix} 1 \\ 0.414 \end{Bmatrix}, \quad \{\phi_2\} = \begin{Bmatrix} 1 \\ \alpha \end{Bmatrix}$$

where α is an unspecified value. The system was initially at rest in the static equilibrium position. Determine the response.

4.35 The mass matrix, natural frequencies, and normal mode matrix of a two-degree-of-freedom system are

$$[m] = \begin{bmatrix} 2/3 & 0 \\ 0 & 1/3 \end{bmatrix} \text{kg}, \quad \omega_1 = 2 \text{ rad/s},$$

$$\omega_2 = 4 \text{ rad/s}, \quad [\Phi] = \begin{bmatrix} 1/\sqrt{2} & -1 \\ \sqrt{2} & 1 \end{bmatrix}$$

The system is initially at the static equilibrium position with initial velocities $\dot{q}_1 = \dot{q}_2 = 10$ m/s. At $t = 0$, a set of generalized forces $Q_1 = \sin(3t)$ N, $Q_2 = 0.5\cos(3t)$ N is applied to the system. Determine the resulting response.

4.36 In order to isolate a piece of electronic equipment ($m_1 = 2$ kg) from the vertical vibration of the table, it is mounted on spring $k_1 = 100$ N/m. We model the table as a rigid beam having mass $m_2 = 100$ kg that is supported by two springs $k_2 = 400$ N/m. The system was initially at rest when, at $t = 0$, a uniform pulse $F = 200$ N is applied for 2 s and then removed. Determine the response of the equipment.

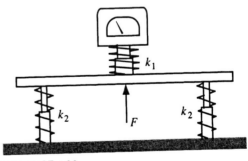

EXERCISE 4.36

4.37 Solve Exercise 4.36 for the case where the excitation is a triangular pulse given by

$F = 400(1 - t/2)[h(t) - h(t - 2)]$, where F has units of newtons and t is measured in seconds.

4.38 Consider the inclined cart with attached mass in Example 4.7 when the ramp angle is $\theta = 36.87°$. At $t = 0$, both bodies are at rest and the spring is stretched to balance the weight of the mass on the incline. At that instant, an exponentially decaying force

$$Q(t) = F_0 \exp[-0.5(k/m)^{1/2}t]$$

acting to the right is applied to the cart. Determine the displacement of each body relative to its initial position in the ensuing motion. Graph the displacement of the cart as a function of the nondimensional time $\hat{t} = (k/m)^{1/2}t$.

4.39 Identical blocks are interconnected by springs as they slide over a smooth horizontal guide bar. The matrix equations of motion are

$$m\begin{bmatrix} 1 & 0 \\ 0 & 1 \end{bmatrix}\begin{Bmatrix} \ddot{x}_1 \\ \ddot{x}_2 \end{Bmatrix} + k\begin{bmatrix} 2 & -1 \\ -1 & 2 \end{bmatrix}\begin{Bmatrix} x_1 \\ x_2 \end{Bmatrix} = \begin{Bmatrix} Q_1 \\ Q_2 \end{Bmatrix}$$

where m and k are basic units of stiffness and mass. When $t = 0$, both blocks pass the static equilibrium position and are moving to the left at speed v. At that instant an impulsive force $F = P\delta(t)$ is applied to block 1, with the result that this block momentarily comes to rest.

(a) Derive an expression for P in terms of v and the properties of the system.

(b) Determine the displacement of each block subsequent to application of the impulsive force. Graph each displacement as a function of the nondimensional time $(k/m)^{1/2}t$.

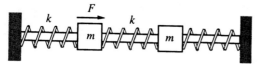

EXERCISE 4.39

4.40 The mass matrix for a two-degree-of-freedom system is $m_{11} = 10$ kg, $m_{12} = 0$, and $m_{22} = 20$ kg. The natural frequencies and modes of free vibration are known to be $\omega_1 = 7.962$ Hz, $\omega_2 = 15.382$ Hz, $\{\phi_1\} = [0.731 \quad 1]^T$, $\{\phi_2\} = [-2.731 \quad 1]^T$. The system is proportionally damped, with proportionality constants $\alpha = 1.0$, $\beta = 0.002$. Determine the free vibration response if the system is released from rest at $q_1 = 60$ mm, $q_2 = 80$ mm.

4.41 The system in Exercise 4.40 is initially at rest in the static equilibrium position when a unit step force is applied to the first generalized coordinate, $Q_1 = h(t)$, $Q_2 = 0$. Determine the resulting response of the generalized coordinates.

4.42 Determine the response of the system in Exercise 4.34 if modal damping ratios are estimated to be $\zeta_1 = 0.05$, $\zeta_2 = 0.10$.

4.43 The following properties are known for a certain three-degree-of-freedom system:

$$[M] = \begin{bmatrix} 600 & 400 & 200 \\ 400 & 1200 & 0 \\ 200 & 0 & 800 \end{bmatrix} \text{kg},$$

$$[K] = \begin{bmatrix} 300 & 0 & -200 \\ 0 & 500 & 300 \\ -200 & 300 & 700 \end{bmatrix} \text{kN/m},$$

$$[C] = \begin{bmatrix} 500 & 300 & -400 \\ 300 & 900 & 600 \\ -400 & 600 & 1300 \end{bmatrix} \text{N-s/m},$$

$$\{Q\} = \begin{Bmatrix} 200\cos(16t) \\ 0 \\ 0 \end{Bmatrix} \text{N}$$

The system was initially at rest at its static equilibrium position. Use the light damping approximation to determine the response. Graph each generalized coordinate as a function of time. From that result, estimate the time required to attain the steady-state condition.

4.44 System parameters for the two-degree-of-freedom system shown are $m_1 = m_2 = 1$ kg, $k_1 = 200$, $k_2 = 300$ N/m, $c_1 = 4$, and $c_2 = 2$ N-s/m. At $t = 0$ the right block is subjected to an exponentially decaying force $F_2 = 10\exp(-2t)$ N, acting to the right. Initial conditions $q_1 = q_2 = \dot{q}_1 = \dot{q}_2 = 0$ at $t = 0$. Determine the response using the light damping approximation.

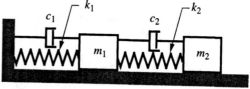

EXERCISE 4.44

4.45 As a model of the earthquake response of buildings, it is desired to study the vibration of the three oscillators when a motion $w(t)$ is imposed on the base. The system parameters are

$$m_1 = 100, \quad m_2 = 200, \quad m_3 = 300 \text{ kg}$$

$$\omega_j = 40, 50, 60 \text{ rad/s}, \quad \text{where } \omega_j = (k_j/m_j)^{1/2}$$

$$\zeta_j = 0.25, 0.15, 0.05, \quad \text{where } \zeta_j = 0.5c_j/(k_jm_j)^{1/2}$$

$$w = 10\exp(-t/\tau) \text{ mm}, \quad \text{where } \tau = 0.05 \text{ s}$$

The system is initially at rest at its static equilibrium position corresponding to $w = 0$. Use the light damping approximation to determine and graph the displacement of each mass as a function of time for $0 < t < 0.4$ s.

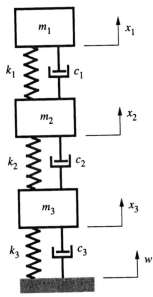

EXERCISE 4.45

4.46 The mass and damping matrices, natural frequencies, and nonnormalized modes for a two-degree-of-freedom system are

$$[M] = \begin{bmatrix} 0.4 & 0.15 \\ 0.15 & 0.30 \end{bmatrix} \text{ kg},$$

$$[C] = \begin{bmatrix} 0.05 & 0 \\ 0 & 0.10 \end{bmatrix} \text{ N-s/m}$$

$$\omega_1 = 0.25 \text{ Hz}, \quad \omega_2 = 0.40 \text{ Hz},$$

$$\{\phi_1\} = \begin{Bmatrix} 1 \\ 1.403 \end{Bmatrix}, \quad \{\phi_2\} = \begin{Bmatrix} 1 \\ -1.069 \end{Bmatrix}$$

(a) Estimate modal damping ratios for this system. Do you expect a response calculation based on a light damping approximation to be accurate?
(b) At $t = 0$ the system was at rest at $q_1 = q_2 = 0$ when it was subjected to the generalized forces $\{Q\} = [F \ \ 0]^T$, where $F(t)$ is as plotted below. Use the light damping approximation and the

superposition method, in conjunction with Appendix B, to derive an expression for the time dependence of each generalized coordinate. Graph q_1/F and q_2/F as functions of t for the case where $T = 2\pi/\omega_2$.

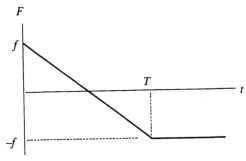

EXERCISE 4.46

4.47 The mass matrix, natural frequencies, and unnormalized vibration modes for a two-degree-of-freedom system are

$$[M] = \begin{bmatrix} 5 & -3 \\ -3 & 4 \end{bmatrix} \text{ kg}$$

$$\omega_1 = 15.68 \text{ rad/s}, \quad \omega_2 = 40.78 \text{ rad/s}$$

$$\{\phi_1\} = \begin{Bmatrix} 1 \\ 1.366 \end{Bmatrix}, \quad \{\phi_2\} = \begin{Bmatrix} 1 \\ -0.366 \end{Bmatrix}$$

The modal damping ratios are estimated to be $\zeta_1 = \zeta_2 = 0.08$. The system is subjected to a harmonic excitation for which the generalized forces are $Q_1 = 50 \sin(20t)$, $Q_2 = 100 \cos(20t)$ N. Determine the steady-state response of the generalized coordinates. Express the result in the form $q_n = \text{Re}[Y_n \exp(i20t)]$ and give the values of Y_1 and Y_2.

4.48 The mass matrix and two normalized modes of an undamped three-degree-of-freedom system are

$$[M] = \begin{bmatrix} 4 & 0 & 0 \\ 0 & 6 & 0 \\ 0 & 0 & 5 \end{bmatrix} \text{ kg}, \quad \{\Phi_1\} = \begin{Bmatrix} 0.15592 \\ -0.30427 \\ -0.26162 \end{Bmatrix},$$

$$\{\Phi_2\} = \begin{Bmatrix} 0.43038 \\ 0.19865 \\ -0.06679 \end{Bmatrix}$$

The corresponding natural frequencies are 3, 4, and 5 Hz. Consider a set of harmonic generalized forces, $Q_j = F_j \cos(\omega t)$, where F_j are constants.
(a) Determine the third normalized mode of the system and the stiffness matrix.

(b) It is observed that at a certain excitation frequency ω the steady-state response is such that $\{q\} = \{\Phi_2\} \cos(\omega t)$. What conditions must exist for this type of response to occur? Will only the second mode participate in the response if the frequency is changed?

(c) It is observed that at a certain excitation frequency ω, the steady-state response is such that $q_1 = 0$, $q_2 \neq 0$, and $q_3 \neq 0$. What condition between the mode coefficients Φ_{nj}, natural frequencies ω_n, and force amplitudes must exist for this type of response to occur? Will q_1 remain zero if the frequency is changed?

(d) It is observed that at a certain excitation frequency ω, the steady-state response is such that $q_1 = q_2 = q_3 = 0$. What condition between the mode coefficients and force amplitudes must exist for this type of response to occur? Will the response remain zero if the frequency is changed?

4.49 A small mass m is supported in the vertical plane by three concurrent springs, such that its static equilibrium position is as shown. Generalized coordinates for the system are the horizontal and vertical displacements u and w, respectively, relative to this position. The mass is loaded by a force $F \sin(\omega t)$ at a constant angle of inclination θ above the horizontal axis.

(a) For some set of values of k_1, k_2, and k_3, the normal mode matrix is known to be

$$[\Phi] = \begin{bmatrix} a & b \\ 1.5a & c \end{bmatrix}$$

Determine the corresponding values of a, b, and c.

(b) For the normal mode matrix determined in part (a), determine the angle θ for which the ratio of the steady-state amplitudes $|w|/|u|$ will be identical to the corresponding amplitude ratio in a free vibration in the second mode.

4.50 A small disk $m = 0.5$ kg is supported in the vertical plane by three concurrent springs, $k_1 = 200$ N/mm, $k_2 = 160$ N/mm, and $k_3 = 240$ N/mm, such that its static equilibrium position is as shown. Generalized coordinates for the system are the horizontal and vertical displacements u and w, respectively, relative to this position. The mass is loaded by a force $F \sin(\omega t)$ at a constant angle of inclination $\theta = 60°$ above the horizontal axis. Because u and w are either in phase or $180°$ out of phase from the excitation, the steady-state displacement is a vector having magnitude D at angle ψ from the horizontal axis, such that $u = D \cos(\psi) \sin(\omega t)$ and $w = D \sin(\psi) \sin(\omega t)$. Determine the magnitude of the displacement D and ψ when $\omega = 0.99\omega_1$, $1.01\omega_1$, $0.5(\omega_1 + \omega_2)$, $0.99\omega_2$, and $1.01\omega_2$, where ω_1 and ω_2 are the natural frequencies.

4.51 Solve Example 4.11 in the situation where the system is modified by inserting identical dashpots parallel to each spring. Their dashpot constant is $c = 0.04(km)^{1/2}$.

4.52 Consider the two-degree-of-freedom system in the sketch. A harmonic force $F \cos(\omega t)$ is applied to the lower mass m_1. Consider two cases: (a) $m_1 = m$, $m_2 = 0.1m$, $k_1 = k_2 = k$; (b) $m_1 = m$, $m_2 = 0.1m$, $k_1 = k$, $k_2 = 0.1k$, where k and m are basic stiffness and mass units. Determine the amplitude and phase lag of the steady-state response of each generalized coordinate in each case when $\omega = (k/m)^{1/2}$.

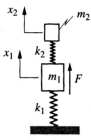

EXERCISE 4.52

4.53 A torque $\Gamma_0 \cos[1.45(g/L)^{1/2}t]$, positive when counterclockwise, is applied to the right bar of the triple pendulum in Exercise 4.30. Determine the steady-state amplitude of the rotation of each bar.

4.54 The base motion of the building model in Exercise 4.45 is $w = 10\sin(50t)$ mm. Determine the displacement amplitude and phase angle rela-

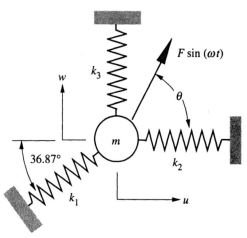

EXERCISES 4.49, 4.50

tive to a pure sine of each mass when the system attains a steady-state condition.

4.55 The sketch repeats the automobile model studied in Example 4.9, in which the chassis is represented as a box of mass m that is translating to the left at constant speed v. The horizontal position of the wheels is $x_1 = vt$, $x_2 = vt - L$, and the ground elevation under a wheel is $w(x) = A\sin(2\pi x/\lambda)$. The equations of motion for this system were shown in the previous exercise to be

$$[M]\begin{Bmatrix} \ddot{y}_1 \\ \ddot{y}_2 \end{Bmatrix} + [K]\begin{Bmatrix} y_1 \\ y_2 \end{Bmatrix} = [K]\begin{Bmatrix} w(vt) \\ w(vt - L) \end{Bmatrix}$$

Evaluate the steady-state amplitudes of y_1 and y_2 in the case where $A = 10$ mm, $L = 5$ m, $m = 1000$ kg, $k = 24$ kN/m, $\lambda = L/3$, and $v = 3$ m/s. Also identify the speeds at which resonances occur.

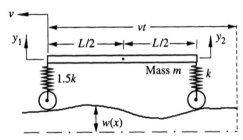

EXERCISES 4.55, 4.56

4.56 Consider the automobile model studied in Example 4.9, in which the chassis is represented as a box of mass m that is translating to the left at constant speed v. In the system of interest, the suspension is modified by inserting dashpots having identical constants c at the front and rear acting parallel to each spring. The horizontal position of the wheels is $x_1 = vt$, $x_2 = vt - L$, and the ground elevation under a wheel is $w(x) = A\sin(2\pi x/\lambda)$.

(a) Show that the equations of motion in this case have the form

$$[M]\begin{Bmatrix} \ddot{y}_1 \\ \ddot{y}_2 \end{Bmatrix} + [C]\begin{Bmatrix} \dot{y}_1 \\ \dot{y}_2 \end{Bmatrix} + [K]\begin{Bmatrix} y_1 \\ y_2 \end{Bmatrix}$$

$$= [K]\begin{Bmatrix} w(vt) \\ w(vt - L) \end{Bmatrix} + [C]\begin{Bmatrix} \dfrac{d}{dt}[w(vt)] \\ \dfrac{d}{dt}[w(vt - L)] \end{Bmatrix}$$

(b) Evaluate the steady-state amplitudes and phase lags of y_1 and y_2 relative to a pure sine function. System parameters are $A = 10$ mm, $L = 5$ m, $m = 1000$ kg, $k = 36$ kN/m, $c = 5$ kN-s/m, $\lambda = L/3$, and $v = 3$ m/s.

(c) For the system parameters specified in part (b), determine and graph the steady-state amplitudes $|y_1|$ and $|y_2|$ as a function of v for $0 < v < 20$ m/s.

4.57 The building in Exercise 4.21 is subjected to horizontal forces at floors 2 and 4 given by

$$F_2 = 2000 \cos(\omega t) \text{ N}$$
$$F_4 = 1500 \sin(\omega t - 2) \text{ N}$$

Determine the steady-state amplitude of the displacement of each floor when $\omega = 0.95\omega_1$ and $\omega = 1.05\omega_1$, where ω_1 is the fundamental natural frequency of the building.

4.58 Consider the excitation of the bladed disk assembly in Example 4.12. The system in this case has 10 blades, and there is strong interblade coupling, $R \equiv \omega_c/\omega_b = 0.5$. Damping is light, $\zeta = 0.005$, and the system has ideal properties, such that the statistical deviations $\Delta f_j = 0$. Determine the steady-state amplitudes $|q_j|$. The frequencies for this evaluation are $\omega = 0.99\omega_2$ and $1.01\omega_2$, where ω_2 is the natural frequency of the second mode.

4.59 Unit impulsive forces are applied periodically at $t = 0, T, 2T, 3T, \ldots$ to a two-degree-of freedom system in which dissipation is negligible. Use a Fourier series to describe the excitation. Then use superposition of harmonic responses to describe the corresponding steady-state response of the system in terms of the natural frequencies and normal modes of free vibration. For what values of the period T will the system resonate?

4.60 Consider the system in Exercise 4.52 in the case where $m_1 = m$, $m_2 = 0.1m$, $k_1 = k$, $k_2 = 0.1k$. The force acting on the lower mass is a periodic square wave, $F(t) = F_0$ if $0 < t < T/2$, $F(t) = -F_0$ if $T/2 < t < T$, $F(t + T) = F(t)$ if $t > T$. Represent this excitation as a Fourier series. Then use superposition of harmonic responses to determine the periodic steady-state displacement of both masses when $T = 2\pi(m/k)^{1/2}$. Graph the results over a time interval of $2T$.

CHAPTER 5

HARMONIC EXCITATION OF MULTI-DEGREE-OF-FREEDOM SYSTEMS

The modal equations showed that steady-state response to harmonic excitation occurs at the frequency of the excitation. (We exclude the case of resonant excitation of an undamped system.) This leads us to an alternative to modal analysis, in which the time dependence is factored out by representing all variables as complex exponentials. The independent variable becomes the excitation frequency, so this is a *frequency domain formulation*. For the same reason, the method often is referred to as *harmonic analysis*. The dependent variables in this viewpoint are the complex amplitudes of the generalized coordinates, which are obtained directly by solving coupled algebraic equations.

By avoiding solution of the eigenvalue problem, frequency domain analysis proves to be simpler for situations where we wish to know the response for a single specified excitation frequency. The method also is particularly useful for non-proportionally damped systems when the amount of damping is large. In addition, by being more direct, we are more likely to obtain an algebraic solution if we employ a frequency domain analysis. We can interrogate such solutions to ascertain the effect of the fundamental physical properties of the system. Another asset is that focusing on harmonic behavior makes the method well suited for application of FFT concepts, which lead, in turn, to some powerful experimental techniques.

On the other hand, harmonic analysis does not provide us with the insight into system characteristics that is afforded by modal analysis. Also, a frequency domain formulation requires individual solution of the governing equations for each frequency (unless we are able to obtain an algebraic solution of the equations). Hence, if we wish to evaluate the response over a range of frequencies, modal analysis is likely to be more efficient.

5.1 FREQUENCY DOMAIN TRANSFER FUNCTION

Harmonic analysis is quite straightforward. We consider the standard equation of motion, eq. (4.1.2), including an arbitrary damping matrix $[C]$. The harmonic excitation is represented as $\{Q\} = \mathrm{Re}[\{F\}\exp(i\omega t)]$. Because the steady-state response occurs at the frequency of the excitation, its form is

$$\{q\} = \mathrm{Re}[\{X\}\exp(i\omega t)] \tag{5.1.1}$$

To obtain an expression for the unknown complex amplitudes X_n, we substitute this expression and the complex representation of $\{Q\}$ into the equations of motion. Each derivative of $\{q\}$ with respect to time leads to a factor $i\omega$. We desire that the substituted form be satisfied at any instant t, so we cancel the exponential time function. Thus, we have

$$[[K] + i\omega[C] - \omega^2[M]]\{X\} = \{F\} \qquad (5.1.2)$$

When damping is present, $[C] \neq [0]$, the above constitutes a set of N complex algebraic equations for the N unknown complex amplitudes X_n. The coefficient matrix $[[K] + i\omega[C] - \omega^2[M]]$ is the *dynamic stiffness*. The inertial and elastic effects in the dynamic stiffness are real and mutually 180° out of phase, while the damping effect is 90° out of phase from both. When the excitation frequency approaches a natural frequency, we know that $|[K] - \omega^2[M]|$ approaches zero. We may interpret this as meaning that the stiffness and inertial effects are canceling in some sense, thereby lessening the dynamic stiffness. At a natural frequency, only damping can resist the exciting forces. If damping is light, there will be significant enhancements in the magnitude of the steady-state response.

In the special case of harmonic excitation of a truly undamped system, $[C] = [0]$, the determinant of the dynamic stiffness actually vanishes when ω matches any natural frequency, $|[K] - \omega_j^2[M]| = 0$. Consequently, eq. (5.1.2) cannot be solved for the value of $\{X\}$ corresponding to specified force amplitudes $\{F\}$. This means that the harmonic steady-state solution is not valid at an undamped resonance. Rather, one needs to introduce a t factor, just like the resonant case for a one-degree-of-freedom system. Except in this special circumstance, it is possible to solve eq. (5.1.2) for $\{X\}$ at any frequency. We shall use the inverse of the coefficient matrix to describe the solution. Hence, the steady-state response amplitudes are given by

$$\{X\} = [[K] + i\omega[C] - \omega^2[M]]^{-1}\{F\} \qquad (5.1.3)$$

The inverted matrix is the *frequency domain transfer function*, which we denote as

$$[G(\omega)] = [[K] + i\omega[C] - \omega^2[M]]^{-1} \qquad (5.1.4)$$

The system's steady-state response is thereby found to be

$$\{q\} = \text{Re}[[G(\omega)]\{F\} \exp(i\omega t)] \qquad (5.1.5)$$

Note that if we merely wish to know $\{X\}$ for a single set of forces, we would employ a standard method, such as Gauss elimination, to solve the simultaneous equations represented by eqs. (5.1.2). Doing so is more efficient than actually finding $[G(\omega)]$ as an inverse. Most mathematics software packages and subroutine libraries are quite capable of solving a very large number of complex algebraic equations.

The term "transfer function" stems from the fact that by knowing $[G(\omega)]$ we may transfer any set of force amplitudes $\{F\}$ into a response. Because of the similarity of eq. (5.1.3) to the relation $X = D(r, \zeta)F/k$ for the steady-state response of a one-degree-of-freedom system, $[G(\omega)]$ may also be called the *complex frequency response matrix*.

To understand the physical significance of $[G(\omega)]$, suppose the harmonic forces are such that $Q_n = \cos(\omega t)$, where n is a specific value and the other generalized forces are zero, that is, $Q_j = \text{Re}[\exp(i\omega t)]\delta_{jn}$, $j = 1, ..., N$. The corresponding elements of $\{F\}$ are zero, except for a unit value in row n. The result of postmultiplying $[G(\omega)]$ by this $\{F\}$ is column n of $[G(\omega)]$. It follows from eq. (5.1.3) that $[G_{jn}(\omega)]$ represents the complex amplitude of generalized coordinate j when the only nonzero generalized force amplitude is $F_n = 1$. From this viewpoint, $[G(\omega)]$ is the dynamic analog of the flexibility matrix for statics. For this reason, some vibration engineers refer to $[G(\omega)]$ as a *mobility matrix*, and mathematicians refer to it as a *Green's function matrix*.

A corollary of the fact that $[M]$, $[C]$, and $[K]$ are symmetric is that $[G(\omega)]$ is symmetric, $[G_{jn}(\omega)] = [G_{nj}(\omega)]$. This symmetry means that the complex amplitude of generalized coordinate j resulting from the nth generalized force having any amplitude F is the same as the complex amplitude of generalized coordinate n when the jth generalized force has the same amplitude. This is the principle of *dynamic reciprocity*, which has several applications, such as acoustic transducer calibration; see Pierce (1981).

EXAMPLE 5.1

The equations of motion for a system are

$$\begin{bmatrix} 200 & 0 & 400 \\ 0 & 600 & 0 \\ 400 & 0 & 1200 \end{bmatrix} \{\ddot{q}\} + \begin{bmatrix} 500 & 300 & -700 \\ 300 & 900 & 600 \\ -700 & 600 & 1300 \end{bmatrix} \{\dot{q}\}$$

$$+ \begin{bmatrix} 50 & -20 & 0 \\ -20 & 40 & 0 \\ 0 & 0 & 30 \end{bmatrix} \{q\} = \begin{Bmatrix} 400 \sin(50t + \pi/3) \\ 0 \\ 200 \cos(50t) \end{Bmatrix} \text{N}$$

where the units of $\{q\}$ and t are meters and seconds, respectively. Use the frequency domain transfer function to determine the steady-state response. Compare the results to those obtained from the light damping approximation.

Solution This example serves to illustrate the ease with which the frequency domain transfer function may be employed to determine steady-state response. It also will provide another baseline for judging the accuracy of the light damping approximation. The generalized forces are specified to be $Q_1 = 400 \sin(50t + \pi/3)$ N, $Q_2 = 0$, and $Q_3 = 200 \cos(50t)$ N. We use the respective complex amplitudes to form the excitation term $\{F\}$, so we have

$$\{Q\} = \text{Re}[\{F\} \exp(i50t)], \qquad \{F\} = \begin{Bmatrix} \dfrac{400}{i} \exp\left(i\dfrac{\pi}{3}\right) \\ 0 \\ 200 \end{Bmatrix}$$

The steady-state response is described by $\{q\} = \text{Re}[\{X\} \exp(i\omega t)]$, where $\omega = 50$ rad/s. Satisfying the equations of motion yields

$$[[K] + i\omega[C] - \omega^2[M]]\{X\} = \{F\}$$

Substitution of the system matrices and $\omega = 50$ rad/s leads to three simultaneous equations,

$$1000 \begin{bmatrix} (-450 + 25i) & (-20 + 15i) & (-1000 - 35i) \\ (-20 + 15i) & (-1460 + 45i) & 30i \\ (-1000 - 35i) & 30i & (-2970 + 65i) \end{bmatrix} \{X\} = \begin{Bmatrix} 346.4 - 200i \\ 0 \\ 200 \end{Bmatrix}$$

The solution of these equations is

$$\{X\} = 10^{-4} \begin{bmatrix} (-26.79 + 4.77i) & (0.35 - 0.16i) & (8.43 - 1.10i) \end{bmatrix}^\text{T} \text{m}$$

Implementation of the light damping approximation for modal damping begins with solution of the free vibration eigenvalue problem, $[[K] - \omega^2[M]]\{\phi\} = \{0\}$. The natural frequencies and normal mode matrix that result are

$$\omega_1 = 4.767, \qquad \omega_2 = 7.348, \qquad \omega_3 = 28.547 \text{ rad/s}$$

$$[\Phi] = 10^{-3} \begin{bmatrix} 7.779 & 15.219 & 121.276 \\ 5.901 & 40.033 & -5.402 \\ 25.913 & -9.447 & -41.705 \end{bmatrix}$$

The next step is to apply the modal transformation to the damping matrix, which leads to

$$[\Phi]^T[C][\Phi] = \begin{bmatrix} 0.863 & 0.438 & -2.964 \\ 0.438 & 1.787 & 2.497 \\ -2.964 & 2.497 & 16.599 \end{bmatrix}$$

The largest off-diagonal term in this product is quite substantial relative to the diagonal terms, which suggests that the light damping approximation might be in error. Nevertheless, we use the diagonal terms of the product to estimate modal parameters, which are found to be

$$\zeta_1 = 0.091, \qquad \zeta_2 = 0.122, \qquad \zeta_3 = 0.291$$

Although the largest modal damping factor is less than unity, it is quite large. This is another indicator that the modal damping approximation might not be accurate.

With the modal properties now known, we may form the solution corresponding to the modal transformation $\{q\} = [\Phi]\{\eta\}$, which leads to the standard modal differential equations,

$$\ddot{\eta}_j + 2\zeta_j\omega_j\dot{\eta}_j + \omega_j^2\eta_j = \{\Phi_j\}^T\text{Re}[\{F\}\exp(i50t)]$$

For the steady-state solution, we set each $\eta_j = \text{Re}[Y_j \exp(i50t)]$, which we substitute into each modal equation. The resulting complex amplitudes of the modal coordinates are

$$Y_1 = (-3.190 + 0.572i)(10^{-3}), \qquad Y_2 = (-1.427 + 1.192i)(10^{-3})$$
$$Y_3 = (-21.786 + 3.664i)(10^{-3})$$

The corresponding complex amplitudes of the generalized coordinates are found from the modal transformation to be

$$\{X\} = [\Phi]\{Y\} = 10^{-4}\begin{bmatrix}(-26.89 + 4.70i) & (0.42 + 3.13i) & (8.39 - 1.49i)\end{bmatrix}^T \text{ m}$$

These values compare well with those obtained directly in the frequency domain. The value of X_2 shows the greatest disagreement, with the wrong sign for its imaginary part. Although this variable has the smallest magnitude of the three generalized coordinates, the error in its predicted value would be very important if q_2 is the generalized coordinate of primary interest.

5.2 STRUCTURAL DAMPING AND MODAL ANALYSIS

We examine the structural damping model here, rather than in the development of modal analysis, because structural damping is valid only in situations where a system vibrates harmonically. A simple modification of the transfer function $[G(\omega)]$ will enable us to account for structural damping. Conversion of the description to a modal form will enable us to identify the implication of structural damping in regard to modal damping ratios.

According to eq. (3.2.23), structural damping is associated with a damping force that is proportional to the magnitude of the elastic force. The structural damping loss factor γ is the factor of proportionality between the elastic and damping forces. We generalize that model to the multiple-degree-of-freedom system by introducing a *complex stiffness*,

$$[K'] = (1 + i\gamma)[K] \tag{5.2.1}$$

Because a dynamic stiffness term that has a factor of i corresponds to damping, we replace $\omega[C]$ in the amplitude relations, eq. (5.1.2), by $\gamma[K]$. This shows that the effective damping constants in a structural damping model decrease in inverse proportion to the response's frequency. The resulting transfer function is

$$[G(\omega)] = [[K](1 + 1\gamma) - \omega^2[M]]^{-1} \tag{5.2.2}$$

Suppose we have determined the system's natural frequencies and normal modes based on ignoring damping. We may employ that knowledge to derive an explicit expression for $[G(\omega)]$ at any frequency that does not require evaluation of a matrix inverse. The simplest derivation applies the modal orthogonality properties to the complex stiffness. A modal expansion of the steady-state amplitudes is

$$\{Y\} = [\Phi]\{X\} \tag{5.2.3}$$

where $[\Phi]$ is the normal mode matrix for the undamped system. Substituting this expression and the complex stiffness into the steady-state amplitude relations, eq. (5.1.2) gives

$$[[K](1 + i\gamma) - \omega^2[M]][\Phi]\{X\} = \{F\} \tag{5.2.4}$$

We premultiply this equation by $[\Phi]^T$ and invoke the modal orthogonality properties, from which we obtain

$$[[\omega_{nat}^2](1 + i\gamma) - \omega^2[I]]\{X\} = [\Phi]^T\{F\} \tag{5.2.5}$$

Because $[\omega_{nat}^2]$ is diagonal, these equations are an uncoupled set, whose solution is

$$X_j = \frac{\{\Phi_j\}^T\{F\}}{\omega_{nat}^2(1 + i\gamma) - \omega^2} \tag{5.2.6}$$

The partitioned expansion of the modal transformation in eq. (5.2.3) then yields

$$\{Y\} = \sum_{j=1}^{N} \{\Phi_j\}X_j = \sum_{j=1}^{N} \frac{\{\Phi_j\}\{\Phi_j\}^T\{F\}}{\omega_j^2(1 + i\gamma) - \omega^2} \tag{5.2.7}$$

The transfer function is the coefficient of $\{F\}$ in this relation, so we find that the transfer function for a structurally damped system is

$$[G(\omega)] = \sum_{j=1}^{N} \frac{\{\Phi_j\}\{\Phi_j\}^T}{\omega_j^2(1 + i\gamma) - \omega^2} \tag{5.2.8}$$

This description for $[G(\omega)]$ resembles a term that appeared in the modal analysis of steady-state response. Specifically, eq. (4.3.42) may be written as $\{q\} = [G(\omega)]\{F\}$, where

$$[G(\omega)] = \sum_{j=1}^{N} \frac{\{\Phi_j\}\{\Phi_j\}^T}{\omega_j^2 + 2i\zeta_j\omega\omega_j - \omega^2} \tag{5.2.9}$$

Thus the structural damping model replaces $2\zeta_j\omega_j\omega$ in the denominator with $\omega_j^2\gamma$. This may be interpreted as stating that the modal damping ratios are

$$\boxed{\zeta_j = \frac{1}{2}\gamma\frac{\omega_j}{\omega}}$$

(5.2.10)

In other words, rather than being independent of the excitation frequency, the effective damping ratio for each mode decreases inversely to the ratio of the excitation frequency to the mode's natural frequency. When ω is very close to a specific natural frequency, the effective damping ratio for the resonating mode will be $\zeta_j = \gamma/2$. This fits the empiric observation that, in many systems, damping ratios derived from experimental measurements seem to be nearly constant from mode to mode.

EXAMPLE 5.2

Extensional vibration of a bar was described by a lumped mass model in Example 4.4. A 100 N harmonic force is applied axially to the bar at its midpoint. The bar is composed of a viscoelastic polymer. Experiments measuring the strain and stress in an oscillatory motion indicate that the material fits the *standard solid model*, for which the Young's modulus is a frequency-dependent quantity given by

$$E(\omega) = E_0 + (E_\infty - E_0)\frac{\omega^2\tau^2}{\omega^2\tau^2 + 1}\left(1 + \frac{i}{\omega\tau}\right)$$

In this expression E_0 is the static Young's modulus, corresponding to very low values of the frequency ω; E_∞ is the Young's modulus for very large ω; and $\tau > 0$ is the relaxation time. It is possible by adjusting the polymer mix to fabricate materials having different values of E_0, E_∞, and τ. Consider a situation where E_0 and E_∞ are set. It is desired to find the value of τ that will minimize the amplitude of vibration when ω is identical to the fundamental natural frequency corresponding to an ideal elastic material whose Young's modulus is E_0 and whose density ρ matches the viscoelastic material properties. Parameters are $E_0 = 8(10^7)$ Pa, $E_\infty = 12(10^7)$ Pa, $\rho = 2000$ kg/m^3, $A = 0.001$ m^2, and $L = 1$ m. Use a five-element lumped mass model ($N = 5$) to determine the amplitude of the displacement at the midpoint as a function of τ. Graph the result, and use it to determine the optimal value of τ for minimizing that amplitude.

Solution   We will see in this example that the frequency domain formulation is very amenable to performing parameter studies for design. Polymeric materials are finding increasing usage in engineering. The frequency domain transfer function offers a straightforward method for analyzing the response of systems composed of such materials. Generalized coordinates for the bar are the axial displacement, positive to the right, of each of the lumped mass elements. The stiffness and mass matrix were derived in Example 4.4. For the case $N = 5$ we have

$$[M] = \frac{m}{N}[I], \qquad [K] = Nk[\hat{K}], \qquad [\hat{K}] = \begin{bmatrix} 3 & -1 & 0 & 0 & 0 \\ -1 & 2 & -1 & 0 & 0 \\ 0 & -1 & 2 & -1 & 0 \\ 0 & 0 & -1 & 2 & -1 \\ 0 & 0 & 0 & -1 & 3 \end{bmatrix}$$

In these expressions m is the total mass of the bar, which we compute from the density, length, and cross-sectional area to be $m = \rho AL = 2$ kg. The parameter k is the stiffness unit $k = EA/L$. Because E is not constant, we have

$$[K] = N\frac{E_0 A}{L}\chi(\omega\tau)[\hat{K}]$$

where $\chi(\omega\tau)$ denotes the dependence of the Young's modulus on the nondimensional combination $\omega\tau$,

$$\chi(\omega\tau) = 1 + \left(\frac{E_\infty}{E_0} - 1\right)\frac{\omega^2\tau^2}{\omega^2\tau^2 + 1}\left(1 + \frac{i}{\omega\tau}\right)$$

The excitation is specified to be a 100 N amplitude harmonic force acting at the midpoint of the bar. For $N = 5$, this force acts at the third lumped mass, so the generalized force may be written as

$$\{Q\} = F_0\text{Re}[\{F\}\exp(i\omega t)], \qquad \{F\} = \begin{bmatrix} 0 & 0 & 1 & 0 & 0 \end{bmatrix}^T, \qquad F_0 = 100 \text{ N}$$

The frequency domain amplitude equations become

$$\left[N\frac{E_0A}{L}\chi(\omega\tau)[\hat{K}] - \omega^2\frac{m}{N}[I]\right]\{X\} = F_0\{F\}$$

A comparison of the stiffness term in this expression with $[K]$ in eq. (5.2.1) shows that the real stiffness and loss factor are

$$[K] = N\frac{E_0A}{L}\text{Re}[\chi(\omega\tau)][\hat{K}], \qquad \gamma = \frac{\text{Im}[\chi(\omega\tau)]}{\text{Re}[\chi(\omega\tau)]}$$

The expression for γ does not fit the structural damping model, in which the loss factor is independent of ω. The graph shows that the maximum γ occurs near $\omega\tau = 0.8$. Another difference is that $[K]$ is not constant.

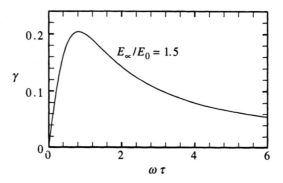

The excitation frequency is specified to be the fundamental natural frequency of the system when the Young's modulus is E_0. The previous analysis expressed the natural frequencies in terms of nondimensional eigenvalues as

$$\omega_n^2 = \frac{EA}{mL}\lambda_n$$

Thus, we set the frequency for the present study at $\omega^2 = (E_0A/mL)\lambda_1$. The value of λ_1 was previously found in the $N = 5$ case to be $\lambda_1 = 9.549$, which corresponds to a fundamental frequency $\omega_1 = 618$ rad/s. When we substitute this description of ω into the frequency domain equations, we find that we obtain a nondimensional set of equations given by

$$\left[N\chi(\omega\tau)[\hat{K}] - \lambda_1\frac{1}{N}[I]\right]\{X\} = \frac{F_0L}{E_0A}\{F\}$$

We will solve these equations by setting F_0L/E_0A to unity, after which we may determine the dimensional displacement by substituting the actual value of this factor.

We need to solve the preceding amplitude equations for a range of $\omega\tau$. When $\omega\tau = 0$, we have $\chi(\omega\tau) = 1$, which corresponds to the resonance. It is not stated what range of $\omega\tau$ we should consider, but an examination of $\chi(\omega\tau)$ suggests that $\omega\tau = 1$ is an important value. Thus, we select the range $0 < \omega\tau \le 10$. For each value of $\omega\tau$, we use mathematical software to solve for $\{X\}$. We save the value of X_3 for each value of $\omega\tau$, because it is requested that we evaluate the midpoint displacement. In order to see more clearly the behavior over the full range of amplitudes and relaxation times, the following graphs use log scales to display $|X_3|$ and arg (X_3) as functions of τ. Also, in order to get a good number of data points in the region where $\omega\tau$ is small, which the log scale expands, the values of $\omega\tau$ for which we perform the computation were taken to be $\omega\tau_p = 10 \log_{10}(p/100)$, $p = 1, ..., 100$.

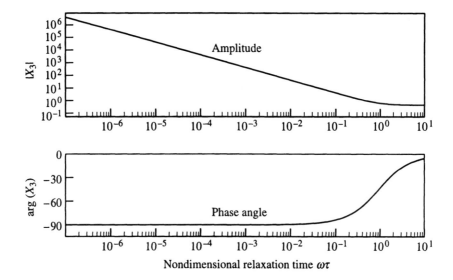

The results show an interesting behavior. For small values of $\omega\tau$, we see that the amplitude decreases steadily, while the phase angle for the response lags the excitation by almost 90°. In other words, the excitation is essentially resonant, with the increasing value of τ for fixed ω increasing the material's damping effect. Beyond $\omega\tau = 1$, the amplitude of X_3 ceases to decrease, and the phase lag of X_3 approaches zero. The phase lag being zero tells us that the material is not providing much damping, while the amplitude behavior indicates that the stiffness is approaching a constant value. This constant stiffness differs significantly from the static value, because E_∞ is 50% greater than E_0. In effect, the fixed frequency $\omega = 618$ rad/s is not close to the fundamental resonance when $\tau > 1/\omega_1$. The dimensional magnitude that X_3 approaches is $(F_0L/E_0A)|0.449 - 0.14i| = 4.49$ mm.

5.3 VIBRATION ABSORBER

The system we shall consider here is remarkably simple, in that it has only two degrees of freedom. All of its modal properties are easy to determine, and its steady-state response follows directly from the developments of the preceding section. However, the specific features of its response will lead us to a concept by which a system's response to excitation at a specified frequency may be significantly reduced.

The basic concept concerns a preexisting system, which we represent as a one-degree-of-freedom system having mass m_1, stiffness k_1, and dashpot constant c_1. Suppose it is necessary to subject this system to a resonant harmonic excitation—for example, because the system is supporting an unbalanced motor that must rotate at the natural frequency. Such excitation will lead to excessive vibration amplitudes, so we must redesign the system.

We could merely alter the system's mass or stiffness, or greatly increase the amount of damping, but doing so might be very expensive. We consider here an alternative that can be much more cost effective. Let us attach to the original system another spring-mass-dashpot combination. The result is the two-degree-of-freedom system depicted in Figure 5.1. The equations of motion for the modified system are

$$m_1\ddot{y}_1 + (c_1 + c_2)\dot{y}_1 - c_2\dot{y}_2 + (k_1 + k_2)y_1 - k_2y_2 = \text{Re}[F\exp(i\omega t)]$$
$$m_2\ddot{y}_2 + c_2\dot{y}_2 - c_2\dot{y}_1 + k_2y_2 - k_2y_1 = 0 \qquad (5.3.1)$$

We set $y_j = \text{Re}[Y_j\exp(i\omega t)]$ for steady-state response, which leads to the following complex amplitude-frequency equations,

$$\begin{bmatrix} [(k_1 + k_2) + i\omega(c_1 + c_2) - \omega^2 m_1] & -(i\omega c_2 + k_2) \\ -(i\omega c_2 + k_2) & k_2 + i\omega c_2 - \omega^2 m_2 \end{bmatrix} \begin{Bmatrix} Y_1 \\ Y_2 \end{Bmatrix} = \begin{Bmatrix} F \\ 0 \end{Bmatrix} \qquad (5.3.2)$$

It is not too difficult to solve these equations algebraically for the amplitudes. Let us do so for the case where damping is negligible, $c_1 = c_2 = 0$. We then find that

$$Y_1 = F\frac{(k_2 - \omega^2 m_2)}{\Delta}, \qquad Y_2 = F\frac{k_2}{\Delta}$$
$$\Delta = [(k_1 + k_2) - \omega^2 m_2](k_2 - \omega^2 m_2) - k_2^2 \qquad (5.3.3)$$

The denominator Δ is the determinant of the system of coefficients in eq. (5.3.2). This quantity is identical to the characteristic equation for the undamped system, which is an aspect that we will discuss later.

Recall that our objective is to reduce the value of Y_1 when ω matches the natural frequency $(k_1/m_1)^{1/2}$ of the lower spring and mass. The numerator for Y_1 vanishes at $\omega = (k_2/m_2)^{1/2}$, and Δ is nonzero at that frequency. It follows that if we select the upper system such that $k_2/m_2 = k_1/m_1$, then the lower mass will not oscillate when $\omega = (k_1/m_1)^{1/2}$. Now note that $(k_1/m_1)^{1/2}$ is the natural frequency of the original system prior to attachment of the upper spring and mass. In other words, rather than merely reducing the troublesome vibration at the original resonance, we can eliminate it. Also, note that the corresponding amplitude for the upper mass is $Y_2 = -F/k_2$, which resembles the static deformation of k_2 resulting from a compressive force F.

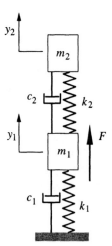

FIGURE 5.1 Two-degree-of-freedom vibration absorber with damping.

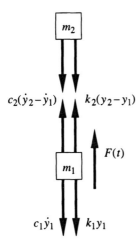

$c_2(\dot{y}_2 - \dot{y}_1)$ $k_2(y_2 - y_1)$

$F(t)$

m_1

$c_1\dot{y}_1$ $k_1 y_1$

FIGURE 5.2 Free body diagrams of the vibration absorber elements.

Both results are readily explained by considering the free body diagrams in Figure 5.2, with the condition that $y_1 = 0$. Because m_1 is not moving, the forces acting on it must equilibrate at every instant, which is satisfied (when $y_1 \equiv 0$) if $k_2 y_2 + \text{Re}[F \exp(i\omega t)] = 0$. Thus we have $y_2 = -\text{Re}[(F/k_2) \exp(i\omega t)]$. Next we observe that if $y_1 = 0$, then the upper system must be executing a fixed-base free vibration, which requires that $\omega = (k_2/m_2)^{1/2}$. We select k_2/m_2 to match k_1/m_1 in order to make this special condition occur at the frequency where the lower system would resonate if the upper system were not present.

The upper system is referred to as a *vibration absorber*. We select its stiffness and mass so as to tune its isolated (that is, fixed-base) natural frequency $(k_2/m_2)^{1/2}$ to match the natural frequency $(k_1/m_1)^{1/2}$ of the one-degree-of-freedom system that must be subjected to a resonant excitation. At that frequency, the spring force the vibration absorber applies to the original system balances the excitation force, so the original system does not move.

The ability of the vibration absorber to quiet y_1 depends only on the ratio k_2/m_2, but not on the actual size of m_2 and k_2. For that reason, it might seem that the upper system can be extremely lightweight. However, two factors can make the size of m_2 significant. Small m_2 leads to small k_2, in which case the amplitude $Y_2 = -F/k_2$ will be very large. If the vibration amplitude of the absorber is excessive, then the spring is likely to fail due to fatigue.

The other consideration limiting the smallness of m_2 and k_2 pertains to situations where the excitation frequency might drift around the nominal value $\omega = (k_1/m_1)^{1/2}$. Let us plot the amplitudes Y_1 and Y_2 as a function of ω. Because we are ignoring damping, we shall consider the amplitudes to be real values that may be positive or negative, rather than representing them as a magnitude and phase angle. The result is shown in Figure 5.3 for a case where $k_2 = 0.1 k_1$ and $m_2 = 0.1 m_1$. There are two resonances corresponding to $\Delta = 0$, whereas the original system had only one. Clearly, this is a consequence of adding a degree of freedom. When we set $k_2/m_2 = k_1/m_1$ in order to nullify Y_1 at the original natural frequency, the natural frequencies of the new system obtained by solving $\Delta = 0$ are

$$(\omega_{1,2})^2 = \frac{k_1}{m_1}\left\{1 + \frac{1}{2}\frac{m_2}{m_1} \mp \left[\frac{m_2}{m_1} + \frac{1}{4}\left(\frac{m_2}{m_1}\right)^2\right]^{1/2}\right\} \qquad (5.3.4)$$

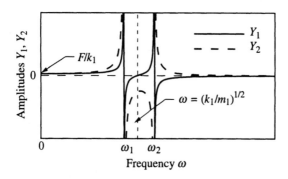

FIGURE 5.3 Frequency response of an undamped vibration absorber, $m_2/m_1 = k_2/k_1 = 0.1$, $\gamma = 0$.

It is evident from this that the natural frequencies become increasingly close as the ratio m_2/m_1 decreases. Hence, if we are not certain that ω will exactly equal $(k_1/m_1)^{1/2}$, we would want to select m_2/m_1 to be sufficiently large to place the two natural frequencies well outside the range in which ω might fall.

To fully understand the effect of damping, we need to consider first the overall frequency dependence of the undamped system's response. We see in Figure 5.3 that both amplitudes have the value F/k_1 at $\omega = 0$, which is the quasi-static deformation of the lower spring due the force. As the frequency approaches the fundamental natural frequency ω_1 from below, both amplitudes become positively infinite. Increasing the frequency slightly beyond ω_1 results in a 180° phase shift in both responses. This is like the behavior of a one-degree-of-freedom system. Below resonance, the elastic effects, which are in phase with the excitation, dominate. Beyond the first natural frequency, the inertial effects, which are 180° out of phase, take over. When ω is increased beyond ω_1, both amplitudes decrease in magnitude. As ω approaches $(k_1/m_1)^{1/2}$, Y_1 becomes zero, and then increases positively. In contrast, Y_2 attains a minimum negative value in the vicinity of $\omega = (k_1/m_1)^{1/2}$, and then increases negatively. Further increase of ω toward the second natural frequency ω_2 causes both amplitudes to approach infinity, but Y_1 is positive, while Y_2 is negative. Passing the resonance results in a 180° phase shift for both amplitudes, as it did for the fundamental resonance. Increasing the frequency beyond ω_2 causes both amplitudes to decrease in magnitude because the inertial effect, which is proportional to the square of the frequency, becomes the dominant effect.

We shall use a structural damping model to illustrate the effects of dissipation. Correspondingly, we set $c_j = \gamma k_j/\omega$ for each dashpot, so that $[C] = (\gamma/\omega)[K]$, as required by eq. (5.2.1). The amplitude dependence in Figure 5.4 and the phase angle dependence in Figure 5.5 correspond to a 5% loss factor ($\gamma = 0.05$). Because the amplitudes in this case are complex, the results are based on representing the ampli-

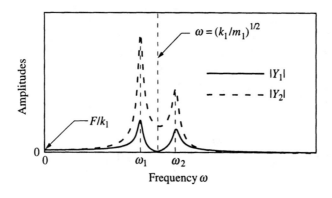

FIGURE 5.4 Amplitude-frequency dependence for a vibration absorber, $m_2/m_1 = k_2/k_1 = 0.1$, $\gamma = 0.05$.

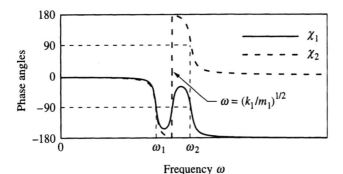

FIGURE 5.5 Phase angle–frequency dependence for a vibration absorber, $\gamma = 0.05$.

tudes in polar form as $Y_j = |Y_j| \exp(i\chi_j)$. The phase angles are selected in the range $-180° < \chi_j \leq 180°$, and we take F to be real. Consequently, positive phase angles correspond to a response that leads the excitation, while negative angles signify a lag.

We see in Figure 5.4 that the resonant peaks are finite, and that both amplitudes have lower peak values at the second natural frequency. Although it is difficult to tell from the figure, $|Y_1|$ is not quite zero at $\omega = (k_1/m_1)^{1/2}$. This small displacement results because there is a damping force that is $90°$ out of phase from the inertial resistance and elastic force, which essentially cancel each other. In all other respects, a small amount of damping has little effect on the frequency dependence of the amplitudes.

In Figure 5.5, both χ_1 and χ_2 decrease from zero as ω is increased from zero, which represents a lag relative to the excitation. When ω approaches ω_1, both angles undergo a rapid change, crossing $\chi_1 = \chi_2 = -90°$ at $\omega = \omega_1$. Beyond that frequency, χ_1 attains a minimum close to $-180°$, then increases to $-90°$ at $\omega = (k_1/m_1)^{1/2}$. From there, it increases almost to zero, and then decreases to $-90°$ at $\omega = \omega_2$. Beyond the second natural frequency, we see that χ_1 is essentially $-180°$. The phase angle χ_2 shows an interesting feature. Up to $\omega = \omega_1$, we see that $\Delta_2 \approx \Delta_1$. Beyond that frequency Δ_2 continues to decrease. The $360°$ shift of χ_2 at $\omega = (k_1/m_1)^{1/2}$ is merely a consequence of limiting the angle to $-180° < \chi_2 \leq 180°$. If we were to subtract $360°$ from the values beyond $\omega = (k_1/m_1)^{1/2}$, we would see that the curve is smooth. Unwrapping the phase angle in this manner leads us to recognize that $\chi_2 = +90° = -270°$ at $\omega = \omega_2$. Further increase of ω beyond ω_2 brings χ_2 close to zero.

The present results show that adding damping to a vibration absorber lessens vibrational amplitudes at the combined system's natural frequencies. That improvement comes at the penalty of increasing $|Y_1|$ at the original natural frequency, $\omega = (k_1/m_1)^{1/2}$. This leads to the concept of an *untuned vibration absorber*, which is used to reduce vibrational amplitudes over a range of frequencies. The concept is explained by Den Hartog (1956), and illustrated in Exercise 5.20.

Several of the response features exhibited by the vibration absorber are common to all lightly damped systems. The damping effect is most obvious at resonances, because the ability of the elastic and inertial effects to carry the excitation tend to cancel when a system oscillates close to its undamped natural frequency. Because the damping forces are proportional to velocities, *all* phase angles at resonance will essentially be $\pm 90°$ relative to the excitation. In contrast, at other frequencies, such as $\omega = (k_1/m_1)^{1/2}$ in Figure 5.5, some *but not all* of the phase angles might be $\pm 90°$. Furthermore, changing ω across a resonance generally shifts the phase of all generalized coordinates by approximately $180°$. These observations can be quite useful in experimental applications.

5.4 FFT TECHNIQUES

It is possible to employ the frequency domain transfer function $[G(\omega)]$ to determine transient, as well as steady-state, response. The concept is an extension of Section 3.8.2, where FFT techniques were developed to evaluate a convolution integral for one-degree-of-freedom systems. One of the primary motivations for extending the formulation to multi-degree-of-freedom systems is that it leads to extremely useful techniques for experimentally identifying modal properties. In turn, those concepts form the heart of a number of techniques for diagnosing and correcting vibration problems.

Another virtue of using FFTs to address transient vibrations is that the formulation is valid for any amount of damping. High levels of damping are encountered in systems built with viscoelastic materials, such as polymers. The constitutive equations for such media depend on the time history of stress and strain, rather than merely the instantaneous values of these quantities. (We encountered one such case in Example 5.2.) In the frequency domain, the relation between stress and strain appears to be like that of an elastic material, except that the elastic constants are functions of frequency. It is significantly easier to measure these frequency-dependent properties than it is to determine stress-strain relations in the time domain. Hence, if we were to formulate the stiffness matrix for a system whose materials are viscoelastic, we would obtain a complex stiffness matrix that depends on frequency. Another area in which a frequency domain formulation is preferable is vibration of submerged structures. In such cases, a portion of the resistance of the water to vibration of the wetted surface is like an added mass, corresponding to moving a mass of water around the structure. The fluid also resists vibration of the surface like a system of dashpots, corresponding to power that is dissipated from the structure in the form of sound radiated into the water. These effects are manifested in the equations of motion as frequency-dependent, complex contributions to the dynamic stiffness (Junger and Feit, 1986). Thus, in both viscoelastic and submerged acoustic systems, one is likely to have equations of motion that have been formulated directly as a frequency domain transfer function.

5.4.1 Analysis of Transient Response

The first step in applying the FFT technique is to sample the time-dependent set of generalized forces. This requires selection of a sampling interval Δ and a time interval $0 < t < T'$ within which we will view the response. (Criteria for these parameters will be discussed later.) We sample the transient excitation $\{Q(t)\}$ at the discrete instants $t_j = j\Delta$, so the number of samples is $J' = T'/\Delta$. Let us define $\{Q_n\}$ to be a column of time values for the nth generalized force. We use each column as a partition of a rectangular array $[Q]$, that is,

$$[Q] = \left[\{Q_1\} \ \{Q_2\} \ \cdots \ \{Q_N\} \right] \tag{5.4.1}$$

where N is the number of degrees of freedom.

Because we will be performing convolution in the frequency domain, we zero pad each column in order to defeat wraparound error. If we hold Δ fixed, the procedure defines the total length of the time data to be $J = 2^p \geq 2J'$. This makes $T = J\Delta$ the period for the DFT. Correspondingly, $J - J'$ zeroes are inserted at the end of each $\{Q_n\}$, so that $[Q]$ has J rows and N columns. The mathematical description of these operations is

$$Q_{jn} = \begin{cases} Q_n(j\Delta), & j = 0, \ldots, J' - 1 \\ 0, & j = J', \ldots, J - 1 \end{cases} \tag{5.4.2}$$

The next operation entails converting each generalized force to the frequency domain. To do so, starting from $n = 1$, we successively take the discrete Fourier transform of each $\{Q_n\}$ data set. The DFT operation was defined in eq. (3.7.18). Although we would carry out this operation with an FFT computer program, let us write it in matrix notation. Doing so will enable us to keep track of the manner in which data is stored. Thus, we have

$$[F] \equiv \Big[\{F_1\} \{F_2\} \cdots \{F_N\}\Big] = [E_{\mathrm{DFT}}]\Big[\{Q_1\} \{Q_2\} \cdots \{Q_N\}\Big] \qquad (5.4.3)$$

where $[E_{\mathrm{DFT}}]$ is a matrix of exponential factors; see eq. (3.7.22). Note that the above form corresponds to sequentially inputting a *column* of $[Q]$ to the computer routine, and storing the resulting output as the corresponding *column* of $[F]$. Thus, column $\{F_n\}$ contains the DFT coefficients associated with generalized force Q_n.

As a consequence of zero padding, the DFT time duration is $T = J\Delta$. The fundamental frequency is $2\pi/T$, so the discrete frequencies associated with the DFT are $\omega_k = 2\pi k/T$, $k = 0, \ldots, J/2$. (Because the time data are real, we only need to consider $\omega_k \geq 0$. The DFT response data for negative frequencies will be the complex conjugates of those for positive frequencies.) A different perspective for the DFT data contained in $[F]$ is that row k holds the DFT coefficients for all generalized forces at discrete frequency ω_k. Let us denote the kth row of $[F]$ as $[F_k]$, where $[]$ serves to distinguish this arrangement from the column $\{F_n\}$ associated with a specific generalized force.

Because $[F_k]$ represents the set of complex generalized force amplitudes at frequency ω_k, we may use the complex frequency response to determine the corresponding complex amplitudes of the generalized coordinates. Hence, we apply eq. (5.1.5), using $[F_k]^T$ as the harmonic force coefficients and $[G(\omega_k)]$ as the complex frequency response. Specifically, we have

$$\{X_k\} = [G(\omega_k)][F_k]^T, \quad k = 0, 1, \ldots, J/2 \qquad (5.4.4)$$

The implication of the notation $\{X_k\}$ is that it forms the kth column of a rectangular array $[X]$. The complex frequency response calculation is performed only for the first $J/2 + 1$ rows of $[F]$. We use the complex conjugate mirroring property in eq. (3.7.33) to fill in the remaining columns of $[X]$, that is,

$$\{X_k\} = \{X_{J-k}\}^*, \quad k = \frac{J}{2} + 1, \ldots, J - 1 \qquad (5.4.5)$$

If we view $[X]$ row-wise, we see that $[X_n]$ contains the DFT coefficients of generalized coordinate q_n at each DFT frequency. The corresponding discretized time response of q_n is obtained by applying the inverse DFT successively to each of these rows. The matrix notation for this operation is

$$\{q_n\} = [E_{\mathrm{IFFT}}][X_n]^T, \quad n = 1, 2, \ldots, N \qquad (5.4.6)$$

Once again, the matrix notation is for presentation purposes only, because we actually would carry out the computation by calling an IFFT computer routine. This would entail successively inputting row n of $[X]$. The output would be the corresponding column q_n, which is the time data for that generalized coordinate. An interesting aspect of the procedure is that we may selectively apply the inverse FFT to only those generalized coordinates in which we are interested.

Proper selection of the FFT window T is crucial to performing this procedure accurately. In order to establish a guideline, recall from Section 3.8.2 that the Fourier transform of a unit impulse force at $t = 0$ is a force amplitude of unity at all frequencies, that is, the transform pair is $Q(t) = \delta(t) \Leftrightarrow F(\omega) = 1$. Consider a situation where the nth generalized force is a unit impulse, $Q_n = \delta(t)$, and the other generalized forces are zero. The Fourier transform of this force is a harmonic generalized force $Q_n = \text{Re}\,[\exp(i\omega t)]$ at any ω. Hence, the Fourier transform of the system's response to an impulsive generalized force will be identical to the system's complex frequency response to that generalized force being harmonic, with unit complex amplitude at all frequencies.

In eq. (3.8.30), we established the guideline $T \geq 2T' \geq 16/\Delta\omega$ for selection of the time window in a convolution, where $\Delta\omega$ is the bandwidth of the resonance for a lightly damped system. This criterion assured that T' will be sufficient to observe the impulse response decay to negligible values, and also that at least two DFT frequencies fall in the range of a resonance. Now consider plotting as a function of ω any element of the complex frequency response $[G(\omega)]$. If the damping is not too heavy, it will exhibit a peak magnitude at each natural frequency. To select a suitably large T, we should examine each resonance peak to identify the minimum bandwidth, $\min(\Delta\omega)$. We follow the criterion developed for one-degree-of-freedom systems, by requiring that $\omega_1 = 2\pi/T \leq (\pi/8)\min(\Delta\omega)$, so

$$\boxed{T \geq 2T' \geq \frac{16}{\min(\Delta\omega)}} \qquad (5.4.7)$$

The other discretionary aspect of sampling is the interval Δ, which must satisfy the Nyquist sampling criteria for the excitation. We may estimate the appropriate choice by identifying the most rapid fluctuation in time exhibited by any of the generalized forces. We then select Δ to be no more than one-half the period of this fluctuation. We may verify that Δ was adequately small by examining the excitation's DFT coefficients $[F]$. Row n of $[F]$ consists of the DFT coefficients for all generalized forces at frequency ω_n. The Nyquist critical frequency corresponding to the selected sampling interval is π/Δ rad/s, which equals $\omega_{J/2}$. Near this frequency the DFT data should be small. Hence, we should inspect the rows of $[F]$ around $J/2$. If these data are not small relative to the magnitude of the coefficients at lower frequencies, we should suspect that aliasing has occurred.

Another issue to consider in selecting Δ is the range of natural frequencies of the system. By definition the excitation is negligible for frequencies above the Nyquist critical value, which we shall denote as $\omega_{\max}$. If the excitation is *narrowband*, then $\omega_{\max}$ will be lower than the highest natural frequency, $(\omega_{\text{nat}})_N$. In that case the higher modes of the system will have little excitation, which suggests that selecting Δ such that $\omega_{\max} = \pi/\Delta$ will be sufficient. The case of wide-band excitation, such as that encountered in impulsive loading situations, is one in which $\omega_{\max}$ extends beyond the highest natural frequency, $\omega_{\max} \gg (\omega_{\text{nat}})_N$. It would not be wrong to set Δ based on $\omega_{\max}$, but doing so would be inefficient. In lightly damped systems the contributions to the convolution products in eq. (5.4.4) will be dominated by the DFT frequencies in the vicinity of natural frequencies, where $[G(\omega)]$ has its largest values. Setting the Nyquist critical frequency π/Δ to be some multiple α of the highest natural frequency, for example by setting

$$\Delta = \frac{\pi}{\alpha(\omega_{\text{nat}})_N} \qquad (5.4.8)$$

should yield adequate sampling. Values of α ranging between two and four are commonly used. The required value of α goes up with increasing damping, because the peaks in the complex frequency response become less dominant and the width of the peaks becomes broader.

EXAMPLE 5.3

Fuzzy structure is a term that has been used to describe a system in which a master structure has numerous attachments whose characteristics are not fully known; see Soize (1993). Physical phenomena encountered in such systems may be explored by studying the system in the diagram. The master structure is a translating block m_0 supported by springs whose combined stiffness is k_0. The attachments to the master structure are five damped one-degree-of-freedom systems having mass m_j, stiffness k_j, and damping c_j. The attached masses are all identical and small relative to the mass of the master structure, being given by $m_j = 0.01m_0$. The spring stiffnesses are set to be close, with an average that would make them vibration absorbers for the master structure, specifically $k_1 = 0.009k_0$, $k_2 = 0.0095k$, $k_3 = 0.010k_0$, $k_4 = 0.0105k_0$, and $k_5 = 0.011k_0$. The dashpots for each attached subsystem are such that, if the subsystem were instead fastened to a fixed base, its damping would be 0.5% of critical, which corresponds to $c_j = 0.01(k_j m_j)^{1/2}$. Use FFT techniques to determine the manner in which this system responds to an impulse force $F = \delta(t)$ applied to the master structure.

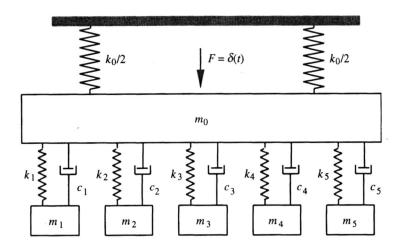

Solution All of the techniques we developed for using FFTs to analyze one-degree-of-freedom systems will also be needed here. In addition to illustrating the application of the frequency domain transfer function, the system response will demonstrate that small attachments can have a significant effect on a system. The first task is to derive the equations of motion. Because MATLAB requires that the base subscript for arrays be 1, we shall denote the generalized coordinates for the attachments as $y_1, \ldots, y_5$, and denote the vertical displacement of the master structure as y_6. We define each to be positive downward, measured from the static equilibrium position. The kinetic and potential energies are

$$T = \frac{1}{2}m_0\dot{y}_6^2 + \frac{1}{2}\sum_{j=1}^{5} m_j\dot{y}_j^2, \qquad V = \frac{1}{2}k_0 y_6^2 + \frac{1}{2}\sum_{j=1}^{5} k_j(y_j - y_0)^2$$

It follows from these expressions that the inertia matrix is diagonal, and that the stiffness coupling is solely between y_6 and each of the attachments, which corresponds to row and column six of $[K]$ being the only nonzero terms off the diagonal. All of the inertia terms are proportional to m_0 and all of the stiffness terms are proportional to k_0, so we factor these out of the respective matrices as basic units. This corresponds to representing the inertia and stiffness matrices as $m_0[M']$ and $k_0[K']$, whose nonzero elements are

$$M'_{jj} = m_j/m_0 = 0.01, \quad j = 1, \dots, 5; \qquad M'_{66} = 1$$

$$K'_{jj} = k_j/k_0 = 0.0090, 0.0095, 0.010, 0.0105, 0.0110$$

$$K'_{j6} = K'_{6j} = -k_j/k_0, j = 1, \dots, 5; \qquad K'_{66} = 1 + \sum_{j=1}^{5} k_j/k_0 = 1.05$$

The power dissipation resembles V, except that there is no dashpot parallel to k_0, so we have

$$\mathcal{P}_{dis} = \frac{1}{2}\sum_{j=1}^{5} c_j(\dot{y}_j - \dot{y}_0)^2$$

It is specified that $c_j = 0.01(k_j m_j)^{1/2}$, so that the damping matrix will be proportional to $(k_0 m_0)^{1/2}$. We factor this combination out, so that the damping matrix is $(k_0 m_0)^{1/2}[C']$, where

$$C'_{jj} = \frac{c_j}{(k_0 m_0)^{1/2}}$$

$$= 0.949(10^{-4}), 0.975(10^{-4}), 1.000(10^{-4}), 1.025(10^{-4}), 1.049(10^{-4})$$

$$C'_{j6} = C'_{6j} = -\frac{c_j}{(k_0 m_0)^{1/2}}, \quad j = 1, \dots, 5$$

$$C'_{66} = \sum_{j=1}^{5} \frac{c_j}{(k_0 m_0)^{1/2}} = 4.997(10^{-4})$$

The equations of motion that result are

$$m_0[M']\{\ddot{q}\} + (k_0 m_0)^{1/2}[C']\{\dot{q}\} + k_0[K'] = \{Q\}$$

The scaling factors may be eliminated by setting $m_0 = 1$ and $k_0 = 1$, which corresponds to using $(k_0/m_0)^{1/2}t$ as a nondimensional time parameter. Then the nondimensional form of the frequency domain transfer function is

$$[G(\omega)] = \left[[K'] + i\omega[C'] - \omega^2[M'] \right]^{-1}$$

where $\omega(k_0/m_0)^{1/2}$ is the dimensional frequency.

The excitation is an impulse, for which the only nonzero generalized force is $Q_6 = \delta(t)$. The Fourier transform of this force is unity at all ω, and the relationship between Fourier transforms and DFTs is described by eq. (3.8.28). Thus, the DFT of the generalized force Q_6 at any frequency ω_n is $1/\Delta$. It follows that the rectangular array $[F]$ holding the DFT data, which is arranged such that column j holds the DFT data for generalized force Q_j, is

$$F_{nj} = \begin{cases} 0, & j = 1, \dots, 5 \\ 1/\Delta, & j = 6 \end{cases}$$

The frequencies associated with the DFT are $n\omega_1$, where $\omega_1 \equiv 2\pi/T$ is the fundamental frequency corresponding to a time window T.

This excitation is broadband, because its DFT coefficients are constant across any frequency interval. Such a situation requires that we select the time increment Δ on the basis of the natural frequencies, as described by eq. (5.4.8). If each of the spring-mass systems were isolated from the others, the natural frequency of each, nondimensionalized by $(k_0/m_0)^{1/2}$, would be approximately 1. Thus, we anticipate that the frequency of the free response following application of the impulsive force will be approximately 1. If we take $\beta = 2.5$ in eq.

(5.4.8), we obtain $\Delta = 0.4\pi$. In view of the qualitative nature of the argument, let us halve this suggested value to $\Delta = 0.2\pi$. The estimate that the highest natural frequency is unity would suggest that the shortest period exhibited by the response will be 2π. Thus, our choice for Δ will place approximately 10 sampling intervals in the space of the shortest period, which far exceeds Nyquist criteria.

To select T we use eq. (5.4.7). We do not know the precise bandwidths of the various resonances, so we invoke a qualitative argument. If the individual attached subsystems were instead fastened to a fixed base, their (nondimensional) natural frequency would be close to one and their damping ratio would be $\zeta = 0.005$. Thus, we estimate that all bandwidths are $2\zeta\omega_{nat} = 0.01$. Equation (5.4.7) then indicates that $T = 16/0.01 = 1600$ is the minimum acceptable value. The corresponding number of sampling intervals is $J' = T/\Delta = 600/(0.2\pi) = 2546.5$. FFTs are done most efficiently when the sample size is a power of 2, so we select $J = 2^{12} = 4096$ as the smallest power of 2 that is greater than the initial estimate J'. The corresponding time interval is $\Delta = T/J = 0.39063$. The nonnegative DFT frequencies are

$$\omega_k = k\frac{2\pi}{T}, \quad k = 0, 1, ..., \frac{J}{2}$$

The next operation implements eq. (5.4.4). Because $[F]$ contains $1/\Delta$ in every element of column 6, and zeroes everywhere else, the only nonzero element of each $[F_k]^T$ is $1/\Delta$ in row 6. To evaluate $\{X_k\} = [G(\omega_k)]\{F_k\}^T$ in MATLAB, we recognize that solving a linear equation is more efficient than performing an inverse. Thus, we define a constant vector F and use the left matrix divide in conjunction with the dynamic stiffness. These operations are implemented within a loop ranging over the DFT indices, specifically,

MATLAB: $F = [(1:5) * 0, 1/delta]'; for\ k = 0: J/2, w = k * 2 * pi/T;$
      $X(:, k + 1) = (K + i * w * C - w^2 * M)\backslash F;\ end$

The analogous operations in Mathcad use the *lsolve* function,

Mathcad: $n := 1;6 \quad F_n := (n = 6) \quad k := 0;J/2 \quad \omega_{k+1} := \dfrac{k * 2 * \pi}{T}$

      $X^{<k+1>} := lsolve((K + i * w * C - \omega_k^2 * M), F)$

Note that the preceding assumes that Mathcad has been set up such that 1 is the lowest subscript for an array.

The procedure for returning to the time domain in MATLAB follows fairly closely those we outlined in the formal development. To use the *IFFT* routine in MATLAB, we must supplement the data as prescribed by eq. (5.4.5). This operation may be performed in a single step. Recall that the index for the positive frequency data in MATLAB ranges from 1 to $N/2 + 1$. Adjusting eq. (5.4.5) to have indices in this range leads to $X = [X, conj(fliplr(X(:, 2:N/2)))]$. In this expression, X initially is an array having 6 rows and $N/2 + 1$ columns, so $X(:, 2:N/2)$ denotes the submatrix formed from columns 2 to $N/2$ of X. The *fliplr* function swaps the columns of this submatrix from left to right, after which the *conj* function returns the complex conjugate. The result is a matrix consisting of the original X augmented to the right by the processed submatrix. Equation (5.4.6), which applies the inverse FFT, may be implemented with a loop:

MATLAB: $for\ n = 1:6;\ q(:, n) = IFFT(X(n, :));\ end$

Element $q(j,n)$ then gives the value of generalized coordinate q_n at time $t_j = j\Delta$.

If we use the *IFFT* routine in Mathcad, we do not need to implement eq. (5.4.5), because it is implicit in the algorithm. Another difference stems from our earlier identification of the DFT of an impulse as $1/\Delta$. The DFT values returned by Mathcad's *FFT* routine are multiplied by $1/J$ relative to values associated with our definition of the DFT. The simplest way of compensating for this is to proceed with all computations as described thus far, and then to introduce the $1/J$ factor, specifically

Mathcad: $$q^{<n>} = \frac{1}{J} * IFFT[(X^T)^{<n>}]$$

where $(X^T)^{<n>}$ is the nth column of the transpose of X and $q^{<n>}$ contains the values of generalized coordinate q_n at the succession of time instants.

The first graph depicts the master response, q_6, over the full interval T. The oscillation is sufficiently rapid that we cannot distinguish much detail in a plot over this long time interval. However, it is apparent that the response decays. The fact that the values at the largest t are quite small indicates that the value of T we selected is sufficiently large.

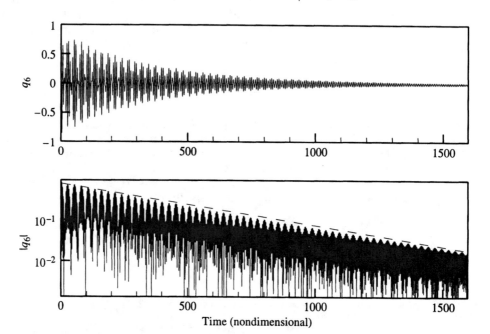

The log decrement is useful for identifying the damping ratio for a one-degree-of-freedom system. The dashed line in the second graph displays the analogous property for the present system. That graph plots $|q_6|$ on a log scale as a function of t. (Any points at which q_6 actually is zero may be eliminated from the plotted data.) The straight dashed line corresponds to an exponential decay. Rather than measuring the slope, we evaluate the equivalent damping constant from the earliest and latest maxima, which are $|q_6|_{max} = 0.967$ at $t = 1.5$ and $|q_6|_{max} = 0.019$ at $t = 1599.2$. Fitting this to the envelope function for an underdamped one-degree-of-freedom system gives $0.019 = 0.967 \exp[-\zeta_{eq}\omega_{nat}(1599.2 - 1.5)]$. We have not computed the undamped natural frequencies, so let us estimate $\omega_{nat} = 1$, based on the properties of the master spring-mass system. This leads to $\zeta_{eq} = 0.0025$, which is half the value of ζ for an isolated attachment. Plotting the response of the attachments in the same manner would show that all the generalized coordinates are attenuated at essentially the same rate.

To see the details of the response we must examine it over a shorter interval, as is done in the next set of graphs. The vibration amplitude for the attachments is larger than that of the master system, but the ratio of amplitudes is much smaller than we might anticipate on the basis of mass or stiffness ratios for the attachments and the master. A primary feature of these graphs is the beating exhibited by each response. This is much like the situation for the coupled pendulums in Example 4.8. In both that system and the present one, the individual subsystems have similar natural frequencies. Unlike the coupled pendulums, in which the two parts have equal mass, the total mass of the attachments is only 5% of the master system's mass. Nevertheless, we see that the attachments have a major influence on the master system.

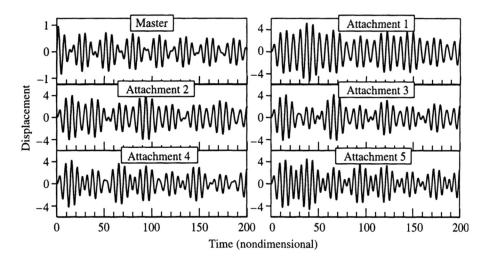

5.4.2 Experimental Modal Analysis

The preceding section focused on using FFTs as an analytical tool for determining transient response. Its developments also provide the basis for experimental procedures. The ultimate objective here is to construct from measured response data an analytical model of an existing system. Such a model could be used for a variety of purposes, such as predicting the system's response to a wide range of excitations, or suggesting redesigns aimed at controlling vibration. We shall first consider an approach that directly obtains the frequency domain transfer function. This procedure is quite general, in that it is not restricted to situations where modal damping is a good approximation. The second procedure, called *experimental modal analysis*, deduces a system's natural frequencies and mode shapes from measured response data. It is limited to systems in which a modal damping model is suitable, but it requires much less effort to implement.

For our discussion, we make several assumptions of ideal conditions:

1. We have complete access to the system, so that we may measure the response at any location.

2. Errors associated with all measurements are negligible. That includes being able to sample time responses sufficiently finely over a sufficiently long interval to obtain accurate FFTs.

3. We have perfect knowledge of the degrees of freedom, so that the generalized coordinates we measure fully characterize the system's motion. As a corollary, the associated generalized forces capture all relevant aspects of any significant excitation to which the system will be subjected.

In practice, any of these conditions might not be met. The effects of each are topics for past and current research. A good starting point for learning about these nonideal circumstances is the text by Ewins (1986).

What measurements are required depends strongly on the type of information we seek. Let us first consider measuring a specific element of $[G(\omega)]$ at a specific frequency. We established earlier that $G_{nj}(\omega)$ is the complex amplitude of q_n when the only nonzero generalized force is $Q_j = \cos(\omega t)$. We saw in Section 3.8.2 that such measurements may be obtained by either a shaker test or by generating an impulsive excitation using a hammer. If we use a shaker to generate a harmonic excitation, a motion sensor can be used to measure the amplitude and relative phase angle of the response. (For the purposes of this discussion we shall assume that displacement is measured. If one uses an accelerometer, displacement coefficients in the frequency domain may be obtained from $X = -A/\omega^2$, where A is the accelerometer output.) To carry out this test, we must mount the shaker such that only one generalized force Q_j is nonzero. If the shaker's frequency is ω, then the magnitude of the shaker force will be $\varepsilon m \omega^2$, where εm is the shaker's imbalance. Because the shaker force is the sole excitation, the associated generalized force will be $Q_j = \beta_j \varepsilon m \omega^2$, where β_j is a constant factor of proportionality that depends on the definition of Q_j. The frequency of the excitation is set by controlling the motor. Dividing the complex amplitude X_n of each measured generalized coordinate by $\beta_j \varepsilon m \omega^2$ leads to the value of G_{nj} at that frequency. If we need to know G_{nj} over a frequency range, we merely step through the frequencies of interest.

In the impulsive excitation method discussed in Section 3.8.2, a hammer generates an impulsive force $f(t)$. A load cell on the hammer measures the instantaneous force, which is sampled at a set of discrete instants. As is true for the shaker test, care must be taken that the impulsive force only gives rise to the relevant generalized force. If this is true, then the measurement of the hammer force yields a set of time data for the jth generalized force, $Q_j(k\Delta) = \beta_j f(k\Delta)$. Simultaneously, we measure instantaneous generalized coordinate values, from which we construct a set of time data $q_n(k\Delta)$. The processing requires ratioing DFT coefficients, which is a deconvolution process. FFT evaluations of the padded force and position yield the DFT coefficients F_{mj} of the jth generalized force, and the DFT coefficients X_{mn} for generalized coordinate q_n. The ratio X_{mn}/F_{mj} gives the value of G_{nj} at discrete frequency ω_m. These operations may be described as

$$\{F_j\} = [E_{\text{DFT}}]\{Q_j\}, \quad \{X_n\} = [E_{\text{DFT}}]\{q_n\}$$

$$G_{nj}(\omega_m) = \frac{X_{mn}}{F_{mj}}, \quad m = 0, 1, ..., N/2 \tag{5.4.9}$$

Now let us consider measuring many elements of $[G(\omega)]$. Application of a single generalized force induces a response in every generalized coordinate. Thus, if N motion sensors are available, we may simultaneously capture the values of G_{nj} for $n = 1, ..., N$ using either a shaker or an impulse hammer to create only the jth generalized force. (If the number of sensors is limited, then the test must be repeated with the sensors repositioned appropriately.) In other words, generating the jth generalized force enables us to fill in the jth column of $[G(\omega)]$.

If we repeat this series of measurements, with a different generalized force being excited in each series, we will obtain the full transfer function $[G(\omega_k)]$ at all frequencies ω_k. From this data, we may invoke eq. (5.1.5) to determine the response to any set of harmonically varying forces. To predict transient response, we would follow the DFT inversion steps outlined in the previous section to perform frequency domain convolution.

An interesting aspect of this procedure is that it is directly analogous to the short-window impulse response method outlined in Section 3.8.2. It was shown there that the DFT of the impulse response could be treated as the complex frequency response, even though such a property is actually correct only if the time window is sufficiently long. The same is true here. Specifically, because of the principle of causality, we may follow the preceding procedure, regardless of the duration of T. If T is too small, the experimentally derived $[G(\omega_k)]$ will not be the true transfer function due to leakage error. Nevertheless, we may use it to perform frequency domain convolution with an excitation, provided that the DFT of the excitation is consistent in the frequency increment and sample size of the $[G(\omega_k)]$ data.

The fundamental difficulty with the preceding procedure is that it requires measurement of the full set of generalized coordinates resulting from generation of each generalized force. If the number of degrees of freedom is large, many displacements will need to be measured, and many repetitions of the experiment, successively generating each generalized force, will be required.

Experimental modal analysis (EMA) offers a much more efficient procedure. Essentially, rather than trying to measure the full $[G(\omega)]$ directly, in what might be called a "brute force" approach, it exploits fundamental knowledge of the features of modal response to substantially lessen the experimental effort. As a corollary, EMA in its standard form is only suitable for systems in which damping is not sufficiently strong to cause individual resonant peaks to merge. This limits EMA to systems that are relatively lightly damped, or else well represented by structural damping. (Several techniques to address this limitation have been developed; see Inman (1989)).

We begin by assuming that the frequency dependence of a *single* column n of $[G(\omega)]$ has been measured. The concept is to match this measured data to the modal representation of $[G(\omega)]$, eq. (5.2.9), which is repeated below in indicial form,

$$G_{nj}(\omega) = \sum_{k=1}^{N} \frac{\Phi_{nk}\Phi_{jk}}{\Omega_k^2 + 2i\zeta_k\Omega_k\omega - \omega^2} \tag{5.4.10}$$

where Ω_k denotes the natural frequencies. This expression contains N^2 parameters that are the elements of $[\Phi]$, as well as N unknown Ω_k, and N unknown modal damping ratios ζ_k. Thus, it is conceivable that, given a sufficient number of measured values of G_{kn}, we may ascertain these modal properties.

A simplistic scheme demonstrates the feasibility of the process. It entails focusing our attention on the behavior of $G_{nj}(\omega)$ in the vicinity of resonances. We assume that the system is lightly damped and the frequencies are far apart. In that case, when ω is close to one of the natural frequencies, the resonant contribution of that mode will be dominant. If p is the number of the resonating mode, then we have

$$G_{nj}(\omega) = \frac{\Phi_{np}\Phi_{jp}}{\Omega_p^2 + 2i\zeta_p\Omega_p\omega - \omega^2} + \varepsilon_{nj}, \quad \omega \approx \Omega_p \tag{5.4.11}$$

The term ε_{nj} represents a residual correction to the leading term. Let us assume that this residual is negligible. Then the above expression is merely the complex frequency response of a single-degree-of-freedom system. Because we have restricted the damping ratio to be small, we can identify the natural frequencies ω_p by scanning a G_{nj} element as a function of ω. The plot for any element's magnitude should show a peak when ω equals any ω_p (assuming that the mode does not have a null at Φ_{np} or Φ_{jp}). Once we have identified the natural frequency, we may use the resonance's bandwidth to determine the damping ratio. According to eq. (3.2.17), for a lightly damped system the amplitude at the half-power points is a factor of $1/\sqrt{2}$ smaller than the peak. Furthermore, eq. (3.2.15) states that the frequency interval between the half-power points is $\Delta\omega = 2\zeta\omega_p$.

Thus, the concept is to scan a plot of $|G_{nj}|$ as a function of ω, identify the frequencies ω_p at which peaks occur as the natural frequencies, and denote the peak amplitudes as $|G_{nj}(\omega_p)|$. We then would draw a horizontal line at $|G_{nj}(\omega_p)|/\sqrt{2}$ in the vicinity of the resonance, and mark as the half-power points the intersections of this line with the $|G_{nj}(\omega)|$ curve. This enables us to measure the bandwidth $\Delta\omega_p$ for each resonance, and then identify the damping ratio according to $\zeta_p = \Delta\omega_p/(2\omega_p)$.

Once we have determined all natural frequencies and modal damping ratios, we proceed to use eq. (5.4.11) to determine the mode coefficients. If we ignore the residuals, setting $\omega = \omega_p$ leads to

$$\Phi_{kp}\Phi_{np} = 2i\zeta_p\omega_p^2 G_{kn}(\omega_p) \tag{5.4.12}$$

The i factor arises from eq. (5.4.11), which indicates that the $G_{kn}(\omega_p)$ values will be imaginary values.

In the assumed scenario, we have measured G_{kn} as a function of ω, with n fixed and $k = 1, \ldots, N$, so we know the $G_{kn}(\omega_p)$ values at each natural frequency. The simplest procedure is to first determine Φ_{np}, which is the generalized coordinate associated with the column number of the $[G(\omega)]$ data. Setting $k = n$ in eq. (5.4.12) gives

$$\Phi_{np} = \left[2\zeta_p|G_{nn}(\omega_p)|\right]^{1/2}\omega_p \tag{5.4.13}$$

Note that the usage of $|G_{nn}(\omega_p)|$ in the preceding is a consequence of the fact that $G_{nn}(\omega_p)$ will be a negative imaginary value, because the numerator in eq. (5.4.11) is positive when $k = n$. Correspondingly, we consider Φ_{np} to be positive. The other elements of $\{\Phi_p\}$ may be determined by taking a ratio of eq. (5.4.12) for any other k to the equation evaluated at $k = n$, which gives

$$\Phi_{kp} = \Phi_{np}\frac{iG_{kn}(\omega_p)}{|G_{nn}(\omega_p)|_p} \tag{5.4.14}$$

It follows that we may fill in the entire normal mode matrix if we know the peak values of a specific column of $[G]$ at each natural frequency. Interestingly, the same argument applies if we instead know the peak values of a single row of $[G]$. This follows from the principle of dynamic reciprocity, according to which $|G_{kn}(\omega_p)| = |G_{nk}(\omega_p)|$; see Section 5.1. Furthermore, once we know the full set of modal properties, we may determine the value of $[G]$ at any frequency by invoking eq. (5.4.10).

The ability to obtain all of the modal properties from the properties of a single column or row of $[G]$ in the vicinity of each of the natural frequencies leads to a significant reduction in effort. As noted earlier, a column of $[G(\omega)]$ constitutes the response of all generalized coordinates to a single unit generalized force having unit amplitude at frequency ω. Because $[G(\omega)]$ is symmetric, the same number column and row are identical. A row of $[G(\omega)]$ is the response of a single generalized coordinate when each of the generalized forces is successively assigned a unit amplitude at frequency ω. Hence, the measurements required for experimental modal analysis can be obtained by instrumenting all of the generalized coordinates and exciting only one generalized force, which gives a column of $[G]$. Alternatively, we can measure only one generalized coordinate while we excite each of the generalized forces sequentially, which fills in a row of $[G]$.

The foregoing is a drastically oversimplified description of the experimental modal analysis process. For example, if the sampling interval Δ is not much smaller than the bandwidth, the resonant peaks will not be represented at high resolution. This makes it difficult to identify accurately the peak values of $|G_{nj}(\omega)|$ and the frequencies at which the peaks occur. The Nyquist plot discussed in Example 3.4 is effective in overcoming that difficulty. Another common difficulty is that the DFT period T

cannot be made sufficiently large because the system is lightly damped. In that case the DFT of the impulse response will not yield a good estimate of $G_{nj}(\omega)$. A shaker test that zooms in on resonant peaks circumvents this difficulty. Also, if impulse response measurement is necessary, inability to extend T can sometimes be overcome by applying the exponential windowing function in Section 3.8.2. In essence, that technique introduces an artificial amount of damping into each modal response, which may be subtracted out from the raw estimate of each ζ_p.

Another common difficulty is proximity of the natural frequencies, which makes it necessary to account for the residual correction in eq. (5.4.11). A remedy here is to use curve-fitting procedures to match eq. (5.4.10) to the measured frequency dependence. Also, experimental errors can originate in the measurement system, the generation of the exciting forces, the FFT evaluation, or the fundamental model on which the analysis is based. Minimizing the effects of each type of error is an important aspect of experimental modal analysis.

In any event, the simple approach outlined here leads to the fundamental realization that we can reconstruct the full transfer function at any frequency from knowledge of the behavior of a single column or row of $[G]$ in the vicinity of resonant peaks. In contrast, without experimental modal analysis, we must generate N distinct excitations, and measure each element of the transfer function directly.

REFERENCES AND SELECTED READINGS

DEN HARTOG, J. P. 1956. *Mechanical Vibration*, 4th ed. McGraw-Hill, New York.

EWINS, D. J., 1986. *Modal Testing: Theory and Practice.* Research Studies Press Ltd; Letchworth, Hertfordshire, England.

INMAN, D. J. 1989. *Vibration.* Prentice Hall, Englewood Cliffs, NJ.

JAMES, M. L., SMITH, G. M., WOLFORD, J. C., & WHALEY, P. W. 1989. *Vibration of Mechanical and Structural Systems,* Harper & Row, New York.

JUNGER, M. C. & FEIT, D. 1986. *Sound, Structures, and Their Interaction.* Acoustical Society of America, New York.

MORSE, P. M., 1981. *Vibration and Sound.* Acoustical Society of America, New York.

NEWLAND, D. E. 1989. *Mechanical Vibration Analysis and Computation,* Longman Scientific and Technical, White Plains, NY.

PIERCE, A. D. 1981. *Acoustics.* Acoustical Society of America, New York.

SOIZE, C. 1993. "A Model and Numerical Method in the Medium Frequency Range for Vibroacoustic Prediction Using the Theory of Structural Fuzzy." *Journal of the Acoustical Society of America,* 94, 849–865.

VERNON, J. B. 1967. *Linear Vibration Theory.* John Wiley & Sons, New York.

VIERCK, R. K. 1979. *Vibration Analysis.* 2nd ed. Harper & Row, New York.

ZAVERI, K. 1984. *Modal Analysis of Large Structures— Multiple Exciter Systems.* Bruel and Kjaer Publications, Naerum, Denmark.

EXERCISES

5.1 The inertia, damping, stiffness, and force matrices for a system are

$$[M] = \begin{bmatrix} 20 & -5 \\ -5 & 10 \end{bmatrix} \text{kg}, \quad [C] = \begin{bmatrix} 4 & 1 \\ 1 & 8 \end{bmatrix} \text{N-s/m}$$

$$[K] = \begin{bmatrix} 400 & 0 \\ 0 & 600 \end{bmatrix} \text{N/m}, \quad \{Q\} = \begin{Bmatrix} 30 \cos(20t) \\ 50 \sin(20t) \end{Bmatrix} \text{N}$$

Use the frequency domain formulation to determine the steady-state responses $q_1(t)$ and $q_2(t)$. Express the result in the form $q_j = A_j \sin(20t - \psi_j)$. Compare the values of the amplitudes A_j and phase lags ψ_j to the values that result when the problem is solved using the light damping approximation based on diagonalizing the damping matrix.

5.2 The two-degree-of-freedom system on the next page is subjected to a harmonic excitation, $F(t) = 5 \cos(\omega t)$ N. System parameters are $m_1 = 1$, $m_2 = 4$ kg; $k_1 = 200$, $k_2 = 300$, $k_3 = 100$ N/m; and $c_1 = 2$, $c_2 = 0.5$ N-s/m. Use

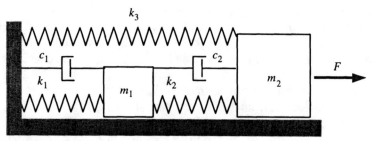

EXERCISE 5.2

frequency domain analysis to derive and graph the corresponding steady-state amplitudes as a function of ω. The maximum value of ω should be sufficiently large to see the amplitude peaks associated with both natural frequencies.

5.3 Use frequency domain analysis to solve Exercise 4.51.

5.4 Use frequency domain analysis to solve Exercise 4.54.

5.5 Consider the automobile model in Exercise 4.56. Use frequency domain analysis to determine the displacement amplitudes at the front and rear as a function of v. Graph the results for $0 < v < 20$ m/s. Compare the results for $c = 5$ kN-s/m to those for $c = 0.5$ kN-s/m.

5.6 A motor is supported by a pair of interconnected bars, which are shown in their static equilibrium position. Each bar has mass m_1 and may be considered to be rigid. The identical shock absorbers consist of a spring k acting in parallel with a dashpot c. The nonrotating parts of the motor have mass m_2, and the rotating mass is m_3.

(a) Define a set of generalized coordinates for the system, and derive the corresponding equations of motion. For this, it is permissible to ignore the effects of rotatory inertia for the motor housing and rotor.

(b) Consider the case where $m_1 = 20$, $m_2 = m_3 = 2$ kg, $k = 50$ kN/m, $c = 200$ N/m, $L = 400$ mm. Determine the steady-state amplitude of each bar's rotation as a function of frequency. Graph the results for a range of ω from zero to 40% above the highest frequency at which these amplitudes attain peak values.

5.7 Consider the three-degree-of freedom model of an automobile in Exercise 4.10. Generalized coordinates for vertical displacement are the upward displacement of the center of mass y_G, pitching rotation θ_y, and rolling rotation θ_x. These motions are superposed onto the forward motion of the vehicle at speed v in the x direction. A dashpot is placed in parallel to each of the springs shown in the sketch for Exercise 4.10. The dashpot constant is $c = 1.0$ kN-s/m. A linearized

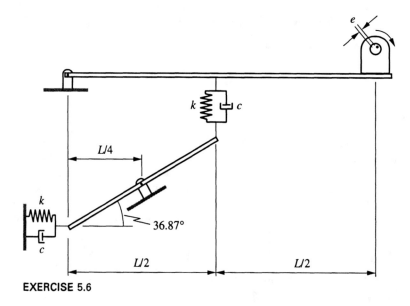

EXERCISE 5.6

approximation of the vertical displacement of the point at which each spring-dashpot pair is attached to the chassis is

$$y_A = y_G - 1.3\theta_y - 0.9\theta_x$$
$$y_B = y_G - 1.3\theta_y + 0.9\theta_x$$
$$y_C = y_G + 1.5\theta_y - 0.9\theta_x$$
$$y_D = y_G + 1.5\theta_y + 0.9\theta_x$$

A corrugated road is approximated as being sinusoidal, such that the ground elevation is $w = 0.05\sin(2X)$ m, where X measures distance perpendicular to the corrugation. If ψ measures the angle in which the vehicle is heading relative to the X direction, then the X value (m) under each wheel is

$$X_A = vt\cos\psi + 0.9\sin\psi$$
$$X_B = vt\cos\psi - 0.9\sin\psi$$
$$X_C = (vt - 2.8)\cos\psi + 0.9\sin\psi$$
$$X_D = (vt - 2.8)\cos\psi - 0.9\sin\psi$$

Because the elongation of each spring-dashpot pair is the difference between the displacements of the attachment point and of the wheel, the effect of the changing ground elevation is equivalent to applying upward forces at wheel P given by

$$F_P = k_P w(X_P) + c_P \dot{w}(X_P)$$
$$= 0.05 k_P \sin(2X_P)$$
$$+ 0.10(v\cos\psi)c_P \cos(2X_P)$$

Determine the steady-state amplitudes of y_G, θ_x, and θ_y as a function of v when $\psi = 20°$ and $\psi = 70°$. Graph the results for $0 < v < 20$ m/s.

5.8 Solve Exercise 4.52 in the case where there is structural damping with a loss factor, $\gamma = 0.001$.

5.9 Solve Exercise 4.54 in the case where, rather than having dashpots, dissipation is represented as structural damping. Consider as the loss factor $\gamma = 0.004$ and $\gamma = 0.04$.

5.10 A torque $\Gamma_0 \cos(\omega t)$, positive when counterclockwise, is applied to the right bar of the triple pendulum in Exercise 4.30. Dissipation due to friction in the pivot of each bar fits the structural damping model, with $\gamma = 0.06$. Determine the steady-state amplitude of the rotation of each bar as a function of ω. The frequency range is $0 < \omega < 1.2\omega_{max}$, where ω_{max} is the highest frequency at which the amplitudes peak.

5.11 In Exercise 4.21, a model of a four-story building was described in terms of measured flexibility data. Consider a situation where horizontal force $F_3 = 200\sin(\omega t - \pi/3)$ N is applied to the third floor and $F_4 = 150\cos(\omega t)$ N is applied to

the fourth floor. The structural damping loss factor is $\gamma = 0.02$. Determine as a function of ω the amplitude of the steady-state response of u_4, and the phase lag of this response relative to a pure cosine. Plot the results as functions of ω covering the interval extending up to 50% higher than the fourth natural frequency.

5.12 Consider the steady-state response of the bladed disk assembly described in Exercise 4.58 when the system has structural damping. Consider the weakly coupled case, $R = 0.05$. Plot the amplitude of the steady-state response of blades 1 and 6 as a function of the frequency ratio ω/ω_b in the range $0.98\omega_b \leq \omega \leq 1.02\omega_b$, which covers the range of the natural frequencies. Consider two loss factor cases, $\gamma = 0.001$ and $\gamma = 0.01$. Does increasing the amount of damping cause modal coupling?

5.13 Solve Exercise 5.12 in the strongly coupled case, $R = 0.5$. The frequency range for the evaluation should be $0.8\omega_b \leq \omega \leq 1.6\omega_b$, which contains all natural frequencies for this value of R.

5.14 An electric motor is mounted on a rigid platform that is supported by four identical springs, each with a spring constant of 50 kN/m. The mass of the motor and its flywheel is 50 kg, and the mass of the platform is 100 kg. The radii of gyration are 60 mm for the motor about the centroid S and 300 mm for the platform about its centroid G. The rotor imbalance is $\varepsilon m = 0.02$ kg-m. Snubbers at the ends of the platform prevent sidesway. The friction effect of the snubbers can be represented as structural damping with a loss factor $\gamma = 0.05$. Use the frequency domain transfer function method to determine the amplitude of the displacement of the motor's shaft as a function of rotation rate Ω. (*Hint:* The platform will rotate

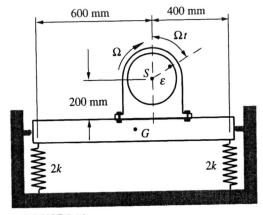

EXERCISE 5.14

because the motor's shaft is not concurrent with the platform's center of mass. Also, because the motor is fastened rigidly to the platform, the displacement of the motor's shaft S is related to the displacement of the center of mass G of the platform by $\Delta \bar{r}_S = \Delta \bar{r}_G + \theta \bar{k} \times \bar{r}_{S/G}$, where θ is the angle of rotation of the platform.)

5.15 A one-degree-of-freedom system whose mass is 50 kg and whose stiffness is 20 kN/m must sustain a harmonic force $500 \sin(\omega t)$ N. The frequency range is $18 < \omega < 22$ rad/s. It is desired to use a vibration absorber to reduce the displacement when ω equals the natural frequency. It also is necessary that the natural frequencies of the new system be outside the range of possible ω values. What is the smallest possible mass and stiffness for the absorber to meet these requirements?

5.16 Consider the vibration absorber model in Figure 5.1 in a situation where $m_1 = 100$ kg and $k_1 = 40$ kN/m. The absorber is tuned to the natural frequency of the isolated lower system, with $m_2 = 0.5$ kg. The damping constants are defined relative to the properties of each isolated system, such that $c_1 = 2\zeta_1(k_1 m_1)^{1/2}$ and $c_2 = 2\zeta_2(k_2 m_2)^{1/2}$. The frequency range for the force applied to m_1 is $1 \text{ Hz} \le \omega \le 5 \text{ Hz}$. Plot $k_1 |Y_1|/|F|$ and $k_1 |Y_2|/|F|$ as a function of ω in this range for two cases: $\zeta_1 = \zeta_2 = 0.0005$ and $\zeta_1 = \zeta_2 = 0.005$.

5.17 The motor is mounted on a block that translates over a smooth horizontal plane. The combined mass of the motor and the block to which it is fastened is 20 kg. The spring constant k_1 is such that if the right spring k_2 were not present, the system would have a natural frequency of 20 Hz. Damping is negligible, $c_1 = 0$.

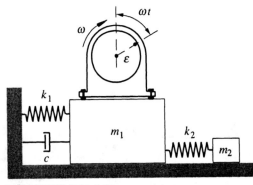

EXERCISES 5.17, 5.18

(a) If $m_2 = 4$ kg, what value of k_2 will minimize the vibration of m_1 when the motor's rotation rate matches the natural frequency of the original system?

(b) If $m_2 = 4$ kg, what value of k_2 will minimize the vibration of m_1 when the motor rotates at 15 Hz?

(c) For the system parameters in case (b), determine the amplitude of the vibration of m_1 when the motor's rotation rate matches the natural frequency of the original system.

5.18 The motor is mounted on a block that translates over a smooth horizontal plane. The combined mass of the motor and the block to which it is fastened is $m_1 = 20$ kg, and $m_2 = 4$ kg. The rotor imbalance is $\varepsilon m = 0.008$ kg-m². The spring constant k_1 and dashpot constant c_1 are such that if the right spring k_2 were not present, the system would have a natural frequency of 20 Hz and the critical damping ratio would be $\zeta = 0.01$. Also k_2 is selected such that the m_2, k_2 system would be a tuned vibration absorber for the m_1, k_1 system. Determine and graph as a function of the motor's rotation rate the amplitude of the displacement of m_1 and m_2. Compare the result to what would be obtained without the vibration absorber.

5.19 The configuration shown in the sketch corresponds to the static equilibrium position of the system in the absence of the harmonic force $F \sin(20t)$, where t is measured in seconds. The masses are 4 kg for the upper bar and 2 kg for the lower bar. Also, $L = 800$ mm and $k_1 = 1500$ N/m.

(a) Suppose damping were not present, $c_1 = c_2 = 0$. For what spring stiffness k_2 will the upper bar not vibrate in the steady-state condition associated with this excitation?

(b) Consider the case where $c_1 = c_2 = 100$ N-s/m. Determine the steady-state amplitude of the angular rotation of the upper bar, and the phase angle of this rotation relative to a pure sine function, when k_2 has the value found in part (a).

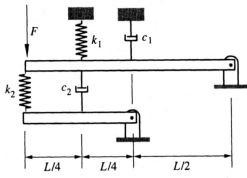

EXERCISE 5.19

5.20 Consider a system like that in Figure 5.1 for which $m_2/m_1 = 0.30$. The added system is mistuned such that $k_2/m_2 = 0.6k_1/m_1$ and it is heavily damped, with $c_2 = 0.6(k_2m_2)^{1/2}$. For the original system $c_1 = 0.01(k_1m_1)^{1/2}$, which corresponds to light damping. The force applied to the lower mass is $F\cos(\omega t)$. Plot as a function of ω the nondimensional amplitude $k_1|Y_1|/F$. Compare these results to what would be obtained without the vibration absorber. Specifically, identify the frequency range over which the absorber is effective in reducing the vibrational amplitude. (This model is essentially the untuned vibration absorber analyzed by Den Hartog (1956)).

5.21 Consider the vibration absorber model in Figure 5.1 in a situation where $m_1 = m$, $m_2 = 0.2m$, $k_1 = k$, and $k_2 = 0.2k$, where m and k are basic units scaling mass and stiffness. The lower mass is subjected to a periodic excitation given by $F = \beta k\{\sin[(k/m)^{1/2}t] + 2\cos[2(k/m)^{1/2}t]\}$, where β is a scaling factor having units of length. Determine the steady-state response of each mass if (a) $c_1 = c_2 = 0$ and (b) $c_1 = 0.2(km)^{1/2}$, $c_2 = 0.5 (km)^{1/2}$.

5.22 Damping in the three-degree-of-freedom system is negligible. Measured vibration amplitudes of masses m_1 and m_3 resulting from application of a force $1000\cos(\omega t)$ N to block 1 are as graphed, but it is not specified whether curve A refers to the amplitude of the left block or the right.

(a) Explain why it must be that curve C describes the motion of the left block.

(b) It is known that $m_1 = m_2 = 2$ kg and $m_3 = 1$ kg. What are the values of k_1, k_2, and k_3?

5.23 The sketch shows a motor that is mounted on a spring-suspended base, with a second platform suspended below. The masses of the upper and lower platforms and the motor are m_1, m_2, and m_3,

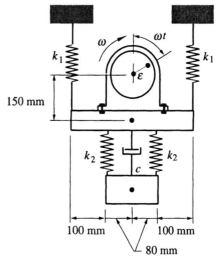

150 mm

100 mm   100 mm

80 mm

EXERCISE 5.23

respectively. Also, the centroid of the motor coincides with its shaft, and the centroidal radii of gyration are κ_1, κ_2, and κ_3 for the upper and lower platforms and the motor, respectively. The center of mass of the rotor is at a distance ε from the axis of rotation. The rotor turns at the constant rate ω clockwise. Although the motor is mounted at the middle of the upper platform, this platform rotates, as well as translates, because the motor's shaft is above the centroid of the platform. This, in turn, induces a rotation of the lower platform. However, sidesway of each platform is prevented by snubbers.

(a) Select as generalized coordinates for this system the vertical translation y_1 of the upper platform's center of mass, the rotation θ_1 of the upper platform, the vertical translation y_2 of the lower platform, and the rotation θ_2 of the lower platform, all of which are measured relative to the system's

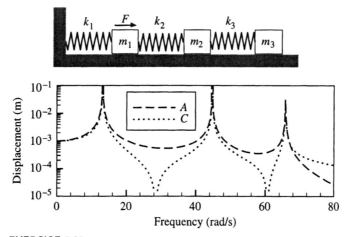

EXERCISE 5.22

static equilibrium position. Derive the equations of motion for this system.

(b) Suppose $c = 0$ and rotation of both platforms were somehow prevented. What combination of k_2 and m_2 values would minimize vibration of the motor's platform at the designated rotation rate?

(c) Consider the case where the system parameters are $m_1 = 5$ kg, $m_3 = 2$ kg, $k_1 = 10$ kN/m, $k_2 = 2$ kN/m, $c = 200$ N-s/m, and the motor's imbalance is $\varepsilon m = 3$ kg-mm. The radii of gyration are $\kappa_1 = 120$ mm, $\kappa_2 = 60$ mm, and $\kappa_3 = 80$ mm. The lower platform's mass fits the criterion established in part (b). Use the frequency domain transfer function for this system to determine the steady-state response as a function of the rotation rate ω.

5.24 Determine the response of the system in Example 5.3 when there are nine substructures whose spring stiffness is evenly distributed from $0.009k_0$ to $0.011k_0$. The damping ratio for each attached subsystem is $\zeta = 0.005$.

5.25 Consider the system in Exercise 5.2 in the case where the exciting force $F(t)$ is a periodic squarewave, $F(t) = F_0$ if $0 < t < T/2$, $F(t) = 0$ if $T/2 < t < T$, $F(t + T) = F(t)$. Use FFTs to derive and graph the corresponding steady-state response of block 2 when $T = 0.40$ s.

5.26 Consider the automobile in Exercise 4.56 in the case where the ground elevation is that of a bump defined by $w(x) = 0.4(x/\lambda - x^2/\lambda^2)$ $[h(x) - h(x - \lambda)]$ m, where $\lambda = 2$ m is the width of the bump. Use FFT techniques to determine the vertical displacement of the front and rear as a function of time.

5.27 Consider the system in Exercise 4.45, in which a three-story building subjected to a base motion from an earthquake is modeled as a string of three oscillators. Use the frequency domain transfer function and FFTs to determine the displacement of each oscillator mass when the damping constants are described by $\zeta_j = 0.5c_j/(k_jm_j)^{1/2} = 0.8, 0.6, 0.4$.

5.28 Axial free vibration of a bar composed of a polymer was analyzed with a lumped mass model in Example 5.2. The situation of interest here is application of an axial force at the midpoint, whose time signature is a uniform pulse: $F(t) = 100$ N, $0 < t < T$; $F(t) = 0$ otherwise. The pulse duration is $T = 10$ meters, which also equals the time

constant τ for the material. Determine the axial displacement of the midpoint resulting from application of this force.

5.29 A two-degree-of-freedom system is loaded by a harmonic force such that the generalized forces are $Q_1 = 0$, $Q_2 = 200 \cos(\omega t)$ N. The steady-state response of the system is measured as the frequency is swept upward from zero to a value much greater than the second natural frequency. The system is known to be sufficiently lightly damped that setting ω close to either natural frequency makes that mode's contribution to the response dominant. Important measured data are

- At $\omega = 5.734$ rad/s, the responses are $q_1 = 0.1019 \sin(\omega t)$ and $q_2 = 0.2467 \sin(\omega t)$, where the unit for all generalized coordinates is meters.
- At $\omega = 5.723$ rad/s, the amplitude of q_1 is $0.1019/\sqrt{2}$.
- At $\omega = 18.478$ rad/s, $q_1 = -0.0927 \sin(\omega t)$, and $q_2 = 0.0622 \sin(\omega t)$.
- At $\omega = 18.515$ rad/s, the amplitude of q_2 is $0.0622/\sqrt{2}$.

These data are sufficient to identify the system's natural frequencies, modal damping ratios, and normal modes.

1. Based on the measured data, what are the system's natural frequencies?
2. Based on the measured data, what are the system's modal damping ratios?
3. Based on the measured data, what are the normal modes?

5.30 A shaker is mounted on the third floor of a three-story building. Accelerometers measured the horizontal displacement of each floor as the shaker's frequency is slowly swept from a very small value to a frequency well above the third resonance. The complex amplitude of the steady-state displacements at each frequency was divided by the corresponding complex amplitude of the shaker force. The result is the accompanying set of plots, which give the displacement amplitude in units of meters/newton, and the relative phase angle as functions of the shaker frequency. In these graphs, X_j denotes the displacement of floor number j. Use these data to estimate the natural frequencies, the modal damping ratios, and the normal modes.

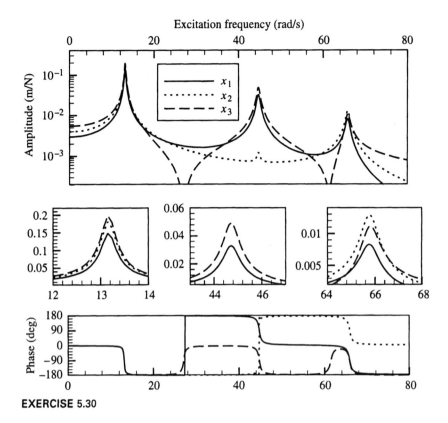

EXERCISE 5.30

CHAPTER 6

VIBRATION OF ELASTIC BARS: THE RITZ METHOD AND THE RAYLEIGH RATIO

In this chapter we turn our attention to continuous systems, in which the motion is characterized by a displacement field whose value is a function of position as well as time. Because a function may assume an arbitrary value at each location (subject to limitations on continuity that we will identify), a continuous system may be considered to have an infinite number of degrees of freedom. A straight bar is the continuum with the simplest dynamic laws. The definition of such a body is that its shape is the locus of cross-sections perpendicular to a straight line, in the case where the centroid of each cross-section is situated on that line and the dimensions of the largest cross-section are much smaller than the axial length. The cross-section can vary along the centroidal axis, but the uniform case is most commonly encountered.

A straight bar undergoes three fundamental types of deformation: extension, torsion, and flexure, each of which occurs independently in linear vibration theory. Each type of deformation is described by a displacement field consisting of a single independent variable defined relative to the bar's longitudinal axis—axial displacement of cross-sections in the case of extension, rotation of cross-sections for torsion, and transverse displacement in the case of flexure. The manner in which a typical cross-section displaces in each type of deformation is depicted in Figure 6.1. Other types of continua commonly used in engineering applications are curved bars, membranes, plates, and shells. In comparison to straight bars, each of those systems is complicated by the spatial dependence of its displacement field and/or coupling of different deformation effects.

For extension, the cross-sections exhibit an axial displacement that depends only on the axial location of the cross-section. This leads to an axial strain that is uniform over the cross-section; an axial force is the stress resultant acting on an exposed cross-section. In torsion, cross-sections undergo a rotation that varies axially. This

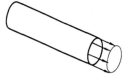
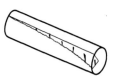
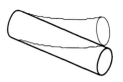

Axial displacement   Torsional displacement   Flexural displacement

FIGURE 6.1 Basic types of deformation of a straight bar.

leads to a shear strain that varies across a cross-sectional face. A couple acting about the bar's axis is the internal stress resultant in this case. In flexure, as in extension, the primary deformation is axial strain. However, flexural deformation depends on the curvature of the deformed axis of the bar, and the axial strain is not constant over a cross-section. The primary internal stress resultants in this case are the bending moment and transverse shear force. Indeed, a synonym for flexure is bending. A bar that carries loads by undergoing flexural deformation is commonly referred to as a *beam*.

A rigorous treatment of vibration of continua leads to *field equations*, which is the term used to refer to motion equations governing a displacement that depends on position and time. These equations express the balance between internal forces and acceleration effects that must exist everywhere within the bar. The field equations are partial differential equations of motion because the inertial effects depend on the accelerations, while the elastic effects depend on spatial derivatives of the displacement field. Most texts begin the study of vibration of continua by discussing the formulation and solution of the associated field equations. Such an approach tends to emphasize the differences between continuous and discrete models, even though the physics of linear vibration theory is similar for both types of models.

The alternative method with which we shall begin our studies is known by several terms. The more descriptive one is the *Ritz series method*. Its origins may be traced back to the work of the German mathematician W. Ritz (1909), who used series expansions in conjunction with variational principles to formulate various problems in mathematics and engineering. Most practicing engineers refer to the procedure as the *method of assumed modes*. As we will see, this is a misnomer, because one does not approach it from the standpoint of guessing mode shapes. We will use the three types of bar deformation, axial extension, torsion, and flexure, as a framework for formulating the Ritz method. One of the primary merits of the Ritz method is that it is readily extended to other types of media, such as plates and shells, although we shall not do so in this text. We will use the method as the theoretical foundation for later developments, especially the finite element and modal synthesis methods, which permit analysis of complicated structures.

The present formulation is limited to situations where the static equilibrium position is a suitable reference state for displacement. (This excludes systems such as rotating shafts.) Furthermore, because the dynamic response may be superposed on the static displacement field, we will consider the bar to be undeformed at its static equilibrium position.

The study of vibration of continua is the dynamic analog to mechanics of materials. The displacement field will depend on spatial position as well as time. Hence, each displacement component contributes to the kinetic and potential energy of the bar, and both energies are distributed over the length of the bar. Correspondingly, bars (and all other continua) are often referred to as *distributed parameter systems*. From this perspective, the essential difference between systems of rigid bodies and continua is that a material element of a continuum simultaneously stores kinetic and potential energy. (Some materials also dissipate a substantial amount of energy when they deform.)

6.1 EQUATIONS OF MOTION

Our framework for the Ritz series method will be the power balance formulation, which is equivalent for linear time-independent systems to (the more general) Lagrange's equations. The basic quantities required to form these equations are the system's kinetic energy T and potential energy V. The essence of the procedure is to use a series expansion to represent the spatial dependence of the displacement field. The coefficients of this series are unknown functions of time, which become the generalized coordinates for the response. We use the series to construct the energy expressions, which will lead us to inertia and stiffness matrices. If dissipation is present, we obtain the damping matrix by constructing the power dissipation corresponding to the displacement series. We also use the displacement series to construct the power input, from which we obtain the generalized forces. The result of these operations is a description that matches the standard matrix equation of motion for discrete systems.

Once these equations have been determined, we may solve them by the methods established in the previous chapters. An important additional step required of continuous systems is to synthesize the physical displacement field associated with the displacement series. Our focus in the initial development is determination of the equations of motion. After we have developed the basic modeling procedure, we will implement modal analysis, which will provide many insights into the underlying physical phenomena and mathematical properties of the Ritz series method.

Before we proceed to the specific steps, it is useful to consider the theoretical justification for the procedure. When we represent the displacement field of a continuous system in terms of a finite-length series, we inherently reduce the number of degrees of freedom of the system. By using the power balance formulation, which is effectively equivalent to Lagrange's equations, to derive the equations of motion for this reduced system, we obtain relations that most closely satisfy the physical laws governing the system, given the limitations of the series. Clearly, such a description is an approximation. One check on the approximation is to verify that the series has converged by extending the length of the series, and comparing the results to those obtained from the shorter series. After we have developed the basic techniques, we will explore principles originating from work by Lord Rayleigh that prove the Ritz series converges, as well as assist us to verify the adequacy of a solution. In some situations, it is possible to extend the series to a large number of terms. Such a development will lead to a solution that is almost as accurate as one obtained by solving the field equations, whose analysis we address in the next chapter.

6.1.1 Axial Displacement

The power balance method requires expressions for the mechanical energy associated with each of the basic displacements encountered in a bar. These are developed in detail in the next chapter. For the present purpose, it is sufficient merely to refer to the final expressions or, alternatively, recall the expressions developed in mechanics of materials. We begin by considering axial motion of a straight bar. Figure 6.2 depicts a bar undergoing an axial displacement in response to a distributed force $f_x(x, t)$. (The physical units of f_x are force per unit length.) All points on a cross-section execute the same axial displacement $u(x, t)$. The cross-sectional area may vary gradually without invalidating the basic mechanics of materials theory, so we denote this area as $A(x)$.

The highlighted segment in Figure 6.2 covers an interval dx. Because of the infinitesimal size of dx, all points in this segment have essentially the same velocity $\dot{u}$ in the x direction, and the mass of the segment is $\rho A\, dx$, where ρ is the density. We integrate over the length L to obtain the kinetic energy of the bar,

$$T_{\text{bar}} = \frac{1}{2}\int_0^L \dot{u}^2 \rho A\, dx \qquad (6.1.1)$$

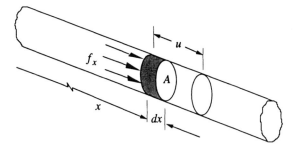

FIGURE 6.2 Axial deformation of a straight bar.

The axial strain is $\varepsilon_{xx} = \partial u/\partial x$. According to linear elasticity, the potential energy per unit volume due to this strain is $\frac{1}{2}E\varepsilon_{xx}^2$, where E is Young's modulus. The corresponding potential energy, which is known as (elastic) *strain energy*, is

$$V_{\text{bar}} = \frac{1}{2}\int_0^L EA\left(\frac{\partial u}{\partial x}\right)^2 dx \tag{6.1.2}$$

In keeping with the restriction stated earlier, $u = 0$ corresponds to a static configuration. We define the axial load f_x to be the disturbance forces causing the bar to displace from this equilibrium configuration. The axial resultant of the external force acting on a segment dx is $f_x\,dx$, and the velocity of this segment is $\dot{u}$, so the power input to the bar by the external forces is

$$\mathscr{P}_{\text{in}} = \int_0^L f_x \dot{u}\,dx \tag{6.1.3}$$

Cases where the axial force is a concentrated force F applied at location x_F may be obtained formally from the above by using a Dirac delta singularity function to represent the force, $f_x = F\delta(x - x_F)$, or more simply by multiplying the force by the velocity of its point of application. Either approach leads to a contribution of $F\dot{u}(x_F, t)$. In the most general case, several concentrated forces, as well as a distributed force, are present, so the total power input is

$$\mathscr{P}_{\text{in}} = \int_0^L f_x \dot{u}\,dx + \sum F\dot{u}(x_F, t) \tag{6.1.4}$$

where a sum that does not contain a subscript should be understood to extend over all similar quantities within that sum.

In real applications, a bar often has other mechanical elements attached to it. In Figure 6.3, a small mass, a spring, and a dashpot are fastened at locations x_m, x_k, and

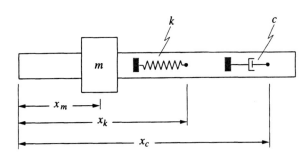

FIGURE 6.3 Attachments to a bar undergoing axial displacement.

x_c, respectively. The power balance formulation requires that we account for the contributions of these elements. The kinetic energy of the block is $\frac{1}{2}m\dot{u}(x_m, t)^2$ and the potential energy of the spring is $\frac{1}{2}ku(x_k, t)^2$. Thus, if several of these elements are present, the total system energies are

$$
T = \frac{1}{2}\int_0^L \dot{u}^2 \rho A\, dx + \frac{1}{2}\sum m\dot{u}(x_m, t)^2
$$

$$
V = \frac{1}{2}\int_0^L EA\left(\frac{\partial u}{\partial x}\right)^2 dx + \frac{1}{2}\sum ku(x_k, t)^2
$$

(6.1.5)

Energy can be dissipated within the bar itself through a variety of mechanisms. A detailed description of such losses is beyond the scope of the present work, but a simple proportional damping model is often used to represent the effect of viscoelasticity. In it, the stress is taken to be related to strain as

$$
\sigma = E(\varepsilon + \gamma\dot{\varepsilon})
$$

(6.1.6)

In the case of harmonically varying strain, γ is the loss factor; see eq. (3.2.23). The portion of σ that depends on $\dot{\varepsilon}$ leads to a dissipation per unit volume of $\gamma E\dot{\varepsilon}^2$, which integrated over the volume of the bar leads to a term resembling the strain energy of the bar. In addition, if there is a dashpot anchored to the ground at x_c, the power dissipation will be $c[\dot{u}(x_c, t)^2]$, so the function for the whole system is

$$
\mathcal{P}_{\text{dis}} = \int_0^L \gamma EA\left(\frac{\partial \dot{u}}{\partial x}\right)^z dx + \sum c\dot{u}(x_c, t)^2
$$

(6.1.7)

Each of the above quantities depends on the axial displacement $u(x, t)$. In the Ritz method we select a set of N functions $\psi_j(x)$ to represent the axial variation of the displacement field. These functions must satisfy certain conditions we discuss later, but an obvious one is that the functions must be linearly independent. For example, if we select $\psi_1 = x$ and $\psi_2 = x^2$, we cannot select $\psi_3 = \alpha x + \beta x^2$, regardless of the values of the constants α and β. We let q_j denote the contribution of ψ_j to the overall displacement field. The amount of this contribution may change with elapsed time. Correspondingly, the Ritz series representation of the displacement is

$$
u(x, t) = \sum_{j=1}^N \psi_j(x)q_j(t)
$$

(6.1.8)

From a mathematical perspective, eq. (6.1.8) maps the continuous function u into an N-dimensional space whose directions are the functions ψ_j, so these functions are called *basis functions*. The series coefficients q_j in this viewpoint represent projections of u in the direction of each basis function. Once the basis functions have been selected, the only free variables remaining are the q_j, so these are the generalized coordinates. Note that it is useful to define the basis functions nondimensionally, so that the generalized coordinates have the dimensions of length.

The power balance method requires expressions for the mechanical energies and power in terms of the generalized coordinates. We obtain these by substituting the Ritz series into the expressions for T, V, $\mathcal{P}_{\text{dis}}$, and $\mathcal{P}_{\text{in}}$. Equations (6.1.5) feature squares of displacement. In order to account for all combinations when the series expansion is substituted into these products, different indices j and n are used to form each term. For the kinetic energy term this leads to

$$T = \frac{1}{2} \int_0^L \left(\sum_{j=1}^N \psi_j(x) \dot{q}_j(t) \right) \left(\sum_{n=1}^N \psi_n(x) \dot{q}_n(t) \right) \rho A \, dx$$

$$+ \frac{1}{2} \sum m \left(\sum_{j=1}^N \psi_j(x_m) \dot{q}_j(t) \right) \left(\sum_{n=1}^N \psi_n(x_m) \dot{q}_n(t) \right) \tag{6.1.9}$$

To simplify this, we collect the coefficients of $\dot{q}_j \dot{q}_n$ and bring the summations over j and n outside the products, which leads to

$$T = \frac{1}{2} \sum_{j=1}^N \sum_{n=1}^N M_{jn} \dot{q}_j \dot{q}_n \tag{6.1.10}$$

where the inertia coefficients are

$$M_{jn} = M_{nj} = \int_0^L \psi_j \psi_n \rho A \, dx + \sum m \psi_j(x_m) \psi_n(x_m) \tag{6.1.11}$$

We follow a similar procedure to describe the potential energy in eqs. (6.1.5), which leads to

$$V = \frac{1}{2} \sum_{j=1}^N \sum_{n=1}^N K_{jn} q_j q_n \tag{6.1.12}$$

where the stiffness terms are

$$K_{jn} = K_{nj} = \int_0^L EA \frac{d\psi_j}{dx} \frac{d\psi_n}{dx} dx + \sum k \psi_j(x_k) \psi_n(x_k) \tag{6.1.13}$$

Application of the same procedure to eq. (6.1.7) leads to

$$\mathcal{P}_{\text{dis}} = \sum_{j=1}^N \sum_{n=1}^N C_{nj} \dot{q}_j \dot{q}_n$$

$$C_{jn} = C_{nj} = \int_0^L \gamma EA \frac{d\psi_j}{dx} \frac{d\psi_n}{dx} dx + \sum c \psi_j(x_c) \psi_n(x_c) \tag{6.1.14}$$

It should be noted in regard to the preceding that the viscoelastic effect for most materials is likely to be negligible if dashpots are attached.

To obtain the generalized forces Q_j we substitute the Ritz series into $\mathcal{P}_{\text{in}}$, eq. (6.1.4), and then collect the coefficients of each $\dot{q}_j$, from which we find

$$\mathcal{P}_{in} = \sum_{j=1}^{N} Q_j \dot{q}_j$$

$$Q_j = \int_0^L f_x \psi_j \, dx + \sum F \psi_j(x_{Fn}, t)$$

(6.1.15)

The coefficients appearing above require the evaluation of definite integrals whose integrand contains the basis functions. The integrals may be evaluated analytically if the basis functions are not too complicated. It is equally valid to employ numerical integration routines in standard mathematical software. In any event, it is evident that the procedures for determining the inertia, stiffness, damping, and generalized forces are quite different from those for a discrete system. However, the associated expressions for T, V, $\mathcal{P}_{dis}$, and $\mathcal{P}_{in}$ have the same form as the corresponding terms for a discrete N-degree-of-freedom system. It follows that the equations of motion governing the generalized coordinates are

$$[M]\{\ddot{q}\} + [C]\{\dot{q}\} + [K]\{q\} = \{Q\}$$

(6.1.16)

Consequently, analysis of the response of a continuous system according to the Ritz method differs from an analysis of discrete systems solely in the manner in which the system coefficients are generated. This is one of the attractive features of the method. Of course, we must use the q_j values in conjunction with the Ritz series, eq. (6.1.8), to evaluate the displacement field u at locations and instants of interest.

A useful application of these concepts applies to situations where we wish to develop a discrete model of a system, but we believe that a spring has sufficient mass that its inertial effects cannot be ignored. We may develop a simple correction by considering the spring to be like an elastic bar undergoing extensional deformation. Such a situation is depicted in Figure 6.4. For the sake of simplicity, we only consider the case where both ends of the spring move axially. We let u_1 and u_2 denote the end displacements. The deformation of the spring is $\Delta = u_2 - u_1$, so the potential energy is $V = \frac{1}{2}k(u_2 - u_1)^2$, as usual. To describe the kinetic energy, we introduce an approximation that the displacement depends linearly on the distance s from the left end, so that

$$u(s) = u_1 + \frac{s}{L}(u_2 - u_1)$$

(6.1.17)

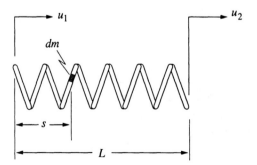

FIGURE 6.4 Evaluation of kinetic energy for an elastic spring.

When we collect the coefficients of u_1 and u_2, we see that this has the form of a Ritz series,

$$u(s) = \psi_1(s)u_1 + \psi_2(s)u_2, \qquad \psi_1(s) = 1 - \frac{s}{L}, \qquad \psi_2(s) = \frac{s}{L} \qquad (6.1.18)$$

The mass per unit length of the spring is its total mass m_{sp} divided by its unstretched length L. Thus, we may evaluate its kinetic energy by employing eqs. (6.1.10) and (6.1.11), with ρA replaced by $(m_{sp}/L)ds$. The approximate kinetic energy of the spring that results is

$$T_{sp} = \tfrac{1}{2}(M_{11}\dot{u}_1^2 + 2M_{12}\dot{u}_1\dot{u}_2 + M_{22}\dot{u}^2) \qquad (6.1.19)$$

where the inertia coefficients are

$$M_{11} = \int_0^L \left(1 - \frac{s}{L}\right)^2 \frac{m_{sp}}{L}\, ds = \frac{1}{3}m_{sp} = M_{22}$$

$$ \qquad (6.1.20)$$

$$M_{12} = \int_0^L \left(\frac{s}{L}\right)\left(1 - \frac{s}{L}\right)\frac{m_{sp}}{L}\, ds = \frac{1}{6}m_{sp}$$

If u_1 and u_2 are actually generalized coordinates for the system, these inertia coefficients may be added directly to the contributions from other bodies.

At this juncture, we return to the central issue in the Ritz series method—selection of the basis functions. In addition to the requirement that the functions be independent, another obvious condition stems from the observation that the displacement u must be a continuous function of x. (A discontinuous u would imply that the bar has been severed.) Thus, *all basis functions used to describe axial displacement must be continuous.*

To identify the last requirement, we consider the manner in which the bar is supported, which we have not yet considered. A support constrains the displacement of the bar. Such motion restrictions are imposed by a corresponding constraint force. Figure 6.5 depicts a bar whose ends are fixed. Correspondingly, (unknown) constraint forces F_1 and F_2 act at the respective ends to prevent displacement, $u(0, t) = u(L, t) = 0$. (These forces are the resultants of the stress distribution acting over the cross-section at each end.) Obviously, the velocity of each end is also zero, so the forces F_1 and F_2 do not contribute to the power input. However, eq. (6.1.15) indicates that both F_1 and F_2 will contribute to $\mathcal{P}_{in}$ if the basis functions are allowed to have arbitrary values at $x = 0$ and $x = L$. We therefore must impose the restriction that the basis functions vanish at any end that is fixed,

$$\boxed{\psi_j = 0 \text{ at any end where axial motion is prevented}} \qquad (6.1.21)$$

Thus, the basis functions we would select to describe the bar in Figure 6.5 must satisfy $\psi_j(0) = \psi_j(L) = 0$.

$x = 0$ $x = L$

FIGURE 6.5 Physical supports and constraint forces for axial motion of a bar.

This result exemplifies a fundamental aspect of Lagrange's equations and the power balance formulation. According to the development in Appendix A, constraint forces do no work in a virtual displacement that is consistent with the kinematical restriction they impose. It also is shown in Appendix A that virtual displacement is analogous to velocity in situations where the generalized coordinates are measured from a stationary reference position. Therefore, the constraint forces should not contribute to the power input, which is a condition that can be attained only if the basis functions satisfy the displacement conditions imposed by the constraint forces. The kinematical restriction on displacement imposed by a constraint force is called a *geometric boundary condition*. Regardless of the type of continuum under consideration, the Ritz series method will always require that

The basis functions must satisfy all geometric boundary conditions.

Satisfaction of this requirement ensures that the Ritz series does not result in power input from the constraint forces.

A corollary of this specification is a caution to select a set of basis functions such that at least some of the basis functions satisfy *only* those geometric boundary conditions that actually apply. In other words, if an end is *not* prevented from moving, we should select some or all basis functions that are nonzero at that location. In part, this recommendation follows from the fact that an essential difference between two bars is the manner in which they are supported. Selecting basis functions that satisfy geometric conditions not associated with the system will mask such differences. Furthermore, it will have the physical effect of making the system seem to be stiffer than it actually is. In addition, if the basis functions unnecessarily satisfy a geometric boundary condition at some location, then the functions are likely to be comparatively small in the vicinity of that location. This would tend to account insufficiently for any external forces acting in the vicinity of that location.

A displacement field that satisfies the geometric boundary conditions is said to be *kinematically admissible*, which refers to the fact that such a displacement could actually be obtained if the external force system were suitably selected. In the next chapter, which develops and solves the field equations, we will encounter another type of boundary condition, which arises when an internal force resultant within the bar is specified at some location. Such criteria constitute *natural boundary conditions*. Basis functions that satisfy all natural boundary conditions, as well as the geometric conditions, are called *comparison functions*. The identification of a set of comparison functions can be difficult. Fortunately, it is not necessary to satisfy the natural conditions when we use the Ritz series method, because their effect will be contained in the potential and kinetic energies.

An important consideration in selecting basis functions is whether rigid body motion is kinematically admissible. If it is, we should include basis functions that describe such a motion. By definition, all strain components are zero in a rigid body motion. For axial displacement, this requires that u is independent of x, which corresponds to a translation in the axial direction. Such a motion is possible if neither end is prevented from moving. The associated basis function to be included in the Ritz series in this case is $\psi_1 = 1$.

Note that we are drawing a distinction here between systems where rigid body motion is merely kinematically admissible and those in which it actually may occur. As we saw for discrete systems, a system that is truly capable of rigid body motion will have one or more natural frequencies that are zero. The potential energy in such systems is only positive semidefinite, meaning that it is possible to move the system without storing energy. If an axially aligned spring is attached to a bar whose ends are unconstrained, rigid body motion would not actually occur, because any tendency of the bar to move as a rigid body will lead to deformation of the spring, which in turn will produce a force on the bar that causes it to deform. Nevertheless, we would use $\psi_1 = 1$ as one of the basis functions for such a system. In part, including $\psi_1 = 1$ assures that at least one basis function does not satisfy a geometric boundary condition that is inappropriate to the system under consideration. There also is a more subtle reason to include a basis function corresponding to rigid body motion when such motion is kinematically admissible. When one begins the analysis, it will not be apparent how stiff any attached springs might be relative to the other effects. Picking one function to be a rigid body mode ensures that there is at least one stiffness coefficient in which the energy stored in attached springs is not overwhelmed by deformation effects within the bar.

For any system, a variety of types of functions fit all criteria required of basis functions. Another consideration stems from solutions of the field equations. In particular, we will see in the next chapter that the lowest frequency modes have the most gradual variation in the axial direction, essentially because they contain the least amount of potential energy for a given amount of kinetic energy. Thus, the basis functions we select should be such that the lowest numbered ones correspond to the most gradual change in the axial direction consistent with the boundary conditions. Consider this criterion in the case of a fixed-free bar, for which there is one geometric boundary condition. If $x = 0$ is the fixed end, then $\psi_j(0) = 0$. Power series terms $(x/L)^r$ fit this boundary condition if r is an integer greater than zero. (We have nondimensionalized the terms relative to the length L for the sake of making all basis functions have the same order of magnitude.) The condition that the lowest-order functions should vary most gradually leads us to select the power r to be j, so that $\psi_j = (x/L)^j$.

The preceding typifies a parameterized family of basis functions, in which the index j of the basis function defines a particular member of the family. Power series terms are used commonly as basis functions when geometric boundary conditions apply at one end only, which can be designated $x = 0$. When there also are conditions to satisfy at $x = L$, we need a family of functions having a sufficient number of free parameters to satisfy all conditions. It is possible to use polynomials for this purpose, but there are potential problems in doing so. Most important is the issue of completeness. Another possible difficulty is that the selected set of functions actually are not independent.

One way of ensuring linear dependence is to select basis functions that have been successful for similar situations. Here Appendix C, which lists modal solutions for some standard problems involving uniform beams, can be invaluable. For axial motion there are only three possibilities regarding geometric boundary conditions: both ends fixed, one end fixed, or neither end fixed. Regardless of the nature and location of devices that are attached to the bar, and whether the bar's cross-sectional properties are constant, we may select the basis functions to match the entry in Appendix C that has the same set of geometric boundary conditions. The entries for axial vibration modes in Appendix C in all cases are either sine or cosine functions, and their period in the axial direction is either $2L$ or $4L$. Thus, these modal solutions suggest basis functions having the form

$$\psi_j = \sin\left(\frac{j\pi x}{L}\right), \cos\left(\frac{j\pi x}{L}\right), \sin\left(\frac{j\pi x}{2L}\right), \text{ or } \cos\left(\frac{j\pi x}{2L}\right) \qquad (6.1.22)$$

Selecting basis functions in this manner is equivalent to using half-range Fourier series to represent the axial displacement.

EXAMPLE 6.1

A concentrated force $F(t)$ is applied at the midpoint of the bar. The bar is aluminum, with a 60 mm diameter circular cross-section. The shock absorber at the right end consists of a spring and dashpot acting in parallel, with $k = 100$ MN/m and $c = 4$ kN-s/m. Derive the coefficient matrices for the equations of motion associated with a Ritz series representation using three sinusoidal basis functions.

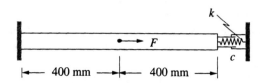

Solution This example addresses the basic operations required to implement a Ritz series analysis of continuous systems, including the considerations entailed in selecting basis functions. We define the left end as $x = 0$ and take axial displacement to be positive rightward. The only geometric boundary condition is that $u = 0$ at $x = 0$. It is stipulated that the basis functions should be sinusoidal, so we shall select a family of functions having the form $\sin(\alpha x)$. Because the displacement is nonzero at $x = L = 0.8$ m, a convenient family of functions sets $\psi_j = 1$ at $x = L$. This leads to the condition that αL should be an odd multiple of $\pi/2$, so we select

$$\psi_j = \sin\left[\frac{(2j - 1)\pi x}{2L}\right], \quad j = 1, 2, 3$$

After we have evaluated the basis function coefficients q_j, which serve as the generalized coordinates, we may reconstruct the displacement at any location from the Ritz series,

$$u(x, t) = \sum_{j=1}^{3} \sin\left[\frac{(2j - 1)\pi x}{2L}\right] q_j(t)$$

No concentrated masses are attached to the bar, so the inertia coefficients are given by eq. (6.1.11) to be

$$M_{jn} = \int_0^L \psi_j \psi_n \rho A \, dx = \rho A \int_0^L \sin\left[\frac{(2j - 1)\pi x}{2L}\right] \sin\left[\frac{(2n - 1)\pi x}{2L}\right] dx$$

$$= \begin{cases} \frac{1}{2}\rho AL & \text{if } j = n \\ 0 & \text{if } j \neq n \end{cases}$$

A spring is attached to the bar at $x = L$, so the stiffness coefficients obtained from eq. (6.1.13) are

$$K_{jn} = \int_0^L EA \frac{d\psi_j}{dx} \frac{d\psi_n}{dx} dx + k\psi_j(L)\psi_n(L)$$

$$= EA \frac{(2j - 1)(2n - 1)\pi^2}{4L^2} \int_0^L \cos\left[\frac{(2j - 1)\pi x}{2L}\right] \cos\left[\frac{(2n - 1)\pi x}{2L}\right] dx$$

$$+ k\sin\left[\frac{(2j - 1)\pi}{2}\right] \sin\left[\frac{(2n - 1)\pi}{2}\right]$$

$$= \begin{cases} \dfrac{1}{8} \dfrac{\pi^2 (2j - 1)^2 EA}{L} + k & \text{if } j = n \\ (-1)^{j+n} k & \text{if } j \neq n \end{cases}$$

The effect of the dashpot at $x = L$ appears in the damping coefficients, which are found from eq. (6.1.14) to be

$$C_{jn} = c\psi_j(L)\psi_n(L) = (-1)^{j+n}c$$

The last step in evaluating the basic terms for the equations of motion is to characterize the generalized forces associated with the concentrated force at $x = L/2$. The power it inputs to the system is $F\dot{u}(L/2,t)$, so we have

$$Q_J = F\psi_j\left(\frac{L}{2}\right) = F\sin\left[\frac{(2j-1)\pi}{4}\right]$$

We next evaluate the various terms by substituting the specific parameters of the system. The density of aluminum is $\rho = 2700$ kg/m^3, and its Young's modulus is $6.9(10^{10})$ Pa. The cross-sectional area corresponding to a 60 mm diameter is 0.002827 m^2. In combination with the given values of L, k, c, and F, the matrices for the set of equations of motion are

$$[M] = \begin{bmatrix} 3.0536 & 0 & 0 \\ 0 & 3.0536 & 0 \\ 0 & 0 & 3.0536 \end{bmatrix}, \quad [K] = \begin{bmatrix} 0.4009 & -0.10 & 0.10 \\ -0.10 & 2.8077 & -0.10 \\ 0.10 & -0.10 & 7.6214 \end{bmatrix}(10^9)$$

$$[C] = \begin{bmatrix} 4000 & -4000 & 4000 \\ -4000 & 4000 & -4000 \\ 4000 & -4000 & 4000 \end{bmatrix}, \quad \{Q\} = 0.7071F\begin{Bmatrix} 1 \\ 1 \\ -1 \end{Bmatrix}$$

The next step after the derivation of the equations of motion would be to solve them, and then use that solution to construct the displacement. These are issues we begin to address in Section 6.1.4.

EXAMPLE 6.2

An elastic bar having cross-sectional area A and density ρ is suspended from block m by spring k_2, and the block is suspended from the ceiling by spring k_1. Force F acts at the bottom of the bar. Develop a three-term Ritz series describing the vertical movement of the system.

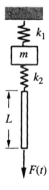

Solution In this exercise we will merge the Ritz series analysis of continuous systems (the bar) with our standard procedure for discrete systems (the block). The first step in the analysis is to recognize that the block is not rigidly fastened to the bar, so its vertical displacement must be described by a generalized coordinate q_4 separate from the three Ritz series coefficients. Thus, the vertical displacement, which we take to be positive downward, is described by

$$\text{Bar:} \quad u = \sum_{j=1}^{3} \psi_j q_j; \qquad \text{Block:} \quad u = q_4$$

It is useful to defer selection of the basis functions ψ_j for this system until we have developed general expressions for the inertia and stiffness coefficients.

The derivation of the inertia and stiffness matrices did not address the situation for this system, so we shall retrace those steps. The kinetic energy is the sum of contributions from the bar and the block, so we have

$$T = \frac{1}{2}\int_0^L \dot{u}^2 \rho A \, dx + \frac{1}{2} m \dot{q}_4^2$$

We substitute the Ritz series into the integrand and bring the summations outside the integral, which leads to

$$T = \frac{1}{2}\sum_{j=1}^{3}\sum_{n=1}^{3}\int_0^L [\psi_j(x)\dot{q}_j(t)][\psi_n(x)\dot{q}_n(t)]\rho A \, dx + \frac{1}{2} m \dot{q}_4^2$$

This system has four generalized coordinates, so the standard form of the kinetic energy is

$$T = \frac{1}{2}\sum_{j=1}^{4}\sum_{n=1}^{4} M_{jn} \dot{q}_j \dot{q}_n$$

We match this to the form of T specific to this system, which entails collecting the coefficients of each $\dot{q}_j \dot{q}_n$ term in the latter expression. This leads to

$$M_{jn} = \int_0^L \psi_j \psi_n \rho A \, dx \text{ if } j, n = 1, 2, \text{ or } 3$$

$$M_{44} = m$$

$$M_{jn} = 0 \text{ otherwise}$$

To obtain the stiffness coefficients, we construct the potential energy as the sum of the contributions of the bar's deformation and those from each of the springs. The elongation of k_1 is the vertical displacement of the block, while the elongation of k_2 is the difference between the downward displacement of the upper end of the bar and of the block,

$$\Delta_1 = q_4, \qquad \Delta_2 = u(x = 0, t) - q_4$$

Note that the role of gravity is to induce a static displacement within the bar and the springs. We take the static equilibrium position to be the reference state, so we ignore the dynamic effect of gravity. Therefore, the potential energy is

$$V = V_{\text{bar}} + V_1 + V_2$$

$$= \frac{1}{2}\int_0^L EA\left(\frac{\partial u}{\partial x}\right)^2 dx + \frac{1}{2}k_1 q_4^2 + \frac{1}{2}k_2[u(0, t) - q_4]^2$$

$$= \frac{1}{2}\int_0^L EA\left(\frac{\partial u}{\partial x}\right)^2 dx + \frac{1}{2}(k_1 + k_2)q_4^2 + \frac{1}{2}k_2 u(0, t)^2 - k_2 u(0, t)q_4$$

We use the Ritz series to represent u anywhere it occurs in this expression. Bringing the summations outside the integral leads to

$$V = \frac{1}{2}EA\sum_{j=1}^{3}\sum_{n=1}^{3}\int_0^L \left(\frac{d\psi_j}{dx}q_j\right)\left(\frac{d\psi_n}{dx}q_n\right)dx + \frac{1}{2}(k_1 + k_2)q_4^2$$

$$+ \frac{1}{2}k_2\sum_{j=1}^{3}\sum_{n=1}^{3}(\psi_j(0)q_n)(\psi_n(0)q_j) - k_2\left(\sum_{j=1}^{3}\psi_j(0)q_j\right)q_4$$

An important aspect of the above step is the use of different subscripts to form each term in a square, which is done in order to ensure that we account for all combinations of terms. This description of V must match the standard form

$$V = \frac{1}{2}\sum_{j=1}^{4}\sum_{n=1}^{4} K_{jn}q_j q_n$$

The matching process, as always, must recognize that the terms for which $j \neq n$ occur twice, which leads to

$$K_{jn} = \int_0^L EA \frac{d\psi_j}{dx}\frac{d\psi_n}{dx} dx + k_2\psi_j(0)\psi_n(0), \quad j, n = 1, 2, 3$$

$$K_{44} = k_1 + k_2$$

$$K_{j4} = -k_2\psi_j(0), \quad j = 1, 2, 3$$

The remaining terms required to form the equations of motion are simpler to obtain. Because the system has no energy dissipation devices, we have $[C] = [0]$. The generalized force terms come from the power input to the system by the force F, which acts at the bottom end. The displacement at that location is $u(L, t) = \sum \psi_j(L)q_j$. The corresponding expression for the power is

$$\mathcal{P}_{in} = F\dot{u}(L, t) = F\sum_{j=1}^{3}\psi_j(L)\dot{q}_j = \sum_{j=1}^{4}Q_j\dot{q}_j$$

We match these alternative descriptions of $\mathcal{P}_{in}$, which leads to

$$Q_j = F\psi_j(L), \quad j = 1, 2, 3; \quad Q_4 = 0$$

To select the basis functions ψ_j of the Ritz series, we observe that neither end of the bar is prevented from moving, so rigid body translation is kinematically admissible. For that reason we select the first function to be $\psi_1 = 1$. Because there are no geometric boundary conditions to satisfy, we may select any set of basis functions to represent the deformation of the bar. Let us select power series terms. Specifically, with the upper end of the bar defined as $x = 0$, we take $\psi_j = (x/L)^{j-1}, j = 1, 2, 3$. Although ψ_2 and ψ_3 satisfy the geometric boundary condition $u = 0$ at $x = 0$, which does not describe this system, this choice is acceptable, because ψ_1 does not satisfy that inappropriate condition.

The integrals for the inertia and stiffness coefficients associated with our choice of basis functions are readily evaluated. To write the results in a more compact form, we let $\mu = \rho AL$ and $\kappa = EA/L$, which leads to

$$[M] = \begin{bmatrix} \mu & 0.5\mu & 0.333\mu & 0 \\ 0.5\mu & 0.333\mu & 0.25\mu & 0 \\ 0.333\mu & 0.25\mu & 0.2\mu & 0 \\ 0 & 0 & 0 & m \end{bmatrix}$$

$$[K] = \begin{bmatrix} k_2 & 0 & 0 & -k_2 \\ 0 & \kappa & \kappa & 0 \\ 0 & \kappa & 1.333\kappa & 0 \\ -k_2 & 0 & 0 & k_1 + k_2 \end{bmatrix}$$

$$\{Q\} = \begin{bmatrix} F & F & F & 0 \end{bmatrix}^T$$

Suppose we had omitted $\psi_1 = 1$ by selecting the basis function set as $(x/L)^j$, $j = 1, 2, 3$. We would have found that all $K_{j4} = K_{4j}$ terms are zero. The expressions for M_{jn} and K_{jn} derived above indicate that the only terms coupling q_4 to q_1, q_2, and q_3 are the off-diagonal terms $K_{j4} = K_{4j}$. Thus, if these stiffness terms were zero, the equations of motion would indicate that the block is not excited by motion of the bar, which obviously is not correct.

6.1.2 Torsion of Circular Bars

Shafts having a circular cross-section are a common feature of rotating machinery in which torsional loads are transferred between system components. We shall limit our attention to circular cross-sections in order to avoid the complications of the theory of torsion for noncircular cross-sections. (See Timoshenko and Goodier, 1970, for a discussion of this theory.)

Figure 6.6 depicts a bar cross-section that has rotated through angle θ about the bar's axis. The excitation consists of a distributed torsional moment τ, whose units are moment per unit length. The cross-sections of a circular bar rotate as a rigid body, but that rotation will depend on the axial location x. The highlighted segment in Figure 6.6 covers a distance dx, over which the rotations differ infinitesimally. Furthermore, because the segment is a thin disk, its centroidal mass moment of inertia is $\rho J dx$, where $J = \frac{1}{2}\pi R^4$ is the area polar moment of inertia of the cross-section. Then the kinetic energy of the segment is $\frac{1}{2}\dot{\theta}^2 \rho J \, dx$. We integrate over the length of the beam to obtain the total kinetic energy of the bar, to which we add the rotational kinetic energy of any attached rigid bodies. Thus,

$$T = \frac{1}{2}\int_0^L \dot{\theta}^2 \rho J \, dx + \frac{1}{2}\sum I_{xx}\dot{\theta}^2(x_I, t)^2 \tag{6.1.23}$$

where I_{xx} is the mass moment of inertia of a body situated at x_I.

To derive the strain energy we use cylindrical coordinates defined relative to the x axis. Each cross-section rotates as a rigid body, so the circumferential displacement of a point at distance r from the x axis is $r\theta$. This gives rise to a circumferential shear strain on the cross-section given by $\varepsilon_{x\theta} = \frac{1}{2}r\,\partial\theta/\partial x$. To obtain the contribution of this strain to the potential energy, we integrate the strain energy per unit volume $2G\varepsilon_{x\theta}^2$, where G is the elastic shear modulus, over the cross-section and the length of the bar. In addition, torsional springs κ (units are moment per radian) exerting moments about the x axis might be attached at a variety of locations x_κ, so the total potential energy is

$$V = \frac{1}{2}\int_0^L GJ\left(\frac{\partial\theta}{\partial x}\right)^2 dx + \frac{1}{2}\sum \kappa\theta(x_\kappa, t)^2 \tag{6.1.24}$$

Dissipation has a similar form to V, with a viscoelastic coefficient γG replacing G and torsional damping constants replacing spring stiffnesses. Hence, the power dissipation function is

$$\mathcal{P}_{\text{dis}} = \frac{1}{2}\int_0^L \gamma GJ\left(\frac{\partial\dot{\theta}}{\partial x}\right)^2 dx + \frac{1}{2}\sum \chi\dot{\theta}(x_\chi, t)^2 \tag{6.1.25}$$

where χ are torsional damping constants (dimensions of moment per unit angular velocity) for dashpots whose locations are x_χ.

To form the power input by the external forces we observe that τdx is the moment about the x axis exerted by the distributed torsional moment on a segment dx. Because the angular velocity of the segment about this axis is $\dot{\theta}$, the contribution of this segment to $\mathcal{P}_{\text{in}}$ is $\dot{\theta}\tau dx$. To this we add the contribution of all concentrated torsional moments Γ acting about the x axis at locations x_γ, which leads to

$$\mathcal{P}_{\text{in}} = \int_0^L \tau\dot{\theta}\, dx + \sum \Gamma\dot{\theta}(x_\gamma, t) \tag{6.1.26}$$

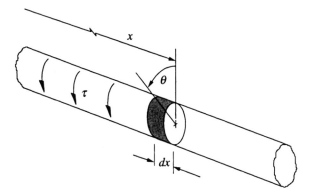

FIGURE 6.6 Torsional deformation of a straight bar.

The Ritz series representation of the cross-sectional rotation field has the standard form of eq. (6.1.8), in which the individual terms are products of basis functions $\psi_j(x)$ and coefficients $q_j(t)$ that serve as generalized coordinates,

$$\theta(x, t) = \sum_{j=1}^{N} \psi_j(x)q_j(t) \tag{6.1.27}$$

(Here, defining the basis functions nondimensionally leads to q_j having units of radians.)

The important aspect of the preceding expressions for T, V, $\mathcal{P}_{\text{dis}}$, $\mathcal{P}_{\text{in}}$, and the Ritz series is their similarity to the counterparts for axial displacement. Specifically, θ for torsion is analogous to u for axial displacement, with G and J replacing E and A, respectively. The inertial properties of attached masses for torsional motion are I, rather than m, and the torsional spring constant κ replaces k. For external power input, τ replaces f_x and Γ replaces F.

The similarity between extension and torsion extends to the geometric boundary condition. If an end is fixed, for example by welding it to a wall, a torsional couple enforces the constraint against rotation. This couple should not contribute to $\mathcal{P}_{\text{in}}$. Hence, it is necessary that all basis functions satisfy the geometric boundary conditions,

$$\psi_j = 0 \text{ at any end where torsional rotation is prevented} \tag{6.1.28}$$

As a result of the similarity between axial displacement and torsional rotation, eq. (6.1.16) also represents the equations of motion for torsional motion. We may exploit the similarity to express, without derivation, the inertia, stiffness, and dissipation constants and the generalized forces,

$$
\begin{aligned}
M_{jn} = M_{nj} &= \int_0^L \psi_j \psi_n \rho J \, dx + \sum I \psi_j(x_I) \psi_n(x_I) \\
K_{jn} = K_{nj} &= \int_0^L GJ \frac{d\psi_j}{dx} \frac{d\psi_n}{dx} \, dx + \sum \kappa \psi_j(x_\kappa) \psi_n(x_\kappa) \\
C_{jn} = C_{nj} &= \int_0^L \gamma GJ \frac{d\psi_j}{dx} \frac{d\psi_n}{dx} \, dx + \sum \chi \psi_j(x_\chi) \psi_n(x_\chi) \\
Q_j &= \int_0^L \tau \psi_j \, dx + \sum \Gamma \psi_j(x_\gamma)
\end{aligned}
\tag{6.1.29}
$$

The corresponding equations of motion have the standard form $[M]\{\ddot{q}\} + [C]\{\dot{q}\} + [K]\{q\} = \{Q\} = [K]\{q\} = \{Q\}$.

The case of rigid body motion in torsion corresponds to a spinning motion about the longitudinal axis. In that case, θ is independent of x, as it would be for a rotating shaft. As before, rigid body motion is kinematically admissible if neither end is constrained. The basis function to be included in such systems is $\psi_1 = 1$, even if attached torsional springs actually prevent the bar from rotating as a rigid body.

It is interesting that another vibratory system has analogous behavior to that of axial displacement of bars. Exercise 6.10 and Example 11.6 discuss the transverse vibration of highly tensioned cables. The analysis for each reveals that the displacement w perpendicular to the cable is governed by mathematical forms similar to those governing u. To use the analogy one merely needs to replace the extensional rigidity EA with the cable tension F.

EXAMPLE 6.3

A circular rod is supported by ball bearings that do not hinder its torsional rotation. At $x = 0$, there is a torsional spring $\kappa = 2GJ/L$, and a specified couple Γ is applied at $x = L$. A flywheel is attached to the bar at the midpoint, $x = L/2$. The mass of the flywheel equals the total mass of the rod, and the radius of gyration of the flywheel about the centerline of the rod is three times that of the rod. Evaluate the coefficient matrices associated with a four-term Ritz series describing the torsional motion of the bar.

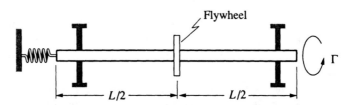

Solution This example will shed further light on the selection of basis functions and the process by which the equations of motion are formulated. We shall measure x from the left end. Both ends of the bar may rotate, so there are no geometric boundary conditions to satisfy. Thus, any power of x is kinematically admissible, and therefore a suitable basis function. In order to be sure that we account for the effect of the torsional spring at the left end, we select rigid body rotation as the first function, $\psi_1 = 1$. Hence, our choice for the four basis functions is $\psi_j = (x/L)^{j-1}, j = 1, 2, 3, 4$. The corresponding Ritz series for the torsional rotation is

$$\theta = \sum_{j=1}^{4} \left(\frac{x}{L}\right)^{j-1} q_j$$

The bar and the flywheel store kinetic energy,

$$T = \frac{1}{2}\int_0^L \dot{\theta}^2 \rho J \, dx + \frac{1}{2}I_{xx}\dot{\theta}(L/2, t)^2$$

while potential energy stored in the bar and the torsional spring is

$$V = \frac{1}{2}\int_0^L GJ\left(\frac{\partial\theta}{\partial x}\right)^2 dx + \frac{1}{2}\kappa\theta(0, t)^2$$

It was stated that the bearings do not impede the rotation of the bar, so damping is negligible, $[C] = [0]$. Power is input to the system at the right end, where the external torque is Γ and the rotation rate is $\dot{\theta}(L, t)$, so that

$$\mathscr{P}_{\text{in}} = \Gamma\dot{\theta}(L, t)$$

A comparison of these energy and power expressions to the standard ones enables us to identify the specific terms in eqs. (6.1.29) contributing to the system matrices. The inertia coefficients are

$$M_{jn} = \int_0^L \psi_j \psi_n \rho J \, dx + I \psi_j(L/2) \psi_n(L/2) = \rho J \int_0^L \left(\frac{x}{L}\right)^{j+n-2} dx + I_{xx}\left(\frac{1}{2}\right)^{j+n-2}$$

For a circular rod, the area moment of inertia is $J = \frac{1}{2}\pi R^4$, the mass is $m = \rho\pi R^2 L$, and the mass moment of inertia is $\frac{1}{2}mR^2$. The radius of gyration of the flywheel is three times that of the bar, so $I_{xx} = \frac{9}{2}mR^2$. Thus, we have

$$M_{jn} = \frac{\pi}{2}\left[\left(\frac{1}{j+n-1}\right) + 9\left(\frac{1}{2}\right)^{j+n-2}\right]\rho R^4 L, \quad j, n = 1, 2, 3, 4$$

The torsional spring stiffness is $\kappa = 2GJ/L$, so

$$K_{jn} = \int_0^L GJ \frac{d\psi_j}{dx}\frac{d\psi_n}{dx} dx + \kappa \psi_j(0)\psi_n(0)$$

$$= \begin{cases} \kappa & \text{if } j = n = 1 \\ \dfrac{GJ}{L^2}\displaystyle\int_0^L (j-1)(n-1)\left(\frac{x}{L}\right)^{j+n-4} dx & \text{if } j \geq 2 \text{ and } n \geq 2 \\ 0 & \text{otherwise} \end{cases}$$

$$= \begin{cases} \pi\dfrac{GR^4}{L} & \text{if } j = n = 1 \\ \dfrac{\pi(j-1)(n-1)}{2(j+n-3)}\dfrac{GR^4}{L} & \text{if } j \geq 2 \text{ and } n \geq 2 \\ 0 & \text{otherwise} \end{cases}$$

As an aside, note that the only term in which the effect of the torsion spring appears is K_{11}, which is associated with ψ_1. This demonstrates why one needs to include a basis function describing rigid body motion whenever such motion is kinematically admissible. The generalized forces describe the power input at $x = L$ by Γ, specifically

$$Q_j = \Gamma \psi_j(L) = \Gamma$$

Because the basic system parameters are not given, it is useful to factor them out of the coefficient matrices. The corresponding equations of motion are

$$\rho R^4 L[M']\{\ddot{q}\} + \frac{GR^4}{L}[K']\{q\} = \{Q\}$$

where $[M']$ and $[K']$ are the numerical factors of the respective matrices,

$$[M'] = \begin{bmatrix} 15.708 & 7.854 & 4.058 & 2.160 \\ 7.854 & 4.058 & 2.160 & 1.198 \\ 4.058 & 2.160 & 1.198 & 0.704 \\ 2.160 & 1.198 & 0.704 & 0.445 \end{bmatrix}, \quad [K'] = \begin{bmatrix} 3.142 & 0 & 0 & 0 \\ 0 & 1.571 & 1.571 & 1.571 \\ 0 & 1.571 & 2.094 & 2.356 \\ 0 & 1.571 & 2.356 & 2.827 \end{bmatrix}$$

6.1.3 Flexural Displacement

Flexure of a bar gives rise to greater complexity in the fundamental energy expressions. Our study of this important type of motion will lead us to a broader view of geometric boundary conditions. The flexure of a beam is depicted in Figure 6.7. The

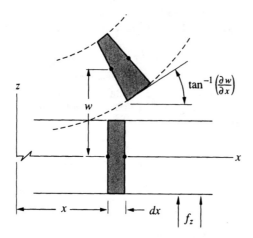

FIGURE 6.7 Flexural deformation of a bar.

figure describes displacement $w(x, t)$ of points on the x axis in response to a transverse distributed force $f_z(x, t)$ acting in the positive z direction. If the bar is symmetric relative to the xz plane, meaning that cross-sections on either side of the xz plane are mirror images, then the displacement w lies in that plane. Each cross-section rotates so as to remain perpendicular to the deformed position of the centroidal x axis. In the linear approximation, the displacement w of centroidal points is very small relative to the length of the bar. Correspondingly the slope $\partial w/\partial x$ is essentially the rotation of the cross-section, so the axial displacement at distance z from the x axis is $-z(\partial w/\partial x)$. The fact that $\partial w/\partial x$ itself may vary from one location to another leads to an axial strain $\varepsilon_{xx} = -z(\partial^2 w/\partial x^2)$.

The strain energy of the bar is obtained by integrating $\frac{1}{2}\sigma_{xx}\varepsilon_{xx}$ over a cross-section and over the length, which gives

$$V_{\text{bar}} = \frac{1}{2}\int_0^L EI\left(\frac{\partial^2 w}{\partial x^2}\right)^2 dx \qquad (6.1.30)$$

where I is the second moment of area about an axis perpendicular to the xz plane and intersecting the centroid of the cross-section.

In order to express the kinetic energy, we observe in Figure 6.7 that the difference between the rotations at the two ends of the dx segment is infinitesimal. Hence, in a first-order approximation the segment, whose mass is $\rho A\, dx$, moves as a rigid body undergoing transverse displacement w and rotation $\partial w/\partial x$. The rotational kinetic energy generally is unimportant, unless one considers very high-frequency motions. As long as we limit the frequencies to a certain upper limit, we may ignore rotatory inertia. (The theory of flexure employed here becomes invalid at high frequencies. Timoshenko beam theory, developed in Section 7.9, addresses such situations.) Consequently, the kinetic energy of the bar is taken to be

$$T_{\text{bar}} = \frac{1}{2}\int_0^L \dot{w}^2\rho A\ dx \qquad (6.1.31)$$

The power input to the bar by the distributed force is like the corresponding quantity for axial motion. The resultant force acting on the differential segment of the bar is $f_z\, dx$ acting in the z direction, and the overall velocity of the segment is $\dot{w}$, which leads to $\dot{w}f_z\, dx$ as the contribution of this segment to $\mathcal{P}_{\text{in}}$. The force resultants acting on an exposed cross-section in flexure are a bending

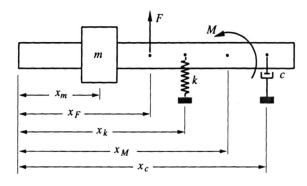

FIGURE 6.8 Discrete attachments and concentrated loads acting on a bar undergoing flexural displacement.

moment about the y axis coinciding with the cross-section and a shear force acting in the xz plane. The analogous loadings acting on the bar are concentrated transverse forces F at locations x_F and couples M at locations x_M, as shown in Figure 6.8. These effects are included in the formulation by adding them to $\mathcal{P}_{\text{in}}$ for the distributed force. By convention, a transverse force is positive if it acts in the sense of positive w, and a couple is positive if it is in the same sense as a positive rotation $\partial w/\partial x$. The corresponding linear and angular velocity components are $\dot{w}$ and $\partial(\partial w/\partial x)/\partial t \equiv \partial \dot{w}/\partial x$, respectively, so the total power input to the system is

$$\mathcal{P}_{\text{in}} = \int_0^L f_z \dot{w}\, dx + \sum F\dot{w}(x_F, t) + \sum M\frac{\partial \dot{w}}{\partial x}(x_M, t) \qquad (6.1.32)$$

Attached masses, springs, and dashpots, which are shown in Figure 6.8, also exert transverse forces and couples on the bar. We describe these effects in terms of contributions to T, V, and $\mathcal{P}_{\text{dis}}$, respectively. A transverse force is associated with the translational motion of a mass, and a couple is associated with rotation. However, unless the mass is very large, we may neglect the rotational inertia effect. As a result, we shall model attached masses as particles m at location x_m that move at speed $\dot{w}(x_m, t)$ in unison with the bar. The corresponding total kinetic energy of the system is

$$T = \frac{1}{2}\int_0^L \dot{w}^2 \rho A \, dx + \frac{1}{2}\sum m\dot{w}(x_m, t)^2 \qquad (6.1.33)$$

Springs and dashpots may be of the extensional or torsional type. We obtain the total potential energy by adding the energy stored in each type of spring to the strain energy, which leads to

$$V = \frac{1}{2}\int_0^L EI\left(\frac{\partial^2 w}{\partial x^2}\right)^2 dx + \frac{1}{2}\sum kw(x_k, t)^2 + \frac{1}{2}\sum \kappa\left[\frac{\partial w}{\partial x}(x_\kappa, t)\right]^2 \qquad (6.1.34)$$

In flexure, the transverse motion of the bar can generate air resistance. A simple model for the fluid-resisting force is to consider it to be proportional to the transverse velocity, with factor of proportionality c_v. In addition, viscoelastic effects are represented by a term like the strain energy of the bar, with $\partial^2 w/\partial x^2$ replaced by $\partial^2 \dot{w}/\partial x^2$.

Adding these effects to the power dissipated by extensional dashpots c and torsional dashpots χ gives

$$
\mathcal{P}_{dis} = \int_0^L \left[\gamma EI \left(\frac{\partial^2 \dot{w}}{\partial x^2} \right)^2 + c_v \dot{w}^2 \right] dx + \sum c [\dot{w}(x_c, t)]^2
$$
$$
+ \sum \chi \left[\frac{\partial \dot{w}}{\partial x}(x_\chi, t) \right]^2 \tag{6.1.35}
$$

Once the mechanical energies, dissipation function, and power input have been characterized in terms of the displacement, the implementation of the Ritz method for flexure proceeds in the same manner as the previous cases. We substitute the basic series expansion, which has the same form as eq. (6.1.8),

$$
w(x, t) = \sum_{j=1}^{N} \psi_j(x) q_j(t) \tag{6.1.36}
$$

into each of the above expressions, and then collect like coefficients of the generalized coordinates q_j or generalized velocities $\dot{q}_j$. The resulting expressions for T, V, $\mathcal{P}_{dis}$, and $\mathcal{P}_{in}$ are like the earlier forms, except that the coefficients are now

$$
M_{jn} = M_{nj} = \int_0^L \psi_j \psi_n \rho A \, dx + \sum m \psi_j(x_m) \psi_n(x_m)
$$

$$
K_{jn} = K_{nj} = \int_0^L EI \frac{d\psi_j}{dx^2} \frac{d^2\psi_j}{dx^2} dx + \sum k \psi_j(x_k) \psi_n(x_k)
$$
$$
+ \sum \kappa \frac{d\psi_j}{dx}(x_\kappa) \frac{d\psi_n}{dx}(x_\kappa)
$$

$$
C_{jn} = C_{nj} = \int_0^L \left[\gamma EI \frac{d^2\psi_j}{dx^2} \frac{d^2\psi_j}{dx^2} + c_v \psi_j \psi_n \right] dx + \sum c \psi_j(x_c) \psi_n(x_c) \tag{6.1.37}
$$
$$
+ \sum \chi \frac{d\psi_j}{dx}(x_\chi) \frac{d\psi_n}{dx}(x_\chi)
$$

$$
Q_j = \int_0^L f_z \psi_j \, dx + \sum F \psi_j(x_F) + \sum M \frac{d\psi_j}{dx}(x_M)
$$

The two primary differences between the above coefficients and those for axial displacement are the presence of second-order derivatives and the fact that force and moment effects must both be considered. Once we have evaluated these coefficients, the response is governed by the standard differential equations of motion, $[M]\{\ddot{q}\} + [C]\{\dot{q}\} + [K]\{q\} = \{Q\}$.

As we did for axial and torsional motions, we identify the geometric boundary conditions by examining $\mathcal{P}_{in}$. We see in eq. (6.1.32) that two types of concentrated external loadings input power to a flexural motion: transverse forces and couples. At an end, the transverse force is equivalent to the shear force acting on a cross–section, and the couple is equivalent to the bending moment. Any support that prevents transverse displacement, $w = 0$, does so by exerting an unknown transverse force. It is also possible to prevent rotation at an end, $\partial w / \partial x = 0$, in which case the support exerts an unknown couple. Although either type of constraint force might be present, neither contributes to the power input because the corresponding motion term is zero. In order to assure that the formulation does not mistakenly lead to contributions of constraint forces to $\mathcal{P}_{\text{in}}$, we must impose the following geometric boundary conditions:

$$\psi_j = 0 \text{ at any location where displacement is prevented}$$

$$\frac{\partial \psi_j}{\partial x} = 0 \text{ at any location where rotation is prevented}$$

(6.1.38)

Four basic types of end conditions appear in Figure 6.9. A clamped (or fixed) end prevents both displacement and rotation, and a hinged (or pinned) end prevents displacement, but permits rotation. (A roller support imposes the same constraint as a hinge in linear theory. It is used to permit axial displacement resulting from reduction of the distance between the ends, which can have a substantial nonlinear effect when the displacements are large.) A guided end, which is not often encountered, may be obtained by welding the bar to a collar that is free to slide over a fixed guide. At such an end, the bar is prevented from rotating, but displacement is possible. At a free end, there is no constraint on either type of motion. In any situation, one should select basis functions that satisfy only those geometric boundary conditions that actually apply, as was true for axial displacement and torsional rotation.

Once again, our selection of basis functions must address whether rigid body motion is kinematically admissible. Rigid body motion leading to transverse displacement can be manifested as a translation or a rotation about some location $x = x_0$. In the former case, the displacement w is independent of x; the translational basis function is $\psi_{1,\text{tr}} = 1$. In the small displacement approximation of linear theory, a rigid body rotation leads to a transverse displacement that is proportional to the radial distance from the center of rotation. The basis function for rigid body rotation, written nondimensionally, is $\psi_{1,\text{rot}} = (x - x_0)/L$. Several configurations necessitate the inclusion of either or both functions. If a bar is supported only by a hinge at $x = x_0$, then rotation, but not translation, can occur, so only $\psi_{1,\text{rot}}$ is admissible. If only guided supports are present, then translation, but not rotation can occur, so only $\psi_{1,\text{tr}}$ is admissible. Both types of rigid body motion are kinematically admissible if the bar is not supported. In the last case, it is best to consider the rotation to occur relative to the center of mass, whose location we denote as x_G. Hence, we would use $\psi_{1,\text{tr}} = 1$ and

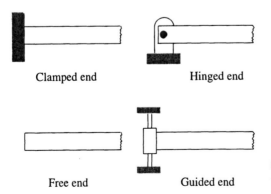

Clamped end   Hinged end

Free end   Guided end

FIGURE 6.9 Basic types of supports for a bar undergoing flexural displacement.

$\psi_{1,\text{rot}} = (x - x_G)L$ to model an unconstrained bar. As before, the decision as to which rigid body modes should be included ignores the presence of attached springs.

Because there are two types of constraint forces, it is possible to have a support placed between the ends of a bar without isolating the segments on either side. For example, a pin placed at the midpoint prevents that point from moving, but it does not prevent rotation. The rotation to the left of the pin must match the rotation to the right, for otherwise the bar would form a kink at which $\partial^2 w/\partial x^2$ is infinite, corresponding to an infinite bending moment. This feature is distinctly different from the axial and torsional cases considered earlier, in which constraining the motion at a location interior to the bar isolates each segment. We shall reserve consideration of such situations until Chapter 9, where we treat it in the context of the general topic of coupling of subsystems. However, one aspect of the consideration of intermediate supports is applicable to any situation, for it shows that *the basis functions for flexure must be continuously differentiable, rather than merely being continuous*. Any basis function that has a discontinuous first derivative would lead to an infinite flexural strain ε_{xx}.

The necessity to satisfy as many as two geometric boundary conditions at each end can complicate the task of selecting basis functions. Let us consider a beam having a free end, which we define to be $x = L$. Suppose we wish to select basis functions from power series terms. If $x = 0$ is clamped, (that is, a cantilevered beam), the geometric boundary conditions are $w = \partial w/\partial x = 0$ at $x = 0$, which is satisfied by defining $\psi_j = (x/L)^{j+1}$, $j = 1, 2, \dots$. A pinned end requires only $w = 0$ at $x = 0$, which leads to $\psi_j = (x/L)^j$. The first basis function in this case represents rigid body rotation. For a guided end, in which $\partial w/\partial x = 0$ at $x = 0$, all power series terms except the first order one identically satisfy the geometric boundary condition, so we may select $\psi_1 = 1$, $\psi_j = (x/L)^j$, $j = 2, 3, \dots$. Here the first basis function represents rigid body translation. The remaining possibility is that both ends are free. There are no geometric boundary conditions to satisfy in this case, so any power series term is kinematically admissible. Hence, in the case of a free-free beam, the first two power series basis functions we would use are the rigid body translational and rotational modes. Setting $\psi_j = [(x - x_G)L]^{j-1}$ references all modes to the center of mass, which can have certain advantages.

When geometric boundary conditions must be satisfied at both ends, the best practice is to take $x = 0$ as the end that has more geometric conditions. We could endeavor to construct a set of power terms that simultaneously satisfy the geometric boundary conditions at both ends. However, doing so might inadvertently lead to selection of a set of functions that are not linearly independent. A more robust alternative is to base the selection on half-range Fourier series like those identified in eq. (6.1.22) for axial displacement. However, these functions might need to be modified to account for the greater number and variety of geometric conditions in flexure.

To explore these issues let us begin with the case of a simply supported (that is, hinged-hinged) beam. Any sinusoidal term whose zeroes occur at $x = 0$ and $x = L$ matches the geometric boundary conditions. Furthermore, the slope of a sinusoidal term at its zeroes is not zero, so such terms would not inappropriately satisfy the geometric boundary condition of zero rotation. The sinusoidal term that varies most gradually, while fitting these specifications, is $\sin(\pi x/L)$, so we would select $\psi_j = \sin(j\pi x/L)$ as the basis functions for a beam that is pinned at both ends. Interestingly, in the special case where the beam has constant cross-sectional properties and no attached masses, springs, or dashpots, these basis functions actually satisfy the field equations.

Now suppose the beam of interest has one clamped end and one hinged end. When there is more than one geometric boundary condition to satisfy at an end, it is useful to form the basis functions from products of simpler functions. The derivative of a product is $d(fg)/dx \equiv (df/dx)g + f(dg/dx)$. Thus, if $f = g = 0$ at an end, then the derivative of the product will also be zero at that end. With this in mind, we adapt the basis functions for a simply supported beam by multiplying them by a factor x/L. Hence, a suitable set of basis functions for a clamped-pinned beam is $\psi_j = (x/L)\sin(j\pi x/L)$. For a clamped-clamped beam, we multiply this set by an additional factor that vanishes at $x = L$, which leads to $\psi_j = (x/L)(1 - x/L)\sin(j\pi x/L)$.

A different modification of a Fourier series arises in the case of a clamped-guided bar. Here, we seek basis functions that have zero value at $x = 0$, but not $x = L$, while simultaneously having zero slope at $x = 0$ and $x = L$. Cosine functions, $\cos(j\pi x/L)$, fit the slope conditions, but they are nonzero at $x = 0$. We may adjust this by offsetting the function relative to its value at $x = 0$, which leads to the basis functions $\psi_j = 1 - \cos(j\pi x/L)$. An interesting aspect of this set of functions is that some of them satisfy a geometric boundary condition that does not apply to the clamped-guided beam. Specifically, it is apparent that when j is even, these functions satisfy $\psi_j = 0$ at $x = L$. This is permissible, in that some of the basis functions, those for odd j, lead to displacement at $x = L$. As an aid to formulating problems, Table 1 lists basis functions that may be used for each of the 10 independent combinations of end conditions. Of course, the tabulated functions are one of many possibilities.

Another technique for selecting basis functions uses mode functions for standard configurations, which are listed in Appendix C. Ten independent configurations of geometric boundary conditions for flexural displacement appear there. The tabulated case we would select matches the geometric conditions for the beam of interest. (Cases where $x = L$, rather than $x = 0$, has the support condition of interest may be obtained by swapping the definition of $x = 0$ or, alternatively, by defining $x' = L - x$.) Note that the selection should be based solely on matching the geometric conditions;

TABLE 1 Suitable Basis Functions for Flexure

Boundary conditions

$x = 0$	$x = L$	ψ_j
Clamped	Clamped	$\frac{x}{L}\left(1 - \frac{x}{L}\right)\sin\left(\frac{j\pi x}{L}\right)$
	Guided	$1 - \cos\left(\frac{j\pi x}{L}\right)$
	Pinned	$\frac{x}{L}\sin\left(\frac{j\pi x}{L}\right)$
	Free	$\left(\frac{x}{L}\right)^{j+1}$
Pinned	Pinned	$\sin\left(\frac{j\pi x}{L}\right)$
	Guided	$\sin\left(\frac{(2j-1)\pi x}{2L}\right)$
	Free	$\left(\frac{x}{L}\right)^{j}$
Guided	Guided	$\cos\left(\frac{j\pi x}{L}\right)$
	Free	$\cos\left(\frac{(j-1)\pi x}{2L}\right)$
Free	Free	$\left(\frac{x - x_G}{L}\right)^{j-1}$

whether masses, springs, or dashpots are attached at an end is irrelevant. The difficulty with using the modal properties in Appendix C is that most of the entries feature formulas that depend on a parameter α whose value is a root of a complicated algebraic equation. At the same time, selecting ψ_j to be the appropriate mode functions often leads to rapid convergence of the Ritz series.

EXAMPLE 6.4

The right end, $x = L$, of a reinforced concrete beam is clamped and the left end, $x = 0$, is guided. The depth of the beam varies linearly, so that the composite cross-sectional properties (including the effect of the reinforcing rods) are $\rho A = \mu(1 + x/L)$ and $EI = \sigma(1 + x/L)^3$, where μ and σ are constants. The excitation is a uniform load per unit length $f_z(x, t) = F(t)$. Select a suitable set of basis functions to use in a four-term Ritz series description of the displacement. Then derive the corresponding equations governing the motion.

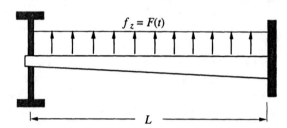

Solution In addition to illustrating the application of the Ritz series method for flexural displacement, we will see in this example that the ability of mathematical software to perform numerical integration can be extremely useful. The preceding discussion identified a set of basis functions suitable to a clamped-guided beam. Those functions were defined on the basis of the clamped end being $x = 0$, which is the opposite of the present definition of x. We therefore introduce the transformation $x' = L - x$, so that the clamped end corresponds to $x' = 0$. Thus the basis functions we shall employ are

$$\psi_j = 1 - \cos\left(\frac{j\pi x'}{L}\right), \quad j = 1, \ldots, 4$$

This satisfies the geometric boundary conditions $\psi_j = \partial\psi_j/\partial x' = 0$ at $x' = 0$ and $\partial\psi_j/\partial x' = 0$ at $x' = L$. Also, as noted earlier, $\psi_j \neq 0$ at $x' = L$ when j is odd, so half the basis functions fit the criterion that some basis functions should be nonzero if a location is permitted to displace.

There are no concentrated mass or spring attachments, so we may employ eqs. (6.1.37) with only the integral terms. In order to use those expressions we must use $x'/L = 1 - x/L$ to transform the mass per unit length and flexural rigidity functions to the new definition of position, $\rho A = \mu(2 - x'/L)$ and $EI = \sigma(2 - x'/L)^3$. This leads to

$$M_{jn} = \int_0^L \psi_j \psi_n \rho A \, dx'$$

$$= \mu \int_0^L \left[1 - \cos\left(\frac{j\pi x'}{L}\right)\right]\left[1 - \cos\left(\frac{n\pi x'}{L}\right)\right]\left(2 - \frac{x'}{L}\right) dx$$

$$K_{jn} = \int_0^L EI \frac{d^2\psi_j}{(dx')^2} \frac{d^2\psi_n}{(dx')^2} \, dx'$$

$$= \sigma \left(\frac{j\pi}{L}\right)^2\left(\frac{n\pi}{L}\right)^2 \int_0^L \cos\left(\frac{j\pi x'}{L}\right)\cos\left(\frac{n\pi x'}{L}\right)\left(2 - \frac{x'}{L}\right)^3 dx'$$

These integrands are composed of combinations of elementary functions, so we could evaluate the integrals analytically. The symbolic capabilities of mathematical software are useful aids for that task. In any event analytical integration would lead to complicated formulas for the coefficients, which we would evaluate for $1 \le j, n \le 4$. A simpler alternative comes from recognizing that we ultimately require only numerical values, so we can arrive at the desired results directly by evaluating the integrals numerically. Rather than assuming a numerical value for L in order to carry out these operations, we introduce $y = x'/L$ into each integrand. The integration then is over $0 \le y \le 1$. Also, because $dx' = L\,dy$, this substitution leads to an additional L factor in each expression.

Numerical integration may be performed in MATLAB by using the *quad8* function. To achieve this for the present situation we need to define two function M-files, which we shall call *m_cg.m* and *k_cg.m*. These function files respectively evaluate the integrands for M_{jn} and K_{jn}. The first line of these files indicates that the M-file is a function of x, j, and n—specifically, *function f = m_cg(y,j,n)* and *function f = k_cg(y,j,n)*—where f in each file is the respective integrand. An important aspect of these functions is that they must be constructed such that f is a column of values corresponding to a column of y values. Thus, we need to be sure to use the scalar operation notation, which precedes operands with a dot. For example, in *m_cg.m*, we would write $f = (2 - y).*(1 - \cos(j*pi*y)).*(1 - \cos(n*pi*y))$. To evaluate an M_{jn} coefficient for a specific j, n pair, we then write (in the MATLAB workspace or a script M-file) $M(j,n) = quad8('m\_cg', 0, 1, [], [], j, n)$, where the first argument is the name of the function that evaluates the integrand, the second and third arguments are the lower and upper limits for the integral, the fourth and fifth arguments are error tolerance and trace parameters that we need not specify, and all arguments after the fifth are values that $'m\_cg'$ requires. Similar considerations apply to the evaluation of a K_{jn} value.

Because $[M]$ and $[K]$ in the present situation are only 4×4 arrays, the integrations will be performed quickly. If we had many terms to compute, $N \gg 4$, we would exploit the symmetry property of these arrays in order to reduce the number of evaluations. We could implement this by evaluating $M(j,n)$ inside loops over j from 1 to N, and over n from 1 to j, with the additional specification that $M(n,j) = M(j,n)$.

A simpler MATLAB alternative is available to those who have installed the Symbolic Toolbox. To do so one would begin by defining x to be a symbolic real variable, and defining basis functions symbolically within a loop over the function's index. The integrals would be evaluated symbolically within a pair of loops extending over the row and column numbers. The *double* function then would convert the coefficients to numerical values. A program fragment to implement this is

MATLAB: *syms y real; N = 4; H = 2 − y;*
           *for j = 1 : N; psi(j) = (1 − cos(j * pi * y))'; end*
           for j = 1 : N; for n = 1 : N
           *MM(j, n) = int(psi(j) * psi(n) * H, y, 0, 1);*
           *KK(j, n) = int(diff(psi(j), 2) * diff(psi(n), 2) * H^3, y, 0, 1);*
           end; end; M = double(MM); K = double(KK);

Mathcad merely requires that we write M_{jn} and K_{jn} as integrals, just as they appear above. If we wish to use symmetry, we may define functions of j and n whose right sides are the integral expressions, that is, $m\_cg(j, n) := \ldots$ and $k\_cg(j, n) := \ldots$. Then, with j and n defined as indices ranging from 1 to N, we would write $M_{j,n} = if(j \ge n, m\_cg(j, n), 0)$, which would fill in terms along and below the diagonal. To invoke symmetry for the terms above the diagonal, we would then write $M_{j,n} := if(n > j, M_{n,j}, M_{j,n})$.

In view of the extra L factor resulting from the substitution $x' = Ly$ in the integrals, the results are

$$[M] = \mu L \begin{bmatrix} 1.8447 & 1.4099 & 1.2748 & 1.3127 \\ 1.4099 & 2.25 & 1.5829 & 1.5 \\ 1.2748 & 1.5829 & 2.205 & 1.5809 \\ 1.3127 & 1.5 & 1.5809 & 2.25 \end{bmatrix}$$

$$[K] = \frac{\sigma}{L^3} \begin{bmatrix} 193.75 & 304.69 & 124.91 & 177.63 \\ 304.69 & 2966.7 & 2555 & 789.57 \\ 124.91 & 2555 & 14894 & 10012 \\ 177.63 & 789.57 & 10012 & 46934 \end{bmatrix}$$

No dashpots are present in this system, so $[C] = [0]$. We evaluate the generalized force expressions according to the last of eqs. (6.1.37). The only excitation is the distributed force f_z, which is independent of x, so we have

$$Q_j = \int_0^L f_z \psi_j \, dx = F(t) \int_0^L \left[1 - \cos\left(\frac{j\pi x'}{L}\right)\right] dx' = FL$$

The corresponding equations of motion for the Ritz series coefficients are

$$[M]\{\ddot{q}\} + [K]\{q\} = FL\begin{bmatrix} 1 & 1 & 1 & 1 \end{bmatrix}^{\mathrm{T}}$$

EXAMPLE 6.5

A simple model for vertical displacement of the wings and fuselage of an airplane considers the wings to be beams attached to a central mass. This is illustrated in the figure, where the beams are considered to have constant cross-sectional properties ρA and EI and the fuselage, whose mass is m, is considered to have negligible rotatory inertia. Identify a Ritz series suitable for analyzing small vertical displacements of this system. Then describe the expressions for the stiffness and mass coefficients.

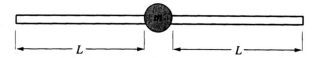

Solution In the process of formulating this solution, we will gain additional insight into the incorporation of basis functions representing rigid body rotation. We also will see how considerations of symmetry may be used to reduce the length of the Ritz series. Rather than modeling this system as two identical beams of length L, it is advantageous to consider it to consist of one beam of length $2L$ with an attached mass at its midpoint. (This viewpoint is allowable if the fuselage diameter is much less than $2L$.) The system is unsupported, which means that it is capable of rigid body translation and rotation. The center of mass of the system is the center of the fuselage. Translation in the vertical direction corresponds to a uniform displacement, so we take $\psi_1 = 1$, while small rotations about the center of mass correspond to vertical displacement that increases linearly with distance from the center of mass. Representation of such a motion is expedited by taking the center of mass to be $x = 0$, so we select the basis function to be $\psi_2 = x/L$. Because it is specified that the fuselage's moment of inertia is negligible, we may correspondingly consider the beam to cover $-L \leq x \leq L$. There are no geometric boundary conditions to satisfy. This enables us to select power series terms as the basis functions, $\psi_j = (x/L)^{j-1}$, $j \geq 3$, representing deformation of the beam. An advantage of this selection is the fact that the rigid body basis functions are part of this family, which expedites evaluation of the inertia and stiffness coefficients.

$$\psi_j = \left(\frac{x}{L}\right)^{j-1}, \; j = 1, 2, \ldots$$

Because the fuselage represents an attached mass, the inertia and stiffness coefficients obtained from eqs. (6.1.37) are

$$M_{jn} = \int_{-L}^{L} \psi_j \psi_n \rho A \, dx + m \psi_j(0) \psi_n(0)$$

$$K_{jn} = \int_{-L}^{L} EI \frac{d^2 \psi_j}{dx^2} \frac{d^2 \psi_n}{dx^2} dx$$

We next observe that the basis functions for odd values of j are even functions of x, that is, $\psi_j(-x) = \psi_j(-x)$, while the even values of j correspond to odd functions of x, $\psi_j(-x) = -\psi_j(-x)$. This observation is significant for several reasons. When j and n have different parity, that is, one is odd and the other is even, both integrands for negative $-|x|$ have the opposite sign as those at $+|x|$. This means that the associated integrals will vanish. Furthermore, $\psi_j(0) = 0$ except for $j = 1$. As a result, we find that M_{jn} and K_{jn} are zero unless j and n are both even or odd. A minor additional saving arising from the basis functions being either even or odd functions is that we may evaluate the integrals as twice the integral over $0 \le x \le L$. These considerations lead to

$$M_{j,n} = M_{n,j} = \begin{cases} 2\rho AL + m & \text{if } j = n = 1 \\[2mm] 0 & \text{if } \begin{cases} j \text{ is odd and } n \text{ is even} \\ j \text{ is even and } n \text{ is odd} \end{cases} \\[4mm] \dfrac{2}{j+n-1}\rho AL & \text{otherwise} \end{cases}$$

$$K_{j,n} = K_{n,j} = \begin{cases} 0 & \text{if } \begin{cases} j < 3 \\ n < 3 \\ j \text{ is odd and } n \text{ is even} \\ j \text{ is even and } n \text{ is odd} \end{cases} \\[6mm] 2\dfrac{(j-1)(j-2)(n-1)(n-2)}{(j+n-5)}\left(\dfrac{EI}{L^3}\right) & \text{otherwise} \end{cases}$$

It is convenient to define a parameter $\varepsilon = m/(2\rho AL)$, which is the ratio of the fuselage mass to the mass of the wings. For illustrative purposes, let us consider a six-term Ritz series, in which case we find

$$[M] = \rho AL \begin{bmatrix} 2+2\varepsilon & 0 & 2/3 & 0 & 2/5 & 0 \\ 0 & 2/3 & 0 & 2/5 & 0 & 2/7 \\ 2/3 & 0 & 2/5 & 0 & 2/7 & 0 \\ 0 & 2/5 & 0 & 2/7 & 0 & 2/9 \\ 2/5 & 0 & 2/7 & 0 & 2/9 & 0 \\ 0 & 2/7 & 0 & 2/9 & 0 & 2/11 \end{bmatrix}$$

$$[K] = \frac{EI}{L^3} \begin{bmatrix} 0 & 0 & 0 & 0 & 0 & 0 \\ 0 & 0 & 0 & 0 & 0 & 0 \\ 0 & 0 & 8 & 0 & 16 & 0 \\ 0 & 0 & 0 & 24 & 0 & 48 \\ 0 & 0 & 16 & 0 & 56.6 & 0 \\ 0 & 0 & 0 & 48 & 0 & 114.29 \end{bmatrix}$$

The feature that both matrices have alternating zeroes has an important consequence when we recall that they are multiplied by $\{\ddot{q}\}$ or $\{q\}$ in the equation of motion. Each row of the matrix product represents terms in a scalar differential equation. Hence, the differential equations for the odd-numbered rows have terms that depend on q_1, q_3, and q_5, while the even-numbered rows yield equations that depend on q_2, q_4, and q_6. In other words, the odd-numbered Ritz coefficients are uncoupled from the even-numbered coefficients. We now observe that ψ_j for $j = 1, 3, 5, \ldots$ are even functions of x, so they represent flexural displacements that are in the same direction for $x > 0$ and $x < 0$. Such functions represent a *symmetric displacement field* relative to $x = 0$. (Symmetry refers to features on one side that are the mirror image of what is on the other side.) For $j = 2, 4, 6, \ldots$, the functions ψ_j represent flexural displacements for $x < 0$ that are in the opposite direction from those for $x > 0$. Such functions represent an *antisymmetric displacement field* relative to $x = 0$. The Ritz series involves multiplication of each ψ_j by the corresponding q_j, which means that the symmetric and antisymmetric displacement fields are neither inertially nor elastically coupled. This is a common feature shared by all systems that have a plane of symmetry, a fact that is proven in the next chapter.

Whether the response is actually symmetric or antisymmetric depends on the excitation. If we know that we are interested solely in either case, we can omit the other type of terms, which reduces by a factor of 2 the length of the Ritz series required to attain a specified level of accuracy. In the general case the excitation can be split into symmetric and antisymmetric parts. (The procedure for doing this is discussed in Section 7.5.) Then, instead of using a Ritz series of length N to analyze the general displacement field, we can use different Ritz series of length $N/2$ to analyze individually the symmetric and antisymmetric parts of the displacement. Even though this process would require solving the equations of motion twice, it actually is more efficient numerically because each analysis requires only half the number of terms as the general analysis. This capability can be very useful when there are many variables, which is usually the case for finite element analyses (see Chapter 8).

Recognition of symmetry properties can also assist us in the selection of basis functions. Some individuals prefer to use sinusoidal functions to form the ψ_j. An important consideration for the wing-fuselage system is that some or all of such functions should have nonzero values and nonzero first derivatives at the ends. Symmetric displacement corresponds to even functions of x, which suggests cosines whose argument is proportional to x. Antisymmetric displacement corresponds to odd functions of x, which is a criterion met by sine functions whose argument is proportional to x. Suitable families of functions fitting the requirements at $x = \pm L$ are

$$\text{Symmetric displacement:} \quad \psi_j = \cos\left(\frac{j\pi x}{2L}\right), \quad j = 0, 1, 2\ldots$$

$$\text{Antisymmetric displacement:} \quad \psi_j = \sin\left(\frac{j\pi x}{2L}\right), \quad j = 0, 1, 2\ldots$$

Note that for symmetric displacement, this definition gives $\psi_0 = 1$, which matches a rigid body translation. However, none of the ψ_j for antisymmetric displacement match a rigid body rotation about the center. Nevertheless, this is a valid set. In fact, extending it to $j = \infty$ leads to the Ritz series being a half-range Fourier series that can represent any odd function of x defined over $-L < x < L$.

6.1.4 Implementation and Solution

A Ritz series analysis of displacement begins with selection of the basis functions appropriate to the specific type of displacement and the manner in which the bar is supported. The guidelines for selection of these basis functions are quite loose, in that they only require that the functions be linearly independent, satisfy the appropriate continuity conditions, and satisfy all geometric boundary conditions. The development outlined some general concepts to further assist us in this task. The standard expressions for $[M]$, $[K]$, $[C]$, and $\{Q\}$ for each type of displacement feature integrals containing products of the basis functions or their derivatives. Such integrals may be evaluated analytically, or numerically using mathematical software. If the latter approach is taken, one must take precautions to set the tolerance parameter for the integration routine sufficiently small. Otherwise, numerical errors might contaminate the equations. (This tends to be an issue only when N is large.)

Regardless of the type of motion, the Ritz method leads to ordinary differential equations of motion, $[M]\{\ddot{q}\} + [C]\{\dot{q}\} + [K]\{q\} = \{Q\}$, whose form is the same as those for discrete systems. As we saw in previous chapters, such equations may be solved by a variety of techniques. If the excitation is harmonic, such that $\{Q\} = \text{Re}[\{F\}\exp(i\omega t)]$, frequency domain analysis may be employed directly to find the steady-state response. Specifically, the transfer function associated with eq. (6.1.16) is

$$[G(\omega)] = [[K] + i\omega[C] - \omega^2[M]]^{-1} \qquad (6.1.39)$$

The corresponding steady-state response of the Ritz series coefficients is

$$\{q\} = \text{Re}\{[G(\omega)]\{F\}\exp(i\omega t)\} \qquad (6.1.40)$$

We may extend frequency domain analysis to cases of transient excitation by introducing FFT concepts. However, modal analysis provides a more efficient alternative because it uncouples the equations of motion. In addition, for free vibration problems where we know the initial displacement or velocity field, it is not yet apparent how we may identify the appropriate initial conditions for the q_j variables. Modal analysis will also provide that capability. Rather than implementing modal analysis at the present juncture, we will wait until we have examined the nature of modes as displacement functions.

Regardless of the method we employ to solve the equations of motion, it is important to recognize that the solution for the basis function factors q_j are not the quantities we need. Rather, they must be used in conjunction with the basic Ritz series, eqs. (6.1.8), (6.1.27), or (6.1.36), to determine the physical displacement at the locations and instants of interest.

EXAMPLE 6.6

Consider the bar in Example 6.1 when the force applied at the midpoint is $F = 25\cos(\omega t)$ kN. Use frequency domain analysis to solve those equations for the steady-state response as a function of the excitation frequency ω. In particular, evaluate the displacement at the right end, and at the location where F is applied.

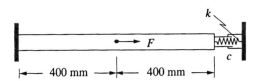

Solution We shall examine here the procedures for using frequency domain analysis to solve the equations of motion associated with a Ritz series. The computational procedure for evaluating the Ritz series as a function of x corresponding to the solution for the q_j coefficients is a primary focus of the discussion. The Ritz series employed in Example 6.1 was

$$u(x, t) = \sum_{j=1}^{3} \sin\left[\frac{(2j-1)\pi x}{2L}\right] q_j(t)$$

where $L = 0.8$ m. The equations of motion are the standard

$$[M]\{\ddot{q}\} + [C]\{\dot{q}\} + [K]\{q\} = \{Q\}$$

where the values of $[M]$, $[C]$, and $[K]$ are listed in the solution to Example 6.1, and the generalized forces are

$$\{Q\} = 0.7071 F \begin{Bmatrix} 1 \\ 1 \\ -1 \end{Bmatrix} = \text{Re}[\{F\} \exp(i\omega t)], \qquad \{F\} = 17.678(10^3) \begin{Bmatrix} 1 \\ 1 \\ -1 \end{Bmatrix} \text{N}$$

Frequency domain analysis offers a direct method for determining the steady-state response over a range of frequencies. However, the range of frequencies for the sweep was not specified. To determine this we recognize that we have constructed a three-degree-of-freedom model of the bar. Accordingly, the range of frequencies should capture three resonances. We use our mathematical software to evaluate the natural frequencies associated with $[[K] - \omega^2[M]]\{\phi\} = \{0\}$, which gives

$$\omega_1 = 1.138(10^4), \qquad \omega_2 = 3.033(10^4), \qquad \omega_3 = 4.997(10^4) \text{ rad/s}$$

Going 25% beyond the highest natural frequency should be adequate to see the last peak in a transfer function, so we shall compute the response for $0 \le \omega \le 6.25(10^4)$ rad/s. We also guess that a frequency increment of 100 rad/s will be adequate to see the peaks in detail. (The computed results will indicate whether either guess was inadequate.)

Setting $\{q\} = \text{Re}[\{X\} \exp(i\omega t)]$ to determine the steady-state response leads to

$$\{X\} = [[K] + i\omega[C] - \omega^2[M]]^{-1} \{F\}$$

This relation must be solved for each value of ω in the sweep range; we denote these values as Ω_p, $p = 1, 2, \ldots$. It will be instructive to examine the solution for each element of $\{X\}$ as a function of frequency, so we arrange the computations to store $\{X\}$ for each Ω_p. One scheme for doing this is to successively append the solutions for $\{X\}$, thereby creating a rectangular array $[XX]$ whose rows hold X_1, X_2, and X_3 as a function of ω. A MATLAB program fragment is

MATLAB: $\quad$ *for* $p = 1 : P + 1$; $\quad om = (p - 1) * delta$;
$\qquad\qquad XX(:, p) = (K + i * om * C - om\wedge 2 * M) \backslash \{F\}$; $\quad end$

where P is the number of frequencies of interest and $delta = 6.25(10^4)/P$ rad/s. In Mathcad, these operations may be implemented with a range index p, such that

Mathcad: $\quad p := 0; P \quad \Omega_p = p * delta$
$\qquad\qquad XX^{<p>} = [K + i * \Omega_p * C - (\Omega_p)^2 * M]^{-1} * \{F\}$

Row j of $[XX]$ contains the X_j values at each frequency Ω_p. The first graph displays the magnitude of each X_j; there are indeed three resonances in the frequency range we selected.

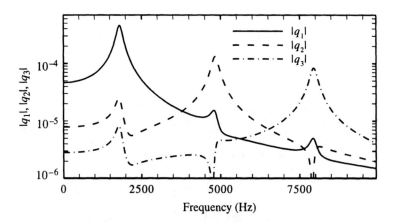

Note that we have used a log scale to plot the amplitudes because the damping is quite light. The fact that the peaks seem to be well resolved suggests that the frequency increment of

100 rad/s was adequate. The maxima occur at 11,350, 30,300, and 49,900 rad/s, which differ from the three undamped natural frequencies by less than the frequency increment for the sweep. We also see in the figure that X_1 is the largest coefficient at the first resonance, X_2 is the largest at the second resonance, and X_3 is the largest at the third. Such a situation suggests that the associated term in the Ritz series is the dominant contributor to the displacement field at each resonance. (Usually, several basis functions will have comparable magnitudes. The underlying reason for the dominance of one term in the present case lies in the properties of vibration mode functions, which are treated in Sections 6.2 and 6.3.)

The generalized coordinates are the variables of interest for a system composed of rigid bodies, but their evaluation is only an intermediate step in Ritz series analysis of a deformable continuum, where the associated physical displacement must be determined. For steady-state response, we set $\{q\} = \mathrm{Re}[\{X\} \exp(i\omega t)]$. Thus, the Ritz series gives

$$u(x, t) = \sum_{j=1}^{3} \sin\left[\frac{(2j - 1)\pi x}{2L}\right] \mathrm{Re}[X_j \exp(i\omega t)]$$

The collected coefficient of $\exp(i\omega t)$ is the complex amplitude of the displacement at the specified x, which we shall denote as $U(x)$,

$$u(x, t) = \mathrm{Re}[U(x) \exp(i\omega t)], \quad U(x) = \sum_{j=1}^{3} \sin\left[\frac{(2j - 1)\pi x}{2L}\right] X_j$$

To evaluate the displacement we recall that the columns of the $[XX]$ array hold the value of $\{X\}$ at each ω in the swept range. If we define a 1×3 row array $[psi\_x]$ whose column j is the value $\psi_j(x)$, where x is the location of interest, the product of $[psi\_x]$ and a column of $[XX]$ will be the corresponding value of $U(x)$ at the frequency associated with the selected column of $[XX]$. Thus, the product $[psi\_x][XX]$ gives a row array that holds the value of $U(x)$ at each frequency. To carry out these steps in MATLAB and Mathcad, we write

MATLAB: $\quad j = 1 : 3; \quad psi\_x = \sin((2 * j - 1) * pi * x/(2 * L)); \quad U\_x = psi\_x * XX$

Mathcad: $\quad j := 1;3 \quad psi\_x_j := \sin((2 * j - 1) * pi * x/(2 * L)) \quad U\_x := psi\_x^T * XX$

The result of these steps for $x = L/2$ and $x = L$, which are the requested locations, is depicted in the next set of graphs. The data displayed there are the amplitude and phase lag of the complex function $U(x)$.

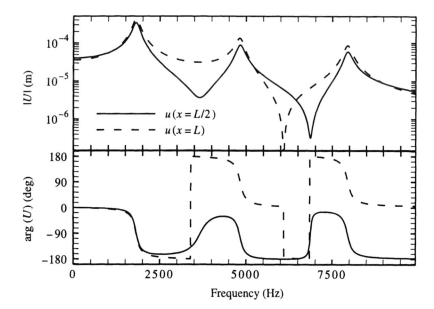

Recall that the excitation force varies as $\cos(\omega t)$, so a negative argument for a $U(x)$ value corresponds to a displacement that lags behind the force. Consider the situation at the resonances. The displacements at $x = L/2$ and $x = L$ are comparable in magnitude. The decreasing nature of the successive peaks stems from the fact that the force exerted by the dashpot is $c\dot{u}(x = L) = \mathrm{Re}[i\omega u(x = L)]$, so the dashpot becomes more effective as the frequency increases. Note that at each resonance $u(x = L/2)$ lags by approximately 90° relative to the excitation. As usual, this characteristic arises because the inertial and elastic contributions to the overall impedance essentially cancel at a resonance, which leaves only the velocity-dependent damping effect to counter the force. Interestingly, the displacement at $x = L$ *leads* the excitation by 90° at the second and third resonances. The logarithmic scale exaggerates the width of the resonance peaks, but it is obvious that the system is very lightly damped. The 360° phase shifts in $u(x = L)$ are merely a result of limiting the values to $-180° < \arg(U) \leq 180°$. An interesting situation occurs near 6100 Hz, where $u(x = L)$ shifts from being in phase with the force to being 180° out of phase, and the value of $|U(x = L)|$ shows an extreme dip. This behavior arises because $|U(x = L)|$ actually vanishes at approximately 6090 Hz. Thus, $x = L$ is a *node* when the frequency is 6090 Hz.

The occurrence of nodes is not unusual. To visualize them it is necessary to examine the complex amplitude function $U(x)$ associated with a specific frequency. The procedure for computing this quantity at any x is comparable to the manner in which the frequency sweep was evaluated, except that now the frequency is held constant and the locations are varied. Thus, we form a rectangular matrix $[psi\_vals]$ by stacking the previous $[psi\_x]$, so that the successive rows hold the value of all basis functions at a specified x in a discretized range $0 \leq x \leq L$. The product of row n of $[psi\_vals]$ and column s of $[XX]$ is $U(x_n)$ when $\omega = \Omega_s$,

MATLAB: $delta = L/N$; $x = [0 : delta : L]'$;
                   for $j = 1 : 3$; $psi\_vals(:, j) = \sin((2 * j - 1) * pi * x/(2 * L))$;   end;
                   $U\_vals = psi\_vals * XX(:, s)$;

Obviously, the value of s corresponding to $\omega = \Omega_s$ must be specified. Then $XX(:, s)$ is the column of basis function amplitudes at that frequency. In this program fragment, N is the number of intervals into which we divide the total length L ($N = 100$ is sufficient). Also, the sine function works element by element when its argument is a matrix. Because we have defined x to be a column, the evaluation of the sine function yields a column at each j. The analogous operations may be implemented in Mathcad by writing

Mathcad: $j := 1;3$ $n := 1; N + 1$ $x_n := (n - 1) * L/N$
                 $\psi_{n, j} := \sin((2 * j - 1) * \pi * x_n/(2 * L))$ $U\_vals = \psi * XX^{<s>}$

The magnitude $|U(x)|$ is the amplitude of the harmonic vibration that will be seen at the various axial locations, while $\arg(U(x))$ indicates the phase of the vibration at each x relative to a pure cosine function. This information is displayed in the last pair of graphs for the three resonant frequencies at which the displacement amplitudes are maximized. The locations at which $|U|$ is very small and $\arg(U)$ undergoes a 180° phase shift correspond to nodes. Thus, there are no nodes at the first resonance, one node at the second, and two nodes at the third resonance. These are essentially the patterns of $\psi_1 = \sin(\pi x/2L)$, $\psi_2 = \sin(3\pi x/2L)$, and $\psi_3 = \sin(5\pi x/2L)$, respectively, which is a result we anticipated when we observed that the term associated with a different X_j coefficient is dominant at each resonance.

As a closing note, one should remember that the present solution is an approximation of the actual response. Changing the length of the Ritz series or the selection of the basis functions might alter the displacement function $U(x)$ at any frequency. This issue is the topic of Exercise 6.16, where a different set of basis functions is specified for the present system. In fact, it turns out that the function $U(x)$ obtained here is exceptionally good in a frequency range extending somewhat beyond the first resonance. This behavior is a consequence of the convergence properties discussed in Section 6.4.

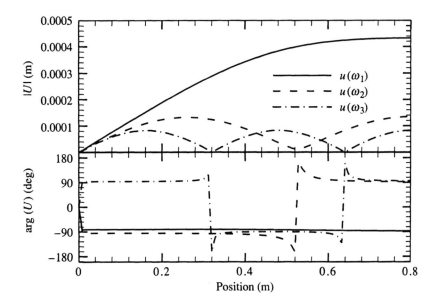

6.2 EIGENSOLUTIONS

A fundamental property of continuous systems is the fact that their vibration modes are functions of position, as opposed to discrete sets of numbers. In this section we will see how these modes may be determined by the Ritz method. The methods developed in the next chapter lead to an alternative solution for the modal properties based on solving partial differential equations of motion. Comparing the two types of eigensolutions will enable us to explore the actualities associated with implementing the Ritz method, especially the influence of the selection of basis functions on the solution.

Regardless of what method one uses to determine the modal properties, one of the primary reasons for such an analysis is decoupling the equations of motion associated with a modal transformation. Such a development will enable us to determine the response for nonzero initial conditions, which the developments thus far do not permit. Furthermore, as is true for discrete systems, decoupling the equations of motion often greatly enhances the efficiency of any response calculation.

The first phase of a modal analysis is evaluation of the natural frequencies and modes associated with the q_j variables for a Ritz series. This eigensolution proceeds in a straightforward manner. The only discretionary aspect is the beginning, where we must select the type and number of basis functions to use in forming the fundamental series. For the purpose of obtaining the free vibration modes we only need to evaluate the inertia coefficients M_{jn} and stiffness coefficients K_{jn}. However, if we also are concerned with forced response or dissipation, we might as well evaluate the generalized forces Q_j and damping coefficients C_{jn} at this stage. The next step entails solution of the general eigenvalue problem. We use complex exponentials to represent the homogeneous solution of the undamped equations of motion, eq. (6.1.16), so

$$\{q\} = \mathrm{Re}[B\{\phi\} \exp(i\omega t)] \tag{6.2.1}$$

where B is an arbitrary complex constant. We require that this satisfy eq. (6.1.16) with $[C] = [0]$ and $\{Q\} = \{0\}$, which leads us to the general eigenvalue problem

$$\boxed{[[K] - \omega^2[M]]\{\phi\} = \{0\}} \tag{6.2.2}$$

Because the Ritz series has N terms, solution of the foregoing leads to N natural frequencies ω_n and corresponding orthogonal modes $\{\phi_n\}$.

We normalize the modes in the usual manner by scaling the computed modes to have a unit modal mass,

$$\{\Phi_j\} = \frac{\{\phi_j\}}{[\{\phi_j\}^T[M]\{\phi_j\}]^{1/2}} \qquad (6.2.3)$$

The corresponding first orthogonality property of the normal modes is

$$\{\Phi_j\}^T[M]\{\Phi_n\} = \delta_{jn} \qquad (6.2.4)$$

and the second orthogonality condition associated with this normalization is

$$\{\Phi_j\}^T[K]\{\Phi_n\} = \omega_n^2 \delta_{jn} \qquad (6.2.5)$$

In most situations, the potential energy is positive definite, in which case all $\omega_m^2 > 0$. However, if rigid body motion is possible, then the potential energy is only positive semidefinite and one or more natural frequencies are zero. In any event, we sequence the eigensolutions such that $0 \leq \omega_1 \leq \omega_2 \leq \cdots \leq \omega_N$.

The mode shapes $\{\Phi_j\}$ are the proportions by which the various q_j contribute to the Ritz series. To determine the modal displacement field, we use eq. (6.2.1) to represent the generalized coordinates in the Ritz series. We shall use axial displacement for the discussion, but all aspects apply equally to torsional and flexural motion. We replace $\{\phi\}$ with $\{\Phi_n\}$ and ω with ω_n in eq. (6.2.1), where n is the mode number. Substitution of the resulting $\{q\}$ into eq. (6.1.8) yields

$$u(x, t) = \sum_{j=1}^{N} \text{Re}[B\Phi_{jn}\exp(i\omega_n t)]\psi_j(x) \qquad (6.2.6)$$

where Φ_{jn} is the jth element of mode vector n. Because B and the complex exponential do not contain the index j for the basis function, they may be brought out of the summation. The factor of these terms is the *mode function* $\Psi_n(x)$, specifically

$$u(x, t) = \text{Re}[\Psi_n(x)B\exp(i\omega_n t)] \equiv \Psi_n(x)|B|\cos[\omega_n t + \arg(B)]$$

$$\Psi_n(x) = \sum_{j=1}^{N} \Phi_{jn}\psi_j(x) \qquad (6.2.7)$$

Note that the mode function is a real quantity, and that the above relation applies equally to torsional rotation θ and flexural displacement w. In any case, a complete eigensolution entails evaluation of the natural frequencies ω_n and mode functions $\Psi_n(x)$.

The mode functions for most systems share a common property. Consider the case of a hinged-clamped beam whose cross-sectional properties are constant. Figure 6.10 displays the first five mode functions. The numbered labels in the figure correspond to the sequential ordering of the corresponding natural frequency. It is evident that the mode functions show more rapid spatial variation as their number goes up. According to eq. (6.2.7), a vibration following a single mode multiplies that mode function by a harmonic time function. Thus, if we were to take a snapshot of the displacement pattern at any instant, we would see the same proportions, with the vertical scale changing from instant to instant. A corollary of this behavior is the fact that the locations where the mode function is zero do not displace; they are called *nodes*. The

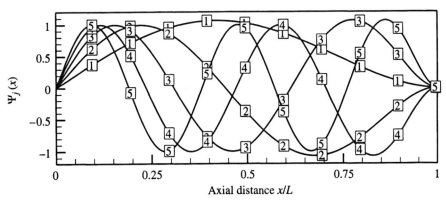

FIGURE 6.10 Flexural mode functions for a hinged-clamped bar.

number of nodes increases by one as we go up in mode number. This behavior is exhibited by the extensional, torsional, and flexural modes for most bars, regardless of whether springs and masses are attached to them.

A corollary of the nodal feature is the increased number of lobes (that is, maxima and minima) exhibited by the mode functions with increasing mode number. A simple explanation is that because of the manner in which $[\Phi]$ is normalized, the equivalent mass associated with each mode is unity. We will see in the next section that ω_n^2 is the ratio of the equivalent spring constant for the mode to its equivalent mass. Because ω_n increases with increasing n, the spring effect must also increase with increasing n. A higher spring constant corresponds to more potential energy, which is manifested by the successive mode functions showing increasing amounts of deformation.

The decision to implement a Ritz solution may be approached from two viewpoints, depending on our requirements. In some cases, we might seek merely a qualitatively correct answer, as would be the case if we wished to check results obtained from a finite element program. We need not be very thoughtful in selecting basis functions in this case, except for satisfying linear independence and the geometric boundary conditions. In such an environment, one aspect of verifying the general correctness of the Ritz solution entails adding a few terms to the Ritz series, and then ascertaining that the response parameters of interest do not change much with the longer series. If the change is unacceptably large, we would insert additional terms and repeat the comparison.

The other viewpoint extends the foregoing approach by examining all aspects of the solution, with the objective of obtaining a convergent answer. (We will consider a proof of convergence in Section 6.4.) Without verification that the Ritz series has converged, the natural frequencies and mode functions should be understood to be approximations that depend on the number of terms N used to obtain the result. In contrast, convergent results are the true eigensolutions (to the precision associated with the convergence criteria used) associated with the basic assumptions used to represent the system. This eigensolution will match results obtained by solving field equations, as in the next chapter.

Performing an analysis that leads to convergent results requires that we select basis functions in an orderly manner so as to ensure independence as the number of functions is increased. One way of doing so is to select the functions from parameterized families, for example, power series terms or Fourier series, as discussed in previous sections. Many practitioners find it convenient to implement the procedure using a specific class of special functions, such as Bessel functions. This will be demonstrated in Example 6.12.

The number of modes that are evaluated equals the length of the Ritz series. We will see in Section 6.4 that the largest errors in the modal properties obtained from a Ritz series analysis are encountered in the modes having the highest natural frequencies. A continuous system has an infinite number of eigenstates because it has an infinite number of degrees of freedom. However, there usually is an upper limit to the frequencies of interest, either because the frequency spectrum of the excitation is limited, or else because the basic mechanics of materials assumptions inherent to bar theory became invalid. Thus, a reasonable criterion for the length of the Ritz series is that convergence should be attained for the modes whose natural frequencies are below that upper limit. We will consider this issue in more detail when we consider different types of excitations.

EXAMPLE 6.7

A block is attached to two bars composed of the same material but having different diameters, as shown. The mass of the block is $m = \pi \rho D^2 L$, where ρ is the material density of the bars. Determine the natural frequencies and mode functions for axial displacement corresponding to Ritz series using two and four basis functions.

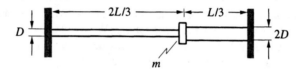

Solution This example demonstrates the basic operations leading to mode functions. It also serves as a precursor to our later study of the convergence properties of Ritz series. Rather than considering this system to consist of two bars, we may implement an analysis of a single bar by considering its diameter to change discontinuously at the location where the block is attached. Thus, if we consider the left end to be $x = 0$, we take the cross-sectional area to be $A = \pi D^2/4$ for $x < 2L/3$, and $A = \pi D^2$ for $x > 2L/3$. The geometric boundary conditions require that $u = 0$ at $x = 0$ and $x = L$. Sinusoidal functions fitting these conditions are

$$\psi_j = \sin\left(\frac{j\pi x}{L}\right), \quad j = 1, \ldots, N$$

where we are interested in the cases $N = 2$ and $N = 4$. The integrals for the inertia coefficients described by eq. (6.1.11) extend over $0 < x < L$, which we break into two parts in order to account for the differences in the cross-sectional area. The block is attached at $x = 2L/3$, so we have

$$M_{jn} = \int_0^L \psi_j \psi_n \rho A \, dx + m \psi_j(2L/3) \psi_n(2L/3)$$

$$= \frac{\pi \rho D^2}{4} \int_0^{2L/3} \sin\left(\frac{j\pi x}{L}\right) \sin\left(\frac{n\pi x}{L}\right) dx + \pi \rho D^2 \int_{2L/3}^L \sin\left(\frac{j\pi x}{L}\right) \sin\left(\frac{n\pi x}{L}\right) dx$$

$$+ \pi \rho D^2 L \sin\left(\frac{2j\pi}{3}\right) \sin\left(\frac{2n\pi}{3}\right)$$

We obtain the stiffness coefficients from eq. (6.1.13), with the integral broken into two parts as above,

$$K_{jn} = \int_0^L EA \frac{d\psi_j}{dx} \frac{d\psi_n}{dx} dx$$

$$= \frac{\pi E D^2}{4}\left(\frac{\pi^2 jn}{L^2}\right)\int_0^{2L/3} \cos\left(\frac{j\pi x}{L}\right)\cos\left(\frac{n\pi x}{L}\right) dx$$

$$+ \pi E D^2\left(\frac{\pi^2 jn}{L^2}\right)\int_{2L/3}^L \cos\left(\frac{j\pi x}{L}\right)\cos\left(\frac{n\pi x}{L}\right) dx$$

We could evaluate these integrals analytically, but in the end we will need numerical values, so we shall use the numerical procedure outlined earlier. In order to eliminate L from the integrands, we substitute $x' = x/L$, which leads to an additional L factor in each coefficient. The result for $N = 4$ is

$$[M] = \rho D^2 L \begin{bmatrix} 2.979 & -2.681 & 0.244 & 2.291 \\ -2.681 & 3.223 & -0.390 & -2.194 \\ 0.244 & -0.390 & 0.785 & -0.278 \\ 2.291 & -2.194 & -0.278 & 3.101 \end{bmatrix}$$

$$[K] = \frac{ED^2}{L} \begin{bmatrix} 9.354 & -6.410 & 2.404 & 2.564 \\ -6.410 & 27.801 & -15.385 & 12.821 \\ 2.404 & -15.385 & 69.764 & -43.598 \\ 2.564 & 12.821 & -43.598 & 120.436 \end{bmatrix}$$

A useful aspect of the formulation is that once we have the system coefficients for the largest series length N, we may obtain the matrices for any smaller N by extracting the appropriate submatrices from the upper left corner. In other words, to analyze the $N = 2$ case, we form the associated $[M]$ and $[K]$ from the upper left 2×2 submatrix of the above terms. This feature, which is the *embedding property* addressed in further detail in Section 6.4, enables us to avoid recomputing coefficients. We shall use the $N = 4$ case to illustrate the procedures, and then list the results obtained for $N = 2$.

The next step is to obtain the eigensolution for $[[K] - \lambda[M]]\{\phi\}$. The eigenvalue parameter λ is defined so as to eliminate the factors E, D, and L, which is achieved by setting

$$\rho D^2 L \omega^2 = \frac{ED^2}{L} \lambda \quad \Rightarrow \quad \omega_j = \left(\frac{E}{\rho L^2} \right)^{1/2} \lambda_j$$

The numerical analysis should also normalize the mode associated with each of the eigenvalues λ_j. The results listed below are sorted in ascending order of the natural frequency.

$$\left(\frac{\rho L^2}{E} \right)^{1/2} \omega_j = 1.723, \, 5.060, \, 9.517, \, 10.556$$

$$[\Phi] = \begin{bmatrix} 0.489 & 1.121 & 0.082 & 0.394 \\ -0.078 & 1.073 & -0.110 & -0.469 \\ -0.013 & 0.090 & -1.227 & -0.568 \\ 0.025 & -0.153 & -0.113 & -0.990 \end{bmatrix}$$

It should be noted that the embedding property does not extend to the mode matrix. Specifically, to carry out the analysis for $N = 2$, one must compute the eigensolution associated with the 2×2 $[M]$ and $[K]$ matrices.

The columns of $[\Phi]$ are the respective modes associated with the Ritz coefficients q_j. The quantities we seek are the mode functions $\Psi_j(x)$ defined by eq. (6.2.7). Plotting any of these functions requires that we evaluate it at many values of x. This task can be expedited by adapting the procedure in Example 6.1. We may write eq. (6.2.7) as

$$\Psi_n(x_p) = \sum_{j=1}^{N} \Phi_{jn} \psi_j(x_p) = \{\Phi_n\}^T \{\psi(x_p)\} \equiv \{\psi(x_p)\}^T \{\Phi_n\}$$

where $\{\psi(x_p)\}$ represents a column of N values formed by evaluating the respective basis functions at a specified location x_p. We are interested in many values of x_p, so we form a rectangular array $[psi\_vals]$ in which row p holds the successive values of ψ_n at a specific x_p. According to the preceding equation, the product of $[psi\_vals]$ and column n of $[\Phi]$ gives the values of $\{\Psi_n\}$ at each x_p location. The partitioning concept for matrices then leads us to an array whose columns are the successive mode functions at each x location,

$$[\Psi(x)] \equiv \left[\{\Psi_1(x)\} \, \{\Psi_2(x)\} \, \cdots \, \{\Psi_N(x)\} \right] = [\psi(x)][\Phi]$$

The elements of each column of $[\Psi(x)]$ may be plotted against the corresponding x_p value in order to visualize the mode functions. The plot of the result for $N = 4$, which gives four mode functions, shows several interesting features.

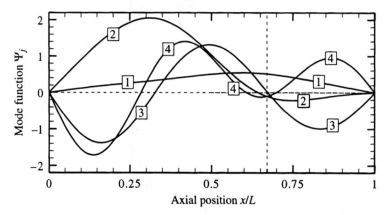

In assessing these curves, it is important to recognize that their magnitude is unimportant. The vertical scale for each would be arbitrary if we did not normalize $[\Phi]$ in computing the functions. Note that the number of maxima and minima in a mode matches the mode number. This feature of mode functions is one of the reasons for suggesting that basis function sets be selected such that they become more curved as the function number increases.

The vertical dashed line marks the position of the block. The first mode seems to be almost straight away from the vicinity of the block, which means that the strain in those regions is almost independent of x. In contrast, the second, third, and fourth modes are highly curved, and almost zero at the location of the block. Hence, the block seems to be situated at a node for these higher mode functions. These features may be explained physically by noting that the total mass of the beam is $\frac{1}{2}\pi\rho D^2 L$, which is one-half the mass of the block. In the first mode, whose natural frequency is the lowest, the primary contribution to the kinetic energy comes from the block, and the bar segment on each side of the block acts like an elastic spring. In fact, we can construct a lumped mass estimate of the fundamental natural frequency by recalling that the spring constant for static elongation of a bar whose length is ℓ is $k = EA/\ell$. These values for the segments to the left and right of the mass are

$$k_1 = \frac{3\pi ED^2}{8L}, \qquad k_2 = \frac{3\pi ED^2}{L}$$

Considering these springs to restrain the block leads to an estimate for the fundamental natural frequency as

$$(\omega_1)_{\text{est}} = \sqrt{\frac{k_1 + k_2}{m_{\text{block}}}} = 1.94\sqrt{\frac{E}{\rho L^2}}$$

which is close to, but greater than, the computed value of ω_1. The reason that this computation overestimated the fundamental natural frequency lies in the Rayleigh ratio, which is discussed in Section 6.4. The relative size of the mass of the block and of the bar also explains why the block location is almost a node for all mode functions except the first. In those modes the oscillation frequency is very high, which means that the acceleration of the block would be very large if the amplitude of the oscillation were large. In combination with the large value of the block's mass, it follows that a very large force would be required to move the block appreciably. Thus, the block represents a substantial resistance to displacement at the second and higher natural frequencies.

To close out this exercise, we consider the results obtained from a two-term Ritz series, $N = 2$. As described earlier, we extract the corresponding $[M]$ and $[K]$, then redo each ensuing step of the analysis. The natural frequencies obtained from this analysis are

$$\left(\frac{\rho L^2}{E}\right)^{1/2} \omega_j = 1.754, 5.430$$

which are close to, but greater than, ω_1 and ω_2 obtained from the $N = 4$ analysis. (Here also, the Rayleigh ratio will explain why $N = 2$ gives larger estimates than $N = 4$.) A plot of the two mode functions obtained from this analysis shows that they are reasonably close to the first two mode functions obtained from the $N = 4$ analysis.

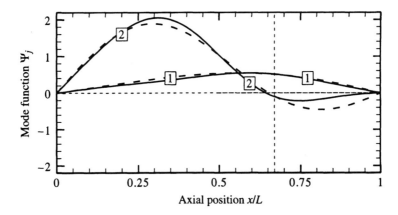

EXAMPLE 6.8

Vertical displacement of a wing and fuselage was modeled in Example 6.5. Consider the fuselage mass to equal the total mass of the wings. Determine the natural frequencies and mode shapes associated with the inertia and stiffness matrices determined in that example.

Solution This exercise will illustrate the treatment of rigid body motion in the analysis of mode shapes. The basis functions used to formulate Example 6.5 were $\psi_j = (x/L)^{j-1}$, with $x = 0$ at the midpoint where the fuselage is located. Although we could split these functions into sets describing displacements that are symmetric or antisymmetric about $x = 0$, we shall keep them together and thereby obtain all natural frequencies and modes in a single analysis. The parameter ε in the previous exercise is the ratio of the fuselage and wing masses, so we set $\varepsilon = 1$ for the prescribed case.

To evaluate the natural frequencies and mode matrix associated with the Ritz coefficients q_j, we define a nondimensional eigenvalue parameter to be $\lambda = \rho A L^4 \omega^2 / EI$. This allows us to solve the eigenvalue problem $[[K] - \lambda[M]]\{\phi\} = \{0\}$, where $[M]$ and $[K]$ are the matrices listed in Exercise 6.5, with EI/L^3 and ρAL equated to unity. There are six basis functions, so the numerical analysis yields six eigenvalues and six eigenvectors. After these are sorted in ascending order of their eigenvalues and normalized, the result is

$$\left(\frac{EI}{\rho AL^4}\right)^{1/2} \{\omega\} = [0 \quad 0 \quad 4.31 \quad 15.88 \quad 48.59 \quad 113.92]^T$$

$$[\Phi] = \begin{bmatrix} 0.5 & 0 & 0.322 & 0 & 0.188 & 0 \\ 0 & 1.225 & 0 & -3.420 & 0 & -3.939 \\ 0 & 0 & -2.229 & 0 & -5.843 & 0 \\ 0 & 0 & 0 & 7.631 & 0 & 19.615 \\ 0 & 0 & 0.490 & 0 & 7.858 & 0 \\ 0 & 0 & 0 & -2.704 & 0 & -18.270 \end{bmatrix}$$

Two natural frequencies are zero, which means that they correspond to rigid body modes. The first column of $[\Phi]$ is the first mode vector. Because only the first element in this column is nonzero, only the first basis function, $\psi_1 = 1$, contributes to the motion, which shows that this mode represents rigid body translation. Similarly, the second column of $[\Phi]$, which is the second mode vector, only has a nonzero value in the second row. Hence, it represents a displacement that varies as $\psi_2(x) = x/L$. This is the small displacement approximation of rigid body rotation. Also, note that the other columns of $[\Phi]$ have zeroes alternately in the odd and even rows. Such behavior is a consequence of the uncoupling of symmetric and antisymmetric displacement effects, as explained in Example 6.5.

To compute values of the mode functions at a number of points along the beam, we employ the procedure in the previous example. Thus, we define a set of equally spaced locations x_m and form a matrix of basis function values $[psi\_vals]$, whose element in the mth row and jth column is $\psi_j(x_m)$. The nth mode function evaluated at these locations is then found from the nth column of the product $[\Psi(x)] = [psi\_vals][\Phi]$.

The graph of these results shows several interesting features. It is evident that the first mode is constant and the second mode is proportional to x, as we anticipated. We also can see that the first, third, and fifth modes are even functions of x, and therefore symmetric about the midpoint, while the second, fourth, and sixth modes are odd functions of x representing antisymmetric motion. As was true for the previous example, the number of nodes in a mode function increases with increasing mode number.

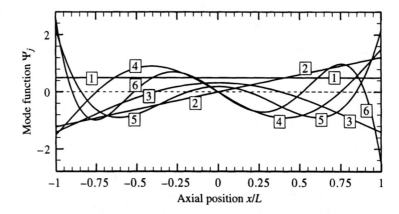

EXAMPLE 6.9

The mass of the block is one-half of the mass of the pinned-clamped beam, $m = \frac{1}{2}\rho AL$, and the spring stiffness k is selected such that $k/m = \Omega_1^2$, where Ω_1 is the fundamental natural frequency of the beam when nothing is suspended from it. Use a three-term Ritz series to determine the natural frequencies and mode functions of flexural motion.

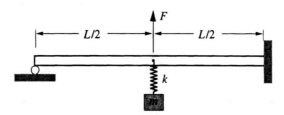

Solution This example is intended to emphasize the ease with which the Ritz series method can treat systems in which discrete and continuous parts respond jointly. The derivation of the inertia and stiffness coefficients shares many of the features of Example 6.2, except that we now will go on to compute the mode functions. It is specified that we should use a three-term Ritz series, and the displacement of interest is the transverse component w, so we have

$$w = \sum_{j=1}^{3} \psi_j q_j$$

Because we are concerned with vertical displacement, we assume that the block only moves in that direction; we let q_4 denote its displacement from the equilibrium position.

The basis functions must satisfy three geometric boundary conditions, because there can be no displacement at the left end of the beam, while the right end can neither displace nor rotate. Because there are more geometric conditions to satisfy at the right end, we define that location to be $x = 0$. Thus, we must select basis functions that satisfy the conditions that $\psi_j = d\psi_j/dx = 0$ at $x = 0$, $\psi_j = 0$ at $x = L$. A set of functions satisfying these conditions was identified in Section 6.1.3 as being

$$\psi_j = \frac{x}{L} \sin\left(\frac{j\pi x}{L}\right), \quad j = 1, 2, 3$$

Because the block is not attached to the bar, we rederive expressions for the inertia and stiffness coefficients by constructing the kinetic and potential energy expressions. When w and q_4 are both considered to be positive in the upward direction, we have

$$T = \frac{1}{2}\int_0^L \dot{w}^2 \rho A \, dx + \frac{1}{2} m \dot{q}_4^2$$

$$V = \frac{1}{2}\int_0^L EI\left(\frac{\partial^2 w}{\partial x^2}\right)^2 dx + \frac{1}{2} k[w(L/2, t) - q_4]^2$$

Into these expressions we substitute the Ritz series, $w = \sum \psi_j q_j$, and collect like terms. The result is

$$T = \frac{1}{2}\sum_{j=1}^{3}\sum_{n=1}^{3}\left(\rho A \int_0^L \psi_j \psi_n \, dx\right)\dot{q}_j \dot{q}_n + \frac{1}{2} m \dot{q}_4^2$$

$$V = \frac{1}{2}\sum_{j=1}^{3}\sum_{n=1}^{3}\left(EI\int_0^L \frac{d^2\psi_j}{dx^2}\frac{d^2\psi_n}{dx^2}dx\right)q_j q_n + \frac{1}{2}k\sum_{j=1}^{3}\sum_{n=1}^{3}\psi_j(L/2)\psi_n(L/2)q_j q_n$$

$$- k\sum_{j=1}^{3}\psi_j(L/2)q_j q_4 + \frac{1}{2}k q_4^2$$

We match these to the standard quadratic sums for T and V, which leads to

$$M_{jn} = M_{nj} = \rho A\int_0^L \psi_j \psi_n \, dx, \quad j, n = 1, 2, 3$$

$$M_{44} = m, \quad M_{jn} = M_{nj} = 0, \quad j = 1, 2, 3, \quad n = 4$$

$$K_{jn} = K_{nj} = EI\int_0^L \frac{d^2\psi_j}{dx^2}\frac{d^2\psi_n}{dx^2}dx + k\psi_j(L/2)\psi_n(L/2), \quad j, n = 1, 2, 3$$

$$K_{j4} = K_{4j} = -k\psi_j(L/2), \quad K_{44} = k$$

We use numerical integration to evaluate these coefficients, which entails introducing the nondimensional position $x' = x/L$ into the integrands. The result is that the integrals for inertia coefficients have the common factor ρAL, and the factor for the stiffness integrals is EI/L^3. We know that $m = 0.5\rho AL$, but we will need to determine the natural frequency of the beam when $m = k = 0$ in order to determine k, so we define

$$\mu = \frac{m}{\rho AL}, \quad \kappa = \frac{kL^3}{EI}$$

This enables us to write the computed matrices as

$$[M] = \rho AL \begin{bmatrix} 0.413 & -0.0901 & 0.0901 & 0 \\ -0.0901 & 0.1603 & -0.0973 & 0 \\ 0.0901 & -0.0973 & 0.1639 & 0 \\ 0 & 0 & 0 & \mu \end{bmatrix}$$

$$[K] = \frac{EI}{L^3} \begin{bmatrix} 222 & 715 & 267 & -0.5\kappa \\ 715 & 2479 & 911 & 0 \\ 267 & 911 & 521 & 0.5\kappa \\ -0.5\kappa & 0 & 0.5\kappa & \kappa \end{bmatrix}$$

The spring stiffness is defined in terms of the fundamental frequency of the beam by itself. To evaluate this quantity, we solve the eigenvalue problem associated with the upper left 3×3 submatrices evaluated at $k = 0$. We denote these as $\rho AL[\hat{M}]$ and $(EI/L^3)[\hat{K}]$, where $[\hat{M}]$ and $[\hat{K}]$ represent the numerical factors of the respective matrices. Thus, this task is to solve

$$[[\hat{K}] - (\rho AL^2/EI)\Omega^2[\hat{M}]]\{\phi\} = \{0\}$$

This yields

$$\left(\frac{\rho AL^4}{EI}\right)^{1/2} \Omega_j = 8.72, 29.51, 288.6$$

From this, we may determine κ because $k = \Omega_1^2 m$ corresponds to

$$\kappa = k\left(\frac{L^3}{EI}\right) = \left(\frac{L^3}{EI}\right)\Omega_1^2 m = \left(\frac{L^3}{EI}\right)\Omega_1^2(\mu\rho AL) = (8.72)^2\mu = 38.02$$

The next step is to form the 4×4 matrices corresponding to nonzero values of m and k. We then solve the eigenvalue problem $[[K'] - (\rho AL^4/EI)\Omega^2[M']]\{\phi\} = \{0\}$. After the modes are normalized, the resulting eigensolution is

$$\left(\frac{\rho AL^4}{EI}\right)^{1/2} \omega_j = 6.70, 16.69, 50.63, 109.95$$

$$[\Phi] = \begin{bmatrix} 1.071 & 2.718 & 1.785 & 0.997 \\ 0.196 & 0.412 & 3.534 & 2.132 \\ -0.010 & -0.003 & 0.610 & 3.261 \\ 1.319 & -0.511 & -0.018 & 0.007 \end{bmatrix}$$

We wish to evaluate the mode functions at a sequence of P points. This evaluation proceeds almost as it did in the previous examples. The only difference is that in this system q_1, q_2, and q_3 are the Ritz series coefficients, while q_4 is the displacement of the block. It follows that the last row of $[\Phi]$ represents the displacement of the block in each mode. To compute the beam displacement, we define $[psi\_vals]$ to be a $P \times 3$ matrix, such that row p has the values $\psi_1(x_p)$, $\psi_2(x_p)$, and $\psi_3(x_p)$. We then define $[\Phi]_{3\times4}$ to be the first three rows of $[\Phi]$. The mode function values for the beam then are the columns of

$$[\Psi(x)] = [\psi(x)][\Phi]_{3\times4}$$

In addition to plotting each column of $[\Psi(x)]$ against the x_p values, we describe the motion of the block in each mode as a point value $\Phi_{4,j}$ which we plot at $x = L/2$. The result is as shown.

An interesting aspect of these mode functions is the similarity between the displacement pattern of the beam in the first two modes. In fact, if these functions were scaled such that their maximum value were one, we would see that the two functions are almost identical. The main difference between the first and second modes is that the block and beam displacements are in phase in the first mode but are 180° out of phase in the second mode ($\Psi_2 > 0$ at $x = L/2$, while $\Phi_{4,2} < 0$). In essence, this behavior arises because the suspended spring-mass system was tuned to match the fundamental natural frequency of the isolated beam, so it behaves like a

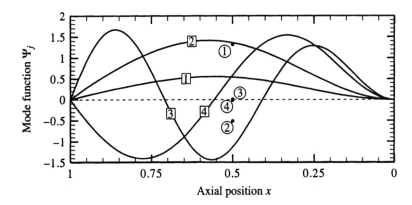

vibration absorber, in which the fundamental mode of the isolated beam splits into two modes. We also see from this figure, as well as the values of Φ_{43} and Φ_{44}, that the block displaces very little in the third and fourth modes.

6.3 MODAL ANALYSIS

6.3.1 Orthogonality of the Mode Functions

An alternative to writing the mode functions in summation form, as is done in eq. (6.2.7), is to use matrix notation, such that

$$\Psi_n(x) = [\psi]\{\Phi_n\} \tag{6.3.1}$$

where $[\psi]$ is a row formed from the basis functions,

$$[\psi(x)] = \left[\psi_1(x) \ \psi_2(x) \ \dots \ \psi_N(x)\right] \tag{6.3.2}$$

We previously used a discrete version of $[\psi(x)]$ to compute displacements. One virtue of expressing mode functions in this manner is that it tends to emphasize the similarity between continuous and discrete systems. For example, the matrix form of a Ritz series like eq. (6.1.8) is

$$u(x, t) = [\psi]\{q\} \tag{6.3.3}$$

This relation is analogous to the linear transformation by which a set of generalized coordinates is used to represent small physical displacements of a discrete system. For example, consider displacement of the rigid bar that is pinned to a translating collar in Figure 6.11. This system has two degrees of freedom. When the displacement s of the

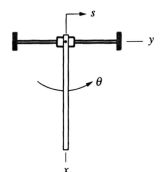

FIGURE 6.11 A two-degree-of-freedom system consisting of a rigid bar suspended from a slider.

collar and the rotation θ of the bar are used as generalized coordinates, the linearized transverse (that is, horizontal) displacement of any point on the bar is $w = s + x\theta$. Viewed as a Ritz series, this corresponds to $\psi_1(x) = 1$ and $\psi_2(x) = x$.

To continue with this analogy, observe that eqs. (6.1.10), (6.1.12), (6.1.14), and (6.1.15) respectively describe T, V, $\mathcal{P}_{\text{dis}}$, and $\mathcal{P}_{\text{in}}$ for the bar in the same manner as the terms for a discrete system. The first of eqs. (6.2.7) represents the physical displacement field corresponding to the normal mode vector $\{\Phi_n\}$ for the generalized coordinates. The fact that the mode functions $\Psi_n(x)$ are real indicates that all points vibrate at frequency ω_n either in phase or 180° out of phase from each other. The mode function gives the proportions between the displacements of points in the bar associated with this mode. This is the same kind of information as what is provided by modal vectors for discrete systems.

The analogy between the Ritz series representation and discrete system properties may be extended to the description of general free vibration. We observe that the first of eqs. (6.2.7) is one of N possible modal free vibrations. The most general free vibration is a linear combination,

$$u(x, t) = \sum_{n=1}^{N} \Psi_n(x)|B_n| \cos[\omega_n t + \arg(B_n)] \tag{6.3.4}$$

The coefficients B_n are dictated by the initial conditions for the free vibration. (How we determine these coefficients is a task we will address later in this section.) Thus, we see that any free vibration consists of a superposition of displacements at each natural frequency; the spatial distribution at each frequency is given by the shape of the eigenmode. This obviously is the same as the behavior of discrete systems.

In view of the foregoing, it is logical for us to exploit the analogy to study forced response. We do so by replacing the harmonically varying terms appearing in eq. (6.3.4) with arbitrary time functions $\eta_n(t)$. This leads to a *modal series* for the displacement field

$$u(x, t) = \sum_{n=1}^{N} \Psi_n(x)\eta_n(t) \tag{6.3.5}$$

The η_n variables give the contribution of each mode function to the displacement, so they are *modal coordinates*. Note that although the foregoing describes u, it is equally applicable to w or θ.

Equation (6.3.5) is a Ritz series like eq. (6.1.8), except that the original basis functions we selected are replaced by the mode functions. In this viewpoint, the η_n are merely a new set of generalized coordinates. However, the modal series has a special significance because it will lead to uncoupled ordinary differential equations for the modal coordinates, just as it did for discrete systems. In the course of proving that the equations for the modal coordinates are uncoupled, we will derive orthogonality relations that are satisfied by the mode functions. The particular form of these orthogonality relations depends on the type of displacement we are analyzing, but the equations for the modal coordinates will be identical to those for discrete systems.

We begin with the case of axial motion by using eq. (6.3.5) to form the kinetic energy. The similarity of eq. (6.3.5) to the fundamental Ritz series, eq. (6.1.8), enables us to write the kinetic energy in the manner of eq. (6.1.1),

$$T = \frac{1}{2}\sum_{j=1}^{N}\sum_{n=1}^{N} \hat{M}_{jn}\dot{\eta}_j\dot{\eta}_n \tag{6.3.6}$$

where the inertia coefficients are given by eq. (6.1.11) with ψ_j replaced by Ψ_j,

$$\hat{M}_{jn} = \int_0^L \Psi_j \Psi_n \rho A\, dx + \sum m \Psi_j(x_m) \Psi_n(x_m) \tag{6.3.7}$$

When we use the second of eqs. (6.2.7) to represent the mode functions, we find that

$$\hat{M}_{jn} = \sum_{r=1}^N \sum_{s=1}^N \Phi_{rj} \Phi_{sn} \left[\int_0^L \psi_r(x) \psi_s(x) \rho A\, dx + \sum m \psi_r(x_m) \psi_s(x_m) \right] \tag{6.3.8}$$

Equation (6.1.11) defines the bracketed term in the above as the inertia coefficient M_{rs}. Therefore, the right side of the equation is the indicial form of the matrix product $\{\Phi_j\}^T[M]\{\Phi_n\}$. In view of the orthogonality of the modal vectors, eq. (6.2.4), this term is zero unless the indices j and n are the same. We thereby obtain the first orthogonality condition for the mode functions of a bar undergoing axial displacement,

$$\boxed{\hat{M}_{jn} \equiv \int_0^L \Psi_j \Psi_n \rho A\, dx + \sum m \Psi_j(x_m) \Psi_n(x_m) = \delta_{jn}} \tag{6.3.9}$$

In other words, the first orthogonality relation is the condition for which the inertia matrix associated with a modal series is the identity matrix. (If we had not normalized the mode vectors $\{\Phi_n\}$ according to eq. (6.2.4), this inertia matrix would be diagonal, but its elements would not have unit values.) A corollary of this result is reduction of the kinetic energy to a sum of squares given by

$$T = \frac{1}{2} \sum_{j=1}^N \dot{\eta}_j^2 \tag{6.3.10}$$

This result should not be surprising, for eq. (6.3.9) is very much like the orthogonality condition for the eigenmodes of a discrete system. The integrand is an inner product of the modal displacements at a common location, weighted by the differential element of mass $\rho A\, dx$. Thus, the integral is the analogy for continuous systems of the summation process entailed with evaluating the modal orthogonality $\{\phi_j\}^T[M]\{\phi_n\}$ when $[M]$ is diagonal.

We obtain the second orthogonality relation for the mode functions by substituting the modal series, eq. (6.3.5), into the potential energy given by eq. (6.1.2). The similar forms of eqs. (6.3.5) and (6.1.8) lead to

$$V = \frac{1}{2} \sum_{j=1}^N \sum_{n=1}^N \hat{K}_{jn} \eta_j \eta_n \tag{6.3.11}$$

where the stiffness coefficients are found from eq. (6.1.13) to be

$$\hat{K}_{jn} = \int_0^L EA \frac{d\Psi_j}{dx} \frac{d\Psi_n}{dx} dx + \sum k \Psi_j(x_k) \Psi_n(x_k) \tag{6.3.12}$$

We substitute the representation of the mode functions given by eq. (6.2.7) to find

$$\hat{K}_{jn} = \sum_{r=1}^N \sum_{s=1}^N \Phi_{rj} \Phi_{sn} \left[\int_0^L EA \frac{d\Psi_r}{dx} \frac{d\Psi_s}{dx} dx + \sum k \Psi_r(x_k) \Psi_s(x_k) \right] \tag{6.3.13}$$

According to eq. (6.1.13), the term within the brackets is the stiffness coefficient K_{rs}. Hence, the term within the double sum is the indicial form of $\{\Phi_j\}^T[K]\{\Phi_n\}$. The second orthogonality relation, eq. (6.2.5), for the mode vectors leads to the second orthogonality relation for the mode functions in axial displacement,

$$\hat{K}_{jn} \equiv \int_0^L EA\frac{d\Psi_j}{dx}\frac{d\Psi_n}{dx}dx + \sum k\Psi_j(x_k)\Psi_n(x_k) = \omega_j^2\delta_{jn} \qquad (6.3.14)$$

Hence, the stiffness matrix associated with the modal coordinates is diagonal, with elements equal to ω_j^2. Substitution of this relation into eq. (6.3.11) shows that the potential energy expressed in terms of the η_j variables is a sum of squares, given by

$$V = \frac{1}{2}\sum_{j=1}^N \omega_j^2\eta_j^2 \qquad (6.3.15)$$

The orthogonality relations for the case of torsional motion have a similar appearance to eqs. (6.3.9) and (6.3.14). We form the mode functions according to eq. (6.3.1), with $[\Phi]$ now taken to be the normal mode matrix associated with the stiffness and mass matrices for a Ritz series representing torsional displacement. We use these mode functions to form the displacement field in the manner of eq. (6.3.5),

$$\theta(x, t) = \sum_{n=1}^N \Psi_n(x)\eta_n(t) \qquad (6.3.16)$$

Because of the similarity in the form of the expressions for T and V in the case of axial and torsional motion, it follows that using this modal series for θ leads to

$$\hat{M}_{jn} = \int_0^L \Psi_j\Psi_n\rho J\, dx + \sum I\Psi_j(x_I)\Psi_n(x_I) = \delta_{jn} \qquad (6.3.17)$$

$$\hat{K}_{jn} = \int_0^L GJ\frac{d\Psi_j}{dx}\frac{d\Psi_n}{dx}dx + \sum \kappa\Psi_j(x_\kappa)\Psi_n(x_\kappa) = \omega_j^2\delta_{jn} \qquad (6.3.18)$$

The same procedure may be used to derive the orthogonality relations for flexural mode functions. After the eigenvalue problem associated with the stiffness and mass coefficients of the Ritz series has been solved, we form the modal series for flexural displacement,

$$w(x, t) = \sum_{n=1}^N \Psi_n(x)\eta_n(t) \qquad (6.3.19)$$

Substitution of this representation into the expression for T in flexural motion, followed by parallel operations to those for axial displacement, leads to the first orthogonality condition for flexural mode functions,

$$\hat{M}_{jn} = \int_0^L \Psi_j\Psi_n\rho A\, dx + \sum m\Psi_j(x_m)\Psi_n(x_m) = \delta_{jn} \qquad (6.3.20)$$

This condition is like those for axial and torsional displacement because we neglected the rotational kinetic energy of attached masses. Similar treatment of the strain energy V leads to the second orthogonality condition for flexural mode functions,

$$\hat{K}_{jn} = \int_0^L EI \frac{d^2\Psi_j}{dx^2} \frac{d^2\Psi_n}{dx^2} dx + \sum k\Psi_j(x_k)\Psi_n(x_k)$$

$$+ \sum \kappa \frac{d\Psi_j}{dx}(x_k) \frac{d\Psi_n}{dx}(x_k) = \omega_j^2 \delta_{jn} \qquad (6.3.21)$$

The primary difference between this relation and those for axial and torsional modes is the presence of second derivatives and the contribution of the first derivative of the mode function when torsional springs are present.

6.3.2 Modal Response

Because the modal versions of the kinetic and potential energies, eqs. (6.3.10) and (6.3.15), respectively, are sums of squares, the equations of motion for the modal amplitudes are uncoupled inertially and elastically. However, as we found for discrete systems, the modal equations of motion are coupled if there is arbitrary damping. To compute modal damping coefficients associated with the light damping approximation directly from the original $[C]$ values, we use eq. (6.3.5) to represent u in the power dissipation function, eq. (6.1.7), which gives

$$\mathscr{P}_{dis} = \sum_{j=1}^N \sum_{j=n}^N \left[\int_0^L \gamma EA \left(\frac{d\Psi_j}{dx}\right)\left(\frac{d\Psi_n}{dx}\right) dx + \sum c\Psi_j(x_c)\Psi_n(x_c) \right] \dot{\eta}_j \dot{\eta}_n$$

$$\mathscr{P}_{dis} = \sum_{j=1}^N \sum_{j=n}^N \hat{C}_{jn} \dot{\eta}_j \dot{\eta}_n \qquad (6.3.22)$$

Substitution of eq. (6.2.7) shows that the modal damping coefficients are

$$\hat{C}_{jn} = \hat{C}_{nj} = \int_0^L \gamma EA \left(\frac{d\Psi_j}{dx}\right)\left(\frac{d\Psi_n}{dx}\right) dx + \sum c\Psi_j(x_c)\Psi_n(x_c)$$

$$= \sum_{r=1}^N \sum_{s=1}^N \Phi_{rj}\Phi_{sn} \left[\int_0^L EA \left(\frac{d\psi_r}{dx}\right)\left(\frac{d\psi_s}{dx}\right) dx + \sum_{s=1}^N c\psi_r(x_k)\psi_s(x_k) \right] \qquad (6.3.23)$$

$$= \sum_{r=1}^N \sum_{s=1}^N \Phi_{rj}\Phi_{sn} C_{rs}$$

Thus, $\hat{C}_{jn}$ are the elements of the damping matrix associated with the modal coordinate transformation

$$[\hat{C}] = [\Phi]^T[C][\Phi] \qquad (6.3.24)$$

The last quantities we require to form the equations of motion for the modal coordinates are the modal generalized forces. We obtain these by replacing $\psi_j(x)$ in the power input, so that

$$\mathscr{P}_{in} = \sum_{j=1}^N \hat{Q}_j \dot{\eta}_j \qquad (6.3.25)$$

where

$$\hat{Q}_j = \int_0^L f_x \Psi_j \, dx + \sum F \Psi_j(x_F) \tag{6.3.26}$$

We substitute eq. (6.2.7) for the mode functions and compare the resulting expressions to eqs. (6.1.15), which leads to a relation between the modal generalized forces and those associated with the original basis functions,

$$\hat{Q}_j = \sum_{r=1}^N \Phi_{rj} \left[\int_0^L f_x \psi_r \, dx + \sum F \psi_j(x_F) \right] = \sum_{r=1}^N \Phi_{rj} Q_r \tag{6.3.27}$$

The matrix form of this relation is

$$\boxed{\{\hat{Q}\} = [\Phi]^T \{Q\}} \tag{6.3.28}$$

We see from the preceding that the key step required to implement modal analysis in conjunction with the Ritz series formulation is solution of the eigenvalue problem for the original generalized coordinates q_j. That analysis yields the natural frequencies ω_j and eigenvectors $\{\Phi_j\}$. From those properties, we can form differential equations for the modal coordinates η_j with a few simple computations. In view of the uncoupled nature of the inertia and stiffness matrices associated with η_j, these equations are

$$\{\ddot{\eta}\} + [\Phi]^T [C][\Phi]\{\dot{\eta}\} + [\text{diag}(\omega^2)]\{\eta\} = [\Phi]^T\{Q\} \tag{6.3.29}$$

If $[C]$ represents proportional damping, or if the damping is light, we may drop the off-diagonal damping elements of $[\Phi]^T[C][\Phi]$; see Section 3.3.7. The modal damping constants are then obtained from

$$\zeta_j = \frac{1}{2\omega_j} \hat{C}_{jj} \tag{6.3.30}$$

Alternatively, in many situations we may uncouple these equations based on empirical and/or experimental estimates of modal damping ratios. In any event, modal damping uncouples the equations for the principal coordinates, such that

$$\boxed{\ddot{\eta}_j + 2\zeta_j\omega_j\dot{\eta}_j + \omega_j^2\eta_j = \{\Phi_j\}^T\{Q\}, \quad j = 1, \ldots, N} \tag{6.3.31}$$

If the damping matrix $[C]$ is too large to consider the modal equations to be uncoupled, there is little merit in implementing eqs. (6.3.29). Such cases call for solution of the original equations of motion, eq. (6.1.16). One possible method in that case is frequency domain analysis, supplemented by FFT techniques in the case of a transient problem, which is the subject of the next chapter. We also can employ state-space damped modal analysis (see Chapter 10), which will lead to uncoupled equations without approximation. Another alternative for large $[C]$ is to use numerical integration techniques, such as Runge-Kutta.

The equations for the modal coordinates for a specified set of generalized forces may be solved by standard methods. After we have solved these equations, we must synthesize the displacement field. It is simpler to use the original basis functions ψ_n, rather than the mode functions Ψ_n, for this purpose. Hence, we substitute the mode functions given by eq. (6.2.7) into the modal series, eq. (6.3.5), from which we obtain

$$\boxed{u(x, t) = \sum_{j=1}^N \sum_{n=1}^N \Phi_{nj}\psi_n(x)\eta_j(t) \equiv [\psi(x)][\Phi]\{\eta(t)\}} \tag{6.3.32}$$

The matrix form is particularly useful for computational purposes. Suppose we wish to evaluate $u(x, t)$ at a succession of P points, $x = x_p$. We then may consider $[\psi]$ to be a rectangular array having P rows and N columns. Row p of $[\psi]$ would consist of the sequence of basis function values $\psi_n(x_p)$. Computation of the matrix product $[\psi][\Phi]\{\eta\}$ would yield a column whose elements are the displacements at the various x_p corresponding to the instant at which $\{\eta\}$ is evaluated. This idea may be extended to evaluate time histories at the various x_p locations. Let $[u]$ be an array whose elements are $u_{ps} = u(x_p, t_s)$, where t_s are a succession of time instants. Each column of $[u]$ is the product $[\psi][\Phi]\{\eta(t_s)\}$. In other words,

$$[u] = \Big[\{[\psi][\Phi]\{\eta(t_1)\}\} \ \{[\psi][\Phi]\{\eta(t_2)\}\} \ \ldots\Big] = [\psi][\Phi][\eta] \qquad (6.3.33)$$

where the columns of $[\eta]$ hold the modal coordinates at each time instant. Once $[u]$ is computed in this manner, plots of displacement as a function of time at a specific location may be obtained by extracting the appropriate row of $[u]$, while plots of displacement as a function of position at a specific instant may be obtained by extracting the appropriate column of $[u]$.

6.3.3 Initial Conditions: The Expansion Theorem

We could have obtained eq. (6.3.29) much more directly. To do so we would start from the original equations, $[M]\{\ddot{q}\} + [C]\{\dot{q}\} + [K]\{q\} = \{Q\}$, into which we would substitute the modal transformation, $\{q\} = [\Phi]\{\eta\}$. The orthogonality of conditions for $[\Phi]$ then leads to eq. (6.3.29). Given this observation it is fair to ponder why it was necessary to identify the mode functions. A primary reason is that their properties lead us to a way in which we can address nonzero initial conditions. Of course, if the system is initially at rest in the undeformed position, we have $\eta_j = \dot{\eta}_j = 0$ and may proceed directly to the solution. The case of nonzero initial conditions presents a new challenge because the Ritz series has a finite number of coefficients, while the displacement field is a function. The resolution to this dilemma lies in the *expansion theorem*.

Let us consider the general task of representing an arbitrary admissible function $h(x)$ in terms of a modal series like eq. (6.3.5), in which the coefficients are H_n,

$$h(x) \approx \sum_{j=1}^{N} H_n \Psi_n(x) \qquad (6.3.34)$$

(If $h(x)$ is not an admissible function, this series is not valid, because the left side will lead to a function that satisfies the geometric boundary conditions, while $h(x)$ does not.) The approximate equality results from using a finite length series to represent a continuous function. The error $\varepsilon(x)$ is the difference between the true value of the function and the value obtained from its series representation,

$$\varepsilon(x) = h(x) - \sum_{n=1}^{N} H_n \Psi_n(x) \qquad (6.3.35)$$

The "best fit" of the series is that for which ε is orthogonal to the N mode functions that we have determined; in that case the discrepancy will be associated only with mode functions that have not been addressed in the analysis. We shall use the first orthogonality condition to formulate this best fit requirement. (Recall that this relation has essentially the same form for axial, torsional, and flexural motion.) We replace mode function Ψ_n in eq. (6.3.9) with ε, and correspondingly require that

$$\int_0^L \varepsilon \Psi_j \rho A \, dx + \sum m \varepsilon(x_m) \Psi_j(x_m) = 0, \quad j = 1, \ldots, N \tag{6.3.36}$$

We substitute eq. (6.3.35) into these conditions, which yields

$$\int_0^L \left(h - \sum_{n=1}^N H_n \Psi_n(x) \right) \Psi_j \rho A \, dx$$

$$+ \sum m \left[h(x_m) - \sum_{n=1}^N H_n \Psi_n(x_m) \right] \Psi_j(x_m) = 0, \quad j = 1, \ldots, N \tag{6.3.37}$$

A minor rearrangement of terms, based on the fact that H_n is a constant, converts this to

$$\sum_{n=1}^N H_n \left[\int_0^L \Psi_n \Psi_j \rho A \, dx + \sum m \Psi_n(x_m) \Psi_j(x_m) \right]$$

$$= \int_0^L h \Psi_j \rho A \, dx + \sum m \Psi_j(x_m) h(x_m), \quad j = 1, \ldots, N \tag{6.3.38}$$

According to eq. (6.3.9), the term within the brackets is zero for all j, n pairs except for $n = j$, in which case its value is unity. Hence, we find that

$$H_j = \int_0^L h(x) \Psi_j \rho A \, dx + \sum m \Psi_j(x_m) h(x_m), \quad j = 1, \ldots, N \tag{6.3.39}$$

It usually is more convenient to use the original basis functions to evaluate the coefficients G_j. When we use eq. (6.2.7) to describe the mode functions, we obtain

$$H_j = \sum_{n=1}^N \Phi_{nj} \left[\int_0^L h(x) \psi_n \rho A \, dx + \sum m \psi_n(x_m) h(x_m) \right] \tag{6.3.40}$$

The matrix form of this relation is

$$\boxed{\{H\} = [\Phi]^T \{h\}, \qquad h_n = \int_0^L h(x) \psi_n \rho A \, dx + \sum m \psi_n(x_m) h(x_m)} \tag{6.3.41}$$

We may apply the expansion theorem given by eqs. (6.3.34) and (6.3.41) directly to the task of determining the initial values of η_j and $\dot{\eta}_j$. To do so, we merely replace $h(x)$ by $u_0(x)$ and $\dot{u}_0(x)$, respectively. Note that the initial displacement and velocity functions must be kinematically admissible, that is, they must satisfy all geometric boundary conditions. It is interesting to observe at this juncture that the analysis of free and transient vibration with nonzero initial conditions constitutes the only situation that actually requires knowing the specific orthogonality condition for the mode functions. In the case of transient excitation with the system initially at rest in the static equilibrium condition, the initial values are identically zero. For an analysis of the steady-state response to harmonic excitation, the initial conditions are irrelevant.

As a closure for this section, let us summarize the steps needed to implement a modal analysis of response. We begin by selecting a set of basis functions and determining $[M]$, $[K]$, $[C]$, and $\{Q\}$ appropriate to the type of motion of interest. We then determine the natural frequencies ω_j and normalized mode matrix $[\Phi]$ by solving the eigenvalue problem, eq. (6.2.2), associated with the Ritz series coefficients. We then solve the uncoupled differential equations for the modal coordinates, eq. (6.3.31). If necessary, we find the associated initial conditions by employing the expansion theorem, eq. (6.3.41). Finally, we construct the displacement field at any locations and instants of interest according to eq. (6.3.32).

6.3.4 Free Vibration

We consider here situations where there are no external forces exciting the bar. If the initial displacement field $u(x, 0) = u_0(x)$ and the initial velocity field $\dot{u}(x, 0) = \dot{u}_0(x)$ are not both zero, the bar will execute a free vibration. (Recall that we are using u as a generic symbol representing any of the types of displacement that a bar may undergo.) Free vibration is described by the complementary solution for the modal coordinates. The initial conditions for these variables may be found by applying the expansion theorem, eq. (6.3.41), which gives

$$
\eta_j(0) = \sum_{n=1}^{N} \Phi_{nj} \left[\int_0^L u_0(x)\psi_n \rho A\, dx + \sum m\psi_n(x_m)u_0(x_m) \right]
$$

$$
\dot{\eta}_j(0) = \sum_{n=1}^{N} \Phi_{nj} \left[\int_0^L \dot{u}_0(x)\psi_n \rho A\, dx + \sum m\psi_n(x_m)\dot{u}_0(x_m) \right]
$$

(6.3.42)

It is likely that the damping will be light, so we shall consider the damping ratios ζ_j to be less than critical. The corresponding homogeneous solution of the modal coordinate equations (6.3.31) is

$$
\eta_j = \exp(-\zeta_j\omega_j t)\Big\{ \eta_j(0)\cos[(1 - \zeta_j^2)^{1/2}\omega_j t]
$$

$$
+ \frac{\dot{\eta}_j(0) + \zeta_j\omega_j\eta_j(0)}{(1 - \zeta_j^2)^{1/2}\omega_j}\sin[(1 - \zeta_j^2)^{1/2}\omega_j t] \Big\}
$$

(6.3.43)

These expressions are readily evaluated at any instant. The corresponding displacement field may be determined from the modal series using either eq. (6.3.5), which uses the computed mode functions, or eq. (6.3.32), which uses the original basis function. The result is like the general free vibration described by eq. (6.3.4), except for the inclusion of damping. A comparison of the two shows that the previous complex amplitude factor is

$$
B_j = \eta_j(0) - i\left[\frac{\dot{\eta}_j(0) + \zeta_j\omega_j\eta_j(0)}{(1 - \zeta_j^2)^{1/2}\omega_j} \right]
$$

(6.3.44)

The vibration modes whose spatial dependence is most similar to the spatial distribution of the initial conditions will have the largest values of $|B_j|$. In the early stages of a free vibration, these will be the strongest participants in the ensuing free vibration. Thus, we need to be certain that the number of terms used to form the modal series is sufficiently large that the B_j values for the highest j in the series are small compared to the largest terms. However, with increasing elapsed time, modes with the largest values of $\zeta_j \omega_j$ will decay most rapidly. If all modal damping ratios have comparable magnitude, the low-frequency modes will be most persistent, and fewer terms in the modal series will be required as t increases.

6.3.5 Impulse Response

Our interest here is in situations where some or all of the generalized forces are well modeled as impulses. The magnitude of an impulsive force is its impulse $G = \int F \, dt$. Thus, we write the modal equations of motion in this case as

$$\ddot{\eta}_j + 2\zeta_j \omega_j \dot{\eta}_j + \omega_j^2 \eta_j = \{\Phi_j\}^{\mathrm{T}} \{G\} \delta(t) \tag{6.3.45}$$

We assume that the initial conditions are zero, $\eta_j(0) = \dot{\eta}_j(0) = 0$.

The modal differential equation represents a one-degree-of-freedom system having a unit modal mass. The impulse of the excitation acting on this system is $\{\Phi_j\}^{\mathrm{T}}\{G\}$. The impulse equals the momentum change, which occurs instantaneously in the impulse function approximation. In view of the unit value of the modal mass, this leads to initial conditions at $t = 0^+$ given by $\{\eta_j\} = 0$ and $\dot{\eta}_j = \{\Phi_j\}^{\mathrm{T}}\{G\}$. After the impulsive force acts, which corresponds to $t > 0^+$, the system executes a free vibration. It follows from eq. (6.3.43) that the modal impulse response is

$$\boxed{\eta_j = \frac{\{\Phi_j\}^{\mathrm{T}}\{G\}}{(1 - \zeta_j^2)^{1/2}\omega_j} \exp(-\zeta_j \omega_j t) \sin[(1 - \zeta_j^2)^{1/2} \omega_j t]} \tag{6.3.46}$$

This expression shows that an impulsive force excites every mode for which $\{\Phi_j\}^{\mathrm{T}}\{G\}$ is nonzero. The contribution of each modal coordinate to the displacement field is $\Psi_j(x)\eta_j$, but all Ψ_j have comparable orders of magnitude. Also, in view of the oscillatory nature of $\Psi_j(x)$, it is likely that $\{\Phi_j\}^{\mathrm{T}}\{G\}$ will not become progressively smaller with increasing j. In such cases, convergence of the modal series will be a consequence of the presence of ω_j in the denominator of eq. (6.3.46). Thus, many modes will be required to attain convergence for early values of t. However, the instantaneous contribution of a mode is attenuated by the exponential decay factor $\zeta_j \omega_j$. As in the case of free vibration, if the ζ_j values are comparable in magnitude, then the high-frequency modes will be attenuated more rapidly.

6.3.6 Harmonic Steady-State Response

As usual, we represent a harmonic excitation in complex notation. If the distributed and concentrated forces exciting the bar all vary at frequency ω, we will find that the generalized forces are

$$Q_j = \mathrm{Re}[F_j \exp(i\omega t)] \tag{6.3.47}$$

The steady-state solution for the modal amplitudes obtained from eq. (6.3.31) is

$$
\eta_j = \mathrm{Re}\left[\frac{\{\Phi_j\}^{\mathrm{T}}\{F\}}{\omega_j^2 + 2i\zeta_j\omega_j\omega - \omega^2}\exp(i\omega t)\right]
\tag{6.3.48}
$$

We form the displacement field associated with this set of modal amplitudes according to the modal series, eq. (6.3.32). It is useful to separate the dependencies on x and t in this expression, such that

$$
u(x, t) = \mathrm{Re}[X(x)\exp(i\omega t)]
$$

$$
X(x) = \sum_{j=1}^{N} [\psi]\{\Phi_j\}\frac{\{\Phi_j\}^{\mathrm{T}}\{F\}}{\omega_j^2 + 2i\zeta_j\omega_j\omega - \omega^2}
\tag{6.3.49}
$$

Because $[\psi]\{\Phi_j\} = \{\Psi_j\}$, the fractional factor represents the contribution of each mode to the displacement field. Note that unless all $\zeta_j = 0$ and all elements of $\{F\}$ are in phase, we will find that $X(x)$ is complex. We may write it in polar form as $X(x) = |X(x)|\exp[-i\Theta(x)]$. If we were to simultaneously measure as a function of t the steady-state response at different locations, we would see each response varies harmonically at frequency ω and amplitude $|X(x)|$. The lag of each displacement relative to a cosine would be $\Theta(x)/\omega$. A different way of regarding this response is to take a "snapshot" of the steady-state displacement field at different instants, for example with a strobe light. At each instant we would see different functions of x. Only in the undamped case is $X(x)$ real, in which case the proportions of the displacement at different locations remain constant as the overall amplitude varies.

Equation (6.3.49) enables us to recognize the primary factors influencing the steady-state response. In view of eq. (4.3.18), $\{\Phi_j\}^{\mathrm{T}}\{F\}$ is the jth modal generalized force. Furthermore, the denominator represents the complex stiffness of mode j. From this, we conclude that the amount by which each mode contributes to the spatial distribution function $X(x)$ depends on two factors. The degree to which the excitation resembles the mode dictates the magnitude of the modal generalized forces, while the proximity of the excitation frequency to the natural frequency associated with that mode affects the magnitude of the complex frequency response.

There are many similarities between eq. (6.3.49) and its analog for discrete systems. One key difference is that for a discrete system, N is set at the number of degrees of freedom. In contrast, for a Ritz series, N is the number of basis functions, which is at our discretion. The properties of eq. (6.3.49) lead to some general criteria regarding the selection of N. Suppose $\omega_{\max}$ is the highest possible excitation frequency. For any $\omega \le \omega_{\max}$, the contribution of the high-frequency modes, whose natural frequency is substantially larger than $\omega_{\max}$, essentially decreases as the inverse square of ω_j. It follows that modes whose natural frequency is some multiple of $\omega_{\max}$, that is, $\omega_j > \beta\omega_{\max}$, may be ignored. Values of β in common usage lie in the range $2 \le \beta \le 4$. This leads to a guideline for selecting N, specifically, that $\omega_N \ge \beta\omega_{\max}$. A warning regarding this guideline pertains to the fact that the modal force $\{\Phi_j\}^{\mathrm{T}}\{Q\}$ describes the significance of the force magnitudes. Although the complex frequency response factor for a mode might seem to be very small above $\omega_{\max}$, a mode for which

$\omega_j > \beta\omega_{max}$ might have a modal force $\{\Phi_j\}^T\{Q\}$ that is very large in comparison to the lower modal forces.

The guideline $\omega_N \geq \beta\omega_{max}$ may be extended to transient excitation. To do so, we need to consider the DFT spectrum of the excitation. By definition, the portion of the spectrum above the Nyquist critical frequency is not significant. Thus, we may use the Nyquist frequency as ω_{max}.

The preceding truncation guideline does not address a basic question. Specifically, the guidelines are expressed in terms of mode functions and natural frequencies. Because these quantities are derived from finite-length Ritz series, they are only approximations of the true quantities. One might question whether truncating the series according to the suggested guidelines will give natural frequencies and mode functions that are sufficiently accurate. We explore this issue in detail in the next section. In essence, we will see that the truncation guideline, $\omega_N > \beta\omega_{max}$, is more than adequate to ensure accurate evaluation of modes whose natural frequency does not exceed ω_{max}.

EXAMPLE 6.10

In Example 6.3, the coefficient matrices associated with a Ritz series analysis of torsional motion were evaluated. Consider a case where, at $t = 0$, the system was at rest in its static equilibrium position, at which instant a step torque $\Gamma h(t)$ was applied to the right end. Use a four-term Ritz series to determine the ensuing rotation as a function of time at the left and right ends of the bar and at the location of the flywheel.

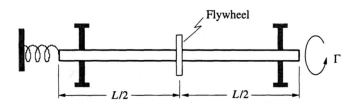

Solution The main objective here is to illustrate the procedures whereby modal analysis for discrete systems may be employed in conjunction with the Ritz series method to analyze the response of bars. The Ritz series employed in Example 6.3 is

$$\theta = \sum_{j=1}^{4}\left(\frac{x}{L}\right)^{j-1}q_j$$

The equations of motion are

$$\rho R^4 L[M']\{\ddot{q}\} + \frac{GR^4}{L}[K']\{q\} = \{Q\}$$

where $[M']$ and $[K']$ are numerical factors listed in Example 6.3 and $Q_j = \Gamma$. In order to remove the dimensional properties from consideration, we define a nondimensional time $\tau = \Omega t$, which leads to $\{\ddot{q}\} = \Omega^2\{d^2q/d\tau^2\}$. Thus, if we set $\rho R^4 L\Omega^2 = (GR^4)/L$, that is

$$\Omega = \left(\frac{G}{\rho L^2}\right)^{1/2}$$

we convert the equations of motion to

$$[M']\left\{\frac{d^2q}{d\tau^2}\right\} + [K']\{q\} = \frac{L}{GR^4}\{Q\}$$

so that the system parameters merely scale the magnitude of the generalized forces. It is specified that there is no initial displacement or velocity; the corresponding initial conditions are $\{q\} = \{dq/d\tau\} = \{0\}$ at $\tau = 0$.

We obtain the solution by modal analysis. The first step is to use mathematical software to evaluate the undamped modal properties by solving

$$[[K'] - \omega^2[M']]\{\phi\} = \{0\}$$

and then to normalize the modes. This result is

$$\omega_1 = 0.359, \quad \omega_2 = 1.080, \quad \omega_3 = 3.716, \quad \omega_4 = 15.916$$

$$[\Phi] = \begin{bmatrix} 0.0241 & -0.0109 & -0.0301 & 0.0398 \\ -0.0708 & -0.0604 & -0.1800 & -0.3507 \\ 0.7069 & 0.5778 & 0.7801 & 0.7910 \\ -0.7034 & -0.8139 & -0.5985 & -0.4997 \end{bmatrix}$$

(The natural frequencies are nondimensional quantities; the corresponding dimensional quantities are $\Omega\omega_j$.) We now introduce the modal transformation, which uncouples the equations of motion,

$$\{q\} = [\Phi]\{\eta\}$$

$$\frac{d^2\eta_j}{d\tau^2} + \omega_j^2\eta_j = \{\Phi_j\}^{\mathrm{T}}\frac{L}{GR^4}\{Q\} \equiv \frac{\Gamma L}{GR^4}F_j h(\tau)$$

$$F_j = \sum_{n=1}^{4} \Phi_{nj}$$

The solution for each modal coordinate is $(\Gamma L F_j/GR^4)$ multiplied by the unit step response for an undamped one-degree-of-freedom system having unit mass and natural frequency ω_j,

$$\eta_j = \left(\frac{\Gamma L}{GR^4}\right)\left(\frac{F_j}{\omega_j^2}\right)\{1 - \cos(\omega_j\tau)\}h(\tau)$$

The last step is to evaluate the torsional displacements. We shall implement eq. (6.3.33). The time increments at which the modal coordinates are evaluated must be sufficiently small to resolve the shortest period of each step response, which corresponds to the highest natural frequency. Hence, we set $\Delta = (2\pi/\omega_4)/8$. There is no damping, so the response does not decay. If we take the largest value of τ to be four times the longest (nondimensional) period of a modal coordinate, we should be able to see the general pattern. We therefore take the largest t of interest to be $8\pi/\omega_1$. Dividing the observation period by the sampling interval leads to $J = \mathrm{int}\,[(8\pi/\omega_1)/\Delta] + 1$ intervals, so the discretized time instants are $\tau_n = (n-1)\Delta$, $n = 1, \ldots, J$.

We store the values of η_j at each discrete instant τ_n as successive columns of a $4 \times J$ array $[Z]$. There are three locations of interest, so we define $[\psi]$ to be a 3×4 matrix whose rows hold the basis function values at $x = 0$, $L/2$, and L. The displacement matrix $[u]$ obtained from the product $[\psi][\Phi][Z]$ will be a matrix whose columns are the values of $\{q\}$ at each instant. We may implement these operations in Mathcad by writing

Mathcad: $\quad \Delta = \dfrac{\pi}{4\omega_4} \quad J := ceil\left(\dfrac{8\pi}{\omega_1\Delta}\right) \quad n := 1;J \quad \tau_n := (n-1)*\Delta$

$$j := 1;4 \quad Z_{j,n} = \left(\frac{1}{\omega_j^2}\right)(1 - \cos(\omega_j\tau_n))$$

$$p := 1;3 \quad x_1 = 0 \quad x_2 = 0.5 \quad x_3 = 1 \quad \psi_{p,j} := (x_p)^{j-1}$$

$$u = \psi * \Phi * Z$$

This computation is based on setting $\Gamma L F_j / G R^4 = 1$ and considering x to be nondimensional. The corresponding operations in MATLAB are

MATLAB: $delta = pi/(4 * omega(4))$; $J = ceil(8 * pi/(omega(1) * delta))$;
$tau = 0 : delta : (J - 1) * delta$;
$for\ j = 1 : 4$; $for\ n = 1 : J$;
$Z(j, n) = (1/omega(j)^2) * (1 - \cos(omega(j) * tau(n)))$; end; end
$x = [0; 0.5; 1]$; $for\ j = 1 : 4, psi(:, j) = x.^(j - 1)$
$u = psi * PHI * Z$;

Before we examine the results, it is useful to consider a one-degree-of-freedom model in which the shaft is rigid, in which case the only elastic effect would be the torsional spring. That solution may be obtained from the present analysis by taking the Ritz series to contain only the rigid body mode, with $\psi_1 = 1$. In that case, q_1 would be the rigid body rotation θ_{rb} of each cross-section. The only inertia and stiffness terms in this model are M_{11} and K_{11}, so the nondimensional equation of motion for this model would be

$$M'_{11} \frac{d^2 \theta_{rb}}{d\tau^2} + K'_{11} \theta_{rb} = \frac{\Gamma L}{G R^4} h(\tau)$$

The result would be the step response of a one-degree-of-freedom system whose natural frequency is $(K'_{11}/M'_{11})^{1/2}$.

The first graph depicts the results for the three specified locations and the prediction obtained from the rigid body model. We see that the rotations at the three locations occur nearly

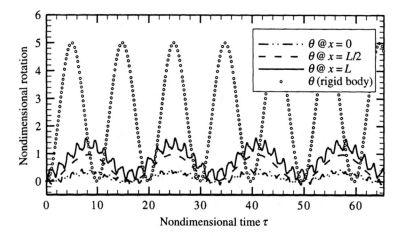

synchronously, but with different amplitudes. The fact that the overall rotation increases substantially with increasing distance from the left end suggests that the torsional spring at the left end is quite stiff relative to the elastic stiffness of the bar. We also see that the rigid body model yields a prediction that oscillates too rapidly, and also overpredicts the rotation. The overprediction can be explained by considering a given rotation at the right end where the torque is applied. In reality, the work done by that torque will be stored partially as kinetic and potential energy associated with deformation of the bar. Because the rigid body model does not address the bar's deformation, the only place where the same amount of work can be deposited in that model is in kinetic energy of a rigid bar and of the flywheel, and in potential energy of the spring, which thereby increases those effects. The higher oscillation rate obtained from the rigid body model has a different explanation, which is addressed in the next section.

As a last step to the analysis, let us examine whether we used a sufficient number of terms to formulate the Ritz series. If we have constructed the worksheet for our software program appropriately, it will be possible to lengthen the series by merely extending the index range over which the elements of $[M']$, $[K']$, and $\{F\}$ are computed. In the next graph the rotation at $x = L$ using eight terms is compared to the four-term result. It is evident that the extended series yields the same overall pattern, with minor differences in the high-frequency fluctuation.

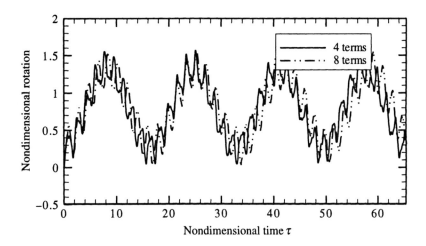

6.4 RAYLEIGH RATIO AND CONVERGENCE PROPERTIES

We thus far have not paid much attention to selecting the number and type of basis functions. This is an important consideration, because without it we are left to wonder whether the results we obtain by the Ritz method are a valid representation of the actual response. The procedure developed in the previous section for evaluating response relies on the eigensolution for the free system, so the behavior of the natural frequencies and modes is the central issue.

We initiate the investigation by considering the Rayleigh ratio, which was suggested by Lord Rayleigh as a simple way of estimating the fundamental natural frequency. (See Rayleigh (1945), which is a reprint of his book, whose first edition was published in 1877.) Suppose a bar is executing a free vibration in the absence of dissipation. With the understanding that it does not matter whether the vibration is extensional, torsional, or flexural, we let $u(x, t)$ denote the displacement field. If this free vibration consists of only mode j, then $u = \mathrm{Re}[\Psi_j(x)\exp(i\omega_j t)]$, where we take the amplitude factor for the mode to be unity, $B_j = 1$. Correspondingly, the velocity field is given by $\dot{u} = \omega_j \mathrm{Re}[i\Psi_j(x)\exp(i\omega_j t)]$. It follows that at intervals of one-half the natural period $2\pi/\omega_j$ the velocity is zero everywhere. At those instants, the magnitude of the displacement at all locations is a maximum, given by $\max(u) = \Psi_j(x)$. Correspondingly, at these instants the kinetic energy is zero and the potential energy has its maximum value

$$\boxed{\max(V) = \Lambda} \tag{6.4.1}$$

We may determine Λ by replacing the displacement field u, θ, or w by Ψ_j in the expression for V given by eq. (6.1.5), (6.1.24), or (6.1.34), respectively. For example, for axial motions, we have

$$\Lambda = \frac{1}{2}\int_0^L EA\left(\frac{\partial \Psi_j}{\partial x}\right)^2 dx + \frac{1}{2}\sum k\Psi_j(x_k)^2 \tag{6.4.2}$$

At intervals of one-quarter of the natural period following occurrence of the maximum displacement, the bar returns to its undeformed position, $u = 0$, so $V = 0$ at those instants. The magnitude of $\dot{u}$ at all locations then is a maximum, given by $\max(\dot{u}) = \omega_j \Psi_j$. Hence, the kinetic energy at these instants has a maximum value given by

$$\max(T) = \omega_j^2 Y \qquad (6.4.3)$$

where we find Y by replacing the velocity field $\dot{u}$, $\dot{\theta}$, or $\dot{w}$ by Ψ_j in the expression for T given by eqs. (6.1.5), (6.1.24), or (6.1.34), respectively. For example, the case of axial motion leads to

$$Y = \frac{1}{2} \int_0^L \Psi_j^2 \rho A \, dx + \frac{1}{2} \sum m \Psi_j(x_m)^2 \qquad (6.4.4)$$

The central fact now is that, in the absence of dissipation, mechanical energy must be conserved at all instants, such that $T + V$ remains constant. Applying this principle to the instants when displacement or velocity is zero leads to $\max(T) = \max(V)$. It follows that

$$\omega_j^2 = \frac{\Lambda}{Y} \qquad (6.4.5)$$

This identity has little direct value, for it relies on knowing the mode function. However, a simple generalization leads to Rayleigh's ratio. Suppose we regard Y and Λ as functionals of a *trial function* $\psi(x)$ that is not necessarily a mode function. (The term *functional* refers to a quantity that is a function of a function.) We use brackets to indicate the function under consideration, so the functionals corresponding to eqs. (6.4.2) and (6.4.4) are

$$\Lambda[\psi] = \frac{1}{2} \int_0^L EA \left(\frac{\partial \psi}{\partial x} \right)^2 dx + \frac{1}{2} \sum k \psi(x_k)^2$$

$$\qquad (6.4.6)$$

$$Y[\psi] = \frac{1}{2} \int_0^L \psi^2 \rho A \, dx + \frac{1}{2} \sum m \psi(x_m)^2$$

The Rayleigh ratio, which we denote as $\mathcal{R}[\psi]$, is

$$\mathcal{R}[\psi] = \frac{\Lambda[\psi]}{Y[\psi]} \qquad (6.4.7)$$

Clearly, $\mathcal{R}[\psi_j] = \omega_j^2$, but what is this ratio when ψ is not a mode function? The only limitation we place on ψ is that it be an admissible function for the system under consideration, that is, it must satisfy the geometric boundary conditions. We may employ the expansion theorem, eq. (6.3.34), to represent any admissible function. (The fact that we do not actually know the mode functions is not important to this development.) We therefore write

$$\psi = \sum_{n=1}^{N} a_n \Psi_n(x) \qquad (6.4.8)$$

where a_n are (unknown) constants. We next substitute this series into $\Upsilon[\psi]$ and $\Lambda[\psi]$. For the former, we observe that $\Upsilon[\psi]$ is the same as T, except that ψ replaces $\dot{u}$, while $\Lambda[\psi]$ is the same as V with u replaced by ψ. Furthermore, eq. (6.4.8) is like the modal series, eq. (6.3.5), which leads to diagonal inertia and stiffness matrices. It follows that

$$\Upsilon[\psi] = \frac{1}{2}\sum_{j=1}^{N} a_j^2$$

$$\Lambda[\psi] = \frac{1}{2}\sum_{j=1}^{N} \omega_j^2 a_j^2$$

(6.4.9)

We substitute these expressions into the definition of $\mathfrak{R}[\psi]$ and factor out the fundamental natural frequency from the term in the numerator, so that

$$\mathfrak{R}[\psi] = \omega_1^2 \frac{\displaystyle\sum_{j=1}^{N} \frac{\omega_j^2}{\omega_1^2}a_j^2}{\displaystyle\sum_{j=1}^{N} a_j^2}$$

(6.4.10)

By definition the natural frequencies are numbered in ascending order, so that $\omega_j/\omega_1 \geq 1$. Hence, every term in the sum contained in the numerator is greater than, or equal to, the corresponding term in the denominator sum. It follows that the fraction has a value greater than or equal to unity, which leads to the conclusion that

$$\boxed{\mathfrak{R}[\psi] \geq \omega_1^2}$$

(6.4.11)

The equality of $\mathfrak{R}[\psi]$ and ω_1^2 corresponds to all coefficients $a_j = 0$ for $j > 1$. In other words,

The square root of the Rayleigh ratio corresponding to an arbitrary admissible trial function is an upper bound for the fundamental frequency of the system, with equality occurring only if the trial function actually is the fundamental mode function.

To gain some insight into how large the error in this bound might be, suppose we believe that our choice for ψ has a small error ε relative to the true fundamental mode function. That is, we believe that $\psi = \Psi_j[1 + \varepsilon f(x)]$, where $f(x)$ is a function having unit order of magnitude. The normalization factor for the mode is irrelevant because we are forming ratios. Thus, to indicate the assumed proximity of the trial function to the first mode, we set $a_1 = 1$ in the modal series, eq. (6.4.8), and consider all other $a_j = O(\varepsilon)$. Equation (6.4.10) then indicates that

$$\mathfrak{R}[\psi] = \omega_1^2 \frac{1 + \displaystyle\sum_{j=2}^{N} \frac{\omega_j^2}{\omega_1^2}O(\varepsilon^2)}{1 + \displaystyle\sum_{j=1}^{N} O(\varepsilon^2)} = \omega_1^2[1 + O(\varepsilon^2)]$$

(6.4.12)

We deduce from this that the upper bound estimate obtained from the Rayleigh ratio should be much better than the trial function used to evaluate the ratio.

When we implement this estimate, it is useful to observe that the fundamental mode function has the least curvature because its only nodes (i.e., zeroes) are those imposed by the geometric boundary conditions. If we select a trial function that has no more nodes than necessary to satisfy the geometric boundary conditions and has the least amount of curvature for that type of function, we should be able to estimate ω_1 to within 40%. For example, a cantilevered uniform beam has the geometric boundary conditions $w = \partial w/\partial x = 0$ at $x = 0$. If we choose to use a power function $\psi = (x/L)^n$ as the trial function, satisfying the geometric boundary conditions requires $n \geq 2$. For $n = 2$, the Rayleigh ratio gives $(\rho AL^4/EI)^{1/4}\omega_1 \leq 4.472$, which is 27% greater than the true value of 3.516. In contrast, because $\psi = (x/L)^4$ exhibits greater relative curvature, it gives a much higher estimate, $(\rho AL^4/EI)^{1/4}\omega_1 \leq 16.100$. Now suppose we choose to use as a trial function $\psi = x \sin(\pi x/L)$, which properly satisfies the geometric boundary conditions at $x = 0$, but also has a zero at $x = L$, which is not a geometric boundary condition for a cantilevered beam. The Rayleigh ratio estimate obtained from this function is $(\rho AL^4/EI)^{1/4}\omega_1 < 17.52$. (In fact, the last function is an excellent trial function for estimating the fundamental frequency of a uniform clamped-hinged beam, whose fundamental frequency is $(\rho AL^4/EI)^{1/4}\omega_1 = 15.418$.)

With very little effort, the Rayleigh ratio can be used to check numerical work. A comparison of the functionals in eq. (6.4.6) to the inertia coefficients, eq. (6.1.11), and the stiffness coefficients, eq. (6.1.36), shows that $\Upsilon[\psi_j] = M_{jj}$ and $\Lambda[\psi_j] = K_{jj}$. Because $\mathscr{R}[\psi_j] \geq \omega_1^2$, it follows that

$$\boxed{\omega_1^2 \leq \min\left(\frac{K_{jj}}{M_{jj}}\right)} \tag{6.4.13}$$

Hence, we can obtain a quick estimate of the fundamental natural frequency by examining the ratios of corresponding diagonal terms in the stiffness and inertia matrices. If we have selected the first basis function to be the one that has the least curvature, then it is likely that K_{11}/M_{11} will be the minimum value.

The Rayleigh ratio has applications beyond merely estimating the natural frequency. An important property is its stationarity. Suppose that the trial function ψ we select is close to any of the mode functions. We express this proximity by setting $a_j = 1$, where j is the number of the mode that ψ is presumed to resemble. We set the other $a_n = \varepsilon a_n'$, where $|\varepsilon| \ll 1$ and a_n' are $O(1)$ quantities. In this case eqs. (6.4.9) become

$$\Upsilon[\psi] = \frac{1}{2}\left[1 + \varepsilon^2 \sum_{\substack{n=1 \\ n \neq j}}^{N} (a_n')^2\right]$$

$$\Lambda[\psi] = \frac{1}{2}\omega_j^2\left[1 + \varepsilon^2 \sum_{\substack{n=1 \\ n \neq j}}^{N} \frac{\omega_n^2}{\omega_j^2}(a_n')^2\right] \tag{6.4.14}$$

Because ε is small, it is sufficient to retain only the leading terms in series expansions, so the Rayleigh ratio corresponding to the above relations is

$$\mathscr{R}[\psi] = \omega_j^2 \frac{1 + \varepsilon^2 \sum_{\substack{n=1 \\ n \neq j}}^{N} \frac{\omega_n^2}{\omega_j^2}(a_n')^2}{1 + \varepsilon^2 \sum_{\substack{n=1 \\ n \neq j}}^{N} (a_n')^2} = \omega_j^2 \left[1 + \varepsilon^2 \sum_{\substack{n=1 \\ n \neq j}}^{N} \left(\frac{\omega_n^2}{\omega_j^2} - 1 \right)(a_n')^2 + O(\varepsilon^4) \right] \quad (6.4.15)$$

Differentiating the ratio with respect to ε leads to

$$\left. \frac{d\mathscr{R}[\psi]}{d\varepsilon} \right|_{\varepsilon=0} = 0 \qquad (6.4.16)$$

It follows that the Rayleigh ratio has an extreme value when the trial function is a mode function, $\varepsilon = 0$. We say that the Rayleigh ratio is stationary to perturbations of the trial function when the trial function is actually a mode function. The quantity within the summation in eq. (6.4.15) may be either positive or negative because ω_n may be above or below the selected ω_j. Thus, it would seem as though the Rayleigh ratio could be a maximum or minimum at $\varepsilon = 0$, but we will soon see that it is a minimum.

The stationarity property has important implications for the representation of the response as a Ritz series. Let us use an N-term Ritz series as the trial function, so that

$$\psi = \sum_{j=1}^{N} c_j \psi_j(x) \qquad (6.4.17)$$

The similarity of $Y[\psi]$ and $\Lambda[\psi]$ to T and V, respectively, enables us to employ eqs. (6.1.10) and (6.1.12) to express the functionals. Because the basis functions are specified, the functionals now depend on the coefficients c_j, so we have

$$Y[c_j] = \frac{1}{2} \sum_{j=1}^{N} \sum_{n=1}^{N} M_{jn} c_j c_n$$

$$\Lambda[c_j] = \frac{1}{2} \sum_{j=1}^{N} \sum_{n=1}^{N} K_{jn} c_j c_n \qquad (6.4.18)$$

Because $\mathscr{R}[c_j] \equiv \Lambda[c_j]/Y[c_j]$ has an extreme value when the c_j values are those for a mode function, we set $d\mathscr{R}/dc_j = 0$ for $j = 1, \ldots, N$, or

$$\frac{d\mathscr{R}[c_j]}{dc_j} = \frac{1}{Y[c_j]} \frac{d\Lambda[c_j]}{dc_j} - \frac{\Lambda[c_j]}{Y[c_j]^2} \frac{dY[c_j]}{dc_j}$$

$$\equiv \frac{1}{Y[c_j]} \left\{ \frac{d\Lambda[c_j]}{dc_j} - (\mathscr{R}[c_j]) \frac{dY[c_j]}{dc_j} \right\} = 0, \quad j = 1, \ldots, N \qquad (6.4.19)$$

The value of $Y[c_j]$ is finite, so the term inside the braces must be zero. The process of evaluating the derivatives of the quadratic sums describing $Y[c_j]$ and $\Lambda[c_j]$ is the same as that used in Section A.1 to obtain the standard Lagrange's equations corresponding to quadratic sums for T and V. Hence, we have

$$\frac{d\Upsilon[c_j]}{dc_j} = \sum_{n=1}^{N} M_{jn}c_n$$

$$\frac{d\Lambda[c_j]}{dc_j} = \sum_{n=1}^{N} K_{jn}c_n$$

(6.4.20)

Correspondingly, satisfying eq. (6.4.19) requires that

$$\sum_{n=1}^{N} K_{jn}c_n - (\mathscr{R}[c_j]) \sum_{n=1}^{N} M_{jn}c_n = 0, \quad j = 1, ..., N$$

(6.4.21)

The matrix form equivalent to the foregoing is

$$[[K] - (\mathscr{R}[c_j])[M]]\{c\} = \{0\}$$

(6.4.22)

This is the same matrix eigenvalue problem as eq. (6.2.2), which leads to the mode functions. We therefore conclude that the values of $\{c\}$ that extremize the Rayleigh ratio are proportional to the mode vectors $\{\Phi_n\}$, and that the corresponding extremized values of the Rayleigh ratio are the values ω_n^2. The process of finding the modal equations by extremizing the Rayleigh ratio according to eqs. (6.4.19) and (6.4.22) is known as the *Rayleigh-Ritz method*. Many texts use it as the fundamental tool for finding modal properties associated with a Ritz series. However, this procedure only addresses the eigensolution, so it is a subset of the formulation in Sections 6.1 to 6.3, which leads to equations of motion that include the effects of external forces and dissipation.

Because the stationary values of the Rayleigh ratio are the eigenvalues ω_n^2 derived from the Ritz method, we may apply two theorems pertaining to $\mathscr{R}[c_j]$ directly to the natural frequencies obtained by solving the Ritz eigenvalue problem, eq. (6.2.2). To discuss these properties we need to distinguish between a true eigensolution for the continuous system and one obtained from an N-term Ritz series. To do so we use a superscript equal to the series length to denote any quantity derived from a Ritz series, while quantities without superscripts denote quantities that are derived in the limit $N \longrightarrow \infty$, or equivalently, by solving the field equations. The first property of interest is the *upper bound theorem*. It states that the extreme values of $\mathscr{R}\{c_j\}$, which equal the various $(\omega_n^{(N)})^2$, constitute upper bounds to the true value of ω_n^2. In other words,

$$\boxed{\omega_n \leq \omega_n^{(N)}, \quad n = 1, ..., N}$$

(6.4.23)

The other property involves two alternative analyses. In the second analysis, the Ritz series is extended by adding one basis function to the set of N functions used in the first analysis. Then, for $n = 1, ..., N$, the natural frequencies $\omega_n^{(N+1)}$ will be less than, or equal to, the corresponding $\omega_n^{(N)}$. Furthermore for $n = 2, ..., N + 1$, the frequencies $\omega_n^{(N+1)}$ will be greater than, or equal to, the preceding $\omega_{n-1}^{(N)}$. This is called the *separation theorem,* which we may express as

$$\boxed{\omega_1^{(N+1)} \leq \omega_1^{(N)} \leq \omega_2^{(N+1)} \leq \omega_2^{(N)} \leq \cdots \leq \omega_N^{(N+1)} \leq \omega_N^{(N)} \leq \omega_{N+1}^{(N+1)}}$$

(6.4.24)

In other words, the approximate natural frequencies obtained by adding one term separate the ranges for the approximate frequencies associated with the shorter series.

Suppose we continue to add terms to the Ritz series without changing the preceding basis functions. We conclude from eq. (6.4.24) that this will lead to decreasing values of the natural frequencies, while eq. (6.4.23) states that the natural frequencies will never be smaller than the true values. Simultaneously satisfying both characteristics leads us to the conclusion that

As the Ritz series length $N \longrightarrow \infty$, the approximate natural frequencies monotonically approach the corresponding true frequencies from above.

To a certain extent this property is of theoretical interest only, because we obviously cannot form an infinite-length series. Furthermore, it does not address the computational difficulties associated with the accumulation of error that we often encounter as we increase N, which is a phenomenon displayed in Example 6.12. Nevertheless, it does constitute a proof that *natural frequencies, and therefore mode functions, obtained by the Ritz method will tend to converge to the true eigensolutions.*

The mathematical proofs of the upper bound and separation theorems are quite sophisticated; the interested reader may find them in the work by Meirovitch (1997). A heuristic argument is available to support both theorems. Suppose that ψ_j constitute a set of admissible basis functions whose number may be extended indefinitely. These functions form a set that completely spans the space of admissible functions, which means that any trial function may be expressed as an infinite series of these basis functions,

$$\psi(x) = \sum_{j=1}^{\infty} c_j \psi_j(x) \tag{6.4.25}$$

The constants c_j are determined by minimizing the Rayleigh ratio. When we limit the trial function to be merely an N-term Ritz series, as in eq. (6.4.17), we are imposing additional constraints that $c_{N+1} = c_{N+2} = c_{N+3} = \cdots = 0$. In general, imposing constraints on a system tends to make the system stiffer, which means that the constrained system will have higher natural frequencies than the unconstrained one. This is the aspect contained in the upper bound theorem, eq. (6.4.23). (This argument is given by Meirovitch (1967).) When we lengthen the series by one term, we effectively remove the constraint that $c_{N+1} = 0$. This makes the system less stiff, thereby reducing the natural frequencies, which tends to explain the separation theorem, eq. (6.4.24).

The computations required to form the equations of motion associated with lengthening the Ritz series may be reduced significantly by exploiting a fundamental feature. Suppose we have already formulated the equations of motion, eq. (6.1.16), for an N-term Ritz series given by

$$u^{(N)} = \sum_{j=1}^{N} \psi_j(x) q_j(t) \tag{6.4.26}$$

Correspondingly, we have evaluated the inertia, stiffness, damping, and generalized forces, which we denote as $[M^{(N)}]$, $[K^{(N)}]$, $[C^{(N)}]$, and $\{Q^{(N)}\}$, respectively. Now let us lengthen the series to N' terms, *with the first N basis functions unchanged.* The new Ritz series therefore may be written as

$$u^{(N')} = \sum_{j=1}^{N} \psi_j(x) q_j(t) + \sum_{j=N+1}^{N'} \psi_j(x) q_j(t) \tag{6.4.27}$$

We need to determine the corresponding values of $[M^{(N')}]$, $[K^{(N')}]$, $[C^{(N')}]$, and $\{Q^{(N')}\}$. Because the first summation above is the same as the one appearing in eq.

(6.4.26), the new matrix elements will be the same as the previous values if their indices are less than or equal to N, that is,

$$
\left.\begin{aligned}
M_{j,n}^{(N')} &= M_{j,n}^{(N)} \\
K_{j,n}^{(N')} &= K_{j,n}^{(N)} \\
C_{j,n}^{(N')} &= C_{j,n}^{(N)} \\
Q_{j}^{(N')} &= Q_{j}^{(N)}
\end{aligned}\right\}, \quad j, n = 1, 2, \ldots, N
\tag{6.4.28}
$$

We say that the old values are *embedded* in the new matrices. This term stems from a partitioned description of the matrices. For example, $[M^{(N)}]$ forms the upper left block of $[M^{(N')}]$, which we depict as

$$
[M^{(N')}] = \begin{bmatrix} [M^{(N)}] & [\times] \\ [\times] & [\times] \end{bmatrix}
\tag{6.4.29}
$$

where $[\times]$ denotes submatrices that must be determined to fill in the $N' \times N'$ matrix. In view of the symmetry of the inertia matrix, without the embedding property, we would need to evaluate $\frac{1}{2}N'(N' + 1)$ elements of $[M^{(N')}]$. The embedding property enables us to compute only $\frac{1}{2}N'(N' + 1) - \frac{1}{2}N(N + 1) \equiv \frac{1}{2}(N' - N)(N' + N + 1)$ additional terms. For example, if $N = 12$ and $N' = 15$, we need to evaluate 42 elements with embedding versus 120 elements without it.

The embedding property may be exploited in an alternative way to check whether the solution to any analysis is sufficiently convergent. If N is the number of terms used to obtain that solution, we can obtain the solution for a smaller length series, $N' < N$, by crossing out of the system matrices $N - N'$ rows at the bottom and $N - N'$ columns to the right. Obviously, this requires no additional evaluations of matrix elements. If the differences between the solution to the resulting reduced order equations and the original solution are sufficiently small, we may deduce that the original solution was adequate.

EXAMPLE 6.11

Consider the clamped-pinned beam in Example 6.9, in the situation where no spring-mass system is attached to the beam. The basis functions used there were $\psi_j = (x/L)\sin(j\pi x/L)$, where $x = 0$ corresponds to the clamped end. Consider using Ritz series ranging from one to four terms to determine the natural frequencies and mode shapes. Examine the resulting natural frequencies in light of the separation theorem. Then compare the mode functions obtained from each series length.

Solution This example will shed a practical light on the theoretical convergence properties associated with the Ritz series formulation. The elements of the evaluation of the natural frequencies and mode functions were laid out in Example 6.9. We follow that analysis with terms that contain the mass of the suspended block and the stiffness of the attached spring equated to zero,

$$
M_{jn} = \rho A \int_0^L \psi_j \psi_n \, dx, \quad K_{jn} = EI \int_0^L \frac{d^2\psi_j}{dx^2} \frac{d^2\psi_n}{dx^2} dx
$$

For each series length N, we solve the eigenvalue problem $[[K'] - (\omega')^2[M']]\{\phi\} = \{0\}$, where $[M']$ and $[K']$ are, respectively, the numerical factors of ρAL and EI/L^3 in $[M]$ and $[K]$. This gives N nondimensional natural frequencies, $\omega_j' = \omega_j(\rho AL^4/EI)^{1/2}$, and N corresponding modes vectors $\{\phi_j\}$. We construct the normal mode matrix in the usual manner, based on scaling each mode vector such that $[\Phi]^T[M'][\Phi] = [I]$. The mode functions at discrete locations are then obtained by forming a $P \times N$ matrix $[\psi(x)]$ such that the element in row p and column j is $\psi_j(x_p)$. The product $[\Psi] = [\psi(x)][\Phi]$ constitutes an array whose jth column holds the normalized mode function $\Psi(x)$ evaluated at each point x_p.

We must implement this procedure for each N. In order to avoid recalculating the matrices for different N, we set up our work to compute the matrices for the longest series, $N = 4$. We may then solve the eigenvalue problem for any $N \le 4$ by using the upper left $N \times N$ submatrices as the input to the solver. Thus, the eigensolution operations at any N may be implemented in MATLAB by writing

$$\text{MATLAB:} \quad [phi, omega] = eig(K(1:N, 1:N), M(1:N, 1:N))$$

Mathcad requires the submatrix function to implement the same operations, specifically

$$\text{Mathcad:} \quad M' := submatrix(M, 1, N, 1, N) \quad K' := submatrix(K, 1, N, 1, N)$$
$$\omega' := \sqrt{genvals(K', M')} \quad \phi := genvecs(K', M')$$

The results in either method should then be sorted in order of increasing value of the natural frequency in order to facilitate comparison of values for different N.

Before we consider the specific results, it is useful to recall that we can use each basis function to form a Rayleigh ratio. According to eq. (6.4.13), the ratio $(K'_{nn}/M'_{nn})^{1/2}$ for each diagonal element of the matrices in the case $N = 4$ will be greater than or equal to the fundamental natural frequency ω_1'. This simple computation yields $\omega_1' \le 17.5, 47.9, 97.6$, and 166.73. Obviously, the first value is the smallest upper bound among the four possible values.

The columns of the following table give the natural frequencies obtained for the various series lengths. As an alternative perspective for assessing the quality of these results, the analytical values of $\omega_j(\rho AL^4/EI)^{1/2}$ obtained from the hinged-pinned entry in Appendix C also appear in the tabulation. The entry for ω_1 at $N = 1$ is the estimate we obtained from the Rayleigh ratio using the diagonal matrix elements. Reading across a row shows that the Ritz eigenvalues for a specific mode do decrease with increasing series length, and each value exceeds the analytical value. This illustrates the upper bound property. A comparison of the columns illustrates the separation theorem, because each natural frequency ω_j for a specific N falls in the interval between ω_j and ω_{j+1} for the next N value. Another aspect of the tabulated values is the fact that the greatest percentage difference between natural frequencies computed from the Ritz series and analytically occurs in the highest frequency. The lower frequencies are much closer to the corresponding analytic values, with the error decreasing with decreasing mode number. In other words, the greatest error resulting from an N-term series arises in ω_N, and the ω_j values become more accurate as j decreases with N fixed. (This is a general trend, but exceptions are possible.)

$\omega_j(\rho AL^4/EI)^{1/2}$ for $\psi_n = (x/L)\sin(n\pi x/L)$

Mode #	Series length N				Analytical
	1	2	3	4	
1	17.519	15.754	15.753	15.595	15.42430
2		53.708	50.626	50.572	49.96488
3			109.943	105.546	104.24770
4				186.095	178.26973

The fact that ω_N for each N shows the greatest discrepancy from the analytical value suggests that $\Psi_N(x)$ also will show the greatest discrepancy. This is borne out by the figures, which compare a specific mode function obtained from each series length to the analytic mode function listed in Appendix C. Note that the right end is $x = 0$ in the figure, consistent with the manner in which the Ritz basis functions were defined.

When $N = 1$, we only can estimate $\Psi_1(x)$. The first figure shows that the shape of this function is close to the analytical result, which explains why the Rayleigh ratio estimate of ω_1 was less than 15% above the analytic value. We also see that the first mode functions obtained from $N > 1$ are quite close to the analytical function. These trends continue for the second and higher mode functions. It is inherent to the Ritz series method that determining the jth modal properties requires $N \geq j$, so increasing the mode number yields progressively fewer Ritz mode functions to compare to the analytical result. In each case, the smallest series length leads to a mode function that differs the most from the analytical solution. We also see that a series that is only one term longer than the mode's number is adequate to get exceptionally good agreement.

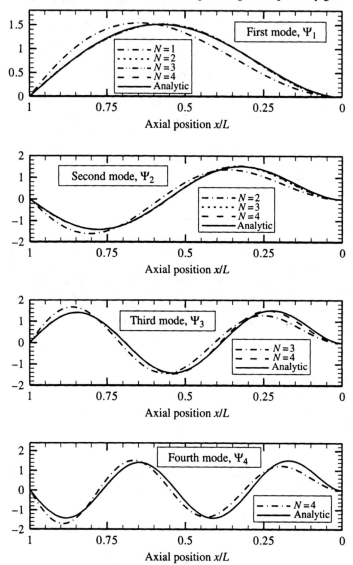

The trends observed in this example regarding the improvements associated with increasing the length of a Ritz series are general. However, the proximity of most results to the corresponding analytical value is uncommon. Particularly, the tabulated natural frequencies indicated a relatively small error in ω_N at each N. In other systems, or with other basis functions in this system, it is possible that these values will be as much as 40% or more above the analytical value. Similarly, the highest mode $\Psi_N(x)$ at any N will usually change significantly when terms are added to the Ritz series.

EXAMPLE 6.12

The rate at which the Ritz formulation converges depends on the choice of the set of basis functions. Alternative basis functions for flexural vibration of a cantilevered (clamped-free) beam having length L and a uniform cross-section are $\psi_n = (x/L)^{n+1}$ and $\psi_n = (x/L)\sin(n\pi x/2L)$. Formulate free vibration analyses of flexural vibration according to the Ritz method for each set of functions. Evaluate the natural frequencies as a function of series lengths in the range $1 \le N \le 10$. Discuss the behavior of each formulation relative to the separation theorem.

Solution This example will explore a question that naturally arises in some situations: If alternative sets of basis functions satisfy the geometric boundary conditions, is one set better than the other? The separation theorem provides a powerful tool for answering this question. We will see that there are indeed "good" and "bad" functions, but not because some give more accurate results than others. The procedure we follow is to construct $[M]$ and $[K]$ for each set of basis functions. Because there are no attachments to the beam, this task entails evaluating

$$M_{jn} = \int_0^L \rho A \psi_j \psi_n \, dx, \qquad K_{jn} = \int_0^L EI \frac{d^2\psi_j}{dx^2} \frac{d^2\psi_n}{dx^2} \, dx$$

for each set of basis functions. Because either set of mode functions actually only depends on x/L, it is convenient to nondimensionalize the calculation by changing variables such that $x = yL$. This leads to the inertia coefficients having a common dimensionalizing factor ρAL and the stiffness coefficients having the factor EI/L^3, as was the case in the previous example. We define M'_{jn} and K'_{jn} as the numerical parts of the coefficients, such that

$$M'_{jn} = \int_0^1 \psi_j \psi_n \, dy, \qquad K'_{jn} = \int_0^1 \frac{d^2\psi_j}{dy^2} \frac{d^2\psi_n}{dy^2} \, dy$$

where the mode functions to be considered are

$$\psi_n = y^{n+1} \quad \text{or} \quad \psi_n = y\sin\left(\frac{n\pi y}{2}\right)$$

Note that half of the second set of functions inappropriately satisfy the geometric boundary condition $w = 0$ at $x = L$, while the other half satisfy $\partial w/\partial x = 0$ at $x = L$, which also does not apply. This is acceptable because the full set does not identically satisfy both conditions.

It is a simple matter to evaluate the inertia and stiffness integrals analytically in the case of the power functions, while the integrals for the sinusoidal-like functions are most easily done using numerical methods. In order to avoid recalculating the matrices for different N, we set up our worksheet to compute the matrices for the longest series, $N = 10$. The MATLAB and Mathcad procedures for using submatrices of the full $[M']$ and $[K']$ in order to solve the eigenvalue problem at lesser N were explained in the previous example.

As a preliminary step, we compute the ω values obtained from the full 10×10 matrix. The values returned by Mathcad are

$$(\omega')^\mathrm{T} = \begin{bmatrix} 3.516 & 22.34 \end{bmatrix}$$

This is not a mistake—the program only returns two eigenvalues when it should give 10. In contrast, when we carry out the same evaluation in MATLAB, we obtain 10 eigenvalues. Why should the two programs give such different results?

A clue to the difficulty comes from examining the condition number of $[M']$ and $[K']$. The condition number of a matrix is closely related to the singular value decomposition of a matrix, which is a linear algebra operation that serves a number of purposes, including identification of the rank of a matrix; see Press et al. (1992). The condition number, which is greater than or equal to 1, measures the degree to which the matrix might be difficult to invert. A condition number that is the reciprocal of the precision with which computations are done, that is, 10^{12} for double precision, is definitely ill-conditioned. However, lower values might still indicate ill-conditioning. The condition numbers for $[M']$ and $[K']$ at $N = 10$ are $3.0(10^{15})$ and $5.6(10)^{12}$, respectively. This situation arises because the basis functions $(x/L)^{j+1}$ at large j are too much alike for numerical work. They are *computationally linearly dependent*, even though they are mathematically linearly independent. This aspect becomes evident when the functions are graphed against x, which would show that the functions for large j are extremely small except near $x = L$.

In order to overcome this ill-conditioning problem we examine the error tolerance parameter *TOL*, which Mathcad uses to test convergence in various contexts. The two values of ω' listed above corresponded to the default value $TOL = 0.001$. We increase this parameter progressively. Ultimately, at $TOL = 10^{-8}$ Mathcad gives 10 eigenvalues for $N = 10$. These values agree with those obtained from MATLAB. We will return to the issue of ill-conditioning later.

We may now implement the eigensolution for $N \times N$ submatrices of $[M]$ and $[K]$ for $N = 1, 2, \ldots, 10$. Each N gives a vector of N eigenvalues in ascending order. The clamped-free beam is also a configuration described in Appendix C, so we can evaluate the analytical prediction for the natural frequencies. These values appear in the following tabulation.

$\omega_j(\rho A L^4/EI)^{1/2}$ for $\psi_n = (x/L)^{n+1}$

	Series length N					
Mode # j	1	2	3	4	5	Analytical
1	4.4721	3.5327	3.5171	3.5160	3.5160	3.5160
2		34.807	22.233	22.158	22.035	22.034
3			118.14	63.347	63.24	61.697
4				281.60	128.52	120.90
5					562.71	199.86

$\omega_j(\rho A L^4/EI)^{1/2}$ for $\psi_n = (x/L)^{n+1}$

	Series length N					
Mode # j	6	7	8	9	10	Analytical
1	3.5160	3.5160	3.5160	3.5160	3.5160	3.5160
2	22.035	22.034	22.034	22.034	22.034	22.034
3	61.716	61.715	61.697	61.697	61.697	61.697
4	128.39	121.12	121.12	121.09	121.09	120.90
5	223.55	223.38	201.09	201.09	199.89	199.86
6	1006.0	356.21	355.98	303.16	303.16	298.56
7		1662.8	536.13	535.8	429.94	416.99
8			2591.3	774.30	773.76	555.17
9				3856.0	1080.6	713.08
10					5521.5	890.73

Several aspects of this tabulation are noteworthy. The separation theorem is verified for every entry obtained from a Ritz series, because the data for any $N > 1$ is less than or equal to the value to its immediate left, but greater than the value to the left and above. The tabulation also is consistent with the upper bound theorem, because each Ritz natural frequency is greater than, or equal to, the corresponding analytical natural frequency. It also is apparent that the error relative to the analytical value increases with increasing mode number for fixed N, with the last value, ω'_N, being significantly in error.

We now repeat the computations using the alternative set of basis functions. Because we encountered numerical problems due to ill-conditioning for the first set of functions, we evaluate the condition number for the 10×10 inertia and stiffness matrices for this second set of functions. These are $4.0(10^{12})$ and $2.1(10^{10})$ for $[M]$ and $[K]$, respectively. These values still are quite large. (The ill-conditioning here apparently stems from the highly oscillatory nature of the ψ_j functions at large j, which leads to strong cancellations when their products are integrated.) In view of the fact that we were able to evaluate the natural frequencies using the first set, even though the condition numbers were larger, we proceed to carry out the eigenvalue computations using the new set of functions. The results for N values from 1 to 5, which form the next tabulation, are representative of the trends for larger N.

$\omega_j(\rho AL^4/EI)^{1/2}$ for $\psi_n = (x/L)\sin(n\pi x/2L)$

Mode # j	Series length N					Analytical
	1	2	3	4	5	
1	3.9109	3.7489	3.5239	3.5222	3.5176	3.516
2		27.666	23.044	22.270	22.188	22.034
3			112.53	67.411	62.824	61.697
4				309.99	142.69	120.9
5					691.22	199.86

Once again, the data agrees with the upper bound and separation theorems. The new values for the highest natural frequency ω'_N are smaller than the corresponding values obtained with the first set of functions. In contrast, the lower natural frequencies at each N are mostly larger than the corresponding first set of data. Because we know that the natural frequencies obtained from the Ritz series constitute upper bounds for the analytical values, the best estimates are the lowest values. Thus, there is no clear choice as to whether the first or second set of functions gives better results.

One issue that is waiting below the surface is ill-conditioning. If we take either set of functions beyond $N = 10$, we ultimately will find that either the solver fails to find N eigenvalues, or else that some ω'_j values are complex. Reducing the error tolerance only slightly raises the value of N at which the failure occurs. (The error tolerance does not seem to be readily adjustable in MATLAB.) This causes concern that such numerical difficulties are inherent to the Ritz series formulation. That this is not the case is demonstrated by a set of basis functions whose definition is not obvious. Bessel functions $J_n(z)$ (see Abramowitz and Stegun, 1965), which arise in the solution of differential equations in polar coordinates, have the property of being oscillatory for sufficiently large z, but being like power functions for small z. The subscript n is the order of the Bessel function. For $n \geq 2$, these functions have the property that $J_n(z) = dJ_n(z)/dz = 0$ at $z = 0$, which are the geometric boundary conditions for the cantilevered beam. We shall use the second-order Bessel functions. To generate a family of basis functions we adjust the scale of z to map different segments of the Bessel function onto the beam's length. Hence, we define $z = \mu x/L$, where μ is a parameter that we can adjust. We impose an additional condition in order to set μ. We know neither the basis function nor its first derivative should be zero at $x = L$. Thus, we impose the condition that $d^2 J_2(\mu x/L)/dx^2 = 0$ at $x = L$. It is conventional to use a prime to denote differentiation of a function with respect to its argument, so the third set of basis functions we shall consider is

$$\psi_n(x) = J_2\left(\frac{\mu_n x}{L}\right), \qquad J''_2(\mu_n) = 0$$

Because of the oscillatory nature of the Bessel functions, the equation for μ_n has an infinite number of roots. Numerical methods may be used to obtain the lowest N roots required to form an N-length Ritz series.

Both MATLAB and Mathcad have built-in functions to evaluate $J_2(z)$. We compute the integrals for the inertia and stiffness coefficients numerically. The condition numbers for $[M']$ and $[K']$ are now 310 and $1.6(10^5)$, respectively, at $N = 10$. The fact that these numbers are significantly smaller than the previous values strongly suggests that the Bessel basis functions will work much better. This conjecture may be verified by determining the natural frequencies using $N = 30$, which would lead to 30 real values with condition numbers of $1.3(10^3)$ and $1.3(10^7)$ for $[M']$ and $[K']$.

Not only are the Bessel functions better behaved in terms of extending the series length, they give exceptionally good results at any N. This is exhibited by the last tabulation. Even the last eigenvalue at each N, that is, ω'_N, is quite close to the analytical value. Furthermore, each natural frequency obtained from the Bessel basis functions is lower than the corresponding value obtained with either of the other basis functions.

$\omega_j(\rho AL^4/EI)^{1/2}$ for $\psi_n = J_2(\mu_n x/L)$

	Series length N					
Mode # j	1	2	3	4	5	Analytical
1	3.6671	3.5232	3.5199	3.5172	3.5168	3.516
2		23.991	22.211	22.175	22.081	22.034
3			66.761	62.348	62.286	61.697
4				130.01	122.36	120.90
5					213.74	199.86

The message to carry away from this example is the adage "Don't change horses in mid-stream." This applies to any usage of Ritz series in vibrations, not just evaluation of modal properties. Some basis functions do work better than others, but it seldom is possible to anticipate which is preferable prior to performing calculations. Once we select a set of basis functions, we should extend the length of the Ritz series until we attain a satisfactory level of convergence. In this scenario, we would change to a different basis function set only if we encounter a numerical difficulty, such as ill-conditioning or exhausting the available computer resources, prior to reaching the desired level of convergence.

6.5   TIME-DEPENDENT BOUNDARY CONDITIONS

In the situations we have addressed thus far, a bar was supported by connecting it to the ground in some manner. This led to geometric boundary conditions expressing the fact that some aspect of the bar's motion is prevented. For example, a clamped end prevents displacement and rotation. However, there are important situations where the motion is constrained, but not as a zero value. For example, suppose one end of a shaft is attached to a strong motor. If the impedance of the bar is sufficiently small, the motor will turn as it would with no attachment, and the shaft will be forced to follow the motor's rotation. The shaft is therefore constrained to execute a specified rotation at the attachment point, which leads to a *time-dependent boundary condition*. The corresponding constraint force is the torque applied by the motor. As is true for immobile supports, the constraint force is an unknown quantity that can only be ascertained in the context of an analysis of the system response.

The fact that constraint forces are not fully specified makes it difficult to treat them as standard nonconservative loadings. Instead, the Ritz series formulation is modified by decomposing the displacement field into two parts. One term is used solely to account for the time-dependent aspects of the boundary conditions. The other term is a standard Ritz series that accounts for any known excitation forces, as well as compensating for artifices introduced by the fact that the first term does not satisfy the equations of motion. With the displacement field represented by the generic symbol w, the decomposition is

$$
\begin{aligned}
w(x, t) &= w_{bc}(x, t) + w_{Ritz}(x, t) \\
w_{Ritz}(x, t) &= \sum_{j=1}^{N} \psi_j(x) q_j(t)
\end{aligned}
\tag{6.5.1}
$$

By definition, w_{Ritz} satisfies the homogeneous geometric boundary conditions associated with immobile supports. This requires that

The basis functions ψ_j are the same as the set of admissible functions we would use if the boundary conditions did not depend on t.

For example, consider the simply supported bar in Figure 6.12, where the right end undergoes a specified rotation $\theta(t)$ due to the unknown couple Γ. Requiring that the rotation at the right end be zero is the time-invariant analog of the requirement that the rotation be a specified function of time. Thus, the basis functions for the Ritz series are those for a bar whose left end is pinned and whose *right end is clamped*, even though the physical nature of the right support is a pin.

The first part of the decomposition requires identification of a suitable functional form for $w_{bc}(x, t)$ solely on the basis of satisfying the time-dependent terms in the boundary conditions. Because this term is found without regard for the equations of motion, it could not exist by itself as a response. For the sake of simplicity, we shall begin by considering how to determine w_{bc} in the case where one time-dependent geometric boundary condition is imposed at $x = L$. For axial, torsional, or flexural displacement, such a boundary condition could be a statement that the displacement at an end $x = L$ must be a specified function of time. In flexure, where rotation represents another possible geometric boundary condition, it could be required that the gradient $\partial w/\partial x$ at $x = L$ be a specified time function, so the time-dependent boundary condition may be

$$
w(L, t) = f(t) \quad \text{or} \quad \frac{\partial w}{\partial x}(L, t) = f(t)
\tag{6.5.2}
$$

By definition, w_{Ritz} gives zero when it is substituted into the above. Hence, the representation of w given by eq. (6.5.1) will satisfy this boundary condition only if w_{bc} satisfies it. Furthermore, we observe that all other boundary conditions, being zero, will be satisfied by w_{Ritz}. Hence, w_{bc} must evaluate to zero when substituted into any

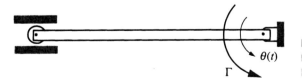

FIGURE 6.12 A simply supported bar with a time-dependent boundary condition.

time-invariant geometric boundary conditions. Let J denote the total number of geometric boundary conditions, with the boundary condition numbered J designated to be the time-dependent one. Furthermore, let the operator, $\mathcal{B}_n = 1$ or $\partial/\partial x$, as appropriate to geometric boundary condition number n at $x = x_n$. We therefore must find a function $w_{bc}(x, t)$ that satisfies

$$\boxed{\mathcal{B}_n(w_{bc})\big|_{x_n} = f(t)\delta_{nJ}, \quad n = 1, \dots, J} \tag{6.5.3}$$

Any function w_{bc} satisfying the above would be acceptable. We shall use an approach that resembles the Ritz series. We select a set of J independent functions $g_j(x)$, which are used to represent w_{bc} as a series,

$$\boxed{w_{bc} = \sum_{j=1}^{J} g_j(x)\alpha_j(t)} \tag{6.5.4}$$

In practice, the g_j functions are usually selected from power series or harmonic functions.

The $\alpha_j(t)$ are determined by satisfying the geometric boundary conditions. When eq. (6.5.4) is substituted into eq. (6.5.3), the dependence on x is replaced by an evaluation at the location of the boundary. We thereby obtain J algebraic equations for the α_j coefficients,

$$\sum_{j=1}^{J} \alpha_j(t)\mathcal{B}_n(g_j)\big|_{x_n} = f(t)\delta_{nJ}, \quad n = 1, \dots, J \tag{6.5.5}$$

In order that this relation be solvable, the terms $\mathcal{B}_n(g_j)$ evaluated at x_n cannot all vanish. This condition will be met if the g_j functions are not admissible for the case where the supports are stationary. In that case, eq. (6.5.5) represents J simultaneous equations for the coefficients α_j. For example, suppose we are interested in flexure of a bar whose end $x = 0$ is clamped and whose end $x = L$ is required to execute a rotation $\partial w/\partial x = f(t)$, but displacement is not constrained at $x = L$. The geometric boundary conditions are then $w = \partial w/\partial x = 0$ at $x = 0$, $\partial w/\partial x = f(t)$ at $x = L$, so $J = 3$. If we select $g_j = (x/L)^{j-1}$, which obviously are not admissible functions when $f(t) = 0$, satisfying the boundary conditions leads to $\alpha_0 = \alpha_1 = 0$, $\alpha_2 = \frac{1}{2}f(t)L$.

This procedure is readily extended to cases where more than one boundary geometric boundary condition is time dependent. To do so, we would require that the right side of eq. (6.5.3) match the time dependency or zero, according to whether each boundary condition is time dependent.

The foregoing specifies w_{bc}, so we now must find the associated basis function coefficients q_j appearing in w_{Ritz}. A time-dependent system is one in which the position depends explicitly on time, as well as the current values of the generalized coordinates. This is precisely the situation described by eq. (6.5.1). The standard power balance equations of motion, given by eqs. (6.1.16), apply only for time-invariant systems. Consequently, we formulate the equations of motion by turning to Lagrange's equations, modified by insertion of a term for the Rayleigh dissipation function $D \equiv \frac{1}{2}\mathcal{P}_{dis}$. (The Lagrange equation formulation is addressed in Appendix A.)

$$\frac{d}{dt}\left(\frac{\partial T}{\partial \dot{q}_j}\right) - \frac{\partial T}{\partial q_j} + \frac{\partial D}{\partial \dot{q}_j} + \frac{\partial V}{\partial q_j} = Q_j, \quad j = 1, \dots, N \tag{6.5.6}$$

The first quantities we shall evaluate are the generalized forces. This requires consideration of the virtual work. In view of eqs. (6.5.1), taking the time derivative of w in order to form the velocity field gives

$$\dot{w}(x, t) = \dot{w}_{bc} + \sum_{j=1}^{N} \psi_j(x)\dot{q}_j(t) \tag{6.5.7}$$

In contrast, a virtual displacement corresponds to assigning infinitesimal increments δq_j to the generalized coordinates in eqs. (6.5.1). Time is held constant in a virtual displacement, so w_{bc} is not altered in this operation. As a result, the virtual displacement at any x is

$$\delta w(x, t) = \sum_{j=1}^{N} \psi_j(x)\delta q_j(t) \tag{6.5.8}$$

At this juncture, we could form the virtual work done by the external forces. However, comparing eqs. (6.5.7) and (6.5.8) leads us to a simpler procedure. Note that the coefficients of δq_j in the second equation match the corresponding coefficients of $\dot{q}_j$ in the first equation, so that δw and $\dot{w}$ have similar forms when $w_{bc} \equiv 0$. However, $w_{bc} \equiv 0$ corresponds to the time-independent situation considered in the preceding sections. It follows that *the generalized forces for the system with time-dependent boundary conditions are the same as those for a system whose constraints are invariant in time.* Hence, the generalized forces are unchanged from the expressions in eqs. (6.1.15), (6.1.29), or (6.1.37). Note that constraint forces, including those imposing motion at boundaries, do no virtual work, and therefore do not appear in the generalized forces. This is a fundamental property of constraint forces mentioned previously.

The case of flexural displacement serves to illustrate the construction of the mechanical energies and Rayleigh dissipation function. We form the kinetic energy by substituting the first of eqs. (6.5.1) into eq. (6.1.33),

$$T = \frac{1}{2}\int_0^L [\dot{w}_{bc} + \dot{w}_{Ritz}]^2 \rho A\, dx + \frac{1}{2}\sum m[\dot{w}_{bc}(x_m, t) + \dot{w}_{Ritz}(x_m, t)]^2 \tag{6.5.9}$$

We collect terms as follows: T_2 is quadratic in $\dot{w}_{Ritz}$, T_1 is linear in $\dot{w}_{Ritz}$, and T_0 does not contain $\dot{w}_{Ritz}$. Thus,

$$\boxed{T = T_2 + T_1 + T_0}$$

$$T_2 = \frac{1}{2}\int_0^L \dot{w}_{Ritz}^2 \rho A\, dx + \frac{1}{2}\sum m\dot{w}_{Ritz}(x_m, t)^2$$

$$T_1 = \int_0^L \dot{w}_{Ritz}\dot{w}_{bc}\rho A\, dx + \sum m\dot{w}_{Ritz}(x_m, t)\dot{w}_{bc}(x_m, t) \tag{6.5.10}$$

$$T_0 = \frac{1}{2}\int_0^L \dot{w}_{bc}^2 \rho A\, dx + \frac{1}{2}\sum m\dot{w}_{bc}(x_m, t)^2$$

Consider the effect of representing w_{Ritz} by the series in eqs. (6.5.1). Because T_0 does not contain any q_j terms, it will not contribute to the Lagrange's equations. We therefore may ignore this term. Next, we observe that T_2 depends on w_{Ritz} in the same way that the total kinetic energy, eq. (6.1.33), depends on w. It follows that T_2 will be the same as the standard quadratic sum, eq. (6.1.10),

$$\boxed{T_2 = \frac{1}{2}\sum_{j=1}^{N}\sum_{n=1}^{N} M_{jn}\dot{q}_j\dot{q}_n} \tag{6.5.11}$$

The boundary correction w_{bc} affects T_1. After we substitute the series for w_{Ritz} into the second of eqs. (6.5.10), we collect like coefficients of each $\dot{q}_j$ and integrate over x. The result is

$$T_1 = \sum_{j=1}^{N}\left[\int_0^L \psi_j\dot{w}_{\text{bc}}\rho A\,dx + \sum m\psi_j(x_m)\dot{w}_{\text{bc}}(x_m, t)\right]\dot{q}_j \tag{6.5.12}$$

We represented w_{bc} as a series in eq. (6.5.4). We substitute the time derivative of that series into the preceding, which leads to

$$\boxed{\begin{aligned} T_1 &= \sum_{j=1}^{N}\sum_{n=1}^{N} \nu_{jn}\dot{q}_j\dot{\alpha}_n \\ \nu_{jn} &= \int_0^L \rho A\,\psi_j g_n\,dx + \sum m\psi_j(x_m)g_n(x_m) \end{aligned}} \tag{6.5.13}$$

Note that the ν_{jn} coefficients are constants.

The procedure for potential energy is the same. We substitute the decomposition of eq. (6.5.1) into eq. (6.1.34) and collect terms that are quadratic, linear, and independent of w_{Ritz}, which gives

$$\boxed{V = V_2 + V_1 + V_0}$$

$$V_2 = \frac{1}{2}\int_0^L EI\left(\frac{\partial^2 w_{\text{Ritz}}}{\partial x^2}\right)^2 dx + \frac{1}{2}\sum k w_{\text{Ritz}}(x_k, t)^2$$

$$+ \frac{1}{2}\sum \kappa\left[\frac{\partial w_{\text{Ritz}}}{\partial x}(x_\kappa, t)\right]^2 \tag{6.5.14}$$

$$V_1 = \int_0^L EI\frac{\partial^2 w_{\text{Ritz}}}{\partial x^2}\frac{\partial^2 w_{\text{bc}}}{\partial x^2}dx + \sum k w_{\text{Ritz}}(x_k, t)w_{\text{bc}}(x_k, t)$$

$$+ \sum \kappa\left(\frac{\partial w_{\text{Ritz}}}{\partial x}\frac{\partial w_{\text{bc}}}{\partial x}\right)\bigg|_{x=x_\kappa}$$

There is no need to explicitly list V_0 because its independence of w_{Ritz} means that it will be independent of the q_j variables, and therefore give no contribution to Lagrange's equations. The quadratic terms have the same appearance as the corresponding terms for the case of immobile supports, so we find that

$$V_2 = \frac{1}{2} \sum_{j=1}^{N} \sum_{n=1}^{N} K_{jn} q_j q_n \qquad (6.5.15)$$

To describe V_1 we substitute the Ritz series for w_{Ritz} and collect coefficients of the terms that are linear in q_j. This yields

$$V_1 = \sum_{j=1}^{N} \sum_{n=1}^{N} \gamma_{jn} q_j \alpha_n \qquad (6.5.16)$$

where the γ_{jn} coefficients are

$$\gamma_{jn} = \int_0^L EI \frac{d^2 \psi_j}{dx^2} \frac{d^2 g_n}{dx^2} dx + \sum k \psi_j(x_k) g_n(x_k) + \sum \kappa \left(\frac{d\psi_j}{dx} \frac{dg_n}{dx} \right) \Bigg|_{x=x_\kappa} \qquad (6.5.17)$$

The procedure for the Rayleigh dissipation repeats the foregoing operations, except that we use the decomposition of velocity in eq. (6.5.7). The result is

$$\mathscr{D} = \mathscr{D}_2 + \mathscr{D}_1 + \mathscr{D}_0 \qquad (6.5.18)$$

where $\mathscr{D}_0$ is discarded because it is independent of $\dot{q}_j$, while

$$\mathscr{D}_2 = \frac{1}{2} \sum_{j=1}^{N} \sum_{n=1}^{N} C_{nj} \dot{q}_j \dot{q}_n$$

$$\mathscr{D}_1 = \sum_{j=1}^{N} \sum_{n=1}^{J} d_{jn} \dot{q}_j \dot{\alpha}_n \qquad (6.5.19)$$

The C_{nj} coefficients are the damping coefficients for the case where all boundaries are immobile, while the d_{jn} quantities are constants given by

$$d_{jn} = \sum c \psi_j(x_c) g_n(x_c) + \sum \chi \left(\frac{d\psi_j}{dx} \frac{dg_n}{dx} \right) \Bigg|_{x=x_\chi} \qquad (6.5.20)$$

We obtain the equations of motion by substituting the sums $T_2 + T_1$, $\mathscr{D}_2 + \mathscr{D}_1$, and $V_2 + V_1$ into Lagrange's equations. The representations of each quantity in terms of the q_j variables, as described above, leads to the following equations of motion,

$$[M]\{\ddot{q}\} + [C]\{\dot{q}\} + [K]\{q\} = \{Q\} - [v]\{\ddot{\alpha}\} - [\gamma]\{\alpha\} - [d]\{\dot{\alpha}\}$$

$$= \{Q_{eff}\} \qquad (6.5.21)$$

Note that we have brought to the right side all terms that do not depend on the q_j variables, and labeled those terms as the effective generalized force, $\{Q_{eff}\}$. This term consists of the actual generalized force plus correction terms that depend on w_{bc}. These corrections represent inertial, dissipative, and elastic effects resulting from the fact that the function w_{bc} we constructed does not satisfy the equations of motion.

For transient responses, the initial conditions to be satisfied by $\{q\}$ are obtained by using eq. (6.5.1) to determine the initial values of u_{Ritz} and $\dot{u}_{\text{Ritz}}$. In particular, the specific form of w_{bc} in eq. (6.5.4) leads to

$$\sum_{j=1}^{N} \psi_j(x) q_j(0) = w(x, 0) - \sum_{j=0}^{J} \alpha_j(0) \chi_j(x) = w_0(x)$$

$$\sum_{j=1}^{N} \psi_j(x) \dot{q}_j(0) = \dot{w}(x, 0) - \sum_{j=0}^{J} \dot{\alpha}_j(0) \chi_j(x) = \dot{w}_0(x)$$

(6.5.22)

The terms $w_0(x)$ and $\dot{w}_0(x)$ represent the net contribution of the actual initial conditions. The corresponding values of $q_j(0)$ and $\dot{q}_j(0)$ may be determined directly from the expansion theorem, eq. (6.3.41).

It is not difficult to recognize that although flexural motion was used to develop the preceding procedure, the technique applies equally to cases of axial and torsional displacement. Of course, the definitions of $[\nu]$, $[\gamma]$, and $[d]$ will need to be adjusted based on the specific forms of T, V, and D. The significant feature of these equations of motion is that they have the same form as the standard equations of motion, eq. (6.1.16). Hence, solving them presents no unfamiliar tasks. Once we have determined the variables q_j as functions of t, we may synthesize the displacement field according to eqs. (6.5.1).

EXAMPLE 6.13

A constant cross-section bar, free at $x = L$, is initially at rest in the undeformed state. At $t = 0$, a constant axial velocity v is imparted to the left end, $x = 0$. Determine the resulting axial displacement.

Solution   This example invokes most of the steps required to adapt the Ritz series method to time-dependent boundary conditions. In addition, the results will shed light on an interesting wave propagation phenomenon, as well as some analytical issues that might cause difficulties. The geometric boundary condition is $u = (vt)h(t)$ at $x = 0$, where $h(t)$ is the step function. There is no geometric condition at $x = L$, so we let $u_{\text{bc}} = (vt)h(t)$, which in the terminology of eq. (6.5.4) corresponds to $g_1(x) = 1$ and $\alpha_1 = (vt)h(t)$. The corresponding representation of the displacement field is

$$u = (vt)h(t) + \sum_{j=1}^{N} \psi_j q_j$$

Satisfying the geometric boundary condition with this expression leads to $\psi_j = 0$ at $x = 0$. Thus, the basis functions for the analysis are what we would use to analyze axial vibration of a fixed-free bar. Suitable sets of basis functions are $\psi_j = (x/L)^j$ or $\psi_j = \sin[(2j - 1)\pi x/2L]$. According to Appendix C, the latter are the actual mode functions for a fixed-free bar, which makes them somewhat easier to implement. In order to illustrate the general procedure, we shall select $\psi_j = (x/L)^j$.

The only external force is the one that imposes the motion at $x = 0$. That end remains stationary in a virtual displacement because t is held constant. Thus, the generalized forces are zero. Also, dissipation is assumed to be negligible, so $[C] = [0]$. The inertia coefficients M_{jn} are the same as those for a fixed-free bar, and eq. (6.5.13) describes the inertial boundary correction coefficients v_{j1}. Because the correction term is $g_1 = 1$, we have

$$M_{jn} = M_{jn} = \int_0^L \rho A \left(\frac{x}{L}\right)^j \left(\frac{x}{L}\right)^n dx = \rho AL \left(\frac{1}{j+n+1}\right)$$

$$v_{j1} = \int_0^L \rho A \left(\frac{x}{L}\right)^j (1) dx = \rho AL \left(\frac{1}{j+1}\right)$$

The stiffness coefficients are

$$K_{jn} = \int_0^L EA \frac{d}{dx}\left[\left(\frac{x}{L}\right)^j\right] \frac{d}{dx}\left[\left(\frac{x}{L}\right)^n\right] dx = \frac{EA}{L} \frac{jn}{j+n-1}$$

The correction term g_1 is a constant, so the stiffness boundary correction coefficients obtained from eq. (6.5.17) vanish,

$$\gamma_{jn} = \int_0^L EA \frac{d}{dx}\left[\left(\frac{x}{L}\right)^j\right] (0) \ dx = 0$$

The standard equation of motion, eq. (6.5.21), contains $\ddot{\alpha}_1$. The first derivative of α_1 is a step function, so we have $\ddot{\alpha}_1 = v\delta(t)$. From this, we find that the equations of motion for the Ritz series coefficients are

$$[M]\{\ddot{q}\} + [K]\{q\} = -\{v\}v\delta(t)$$

The initial conditions apply to the system before the left end begins to move. Correspondingly, we take $h(0^-) = 0$, so that

$$u(x, 0) = 0 = \left[vth(t) + \sum_{j=1}^N \left(\frac{x}{L}\right)^j q_j\right]_{t=0^-} = \sum_{j=1}^N \left(\frac{x}{L}\right)^j q_j(0)$$

$$\dot{u}(x, 0) = 0 = \left[vh(t) + \sum_{j=1}^N \left(\frac{x}{L}\right)^j \dot{q}_j\right]_{t=0^-} = \sum_{j=1}^N \left(\frac{x}{L}\right)^j \dot{q}_j(0)$$

from which it follows that $q_j = \dot{q}_j = 0$ at $t = 0$.

We use modal analysis to solve for q_j. It is not clear how large N should be. We shall begin with $N = 5$, and then check for convergence. The first step is to solve for the nondimensional natural frequencies $\omega'_j = (\rho L^2/E)^{1/2}\omega_j$ by solving $[[K'] - (\omega')^2[M']]\{\phi\} = \{0\}$, where $[M']$ and $[K']$ are the numerical factors of the inertia and stiffness matrices. The natural frequencies are thereby found to be

$$\omega_j = 1.5708, \ 4.7132, \ 7.9390, \ 12.1740, \ 23.3614$$

We normalize the modes according to the usual procedures, so as to satisfy

$$\{\Phi_j\}^T[M]\{\Phi_n\} = \delta_{jn}$$

To determine the response we introduce the mode transformation $\{q\} = [\Phi]\{\eta\}$. The corresponding modal equations of motion are

$$\rho AL \ddot{\eta}_j + \frac{EA}{L}(\omega'_j)^2 \eta_j = -\rho AL\{\Phi_j\}^T\{v'\}v\delta(t)$$

where $\{v'\} = \{v\}/\rho AL$. The modal solutions are impulse responses given by

$$\eta_j = -vL\left(\frac{\rho}{E}\right)\frac{\{\Phi_j\}^T\{v'\}}{\omega'_j} \sin(\omega'_j \tau)h(\tau)$$

where τ is a nondimensional time parameter,

$$\tau = \left(\frac{E}{\rho L^2}\right)^{1/2} t$$

We obtain the displacement field by forming the modal transformation $\{q\} = [\Phi]\{\eta\}$, and then introducing those expressions into the Ritz series. To this, we add the contribution of u_{bc}. Introducing the definition of τ into u_{bc} leads to

$$u(x, \tau) = vL\left(\frac{\rho}{E}\right)^{1/2} \tau h(\tau) + \sum_{n=1}^{N} \left(\frac{x}{L}\right)^n \sum_{j=1}^{N} \Phi_{nj} \eta_j$$

Example 6.10 developed an efficient way to evaluate a Ritz series at a succession of instants and spatial locations. To implement that method here, we define a set of S locations x_s/L and a set of P (nondimensional) time instants τ_p. We then compute an $N \times P$ array [eta_vals] such that $(eta\_vals)_{jp} = \eta_j(\tau_p)$, with the dimensional amplitude factor $vL(\rho/A)^{1/2}$ set to unity. We also compute an $S \times N$ array [psi_vals], such that $(psi\_vals)_{sn} = \psi_n(x_s)$. The Ritz series portion of the displacement may then be computed according to

$$[u\_Ritz] = [psi\_vals][\Phi][eta\_vals]$$

The total displacement is then obtained by adding to the pth column of the displacement array [u_vals] the value $u_{bc}/[vL(\rho/A)^{1/2}] = \tau_p$ associated with that column.

The physical significance of the nondimensional time rests on a development in the next chapter. The quantity $(E/\rho)^{1/2}$ is called the bar speed. It will be seen to be the speed at which an axial displacement wave travels through a bar. (This is equivalent to the speed of sound in a fluid.) Thus, $L(\rho/E)^{1/2}$ is the time it takes such a wave to travel from one end of the bar to the other. Such an interval corresponds to $\Delta\tau = 1$. One way in which to display the response is to graph u, scaled by $vL(\rho/E)^{1/2}$, against x/L for intervals spaced by $\Delta\tau = 0.25$. We then may verify that the Ritz series was sufficiently long by comparing these plots to those obtained for a larger N. The plots for $N = 5$ are very much like those for $N = 2$, except for some additional waviness in each curve for $N = 2$. For the sake of brevity, we only display the results for $N = 10$.

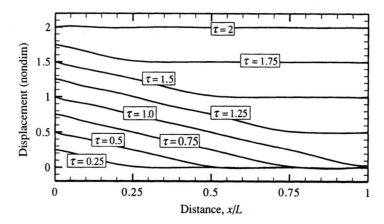

The graphed responses are quite close to those obtained from an analytical solution, which entails applying the d'Alembert solution (Graff, 1975) for the wave equation to the present boundary conditions.

The curves for $\tau < 1$ describe a wave that propagates in the direction of increasing x at the bar speed. The position $x/L = \tau$ is the wavefront. Ahead of the wavefront, there is no displacement, while the displacement increases linearly with distance behind the wavefront. After $\tau = 1$, the wavefront reflects from the free end, $x = L$. The wavefront of the reflected wave is at $x/L = 2 - \tau$, which corresponds to movement in the negative x direction. Behind the wavefront, that is, for $x/L > 2 - \tau$, the combination of the reflected and incident waves creates a displacement field that is constant in x and grows linearly with time. At $\tau = 2$, the wavefront has returned to the left end, and the entire bar has undergone the same displacement. For larger τ beyond what is displayed in the graph, reflections occur at each end after arrival of the wavefront.

A different way in which we may view the response is to plot u as a function of τ at fixed locations. This is the second set of graphs, again using $N = 10$.

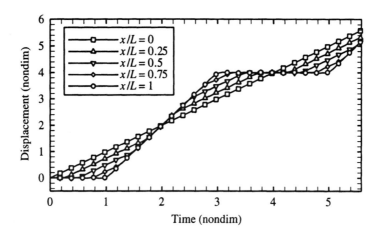

Interpreting this graph requires recognizing that, because of the similar nature of the nondimensionalizations of displacement and time, the slope of the graph is $\dot{u}/v$. At $x = 0$, the imposed displacement is $u = vt$, so the slope is unity. At $x = L$, the displacement is zero prior to $\tau = 1$, which is the instant when the initial wave arrives at that end. Until $\tau = 3$, the slope of u versus τ at that end is 2, which means that the velocity at $x = L$ is $2v$ in that interval. Doubling of v comes about because the reflected wave's velocity reinforces that of the incident wave. The time $\tau = 3$ is the instant when the wavefront that left $x = 0$ at $\tau = 0$ has traveled to $x = L$, then back to $x = 0$, then returned to $x = L$. The displacement at $x = L$ then remains constant until $\tau = 5$, at which time the wavefront returns and displacement again begins to increase. At locations between the ends, $\dot{u}$ is zero, v, or $2v$, according to where the wavefront is relative to that location. Consider viewing a graph such as this plotted over a long interval. A macroscopic view would see a line with unit slope, with small fluctuations corresponding to the various locations along the bar. This macroscopic view is the rigid body motion of the bar.

In closing, it is useful to note that the graphed responses are not exactly straight lines, but the analytical solution would show that they are. It would be logical to attempt to improve the Ritz solution further by going beyond $N = 10$. Unfortunately, such an attempt would fail because of the ill-conditioning phenomenon described in Example 6.12. It is situations like this that emphasize the need for analytical solutions, which is the topic we address in the next chapter.

6.6 RITZ METHOD FOR DISCRETE SYSTEMS

The ability of the Ritz series method to address various types of vibration of bars merely hints at its overall generality. It may be used to describe a variety of continua ranging from plates, which are thin flat sheets, to systems of simple structural elements. Further, it provides the theoretical foundation for the finite element method. A comparison of the three displacement cases for bars—axial, torsional, and flexural—shows that the only substantial change for each is the details leading to evaluation of $[M]$, $[K]$, $[C]$, and $\{Q\}$. The subsequent developments, such as mode functions and their orthogonality properties, the expansion theorem, and the Rayleigh ratio theorems, only differ in details associated with the specifics of the energy and power expressions. This similarity is shared by all continua when their displacement field is linearized.

One historically important use of Ritz series concepts pertains to discrete systems having several degrees of freedom. Prior to the availability of digital computers, solving a large number of equations of motion was quite labor intensive. The Rayleigh ratio and the Rayleigh-Ritz method, which we have seen is essentially the Ritz series method restricted to free vibration, were often used to obtain reasonable, yet readily computed, estimates of the fundamental, or first few, natural frequencies.

We consider a system having N' degrees of freedom described by generalized coordinates q'_j. We wish to approximate the response by a reduced-order Ritz series consisting of $N < N'$ terms. In view of the discrete nature of the system, the basis functions become column vectors $\{\psi_j\}$. The number of elements of each $\{\psi_j\}$ is N', corresponding to each generalized coordinate. In other words, we select N basis vectors $\{\psi_j\}$ whose elements represent different combinations of N' numbers. We will consider selection of these values at the conclusion of the development. The resulting discrete Ritz series has the form

$$\{q'\} = \sum_{j=1}^{N} \{\psi_j\} q_j(t) \equiv [\psi]\{q\} \tag{6.6.1}$$

The q_j represent generalized coordinates for the *reduced-order system*, and $[\psi]$ is an $N' \times N$ array whose columns are the respective basis vectors,

$$[\psi] = [\{\psi_1\}\ \{\psi_2\}\ \cdots\ \{\psi_N\}] \tag{6.6.2}$$

In the postulated situation, we know the full set of equations of motion, which is equivalent to saying that we know $[M']$, $[K']$, $[C']$, and $\{Q'\}$ associated with the original generalized coordinates $\{q'\}$. This enables us to form the kinetic and potential energies, the power dissipation, and the power input for the original system. The matrix forms of these quantities, which we mark with a prime to denote that they are associated with the full system, are

$$T = \tfrac{1}{2}\{\dot{q}'\}^{\mathrm{T}}[M']\{\dot{q}'\}$$

$$V = \tfrac{1}{2}\{q'\}^{\mathrm{T}}[K']\{q'\}$$

$$D = \tfrac{1}{2}\{\dot{q}'\}^{\mathrm{T}}[C']\{\dot{q}'\} \tag{6.6.3}$$

$$\mathscr{P}_{\mathrm{in}} = \{\dot{q}'\}^{\mathrm{T}}\{Q'\}$$

To derive the equations of motion associated with the Ritz series approximation, we substitute the discrete Ritz series, eq. (6.6.1), into the preceding, which yields

$$T = \tfrac{1}{2}\{\dot{q}\}^{\mathrm{T}}[M]\{\dot{q}\}$$

$$V = \tfrac{1}{2}\{q\}^{\mathrm{T}}[K]\{q\}$$

$$\mathscr{P}_{\mathrm{dis}} = \{\dot{q}\}^{\mathrm{T}}[C]\{\dot{q}\} \tag{6.6.4}$$

$$\mathscr{P}_{\mathrm{in}} = \{\dot{q}\}^{\mathrm{T}}\{Q\}$$

where the properties of the reduced system are

$$
\begin{aligned}
[M] &= [\psi]^T[M'][\psi], \quad [K] = [\psi]^T[K'][\psi] \\
[C] &= [\psi]^T[C'][\psi], \quad \{Q\} = [\psi]^T\{Q'\}
\end{aligned}
\tag{6.6.5}
$$

Note that these are $N \times N$ matrices, and therefore smaller than the original arrays. The equations of motion associated with eqs. (6.6.4) are

$$
[M]\{\ddot{q}\} + [C]\{\dot{q}\} + [K]\{q\} = \{Q\}
\tag{6.6.6}
$$

The reduced-order system has only N natural frequencies. To ascertain the quality of these quantities, let us examine the Rayleigh ratio. When the original system executes a free vibration at frequency ω, the modal response is $\{q'\} = \text{Re}[\{\phi'\}\exp(i\omega t)]$. Correspondingly, we have max $\{q'\} = \{\phi'\}$ and max $\{\dot{q}'\} = \omega\{\phi'\}$, so we find from eqs. (6.6.4) that

$$
\begin{aligned}
T_{\max} &= \tfrac{1}{2}\omega^2\{\phi'\}^T[M']\{\phi'\} \\
V_{\max} &= \tfrac{1}{2}\{\phi'\}^T[K']\{\phi'\}
\end{aligned}
\tag{6.6.7}
$$

Equating $T_{\max}$ and $V_{\max}$ leads to the Rayleigh ratio

$$
\mathscr{R}\,[\{\phi\}] = \frac{\{\phi'\}^T[K']\{\phi'\}}{\{\phi'\}^T[M']\{\phi'\}}
\tag{6.6.8}
$$

When $\{\phi'\}$ is actually a mode of the original system, $\{\phi'\} = \{\phi_j'\}$, we have $\mathscr{R} = (\omega_j')^2$. To use the Ritz series in eq. (6.6.1) to determine an approximate natural frequency and mode vector, we now consider $\{\phi'\}$ to be a *trial vector* defined by

$$
\{\phi'\} = [\psi]\{a\}
\tag{6.6.9}
$$

where the elements of $\{c\}$ are to be determined. Substitution of this trial vector into eq. (6.6.8) leads to

$$
\mathscr{R}[\{c\}] = \frac{\{a\}^T[K]\{a\}}{\{a\}^T[M]\{a\}} \equiv \frac{\displaystyle\sum_{j=1}^{N}\sum_{n=1}^{N} K_{jn}a_j a_n}{\displaystyle\sum_{j=1}^{N}\sum_{n=1}^{N} M_{jn}a_j a_n}
\tag{6.6.10}
$$

where $[K]$ and $[M]$ are as described in eqs. (6.6.5).

This summation form of $\mathscr{R}$ given above is equivalent to the representation of $\mathscr{R}$ associated with eqs. (6.4.18) for a bar. Therefore, any conclusions regarding the properties of the Rayleigh ratio derived earlier apply equally to discrete systems. For example, if we employ a single-term approximation, $N = 1$, factors a_1^2 in the numerator and denominator of eq. (6.6.10) cancel, in which case the Rayleigh ratio gives $\omega_1 \geq (K_{11}/M_{11})^{1/2}$. The earlier observations regarding the separation theorem for the approximate natural frequencies apply when $N > 1$, except that it is limited to $N < N'$. Note that if we let $N = N'$, which means that we have selected N' independent basis vectors, then the Ritz series, eq. (6.6.1), no longer represents an approximation. Instead, it is a one-to-one transformation defining a new set of generalized coordinates for the original system.

The key issue associated with this approach is the selection of the basis vectors $\{\psi_j\}$. There are no geometric boundary conditions to satisfy, so these vectors may, in principle, be selected arbitrarily. In the case where we use only one term to construct the Rayleigh ratio, the upper bound for the fundamental natural frequency might be excessively large compared to the corresponding true value, unless the basis function has been judiciously selected. For this reason, the method is best used if we can, through intuition and prior experience, select $\{\psi_1\}$ based on an estimate of the fundamental mode of free vibration. Selection of the basis functions is less critical for $N > 1$, but selecting the $\{\psi_j\}$ vectors based on guesses of the mode shapes is likely to improve the derived modal properties. In this regard it is useful to observe that the vibration modes generally have more sign reversals as the mode number increases. Thus, for a four-degree-of-freedom system, a reasonable set of basis vectors would be $\{\psi_1\} = [1\ 1\ 1\ 1]^T$, $\{\psi_2\} = [2\ 1\ -1\ -2]^T$, and $\{\psi_3\} = [1\ -1\ -1\ 2]^T$. (It is reasonable to conjecture that these considerations led to the usage of the term "method of assumed modes" as an alternative to "Ritz series method.")

REFERENCES AND SELECTED READINGS

ABRAMOWITZ, M., & STEGUN, I.A. 1965. *Handbook of Mathematical Functions.* Dover, New York.

CRAIG, R. R., JR. 1981. *Structural Dynamics.* John Wiley & Sons, New York.

GRAFF, K. F. 1975. *Wave Motion in Elastic Solids.* Ohio State University Press, Columbus.

INMAN, D. J. 1989. *Vibration.* Prentice Hall, Englewood Cliffs, NJ.

MEIROVITCH, L. 1967. *Analytical Methods in Vibrations.* Macmillan, New York.

MEIROVITCH, L. 1997. *Principles and Techniques of Vibrations.* Prentice Hall, Englewood Cliffs, NJ.

NEWLAND, D. E. 1989. *Mechanical Vibration Analysis and Computation.* Longman Scientific and Technical, White Plains, NY.

PRESS, W. H., TEUKOLSKY, S. A., VETTERLING, W. T., & FLANNERY, B. P. 1992. *Numerical Recipes,* 2nd ed. Cambridge University Press, Cambridge, England.

RAYLEIGH, J. W. S. 1945. *Theory of Sound,* Vol. 1. Dover, New York.

RITZ, W. 1909. "Über eine neue Methode zür Lösung gewisser Variationsprobleme die mathematischen Physik." *Journal für die reine und angewandte Mathematik,* 135, 1–61.

SHAMES, I. H., & DYM, C. L. 1985. *Energy and Finite Element Methods in Structural Mechanics.* Hemisphere Publishing, Washington, DC.

TIMOSHENKO, S. P., & GOODIER, J. N. 1970. *Theory of Elasticity,* 3rd ed. McGraw-Hill, New York.

TONGUE, B. H. 1996. *Principles of Vibrations.* Oxford University Press, New York.

WEINSTOCK, R. 1974. *Calculus of Variations.* Dover, New York.

EXERCISES

6.1 Use three power series terms as the basis functions for solving Example 6.1.

6.2 A block whose mass is $m = \rho AL$ is attached at the midpoint of a uniform bar. The bar is anchored at both ends. A sleeve applies a distributed axial load $f_x = f_0 \exp(-\mu t)$ over the region $L/2 \le x \le 3L/4$, as shown. Derive the equations governing axial motion when a four-term Ritz series is used to represent the displacement.

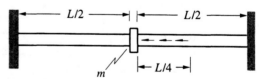

EXERCISE 6.2

6.3 A uniform bar is suspended from the ceiling by a pair of springs. The system is in static equilibrium in the position shown. A dynamic excitation $F(t)$ is applied at the lower end. Find the equations of axial motion corresponding to a three-term Ritz series.

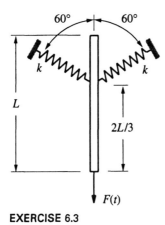

EXERCISE 6.3

6.4 Axial force F pushes the circular bar through a hole in the wall. The wall is lubricated with a viscous fluid that exerts a frictional resistance per unit length $f_x = -\nu \dot{u}(x, t)$. Construct the equations of axial motion associated with a four-term Ritz series.

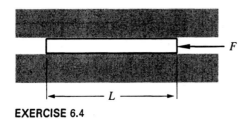

EXERCISE 6.4

6.5 A slender bar has length L and a tapered cross-section for which $A = A_0(1 + x/L)$. Both ends are fixed. Determine the inertia and stiffness matrices associated with a four-term Ritz series for axial displacement.

6.6 A hollow steel circular tube, whose outer diameter is $2R = 100$ mm, and whose wall thickness is $h = 5$ mm, is attached to a wall at its left end, while a disk is welded to the right end. A pis-

ton supported by a spring slides inside this tube. Friction between the disk and the tube is negligible. The mass of the disk, and of the piston, each equal the mass of the tube, and the stiffness of the spring is EA/L, where $A = \pi[R^2 - (R - h)^2]$ is the cross-sectional area of the tube, and $L = 0.8$ m is the length of the tube. Identify a suitable set of three basis functions describing the axial displacement of the tube. Then derive the coefficient matrices of the associated equations of motion for axial free vibration of this system.

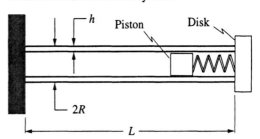

EXERCISE 6.6

6.7 Use a Ritz series consisting of four sinusoidal basis functions to formulate Example 6.3.

6.8 A 25 mm–diameter rod is welded to a wall at its left end, while a flywheel is attached concentrically to its right end. The rod is composed of aluminum, and its length is 500 mm. The flywheel's moment of inertia about the longitudinal axis is 60 kg-mm^2. Identify a three-term Ritz series that is descriptive of the torsional motion of this system, and evaluate the coefficient matrices for the equations of motion associated with this series.

6.9 A circular shaft supported by sleeve bearings at the 1/3 and 2/3 span locations has a flywheel mounted at the left end, where a couple $\Gamma(t)$ is applied. The other end is fixed. The bearings are lubricated by an oil, which may be modeled by considering each bearing to have a damping constant χ. Construct the equations of torsional motion corresponding to a three-term Ritz series.

6.10 A thin cable of length L tensioned by a force F is subjected to a transverse distributed force,

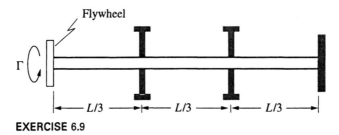

EXERCISE 6.9

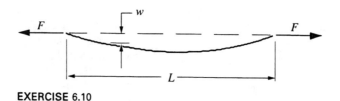

EXERCISE 6.10

$f_z(x, t)$, which results in a transverse displacement $w(x, t)$. In the linear approximation the cable tension is constant along the length and unaltered by the displacement. The displaced arclength between the ends is

$$S = \int_0^L \left[1 + \left(\frac{\partial w}{\partial x} \right)^2 \right]^{1/2} dx$$

The total elongation of the cable is $S - L$, so the work done by the tensile force is $F(S - L)$. This work is stored as potential energy in the cable.

(a) Use the small displacement approximation, in which $\partial w / \partial x \ll 1$, to derive an expression for the potential energy of the cable that is comparable in form to the strain energy expressions for axial and torsional motion of bars. Also show that the kinetic energy expression for the cable is analogous to the expression for a bar. (*Hint*: Use a two-term series expansion to simplify the square root term in the expression for S.)

(b) What are the geometric boundary conditions for the cable?

(c) What are the expressions for the inertia and stiffness matrices corresponding to a Ritz series representation of the displacement $w(x, t)$?

6.11 A clamped-free bar has a spring and dashpot mounted transversely at its midpoint. A transverse exciting force $F(t)$ acts at the right end. Use a three-term Ritz series whose basis functions are power series terms to model the transverse motion. Derive the corresponding equations of motion.

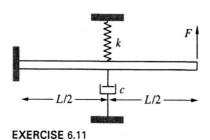

EXERCISE 6.11

6.12 The bar appearing in the figure is supported at its right end by a hinge, where a couple excitation $\Gamma(t)$ is applied. Spring k supports the left end.

The system is constructed such that the bar is horizontal when it is in static equilibrium under the influence of gravity. Find the equations of motion corresponding to a three-term Ritz series for transverse displacement.

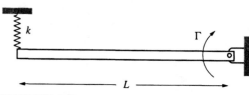

EXERCISE 6.12

6.13 A small block having mass m is supported by a simply supported beam, as shown. Derive the equations governing flexural displacement corresponding to a three-term Ritz series.

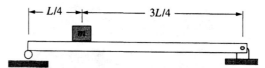

EXERCISE 6.13

6.14 A small block having mass m is attached to the midpoint of a clamped-clamped beam.

(a) Identify three functions ψ_j for a Ritz series describing motions that are symmetric about the midpoint. Then evaluate the corresponding inertia and stiffness matrices.

(b) Identify three functions ψ_j for a Ritz series describing motions that are antisymmetric about the midpoint.

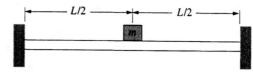

EXERCISE 6.14

6.15 A small block having mass $m = 0.25 \rho A L$ is suspended from the clamped-clamped bar by a spring having stiffness $k = 60 EI/L^3$. Derive the equations governing flexural displacement corresponding to a three-term Ritz series

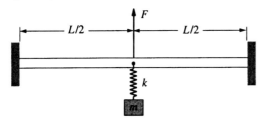

EXERCISE 6.15

6.16 Use three power series terms as the basis functions for solving Example 6.6. Compare the results to those obtained in Example 6.6. What conclusions regarding the influence of the choice of basis functions can one draw from such a comparison?

6.17 The bearings in Exercise 6.9 each have a torsional damping constant of 1 N-m/s. The shaft is composed of steel, its diameter is 40 mm, and its length is 600 mm. The flywheel's mass is 5 kg and its radius of gyration is 80 mm. Use a four-term Ritz series to determine the steady-state response when the torque is $\Gamma = \Gamma_0 \sin(\omega t)$. For the cases where $\omega = 85\%$, 95%, and 105% of the fundamental deformational natural frequency, plot the steady-state amplitude of the cross-sectional rotation as a function of x. Also plot as a function of x the phase angle of the rotation relative to the sinusoidal dependence of $\Gamma(t)$.

6.18 In Exercise 6.8, a sinusoidally varying torque $\Gamma = 20\sin(\omega t)$ N-m is applied to the flywheel. Damping is negligible. It is necessary that the amplitude of the steady-state torsional rotation of the flywheel not exceed 0.10 rad. If the frequency ω is increased very slowly, what is the maximum ω at which this criterion will be met? Use a three-term Ritz series to formulate the solution.

6.19 A harmonically varying distributed force $f_z(x) = 2\sin(\omega t)$ kN/m is applied transversely to a clamped-guided bar. The bar is steel with a 100 mm square cross-section. A dashpot whose constant is $c = 10$ kN-m/s is attached at midspan. Use a three-term Ritz series to model the transverse motion. Use the frequency domain transfer function to determine the steady-state amplitude of the transverse displacement at midspan.

6.20 Consider the pinned-clamped beam with attached mass in Example 6.9. The force at the midpoint is $F\sin(\omega t)$. It is desired to explore whether the spring and mass system can act as a vibration absorber. Toward that end, the suspended mass is taken to be one-tenth the beam's mass, $m = 0.1\rho AL$. The stiffness of the spring is

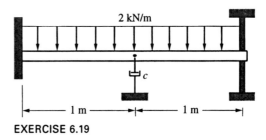

EXERCISE 6.19

selected to be $k = \Omega_1^2 m$, where Ω_1 is the fundamental frequency of the beam if the spring and mass were not attached. (This value may be determined by solving the eigenvalue problem in the case where $[M]$ and $[K]$ contain only terms associated with the Ritz series for the beam.)

(a) Compute the amplitude of the displacement as a function of distance along the bar when $\omega = \Omega_1$.

(b) Compute the amplitude of the displacement at midspan for the frequency range $0 < \omega < 1.2\Omega_1$.

(c) On the basis of the computed results, assess whether the suspended spring-mass system represents a useful vibration absorber.

6.21 A uniform bar is suspended from the ceiling by a pair of springs whose stiffness is $k = EA/L$, where EA is the extensional rigidity of the bar. The system is in static equilibrium in the position shown. Use a three-term Ritz series to determine the natural frequencies and mode functions for axial vibration.

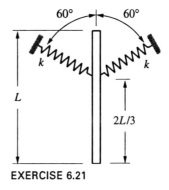

EXERCISE 6.21

6.22 A circular shaft supports a flywheel at its left end, while its right end is fixed. The radius of the flywheel is $4R$, where R is the radius of the shaft's cross-section. Use a two-term Ritz series to estimate the ratio of the flywheel's mass to the shaft's mass for which the fundamental torsional natural frequency will be half what it would be in the absence of the flywheel.

6.23 A bar of length L and uniform cross-section has a block whose mass is $m = \rho AL$ attached at its right end. The left end of the bar is fixed. Determine the natural frequencies and mode functions associated with a three-term Ritz series representation of its axial displacement.

6.24 Consider torsional vibration of the elastic bar in Example 6.3. Use Ritz series consisting of two, four, and six basis functions to determine the natural frequencies and mode shapes. Compare the results for the different series lengths

6.25 A bar has length L and constant cross-sectional properties EI and ρA. The left end of the bar is hinged, and the right end is supported by a transverse spring whose stiffness is $k = 12EI/L^3$. Use four basis functions having the form $\sin(n\pi x/2L)$ to determine the natural frequencies and mode shapes for flexural displacement. Compare the results to those obtained from an eight-term Ritz series.

6.26 A simply supported beam has a variable depth rectangular cross-section for which $h = h_0[1 + \sin(\pi x/L)]$. The cross-sectional width b is constant.

(a) Use a three-term Ritz series to determine the fundamental natural frequency.

(b) Use Appendix C to determine the depth of a uniform beam having a rectangular cross-section of width b, such that the fundamental frequency equals the value determined in part (a).

6.27 Determine the natural frequencies and mode functions associated with a six-term Ritz series representation of the system in Exercise 6.11. Consider cases where $k = 20EI/L^3$ and $k = 200EI/L^3$.

6.28 Determine the natural frequencies and mode functions associated with a three-term Ritz series representation of the cantilevered beam with a spring and block attached at its right end. The spring stiffness is $k = 10EI/L^3$ and the mass m is one-half the mass of the bar, $m = \frac{1}{2}\rho AL$. Consider $b/L = 0$.

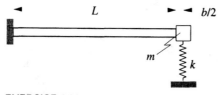

EXERCISE 6.28

6.29 For the system in Exercise 6.14, determine the natural frequencies and mode functions associated with a four-term Ritz series representing the transverse displacement. Consider cases where the

block's mass is $2\rho AL$ and $0.2\rho AL$. Compare the fundamental frequency in each case to what would be obtained from a one-degree-of-freedom system model that ignores the beam's mass.

6.30 In Exercise 6.6 the mass of the piston and of the disk each equal the tube's mass, ρAL. The spring's stiffness is $k = EA/L$. Determine the natural frequencies and mode shapes associated with a three-term Ritz series for the axial displacement.

6.31 Consider the shaft in Exercise 6.9. The excitation is a harmonically varying torque $\Gamma = \Gamma_0 \sin(\omega t)$. The shaft is composed of steel, its diameter is 40 mm, and its length is 600 mm. The flywheel's mass is 5 kg and its radius of gyration is 80 mm. Dissipation in the bearings is negligible. Use a four-term Ritz series and modal analysis to determine the steady-state response. For cases where ω is 95% of the fundamental, second, and third natural frequencies, plot the rotational amplitude as a function of x.

6.32 The spring stiffness in Exercise 6.12 is $k = 2EI/L^3$. The torque varies harmonically at $\Gamma(t) = \Gamma_0 \cos(\omega t)$. Use a four-term Ritz series and modal analysis to determine the steady-state response. Graph the displacement amplitude as a function of x for cases where $\omega = 80\%$ and 99% of the fundamental natural frequency.

6.33 Consider the clamped-guided bar in Exercise 6.19. Suppose the 2 kN/m distributed force is applied suddenly at $t = 0$, after which the force remains constant. At the instant of application, the bar is at rest in its static equilibrium position. The bar is steel with a 100 mm square cross-section. A dashpot whose constant is $c = 4$ kN-s/m is attached at midspan. Use a three-term Ritz series to model the transverse motion. Solve the corresponding differential equations of motion using modal analysis. Graph the displacement at the midpoint as a function of t. From that solution estimate the constant steady-state displacement of the midpoint and the approximate time required to attain that state.

6.34 A clamped-hinged bar is initially at rest in a deformed position in which the transverse displacement is $\omega_0 \sin[\pi(x/L)^2]$. Damping is negligible. It is desired to use a three-term Ritz series.

(a) Use the expansion theorem to evaluate a series representation of the initial displacement. Compare the result of evaluating that series to the actual function.

(b) Determine the free vibration response.

6.35 Consider the axial motion of a tapered bar welded to a wall at $x = 0$, while $x = L$ is free. The cross-sectional area varies as $A = 2A_0(1 - x/L)$. The viscoelastic loss factor in eq. (6.1.7) is $\gamma = 0.001$. At $t = 0$, the bar is released from rest with initial displacement $u = a(x/L)^2 \sin(3\pi x/L)$.

(a) Determine the initial values of the modal coordinates corresponding to a four-term Ritz series. Evaluate and plot the initial displacement field obtained from the Ritz series. Compare that plot to the given initial displacement function.

(b) Evaluate and plot the displacement field as a function of x when $t = 0.5\pi/\omega_1$, π/ω_1, and $2\pi/\omega_1$, where ω_1 is the fundamental natural frequency of the system.

(c) Repeat the analyses in parts (a) and (b) using a 20-term series.

6.36 The system in Exercise 6.6 was initially at rest in its static equilibrium position when an impulsive compressive force $I\delta(t)$ was applied to the disk at the right end. Use a four-term Ritz series for the axial displacement of the tube to represent the displacement field. Solve the associated equations of motion and use that solution to determine the displacement of the piston and the displacement u of the tube as a function of x at $t = \pi/\omega_1$, $2\pi/\omega_1$, and $4\pi/\omega_1$, where ω_1 is the fundamental frequency derived from the Ritz series.

6.37 The beam in Exercise 6.28 is subjected to a concentrated force F at its midpoint. The force is a uniform time pulse whose magnitude is $0.01EI/L^2$ and whose duration is π/ω_1, where ω_1 is the fundamental frequency. Determine the resulting vertical displacement of the block at the end. Plot the result as a function of the nondimensional time parameter $\tau = (EI/\rho AL^4)^{1/2}t$.

6.38 The sketch depicts the automobile model that was addressed in Exercise 4.56 by considering the chassis to be a rigid bar. A refinement of the model represents the chassis as a beam having a uniform cross-section. There are no geometric boundary conditions to satisfy. Hence, if

$x = 0$ is defined to be the center of the beam, a suitable set of basis functions is $\psi_j = (2x/L)^{j-1}$. The first two of these represent rigid body rotation and translation, so an analysis with $N = 2$ yields the rigid body model, and $N > 2$ accounts for deformation of the chassis. Identical dashpots c act parallel to the front and rear springs. The horizontal position of the wheels is $X_1 = vt$, $X_2 = vt - L$, and the ground elevation under a wheel is $w(X) = A \sin(2\pi X/\lambda)$, where v is the horizontal speed of the vehicle. Parameters to be considered are $m = 1000$ kg, $L = 5$ m, $EI = 10^6$ N-m^2, $k = 24$ kN/m, $c = 1$ kN-s/m, $A = 10$ mm, and $\lambda = L/3$. Use a six-term Ritz series to determine the steady-state amplitudes of the vertical displacement at both ends and the middle as a function of v. Compare this result to the dependence obtained from the rigid body model.

6.39 Consider torsional free vibration of the bar in Example 6.3. Formulate a Ritz evaluation of the natural frequencies for $N = 8$. Then use the embedding property to evaluate the natural frequencies predicted by analyses for series lengths ranging from $N = 1$ to $N = 7$. Discuss the behavior of those results relative to the separation theorem.

6.40 Consider axial motion of the system in Example 6.7. Use the embedding property to evaluate the natural frequencies derived from Ritz series for $N \le 6$. Discuss the behavior of those results relative to the separation theorem.

6.41 Consider axial motion of a tapered bar that is welded to a wall at $x = 0$, while $x = L$ is free. The cross-sectional area varies as $A = 2A_0(1 - x/L)$. Use Ritz series with $1 \le N \le 6$ to analyze the modal properties of this bar.

(a) Compare the natural frequencies derived from each series length.

(b) Compare the first and fourth mode functions derived from each series length.

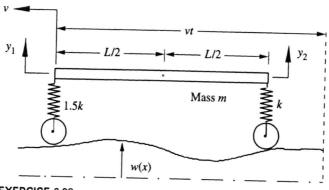

EXERCISE 6.38

6.42 Consider the system in Exercise 6.11 when the spring stiffness is $k = 8EI/L^3$. Use Ritz series and the embedding property to determine the flexural natural frequencies obtained from $N \leq 6$. Assess the results relative to the separation theorem.

6.43 Consider the system in Exercise 6.13 when the block's mass is $m = \rho AL$. Use Ritz series and the embedding property to determine the flexural natural frequencies obtained from $N \leq 10$. Assess the results relative to the separation theorem.

6.44 Consider the system in Exercise 6.12 when the spring stiffness is $k = 30EI/L^3$. Use Ritz series and the embedding property to determine the natural frequencies and mode functions obtained from $N \leq 5$. Assess the natural frequencies relative to the separation theorem, and graphically compare the first and third modes obtained from each N.

6.45 Consider a clamped-guided beam whose cross-section is uniform. Determine the natural frequencies and mode shapes for $N = 4$, 6, and 8. Compare the results for their convergence properties.

6.46 Consider the flexural vibration of a clamped-clamped beam whose depth varies parabolically according to $h = h_0(1 + 4x/L - 4x^2/L^2)$. The cross-sectional area is $A = bh$ and the area moment of inertia is $I = bh^3/12$, where b is a constant. Assess the natural frequencies derived from Ritz series for $N \leq 5$. Also compare the first and third mode functions obtained from each series length.

6.47 Consider the beam with a suspended spring-mass system in Exercise 6.15 when $m = \rho AL$ and $k = 48EI/L^3$. Determine the natural frequencies for Ritz series representations of the transverse displacement of $1 \leq N \leq 8$. Assess the results relative to the separation theorem.

6.48 Solve Example 6.13 in the case where $x = L$ is a fixed end.

6.49 A steel bar whose cross-section is a rectangle 100 mm high by 50 mm wide is supported by a roller and a pin, as shown. The length is $L = 1.5$ m. The right end is connected to a servomotor whose torque Γ imposes a flexural rotation given by $\theta_0 \sin(\omega t)$, where ω is 5% greater than the fundamental natural frequency. The dashpot placed at the midpoint has a constant of 40 N-m/s. Determine the amplitude of the steady-state displacement at the location of the dashpot.

6.50 A cantilevered beam is initially at rest in its static equilibrium position. At $t = 0$, the free end

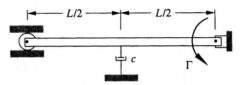

EXERCISE 6.49

is given a constant upward acceleration a_0. Derive an expression for the ensuing flexural displacement. Evaluate and graph this expression at $x = L/4$, $L/2$, and $3L/4$ for the interval $0 \leq t \leq 4\pi/\omega_1$, where ω_1 is the fundamental flexural natural frequency.

6.51 A harmonically varying axial displacement $B\sin(\omega t)$ is imposed on the left end by an unknown force $F(t)$ acting at that end. The stiffness of the spring at the right end is $k = 10EA/L$. Determine and graph the axial displacement amplitude as a function of x/L when $\omega = 1.05\omega_1$, where ω_1 is the fundamental natural frequency of the system.

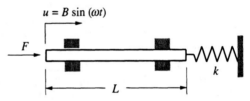

$u = B \sin(\omega t)$

EXERCISE 6.51

6.52 When $t = 0$, the system in Exercise 6.51 was at rest in the undeformed position. Determine and graph as a function of the nondimensional time $\tau = (E/\rho L^2)^{1/2} t$ the ensuing axial displacement at the right end of the bar.

6.53 The left end of a bar is attached to a wall, and a harmonic axial force $F\cos(\omega t)$ acts at the right end. The bar is a 20 mm diameter steel rod, and its length is 500 mm. The force F is such that the axial displacement at the right end is $u(L, t) = 0.02\cos(\omega t)$ mm, and damping is negligible.

(a) Use the decomposition approach to analyze the steady-state displacement field. Implement this analysis with u_{Ritz} described by a six-term Ritz series whose basis functions are polynomials. Then derive an expression for the axial force $F = EA(\partial u/\partial x)$ at the right end corresponding to the series coefficients. Evaluate F as a function of ω.

(b) An alternative analysis to using decomposition is to consider the end force as an excitation. A standard Ritz series analysis can be used to obtain an expression for the displacement, with F as a scale factor. Equating u at the right end to the

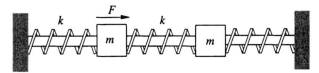

EXERCISE 6.54

required value then yields an equation for F. Determine F as a function of ω by following this procedure using a three-term Ritz series whose basis functions are polynomials.

(c) Compare the level of effort and the results associated with each analysis.

6.54 Identical blocks are interconnected by springs as they slide over a smooth horizontal guide bar. The matrix equations of motion are

$$m\begin{bmatrix} 1 & 0 \\ 0 & 1 \end{bmatrix}\begin{Bmatrix} \ddot{x}_1 \\ \ddot{x}_2 \end{Bmatrix} + k\begin{bmatrix} 2 & -1 \\ -1 & 2 \end{bmatrix}\begin{Bmatrix} x_1 \\ x_2 \end{Bmatrix} = \begin{Bmatrix} Q_1 \\ Q_2 \end{Bmatrix}.$$

Use the Rayleigh ratio to estimate the fundamental natural frequency of this system.

6.55 The inertia and stiffness coefficients for a three-degree-of-freedom system are

$$[M] = \begin{bmatrix} 200 & 0 & 400 \\ 0 & 600 & 0 \\ 400 & 0 & 1200 \end{bmatrix} kg,$$

$$[K] = \begin{bmatrix} 50 & -20 & 0 \\ -20 & 40 & 0 \\ 0 & 0 & 30 \end{bmatrix} kN/m$$

Compare the approximate natural frequencies and mode shapes of this system obtained from a two-term Ritz series to those obtained by solving the full set of equations.

6.56 When a cantilever beam is modeled by a Ritz series whose basis functions are $\psi_j = (x/L)^{j+1}$, $j = 1, \ldots, 8$, the inertia and stiffness matrices are

$$M_{jn} = \frac{1}{j + n + 3}\rho AL,$$

$$K_{jn} = \frac{j(j + 1)n(n + 1)}{j + n - 1}\left(\frac{EI}{L^3}\right)$$

Use a one-term Ritz series whose basis vector elements are $\psi_{j1} = 1/j^2, j = 1, \ldots, 8$, to estimate the fundamental natural frequency of this system. What is the error of this estimate in comparison to the value obtained from the full series?

6.57 Consider the stiffness and mass coefficients defined in Exercise 6.56. Use Ritz series composed of one, two, and three basis vectors to determine an approximation for the fundamental natural frequency and corresponding mode shape. Compare these estimates to the properties obtained from an eigensolution associated with the given $[M]$ and $[K]$ matrices for $N = 8$.

CHAPTER 7

FIELD DESCRIPTIONS FOR VIBRATING BARS

Our objective in this chapter is to derive and solve partial differential equations that model axial, torsional, and flexural displacement in bars. In each case axial position and time will be the independent variables, and a single displacement parameter will be the dependent variable. The Ritz method developed in the preceding chapter is quite versatile in its ability to describe each type of response, and it is readily extended to other types of continuous media. One therefore might wonder why an alternative line of analysis based on partial differential equations is needed.

As we saw in Example 6.12, a basic problem with the Ritz method is the possibility that even though we have selected a set of basis functions that meet all criteria, we might encounter numerical difficulties as we increase the number of terms in order to address higher frequency responses. Another shortcoming of the Ritz method is the fact that the modal properties (natural frequencies and mode functions) are obtained by solving a matrix eigenvalue problem. It is difficult in this approach to determine how any parameter, other than those that set the basic scales, affects the vibrational properties. For example, consider a cantilever beam that supports a small block at its far end. We wish to determine in a qualitative manner how altering the mass of the block relative to the mass of the beam affects the natural frequencies. Addressing this question using a Ritz expansion requires that we solve for the response for a range of mass ratios; in other words, we must perform a quantitative analysis in order to answer a qualitative question.

An equally important reason for determining response parameters by solving field equations is that the solution can shed light on ways in which the Ritz method may be performed more efficiently. This aspect is associated with the fact that modal properties derived by solving the partial differential equations of motion are expressed as analytical functions. It is not too difficult to obtain these functions for a few canonical problems. If we use these solutions to guide us in the selection of basis functions for more complicated systems, rather than selecting the basis functions arbitrarily, we are likely to see a more rapid rate of convergence, and also avoid numerical difficulties associated with extending the Ritz series. A final reason for developing the differential equation approach is that it offers a number of fundamental solutions that may be used as benchmarks for testing results obtained from approximate simulations.

Despite these positive features, models based on partial differential equations are limited in their usefulness. With a few exceptions, it is quite difficult to employ the approach to treat bars whose cross-section is not constant. Also, placing attachments and support constraints anywhere other than the ends of a bar substantially complicates the analysis. Assembling structural networks and mechanical linkages from individual bars leads to further complications. In a similar vein, dissipative effects associated with damping devices (i.e., dashpots) are difficult to incorporate into the formulation. Another negative aspect of the partial differential equation approach is the great enhancement of the complexity of the analysis when one wishes to study curved bars, plates, or shells.

7.1 DERIVATION OF MOTION EQUATIONS

Equations of motion for particles and rigid bodies may be derived by the Newton-Euler momentum-based approach. Alternatively, they may be obtained by the power balance method or Lagrange's equations, both of which are energy based. A comparable dualism exists for deriving the equations of motion governing deformable bodies. The Newton-Euler formulation for continua considers a differential element of the bar. This approach makes us aware of the interplay between the internal and external forces and the motion. In particular, the nature of the internal forces within a bar will be an essential aspect of the identification of appropriate model equations. However, the method becomes increasingly difficult with increasing generality of the displacement field. For plates and shells, the calculus of variations, which uses Hamilton's principle, is far simpler to implement.

7.1.1 Field Equations

Derivations of field equations for extension, torsion, and flexure that use mechanics of materials concepts are founded on a priori kinematical assumptions of the nature of the deformation field. Those assumptions are used in conjunction with strain-displacement, stress-strain, and dynamic equilibrium equations to obtain equations for the displacement field. If one wishes to assess the validity of the basic assumptions, it is necessary to develop an improved model. This might entail refining the initial assumptions, or it might involve developing a solution according to the theory of elasticity.

   Our study of a straight bar is based on the Bernoulli-Euler model for deformation (Shames and Dym, 1985). We shall refer to this description as *classical beam theory*. (Section 7.9 develops a model that addresses some of the approximations in classical beam theory.) It is required that in the undeformed state the centroids of all cross-sections are situated on a common straight line, with distance along this line denoted as x. The cross-sectional dimensions must be small in comparison to the length of the bar in order for the theory to apply. A further restriction is that the cross-sectional shape varies slowly in the x direction. The fundamental kinematical assumption for classical theory is that planar cross-sections maintain their shape and remain perpendicular to the centroidal axis as the bar undergoes deformation. As a result of this assumption, the displacement of any point in the bar is kinematically related to the displacement of the centroid. For our initial work, we restrict our attention to situations where the centroidal points move in a common xz plane, which can occur if all cross-sections are symmetric with respect to the xz plane. The axial displacement of the centroid is u, and the transverse displacement of the centroid is w; both displacement components may depend on x and t.

   We wish to develop a linear theory, which requires that the displacement must be very small compared to the span. Figure 7.1 depicts a side view of a typical cross-section. Because the cross-section must remain perpendicular to the centroidal axis, and its rotation is small in the linearized approximation, the rotation may be approximated by the slope $\partial w/\partial x$ of the centroidal axis. (Note that although all cross-sections move in the x direction, the smallness of the displacement u enables us to consider the

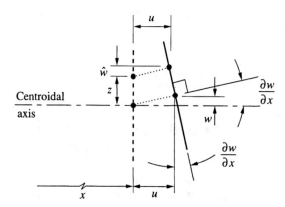

FIGURE 7.1 Displacement of the cross-section of a bar according to Bernoulli-Euler beam theory.

curve locating the centroidal axis *after* deformation to be described by the dependence of w on the original horizontal position x of the centroid.) As a consequence of the invariant shape of the cross-section and the smallness of the rotation, the transverse displacement of any point on the cross-section is essentially w, while the axial displacement varies linearly with the transverse distance z from the point to the centroidal axis. In a linearized description, the rotation of the cross-section is $\partial w/\partial x$, and a point above the centroidal axis displaces backward by $z(\partial w/\partial x)$ relative to the centroidal point. We shall use a caret to denote the displacement components of an arbitrary point in the bar, so we have

$$\hat{u} = u - z \frac{\partial w}{\partial x}, \qquad \hat{w} = w \tag{7.1.1}$$

The axial extensional strain ε_{xx} corresponding to this two-dimensional displacement field is

$$\varepsilon_{xx} = \frac{\partial \hat{u}}{\partial x} = \frac{\partial u}{\partial x} - z \frac{\partial^2 w}{\partial x^2} \tag{7.1.2}$$

The normal stress component acting on the cross-section is the extensional stress $\sigma_{xx} = E\varepsilon_{xx}$. A corollary of the assumption that cross-sections remain planar and perpendicular to the centroidal axis is an assumption that shear stresses acting on the cross-section do not cause any associated strain. (A shear stress distribution must act on the cross-section, but there is no need to characterize it.) In the theory of bars, the direct stress σ_{zz} in the transverse direction also is neglected on the basis that it will be small compared to σ_{xx} if the bar is slender. Hence, the primary stress distribution acting on a cross-section to the right of a transverse cutting plane is as shown in Figure 7.2.

The resultant of the normal stress on a differential element of cross-section area dA is $\sigma_{xx}\, dA$. On the cross-sectional face appearing in Figure 7.2, this resultant exerts a moment $-(\sigma_{xx}\, dA)z$ about an axis perpendicular to the xz plane and intersecting the

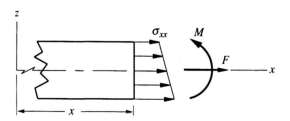

FIGURE 7.2 Stress resultants acting on a cross-section of a bar undergoing axial and flexural deformation.

centroid, with the negative sign stemming from the fact that $\sigma_{xx}\,dA$ exerts a clockwise moment if the stress is tensile. It follows that the resultant of the normal stress distribution on a cross-section is an axial force F and bending moment M,

$$F = \iint_A \sigma_{xx}\,dA = \iint_A E\left(\frac{\partial u}{\partial x} - z\frac{\partial^2 w}{\partial x^2}\right)dA$$

$$M = -\iint_A \sigma_{xx}z\,dA = \iint_A E\left(-\frac{\partial u}{\partial x} + z\frac{\partial^2 w}{\partial x^2}\right)z\,dA$$

(7.1.3)

We next observe that the displacements and E are independent of the transverse distance z. (We will discuss beams of composite material construction later.) When we factor these quantities out of the integral, we are left with integrals of z^0, z^1, and z^2 over the cross-section. These integrals are referred to as *moments of area*. The first reduces to the area, while the third gives the (centroidal) second moment of area about the y axis, which also is known as the area moment of inertia; we denote this quantity as I. Because we have selected $z = 0$ to be the centroid, we find that the first moment of area vanishes, $\iint z\,dA = 0$. As a result, the internal force resultants uncouple, with the axial force depending solely on the axial displacement, and the bending moment depending on the transverse displacement. Specifically, we find that

$$F = EA\frac{\partial u}{\partial x}$$

$$M = EI\frac{\partial^2 w}{\partial x^2}$$

(7.1.4)

A direct consequence of the uncoupling displayed by the above relations is that we can analyze axial and transverse displacements independently. Coupling of different displacement components is an inherent feature of the deformation of other media, such as thin shells.

An interesting generalization of these developments, which considered E to be constant over a cross-section, enables us to apply the present formulation to treat composite bars in which E is a function of z. One example of such a construction is a bar composed of layers arranged in the xy plane. To implement this generalization, we define $z = 0$ as the location at which a first moment of area weighted by E vanishes,

$$\iint_A Ez\,dA = 0$$

(7.1.5)

The locus of points satisfying this condition at each cross-section is referred to as the *neutral axis*, a term that is derived from the fact that σ_{zz} associated with bending in the absence of axial force vanishes at this location. For a homogeneous bar the centroidal axis is the neutral axis, but such a coincidence is not necessarily the case for inhomogeneous bars. Consistent with the weighted first moment of area, we consider EA and EI to be coefficients that are the extensional and flexural rigidities of a composite bar,

$$EA = \iint_A E\,dA, \qquad EI = \iint_A Ez^2\,dA$$

(7.1.6)

These modifications enable us to apply eqs. (7.1.4) in all cases.

The final step in the derivation of the field equations is to eliminate the internal forces. We do so by invoking the Newton-Euler equations for an infinitesimal segment of the beam contained between cross-sections at axial distances x and $x + dx$. Correspondingly, the internal forces at the farther cross-section differ by differential increments from the corresponding values at the nearer location. The free body diagram depicted in Figure 7.3 accounts for the shear force resultant S, even though we have ignored the deformation effects associated with shear. The convention we have used for depicting the shear forces is based on a definition that a positive shear stress acts in the positive z direction on a cross-section whose normal is the positive x direction. The sense of the bending moment is such that a positive curvature, $\partial^2 w/\partial x^2 > 0$, produces compression in the region $z > 0$ in accord with eq. (7.1.2). Note that the isolated beam segment is shown at the undeformed position, which is one of the linearization simplifications arising from the assumption that the displacements are very small. Each cross-section has normal and shear forces and a bending moment. The forces $f_x dx$ and $f_z dx$ represent distributed loads (force per unit length) that may be applied to the bar.

For the formulation of inertial effects, the difference between the rotations of the two cross-sections defining the beam segment is negligible, so the segment may be considered to move as a rigid body. To first order, the displacement of all points is $u\bar{e}_x + w\bar{e}_z$, and the rotation is $\partial w/\partial x$. Hence, the equations of motion for the two-dimensional situation in Figure 7.3 are

$$\sum F_x = (dm)\ddot{u}, \qquad \sum F_z = (dm)\ddot{w}, \qquad \sum M_G = (dI_G)\frac{\partial \ddot{w}}{\partial x} \qquad (7.1.7)$$

where dm is the mass of the element and dI_G is its centroidal moment of inertia. The segment length is dx, so $dm = \rho A dx$, where the mass per unit length ρA may depend on x. A subtlety enters in the characterization of dI_G. Rayleigh (see the 1945 reprint) included the rotatory inertia effect, as above. However, Timoshenko beam theory, which is developed in Section 7.9, reveals that rotatory inertia is unimportant for low-frequency vibration, while at high frequencies the effect of deformation associated with transverse shear stress is comparable to that of rotatory inertia. For this reason, we set $dI_G = 0$ for our basic model.

The force equations derived from the free body diagram in Figure 7.3 are

$$\sum F_x = \frac{\partial F}{\partial x} dx + f_x dx = (\rho A \, dx)\ddot{u}$$

$$\sum F_z = \frac{\partial S}{\partial x} dx + f_z dx = (\rho A \, dx)\ddot{w} \qquad (7.1.8)$$

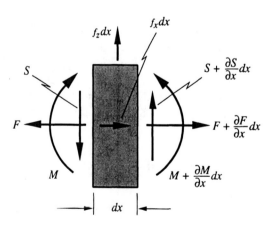

FIGURE 7.3 Free body diagram of a bar segment.

The distributed forces $f_x dx$ and $f_z dx$ are infinitesimal, and their lever arm about the center of mass G is also infinitesimal. Hence, their contribution to the moment equation of motion constitutes second-order differentials, which are unimportant. Thus, we have

$$\sum M_G = S\frac{dx}{2} + \left(S + \frac{\partial S}{\partial x}dx\right)\frac{dx}{2} + \frac{\partial M}{\partial x}dx = 0 \qquad (7.1.9)$$

We drop second-order differentials, which yields the familiar relation between shear and bending moment,

$$S = -\frac{\partial M}{\partial x} \qquad (7.1.10)$$

We substitute this expression into eqs. (7.1.8) and use the force-displacement equations (7.1.4) to eliminate M and F. Canceling the common factor dx and rearranging terms leads to the partial differential equations of motion for extensional and flexural vibration,

$$\boxed{\frac{\partial}{\partial x}\left(EA\frac{\partial u}{\partial x}\right) - \rho A\ddot{u} = -f_x} \qquad (7.1.11)$$

$$\boxed{\frac{\partial^2}{\partial x^2}\left(EI\frac{\partial^2 w}{\partial x^2}\right) + \rho A\ddot{w} = f_z} \qquad (7.1.12)$$

In the above forms the extensional rigidity EA and flexural rigidity EI are allowed to be functions of x, as would be the case for a tapered bar. Obviously, these parameters may be brought out of the derivative in the case of a homogeneous prismatic bar.

The differential equations and force-displacement relations appearing above are the basis for analysis of extensional and flexural motions. The equations for torsional deformation of bars have a similar mathematical structure to those for axial motion. The theory of torsion for bars having a circular cross-section indicates that cross-sections remain planar as they rotate about the centerline. (When the cross-section is noncircular, one needs to address warping of the cross-section; see Timoshenko and Goodier (1970)). A cross-section's rotation is $\theta(x,t)$. In terms of cylindrical coordinates (r, ϕ, x) whose axis is the centerline of the bar, the displacement is $u_\phi = r\theta$, so only the shear strain $\varepsilon_{x\phi} = \varepsilon_{\phi x}$ is nonzero. The corresponding stress component is $\sigma_{x\phi} = \sigma_{\phi x} = Gr(\partial\theta/\partial x)$, where G is the shear modulus. Integrating this stress component over the cross-section yields an expression for the resultant torsional moment Γ carried internally by the bar,

$$\Gamma = GJ\frac{\partial\theta}{\partial x}, \quad J = \iint_A r^2\,dA \qquad (7.1.13)$$

where J is the polar moment of area.

To obtain equations of motion, we consider a differential segment of the bar contained between x and $x + dx$. As shown in Figure 7.4, the net axial moment acting on this segment is the sum of the incremental torsional moment $(\partial\Gamma/\partial x)dx$ and the distributed torsional load $\tau\,dx$. The moment of inertia of the segment is $\rho J\,dx$, and its angular acceleration is $\ddot{\theta}$. In view of eq. (7.1.13), the moment equation of motion for the segment reduces to

$$\boxed{\frac{\partial}{\partial x}\left(GJ\frac{\partial\theta}{\partial x}\right) - \rho J\ddot{\theta} = -\tau} \qquad (7.1.14)$$

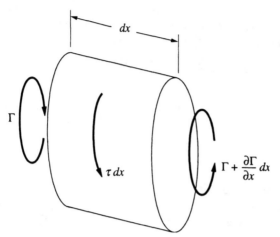

FIGURE 7.4 Segment of a circular rod carrying a torsional load.

Clearly, this differential equation is analogous to eq. (7.1.11), and eq. (7.1.13) is analogous in the same manner to the first of eqs. (7.1.4). Thus, any solution technique appropriate to extensional vibrations may be applied comparably to torsional vibrations.

7.1.2 Boundary Conditions

Solution of field equations (7.1.11), (7.1.12), or (7.1.14) requires specification of the boundary conditions. These are drawn from the geometric conditions, which arise in the Ritz series formulation, and from natural conditions that the internal force resultants must satisfy. Natural boundary conditions apply whenever the internal forces are known a priori. Recall that the internal forces are related to the displacement by

$$F = EA \frac{\partial u}{\partial x}$$

$$\Gamma = GJ \frac{\partial \theta}{\partial x}$$

$$M = EI \frac{\partial^2 w}{\partial x^2}$$ (7.1.15)

$$S = -\frac{\partial}{\partial x}\left(EI \frac{\partial^2 w}{\partial x^2}\right)$$

A fundamental concept of mechanics is that in any system, either the value of a generalized coordinate may be specified, or the value of the associated generalized force may be specified, but not both. A geometric boundary condition, in which either u, θ, w, or $\partial w/\partial x$ is known, requires generation of a constraint force that is sufficient to impose that motion. It follows that solving the field equations for axial and torsional motions requires specification of one boundary condition at each end, whereas flexural motion requires two boundary conditions at each end. The possibilities are

Geometric BC	or	Natural BC
u		F
θ		Γ
w		S
$\dfrac{\partial w}{\partial x}$		M

When the geometric boundary condition is time independent, we say that it is *homogeneous*, because the only variable is the associated displacement. Time dependent boundary conditions may be imposed by large machines or through servo-control.

Geometric boundary conditions were addressed in the previous chapter. As is true for geometric boundary conditions, the natural boundary condition may be time-dependent. This would correspond to a situation where the internal force resultant at a boundary must match a specified external force or couple. However, a time-dependent force may also be considered to be an external load applied very close to the end. The latter viewpoint allows us to consider the external force affecting any natural boundary condition to be zero, so we only need to consider homogeneous natural boundary conditions. The simplest conditions are those in which no springs or masses are attached at an end, in which case the internal force variable must be zero.

The presence of a spring complicates the situation, because displacement of the end leads to deformation of the spring, which, in turn, leads to an internal force. Possible ways in which a spring may act at the end $x = L$ are depicted in Figure 7.5. The basic idea in constructing the boundary condition associated with either case in Figure 7.5 is to characterize the internal force based on the *displacement being positive*, and then to use the internal force-displacement relations, eqs. (7.1.15), to eliminate the force. For example, in the case of an axial spring, an axial displacement u in the x direction at $x = L$ shortens the spring, so the spring force associated with positive u is compressive, $ku = -F$. However, we know that $F = EA(\partial u/\partial x)$, so the homogeneous boundary condition at $x = L$ in this case is $EA(\partial u/\partial x) = -ku$. Torsional motion leads to a similar relation, because a positive rotation θ at $x = L$ results in the application of a restoring torque $\kappa\theta = -\Gamma$, where the negative sign results from the fact that the induced torque would cause a negative $\partial\theta/\partial x$. Thus, the boundary condition would be $GJ(\partial\theta/\partial x) = -\kappa\theta$ at $x = L$. To address flexural displacement situations, we must be consistent with the sign conventions established by Figures 7.1 and 7.3. Thus a positive displacement w at $x = L$ is upward, which results in application of a spring force kw downward. This is the shear force, but a positive S is upward on this cross-section, so $S = -kw$. In view of the force-displacement relation, we are led to the boundary condition $(\partial/\partial x)[EI(\partial^2 w/\partial x^2)] = kw$ at $x = L$. Analysis of the case of a torsional spring exerting a bending moment follows a similar line of reasoning. A positive rotation $\partial w/\partial x$ at the right end is counterclockwise, which results in the spring exerting a clockwise restoring torque $\kappa(\partial w/\partial x)$. However, a positive bending moment on a cross-section whose normal points in the positive x direction is counterclockwise, so the boundary condition is $EI(\partial^2 w/\partial x^2) = -\kappa(\partial w/\partial x)$ at $x = L$.

It is obvious from the preceding development that awareness of the sign conventions for the displacement variable and the internal force is essential. It is left to

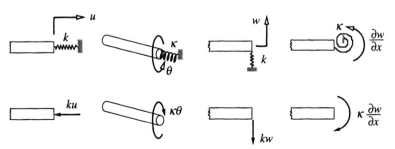

FIGURE 7.5 Spring attachments at the end $x = L$ of a bar.

Exercise 7.1 to examine the corresponding boundary conditions when $x = 0$ is the end at which a spring is attached. In each case the result will be a reversal of one sign.

When a block is attached at an end, the internal forces at that end are exerted by the block. The boundary conditions in such situations may be obtained by isolating the block in a free body diagram. To avoid sign errors, this diagram should be drawn with the internal forces depicted in accord with the sign convention for positive values. Let us consider axial and flexural displacement of the bar in Figure 7.6, where the end is designated as $x = L$. A positive axial force corresponds to elongation of the bar, so $F = EA(\partial u/\partial x)$ is depicted in Figure 7.6 as pulling backward on the block. The axial equation of motion for the block is $-F = m\ddot{u}$, so the boundary condition is $EA(\partial u/\partial x) = -m\ddot{u}$ at $x = L$.

For flexural motion, the complexity of the boundary conditions depends on the relative size of the block. The simpler case corresponds to modeling the block as a particle, according to which the block's displacement may be taken to equal the displacement w of the beam. If S is positive, it is upward when applied to the cross-section of the bar at $x = L$, so the force the bar applies to the block is $S = -(\partial/\partial x)[EI(\partial^2 w/\partial x^2)]$ downward. The equation of motion for transverse motion of the block is $-S = m\ddot{w}$, so we find that $(\partial/\partial x)[EI(\partial^2 w/\partial x^2)] = m\ddot{w}$ at $x = L$. Because rotatory inertia of the attached block is considered to be negligible in this model, the moment M acting on the block must be zero, so the moment boundary condition is $EI(\partial^2 w/\partial x^2) = 0$ at $x = L$.

The block's rotatory inertia may not be ignored if its radius of gyration is a significant fraction of L. The equations of motion of the block in that case must be formulated relative to its center of mass. To develop the associated boundary condition we define I_G to be the centroidal moment of inertia of the block, and let b denote the distance from the end of the bar to the center of mass. Because of the rotation and the finite size of the block, the displacement of the center of mass G is $w + b(\partial w/\partial x)$, so the acceleration of this point is $\ddot{w} + b(\partial \ddot{w}/\partial x)$. The angular acceleration of the block is $\partial \ddot{w}/\partial x$ counterclockwise. The transverse force equation for the block is $-S = m[\ddot{w} + b(\partial \ddot{w}/\partial x)]$, which upon substitution for S becomes the shear boundary condi-

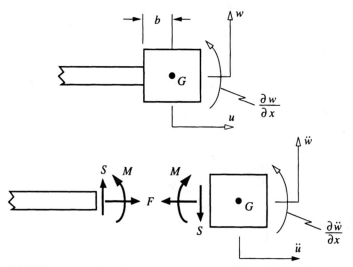

FIGURE 7.6 Axial and flexural boundary conditions for an attached mass at $x = L$.

tion, $(\partial/\partial x)[EI(\partial^2 w/\partial x^2)] = m[\ddot{w} + b(\partial \ddot{w}/\partial x)]$ at $x = L$. Summing moments about the center of mass G leads to $Sb - M = I_G(\partial \ddot{w}/\partial x)$. The corresponding moment boundary condition is $-b(\partial/\partial x)[EI(\partial^2 w/\partial x^2)] - EI(\partial^2 w/\partial x^2) = I_G(\partial \ddot{w}/\partial x)$ at $x = L$. Clearly, these reduce to the preceding set for the particle model when $I_G = 0$ and $b = 0$.

Once again, consistency in following the sign convention for the internal forces is essential to proper development of the boundary conditions. Cases where a block is attached at $x = 0$ are left for exercises. Another modification is to have a spring and a mass at an end. To handle such situations the free body diagram of the mass should show the spring force resulting from a positive displacement variable. Inclusion of this spring force in the equation of motion for the block leads to the appropriate condition.

EXAMPLE 7.1

A disk whose polar moment of inertia about its axis of symmetry is I_P is welded to the circular shaft. A torsional spring κ is attached collinearly to the disk.

(a) Describe the differential equations and boundary conditions that the torsional rotation θ must satisfy if the location where the disk and spring are attached is $x = L$.

(b) Repeat part (a) for the case where the location of the disk and mass is defined to be $x = 0$.

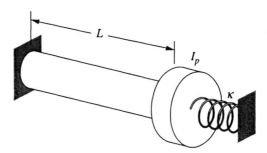

Solution This example will emphasize the need to define consistently the positive sense of various quantities; failing to do so will have serious analytical implications. The internal torque at the right end of the bar results from interaction with the disk, so we draw a free body diagram of the disk. In the first part of the exercise, the disk is at $x = L$, so we define the left end to be $x = 0$. Positive rotation is defined in the free body diagram according to the right hand rule, with the thumb extending in the direction of increasing x.

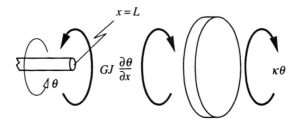

A positive rotation induces an internal torque in the bar and a torque within the spring. When the torsional deformation is such that θ increases with x, the internal torque is $GJ \, \partial \theta/\partial x$. This acts in the positive x direction on the exposed cross-section of the bar, so the torque the bar applies to the disk is in the negative x direction. The torque exerted by the spring is $\kappa\theta$, which tends to return the disk to the position at which $\theta = 0$. Therefore, it also acts in the negative x direction. These torques produce the acceleration of the disk, so we have

$$\sum M_x = -GJ\frac{\partial\theta}{\partial x} - \kappa\theta = I_P\ddot{\theta} \text{ at } x = L$$

The other boundary condition describes the fixed end at $x = 0$,

$$\theta = 0 \text{ at } x = 0$$

The differential equation is the standard one for free torsional vibration, eq. (7.1.14), with GJ being a constant, so that

$$GJ\frac{\partial^2\theta}{\partial x^2} - \rho J\ddot{\theta} = 0$$

When we define the end with the disk to be $x = 0$, the sense of increasing x, and therefore of positive θ, reverses. The internal torque within the bar remains $GJ\,\partial\theta/\partial x$. At the end $x = 0$, a positive $\partial\theta/\partial x$ corresponds to a torque that acts in the negative x direction, so the torque applied by the bar on the disk acts in the positive x direction. The spring's torque still tends to return the disk to $\theta = 0$, so it acts in the negative x direction. Thus, the rotational equation of motion for the disk in this case leads to

$$\sum M_x = GJ\frac{\partial\theta}{\partial x} - \kappa\theta = I_P\ddot{\theta} \text{ at } x = 0$$

The fixed-end boundary condition now is

$$\theta = 0 \text{ at } x = L$$

Also, the differential equation of motion does not depend on how x is defined.

A comparison of the boundary conditions at the disk for either definition of increasing x values shows that there is a sign change. Once we had the condition for $x = L$, we could have obtained the corresponding condition at $x = 0$ directly. To do so, we note that the coordinate system in part (b) replaces x by $-x$ and θ by $-\theta$. This reverses the sign of the terms containing θ and $\ddot{\theta}$, but it leaves intact the sign of $\partial\theta/\partial x$.

7.2 EIGENSOLUTIONS FOR EXTENSION AND TORSION

Eigensolutions of the field equations are the modal properties that are the foundation for evaluating response. The solution process will enable us to recognize the fundamental physical factors affecting a bar's mode functions. This level of insight is not afforded by the Ritz method, which merely describes the mode functions as a linear combination of the basis functions. We begin with the case of extension, whose equations are second order in x, and therefore substantially simpler to solve than the equations for flexure. Because the equations for torsional deformation have the same mathematical structure, the development applies equally well to that case.

To obtain vibration modes, we solve the field equation in the absence of the external loads. Thus, we must find the solution of the homogeneous partial differential equation

$$\frac{\partial}{\partial x}\left(EA\frac{\partial u}{\partial x}\right) - \rho A\ddot{u} = 0 \tag{7.2.1}$$

that also satisfies two boundary conditions drawn from alternative geometric/natural pairs at each end.

Equation (7.2.1) may be solved by the separation of variables method, in which u is decomposed into a product of functions of x and t. The bar represents a conservative system and its undeformed position is a stable static equilibrium configuration. Therefore, the time dependence of u must be sinusoidal, so we set

$$u = \text{Re}[\psi(x)e^{i\omega t}] \tag{7.2.2}$$

The corresponding form of the equation of motion is

$$\frac{d}{dx}\left(EA\frac{d\psi}{dx}\right) + \rho A \omega^2 \psi = 0 \tag{7.2.3}$$

In a nonuniform bar the cross-section varies with axial position, in which case we are confronted with the task of solving a differential equation with variable coefficients. Exercise 7.8 treats a rare practical case where a solution in terms of standard mathematical functions is known. Fortunately, most engineering systems feature uniform bars, in which the extensional rigidity is constant, so that is the situation we shall consider. When we factor a constant EA out of the derivative, we have

$$\boxed{\frac{d^2\psi}{dx^2} + \frac{\rho A}{EA}\omega^2\psi = 0} \tag{7.2.4}$$

The common factor A has not been canceled in the second term in order to accommodate bars of composite construction, for which ρA and EA respectively represent the mass per unit length and extensional rigidity obtained by averaging over a cross-section.

The general solution of eq. (7.2.4) consists of a sine and a cosine. For later work, it is convenient to express the dependence on x nondimensionally, so we have

$$\boxed{\psi = C_1 \sin\left(\alpha\frac{x}{L}\right) + C_2 \cos\left(\alpha\frac{x}{L}\right)} \tag{7.2.5}$$

This form will satisfy eq. (7.2.4) if the parameter α is

$$\boxed{\alpha = \left(\frac{\rho A L^2}{EA}\right)^{1/2}\omega} \tag{7.2.6}$$

It is important to note that eqs. (7.2.5) and (7.2.6) may be adapted to describe torsion of circular rods. To do so, we replace EA with the torsional rigidity GJ, and ρA becomes the moment of inertia per unit length ρJ. Of course, u in eq. (7.2.2) then becomes θ.

At this juncture, we know neither the coefficients C_j nor the parameter α. It still remains to satisfy the boundary conditions. Before we do so, let us consider the general form of these conditions. The displacement at an end might be specified, in which case the homogeneous boundary condition to be satisfied is $u = 0$. In the terminology of the theory of differential equations, this is referred to as a *Dirichlet boundary condition*. It also is possible that the end is free, so that $du/dx = 0$, which is a *Neumann boundary condition*. A third possibility is that the end has an attached spring or mass, in which case we have a boundary condition that relates the value of $\partial u/\partial x$ to u and/or $\ddot{u}$. This is the case of a *mixed (or Robin) boundary condition*.

Now consider the result of using the separated form of u in eq. (7.2.2) to form $\ddot{u}$ or $\partial u/\partial x$ in the boundary condition. The coefficients of ψ or $d\psi/dx$ that arise as a result will contain ω. We consider eq. (7.2.6) to define the frequency in terms of α, so the general solution for ψ must satisfy boundary conditions whose general form is

$$\beta_{11}\psi + \beta_{12}\frac{d\psi}{dx} = 0 \text{ at } x = 0$$

$$\beta_{21}\psi + \beta_{22}\frac{d\psi}{dx} = 0 \text{ at } x = L$$

(7.2.7)

where the β_{jn} coefficients may depend on α. In the case of a Dirichlet condition at an end, $(u = 0)$, we would have $\beta_{j2} = 0$, whereas $\beta_{j1} = 0$ for a Neumann condition $(\partial u/\partial x = 0)$. An example of a Robin boundary condition is an attached mass at $x = L$, for which the boundary condition is

$$\left[EA\frac{\partial u}{\partial x} + m\ddot{u}\right]_{x=L} = 0$$

(7.2.8)

When we substitute eq. (7.2.2) into this condition, we obtain

$$\left[EA\frac{d\psi}{dx} - m\omega^2\psi\right]_{x=L} = 0$$

(7.2.9)

According to eq. (7.2.6), $\omega^2 = \alpha^2(EA/\rho AL^2)$, so the preceding boundary condition corresponds to the second of the general forms in eq. (7.2.7) with $\beta_{21} = -\alpha^2 m/(\rho AL^2)$ and $\beta_{22} = 1$.

The next step is to substitute eq. (7.2.5) into eq. (7.2.7), which leads to two simultaneous equations,

$$\beta_{12}\alpha C_1 + \beta_{11}C_2 = 0$$

$$\left[\beta_{21}\sin(\alpha) + \beta_{22}\frac{\alpha}{L}\cos(\alpha)\right]C_1 + \left[\beta_{21}\cos(\alpha) - \beta_{22}\frac{\alpha}{L}\sin(\alpha)\right]C_2 = 0$$

(7.2.10)

A concise notation for these equations is

$$[D(\alpha)]\begin{Bmatrix}C_1 \\ C_2\end{Bmatrix} = 0$$

(7.2.11)

Suppose we assign an arbitrary value to α. In that case $[D]$ would be invertible, so that the only possible solution would be the trivial one, $C_1 = C_2 = 0$. It follows that the condition for a nontrivial solution is

$$\|[D]\| = 0$$

(7.2.12)

Because the coefficients of $[D]$ depend on α, this is the *characteristic equation*, and the roots α_j of the equation are the eigenvalues. Whereas the characteristic equation for a discrete system is a polynomial, the characteristic equation for all continuous systems will always be transcendental. That is, the terms forming the equation will not merely be polynomials in α. Numerical methods usually will be necessary to determine the roots of such equations. An important general aspect of the transcendental characteristic equations for vibratory systems is that they contain sinusoidally varying terms, as evidenced by the coefficients of C_1 and C_2 in the second of eqs. (7.2.10). Consequently, the characteristic equation will have an infinite number of roots. We number these roots in ascending order, and note that the sign of α is unimportant. As is true for discrete systems, zero eigenvalues are associated with rigid body motion, so we have $0 \le \alpha_1 < \alpha_2 < \cdots$.

The rank of eq. (7.2.11) reduces by 1 when α matches an eigenvalue α_j, which means that the equation for C_1 and C_2 associated with the second row of $[D]$ is the same as the first row's equation. If we select the equation associated with the first row, we have

$$D_{11}(\alpha_j)C_1 + D_{12}(\alpha_j)C_2 = 0 \tag{7.2.13}$$

where the dependence of coefficients of $[D]$ on the eigenvalue is noted explicitly. The above permits a solution for one coefficient in terms of the other,

$$C_2 = -\frac{D_{11}(\alpha_j)}{D_{12}(\alpha_j)}C_1, \quad \text{or} \quad C_1 = 0 \text{ if } D_{12} = 0 \tag{7.2.14}$$

Correspondingly, the eigenfunction derived from eq. (7.2.5) may be written as

$$\psi_j(x) = C_1\left[\sin\left(\alpha_j\frac{x}{L}\right) - \frac{D_{11}(\alpha_j)}{D_{12}(\alpha_j)}\cos\left(\alpha_j\frac{x}{L}\right)\right] \text{ if } D_{12} \ne 0$$

$$\text{or}$$

$$\psi_j(x) = C_2\cos\left(\alpha_j\frac{x}{L}\right) \text{ if } D_{12} = 0$$

$$\tag{7.2.15}$$

These eigenfunctions represent the possible axial displacement fields when the bar is not excited by external forces, so they are the modes of free vibration. The natural frequency of each mode is found from eq. (7.2.6) to be

$$\omega_j = \left(\frac{EA}{\rho A L^2}\right)^{1/2}\alpha_j \tag{7.2.16}$$

The remaining coefficient C_j may be arbitrarily selected. They will be set when we define normal modes later, after we address orthogonality of modes.

An exception to the foregoing development arises in the case of a bar that is neither fixed at some location, nor attached to a fixed foundation by means of a spring. Such a system can translate as a rigid body, in addition to undergoing deformational vibration patterns. Rigid body modes are manifested in discrete systems by the occurrence of one or more modes whose natural frequency is zero. The same is true here. To determine the mode function in this situation, we observe that when $\omega = 0$ for a mode, then $\alpha = 0$ also. In that case, the modal differential equation (7.2.4) becomes $\partial^2 \psi/\partial x^2 = 0$, whose solution is $\psi = C_0 + C_1 x$. However, if the system moves as a rigid body, the strain must be zero everywhere, so that $\partial u/\partial x = 0$, from which it follows that $C_1 = 0$. Correspondingly, a rigid body mode is described by

$$\psi = C_1 \text{ for rigid body motion} \tag{7.2.17}$$

The axial displacement field is the same at all x, so it constitutes a rigid body translation in the axial direction. In general, we can examine the manner in which a bar is supported to identify whether it is free, and therefore capable of rigid body motion. Alternatively, we may test whether eq. (7.2.17) satisfies all boundary conditions when $\alpha = 0$. Because we number the modes in ascending order of their natural frequency, the rigid body mode is ψ_1. In many situations, rigid body motion is the dominant contributor to the system's motion, so it is essential to consider this possibility in the process of obtaining an eigensolution.

EXAMPLE 7.2

The left end of a rod with a constant cross-section is welded to a wall and the right end is free. Consider extensional motions of the bar. Derive expressions for the natural frequencies and mode functions associated with free vibration as a function of the system parameters. How does changing the length affect these free vibration properties? Consider a steel bar. For what range of lengths will the natural frequencies of extensional modes all be above 200 Hz?

Solution   This example treats the simplest situation for determining eigenfunctions, in which there are no mixed boundary conditions. The cross-sectional properties do not depend on x, so the general solution for an eigenfunction is given by eq. (7.2.5),

$$\psi = C_1 \sin\left(\alpha \frac{x}{L}\right) + C_2 \cos\left(\alpha \frac{x}{L}\right)$$

We select $x = 0$ to be the fixed end, whose boundary condition $u = 0$ is somewhat simpler. Satisfying this requires that $C_2 = 0$. The fact that $x = L$ is free to displace requires that $\partial \psi / \partial x = 0$. Because we have already identified that $C_2 = 0$, we now require that

$$\left.\frac{d\psi}{dx}\right|_{x=L} = C_1 \frac{\alpha}{L} \cos(\alpha) = 0$$

We know that $\alpha = 0$ is not a possibility, because the bar cannot displace without being deformed. Furthermore, $C_1 = 0$ leads to the trivial solution $\psi = 0$. Thus, it must be that $\cos(\alpha) = 0$, whose roots are

$$\alpha = \frac{n\pi}{2} \text{ where } n \text{ is odd}$$

Because we use n to count the modes, we can obtain an odd integer from the combination $2n - 1$, $n = 1, 2, \ldots$. The natural frequency corresponding to the nth eigenvalue is found from the definition of α to be

$$\omega_n = \left(\frac{E}{\rho L^2}\right)^{1/2} \alpha_n = \frac{(2n - 1)\pi}{2}\left(\frac{E}{\rho L^2}\right)^{1/2}$$

If we set the arbitrary coefficient C_1 to unity, the mode functions are formed from segments of a sine curve according to

$$\phi = \sin(\alpha_n x) = \sin\left[\frac{(2n - 1)\pi x}{L}\right]$$

The natural frequency decreases as $1/L$ with increasing L, but L merely nondimensionalizes the axial distance for the mode functions. Furthermore, the mode functions are independent of the material properties. We seek the range of L for which the lowest frequency, ω_1, is above 200 Hz. For a steel bar, we set $\rho = 7800$ kg/m³, and $E = 2.07(10^{11})$ Pa, so we want

$$\frac{\pi}{2}\left(\frac{E}{\rho L^2}\right)^{1/2} > 200(2\pi) \quad \Rightarrow \quad L < 6.44 \text{ m}$$

EXAMPLE 7.3

A flywheel is welded to one end of a bar having a uniform circular cross-section, and the other end is welded to a wall. Consider torsional deformation. Let the moment of inertia of the flywheel be $\varepsilon \rho JL$, where ρJ is the mass moment of inertia per unit length for the bar.

(a) Derive the characteristic equation and an expression for the torsional mode functions corresponding to an arbitrary value of ε.

(b) Determine the lowest three natural frequencies and plot the corresponding mode functions in the case of a 1 m long aluminum bar with $\varepsilon = 0.5$.

(c) Consider the limit where the flywheel inertia is very small, but finite, $0 < \varepsilon \ll 1$. Derive expressions for the fundamental natural frequency and mode function in this case. Also discuss the high-frequency modal properties for small ε.

(d) Consider the limit where the flywheel inertia is very large, so that $\varepsilon \gg 1$. Derive expressions for the natural frequencies and mode functions in this case. How does the behavior of the fundamental and high-frequency modes compare to the results in part (c)?

Solution This example demonstrates the solution process when there is a mixed boundary condition. In addition, interpretation of the solution will provide some interesting observations regarding the general nature of modes. We begin the solution by defining the fixed end as $x = 0$, so that the more complicated boundary condition for the flywheel is at $x = L$. We may obtain these boundary conditions by setting the spring stiffness $\kappa = 0$ in Example 7.1, so we have

$$\theta = 0 \text{ at } x = 0, \qquad -GJ\frac{\partial\theta}{\partial x} = \varepsilon\rho JL\ddot{\theta} \text{ at } x = L$$

We set $\theta = \text{Re}\,[\psi \exp(i\omega t)]$ for free vibration. Cancellation of $\exp(i\omega t)$ converts the boundary conditions to

$$\psi = 0 \text{ at } x = 0, \qquad GJ\frac{d\psi}{dx} = \varepsilon\rho JL\omega^2\psi \text{ at } x = L$$

In order to eliminate the frequency ω in favor of the eigenvalue parameter α, we substitute $\omega^2 = (G/\rho L^2)\alpha^2$ in the boundary condition at $x = L$, so that

$$\psi = 0 \text{ at } x = 0, \qquad \frac{d\psi}{dx} = \varepsilon\frac{\alpha^2}{L}\psi \text{ at } x = L$$

The general solution for ψ is

$$\psi = C_1 \sin\left(\alpha\frac{x}{L}\right) + C_2\cos\left(\alpha\frac{x}{L}\right)$$

The condition that $\psi = 0$ at $x = 0$ requires that $C_2 = 0$. We substitute $\psi = C_1 \sin(\alpha x/L)$ into the boundary condition at $x = L$, which leads to

$$\frac{\alpha}{L}[C_1\cos(\alpha)] = \varepsilon\frac{\alpha^2}{L}[C_1\sin(\alpha)]$$

Clearly, $C_1 = 0$ is the trivial solution, and $\alpha = 0$ is not possible because there can be no rigid body motion. Thus, the characteristic equation is

$$\varepsilon\alpha\tan(\alpha) = 1$$

We can use numerical routines that implement techniques such as Newton-Raphson to find the roots of a transcendental equation such as this, but doing so requires starting values. One way of identifying such values is graphically. We express the characteristic equation as an equality of simpler functions by rewriting it as

$$\tan(\alpha) = \frac{1}{\varepsilon\alpha}$$

Although an accurate graph is displayed here, an approximate one is sufficient.

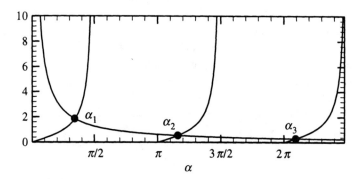

The roots of the characteristic equation correspond to intersections of the functions on either side of the above equation, which are a hyperbola and the tangent function. Increasing or decreasing the value of ε respectively brings the hyperbola closer or farther from the x axis. Regardless of the value of ε, it is clear that the roots must fall in the range $0 < \alpha_1 < \pi/2$, $\pi < \alpha_2 < 3\pi/2$, $2\pi < \alpha_3 < 5\pi/2$, and so on.

To find the roots of the characteristic equation when $\varepsilon = 0.5$, we observe that $\alpha = \pi/4$ corresponds to $\tan(\alpha) = 1$ and $1/(\varepsilon\alpha) = 8/\pi$, so the first root is beyond $\alpha = \pi/4$. For the higher roots, we use $\alpha_j = (j-1)\pi$ as initial guesses. To carry out the analysis in Mathcad, we define a function $F(y) := \tan(y) - 1/(\varepsilon y)$ and then initialize y as a three-element vector containing the guesses for the first three eigenvalues. The *root* function in conjunction with vectorizing then gives the lowest three eigenvalues, with the syntax of the operations being

$$\text{Mathcad:} \qquad \alpha = \overrightarrow{root(F(y), y)}$$

To implement these operations in MATLAB we use a function M-file called "end_mass.m" whose first line is *function F = end_mass(y)*, where $F(y)$ is defined above. A three-element vector y holding the initial guesses is defined and used as the initial guess for the *f zero* function, with the syntax being

$$\text{MATLAB:} \qquad for \ j = 1:3; \quad alpha(j) = fzero('end\_mass', y(j)); \quad end$$

The roots are found to be $\alpha_j = 1.0769, 3.6436$, and 6.5783. The corresponding natural frequencies for an aluminum bar ($E = 6.90(10^{10})$ Pa, $\rho = 2700$ kg/m^3, $L = 1$ m) are

$$\omega_j = \left(\frac{E}{\rho L^2}\right)^{1/2} \alpha_j = 5.44(10^3), 18.42(10^4), 33.26(10^4) \text{ rad/s}$$

Once we have these values it is a simple matter to evaluate the mode shapes as functions of x. The results are depicted on the next page.

As the root number increases for a specified ε, the hyperbolic portion of the characteristic equation comes increasingly close to the horizontal axis. Thus, at a sufficiently high j, the eigenvalues will approach $\alpha_j = (j-1)\pi$. The mode functions for such eigenvalues satisfy

$\psi_j = 0$ at $x = L$. In other words, the high-frequency eigenvalues are the same as those for a bar whose ends are fixed.

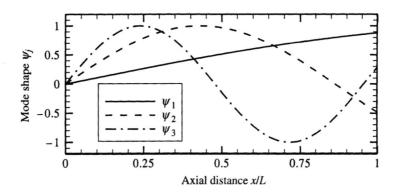

The case where ε is very small at first seems to present the opposite of the preceding situation, for it moves the hyperbola away from both axes. Thus, the first eigenvalue is close to $\alpha_1 = \pi/2$, the second is somewhat less close to $\alpha_2 = 3\pi/2$, and so on. Because $\varepsilon = 0$ is the case where there is no flywheel, we conclude that the lower modes for small ε correspond to those of a fixed-free bar. However, the hyperbola eventually comes close to the horizontal axis as α increases, regardless of the size of ε. Thus, the lower modes for the case of a small disk resemble those of a bar having one end fixed and one end free, while the high-frequency modes are like those of a bar having both ends fixed.

When ε is very large, so that the hyperbola is very close to the axes, the approximation $\alpha_j = (j-1)\pi$ for a fixed-fixed bar is very good for all eigenvalues, except for the first. The approximation $\alpha_1 = 0$ is not adequate because it corresponds to rigid body motion. A good approximation results from recognizing that α_1 will be small when ε is very large, so that $\tan(\alpha_1) \approx \alpha_1$. Thus, the characteristic equation's first root in this case is approximately $(\alpha_1)^2 = 1/\varepsilon$. Eliminating α in favor of ω then leads to

$$\omega_1 \approx \left(\frac{G}{\varepsilon\rho L^2}\right)^{1/2}$$

This approximation has much physical significance. To recognize it we recall that the mass moment of inertia of the flywheel is $I_f = \varepsilon\rho JL$. We also note that $GJ/L = K_T$ is the static torsional stiffness of the bar. It follows from these observations that

$$\omega_1 \approx \left(\frac{K_T}{I_f}\right)^{1/2}$$

This is the natural frequency we would have obtained if we had created a one-degree-of-freedom model based on ignoring the inertia of the bar. Further confirmation of this tendency comes from the first mode function. When α_1 is very small, we have $\psi_1 = \sin(\alpha_1 x/L) \approx \alpha_1 x/L$, which corresponds to a torsional strain that is independent of x, as in the static case.

Although the details vary between systems, the trends we have identified in solving this exercise are general. Specifically, masses present increasing impediments to motion as the natural frequency increases, eventually resulting in their acting to fix the location at which they are fastened. In a similar vein, further exercises will reveal that springs become less effective as the frequency increases. Thus, a very stiff spring might effectively prevent displacement in a low-frequency mode, while the high-frequency modes for the same system will be barely affected by the presence of the spring. The present results for the case of a large mass illustrate the general fact that rigid body models are limited to situations where the oscillation rate is well below the natural frequencies for deformation of the bodies represented as lumped elements.

7.3 EIGENSOLUTIONS FOR FLEXURE

Our approach for determining the modal properties of a bar undergoing flexural motion is much like that for extension and torsion. However, the higher order of the spatial derivatives in this case makes the general functional forms somewhat more complicated, and there are more boundary conditions to be satisfied. We obtain the eigensolution by considering free vibration, $f_z = 0$, so the equation of motion becomes

$$\frac{\partial^2}{\partial x^2}\left(EI\frac{\partial^2 w}{\partial x^2}\right) + \rho A\ddot{w} = 0 \tag{7.3.1}$$

To separate variables we observe that $w = 0$ corresponds to a stable equilibrium position, so the separated form of a free vibration must fit

$$w = \text{Re}\left[\psi(x)e^{i\omega t}\right] \tag{7.3.2}$$

where ω is real and nonnegative. Substitution of this form into the differential equation governing w leads to

$$\frac{d^2}{dx^2}\left(EI\frac{d^2\psi}{dx^2}\right) - \rho A\omega^2\psi = 0 \tag{7.3.3}$$

As is true for axial and torsional vibration, one is not likely to find an analytical solution for flexural vibration of a bar whose cross-sectional properties depend on x. Thus, we shall consider both EI and ρA to be constant for the remainder of this section. The corresponding modal differential equation is

$$\boxed{\frac{d^4\psi}{dx^4} - \frac{\alpha^4}{L^4}\psi = 0} \tag{7.3.4}$$

where the factor L^4 is introduced in order that the eigenvalue α be nondimensional. The value of α is related to the frequency by

$$\boxed{\alpha = \left(\frac{\rho A L^4\omega^2}{EI}\right)^{1/4}} \tag{7.3.5}$$

Because the coefficients of eq. (7.3.4) are constant, the basic form of its solution must be $\psi = \exp(\lambda x/L)$. Substitution of this form into the differential equation leads to $\lambda^4 - \alpha^4 = 0$, so the possible roots are

$$\lambda = +i\alpha, -i\alpha, +\alpha, -\alpha \tag{7.3.6}$$

The first two roots correspond to sine and cosine solutions, whereas the second pair leads to real exponentials, $\exp(\alpha x)$ and $\exp(-\alpha x)$. However, we may simplify the task of satisfying the boundary conditions by noting that the sum and difference of the third and fourth basic solutions are also solutions. This leads to a solution in the form of hyperbolic functions, because

$$\cosh\left(\alpha\frac{x}{L}\right) \equiv \frac{1}{2}\left[\exp\left(\alpha\frac{x}{L}\right) + \exp\left(-\alpha\frac{x}{L}\right)\right]$$

$$\sinh\left(\alpha\frac{x}{L}\right) \equiv \frac{1}{2}\left[\exp\left(\alpha\frac{x}{L}\right) - \exp\left(-\alpha\frac{x}{L}\right)\right]$$

$$\tag{7.3.7}$$

In comparison to exponentials, expressing the general solution for ψ in terms of hyperbolic functions will expedite satisfying the boundary conditions and simplifying the characteristic equation. Each of the properties listed below may be verified directly from the function definitions:

$$\cosh(0) = 1, \qquad \sinh(0) = 0$$

$$\lim_{z \to \infty} \cosh(z) = \infty, \qquad \lim_{z \to \infty} \sinh(z) = \infty, \qquad \lim_{z \to \infty} \tanh(z) = \lim_{z \to \infty} \frac{\sinh(z)}{\cosh(z)} = 1$$

$$\frac{d}{dz}\cosh(z) = \sinh(z), \qquad \frac{d}{dz}\sinh(z) = \cosh(z)$$

$$\cosh^2(z) - \sinh^2(z) = 1$$

$$\cosh^2(z) + \sinh^2(z) = \cosh(2z), \qquad 2\sinh(z)\cosh(z) = \sinh(2z) \qquad (7.3.8)$$

In terms of hyperbolic functions, the general solution of the modal differential equation for flexure is

$$\psi = C_1 \sin\left(\alpha\frac{x}{L}\right) + C_2 \cos\left(\alpha\frac{x}{L}\right) + C_3 \sinh\left(\alpha\frac{x}{L}\right) + C_4 \cosh\left(\alpha\frac{x}{L}\right) \qquad (7.3.9)$$

We shall restrict our attention to situations where spring and mass elements, if present, are attached only at the ends of a bar. For free vibration no motion may be imposed at a boundary. Thus, two homogeneous boundary conditions must be satisfied at each end. These boundary conditions are drawn from the displacement/shear and rotation/bending moment alternatives. In the case where a natural boundary condition applies, it may be that the shear or bending moment is zero. It also is possible that the associated force variable is related to the displacement at that boundary. Such a possibility arises when a spring or inertial body is fastened at the boundary. As we did for a bar, we include all possibilities by allowing each boundary condition to be mixed. According to Section 7.1.2, the boundary conditions may be written as

$$\beta_{11}\psi + \beta_{12}\frac{d^3\psi}{dx^3} = 0 \text{ at } x = 0$$

$$\beta_{21}\frac{d\psi}{dx} + \beta_{22}\frac{d^2\psi}{dx^2} = 0 \text{ at } x = 0$$

$$\beta_{31}\psi + \beta_{32}\frac{d^3\psi}{dx^3} = 0 \text{ at } x = L$$

$$\beta_{41}\frac{d\psi}{dx} + \beta_{42}\frac{d^2\psi}{dx^2} = 0 \text{ at } x = L$$

$$(7.3.10)$$

where the coefficients β_{jk} may be functions of α.

The solution process follows similar steps to those for extension. We substitute the general homogeneous solution into the boundary conditions. After differentiation and evaluation of the trigonometric and hyperbolic functions, the result of substituting eq. (7.3.9) into the above boundary conditions will be

$$[D(\alpha)]\{C\} = \{0\} \qquad (7.3.11)$$

where $\{C\}$ is the vector of modal coefficients, and the elements of $[D]$ contain the boundary coefficients β_{jk}, as well as trigonometric and hyperbolic functions evaluated at $x = L$.

For arbitrary values of α, the only solution of eq. (7.3.11) is the trivial one, $\{C\} = \{0\}$. For a nontrivial solution, $[D]$ must not be invertible, which leads to the characteristic equation

$$\|[D]\| = 0 \qquad (7.3.12)$$

This equation will always be transcendental, having an infinite number of roots, $0 \leq \alpha_1 < \alpha_2 < \cdots$. The natural frequency associated with each eigenvalue is given by eq. (7.3.5) to be

$$\omega_j = \left(\frac{EI}{\rho A L^4} \right)^{1/2} \alpha_j^2 \qquad (7.3.13)$$

We should note that the process of obtaining the characteristic equation for a specific configuration according to eq. (7.3.12) actually entails a 4×4 determinant only if all of the boundary conditions are mixed. In that case, it certainly would be helpful to avail oneself of a symbolic manipulator, such as Maple. In most cases, the equations may be simplified in a two-step process. Satisfying the conditions at $x = 0$ first should enable us to eliminate some coefficients based on the properties of sinusoidal and hyperbolic functions of zero argument.

A clamped-clamped beam represents the greatest degree of constraint that may be imposed by boundary conditions, so its fundamental natural frequency is the largest that can be obtained with any set of boundary conditions. For flexural motion, the smallest root of the characteristic equation for this configuration is $\alpha_1 L = 4.730$ (see Appendix C). The smallest eigenvalue for axial motion of a fixed-fixed bar is $\alpha_1 = \pi$. Given the closeness of these fundamental values, let us compare eqs. (7.3.13) and (7.2.16). For a homogeneous bar, the fundamental natural frequencies are $(E/\rho L)^{1/2} \alpha_1$ for extension, and $(E/\rho L^2)^{1/2} (I/AL^2)^{1/2} \alpha_1^2$ for flexure. Thus, the fundamental flexural frequency is scaled by $(I/AL^2)^{1/2}$ relative to the fundamental frequency for extension. The radius of gyration of an area about an axis is $d = (I/A)^{1/2}$, and the largest possible value of d is the distance from the axis to the farthest point on the cross-section. It follows that the fundamental frequency for flexure is smaller than the fundamental extensional frequency by a factor whose order of magnitude is the ratio of the depth of the beam to the span length L. Physically, this is a consequence of the lesser stiffness of a beam to a transverse load relative to an axial load of the same magnitude. For example, in the case of a cantilevered (that is, clamped-free) bar, a static unit axial force at the free end results in an end displacement of L/EA, while a static unit force applied in the transverse direction at the free end produces an end displacement $L^3/3EI$. The scaling factor for these displacements is $L^2 A/I = (L/d)^2$. The factor d/L is quite small in most engineering applications. Indeed, if the depth to span ratio is much greater than $1/5$, it is likely that classical beam theory will not be accurate; see Section 7.9.

When α is one of the eigenvalues satisfying the characteristic equation, the rank of eq. (7.3.11) is reduced by 1. Hence, one of the corresponding coefficients C_j may be considered to be arbitrary. (The arbitrariness will be removed when we normalize the modes based on their orthogonality properties.) Because the hyperbolic function $\sinh(\alpha x)$ vanishes only at $x = 0$, and $\cosh(\alpha x/L)$ is never zero, there are a few situations where the associated coefficients C_3 and C_4 are both zero. However, there are no situations in which the coefficients of $\sin(\alpha x/L)$ and $\cos(\alpha x/L)$ are both zero, although it usually is the case that either C_1 or C_2 is zero. We therefore consider whichever of C_1 and C_2 is nonzero to be the arbitrary one. Solution of eq. (7.3.11) with $\alpha = \alpha_j$ leads to values of the other coefficients as ratios to the arbitrary one at this eigenvalue. Thus, the eigenfunctions may be written as

$$
\text{either} \quad \psi_j = C_1\left[\sin\left(\alpha_j\frac{x}{L}\right) + (C_2/C_1)\big|_{\alpha_j}\cos\left(\alpha_j\frac{x}{L}\right)\right.
$$

$$
\left. + (C_3/C_1)\big|_{\alpha_j}\sinh\left(\alpha_j\frac{x}{L}\right) + (C_4/C_1)\big|_{\alpha_j}\cosh\left(\alpha_j\frac{x}{L}\right)\right]
$$

$$
\text{or} \quad \psi_J = C_2\left[(C_1/C_2)\big|_{\alpha_j}\sin\left(\alpha_j\frac{x}{L}\right) + \cos\left(\alpha_j\frac{x}{L}\right)\right.
$$

$$
\left. + (C_3/C_2)\big|_{\alpha_j}\sinh\left(\alpha_j\frac{x}{L}\right) + (C_4/C_2)\big|_{\alpha_j}\cosh\left(\alpha_j\frac{x}{L}\right)\right] \tag{7.3.14}
$$

If we restrict our attention to systems that do not have attached masses and springs, there are four alternative types of end conditions: clamped, hinged, guided, and free. This gives rise to 10 unique configurations. For reference purposes, Appendix C lists the characteristic equations and eigenmodes for a uniform bar undergoing flexural motion. The tabulation also gives the corresponding information for axial motion, which has three unique configurations obtained from fixed or free ends.

As was true for extensional and torsional motion, flexural rigid body modes correspond to $\alpha = 0$ having a nontrivial solution for ψ. To address this issue we note that eq. (7.3.4) becomes $d^4\psi/dx^4 = 0$, whose solution is a third-degree polynomial, $\psi = C_0 + C_1x + C_2x^2 + C_3x^3$. However, the condition that the bar move as a rigid body requires that the flexural deformation be zero everywhere. We therefore require that $d^2\psi/dx^2 = 0$, which reduces the possible solution to

$$
\boxed{\psi = C_0 + C_1x \text{ for rigid body motion}} \tag{7.3.15}
$$

If the boundary conditions can be satisfied by this expression without requiring that $C_0 = C_1 = 0$, then rigid body motion is possible. Such a possibility should be obvious by inspection of the physical supports. For example, a free-free beam can undergo rigid body translation and rotation. The boundary conditions in that case are $d^2\psi/dx^2 = d^3\psi/dx^3 = 0$ at $x = 0$ and $x = L$, both of which are satisfied by eq. (7.3.15) for arbitrary C_0 and C_1. The contribution of C_0 to the displacement field is a constant value, so it corresponds to translation in the transverse direction. In contrast, C_1 is associated with a displacement that increases linearly with x. Such a displacement field is characteristic of a rotation about $x = 0$. Orthogonality of the modes, which we shall discuss in the next section, as well as our experience with rigid body dynamics and the Ritz series method, suggests that we should define the center of mass as the reference point for the rotation, at which $\psi = 0$. Because C_0 and C_1 may be selected arbitrarily, we accordingly define the rigid body modes for flexure of a free-free beam as $\psi_{1,\text{tr}} = C_0$ and $\psi_{1,\text{rot}} = C_1(1 - 2x/L)$. A similar analysis for a beam whose only physical support is a hinge would lead to the obvious conclusion that the only rigid body mode is a rotation centered at the pin.

As a way of enhancing our physical understanding, as well as illustrating the general analytical procedure, let us use the foregoing development to compare the modal properties of two beams that differ solely in one boundary condition. Specifically, consider a uniform bar of length L whose left end is hinged and whose right end may be either hinged or clamped. Selecting a hinged end as $x = 0$ enables us to eliminate two coefficients C_j before we consider the other end. At the hinged end, we require $\psi = d^2\psi/dx^2 = 0$. If $x = 0$ at this location then the general solution in eq. (7.3.9) gives

$$C_2 + C_4 = 0 \text{ and } -\alpha^2 C_2 + \alpha^2 C_4 = 0 \tag{7.3.16}$$

Because rigid body motion is not possible, $\alpha \neq 0$. Thus, $C_2 = C_4 = 0$, which means that for either configuration, the mode function must fit $\psi = C_1 \sin(\alpha x/L) + C_3 \sinh(\alpha x/L)$.

A hinged-hinged beam is commonly said to be simply supported. We require that $\psi = d^2\psi/dx^2 = 0$ at $x = L$, which leads to

$$C_1 \sin(\alpha) + C_3 \sinh(\alpha) = 0$$
$$-C_1 \sin(\alpha) + C_3 \sinh(\alpha) = 0 \tag{7.3.17}$$

Because $\sinh(\alpha)$ is nonzero, the preceding requires $C_3 = 0$. Setting $C_1 = 0$ then would lead to a trivial solution, so we find that the characteristic equation is

$$\sin(\alpha) = 0 \tag{7.3.18}$$

The corresponding eigensolution is

$$\psi_j = C_1 \sin\left(\alpha_j \frac{x}{L}\right), \quad \alpha_j = j\pi \tag{7.3.19}$$

The steps for the hinged-clamped case are slightly more involved. Setting $\psi = d\psi/dx = 0$ at $x = L$ leads to

$$C_1 \sin(\alpha) + C_3 \sinh(\alpha) = 0$$
$$C_1 \cos(\alpha) + C_3 \cosh(\alpha) = 0 \tag{7.3.20}$$

The characteristic equation obtained from the determinant of these equations is

$$\sin(\alpha)\cosh(\alpha) - \sinh(\alpha)\cos(\alpha) = 0 \tag{7.3.21}$$

Because the hyperbolic functions become very large for modest values of α, it is convenient to rewrite the characteristic equation as

$$\tan(\alpha) - \tanh(\alpha) = 0 \tag{7.3.22}$$

We must solve this transcendental equation numerically. For each root α_j, the first equation for C_1 and C_3 gives $C_3 = -[\sin(\alpha)/\sinh(\alpha)]C_1$. The corresponding eigenfunction is

$$\psi_j = C_1\left[\sin\left(\alpha_j \frac{x}{L}\right) - \frac{\sin(\alpha_j)}{\sinh(\alpha_j)}\sinh(\alpha_j x)\right] \tag{7.3.23}$$

Numerical solution of the characteristic equation is substantially easier if one has a reasonable initial guess for each eigenvalue. To obtain this value, it is preferable to bring one term in eq. (7.3.22) to the right side. If we graph each term, $\tan(\alpha)$ and $\tanh(\alpha)$, as a function of α, the eigenvalues correspond to intersections of these functions. Such a graph is depicted in Figure 7.7. The behavior of each function is well known, and the graph need not be done carefully. For the hyperbolic tangent it is suf-

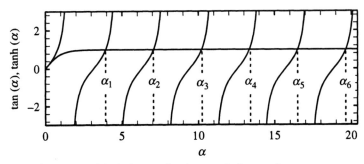

FIGURE 7.7 Graphical picture of a characteristic equation.

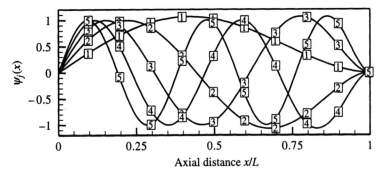

FIGURE 7.8 Flexural mode shapes of a hinged-clamped beam.

ficient to recognize that $\tanh(0) = 0$ and that $\tanh(u)$ is very close to unity beyond $u = \pi$. Correspondingly, we conclude that the roots for the second and higher modes are well approximated by $\alpha_j = (4j + 1)\pi/4$. (The first root is $\alpha_1 = 3.9266$.) Each of these eigenvalues is higher than its counterpart for the hinged-hinged bar, given by eq. (7.3.19). This trend is readily explained by the fact that clamping an end constrains that end more than supporting it with a hinge, so the hinged-clamped bar is stiffer than the hinged-hinged bar.

Figure 7.8 shows the first five mode functions of the hinged-clamped beam. (We set $C_1 = 1$ to draw the figure.) Increasing eigenvalues lead to more nodes, at which the displacement $w = \mathrm{Re}[\psi \exp(i\omega t)]$ is zero for all t. This property is obviously a consequence of the fact that $\alpha_j x/L$ is the argument of the sinusoidal terms forming ψ_j, and α_j increases with increasing j.

EXAMPLE 7.4

Consider flexural vibration of a uniform cantilevered beam. Derive the characteristic equation and an expression for the mode shape corresponding to a root of the characteristic equation.

Solution The objective here is to illustrate the basic operations required to determine flexural modes, including the application of identities involving hyperbolic functions. We know that an eigenfunction ψ has the general form of eq. (7.3.9). The boundary conditions at the clamped end of a cantilevered beam are somewhat simpler than those at the free end, so we define the clamped end as $x = 0$. Thus, the boundary conditions are

$$w = \frac{\partial w}{\partial x} = 0 \text{ at } x = 0$$

$$\frac{\partial^2 w}{\partial x^2} = \frac{\partial^3 w}{\partial x^3} = 0 \text{ and } x = L$$

Because there are no time derivatives in these boundary conditions, ψ must satisfy the same conditions as w. We consider $x = 0$ first in order to simplify the general solution for ψ,

$$\psi(0) = C_2 + C_4 = 0, \quad \left.\frac{\partial \psi}{\partial x}\right|_{x=0} = C_1 + C_3 = 0 \quad \Rightarrow \quad C_3 = -C_1, \quad C_4 = -C_2$$

Note that we know that neither rigid body translation nor rotation is possible, so $\alpha > 0$, which justifies canceling the common α factor in the second boundary condition. Thus, at this stage we have

$$\psi = C_1\left[\sin\left(\alpha\frac{x}{L}\right) - \sinh\left(\alpha\frac{x}{L}\right)\right] + C_2\left[\cos\left(\alpha\frac{x}{L}\right) - \cosh\left(\alpha\frac{x}{L}\right)\right]$$

We now proceed to consider the boundary conditions at $x = L$, which require that

$$C_1[-\sin(\alpha) - \sinh(\alpha)] + C_2[-\cos(\alpha) - \cosh(\alpha)] = 0$$
$$C_1[-\cos(\alpha) - \cosh(\alpha)] + C_2[\sin(\alpha) - \sinh(\alpha)] = 0$$

The matrix form of these equations is

$$\begin{bmatrix} [\sin(\alpha) + \sinh(\alpha)] & [\cos(\alpha) + \cosh(\alpha)] \\ [-\cos(\alpha) - \cosh(\alpha)] & [\sin(\alpha) - \sinh(\alpha)] \end{bmatrix}\begin{Bmatrix} C_1 \\ C_2 \end{Bmatrix} = \begin{Bmatrix} 0 \\ 0 \end{Bmatrix}$$

The characteristic equation is the determinant of the coefficient matrix,

$$[\sin(\alpha) + \sinh(\alpha)][\sin(\alpha) - \sinh(\alpha)] + [\cos(\alpha) + \cosh(\alpha)]^2 = 0$$

Although this equation seems to be quite formidable, it is simplified greatly by applying two identities:

$$\cosh(y)^2 - \sinh(y)^2 = 1, \quad \cos(y)^2 + \sin(y)^2 = 1$$

This transforms the characteristic equation to

$$2\cos(\alpha)\cosh(\alpha) + 2 = 0$$

After we have found a root α_j of the characteristic equation, an expression for the coefficient C_2 in terms of C_1 may be found from either boundary condition at $x = L$. The first one gives

$$C_2 = -C_1\frac{\sin(\alpha_j) + \sinh(\alpha_j)}{\cos(\alpha_j) + \cosh(\alpha_j)}$$

The corresponding mode shape is

$$\psi_j(x) = C_1\left\{\left[\sin\left(\alpha_j\frac{x}{L}\right) - \sinh\left(\alpha_j\frac{x}{L}\right)\right]\right.$$
$$\left. - \frac{\sin(\alpha_j) + \sinh(\alpha_j)}{\cos(\alpha_j) + \cosh(\alpha_j)}\left[\cos\left(\alpha_j\frac{x}{L}\right) - \cosh\left(\alpha_j\frac{x}{L}\right)\right]\right\}$$

The expression for ψ_j exemplifies the general fact that a mode will depend solely on the associated eigenvalue α_j and the nondimensional axial position x/L.

EXAMPLE 7.5

A uniform bar is hinged at its left end, and its right end is supported by a spring of stiffness $k = \kappa EI/L^3$, where κ is a dimensionless parameter.

(a) Determine the lowest three natural frequencies and plot the mode functions for the case where $\kappa = 2$ for a steel bar whose length is 4 m.

(b) Discuss the eigenvalues in the case of a strong spring, $\kappa \gg 1$.

(c) Discuss the eigenvalues in the case of a weak spring, $\kappa \ll 1$.

Solution   This example addresses the intricacies involved in solving the eigenvalue problem for systems having attachments. It also will amplify on the remarks pertaining to the role of springs in the conclusion of Example 7.3. The cross-sectional properties are constant, so the general form of an eigenfunction is given by eq. (7.3.9). The boundary conditions for a hinged end are simpler than those for a spring-supported end, so we place $x = 0$ at the hinge. There is neither displacement nor bending moment at $x = 0$, and the bending moment at $x = L$ also is zero. To determine the other boundary condition at $x = L$, we refer to Figure 7.5. We see there that an upward displacement $w(L, t)$ causes the spring to pull downward. However, positive shear is upward at the right end, so that $S = -EI \, \partial^2 w / \partial x^2 = -kw$ at $x = L$. Because none of the boundary conditions contain time derivatives, the mode functions must satisfy the same conditions as w,

$$\psi = \frac{\partial^2 \psi}{\partial x^2} = 0 \text{ at } x = 0$$

$$\frac{\partial^2 \psi}{\partial x^2} = 0 \text{ and } EI \frac{\partial^3 \psi}{\partial x^3} = k\psi \text{ at } x = L$$

We address the conditions at $x = 0$ in order to remove some coefficients. Specifically, we find that $C_2 = C_4 = 0$, just as we found in eq. (7.3.17). Thus, the mode function satisfying the conditions at $x = 0$ is

$$\psi = C_1 \sin\left(\alpha \frac{x}{L}\right) + C_3 \sinh\left(\alpha \frac{x}{L}\right)$$

Substitution of this expression into the conditions at $x = L$ leads to

$$-\alpha^2 C_1 \sin(\alpha) + \alpha^2 C_3 \sinh(\alpha) = 0$$

$$EI \frac{\alpha^3}{L^3}[-C_1 \cos(\alpha) + C_3 \cosh(\alpha)] = k[C_1 \sin(\alpha) + C_3 \sinh(\alpha)]$$

We know that rigid body motion is not possible because the pin prevents translation and a rigid body rotation of the bar would result in deformation of the spring. We therefore cancel the common α^2 factor in the first equation, regroup the terms to collect like coefficients of C_1 and C_3, and write the result in matrix form. This gives

$$\begin{bmatrix} \sin(\alpha) & -\sinh(\alpha) \\ \dfrac{kL^3}{EI\alpha^3}\sin(\alpha) + \cos(\alpha) & \dfrac{kL^3}{EI\alpha^3}\sinh(\alpha) - \cosh(\alpha) \end{bmatrix} \begin{Bmatrix} C_1 \\ C_3 \end{Bmatrix} = \begin{Bmatrix} 0 \\ 0 \end{Bmatrix}$$

The characteristic equation resulting from setting the determinant of the coefficient matrix to zero is

$$\cos(\alpha)\sinh(\alpha) - \sin(\alpha)\cosh(\alpha) + 2\frac{kL^3}{EI\alpha^3}\sin(\alpha)\sinh(\alpha) = 0$$

This equation may be placed into a more friendly form by dividing by $\sin(\alpha)\sinh(\alpha)$ and introducing the nondimensional factor κ defined in the problem statement. This leads to

$$\cot(\alpha) - \coth(\alpha) + \frac{2\kappa}{\alpha^3} = 0$$

After the roots of the characteristic equation have been determined, the boundary condition equations give

$$C_3 = \frac{\sin(\alpha_j)}{\sinh(\alpha_j)} C_1$$

The corresponding mode function is

$$\psi_j = C_1 \left[\sin\left(\alpha_j \frac{x}{L} \right) + \frac{\sin(\alpha_j)}{\sinh(\alpha_j)} \sinh\left(\alpha_j \frac{x}{L} \right) \right]$$

We generate starting values for numerical solution of the characteristic equation by splitting it into two parts as

$$F_1(\alpha) = \cot(\alpha), \quad F_2(\alpha) = \coth(\alpha) - \frac{2\kappa}{\alpha^3}$$

Note that we have written the characteristic equation in this manner in order to isolate the cyclical term. In addition to being easier to understand, using $F_1(\alpha) - F_2(\alpha) = 0$ based on such decomposition often helps numerical methods converge more rapidly. Both functions are graphed below, with $F_2(\alpha)$ shown for $\kappa = 0.02$ and $\kappa = 200$, as well as $\kappa = 2$, in order to understand the limits where κ is very small and very large.

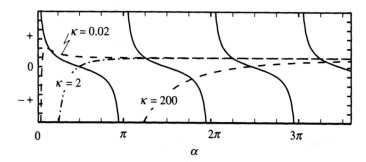

For very small α, $F_2(\alpha)$ is negative. As α increases, $2\kappa/\alpha^3$ becomes progressively smaller, so that $F_2(\alpha) \longrightarrow \coth(\alpha) \approx 1$. Thus, the first root must lie in the range $0 < \alpha_1 < \pi$, and the higher roots are approximately the roots of $\cot(\alpha) = 1$. Reasonable starting values for the numerical root finder are $\alpha_1 = \pi/2$, $\alpha_j = (j - 3/4)\pi$, $j \geq 2$. The lowest three numerical values obtained in the case $\kappa = 2$ are

$$\alpha_1 = 1.55021, \quad \alpha_2 = 3.95991, \quad \alpha_3 = 7.07426$$

To plot the corresponding three mode shapes we set the coefficient $C_1 = 1$, which leads to the second graph.

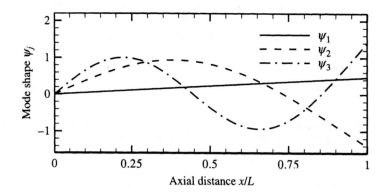

The second and third modes are oscillatory with an increasing number of nodes, but the first mode shape is almost straight, as though there is no flexural deformation. The explanation for the latter feature lies in the case where κ is small.

When $\kappa \gg 1$, the combination $2\kappa/\alpha^3$ is large only if α is not too big. Thus, as shown by the graphs of $F_1(\alpha)$ and $F_2(\alpha)$ for $\kappa = 200$, the first eigenvalue in this case is close to $\alpha_1 = \pi$, which is the lowest eigenvalue for a pinned-pinned beam. However, increasing α further ultimately makes $2\kappa/\alpha^3$ become very small, regardless of the size of κ. Thus, the high-frequency eigenvalues approach those for a system where $\kappa = 0$, which is a pinned-free beam. This demonstrates the loss of effectiveness of springs for high frequencies, which was alluded to in Example 7.3.

In the case where κ is very small, $2\kappa/\alpha^3$ is very small, except for $\alpha \ll 1$. Hence, all modes except the first are essentially the same as what would be obtained for $\kappa = 0$. In other words, these modes are essentially the same as those for a pinned-free beam. The only effect of the spring is in the first mode, because without it the bar could execute a rigid body rotation. The graph of $F_1(\alpha)$ and $F_2(\alpha)$ for $\kappa = 0.02$ shows that the first eigenvalue is very small. Identifying the limiting value of α_1 requires care. Essentially it involves writing $\cot(\alpha_1)$ and $\coth(\alpha_1)$ as series in α, expanding each denominator in powers of α, collecting like terms, then retaining only the lowest powers of α. The limiting form of the characteristic equation is

$$\cot(\alpha_1) - \coth(\alpha_1) + \frac{2\kappa}{\alpha_1^3} \approx -\frac{2}{3}\alpha_1 + \frac{2\kappa}{\alpha_1^3} = 0$$

Substitution of the definition of α into the solution of this equation leads to

$$\alpha_1^4 = 3\kappa \quad \Rightarrow \quad \omega_1 = \left(\frac{3\kappa EI}{\rho AL^4}\right)^{1/2} \equiv \left(\frac{\kappa L^2}{\frac{1}{3}\rho AL^3}\right)^{1/2}$$

This approximation for $\kappa = 2$ gives $\alpha_1 = 1.565$, which is extremely close to the computed value. The significance of this approximation lies in the fact that the mass of the bar is ρAL, so the denominator in the last form is the mass moment of inertia of the bar about its pin. In other words, the fundamental frequency when the spring is very soft is essentially the same as the natural frequency of a one-degree-of-freedom model of a rigid bar pinned at one end and supported by a spring at its other. As we concluded in Example 7.3, models based on considering bodies to be rigid should only be employed in frequency ranges well below the natural frequencies of deformation modes of the bodies.

7.4 WAVE INTERPRETATION OF MODES AND HIGH-FREQUENCY ASYMPTOTICS

The modal properties we have established may be viewed alternatively as wave propagation phenomena. Such a perspective sheds additional light on the underlying phenomena, and it also leads to improvements in the computation of mode functions. We begin with the case of axial (and torsional) vibration. Let us use complex exponentials to represent the sine and cosine functions in the general solution, eq. (7.2.5). This leads to

$$\psi_j = \frac{1}{2i}C_1\left[\exp\left(i\alpha_j\frac{x}{L}\right) - \exp\left(-i\alpha_j\frac{x}{L}\right)\right] + \frac{1}{2}C_2\left[\exp\left(i\alpha_j\frac{x}{L}\right) + \exp\left(i\alpha_j\frac{x}{L}\right)\right] \quad (7.4.1)$$

When we substitute this into the separation of variables form of eq. (7.2.2) and group similar terms, we obtain

$$u = \frac{1}{2}\mathrm{Re}\left\{\left(C_2 - \frac{C_1}{i}\right)\exp\left[i\left(\omega_j t - \alpha_j\frac{x}{L}\right)\right] + \left(C_2 + \frac{C_1}{i}\right)\exp\left[i\left(\omega_j t + \alpha_j\frac{x}{L}\right)\right]\right\} \quad (7.4.2)$$

In the case where EA and ρA are constants, eq. (7.2.1) is known as the wave equation. The d'Alembert general solution of this equation states that

$$u = f\left(t - \frac{x}{c}\right) + g\left(t + \frac{x}{c}\right), \quad c = \left(\frac{E}{\rho}\right)^{1/2} \tag{7.4.3}$$

where f and g denote arbitrary functions. It is clear that eq. (7.4.2) is a special case of the more general solution in eq. (7.4.3). The term $f(t - x/c)$ represents an *extensional wave* propagating in the direction of increasing x, while $g(t + x/c)$ represents an extensional wave propagating in the negative x direction. The propagation speed of each wave is c, because $\Delta x = \pm c \Delta t$ is the amount by which x must change over an interval Δt if we wish to track a constant value of the function f or g. The quantity $t \pm x/c$ is the *phase* of the wave, and c is the *phase speed*. A comparison of eqs. (7.4.2) and (7.4.3) shows that the modal vibration is a sum of sinusoidal waves propagating in opposite directions, as depicted in Figure 7.9. The propagation speed of each wave is

$$c = \frac{\omega_j}{(\alpha_j / L)} = \left(\frac{EA}{\rho A}\right)^{1/2} \tag{7.4.4}$$

If the bar is homogeneous, we have $c = (E/\rho)^{1/2}$, which is called the *bar speed*, c_{bar}. This speed is analogous to the speed of sound in fluids, but its value in metals is quite high, for example, $c_{bar} \approx 5200$ m/s for steel.

The fact that ψ_j is real means that the spatial distribution of modal displacement has a constant shape that oscillates in amplitude as time varies. Such a displacement field is a *standing wave*, which is created when sinusoidal waves of equal amplitude superpose, as indicated by eq. (7.4.2). In essence, each propagating wave that is incident on a boundary reflects, and thereby generates the oppositely traveling wave in order to satisfy the boundary condition at that end.

We employ the same approach to derive the wave description of flexural modes. Our starting point is the general solution for a flexural mode, eq. (7.3.9), in which we express the trigonometric and hyperbolic functions in terms of their equivalent representations as complex and real exponentials, respectively. When we use that form to synthesize the modal displacement field w in eq. (7.3.2), we obtain

$$w = \frac{1}{2}\text{Re}\left\{\left(C_2 - \frac{C_1}{i}\right)\exp\left[i\left(\omega_j t - \alpha_j \frac{x}{L}\right)\right] + \left(C_2 + \frac{C_1}{i}\right)\exp\left[i\left(\omega_j t + \alpha_j \frac{x}{L}\right)\right]\right.$$
$$\left. + (C_4 - C_3)\exp\left(-\alpha_j \frac{x}{L}\right)(i\omega_j t) + (C_4 + C_3)\exp\left(+\alpha_j \frac{x}{L}\right)\exp(i\omega_j t)\right\} \tag{7.4.5}$$

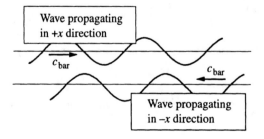

FIGURE 7.9 Decomposition of modal axial displacement into propagating waves.

The first two terms represent sinusoidal waves that respectively propagate in the positive and negative x directions, just like eq. (7.4.2). Their contribution is felt throughout the bar's length. The third and fourth terms represent waves that decay in the positive and negative x directions, respectively, which becomes apparent if we factor out $\exp(\alpha_j)$ from the last term, which converts the expression to

$$w = \frac{1}{2}\mathrm{Re}\left\{(C_2 + iC_1)\exp\left[i\left(\omega_j t - \alpha_j\frac{x}{L}\right)\right] + (C_2 - iC_1)\exp\left[i\left(\omega_j t + \alpha_j\frac{x}{L}\right)\right]\right.$$

$$+ (C_4 - C_3)\exp\left(-\alpha_j\frac{x}{L}\right)\exp(i\omega_j t)$$

$$\left. + [(C_4 + C_3)\exp(\alpha_j)]\exp\left[-\alpha_j\left(\frac{L-x}{L}\right)\right]\exp(i\omega_j t)\right\} \tag{7.4.6}$$

The phase speed of the propagating waves is

$$c_{\mathrm{flex}} = \frac{\omega L}{\alpha} \tag{7.4.7}$$

By definition, $\omega^2 = (EI/\rho AL^4)\alpha^4$, which leads to

$$c_{\mathrm{flex}} = c_{\mathrm{bar}}\frac{d}{L}\alpha \tag{7.4.8}$$

where d is the area radius of gyration, $d = (I/A)^{1/2}$. For either of the sinusoidal waves in eq. (7.4.6), the *wavelength*, which is the spatial distance over which the pattern at fixed t repeats, is $\ell = 2\pi L/\alpha$. For studies of wave propagation, an equation giving the phase speed in terms of wavelength is called a *dispersion relation*. In contrast to extensional waves, which propagate at the constant speed, c_{bar}, the propagation speed of flexural waves increases with decreasing wavelength (increasing α). In other words, the propagation speed of flexural waves forming a mode increases with increasing mode number.

Now let us consider the last two terms in eq. (7.4.6), which contain real exponentials. The factor $\exp(-\alpha_j x/L)$ decreases with increasing x from a maximum value at $x = 0$. In contrast, the factor $\exp[-\alpha_j(L - x)]$ decreases with decreasing x from a maximum value at $x = L$. We say that these are *evanescent waves*, because they disappear with increasing distance from their respective boundaries. The decomposition of a flexural modal vibration into propagating and evanescent waves is depicted in Figure 7.10. When α is reasonably large, these evanescent waves are only significant in regions very close to $x = 0$ and $x = L$, in which case they represent boundary layer-type effects.

The explanation for the occurrence of evanescent waves becomes evident when we consider the fact that there are two boundary conditions at each end, whereas axial displacement requires satisfaction of only one boundary condition. A suitable combination of the propagating flexural waves can be found to satisfy one boundary condition, but satisfying two conditions requires the evanescent wave at that end. In some cases, an evanescent wave will not be generated. For example, if $C_4 = C_2 = 0$, which is the case when $x = 0$ is pinned, we find that there is no evanescent wave in the vicinity of $x = 0$.

Knowledge of the wave representation in eq. (7.4.6) is especially useful for computing high-frequency mode functions. Some mode functions require evaluation

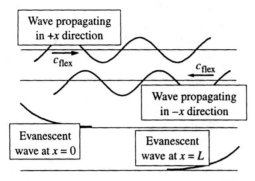

FIGURE 7.10 Decomposition of a flexural modal vibration into propagating and evanescent waves.

of the difference between trigonometric and hyperbolic functions, but the former are quite small compared to the latter when α_j is large. For example, if $\alpha_j > 30$, sinh $(\alpha_j) \approx$ cosh $(\alpha_j) > 5(10^{12})$. This can result in extreme loss of precision, in some cases leading to nonsensical results. For very large values of α_j, it also is possible that underflow or overflow will arise in the calculation. The alternative is to represent the modal displacement in terms of its limiting form for very large α_j, as

$$w = \text{Re } [(\psi_j(x))_{\text{asymptotic}} \exp(i\omega t)] \tag{7.4.9}$$

where $(\psi_j(x))_{\text{asymptotic}}$ is the x dependence associated with the propagating and evanescent waves,

$$\psi_j(x))_{\text{asymptotic}} = \frac{1}{2}\left\{ (B_1)_{\text{asymptotic}} \exp\left(-i\alpha_j \frac{x}{L}\right) + (B_2)_{\text{asymptotic}} \exp\left(i\alpha_j \frac{x}{L}\right) \right.$$
$$\left. + (B_3)_{\text{asymptotic}} \exp\left(-\alpha_j \frac{x}{L}\right) + (B_4)_{\text{asymptotic}} \exp\left[-\alpha_j \left(\frac{L-x}{L}\right)\right] \right\} \tag{7.4.10}$$

As implied by the subscript, the coefficients in this description are the limits as $\alpha_j \rightarrow \infty$. A comparison of the preceding with eq. (7.4.6) shows that

$$B_1 = \lim_{\alpha_j \to \infty} (C_2 + iC_1), \quad B_2 = \lim_{\alpha_j \to \infty} (C_2 - iC_1)$$

$$B_3 = \lim_{\alpha_j \to \infty} (C_4 - C_3), \quad B_4 = \lim_{\alpha_j \to \infty} [(C_4 + C_3)\exp(\alpha_j)] \tag{7.4.11}$$

The outcome of the above evaluations will be a real function $(\psi_j(x))_{\text{asymptotic}}$, based on factoring out the arbitrary coefficient, B_1 or B_2.

EXAMPLE 7.6

Consider the cantilevered beam in Example 7.4. Derive a high-frequency asymptotic representation of the mode function. For the first, second, eighth, and sixteenth modes, evaluate the asymptotic representation, and compare it to the results obtained by a direct evaluation of the analytical eigensolution.

Solution In addition to illustrating the operations leading to a high-frequency asymptotic representation, this example will demonstrate that such representations might be essential. The characteristic equation was found in Example 7.4 to be

$$\cos(\alpha) = -\frac{1}{\cosh(\alpha)}$$

The coefficient C_1 was chosen to be the arbitrary value in the mode function, which was found to be

$$\psi_j(x) = C_1\left\{\left[\sin\left(\alpha_j\frac{x}{L}\right) - \sinh\left(\alpha_j\frac{x}{L}\right)\right]\right.$$

$$\left. - \frac{\sin(\alpha_j) + \sinh(\alpha_j)}{\cos(\alpha_j) + \cosh(\alpha_j)}\left[\cos\left(\alpha_j\frac{x}{L}\right) - \cosh\left(\alpha_j\frac{x}{L}\right)\right]\right\}$$

When α is large, we may approximate the characteristic equation by letting $\cosh(\alpha) \approx 0$. This leads to approximate eigenvalues $\alpha_j = (2j-1)\pi/2$, with the approximation becoming increasingly good as j increases.

To derive an asymptotic approximation of the mode functions for sufficiently large j, we employ eqs. (7.4.10) and (7.4.11). The coefficients are

$$C_2 = -C_4 = -C_1\frac{\sin(\alpha_j) + \sinh(\alpha_j)}{\cos(\alpha_j) + \cosh(\alpha_j)}, \quad C_3 = -C_1$$

When α_j is very large, we have $\sinh(\alpha_j) \longrightarrow \cosh(\alpha_j) \longrightarrow \frac{1}{2}\exp(\alpha_j)$, which is much bigger than unity. Thus, the first three coefficients are

$$B_1 = \lim_{\alpha_j \to \infty}\left[-C_1\frac{\sin(\alpha_j) + \sinh(\alpha_j)}{\cos(\alpha_j) + \cosh(\alpha_j)} + iC_1\right] = (-1+i)C_1$$

$$B_2 = \lim_{\alpha_j \to \infty}\left[\left(-C_1\frac{\sin(\alpha_j) + \sinh(\alpha_j)}{\cos(\alpha_j) + \cosh(\alpha_j)}\right) - iC_1\right] = (-1-i)C_1$$

$$B_3 = \lim_{\alpha_j \to \infty}\left[C_1\frac{\sin(\alpha_j) + \sinh(\alpha_j)}{\cos(\alpha_j) + \cosh(\alpha_j)} + C_1\right] = 2C_1$$

The presence of the $\exp(\alpha_j)$ factor in the definition of B_4 requires a more careful analysis. The definitions of the hyperbolic functions give $\sinh(\alpha_j) - \cosh(\alpha_j) \equiv -\exp(-\alpha_j)$, which leads to

$$B_4 = \lim_{\alpha_j \to \infty}\left\{\left[C_1\frac{\sin(\alpha_j) + \sinh(\alpha_j)}{\cos(\alpha_j) + \cosh(\alpha_j)} - C_1\right]\exp(\alpha_j)\right\}$$

$$= C_1\lim_{\alpha_j \to \infty}\left\{\frac{\sin(\alpha_j) - \cos(\alpha_j) + \sinh(\alpha_j) - \cosh(\alpha_j)}{\cos(\alpha_j) + \cosh(\alpha_j)}\exp(\alpha_j)\right\}$$

$$= C_1\lim_{\alpha_j \to \infty}\left\{\frac{[\sin(\alpha_j) - \cos(\alpha_j)]\exp(\alpha_j) - 1}{\cosh(\alpha_j)}\right\}$$

$$= C_1\lim_{\alpha_j \to \infty}\{2C_1[\sin(\alpha_j) - \cos(\alpha_j)]\} = 2C_1\sin\left[\frac{(2j-1)\pi}{2}\right] = 2(-1)^{j-1}C_1$$

where the last step involved substituting the asymptotic eigenvalues. When we substitute these coefficients into eq. (7.4.10), we find that the asymptotic mode approximation is

$$(\psi_j(x))_{\text{asymptotic}} = \frac{1}{2}C_1\left\{(-1+i)\exp\left(-i\alpha_j\frac{x}{L}\right) + (-1-i)\exp\left(i\alpha_j\frac{x}{L}\right)\right.$$

$$\left. + 2\exp\left(-\alpha_j\frac{x}{L}\right) + 2(-1)^{j-1}\exp\left[-\alpha_j\left(\frac{L-x}{L}\right)\right]\right\}$$

$$= C_1\left\{\sin\left(\alpha_j\frac{x}{L}\right) - \cos\left(\alpha_j\frac{x}{L}\right) + \exp\left(-\alpha_j\frac{x}{L}\right)\right.$$

$$\left. + (-1)^{j-1}\exp\left[-\alpha_j\left(\frac{L-x}{L}\right)\right]\right\}$$

The first set of graphs describes the first and second modes. The actual eigenvalues are $\alpha_1 = 1.8751$ and $\alpha_2 = 4.69409$. The approximate eigenvalues, which were used to evaluate $(\psi_j(x))_{\text{asymptotic}}$, are $\alpha_1 = \pi/2$ and $\alpha_2 = 3\pi/2$. The first value is 16% in error, so it is not surprising that the first mode is not well described by the asymptotic approximation. The second asymptotic eigenvalue is only 0.4% in error, which explains the surprising fact that the high-frequency approximation is quite good for the second mode.

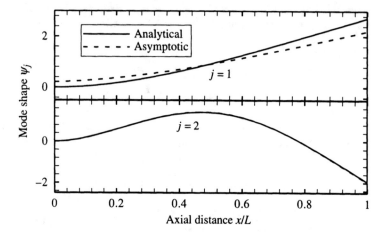

The second set of graphs displays the analytical and asymptotic mode shapes for the eighth and sixteenth modes. The asymptotic approximations $\alpha_8 = 15\pi/2$ and $\alpha_{16} = 31\pi/2$ are indistinguishable from the roots obtained numerically. There is no discernible difference between the analytical and asymptotic results for the eighth mode. The analytical and asymptotic results for the sixteenth mode overlap until $x \approx 0.76L$. At that location, the analytical result suddenly drops to zero, which certainly is not correct. This irregularity is a consequence of loss of precision associated with the large values of the hyperbolic functions when α_j is very large and x/L is not small. Without increasing the precision of the computations, there is no way to compute the analytical mode shape in this region. In contrast, the asymptotic representation has no difficulty.

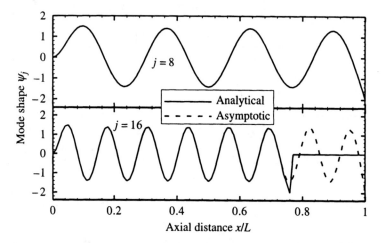

Loss of precision occurs over an increasing range of x as the mode number increases beyond 16. Thus, evaluations of high-frequency modes shapes should use the asymptotic form. At the same time, it should be noted that the mechanics of materials approximations on which our fundamental model is founded lose validity with increasing frequency. Beyond a certain frequency range, one should use Timoshenko beam theory, which we will examine in Section 7.9.

7.5 SYMMETRY AND MODE FUNCTIONS

Physical symmetry can often be used to simplify vibration analysis. Consider a bar that is constructed such that every feature (mass distribution, elastic rigidity, boundary conditions) at position x is duplicated at $L - x$, for $0 \leq x \leq L/2$. We say that such a bar is symmetric with respect to a transverse cutting plane at $x = L/2$. If we were to place a mirror on the symmetry plane, the image appearing in the mirror would look like the segment behind the mirror. Figure 7.11(a) shows symmetric axial and transverse displacements and torsional rotation, while antisymmetric displacements are depicted in Figure 7.11(b). Note that the symmetry we are discussing is physical, not mathematical. That is, a physically symmetric displacement has a mirror image that is the same as that at the matching point behind the mirror. From a mathematical standpoint, symmetric axial displacement is an odd function with respect to $x = L/2$, while symmetric transverse displacement and torsional rotation are even functions,

$$\text{Symmetric:} \begin{cases} u(x) = -u(L - x) \\ w(x) = w(L - x), \quad \theta(x) = \theta(L - x) \end{cases}$$

$$\text{Antisymmetric:} \begin{cases} u(x) = u(L - x) \\ w(x) = -w(L - x), \quad \theta(x) = -\theta(L - x) \end{cases} \tag{7.5.1}$$

An arbitrary displacement field in a physically symmetric system may be decomposed into symmetric and antisymmetric parts,

$$\boxed{\begin{aligned} u(x) &= u_{\text{sym}}(x) + u_{\text{anti}}(x) \\[4pt] w(x) &= w_{\text{sym}}(x) + w_{\text{anti}}(x) \\[4pt] \theta(x) &= \theta_{\text{sym}}(x) + \theta_{\text{anti}}(x) \end{aligned}} \tag{7.5.2}$$

To determine the individual parts we use the decomposition to describe the displacement at the point symmetric to x, then invoke the symmetry definitions appearing in eq. (7.5.1), which gives

$$u(L - x) = u_{\text{sym}}(L - x) + u_{\text{anti}}(L - x) = -u_{\text{sym}}(x) + u_{\text{anti}}(x)$$

$$w(L - x) = w_{\text{sym}}(L - x) + w_{\text{anti}}(L - x) = w_{\text{sym}}(x) - w_{\text{anti}}(x) \tag{7.5.3}$$

$$\theta(L - x) = \theta_{\text{sym}}(L - x) + \theta_{\text{anti}}(L - x) = \theta_{\text{sym}}(x) - \theta_{\text{anti}}(x)$$

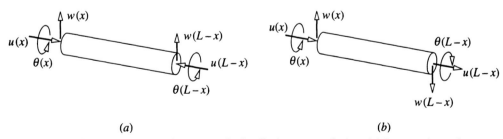

(a)   (b)

FIGURE 7.11 Symmetry and antisymmetry in the displacement of a bar. (a) Symmetric motions. (b) Antisymmetric motions.

We form the sum and difference of eqs. (7.5.2) and (7.5.3) to find the individual parts,

$$
\begin{aligned}
&u_{\text{sym}}(x) = \tfrac{1}{2}[u(x) - u(L - x)], \quad u_{\text{anti}}(x) = \tfrac{1}{2}[u(x) + u(L - x)] \\[2mm]
&w_{\text{sym}}(x) = \tfrac{1}{2}[w(x) + w(L - x)], \quad w_{\text{anti}}(x) = \tfrac{1}{2}[w(x) - w(L - x)] \\[2mm]
&\theta_{\text{sym}}(x) = \tfrac{1}{2}[\theta(x) + \theta(L - x)], \quad \theta_{\text{anti}}(x) = \tfrac{1}{2}[\theta(x) - \theta(L - x)]
\end{aligned}
\tag{7.5.4}
$$

A corollary of this decomposition is the fact that the eigenmodes of a symmetric system correspond to displacement fields that are either symmetric or antisymmetric. Either type may be determined by considering only half the bar, say $0 \le x \le L/2$, with the boundary conditions at the symmetry plane derived by using eq. (7.5.4) to evaluate the displacements and derivatives at the midpoint, $x = L/2$. This gives

$$
\text{Symmetric modes:}
\left\{
\begin{aligned}
&\psi_u = 0 \\[2mm]
&\frac{d\psi_w}{dx} = \frac{d^3\psi_w}{dx^3} = 0 \\[2mm]
&\frac{d\theta}{dx} = 0
\end{aligned}
\right\}
\text{ at the midpoint}
$$

$$
\tag{7.5.5}
$$

$$
\text{Antisymmetric modes:}
\left\{
\begin{aligned}
&\frac{d\psi_u}{dx} = 0 \\[2mm]
&\psi_w = \frac{d^2\psi_w}{dx^2} = 0 \\[2mm]
&\theta = 0
\end{aligned}
\right\}
\text{ at the midpoint}
$$

An interesting observation is that the midpoint boundary conditions for symmetric motions are those of a fixed, free, and guided end for axial, torsional, and flexural displacement, respectively. In contrast, the comparable boundary conditions for antisymmetric motion are free, fixed, and hinged ends.

These equivalent boundary conditions are useful when we are only interested in responses that are either symmetric or antisymmetric. The decomposition also is useful in situations where there are mixed boundary conditions. In that case, we would shift the origin of the coordinate system to the midspan location, so that $-L/2 \le x \le L/2$. The symmetry and antisymmetry boundary conditions, eqs. (7.5.5), then would apply at $x = 0$. Satisfying these conditions first would enable us to eliminate some of the coefficients in the general homogeneous solution $\psi(x)$ prior to addressing the Robin boundary conditions at $x = L/2$.

EXAMPLE 7.7

Identical transverse springs of stiffness k are used to support each end of a bar whose length is L. Decompose the transverse vibration into symmetric and antisymmetric modes. Derive the characteristic equation and an expression for the mode functions corresponding to each type of motion.

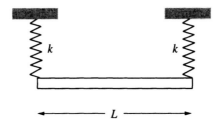

Solution The substantial reduction in difficulty afforded by using symmetry will be evident in this solution. We define the midpoint as $x = 0$, and let the right end be $x = L/2$. An upward displacement at that location results in the spring applying a downward force kw to that end. Our sign convention is that positive shear is upward at that end, so the shear boundary condition there is that $S = -kw$ at $x = L/2$. The right end also is moment free. In combination with the symmetry boundary conditions, eqs. (7.5.5), we then have

$$\text{Symmetric displacement:} \begin{cases} \dfrac{d\psi}{dx} = \dfrac{d^3\psi}{dx^3} = 0 \text{ at } x = 0 \\[2ex] \dfrac{d^2\psi}{dx^2} = 0, \quad EI\dfrac{d^3\psi}{dx^3} = kw \text{ at } x = L/2 \end{cases}$$

$$\text{Antisymmetric displacement:} \begin{cases} \psi = \dfrac{d^2\psi}{dx^2} = 0 \text{ at } x = 0 \\[2ex] \dfrac{d^2\psi}{dx^2} = 0, \quad EI\dfrac{d^3\psi}{dx^3} = kw \text{ at } x = L/2 \end{cases}$$

For either type of motion the general solution for a mode is the standard one in eq. (7.3.9). In the case of symmetric displacement, satisfying the boundary condition at the midpoint, $x = 0$, leads to

$$\psi = C_2 \cos\left(\alpha\frac{x}{L}\right) + C_4 \cosh\left(\alpha\frac{x}{L}\right)$$

where $\alpha_4 = \rho A \omega^2 / EIL^4$. We know that rigid body motion is not possible, so we may cancel any common α factors. Satisfying the boundary conditions at $x = L/2$ then requires that

$$-C_2 \cos\left(\frac{\alpha}{2}\right) + C_4 \cosh\left(\frac{\alpha}{2}\right) = 0$$

$$EI\frac{\alpha^3}{L^3}\left[C_2 \sin\left(\frac{\alpha}{2}\right) + C_4 \sinh\left(\frac{\alpha}{2}\right)\right] = k\left[C_2 \cos\left(\frac{\alpha}{2}\right) + C_4 \cosh\left(\frac{\alpha}{2}\right)\right]$$

To write this in matrix form we define $\lambda = \alpha/2$, which results in

$$\begin{bmatrix} -\cos(\lambda) & \cosh(\lambda) \\ \left[\lambda^3\sin(\lambda) - \dfrac{kL^3}{8EI}\cos(\lambda)\right] & \left[\lambda^3\sinh(\lambda) - \dfrac{kL^3}{8EI}\cosh(\lambda)\right] \end{bmatrix} \begin{Bmatrix} C_2 \\ C_4 \end{Bmatrix} = \begin{Bmatrix} 0 \\ 0 \end{Bmatrix}$$

The characteristic equation is the determinant of the coefficient matrix,

$$\lambda^3[\cos(\lambda)\sinh(\lambda) + \sin(\lambda)\cosh(\lambda)] - \frac{kL^3}{4EI}\cos(\lambda)\cosh(\lambda) = 0$$

which is equivalent to

$$\tan(\lambda) + \tanh(\lambda) = \frac{kL^3}{4EI}\frac{1}{\lambda^3}$$

To determine symmetric modes we would solve this equation for its roots λ_j, from which the corresponding natural frequencies would be determined according to

$$\omega_j = \left(\frac{EI}{\rho A L^4}\right)^{1/2} \alpha_j^2 = 4\left(\frac{EI}{\rho A L^4}\right)^{1/2} \lambda_j^2$$

We use the first of the coefficient equations to determine the ratio C_4/C_2 and substitute $\alpha_j = 2\lambda_j$, from which we find the symmetric mode functions to be

$$\psi_j = C_2\left[\cos\left(\frac{2\lambda_j x}{L}\right) + \frac{\cos(\lambda_j)}{\cosh(\lambda_j)}\cosh\left(\frac{2\lambda_j x}{L}\right)\right], \quad -\frac{L}{2} \le x \le \frac{L}{2}$$

Determination of the antisymmetric mode shapes follows the same steps. Satisfying the boundary conditions at $x = 0$ in this case requires that

$$\psi = C_1\sin\left(\alpha\frac{x}{L}\right) + C_3\sinh\left(\alpha\frac{x}{L}\right)$$

The matrix equation resulting from satisfaction of the boundary conditions at $x = L/2$ is

$$\begin{bmatrix} -\sin(\lambda) & \sinh(\lambda) \\ \left[\lambda^3\sin(\lambda) + \dfrac{kL^3}{8EI}\cos(\lambda)\right] & \left[\lambda^3\sinh(\lambda) - \dfrac{kL^3}{8EI}\cosh(\lambda)\right] \end{bmatrix} \begin{Bmatrix} C_1 \\ C_3 \end{Bmatrix} = \begin{Bmatrix} 0 \\ 0 \end{Bmatrix}$$

After simplification, the characteristic equation for antisymmetric modes is found to be

$$\cot(\lambda) - \coth(\lambda) = -\frac{kL^3}{4EI}\frac{1}{\lambda^3}$$

The antisymmetric mode functions, corresponding to roots λ_j of this equation are

$$\psi_j = C_2\left[\sin\left(\frac{2\lambda_j x}{L}\right) + \frac{\sin(\lambda_j)}{\sinh(\lambda_j)}\sinh\left(\frac{2\lambda_j x}{L}\right)\right], \quad -\frac{L}{2} \le x \le \frac{L}{2}$$

It is interesting to consider the alternative to invoking symmetry arguments to solve this problem. Without symmetry, we would find that we could not eliminate any of the C_j coefficients in the mode function by using one set of boundary conditions. Thus, we would be left with the task of finding the characteristic equation by expanding a 4×4 determinant.

7.6 ORTHOGONALITY

The Ritz series method led to orthogonality conditions that the modes of a continuous system must satisfy. Thus, in some sense it is unnecessary to use the properties of the field equations to rederive those conditions. However, we shall follow other texts by using the properties of the field equations and eigensolutions to rederive the orthogonality conditions. One reason for doing so is to gain confidence in the energy-based derivation of the orthogonality conditions associated with the Ritz series method.

We begin with the axial modes of a bar whose extensional rigidity EA may depend on x. The mode functions satisfy

$$\frac{d}{dx}\left(EA\frac{d\psi_j}{dx}\right) + \rho A\omega_j^2\psi_j = 0 \tag{7.6.1}$$

which applies for any mode number j. We multiply the differential equation for mode j by a specific mode ψ_k and integrate the product over the span of the beam, $0 \le x \le L$. This leads to

$$\int_0^L \psi_k \frac{d}{dx}\left(EA\frac{d\psi_j}{dx}\right)dx + \omega_j^2 \int_0^L \rho A \psi_k \psi_j\, dx = 0 \tag{7.6.2}$$

A comparable relation applies if we switch the k and j indices, so that

$$\int_0^L \psi_j \frac{d}{dx}\left(EA\frac{d\psi_k}{dx}\right)dx + \omega_k^2 \int_0^L \rho A \psi_k \psi_j\, dx = 0 \tag{7.6.3}$$

The second integrals in eqs. (7.6.2) and (7.6.3) are obviously the same. To bring out the similarity of the first integrals, we apply integration by parts to each equation, which gives

$$\psi_k\left(EA\frac{d\psi_j}{dx}\right)\Bigg|_0^L - \int_0^L EA\frac{d\psi_j}{dx}\frac{d\psi_k}{dx}\, dx + \omega_j^2 \int_0^L \rho A \psi_k \psi_j\, dx = 0$$
$$\tag{7.6.4}$$

$$\psi_j\left(EA\frac{d\psi_k}{dx}\right)\Bigg|_0^L - \int_0^L EA\frac{d\psi_j}{dx}\frac{d\psi_k}{dx}\, dx + \omega_k^2 \int_0^L \rho A \psi_k \psi_j\, dx = 0$$

When we subtract the second relation from the first, we obtain

$$\left[\psi_k\left(EA\frac{d\psi_j}{dx}\right) - \psi_j\left(EA\frac{d\psi_k}{dx}\right)\right]\Bigg|_0^L + (\omega_j^2 - \omega_k^2)\int_0^L \rho A \psi_k \psi_j\, dx = 0 \tag{7.6.5}$$

At a fixed end, $\psi_j = \psi_k = 0$, while a free end gives $d\psi_j/dx = d\psi_k/dx = 0$. The boundary terms in eq. (7.6.5) then evaluate to zero, but the same is not true in the case of mixed boundary conditions. Suppose that a mass m_L and spring k_L are fastened at $x = L$. A free body diagram of the block shows the boundary condition to be

$$-EA\frac{\partial u}{\partial x} - k_L u = m_L \ddot{u} \text{ at } x = L \tag{7.6.6}$$

Setting $u = \text{Re}[\psi_n \exp(i\omega_n t)]$ leads to

$$(k_L - m_L \omega_n^2)\psi_n = -EA\frac{d\psi_n}{dx} \text{ at } x = L \tag{7.6.7}$$

If mass m_0 and spring k_0 are attached at end $x = 0$, the mixed boundary condition is like the preceding, except for a reversal of the sign for the derivative,

$$(k_0 - m_0 \omega_n^2)\psi_n = EA\frac{d\psi_n}{dx} \text{ at } x = 0 \tag{7.6.8}$$

Both boundary conditions apply to any mode, so we may use them to eliminate the boundary derivatives in eq. (7.6.5). Doing so for $x = L$ gives

$$\left[\psi_k\left(EA\frac{d\psi_j}{dx}\right) - \psi_j\left(EA\frac{d\psi_k}{dx}\right)\right]\Bigg|_{x=L} = [-\psi_k(k_L - m_L \omega_k^2)\psi_j + \psi_j(k_L - m_L \omega_k^2)\psi_k]|_{x=L}$$

$$= m_L(\omega_j^2 - \omega_k^2)\psi_j\psi_k|_{x=L}$$

Because $x = 0$ is the lower limit for eq. (7.6.5) and the sign of the derivative in eq. (7.6.8) is opposite from the sign of the corresponding term in eq. (7.6.7), a result similar to the preceding is obtained for the $x = 0$ boundary terms. Consequently, mixed boundary conditions at the ends reduce eq. (7.6.5) to

$$m_L(\omega_j^2 - \omega_k^2)\psi_j(L)\psi_k(L) + m_0(\omega_j^2 - \omega_k^2)\psi_j(0)\psi(0)$$

$$+ (\omega_j^2 - \omega_k^2)\int_0^L \rho A \psi_k \psi_j \, dx = 0 \tag{7.6.9}$$

In view of the fact that there are no boundary contributions for fixed or free ends, the preceding expression with m_0 or m_L set to zero also describes those situations.

Two possibilities exist. If $j \neq k$, so that $\omega_j \neq \omega_k$, then eq. (7.6.9) becomes the mass orthogonality condition,

$$\boxed{\int_0^L \rho A \psi_k \psi_j \, dx + m_l \psi_j(L)\psi_k(L) + m_0 \psi_j(0)\psi_k(0) = 0, \quad k \neq j} \tag{7.6.10}$$

In contrast, if $j = k$, so that $\omega_j = \omega_k$, eq. (7.6.9) is satisfied identically. We use this case to remove the arbitrariness of the remaining mode coefficient, C_1 or C_2 in eq. (7.2.15). A *normal mode* Ψ_j is an eigenfunction ψ_j whose arbitrary coefficient has been selected such that eq. (7.6.10) gives unity for $k = j$, that is

$$\int_0^L \rho A \Psi_j^2 \, dx + m_2 \Psi_j(L)^2 + m_1 \Psi_j(0)^2 = 1 \tag{7.6.11}$$

The second orthogonality condition is obtained from the first of eqs. (7.6.4), with the modes now taken to be normalized. We use the boundary conditions, eqs. (7.6.7) and (7.6.8), to eliminate the derivatives at the boundaries. This yields

$$\Psi_k(L)[-(k_L - m_L\omega_j^2)\Psi_j(L)] - \Psi_k(0)[(k_0 - m_0\omega_j^2)\Psi_j(0)]$$

$$- \int_0^L \frac{d\Psi_k}{dx}\left(EA\frac{\partial\Psi_j}{\partial x}\right)dx + \omega_j^2\int_0^L \rho A \Psi_k \Psi_j \, dx = 0 \tag{7.6.12}$$

If $k \neq j$, the mass terms appearing in this relation cancel due to the first orthogonality condition, eq. (7.6.10). Furthermore, if $k = j$, the mass terms add up to ω_j^2 as a consequence of eq. (7.6.11). We thereby obtain the stiffness orthogonality condition,

$$\boxed{\int_0^L EA\frac{d\Psi_k}{dx}\frac{d\Psi_j}{dx}dx + k_2\Psi_k(L)\Psi_j(L) + k_1\Psi_k(0)\Psi_j(0) = \delta_{jk}\omega_j^2} \tag{7.6.13}$$

where δ_{jk} is the Kronecker delta index. Setting k_0 or k_L to zero adapts this relation to cases where the respective ends are fixed or free.

Note that nowhere in the derivation was it assumed that ω_j or ω_k is nonzero. Consequently, eqs. (7.6.10), (7.6.11), and (7.6.13) apply to rigid modes, as well as deformational modes.

Equations (7.6.10) and (7.6.13) are the same as the orthogonality relations for axial motion derived by the Ritz method, eqs. (6.3.9) and (6.3.14), in the special case where springs and masses are attached only at the ends. Some individuals prefer the present derivation because it is more rigorous. However, the present derivation is less general because it is limited to attachments at the ends.

The derivation of the orthogonality relations for flexural modes follows the outline of the preceding derivation. We begin with the differential equation for modes when the flexural rigidity is not constant,

$$\frac{d^2}{dx^2}\left(EI\frac{d^2\psi_j}{dx^2}\right) - \rho A \omega_j^2 \psi_j = 0 \tag{7.6.14}$$

We multiply the modal differential equation for ψ_j by an arbitrarily selected ψ_k, and do likewise for the differential equation governing ψ_k. Both products are integrated by parts over the length of the bar. The primary difference relative to the analysis for axial motion is that the integration by parts must be done twice, specifically

$$\int_0^L \psi_k \frac{d}{dx^2}\left(EI\frac{d^2\psi_j}{dx^2}\right)dx - \omega_j^2 \int_0^L \rho A \psi_k \psi_j \, dx$$

$$\equiv \psi_k \frac{d^2}{dx^2}\left(EI\frac{d^2\psi_j}{dx^2}\right)\Big|_0^L - \frac{d\psi_k}{dx}\left(EI\frac{d^2\psi_j}{dx^2}\right)\Big|_0^L \tag{7.6.15}$$

$$+ \int_0^L \frac{d^2\psi_k}{dx^2} EI \frac{d^2\psi_j}{dx^2} \, dx - \omega_j^2 \int_0^L \rho A \psi_k \psi_j \, dx = 0$$

To describe the boundary terms we consider cases where a transverse spring and a small mass are attached to each end. Figure 7.12 depicts the free body diagrams of each attached mass, with the shear forces depicted according to their positive sign convention. Setting $S = -(\partial/\partial x)[EI(\partial^2 w/\partial x^2)]$ at each end yields the following displacement-shear boundary conditions,

$$-\frac{\partial}{\partial x}\left(EI\frac{\partial^2 w}{\partial x^2}\right) - k_0 w = m_0 \ddot{w} \text{ at } x = 0$$

$$\tag{7.6.16}$$

$$\frac{\partial}{\partial x}\left(EI\frac{\partial^2 w}{\partial x^2}\right) - k_l w = m_l \ddot{w} \text{ at } x = L$$

FIGURE 7.12 Free body diagrams for characterizing shear boundary conditions.

The modal displacement varies harmonically according to $w = \text{Re}[\psi_j \exp(i\omega t)]$, so the boundary conditions satisfied by the mode functions are

$$\frac{\partial}{\partial x}\left(EI\frac{\partial^2 \psi_j}{\partial x^2}\right) + (k_0 - m_0\omega_j^2)\psi_j = 0 \text{ at } x = 0$$

$$\frac{\partial}{\partial x}\left(EI\frac{\partial^2 \psi_j}{\partial x^2}\right) - (k_L - m_L\omega_j^2)\psi_j = 0 \text{ at } x = L \tag{7.6.17}$$

For the sake of brevity, end conditions of a torsional spring or a body having significant rotatory inertia shall be left for an exercise. Hence, the rotation-bending moment boundary condition requires that either there is no rotation, so that $d\psi_j/dx = 0$ at the end, or else there is no bending moment, so that $d^2\psi_j/dx^2 = 0$. In either case, the second set of boundary terms in eq. (7.6.15) vanish. For the first set of limits, we employ the displacement-shear boundary conditions, eqs. (7.6.16), to eliminate derivatives of ψ, which gives

$$\psi_k(0)(k_0 - m_0\omega_j^2)\psi_j(0) + \psi_k(L)(k_L - m_L\omega_j^2)\psi_j(L)$$

$$+ \int_0^L EI\frac{d^2\psi_k}{dx^2}\frac{d^2\psi_j}{dx^2}\,dx - \omega_j^2\int_0^L \rho A\,\psi_k\psi_j\,dx = 0 \tag{7.6.18}$$

This equation obviously holds if we switch j and k, but doing so merely changes ω_j to ω_k. We take the difference between the preceding relation and its mate for the alternative j, k sequence, which gives

$$(\omega_k^2 - \omega_j^2)\left[\int_0^L \rho A\,\psi_k\psi_j\,dx + m_0\psi_j(0)\psi_k(0) + m_L\psi_j(L)\psi_k(L)\right] = 0 \tag{7.6.19}$$

If the mode numbers are different, $\omega_k \neq \omega_j$, then the second factor must be zero. Furthermore, because $\omega_k^2 \equiv \omega_j^2$ when $k = j$, that case may be used to define the factor C_1 or C_2 required to normalize the mode. Doing so leads to the mass orthogonality condition for flexure,

$$\boxed{\int_0^L \rho A\,\Psi_k\Psi_j\,dx + m_0\Psi_k(0)\Psi_j(0) + m_L\Psi_k(L)\Psi_j(L) = \delta_{jk}} \tag{7.6.20}$$

(Additional terms arise when the mass moment of inertia of attached bodies is substantial.) This relation is identical to the second orthogonality condition for flexure obtained by the Ritz method, eq. (6.3.20), in the special case where mass attachments are exclusively at the ends of the bar.

When we use the mass orthogonality relation and the boundary conditions in eqs. (7.6.16) to simplify eq. (7.6.18), we obtain the stiffness orthogonality relation,

$$\boxed{\int_0^L EI\frac{d^2\Psi_k}{dx}\frac{d^2\Psi_j}{dx}\,dx + k_0\Psi_k(0)\Psi_j(0) + k_L\Psi_k(L)\Psi_j(L) = \omega_j^2\delta_{jk}} \tag{7.6.21}$$

This relation is identical to the second orthogonality condition for flexure obtained by the Ritz method, eq. (6.3.21), for spring attachments that are exclusively at the ends of the bar.

An exception to the derivation arises in the case of a free-free beam, which has two rigid body modes. Because $\omega_{1,\text{tr}} = \omega_{1,\text{rot}} = 0$, the argument used to derive eq. (7.6.20) does not apply when j and k refer to the two rigid body modes. The consequence is that the modes are not mutually orthogonal if their undefined coefficients are selected arbitrarily, although they will always be orthogonal to the deformational modes. The rigid body modes can be made to be mutually orthogonal by choosing their coefficients appropriately. The general form of the translational and rotational rigid body modes are

$$\psi_{1,\text{tr}} = C_0, \qquad \psi_{1,\text{rot}} = C_1 + C_2\frac{x}{L} \tag{7.6.22}$$

Making these modes orthogonal according to eq. (7.6.20) requires that

$$\int_0^L \rho A\left(C_1 + C_2\frac{x}{L}\right)dx + m_0 C_1 + m_L(C_1 + C_2) = 0 \tag{7.6.23}$$

from which we obtain

$$\frac{C_1}{C_2} = \frac{\displaystyle\int_0^L \rho A\frac{x}{L}\,dx + m_L}{\left[\displaystyle\int_0^L \rho A\,dx + m_0 + m_L\right]}$$

The denominator in this expression is the total system mass, and the numerator is the first moment of mass with respect to the origin, scaled by $1/L$. Hence, the ratio locates the center of mass of the system,

$$\frac{C_1}{C_2} = -\frac{x_G}{L} \tag{7.6.24}$$

In this relation x_G is the axial position of the center of mass of the system. The corresponding rotational mode is

$$\psi_{1,\text{rot}} = C_1\left(1 - \frac{x}{x_G}\right) = C_1\frac{\hat{x}}{x_G} \tag{7.6.25}$$

where $\hat{x} = x_G - x$ is distance from the center of mass. In other words, the rotational mode that is orthogonal to rigid body translation corresponds to a rotation about the center of mass. This is consistent with the laws of rigid body motion, which state that the motion of the center of mass of a rigid body and rotation about the center of mass are not inertially uncoupled. Interestingly, if we use the above forms to normalize the rigid body modes, we find that

$$C_0 = \frac{1}{m^{1/2}}, \qquad C_1 = \frac{x_G}{I_G^{1/2}} \tag{7.6.26}$$

where m is the total mass of the system and I_G is the centroidal moment of inertia of the system.

$$m = \int_{-x_G}^{L-x_G} \rho A \, dx + m_0 + m_L$$

$$I_G = \int_{-x_G}^{L-x_G} \rho A \hat{x}^2 \, d\hat{x} + m_0 x_G^2 + m_L (L - x_G)^2$$

(7.6.27)

This confirms that the orthogonal rigid body modes represent translation and rotation referred to the center of mass.

7.7 MODAL ANALYSIS OF RESPONSE

In the previous chapter, where we derived the modal response equations associated with a Ritz series, decoupling of the modal coordinates was demonstrated by examining the kinetic and potential energy expressions associated with the modal series. Here we will see a completely different derivation that leads to the same result.

7.7.1 Modal Equations

We begin by using the normal modes as the basis functions for a series expansion of axial displacement,

$$u(x, t) = \sum_{j} \eta_j(t) \Psi_j(x)$$

(7.7.1)

Note that summations should henceforth be understood to extend over all modes, that is, to infinity, unless noted otherwise.

The differential equations for the modal coordinates η_j may be obtained by manipulating eq. (7.1.11). First, we multiply that equation by one of the normal modes ψ_k, and integrate the product over $0 \leq x \leq L$. We then apply integration by parts to the portion of the integral containing derivatives with respect to x, which leads to

$$\Psi_k \left(EA \frac{\partial u}{\partial x} \right) \Big|_0^L - \int_0^L \frac{d\Psi_k}{dx} \left(EA \frac{\partial u}{\partial x} \right) dx - \int_0^L \rho A \Psi_k \ddot{u} \, dx = -\int_0^L f_x \Psi_k \, dx \qquad (7.7.2)$$

The general boundary conditions in eq. (7.6.7) and (7.6.8) enable us to eliminate the derivatives at the boundaries, with the result that

$$-\Psi_k (m_L \ddot{u} + k_L u) \big|_{x=L} - \Psi_k (m_0 \ddot{u} + k_0 u) \big|_{x=0} - \int_0^L \frac{d\Psi_k}{dx} \left(EA \frac{\partial u}{\partial x} \right) dx$$

$$- \int_0^L \rho A \Psi_k \ddot{u} \, dx = -\int_0^L f_x \Psi_k \, dx$$

(7.7.3)

The next step is to substitute the modal series into this relation and to bring the summation outside the integral. Because η_j is independent of x, this gives

$$\sum_j \left[(m_L \ddot{\eta}_j + k_L \eta_j) \Psi_k(L) \Psi_j(L) + (m_0 \ddot{\eta}_j + k_0 \eta_j) \Psi_k(0) \Psi_j(0) \right.$$

$$\left. + \eta_j \int_0^L EA \frac{d\Psi_k}{dx} \frac{d\Psi_j}{dx} + \ddot{\eta} \int_0^L \rho A \Psi_k \Psi_j dx \right] = \int_0^L f_x \Psi_k \, dx \tag{7.7.4}$$

The last step is to collect the coefficients of $\ddot{\eta}_j$ and η_j. These are, respectively, the mass and stiffness orthogonality conditions. Hence, we find that

$$\ddot{\eta}_j + \omega_j^2 \eta_j = Q_j, \quad Q_j = \int_0^L f_x \Psi_j \, dx \tag{7.7.5}$$

In many situations, we have an estimate of modal damping, which we incorporate by letting ζ_j denote the ratio of critical damping for the mode, so that

$$\boxed{\ddot{\eta} + 2\zeta_j \omega_j \dot{\eta}_j + \omega_j^2 \eta_j = Q_j} \tag{7.7.6}$$

We should not be surprised by the sameness of these equations and the modal equations we derived from the Ritz series method, as well as those for discrete systems. The preceding section showed that the orthogonality conditions, eqs. (7.6.9) and (7.6.13) in the case of axial motion, are identical to the orthogonality relations obtained from the Ritz method. Furthermore, the Ritz analysis disclosed that the orthogonality relations reduce the kinetic and potential energy expressions to sums of squares,

$$T = \frac{1}{2} \sum_j \dot{\eta}_j^2, \quad V = \frac{1}{2} \sum_j \omega_j^2 \eta_j^2 \tag{7.7.7}$$

The present development merely shows that the modal series decouples the equations of motion, even though there are an infinite number of modes. We would reach the same conclusion if we were to retrace the derivation's steps in the case of flexure. Thus, we may consider eq. (7.7.6) to be a generic representation for bars, with u representing axial or flexural displacement, or torsional rotation, and Ψ_j representing the associated normal mode functions.

Now that we have proven it once using the field equation properties, we shall henceforth consider decoupling of the modal coordinates to be axiomatic. In that view, the modal differential equations (7.2.4) follow immediately from a modal series representation of displacement. The generalized force Q_j for each type of motion may be obtained directly from the power input. The velocity field is

$$\boxed{\dot{u} = \sum_j \dot{\eta}_j \Psi_j(x)} \tag{7.7.8}$$

If we let f_x denote an axial force that is distributed over along the length of the bar and F_x be one of a set of concentrated axial forces applied at locations x_F, the power input to the system by external forces will be

$$\mathcal{P}_{\text{in}} = \int_0^L f_x \dot{u} \, dx + \sum F_x \dot{u}(x_F, t) = \sum_j Q_j \dot{\eta}_j \tag{7.7.9}$$

Substitution of eq. (7.7.8) into the preceding leads to

$$\text{Axial displacement: } Q_j = \int_0^L f_x \Psi_j \, dx + \sum F_x \Psi_j(x_F) \tag{7.7.10}$$

This is the same as the form that results from eq. (7.7.5) when Dirac delta functions are used to represent the spatial distribution of concentrated forces. The generalized forces for torsion have a similar appearance, except that the distributed torsional load is denoted as τ, and Γ denote couples applied at locations x_γ,

$$\text{Torsional rotation: } Q_j = \int_0^L \tau \Psi_j \, dx + \sum \Gamma \Psi_j(x_\gamma) \tag{7.7.11}$$

For transverse motion, the loading may consist of a transverse distributed force f_z, as well as a set of concentrated transverse forces F at locations x_F. In addition, flexural couples may be applied at locations x_γ. Because the angular velocity is $\partial \dot{w}/\partial x$, the contribution of one such couple to the power input is $\Gamma[\partial \dot{w}/\partial x]$, so the generalized forces for flexural motion are

$$\text{Flexural motion: } Q_j = \int_0^L f_z \Psi_j \, dx + \sum F_z \Psi_j(x_F) + \sum \Gamma \frac{\partial \Psi_j}{\partial x}'(x_\gamma) \tag{7.7.12}$$

Although the present derivation led to differential equations governing the modal coordinates that are the same as those for the Ritz series formulation, there is a subtle distinction between the two approaches. Solution of the field equations leads to a representation of the modes as simple combinations of sinusoidal and hyperbolic functions, rather than as a series. This allows us to characterize the generalized forces more definitively. In some cases, such knowledge will enable us to derive analytical expressions for the response from which we may gain qualitative understanding of the underlying physical phenomena.

7.7.2 Steady-State and Transient Response

We consider harmonic excitation at frequency ω, in which case the generalized forces may be represented as $Q_j = \text{Re}[F_j \exp(i\omega t)]$. Substituting the steady-state form $\eta_j = \text{Re}[X_j(\omega)\exp(i\omega t)]$ into the modal differential equation, eq. (7.7.6), leads to an expression for the amplitude factor X_j. We use the solution for X_j to form η_j, and then synthesize the displacement field, u (or w, or θ). We thereby find that

$$u(x, t) = \text{Re} \sum_j \left[\frac{F_j \Psi_j(x)}{\omega_j^2 + 2i\zeta_j\omega_j\omega - \omega^2} \exp(i\omega t) \right] \tag{7.7.13}$$

Alternatively, if dissipation fits the structural damping model, we may represent the denominator of the preceding expression by $\omega_j^2(1 + i\gamma) - \omega^2$, where γ is the loss factor.

An important issue, first addressed in the previous chapter, is convergence of the modal series. In many applications it is useful to select the series length N in advance so that we may examine the effect of several excitations in a sequence of automated calculations. Let us consider the magnitude of each term contributing to eq. (7.7.13). We have seen for extension and torsion that the modes are sinusoidal and the high-frequency flexural modes are essentially sinusoidal except for evanescent boundary layers near the ends. A corollary of this is that all modes tend to have the same order of magnitude, approximately inversely proportional to the square root of the system mass, $\Psi_j(x) = 0(m^{-1/2})$. If F_j represents concentrated forces, it is likely that their values will not decrease rapidly with increasing j, so we cannot rely on decreasing values of F_j to make series terms negligible.

Hence, if the contribution associated with a particular mode j in eq. (7.7.13) is to be negligible, the denominator must be large compared to the denominator for the other modes. The guideline we identified in Section 6.3.6 relies on the fact that in any vibratory system, the frequency of interest has an upper limit $\omega_{\max}$. In order for the denominator to be large, the excitation frequency ω may not be close to a natural frequency ω_j. However, ω might have any value up to $\omega_{\max}$. It follows that we only can be sure that a modal contribution will be small if its natural frequency is substantially larger than $\omega_{\max}$. The criterion we identified earlier, according to which we select the number of modes N such that $\omega_N > \beta\omega_{\max}$, with $2 \le \beta \le 4$, should be adequate to assure that the omitted modes are not important.

When we wish to analyze transient response we must solve eq. (7.7.6). Without knowing the specific excitation, it is difficult to construct a general guideline. One idea is to perform an FFT decomposition of the generalized forces Q_j, and to set $\omega_{\max}$ based on an inspection of the spectrum. We would then impose the same truncation guideline as that associated with a purely harmonic excitation.

The notion of examining an FFT of the generalized forces needs to be supplemented by an examination of the initial spatial distributions of displacement and velocity. To solve the equations of motion for the η_j we must find the corresponding initial values of η_j and $\dot{\eta}_j$. Of course, if the initial displacement or velocity is zero, then η_j or $\dot{\eta}_j$ will respectively also be zero. To determine nonzero initial values of the principal coordinates we employ the expansion theorem, which was derived in the context of the Ritz series method as eqs. (6.3.34) and (6.3.39). For the sake of completeness we shall summarize the procedure here. We wish to represent a known function $f(x)$ in series form using the normal modes, such that

$$f(x) = \sum_j f_j \Psi_j(x) \tag{7.7.14}$$

We form the right side of the kinetic energy orthogonality condition, replacing one of the normal mode functions in the product with $f(x)$. We form an identity by equating this quantity to itself, except that on the left side of the equality we replace $f(x)$ by its modal series. This yields

$$\int_0^L \rho A\left(\sum_j f_j \Psi_j(x)\right)\Psi_k\,dx + m_0 \Psi_k(0)\left(\sum_j f_j \Psi_j(0)\right) + m_L \Psi_k(L)\left(\sum_j f_j \Psi_j(L)\right)$$

$$= \int_0^L \rho A f(x)\Psi_k\,dx + m_0\Psi_k(0)f(0) + m_L\Psi_k(L)f(L)$$

$$\tag{7.7.15}$$

The orthogonality property of the mode functions then simplifies the identity to

$$f_k = \int_0^L \rho A f(x) \Psi_k(x) \, dx + m_1 f(0) \Psi_k(0) + m_2 f(L) \Psi_k(L) \qquad (7.7.16)$$

We may apply the expansion theorem to evaluate initial conditions for the modal coordinates. If the initial displacement is $w(x, 0)$ and the initial velocity is $\dot{w}(x, 0)$, we have

$$
\begin{aligned}
\eta_j(0) &= \int_0^L \rho A w(x, 0) \Psi_j(x) \, dx + m_1 w(0, 0) \Psi_k(x) + m_2 w(L, 0) \Psi_k(x) \\
\dot{\eta}_j(0) &= \int_0^L \rho A \dot{w}(x, 0) \Psi_j(x) \, dx + m_1 \dot{w}(0, 0) \Psi_k(x) + m_2 \dot{w}(L, 0) \Psi_k(x)
\end{aligned}
\qquad (7.7.17)
$$

Recall that all high-frequency modes tend to depend in a sinusoidal manner on x, with approximately the same order of magnitude. This suggests that the values $\eta_j(0)$ or $\dot{\eta}_j(0)$ might be appreciable for large j if the initial displacement or velocity is a highly oscillatory function of x. In such a case, the modal series must be sufficiently long to include all modes that are excited by the initial conditions.

The issue of the convergence of the modal series affects the accuracy of the solution beyond merely influencing computation of series sums. Suppose that the number of modes required to obtain adequate convergence is very large, which means that many high frequencies modes are significant. However, each simplifying assumption contained in the classical theory for bars breaks down at high frequencies. For example, we already noted that transverse shear deformation and rotatory inertia become important for flexure. For extensional motion the Poisson ratio effect leads to a transverse displacement that may vary through the thickness. In other words, it is very possible that one might obtain a solution that is a proper mathematical solution, but the physical theory that the solution describes is based on approximations that cease to be valid. For this reason it is important to have some idea of the limits of classical bar theory, which is a topic addressed in Section 7.9.

EXAMPLE 7.8

A uniform bar, fixed at its left end and with an attached mass $m = 0.25\rho AL$ at its right end, is subjected to a static compressive force at its right end. At $t = 0$, the force is suddenly removed. Derive an expression for the ensuing axial displacement field. Compare the initial displacement field obtained from this series to the actual initial displacement. Then plot the displacement of the right end as a function of time over an interval lasting six periods of the fundamental mode.

Solution   This example describes the application of the expansion theorem, with which some individuals have difficulty in the case of mixed boundary conditions. The starting point is statement of the full problem. The boundary condition at $x = L$ requires that the axial force accelerate the attached mass, $m\ddot{u} = -EA \, \partial u/\partial x$. Thus, we seek a free vibration response satisfying

$$EA \frac{\partial^2 u}{\partial x^2} - \rho A \ddot{u} = 0$$

subject to the boundary conditions that

$$u = 0 \text{ at } x = 0, \quad EA\frac{\partial u}{\partial x} = -m\ddot{u} \text{ at } x = L$$

The initial axial displacement is due to a static axial force F. The corresponding strain is $\varepsilon_{xx} = du/dx = F/AE$. We integrate this to determine that the initial static displacement is Fx/AE. The bar is initially at rest, so the initial conditions are

$$u = \frac{F}{EA}x, \quad \dot{u} = 0 \text{ at } t = 0$$

The eigenfunction consistent with the fixed boundary condition at $x = 0$ is

$$\psi_j = C_1 \sin\left(\alpha\frac{x}{L}\right)$$

where the eigenvalues α_j are the roots of the characteristic equation

$$\cot(\alpha) = \varepsilon\alpha, \quad \varepsilon = \frac{m}{\rho AL} = 0.25$$

The first 10 eigenvalues are

$$\alpha_j = 1.265, 3.935, 6.814, 9.812, 12.868, 15.954, 19.056, 22.170, 25.290, 28.414$$

Higher roots are well approximated as $\alpha_j = (j-1)\pi$. The natural frequency corresponding to an α_j value is

$$\omega_j = \left(\frac{E}{\rho L^2}\right)^{1/2}\alpha_j$$

Because the mass m is attached to the bar at $x = L$, the first orthogonality condition for normal modes is

$$\int_0^L \rho A \Psi_j \Psi_n \, dx + m\Psi_j(L)\Psi_n(L) = \delta_{jn}$$

Applying this for $j = n$, along with setting $m = 0.25\rho AL$, yields the expression for the normalization coefficient C_{1j} for each mode,

$$C_{1j}^2\left[\int_0^L \rho A \sin\left(\alpha_j\frac{x}{L}\right)^2 dx + m\sin(\alpha_j)^2\right] = 1$$

It follows that

$$C_{1j} = \left[\frac{4\alpha_j}{\rho AL(2\alpha_j - \sin(2\alpha_j) + 4\varepsilon\alpha_j\sin(\alpha_j)^2)}\right]^{1/2}$$

We use the normal modes to form the modal series,

$$u(x, t) = \sum_j \eta_j(t)\Psi_j(x)$$

The system executes a free vibration after removal of the static force, so the modal differential equations are

$$\ddot{\eta}_j + \omega_j^2\eta_j = 0$$

The initial velocity of the bar is zero, so one initial condition is $\dot{\eta}_j = 0$ at $t = 0$. Satisfying this condition leads to

$$\eta_j = B_j \cos(\omega_j t)$$

The coefficient B_j is determined by formulating the expansion theorem. In the first orthogonality condition for the system we replace each occurrence of $\psi_j(x)$ with $u(x, 0)$, which we then represent with the modal series for $u(x, 0)$. This yields

$$\int_0^L \rho A u(x, 0)\Psi_n \, dx + mu(L, 0)\Psi_n(L) = \int_0^L \rho A \left[\sum_j B_j \Psi_j(x)\right]\Psi_n \, dx + m\left[\sum_j B_j \Psi_j(L)\right]\Psi_n(L)$$

We bring the integral on the right side inside the summation. The orthogonality condition serves to filter out of the sum only the term for which $j = n$, so the right side reduces to B_n. The actual initial displacement was found earlier. Substitution of that expression into the left side of the preceding gives

$$B_j = \int_0^L \rho A \left(\frac{F}{EA}x\right)\Psi_n \, dx + m\left(\frac{FL}{EA}\right)\Psi_n(L)$$

$$= \frac{\rho F L^2}{E} C_{1n}\left[\left(\frac{1}{\alpha_n^2} + \varepsilon\right)\sin(\alpha_n) - \frac{\cos(\alpha_n)}{\alpha_n}\right]$$

This expression may be simplified by factoring out $\sin(\alpha_n)$ and using the fact that α_n is a root of the characteristic equation, from which we find that

$$B_n = \frac{\rho F L^2}{E} C_{1n} \frac{\sin(\alpha_n)}{\alpha_n^2}$$

The first six values are

$$B_j = 0.76112, -0.06132, 0.01495, -0.0054, 0.00251, -0.00134$$

We now have all quantities required to synthesize the displacement series, so we form

$$u = \sum_j \Psi_j \eta_j = \sum_j C_{1j} B_j \sin\left(\alpha_j \frac{x}{L}\right)\cos(\omega_j t)$$

In order to verify that we have a dimensionally consistent result, let us substitute the expressions for ω_j, C_{1j}, and B_j. The result is

$$u = \frac{FL}{AE}\sum_j \left[\frac{4\alpha_j}{2\alpha_j - \sin(2\alpha_j) + 4\varepsilon\alpha_j\sin(\alpha_j)^2}\right]\frac{\sin(\alpha_n)}{\alpha_n^2}\sin\left(\alpha_j \frac{x}{L}\right)\cos(\alpha_j \tau)$$

where $\tau = c_{bar}t/L$.

The modal series may be truncated when the B_j coefficients are small because all of the C_{1j} have unit order of magnitude. The list of B_j values suggests that truncation at $N = 5$ would be adequate. An alternative way of deciding the series length is to compare the initial displacement field obtained from the series to the actual initial displacement. The first set of graphs displays $u(x, 0)$ scaled by FL/AE. We see that even $N = 2$ well approximates the straight-line dependence of the actual initial displacement. (This feature is a consequence of the fact that the initial displacement strongly resembles the first mode function. In Exercise 7.32, which replaces the small end mass with a stiff spring, such agreement does not arise. Many more mode functions are required there.)

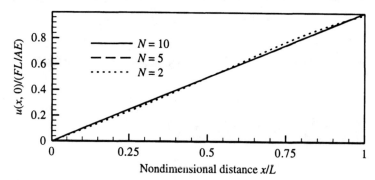

The displacement responses at $x = L$ obtained from series with $N = 2$ and $N = 5$ are displayed in the next graph. We constructed this graph by taking the maximum value of τ to be $6(2\pi/\alpha_1)$, which gives six periods at the fundamental frequency. To select the increment of τ we divide the shortest period of the included modes into 10 intervals, so that $\Delta\tau = 0.1(2\pi/\alpha_5)$. We see little difference between the results from the two series lengths. We also see that response is dominated by the first mode, which we could have anticipated by the fact that B_1 is much larger than the other B_j values.

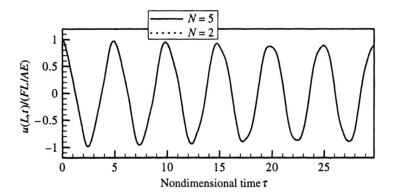

EXAMPLE 7.9

A clamped-clamped beam is initially at rest at its static equilibrium position. At $t = 0$ a concentrated transverse force of constant magnitude F enters at $x = 0$ and continues to move at constant speed v across the beam. Determine the resulting displacement, assuming that damping is negligible. Are there any critical speeds v that should be avoided?

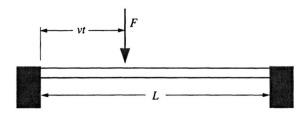

Solution The model of a translating force, which we shall use to illustrate the analysis of transient response, is fundamental to predicting the dynamic response of bridges as vehicles move across them. Most texts only consider the case of a simply supported beam. By considering clamped boundary conditions, we will gain additional perspectives of the formulation and solution of modal equations. For the modal properties of a clamped-clamped beam we turn to Appendix C, from which we find that the characteristic equation is

$$\cos(\alpha)\cosh(\alpha) = 1$$

The first two roots are $\alpha_j = 4.73004, 7.8532$, after which $\alpha_j = (j/2 + 1)\pi$ is sufficiently accurate for any computation. The nonnormalized mode functions are

$$\psi_j = C_{1j}\left\{\left[\sin\left(\alpha_j\frac{x}{L}\right) - \sinh\left(\alpha_j\frac{x}{L}\right)\right] - R_j\left[\cos\left(\alpha_j\frac{x}{L}\right) - \cosh\left(\alpha_j\frac{x}{L}\right)\right]\right\}$$

$$R_j = \frac{\sin(\alpha_j) - \sinh(\alpha_j)}{\cos(\alpha_j) - \cosh(\alpha_j)}$$

The natural frequency corresponding to each eigenvalue is

$$\omega_j = \left(\frac{EI}{\rho A L^4}\right)^{1/2} \alpha_j^2$$

There are no attachments to the beam, so the first orthogonality condition is

$$\int_0^L \rho A \Psi_j \Psi_n \, dx = 1$$

We apply this condition for $n = j$ to find the normalization coefficients for which $\psi_j = C_{1j}\hat{\psi}_j$,

$$C_{1j} = \left(\int_1^L \rho A \hat{\psi}_j^2 \, dx\right)^{-1/2}$$

The first three values are $(\rho A L)^{1/2} C_{1j} = 0.98250, 1.00078, 0.99997$, after which $(\rho A L)^{1/2} C_{1j} = 1$ fits the high-frequency asymptotic approximation of a mode derived from Section 7.4.

The modal series for forced response is

$$w = \sum_j \Psi_j(x)\eta_j(t)$$

The location of the vertical force is $x_F = vt$, so the generalized force obtained from eq. (7.7.12) is $Q_j = F\Psi_j(vt)$. The corresponding modal differential equations of motion are

$$\ddot{\eta}_j + \omega_j^2 \eta_j = F\Psi_j(vt) = FC_{1j}\left\{\left[\sin\left(\frac{\alpha_j v}{L}t\right) - \sinh\left(\frac{\alpha_j v}{L}t\right)\right]\right.$$

$$\left. - R_j\left[\cos\left(\frac{\alpha_j v}{L}t\right)\right] - \cosh\left(\frac{\alpha_j v}{L}t\right)\right)$$

It is given that the initial displacement and velocity of the bar are zero, so the initial conditions for the modal coordinates are $\eta_j = \dot{\eta}_j = 0$ at $t = 0$. We use the method of undetermined coefficients to obtain the particular solution for the modal equations, which is

$$(\eta_j)_p = \frac{FC_{1j}}{\omega_j^2 - \Omega_j^2}[\sin(\Omega_j t) - R_j\cos(\Omega_j t)]$$

$$- \frac{FC_{1j}}{\omega_j^2 + \Omega_j^2}[\sinh(\Omega_j t) - R_j\cosh(\Omega_j t)], \quad \Omega_j = \frac{\alpha_j v}{L}$$

To this we add the complementary solution for each j, and then satisfy the initial conditions. The result is

$$\eta_j = (\eta_j)_p + \frac{2FC_{1j}\Omega_j^2}{\omega_j^4 - \Omega_j^4}\left[R_j\cos(\omega_j t) - \frac{\Omega_j}{\omega_j}\sin(\omega_j t)\right]$$

Now that we have determined the response of the modal coordinates, we may proceed to form the displacement.

As always, an important consideration is the number of modes to use to form the displacement. We observe that the largest magnitudes for η_j will occur whenever $\omega_j \approx \Omega_j$, because this condition will minimize the denominator of several terms forming η_j. In view of the definition of the parameters, this condition occurs when

$$\left(\frac{EI}{\rho A L^4}\right)^{1/2} \alpha_j \approx \frac{v}{L}$$

For a given v, we may evaluate this expression for successive j until we find the index j' that fits it the closest. It should be adequate to truncate the series at three or four terms beyond j', for the denominators of the higher η_j then will be much larger than the denominator for $\eta_{j'}$.

The fact that the equality condition $\omega_j = \Omega_j$ leads to a vanishing denominator causes some to say that it characterizes a critical speed, specifically

$$(v_j)_{cr} = \left(\frac{EI}{\rho A L^2}\right)^{1/2} \alpha_j = c_{bar}\frac{d}{L}\alpha_j$$

where $d = (I/A)^{1/2}$. This critical condition corresponds to a resonant excitation. In the absence of damping, the appropriate particular solution would multiply the sine and cosine terms in $(\eta_j)_p$ by a t factor. Proper interpretation of the significance of this critical speed requires that we recognize that the moving force is a transient excitation that acts on the bar only over the interval $0 \leq t \leq L/v$. After $t = L/v$, the bar executes a free vibration. It follows that a single force cannot cause a true resonance, regardless of the value of v. However, if a continuous sequence of forces is present, as in the case of many vehicles, then a resonant state can be attained. Also, it is important to realize that the critical speed is quite high. For example, in the case of a steel beam with $d/L = 0.005$, which is extremely slender, $(v_1)_{cr} = 122$ m/s.

Let us evaluate the transverse displacement for a few representative cases. In order to nondimensionalize the expressions, we set $v = \mu(v_1)_{cr}$, so that $\mu = 1$ corresponds to the first critical speed. We also nondimensionalize time as $\tau = (EI/\rho AL^4)^{1/2}t$. Because $vt/L = \mu\alpha_1\tau$, $\Omega_j = (EI/\rho AL^4)^{1/2}\mu\alpha_j\alpha_1$, and $\omega_j t = \alpha_j^2\tau$, these substitutions lead to

$$\eta_j = F\frac{\rho AL^4 C_{1j}}{EI}\frac{1}{\alpha_j^2}\left\{\frac{1}{\alpha_j^2 - \mu^2\alpha_1^2}[\sin(\mu\alpha_j\alpha_1\tau) - R_j\cos(\mu\alpha_j\alpha_1\tau)]\right.$$

$$- \frac{1}{\alpha_j^2 + \mu^2\alpha_1^2}[\sinh(\mu\alpha_j\alpha_1\tau) - R_j\cosh(\mu\alpha_j\alpha_1\tau)]$$

$$\left. + \frac{2\mu^2\alpha_1^2}{\alpha_j^4 - \mu^4\alpha_1^4}\left[R_j\cos(\alpha_j^2\tau) - \frac{\mu\alpha_1}{\alpha_j}\sin(\alpha_j^2\tau)\right]\right\}$$

Note that the instant when the force passes $x = L$ corresponds to $vt/L = 1$, or $\tau = 1/\mu\alpha_1$. Hence, this expression applies only for $0 \leq \tau < 1/\mu\alpha_1$. Because C_{1j} is inversely proportional to $(\rho AL)^{1/2}$, and Ψ_j is proportional to C_{1j}, it follows that the displacement obtained from $\sum\Psi_j\eta_j$ will have the common dimensional factor FL^3/EI.

The first graph displays displacement fields when the speed is one-tenth of the lowest critical value.

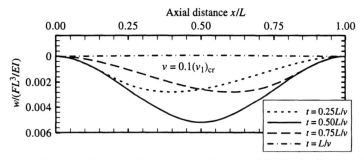

Results are shown for instants when the force is at $x = L/4$, $L/2$, $3L/4$, and L. The speed is sufficiently low in this case that the dynamic aspects of the response are not significant. This is evidenced by the fact that the displacement is essentially zero when $t = L/v$ (the force is at the end of the bar). Indeed, the displacement pattern at each instant is essentially what would be obtained from a static analysis for a concentrated force at the respective instantaneous position. The next case we consider is a speed that is half the lowest critical value.

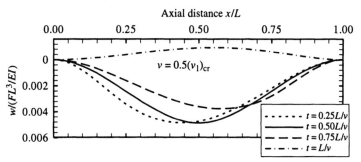

The amplitude of the displacement is essentially the same as it was in the low speed case, but there is evidence of dynamic effects. The displacement when $t = 0.75L/v$ is not the mirror image of the field when $t = 0.25L/v$, and the displacement at $t = L/v$ has rebounded upward due to the inertia of the beam as it returns to the horizontal position. The last case corresponds to a speed that is within 1% of critical.

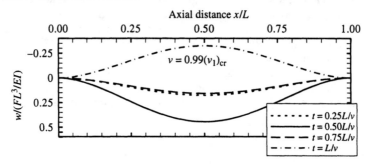

The maximum displacement is approximately one hundred times that encountered at the lower speeds, which confirms that $(v_1)_{cr}$ is indeed a speed to avoid. Interestingly, the displacement when the force leaves the beam is almost as large as, but in the opposite sense from, the displacement when the force is midspan.

EXAMPLE 7.10

Consider a cantilevered beam of length L, with $x = 0$ selected as the clamped end. At $x = 0.8L$ there is a concentrated harmonic force $F \cos(\omega t)$. The excitation frequency is 5% greater than the third natural frequency, $\omega = 1.05\omega_3$. When damping is neglected, the steady-state response will have the form $w = f_w(x)\cos(\omega t)$. The corresponding bending moment and shear force will be $M = f_M(x)\cos(\omega t)$ and $S = f_S(x)\cos(\omega t)$. Evaluate the convergence properties of the modal series for $f_w(x)$, $f_M(x)$, and $f_S(x)$ by considering truncation of the series at differing series lengths. Graph the results for each function at each series length.

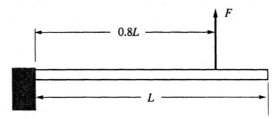

Solution This example will start out as a straightforward application of modal analysis, but the assessment of convergence rates will disclose some interesting computational issues. We worked out the modal properties of a cantilevered beam in Example 7.6. The characteristic equation is

$$\cos(\alpha L) = -\frac{1}{\cosh(\alpha L)}$$

The eigenfunction corresponding to root α_j is

$$\psi_j(x) = C_{1j}\left[\sin\left(\alpha_j\frac{x}{L}\right) - \sinh\left(\alpha_j\frac{x}{L}\right)\right] - R_j\left[\cos\left(\alpha_j\frac{x}{L}\right) - \cosh\left(\alpha_j\frac{x}{L}\right)\right]$$

$$R_j = \frac{\sin(\alpha_j) + \sinh(\alpha_j)}{\cos(\alpha_j) + \cosh(\alpha_j)}$$

There are no attached masses, so the first orthogonality condition for the normal modes is

$$\int_0^L \rho A \Psi_j \Psi_n \, dx = \delta_{jn}$$

We scale the normal mode as $\Psi_j = C_{1j}\psi_j$. Using this to satisfy the preceding when $n = j$ defines the normalization factor,

$$C_{1j} = \left(\int_0^L \rho A \psi_j^2 \, dx\right)^{-1/2}$$

Numerical evaluation of this expression yields $(\rho A L)^{1/2}C_{1j} = 0.9825$, 1.00078, and 0.99997, with $C_{1j} = 1$ for $j \geq 4$.

The modal series for displacement is

$$w = \sum_j \Psi_j(x)\eta_j$$

The upper limit for the sum is not indicated because the appropriate value is an issue we will explore. The modal values are required to evaluate the generalized forces, which are found from eq. (7.7.12) to be $Q_j = \mathrm{Re}[F\Psi_j(0.8L)\exp(i\omega t)]$. The modal equations of motion are

$$\ddot{\eta}_j + \omega_j^2 \eta_j = Q_j$$

We are interested solely in the steady-state response, so we take η_j to be harmonic at frequency ω,

$$\eta_j = \mathrm{Re}[X_j \exp(i\omega t)]$$

where the complex amplitudes are

$$X_j = \frac{F\Psi_j(0.8L)}{\omega_j^2 - \omega^2}$$

It is stipulated that $\omega = 1.05\omega_3$. The relation between the natural frequencies and the eigenvalues α_j gives

$$\omega = 1.05\,\alpha_3^2 \left(\frac{EI}{\rho A L^4}\right)^{1/2}$$

which leads to

$$X_j = \frac{\rho A L^4 F}{EI} \frac{\Psi_j(0.8L)}{\alpha_j^4 - (1.05)^2\alpha_3^4}$$

When we use these expressions to form the modal displacement series, we obtain

$$w = \frac{\rho A L^4 F}{EI} \sum_j \mathrm{Re}\left[\frac{\Psi_j(0.8L)\Psi_j(x)}{\alpha_j^4 - (1.05)^2\alpha_3^4}\exp(i\omega t)\right]$$

All terms except the complex exponential are real. Furthermore, C_{1j} is the proportionality factor for Ψ_j, and C_{1j} is scaled by $(\rho A L)^{-1/2}$. Hence, the displacement obtained from this series may be written as

$$w = f_w(x)\cos(\omega t), \quad f_w(x) = \frac{FL^3}{EI}\sum_j \frac{\rho A L \Psi_j(0.8L)\Psi_j(x)}{[\alpha_j^4 - (1.05)^2\alpha_3^4]}$$

The corresponding bending moment and shear force are

$$M = EI\frac{\partial^2 w}{\partial x^2} = f_M(x)\cos(\omega t), \qquad S = -EI\frac{\partial^3 w}{\partial x^3} = f_S(x)\cos(\omega t)$$

where the functions defining the x dependence of M and S are

$$f_M(x) = FL\sum_j \frac{\rho A L \Psi_j(0.8L)\Psi_j''(x)\alpha_j^2}{[\alpha_j^4 - (1.05)^4\alpha_3^4]}, \qquad f_S(x) = -F\sum_j \frac{\rho A L \Psi_j(0.8L)\Psi_j'''(x)\alpha_j^3}{[\alpha_j^4 - (1.05)^4\alpha_3^4]}$$

In these expressions a prime denotes differentiation of a mode function with respect to the combination $\alpha_j x/L$, so that

$$\Psi_j''(x) = -C_{1j}\left[\sin\left(\alpha_j\frac{x}{L}\right) + \sinh\left(\alpha_j\frac{x}{L}\right)\right] + R_j\left[\cos\left(\alpha_j\frac{x}{L}\right) + \cosh\left(\alpha_j\frac{x}{L}\right)\right]$$

$$\Psi_j'''(x) = -C_{1j}\left[\cos\left(\alpha_j\frac{x}{L}\right) + \cosh\left(\alpha_j\frac{x}{L}\right)\right] + R_j\left[-\sin\left(\alpha_j\frac{x}{L}\right) + \sinh\left(\alpha_j\frac{x}{L}\right)\right]$$

We wish to evaluate $f_w(x)$, $f_M(x)$, and $f_S(x)$ for $0 \le x < L$ corresponding to various series lengths. The excitation is close to the third natural frequency, so we know that we need at least the first three modes. The graphs depict the results when the series length is $N = 3$, 5, and 10.

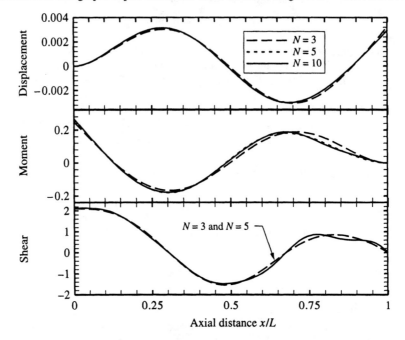

For the displacement and bending moment functions the results for $N = 5$ and $N = 10$ show no perceptible difference, and $N = 3$ gives almost the same result. For the shear there is some difference for $N = 10$ in the vicinity of the location where the force is applied, $x = 0.8L$, which suggests that the series might not have converged in this region. However, if we were to plot the result for $N = 20$, we would find that it coincides with the $N = 10$ curve. Thus, it seems as though $N = 10$ is the convergent result for $f_S(x)$. There is just one difficulty—this result for shear is wrong!

In the analysis of static beams, a common task is to construct a diagram showing the dependence of the internal shear force on the axial distance. In such a diagram, the shear jumps discontinuously across a location where a concentrated force is applied. The same condition applies in the dynamic case. Let S^- and S^+ denote the shear forces immediately to the left and right of $x = 0.8L$. The sketch shows a free body diagram of the beam segment between these two locations.

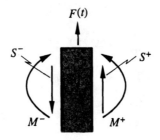

The width of this element is essentially zero, so its inertia is negligible. Consequently, force and moment sums must equilibrate. This leads to the condition that the bending moment and shear forces on either side of $x = 0.8L$ must be related by

$$S^+ = S^- - F, \qquad M^+ = M^-$$

These are called *jump conditions*. Because of the way in which f_S and f_M are defined, these jump conditions require that the value of f_S decrease by unity across $x = 0.8L$, while the bending moment is continuous. (Because $S = -\partial M/\partial x$ the bending moment should have a slope discontinuity at this location.) The question now is, Why does the series solution for f_S fail to display the discontinuity?

The resolution of this issue may be found in Example 7.6. We saw there that a straightforward evaluation of the mode functions gives an erroneous result for the higher frequency modes because of loss of precision associated with hyperbolic functions at large argument. In the present context, the same difficulty arises when we evaluate Ψ_j''' at $x = 0.8L$ in order to extend the length of the series. The computation of these values gives zero for $j > 15$. Example 7.6 showed that the asymptotic approximation gives the correct values for $j > 2$. Hence, we should use the asymptotic representation to evaluate modal values for $j > 10$. We found earlier that $C_{1j} = (\rho AL)^{-1/2}$ for large j, so the previously derived expression is

$$(\Psi_j(x))'''_{\text{asymptotic}} = C_1\left\{\sin\left(\alpha_j\frac{x}{L}\right) - \cos\left(\alpha_j\frac{x}{L}\right) + \exp\left(-\alpha_j\frac{x}{L}\right)\right.$$
$$\left. + (-1)^{j-1}\exp\left[-\alpha_j\left(\frac{L-x}{L}\right)\right]\right\}$$

The third derivative of this function with respect to the argument $\alpha_j x/L$ is

$$(\Psi_j(x))'''_{\text{asymptotic}} = C_1\left\{-\cos\left(\alpha_j\frac{x}{L}\right) - \sin\left(\alpha_j\frac{x}{L}\right) + \exp\left(-\alpha_j\frac{x}{L}\right)\right.$$
$$\left. + (-1)^{j-1}\exp\left[-\alpha_j\left(\frac{L-x}{L}\right)\right]\right\}$$

The resulting function f_S appears on the last set of graphs.

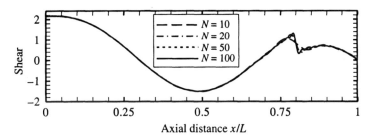

The discontinuity of shear under the force is now evident, and the change across $x = 0.8L$ is $\Delta S = -1$, as required by the jump condition. The results for $N = 50$ and $N = 100$ are essentially the same. As is true when we use a Fourier series to describe a discontinuous function, many series terms are required to get the correct result in the vicinity of a discontinuity.

7.8 TIME-DEPENDENT BOUNDARY CONDITIONS

7.8.1 Mindlin-Goodman Decomposition

We developed a procedure for treating time-dependent geometric boundary conditions in Section 6.5. The approach is based on a superposition of two displacement terms.

The first, u_{bc}, is selected solely on the basis of satisfying boundary conditions, while the second, u_{Ritz}, compensates for the effects in the equations of motion that are not satisfied by u_{bc}. Mindlin and Goodman (1950) applied a similar decomposition to solve field equations. They addressed time-dependent natural, as well as geometric, boundary conditions. The former arise if a known force acts at a boundary. From a practical standpoint, such forces are readily handled as concentrated force or couple excitations, so there is no need to consider that case. Rather than following the Mindlin-Goodman approach directly, we shall adapt it to the procedure in the previous chapter.

We begin with the general procedure for extensional (and torsional) motions. The boundary conditions may be expressed concisely in terms of differential operators $\mathcal{B}_1(u)$ as

$$\mathcal{B}_1(u)\big|_{x=0} = f_1(t), \qquad \mathcal{B}_2(u)\big|_{x=L} = f_2(t) \tag{7.8.1}$$

Any boundary condition that is homogeneous (that is, time independent) may be recovered from the preceding by taking $f_1 = 0$ or $f_2 = 0$, as appropriate.

Linearity of the field equations permits decomposition of the displacement field into two parts,

$$\boxed{u = u_{bc} + u_h} \tag{7.8.2}$$

where the subscripts indicate that the first term will be generated by the boundary conditions, while the second will correspond to homogeneous boundary conditions. When we formulated the Ritz series approach we only required that u_{bc} satisfy the geometric boundary conditions. Now we require that u_{bc} satisfy *all* boundary conditions,

$$\mathcal{B}_1(u_{bc})\big|_{x=0} = f_1(t), \qquad \mathcal{B}_2(u_{bc})\big|_{x=L} = f_2(t) \tag{7.8.3}$$

In our previous work we represented u_{bc} in series form. Mindlin and Goodman provided a recipe for selecting the functions in such a series. They observed that the elasticity terms in the equation of motion represent an ordinary differential equation whose solution has the same number of integration constants as the number of boundary conditions. For constant EA, we have

$$\frac{\partial^2 u_{bc}}{\partial x^2} = 0 \tag{7.8.4}$$

The general solution is a first-order polynomial in x, with coefficients that may have arbitrary dependence on t,

$$\boxed{u_{bc} = a_1(t) + a_2(t)\frac{x}{L}} \tag{7.8.5}$$

We determine the free functions $a_1(t)$ and $a_2(t)$ by using this form to satisfy eq. (7.8.3). Because the boundary operators are linear, this leads to a pair of linear equations,

$$\boxed{\begin{aligned} \left[\mathcal{B}_1(a_1) + \mathcal{B}_1\!\left(a_2\frac{x}{L}\right)\right]\bigg|_{x=0} &= f_1 \\[2ex] \left[\mathcal{B}_2(a_1) + \mathcal{B}_2\!\left(a_2\frac{x}{L}\right)\right]\bigg|_{x=L} &= f_2 \end{aligned}} \tag{7.8.6}$$

Recall that the boundary condition will contain time derivatives when an inertial body is attached to the bar, in which case the above represent linear differential equations for $a_1(t)$ and $a_2(t)$. Otherwise, t arises solely as a parameter, and the equations may be solved algebraically.

Once we have determined u_{bc}, we proceed to determine u_h, for which we use a modal series,

$$u_h(x, t) = \sum_j \eta_j(t) \Psi_j(x) \tag{7.8.7}$$

In essence, u_h is like u_{Ritz} in the previous analysis, with the basis functions now taken to be mode functions. An important aspect of the formulation is that these mode functions are those of the system in the case where no motion is imposed at the supports, so that the boundary conditions are not time dependent.

The differential equations governing the Ritz series generalized coordinates were presented previously in eq. (6.5.21). Because the basis functions are now the normal mode functions, the inertia matrix associated with the modal coordinates is the identity matrix, and the stiffness matrix is the diagonal array of ω_j^2 values. If we assume that modal damping applies, then the modal coordinates are uncoupled. The modal version of eq. (6.5.21) is

$$\ddot{\eta}_j + 2\zeta_j \omega_j \dot{\eta}_j + \omega_j^2 \eta_j = Q_j - \sum_{n=1}^{2} v_{jn} \ddot{\alpha}_n - \sum_{n=1}^{2} d_{jn} a_n \tag{7.8.8}$$

where Q_j represents the modal forces associated with homogeneous boundary conditions.

A comparison of the preceding with eq. (6.5.21) reveals that terms containing the elasticity coefficients γ_{jn} in the previous version do not appear here. This is not an omission. It can be shown through an integration by parts that these coefficients vanish when u_{bc} has been selected to satisfy the elasticity part of the field equations. The v_{jn} and d_{jn} coefficients are associated, respectively, with inertial and dissipative effects generated by u_{bc}. They have two columns, corresponding to the number of a_n terms. To adapt the definitions of those coefficients, eqs. (6.5.13) and (6.5.20), to the present situation, we observe that the previously arbitrary expansion functions used to describe u_{bc} are now $g_1 = 1$, $g_2 = x/L$. In addition, the Ritz basis functions are now specified to be the mode functions. Thus, the coefficient matrices now are

$$v_{jn} = \int_0^L \rho A \left(\frac{x}{L}\right)^{n-1} \Psi_j \, dx + \sum m \left(\frac{x_m}{L}\right)^{n-1} \Psi_j(x_m)$$

$$d_{jn} = \sum c \left(\frac{x}{L}\right)^{n-1} \Psi_j(x_c) \tag{7.8.9}$$

Because a_1 and a_2 were determined at an earlier stage of the analysis, eq. (7.8.8) is ready to solve. The initial conditions for u_h are derived from the superposition representation,

$$u_h(x, 0) = u(x, 0) - u_{bc}(x, 0), \qquad \dot{u}_h(x, 0) = \dot{u}(x, 0) - \dot{u}_{bc}(x, 0) \tag{7.8.10}$$

The right side of each expression represents the effective initial displacement and velocity fields. The corresponding initial values $\eta_j(0)$ and $\dot{\eta}_j(0)$ may then be found from the expansion theorem; see eqs. (7.7.17).

The methodology for handling time-dependent boundary conditions imposed on flexural response is very much the same, except that there are four boundary conditions, any of which may be time dependent. They may be written as

$$\mathcal{B}_1(w)\big|_{x=0} = f_1(t), \qquad \mathcal{B}_2(w)\big|_{x=0} = f_2(t)$$

$$\mathcal{B}_3(w)\big|_{x=L} = f_3(t), \qquad \mathcal{B}_2(w)\big|_{x=L} = f_4(t) \tag{7.8.11}$$

The superposition in eq. (7.8.2) still applies in the case of flexure,

$$\boxed{w = w_{bc} + w_h}$$

but w_{bc} must be altered. In accord with the Mindlin-Goodman procedure, we require that w_{bc} satisfy the elastic part of the field equation, that is,

$$\frac{\partial^4 w_{bc}}{\partial x^4} = 0 \tag{7.8.12}$$

This leads to a cubic polynomial for w_{bc},

$$\boxed{w_{bc} = a_0(t) + a_1(t)\frac{x}{L} + a_2(t)\left(\frac{x}{L}\right)^2 + a_3(t)\left(\frac{x}{L}\right)^3} \tag{7.8.13}$$

The coefficient functions are determined by satisfying the boundary conditions,

$$\left[\mathcal{B}_j(a_0) + \mathcal{B}_j\left(a_1\frac{x}{L}\right) + \mathcal{B}_j\left(a_2\left(\frac{x}{L}\right)^2\right) + \mathcal{B}_j\left(a_3\left(\frac{x}{L}\right)^3\right)\right]\Bigg|_{x=0} = f_j(t), \qquad j=1,2$$

$$\left[\mathcal{B}_j(a_0) + \mathcal{B}_j\left(a_1\frac{x}{L}\right) + \mathcal{B}_j\left(a_2\left(\frac{x}{L}\right)^2\right) + \mathcal{B}_j\left(a_3\left(\frac{x}{L}\right)^3\right)\right]\Bigg|_{x=L} = f_j(t), \qquad j=3,4$$

$$\tag{7.8.14}$$

Simultaneous solution of these equations yields the functions $a_j(t)$, which completes the evaluation of w_{bc}.

The coefficient matrices in the right side of eq. (7.8.8) may then be obtained by noting that for flexure there are four expansion terms $g_j = 1$, x/L, $(x/L)^2$, and $(x/L)^3$ accompanying the a_j terms. Correspondingly, eqs. (7.8.9) apply for n from 1 to 4 so $[\nu]$ and $[d]$ both have four columns, rather than two as in the axial displacement case.

EXAMPLE 7.11

The left end of the simply supported beam is subjected to a rotation, such that $\partial w/\partial x = \theta_0 \sin(\omega t)$, with θ_0 specified as a very small angle. The beam is initially at rest in the undeformed position. It is desired to determine the resulting displacement field.

(a) Determine the function w_{bc} satisfying the time-dependent boundary conditions for this situation.

(b) Determine the full set of equations that the second part of the solution, $w_h \equiv w - w_{bc}$, must satisfy. Identify the type of supports that correspond to the eigenmodes one should use to evaluate w_h; there is no need to actually determine w_h.

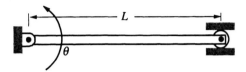

Solution All of the steps required to account for time-dependent boundary conditions will be evident in this example. One of the aspects of the analysis that surprises some individuals is the type of mode functions to be used. We begin by stating the differential equation, boundary conditions, and initial conditions for the system.

$$EI\frac{\partial^4 w}{\partial x^4} + \rho A \ddot{w} = 0$$

$$w = 0, \qquad \frac{\partial w}{\partial x} = \Theta \text{ at } x = 0, \qquad w = \frac{\partial^2 w}{\partial x^2} = 0 \text{ at } x = L$$

$$w = \dot{w} = 0 \text{ at } t = 0$$

where $\Theta = \theta_0 \sin(\omega t)$.

We now introduce the decomposition $w = w_{bc} + w_h$, where eq. (7.8.13) gives

$$w_{bc} = a_1(t) + a_2(t)\frac{x}{L} + a_3(t)\left(\frac{x}{L}\right)^2 + a_4(t)\left(\frac{x}{L}\right)^3$$

This function satisfies the boundary conditions if

$$a_1 = 0, \qquad a_2 = L\Theta, \qquad a_3 = -\frac{3}{2}L\Theta, \qquad a_4 = \frac{1}{2}L\Theta$$

The second part of the superposition, w_h, corrects for the inertial effects, which we determine by modal analysis. The mode functions to be employed satisfy the homogeneous boundary conditions

$$\psi_j\big|_{x=0} = \frac{d\psi_j}{dx}\bigg|_{x=0} = 0, \qquad \frac{d^2\psi_j}{dx^2}\bigg|_{x=L} = \frac{d^3\psi_j}{dx^3}\bigg|_{x=L} = 0$$

These are the conditions for a clamped-pinned beam. A glance at the picture of the system would seem to contradict this conclusion, for the left end is pinned, rather than clamped. However, the picture is deceptive. The time-dependent boundary condition constrains the rotation, and the time-independent analog of an imposed rotation is zero rotation.

With the understanding that the eigenvalues α_j and the mode functions ψ_j are those for a clamped-pinned beam, we proceed to normalize the latter based on the first orthogonality condition,

$$\int_0^L \rho A \Psi_j \Psi_n \, dx = \delta_{jn}$$

The modal series for w_h is

$$w_h = \sum_j \Psi_j \eta_j$$

The equations of motion for the modal coordinates are stated in eq. (7.8.8). There is no damping to consider, so $d_{jn} = 0$. Equation (7.8.9) describes the inertia corrections ν_{jn}. Because there are no attached masses, we have

$$\nu_{jn} = \int_0^L \rho A \Psi_j \left(\frac{x}{L}\right)^{n-1} dx, \qquad n = 1, 2, 3, 4$$

No external forces other than the couple causing the rotation, which is a constraint force, are present. Hence, we set $Q_j = 0$. For the specific a_n coefficients found here, and with $\Theta = \theta_0 \sin(\omega t)$ and damping taken to be zero, we have

$$\ddot{\eta}_j + 2\zeta_j \omega_j \dot{\eta}_j + \omega_j^2 \eta_j = -\sum_{n=1}^4 \nu_{jn} \ddot{\alpha}_n = \left(\nu_{j2} - \frac{3}{2}\nu_{j3} + \frac{1}{2}\nu_{j4}\right) L\omega^2 \theta_0 \sin(\omega t) \qquad (7.8.15)$$

We see that u_{bc} generates a harmonic excitation of u_h. After the modal equations have been solved, the displacement field may be synthesized as

$$u = u_{bc} + u_h = \left[\frac{x}{L} - \frac{3}{2}\left(\frac{x}{L}\right)^2 + \frac{1}{2}\left(\frac{x}{L}\right)^3\right] L\theta_0 \sin(\omega t) + \sum_j \Psi_j \eta_j \qquad (7.8.16)$$

7.8.2 Harmonic Boundary Excitation Wave Solution

The two-part procedure required to solve for the response in the case of arbitrary time dependency of the boundary conditions can be circumvented in situations where the excitation is harmonic and we solely wish to determine the steady-state response. The approach is the analog for continua of the frequency domain solution, in which steady-state response is obtained by factoring out $\exp(i\omega t)$ from all terms in the equation of motion. The formulation is able to account for springs, masses, and dashpots attached anywhere. It also can incorporate the effects of material dissipation. Rather than being rooted in modal analysis, the procedure is founded on wavelike aspects of the displacement field, which is a feature that we first encountered in the course of interpreting the mode functions. The only restrictions we shall impose are that the cross-sectional properties are constant, and that the excitation is restricted to concentrated forces and couples.

The approach we will develop addresses each system by returning to fundamental principles. Much work has gone into systematizing and automating the formulation, in the spirit of the finite element analysis. In fact, some individuals refer to the generalized procedure as *frequency domain finite elements*, but the more traditional name is *transfer matrices*. The text by Pestel and Leckie (1963) is an extensive development of the transfer matrix method. Doyle (1989) describes the application of the transfer matrices in combination with FFTs to evaluate transient, as well as steady-state, response.

Axial Displacement We begin by considering extensional deformation. The steady-state axial displacement at any x must vary at frequency ω, so we set

$$\boxed{u = \text{Re}[\psi(x)\exp(i\omega t)]} \qquad (7.8.17)$$

To incorporate viscoelastic material effects, we consider Young's modulus to be a complex constant in the form $E = \hat{E}(\omega)[1 + i\gamma(\omega)]$, where $\hat{E}(\omega)$ expresses the portion of stress that is in phase with a harmonically varying strain, and $\hat{E}(\omega)\gamma(\omega)$ gives the portion of stress that is 90° out of phase from the strain. The factor $\gamma(\omega)$ is the *viscoelastic loss factor*. Structural damping, in which the loss factor is constant, is obviously a special case. Causality requirements, which stem from the properties of the Fourier transform, dictate that $\gamma \geq 0$ at all frequencies; for most nonpolymeric materials, $\gamma \ll 1$. The frequency dependence of both $\hat{E}$ and γ may be ascertained experimentally.

We substitute u and E into the differential equation governing axial displacement and factor out the complex exponential, which leads to

$$\hat{E}(1 + i\gamma)\frac{d^2\psi}{dx^2} + \rho\omega^2\psi = 0 \tag{7.8.18}$$

We construct a solution of the homogeneous differential equation (7.8.18) as an exponential

$$\psi = B\exp(\lambda x)$$

which satisfies the differential equation if

$$\lambda^2 = -\frac{\rho\omega^2}{\hat{E}(1 + i\gamma)} \tag{7.8.19}$$

Because γ is not negative, the principal root of $(1 + i\gamma)^{1/2}$ lies in the first quadrant of the complex plane. For this reason, we may write the possible values of λ as

$$\lambda = \pm i\hat{k}, \qquad \hat{k} = \left[\frac{\rho\omega^2}{\hat{E}(1 + i\gamma)}\right]^{1/2} = k - i\mu, \qquad k > 0, \qquad \mu \geq 0 \tag{7.8.20}$$

There are two possible values of λ, so the general solution is a linear combination of the two fundamental solutions. Because $\hat{k}$ is complex (unless $\gamma = 0$), we retain the exponential form of the solution,

$$\psi = B_1\exp(i\hat{k}x) + B_2\exp(-i\hat{k}x) \tag{7.8.21}$$

The coefficients B_j are defined by the boundary conditions. At each end, either a geometric or natural boundary condition applies. The former may be homogeneous, or it might impose a harmonically varying displacement. We may express these as

$$u = \text{Re}[F_1\exp(i\omega t)] \text{ at } x = 0, \qquad u = \text{Re}[F_2\exp(i\omega t)] \text{ at } x = L \tag{7.8.22}$$

A natural boundary condition will be time dependent if a known harmonic force acts at an end. The most general case corresponds to attachment of a mass, spring, and dashpot at the end. The corresponding boundary conditions would be

$$k_1 u + c_1\dot{u} + m_1\ddot{u} - EA\frac{\partial u}{\partial x} = \text{Re}[F_1\exp(i\omega t)] \text{ at } x = 0$$

$$k_2 u + c_2\dot{u} + m_2\ddot{u} + EA\frac{\partial u}{\partial x} = \text{Re}[F_2\exp(i\omega t)] \text{ at } x = L \tag{7.8.23}$$

Let us multiply each term containing $\partial u/\partial x$ by a coefficient ε_1 or ε_2. This allows us to use the last set of boundary conditions to represent any possibility. When the natural boundary condition applies, we set $\varepsilon_1 = 1$ or $\varepsilon_2 = 1$ and assign the coefficients k_n, c_n, and m_n as appropriate. To represent a geometric boundary condition, we set $\varepsilon_1 = c_1 = m_1 = 0, k_1 = 1$, or $\varepsilon_2 = c_2 = m_2 = 0, k_2 = 1$.

To find the boundary conditions for the ψ function, we substitute eq. (7.8.17) into eqs. (7.8.23). Factoring out the time dependence then leads to

$$(k_1 + i\omega c_1 - m_1\omega^2)\psi - \varepsilon_1 EA\frac{d\psi}{dx} = F_1 \text{ at } x = 0$$

$$(k_2 + i\omega c_2 - m_2\omega^2)\psi + \varepsilon_2 EA\frac{d\psi}{dx} = F_2 \text{ at } x = L \tag{7.8.24}$$

Substitution of the general solution for ψ, eq. (7.8.21), into the preceding yields a pair of complex linear algebraic equations for B_1 and B_2,

$$
\begin{aligned}
[k_1 + i\omega c_1 &- m_1\omega^2 - i\varepsilon_1 EA\hat{k}]B_1 \\
&+ [k_1 + i\omega c_1 - m_1\omega^2 + i\varepsilon_1 EA\hat{k}]B_2 = F_1 \\
[(k_2 + i\omega c_2 &- m_2\omega^2 + i\varepsilon_1 EA\hat{k})\exp(i\hat{k}L)]B_1 \\
&+ [(k_2 + i\omega c_2 - m_2\omega^2 - i\varepsilon_1 EA\hat{k})\exp(-i\hat{k}L)]B_2 = F_2
\end{aligned}
\tag{7.8.25}
$$

For a specified value of ω, which defines the value of $\hat{k}$, these equations usually may be solved for B_1 and B_2. An exceptional case in which the equations are not solvable occurs when all dissipation effects have been neglected, $c_1 = c_2 = \gamma = 0$. Then the determinant of the system of coefficients will vanish when ω equals any natural frequency. (A few manipulations would show that this determinant is the characteristic equation for the free vibration modes.) Thus, the lack of solvability corresponds to resonance, where the response is not purely harmonic. When dissipation effects have been included, $|B_1|$ and $|B_2|$ will exhibit peaks when ω is close to a natural frequency.

It is instructive to examine the steady-state displacement from a physical viewpoint. We form the displacement by substituting eq. (7.8.21) into eq. (7.8.17), and using eq. (7.8.20) to replace $\hat{k}$. This gives

$$
\begin{aligned}
u = \text{Re}\{B_1 &\exp(\mu x)\exp[i(kx + \omega t)] \\
&+ B_2\exp(-\mu x)\exp[i(-kx + \omega t)]\}
\end{aligned}
\tag{7.8.26}
$$

As we saw in Section 7.4, a term that depends on the combination $kx + \omega t$ represents a wave propagating in the direction of decreasing x, because the function maintains a constant value as t increases and x decreases. Similarly, a term that depends on $-kx + \omega t$ corresponds to a wave propagating in the positive x direction. Thus, B_1 is the amplitude factor of a sinusoidal wave that propagates in the negative x direction. The overall amplitude of this wave decreases *in the direction of propagation* because $\mu > 0$. From this viewpoint, B_2 is the amplitude factor of a sinusoidal wave that propagates in the positive x direction. The overall amplitude of this wave also decreases in the direction of propagation because $\mu > 0$. In other words, harmonic boundary excitations generate waves that travel back and forth within the bar. The wave departing from either end consists of one part that is generated by the excitation at that end, and another part that is a reflection of the oppositely traveling wave that is incident at that end. Dissipation within the bar attenuates both waves as they propagate.

Let us inspect the dissipationless case further. If $\gamma = 0$, we find that $\hat{k} = k = (E/\rho)^{1/2}\omega$. Now recall the wave interpretation of an eigenmode associated with eq. (7.4.2). The modal motions were shown there to be equivalent to a superposition of waves whose phase speed is $c_{\text{bar}} = (E/\rho)^{1/2}$. The phase speed of the waves described by eq. (7.8.26) is $k/\omega = c_{\text{bar}}$ when $\gamma = 0$. The wavelengths in the forced case are $2\pi/k$, whereas the waves forming a free vibration mode have $2\pi/\alpha_j$ as their wavelength, where α_j is an eigenvalue. In other words, the waves generated by harmonic boundary excitation travel at the same speed as the waves forming a free vibration mode. When ω matches a natural frequency and there is no dissipation, the phase speed and wavelength of the forced waves match the properties of waves that can exist without excitation. In essence, the bar then offers no resistance to the excitations, which leads to resonance.

Flexural Displacement As we did for axial displacement, we consider only a uniform bar, and the excitations are considered to originate only at the ends. For the sake of simplicity, we shall consider the bar to be dissipationless, $\gamma = 0$. (Exercise 7.56 addresses the viscoelastic case.) The steady-state displacement is

$$w = \text{Re}[\psi(x)\exp(i\omega t)] \tag{7.8.27}$$

Substitution of this form into the differential equation governing flexural motion in the absence of a distributed force leads to

$$EI\frac{d^4\psi}{dx^4} - \rho A\omega^2\psi = 0 \tag{7.8.28}$$

As a fundamental solution, we let $\psi = B\exp(\lambda x)$, which satisfies the equation of motion if

$$\lambda = \pm ik, \pm k, \quad \text{where } k = \left(\frac{\rho A\omega^2}{EI}\right)^{1/4} \tag{7.8.29}$$

Note that k is real. Because there are four possible values of λ at a specified frequency, the general solution is a linear combination of the corresponding fundamental solutions,

$$\psi = B_1\exp(ikx) + B_2\exp(-ikx) + B_3\exp(kx) + B_4\exp(-kx) \tag{7.8.30}$$

Each end of the beam may have an attached spring, mass, and/or dashpot acting against transverse displacement, as well as torsional analogs resisting rotation. The harmonic excitation may consist of either an imposed displacement or specified shear force and an imposed rotation or specified bending moment. The boundary conditions at each end are alike aside from a change of sign associated with the shear or bending moment term, so we shall explicitly consider only the end $x = 0$. In the first of the shear boundary conditions given by eq. (7.6.16), we insert a term representing a dashpot and bring all terms that depend on w to the left side. These terms must balance a harmonic shear force excitation, which we represent by inserting $\text{Re}[F_1\exp(i\omega t)]$ into the right side of the boundary condition. For the other boundary condition at $x = 0$, we use the bending moment and the restorative torque of a torsional spring to balance an applied rotational excitation, which we represent as $\text{Re}[F_2\exp(i\omega t)]$. Hence, the boundary conditions to be satisfied are

$$-\varepsilon_1\frac{\partial}{\partial x}\left(EI\frac{\partial^2 w}{\partial x^2}\right) - k_1 w - c_1\dot{w} - m_1\ddot{w} = \text{Re}[F_1\exp(i\omega t)] \text{ at } x = 0$$

$$\tag{7.8.31}$$

$$-\varepsilon_2 EI\frac{\partial^2 w}{\partial x^2} - k_2\frac{\partial w}{\partial x} = \text{Re}[F_2\exp(i\omega t)] \text{ at } x = 0$$

As we did for axial motions, the coefficients ε_1 and ε_2 are introduced in order to use one equation to cover geometric and natural boundary conditions. Torsional dashpots and rotatory inertia of attached masses could be included by adding appropriate terms to the second boundary condition.

We substitute the steady-state displacement in eq. (7.8.27) into the boundary conditions above, with ψ expressed according to eq. (7.8.30). This yields two linear algebraic equations for the coefficients B_j. Two more equations result from the pair of boundary conditions at $x = L$. The four equations obtained in this manner are invertible, except when ω matches a natural frequency of the system and dissipation is ignored, in which case there is a resonance.

Once the coefficients B_j have been determined, we may synthesize the displacement field by substituting eq. (7.8.30) into eq. (7.8.27), which leads to

$$
\begin{aligned}
w &= \text{Re}\{B_1 \exp[i(kx + \omega t)] + B_2 \exp[i(-kx + \omega t)]\} \\
&\quad + \exp(kx)\,\text{Re}[B_3 \exp(i\omega t)] + \exp(-kx)\,\text{Re}[B_4 \exp(i\omega t)]
\end{aligned}
\tag{7.8.32}
$$

Clearly, the two complex exponential terms are like those for axial motion, eq. (7.8.26), in the case of a nondissipative material, $\mu = 0$. Their phase speed is

$$
c = \frac{\omega}{k} \equiv \left(\frac{EI}{\rho A}\right)^{1/4} \omega^{1/2}
\tag{7.8.33}
$$

For studies of wave propagation, an equation giving the phase speed in terms of either k or ω is called a *dispersion relation*. In contrast to extensional waves, which propagate at the constant bar speed $(E/\rho)^{1/2}$, the propagation speed of flexural waves increases with increasing frequency. From a different perspective, using eq. (7.8.29) to eliminate ω in the above yields

$$
c = c_{\text{bar}}\left(\frac{2\pi d}{\ell}\right)
\tag{7.8.34}
$$

where $\ell = 2\pi/k$ is the wavelength and $d = (I/A)^{1/2}$ is the radius of gyration of the cross-section; the largest possible value of d is the maximum cross-sectional dimension. This leads to several observations. Most importantly, the flexural phase speed is inversely proportional to the wavelength. (Transverse shear and rotatory inertia effects make this trend invalid when ℓ is comparable to d.) Furthermore, low frequencies correspond to small k according to eq. (7.8.29). Because ℓ for such waves is much greater than d, these waves propagate much more slowly than c_{bar}. For vibration studies at a fixed frequency, the dispersion phenomenon is not a major issue, but the difference between the manner in which dispersive and nondispersive waves propagate has profound implications for studies of transient wave propagation.

We have not addressed the second group of terms in eq. (7.8.32), which vary exponentially in x. We encountered similar terms in the high-frequency asymptotic form of a flexural mode, eq. (7.4.6). The term containing B_4 decays in the direction of increasing x, while the B_3 term decays in the direction of decreasing x. Both terms are waves that evanesce. The displacement field associated with each evanescent wave varies in time with a phase lag that is independent of x, that is, these are standing wave patterns. When k is large, corresponding to high frequencies, these evanescent waves decay very rapidly. Their effect in that case is evident only in the vicinity of the boundaries, just like the evanescent waves contributing to a free vibration mode.

Excitation and Attachments at Intermediate Locations

We shall now modify the procedure in order to handle excitations that consist of one or more concentrated harmonic forces applied along the length of the bar. In the regions between the points where the forces are applied, the wave solution in eqs. (7.8.21) or (7.8.30) is valid. We divide the bar up into segments whose ends are the *nodes* at which the forces are applied. We use different sets of coefficients B_j^n to describe the wave solution in the respective segments n, but the values of ω are the same for each solution. For simplicity, the bar is considered to be homogeneous, so the value of k also is common to all wave solutions. (Inhomogeneous situations require that each segment be described by a different k_n value according to eq. (7.8.20) or (7.8.29).) Equations for the B_j^n coefficients are obtained from the boundary conditions at each end, as well as continuity and jump conditions across the nodes. Identification of the latter conditions is achieved by considering an infinitesimally small segment of the bar surrounding a node at location x_p, as shown in Figure 7.13. In this diagram, we use a minus superscript to denote quantities that are evaluated to the left of x_p, while a plus denotes quantities to the right. The forces F_x and F_z are concentrated axial and transverse harmonic external forces. The segment's width is infinitesimal, but the force and moment resultants are finite. As a consequence, inertial effects are negligible and the equations of static equilibrium apply.

In the case of axial vibration, the displacement on either side of the node must be the same. Equilibrium in the x direction requires that the jump in the internal axial force must balance the applied force. This leads to

$$\boxed{\begin{aligned} u^+ &= u^- \\ \left(EA\frac{\partial u}{\partial x}\right)^+ - \left(EA\frac{\partial u}{\partial x}\right)^- &= -F_x \end{aligned}} \qquad (7.8.35)$$

Similar considerations treat the case where the concentrated force is applied transversely to the beam. The displacement, rotation, and bending moment must be continuous at the location where the force acts, while the jump in the internal shear force must balance the applied force. Thus,

$$\boxed{\begin{aligned} w^+ = w^-, \qquad \left(\frac{\partial w}{\partial x}\right)^+ = \left(\frac{\partial w}{\partial x}\right)^-, \qquad \left(\frac{\partial^2 w}{\partial x^2}\right)^+ = \left(\frac{\partial^2 \omega}{\partial x^2}\right)^- \\ \left(EI\frac{\partial^3 w}{\partial x^3}\right)^+ - \left(EI\frac{\partial^3 w}{\partial x^3}\right)^- = F_z \end{aligned}} \qquad (7.8.36)$$

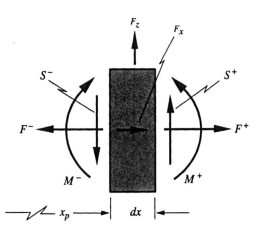

FIGURE 7.13 Jump conditions across a node where an axial or transverse force is applied.

In the case of axial motion, identification of P nodes between the ends breaks the bar into $P + 1$ segments. There are two continuity equations at each node, in addition to one boundary condition at each end. Thus, there will be $2P + 2$ equations to satisfy. Within the nth segment, the solution has the form of eq. (7.8.26), with two associated coefficients B_1^n and B_2^n. There are $P + 1$ segments, so there will be $2P + 2$ simultaneous complex equations for the $2P + 2$ unknown coefficients. The same considerations for the case of transverse displacement shows that there will be $4P + 4$ continuity and boundary conditions if there are P concentrated forces between the ends. Correspondingly, the transverse displacement within the $P + 1$ segments between the forces will be characterized by $4P + 4$ coefficients, B_1^n, B_2^n, B_3^n, and B_4^n.

It is possible to extend the forced wave method further to address springs, masses, and dashpots that are attached elsewhere than the ends of a bar. To do so, we merely add to the nodal external force F_x or F_z, the force applied by the spring, dashpot, or mass expressed in terms of displacement. For example, suppose a spring and dashpot are attached transversely at x_p. The upward displacement at this location is $w(x_p, t) = w^+ = w^-$. The spring opposes the displacement, and the dashpot opposes the velocity. Hence, we would describe the conditions at this node by applying eq. (7.8.36) with $F_z = -kw^+ - c\dot{w}^+$.

EXAMPLE 7.12

Example 7.10 examined the convergence of a modal series. Use the forced wave formulation to determine the displacement amplitude $f_w(x)$, bending moment amplitude $f_M(x)$, and shear amplitude $f_S(x)$.

Solution This exercise demonstrates the formulation of the forced wave method with an interior node point. It exploits the fact that the forced wave formulation can be used in conventional situations where the external forces are known, as well as in situations where the forces constrain motion. Because the wave method does not feature series expressions, questions regarding series convergence do not arise as they do in modal analysis. We formulate the problem by defining $x = 0.8L$ to be a node, so the beam is divided into two segments,

$$0 \leq x < 0.8L, \quad w = \mathrm{Re}[\psi_1(x)\exp(i\omega t)]$$

$$0.8L < x < L, \quad w = \mathrm{Re}[\psi_2(x)\exp(i\omega t)]$$

Within each segment, there are no external forces, so the wave solutions are

$$\psi_j = B_1^j\exp(ikx) + B_2^j\exp(-ikx) + B_3^j\exp(kx) + B_4^j\exp(-kx), \quad j = 1, 2$$

The excitation frequency is specified to be $\omega = 1.05\omega_3$, where ω_3 is the third natural frequency. In terms of the eigenvalue parameter introduced in Example 7.10, this corresponds to

$$\omega = 1.05\alpha_3^2\left(\frac{EI}{\rho AL^4}\right)^{1/2}$$

The wave number for flexural motion is $k^4 = \rho A\omega^2/EI$, so that

$$kL = \sqrt{1.05}\,\alpha_3 = 8.049$$

It will be necessary to evaluate the complex and real exponential functions at $x = 0.8L$ and $x = L$. In order to write the boundary and node conditions more concisely, we define the function values symbolically as

$$\Delta_1 = \exp(ikL), \qquad \Delta_2 = \exp(i0.8kL)$$
$$\varepsilon_1 = \exp(kL), \qquad \varepsilon_2 = \exp(0.8kL)$$
$$\exp(-ikL) = 1/\Delta_1, \qquad \exp(-i0.8kL) = 1/\Delta_2$$
$$\exp(kL) = 1/\varepsilon_1, \qquad \exp(0.8kL) = 1/\varepsilon_2$$

The left end, $x = 0$, is clamped, and the wave function appropriate to that location is ψ_1. Hence, we set $\psi_1 = d\psi_1/dx = 0$ at $x = 0$. This gives

$$B_1^1 + B_2^1 + B_3^1 + B_4^1 = 0$$
$$iB_1^1 - iB_2^1 + B_3^1 + B_4^1 = 0$$

where we have canceled a k factor in the second equation, the rotation condition. The end $x = L$ is free, so we set the shear and bending moment to zero there. The wave function for this location is ψ_2. Thus, the boundary conditions are $d^2\psi_2/dx^2 = d^3\psi_2/dx^3 = 0$ at $x = L$, which gives

$$-B_1^2\Delta_1 - B_2^2/\Delta_1 + B_3^2\varepsilon_1 + B_4^2/\varepsilon_1 = 0$$
$$-iB_1^2\Delta_1 - iB_2^2/\Delta_1 + B_3^2\varepsilon_1 - B_4^2/\varepsilon_1 = 0$$

To complete the analysis we satisfy the node conditions at $x = 0.8L$. At that location the displacement, rotation, and bending moment are continuous. The wave functions are ψ_1 to the left of the node and ψ_2 to the right. Also, the flexural rigidity EI is constant, which allows us to factor that term out of the moment equation. It is convenient for the purpose of solving for the B_n^j coefficients to write the continuity conditions as differences, for example $w^- - w^+ = 0$. Hence, we have

$$(B_1^1\Delta_2 + B_2^1/\Delta_2 + B_3^1\varepsilon_2 + B_4^1/\varepsilon_2) - (B_1^2\Delta_2 + B_2^2/\Delta_2 + B_3^2\varepsilon_2 + B_4^2/\varepsilon_2) = 0$$
$$(iB_1^1\Delta_2 - iB_2^1/\Delta_2 + B_3^1\varepsilon_2 - B_4^1/\varepsilon_2) - (iB_1^2\Delta_2 - iB_2^2/\Delta_2 + B_3^2\varepsilon_2 - B_4^2/\varepsilon_2) = 0$$
$$(-B_1^1\Delta_2 - B_2^1/\Delta_2 + B_3^1\varepsilon_2 + B_4^1/\varepsilon_2) - (-B_1^2\Delta_2 - B_2^2/\Delta_2 + B_3^2\varepsilon_2 + B_4^2/\varepsilon_2) = 0$$

At this juncture, we have seven algebraic equations for the eight unknown coefficients B_n^j. The last equation is the shear jump condition, which requires that $S^+ = S^- - F\cos(\omega t)$. Shear is related to displacement as $S = EI\,\partial^3 w/\partial x^3$. We use ψ_1 to describe the derivative at $x = 0.8L^-$, and ψ_2 is used for $x = 0.8L^+$. Thus, the last node condition is

$$EI\mathrm{Re}\left[\frac{d^3\psi_1}{dx^3}\Big|_{0.8L}\exp(i\omega t)\right] - EI\mathrm{Re}\left[\frac{d^3\psi_2}{dx^3}\Big|_{0.8L}\exp(i\omega t)\right] = -\mathrm{Re}[F\exp(i\omega t)]$$

Matching the coefficients of the complex exponentials leads to the eighth equation for the coefficients,

$$k^3(-iB_1^1\Delta_2 + iB_2^1/\Delta_2 + B_3^1\varepsilon_2 - B_4^1/\varepsilon_2) - k^3(-iB_1^2\Delta_2 + iB_2^2/\Delta_2 + B_3^2\varepsilon_2 - B_4^2/\varepsilon_2) = \frac{F}{EI}$$

Because the quantities Δj and ε_j are functions of kL, the assembled set of equations may be written in matrix form as

$$[D(kL)]\{B\} = \{C\}$$

where all of the coefficients $C_j = 0$ except $C_8 = FL^3/EI$. We substitute $kL = 8.049$, and solve for the coefficients as ratios of the nondimensional force parameter FL^3/EI. Once we have these coefficients, the displacement function $f_w(x)$ is merely $\psi_1(x)$ or $\psi_2(x)$, depending on which side of the node is being evaluated. Similarly, the bending moment function $f_M(x)$ and shear function $f_S(x)$ are obtained by differentiating the ψ_j function appropriate to that x.

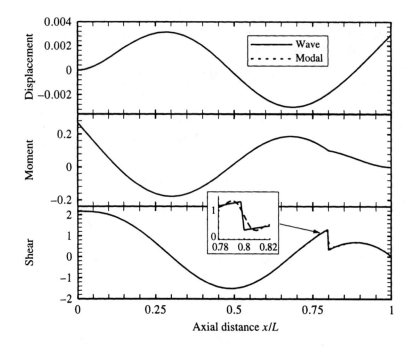

The results appear in the graph, along with the 100-term modal series results.

The agreement is exceptional, but the enlarged view in the vicinity of $x = 0.8L$ shows that only the wave solution truly is discontinuous at that location. A close examination also shows a small slope discontinuity in bending moment, which is associated with $S = -\partial M/\partial x$. Although the results agree, there is little doubt that the forced wave method is easier to implement.

7.9 TIMOSHENKO BEAM THEORY

Our studies thus far were founded on the classical theory of bars, in which cross-sections remain planar and perpendicular to the centroidal axis in axial and flexural deformation. This represents an engineering approximation that substantially simplifies any analysis. The approximation's validity will be examined here by comparing solutions obtained from classical beam theory to those obtained from an improved theory. The Pochammer-Chree solution for waves in bars (Graff, 1975) is a complete solution of the equations of dynamic elasticity for circular bars in which there are no a priori assumptions regarding the nature of the displacement field. Rather than pursuing this highly mathematical formulation, we will develop another engineering approximation that addresses the primary inertia and elastic effects omitted from the classical theory.

7.9.1 Field Equations

One anomaly contained in classical beam theory is the notion that although cross-sections carry a resultant shear force, the associated shear stress distribution is not accompanied by a shear strain. In actuality, the shear stress and strain vary over the cross-section, because the shear stress must be zero at the upper and lower surfaces of the bar. Timoshenko (1922) approximated the effect of shear as an average over the cross-section. This entails allowing each cross-section to rotate independently of the slope of the centroidal axis in the deformed state. As part of the correction to classical beam theory, rotatory inertia of cross-sections is also incorporated into the formulation.

Our concern here is with flexural effects, so we consider the centroid of the cross-section to displace solely in the transverse direction. Further, we assume that y and z are principal axes for the cross-sectional area, $I_{yz} = 0$, so that displacements in the y and z directions are uncoupled. We shall consider the case where displacement occurs in the xz plane.

Figure 7.14 depicts Timoshenko's kinematical approximation of the displacement field. The beam's cross-sections are taken to remain planar in the deformed state. The angle Y measures the rotation of a cross-section relative to the normal to the centroidal axis. It is referred to as the *shear angle* because Y is equated to zero in classical beam theory. The angle χ is the total rotation of a cross-section. As in the classical theory, w is the transverse displacement of the centroidal axis, so $\partial w/\partial x$ is the rotation the cross-section would undergo if there was no shear deformation. It follows from Figure 7.14 that

$$Y = \frac{\partial w}{\partial x} - \chi \qquad (7.9.1)$$

In linear elasticity, the shear strain is one-half the angle by which an infinitesimal rectangular element of material distorts to a parallelogram. Hence, the Timoshenko approximation considers the shear strain to be constant over the cross-section. In our sign convention, positive shear acts in the positive z direction for the cross-section that faces in the positive x direction. Thus, positive Y corresponds to a positive shear strain, such that

$$\varepsilon_{zz} = \tfrac{1}{2}Y \qquad (7.9.2)$$

For the direct stress on the cross-section, we observe from Figure 7.14 that the axial displacement of a point at distance z from the centroidal axis is

$$u = -z\chi \qquad (7.9.3)$$

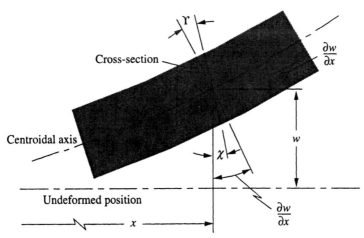

FIGURE 7.14 Kinematics of deformation according to Timoshenko beam theory.

The corresponding axial strain is

$$\varepsilon_{zz} = \frac{\partial u}{\partial x} = -z\frac{\partial \chi}{\partial x} \tag{7.9.4}$$

Expressions for the shear force and bending moment are obtained by evaluating the stress resultants. From Hooke's law we have $\sigma_{xz} = 2G\varepsilon_{xz}$ and $\sigma_{xx} = E\varepsilon_{xx}$. The contribution to the shear force stemming from an element of cross-sectional area dA is $\sigma_{xz}dA$. The contribution to the bending moment of the direct stress is $-z(\sigma_{xx}dA)$, where the minus sign accounts for the fact that positive M represents compression at $z > 0$, whereas positive σ_{xx} is tensile. We integrate these contributions over the entire cross-section to obtain

$$S = \iint_A (\sigma_{xy})\,dA = \iint_A (GY)\,dA$$

$$M = \iint_A (-z\sigma_{xx})\,dA = \iint_A z^2 E\frac{\partial \chi}{\partial x}\,dA \tag{7.9.5}$$

When the constant factors are brought outside, the integrals are the area and the area moment of inertia. However, the first integral is based on an assumption that Y is a constant across the cross-section, which is not correct. For this reason a correction factor κ is introduced into the shear force relation. (We will discuss the value of κ later.) The result is

$$\boxed{\begin{aligned} S &= \kappa GAY = \kappa GA\left(\frac{\partial w}{\partial x} - \chi\right) \\[2mm] M &= EI\frac{\partial \chi}{\partial x} \end{aligned}} \tag{7.9.6}$$

These relations lead to interpretation of the action in Timoshenko beam theory as a superposition of effects. In Figure 7.15(a) the shear force distorts a differential beam element by shear angle Y, without causing the element to rotate. Figure 7.15(b) shows the bending moment inducing a rotation χ with the shear angle unchanged. The result is that the top and bottom surfaces of the beam element have rotated by $\partial w/\partial x$, which is shown in the figure to be Y $+ \chi$, consistent with eq. (7.9.1).

The equations of motion are derived from the free body diagram in Figure 7.3, which was used previously for classical beam theory. Summing forces in the z direction leads to

$$\sum F_x = \left(S + \frac{\partial S}{\partial x}dx\right) - S + f_x dx = (\rho A dx)\ddot{w} \tag{7.9.7}$$

This expression is like the one in classical beam theory. The moment sum introduces the second correction to classical beam theory by accounting for the rotatory inertia. Rotational effects associated with the shear deformation in Figure 7.15(a) are negligible because of the differential width. In contrast, Figure 7.15(b) constitutes a rotation χ about the centroid. The mass moment of inertia about the centroid is $\rho I\,dx$ because the beam element is a waferlike body. Thus, the moment equation of motion is

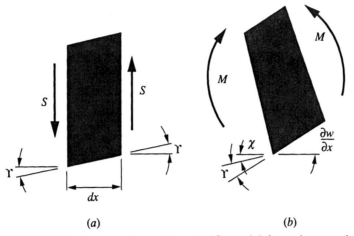

FIGURE 7.15 Superposition of shear and flexural deformation associated with Timoshenko beam theory.

$$\sum M_G = \left(M + \frac{\partial M}{\partial x} dx\right) dx - M\,dx + S\,dx = (\rho I\,dx)\ddot{\chi} \qquad (7.9.8)$$

We substitute the force-displacement relations, eqs. (7.9.6), which results in two coupled partial differential equations of motion,

$$
\kappa GA\left(\frac{\partial^2 w}{\partial x^2} - \frac{\partial \chi}{\partial x}\right) - \rho A\ddot{w} = -f_z
$$

$$
EI\frac{\partial^2 \chi}{\partial x^2} + \kappa GA\left(\frac{\partial w}{\partial x} - \chi\right) - \rho I\ddot{\chi} = 0
$$

$$(7.9.9)$$

Clearly these field equations are substantially more complicated than the single partial differential equation in classical bar theory. Some texts combine them into a single equation by eliminating χ, but doing so is not useful because χ arises in the boundary condition.

The boundary conditions for Timoshenko beam theory are the same alternatives as those for classical beam theory. Specifically, at each end, either the displacement w or the shear force S must be specified, and either the rotation χ or the bending moment M must be specified. In view of the force-displacement relations, this leads to

$$
\text{At each end, specify } \begin{cases} w \text{ or } \kappa GA\left(\dfrac{\partial w}{\partial x} - \chi\right) \\[2ex] \chi \text{ or } EI\dfrac{\partial \chi}{\partial x} \end{cases}
\qquad (7.9.10)
$$

We will not address time-dependent boundary conditions, so the specified values of the boundary conditions will either be zero or functions of w and/or χ in the case of spring or mass attachments at the ends.

For Ritz series formulations, it is important to recognize that the geometric boundary conditions are those that specify the displacement w or rotation χ. Formulating a Ritz series solution now requires selection of kinematically admissible basis functions for w and χ, which are used to form the series for each variable. An interesting aspect is that this selection is somewhat easier than it is in classical beam theory because the geometric boundary conditions pertain to w and χ, but not their derivatives.

An expression for kinetic energy in terms of the displacement variables is required to perform a Ritz series analysis, as well as to establish the orthogonality properties of mode functions. Both translational and rotatory effects contribute to the kinetic energy of the beam, such that

$$T = \frac{1}{2} \int_0^L [\rho A \dot{w}^2 + \rho I \dot{\chi}^2]\, dx \qquad (7.9.11)$$

The strain energy combines the effect of bending and shear deformations, according to

$$V = \frac{1}{2} \int_0^L \left[EI\left(\frac{\partial \chi}{\partial x}\right)^2 + \kappa GA\left(\frac{\partial w}{\partial x} - \chi\right)^2 \right] dx \qquad (7.9.12)$$

EXAMPLE 7.13

A cube having mass m and side dimensions b is welded to the end of the beam, and a spring of stiffness k is attached to the cube. The left end of the bar is clamped. Determine the boundary conditions corresponding to the Timoshenko field equations. Account for both the translational and rotatory inertia of the cube.

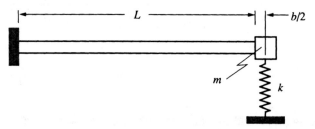

Solution   This example will highlight the differences in the identification of boundary conditions engendered by having two displacement variables to consider. The left end is clamped, which means that there can be no transverse displacement and no rotation of a cross-section. Because these conditions are less intricate than those at the mass-loaded end, we define the left end as $x = 0$, so it must be

$$w = \chi = 0 \text{ at } x = 0$$

We draw a free body diagram of the block to identify the other two boundary conditions.

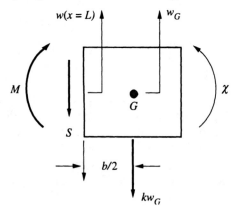

The kinematical features to note are that χ at $x = L$ is the rotation of the cross-section, so that is the rotation of the block. Furthermore, because w at $x = L$ is the displacement of the cross-section, that quantity is also the vertical displacement at the left face of the block. The small displacement approximation then leads to the vertical displacement of the block's center of mass being given by

$$w_G = w(x = L) + \frac{b}{2}\chi(x = L)$$

The shear force acting on the block is positive downward according to our sign convention, so the force equation of motion for the block is

$$\sum F_x = -S - kw_G = m\ddot{w}_G$$

The moment of inertia for the cube about its center of mass is $I_G = (1/6)mb^2$, so the moment equation of motion is

$$\sum M_G = -M + S\frac{b}{2} = \frac{1}{6}mb^2\ddot{\chi}$$

Substitution of the force-displacement relations then leads to

$$\left.\begin{aligned} \kappa GA\left(\frac{\partial w}{\partial x} - \chi\right) + k\left(w + \frac{b}{2}\chi\right) + m\left(\ddot{w} + \frac{b}{2}\ddot{\chi}\right) = 0 \\ EI\frac{\partial \chi}{\partial x} - \kappa GA\frac{b}{2}\left(\frac{\partial w}{\partial x} - \chi\right) + \frac{1}{6}mb^2\ddot{\chi} = 0 \end{aligned}\right\} \text{ at } x = L$$

Notice that these boundary conditions contain the displacement variables and their first x derivative, whereas the shear boundary condition for classical theory would contain $\partial^3 w/\partial x^3$.

7.9.2 Wave Solutions

Our first analysis of the Timoshenko field equations considers flexural waves propagating along an infinitely long bar (Graff, 1975). Because there are no boundary conditions to satisfy, this wave propagation analysis is substantially easier to perform than a modal analysis of a finite-length bar. We consider a harmonic wave propagating in the positive x direction with a wavelength $2\pi/k$. As part of the solution, we must determine the relation between the frequency ω and the wavenumber k. The displacement w and the cross-sectional rotation χ propagate in unison, but we cannot assume that these variables are in phase. Hence, we use complex exponentials to express both variables. For propagation in the positive x direction, the phase variable is $\omega t - kx$, so we write

$$\boxed{w = \text{Re}\{B_w\exp[i(\omega t - kx)]\}, \qquad \chi = \text{Re}\{B_\chi\exp[i(\omega t - kx)]\}} \qquad (7.9.13)$$

We substitute these expressions into eqs. (7.9.9), in which $f_z = 0$. Cancellation of the complex exponential leads to

$$\kappa GA(-k^2B_w + ikB_\chi) + \rho A\omega^2 B_w = 0$$
$$-EIk^2B_\chi + \kappa GA(-ikB_w - B_\chi) + \rho I\omega^2 B_\chi = 0 \qquad (7.9.14)$$

The equivalent matrix form of these relations is

$$\begin{bmatrix} (-\kappa GAk^2 + \rho A\omega^2) & i(\kappa GAk) \\ -i(\kappa GAk) & (-EIk^2 - \kappa GA + \rho I\omega^2) \end{bmatrix}\begin{Bmatrix} B_w \\ B_\chi \end{Bmatrix} = \begin{Bmatrix} 0 \\ 0 \end{Bmatrix} \qquad (7.9.15)$$

The characteristic equation for this set of homogeneous equations is the determinant of the coefficient matrix,

$$(\rho A \omega^2 - \kappa GAk^2)(\rho I \omega^2 - EIk^2 - \kappa GA) - (\kappa GAk)^2 = 0 \qquad (7.9.16)$$

This relation is biquadratic in both ω and k. To simplify its form we introduce the cross-sectional radius of gyration, $d = (I/A)^{1/2}$, which leads to

$$\rho^2 d^2 \omega^4 - \rho[d^2(E + \kappa G)k^2 + \kappa G]\omega^2 + \kappa GEd^2k^4 = 0 \qquad (7.9.17)$$

The key quantity for wave propagation is the phase speed. For the forms in eqs. (7.9.13) this quantity is

$$c = \frac{\omega}{k} \qquad (7.9.18)$$

To obtain the dispersion relation between c and k we substitute $\omega = kc$ into the characteristic equation. Because $G = E/2(1 + \nu)$, this leads to

$$\boxed{\left(\frac{c}{c_{\mathrm{bar}}}\right)^4 - \left[1 + \frac{\kappa}{2(1 + \nu)}\left(1 + \frac{1}{k^2d^2}\right)\right]\left(\frac{c}{c_{\mathrm{bar}}}\right)^2 + \frac{\kappa}{2(1 + \nu)} = 0}$$

where $c_{\mathrm{bar}} = (E/\rho)^{1/2}$ is the phase speed for extensional waves in a bar.

The preceding dispersion equation indicates that, for a specified cross-sectional shape, c/c_{bar} depends only on kd. Furthermore, the order of magnitude of d is the depth h of the cross-section, and k is inversely proportional to the wavelength, $k = 2\pi/\lambda$. Thus, the dispersion relation indicates that c/c_{bar} depends only on the ratio of the beam's depth to the wavelength. For each value of kd, the dispersion equation gives two positive roots for $(c/c_{\mathrm{bar}})^2$. It follows that there are two phase speeds, which we denote as $c_a < c_b$.

The displacements are quite different for c_a and c_b. The wave amplitudes may be determined from either of eqs. (7.9.15), which yields

$$B_\chi = \frac{(\kappa GAk^2 - \rho A \omega^2)}{i(\kappa GAk)}B_w \qquad (7.9.19)$$

When we substitute $\omega = kc_n$, $n = a, b$, we obtain

$$\boxed{(B_\chi)_n = -i\left[1 - 2(1 + \nu)\frac{c_n^2}{(c_{\mathrm{bar}})^2}\right]k(B_w)_n}$$

$$(7.9.20)$$

The amplitude of the centroidal axis' rotation in a wave is $|\partial w/\partial x| = k|(B_w)_n|$. We will see that in the range of kd relevant to conventional engineering applications, $c_a \ll c_{\mathrm{bar}}$ and $c_b \gg c_{\mathrm{bar}}$. Hence, eq. (7.9.20) indicates that in the low-speed wave, $|\partial w/\partial x| \approx |\chi|$, which means that the shear angle Υ is relatively small. In contrast, in the high-speed wave $|\chi|$ is much larger than $|\partial w/\partial x|$, which means that shear deformation is the dominant effect in those waves.

Evaluation of the dispersion relation requires a value for the shear correction factor κ. It was introduced to account for the fact that the average shear stress is not constant across a cross-section. Thus, one approach to selecting it has been to examine the static shear distribution. Another perspective is that κ is a free parameter that may be used to match a phase speed obtained from Timoshenko beam theory to one that is obtained from the dynamic theory of elasticity. These issues are discussed by Shames and

Dym (1985). In either viewpoint, the value of κ depends on the shape of the cross-section. Numerous papers have suggested different values. We shall follow Graff (1975), who used 5/6 and 10/9 for rectangular and circular cross-sections, respectively.

We found in eq. (7.8.34) that classical beam theory leads to a single wave speed, satisfying

$$\frac{c}{c_{\text{bar}}} = kd$$

This relation ignores the effects of transverse shear deformation and rotatory inertia. Prior to Timoshenko's work, Rayleigh (1945) had suggested that classical beam theory should be corrected by including rotatory inertia in the moment equation of motion. One way of obtaining Rayleigh's model is to return to classical theory. The other is to follow Exercise 7.61. There the combination $\partial w/\partial x - \chi$, which equals Y, is eliminated between eqs. (7.9.9), after which $\chi = \partial w/\partial x$ leads to the Rayleigh beam equation,

$$EI\frac{\partial^4 w}{\partial x^4} + \rho A\ddot{w} - \rho I\frac{\partial^2 \ddot{w}}{\partial x^2} = f_z \tag{7.9.21}$$

The characteristic equation obtained by substituting w from eqs. (7.9.13) is

$$EIk^4 - \rho A\omega^2 - \rho Ik^2\omega^2 = 0 \tag{7.9.22}$$

Setting $\omega = kc$ in this relation leads to the dispersion equation for Rayleigh beam theory,

$$\left(\frac{c}{c_{\text{bar}}}\right)^2 = \frac{k^2 d^2}{1 + k^2 d^2} \tag{7.9.23}$$

Figure 7.16 compares the wave speeds obtained from the three alternative theories as a function of kd. Obviously, only Timoshenko theory predicts more than one wave at each value of k. We refer to the curve describing each phase speed as a *branch*. Solutions of the equations of dynamic elasticity show that there actually are an infinite number of branches; see Graff (1975). Such solutions indicate that the c_b branch is quite inaccurate, whereas the c_a branch is quite good for all kd. Both classical beam theory and the Rayleigh theory yield phase speeds very close to c_a in the region $kd < 0.1$. The Rayleigh prediction is somewhat better than classical theory for larger kd. For example, Rayleigh theory at $kd = 0.2$ yields c/c_{bar} that is 4% too high, while the result of classical theory is 6% high. The values diverge more from the correct result with further increase of kd.

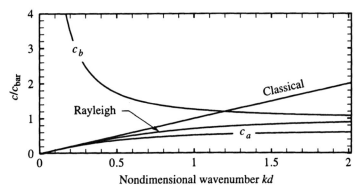

FIGURE 7.16 Dispersion curves for flexural waves in a uniform circular bar, $\nu = 0.3$, $\kappa = 10/9$.

The discrepancy between the phase speeds in Timoshenko and classical beam theories has general implications. For a rectangular bar, $d = h/\sqrt{12}$. The condition $kd < 0.1$ corresponds to an approximate wavelength specification $\lambda > 20h$. Because the wavelength is twice the distance between zeroes, this specification suggests that

> *A flexural mode obtained from classical beam theory will show discrepancies from dynamic elasticity solutions if the distance between nodal points is less than 10 times the depth of the beam.*

This and all subsequent guidelines apply to homogeneous bars. A composite construction that increases the ratio G/E will enhance the discrepancies between classical and Timoshenko beam theories.

7.9.3 Modal Analysis

The analysis in the previous section focused on propagating waves. For finite-length beams, vibration modes consist of a superposition of propagating and evanescent waves. Thus, we consider a displacement field described by

$$w = \text{Re}[B_w \exp(-\mu x) \exp(i\omega t)], \qquad \chi = \text{Re}[B_\chi \exp(-\mu x) \exp(i\omega t)] \quad (7.9.24)$$

where μ may be real, complex, or purely imaginary with any sign.

The equations for the amplitudes B_w and B_χ are like eq. (7.9.15), except that μ replaces ik,

$$\begin{bmatrix} (\kappa G A \mu^2 + \rho A \omega^2) & (\kappa G A \mu) \\ -(\kappa G A \mu) & (EI\mu^2 - \kappa G A + \rho I \omega^2) \end{bmatrix} \begin{Bmatrix} B_w \\ B_\chi \end{Bmatrix} = \begin{Bmatrix} 0 \\ 0 \end{Bmatrix} \quad (7.9.25)$$

The associated characteristic equation is

$$c_{\text{bar}}^2 c_{\text{sh}}^2 \mu^4 + (c_{\text{bar}}^2 + c_{\text{sh}}^2)\omega^2\mu^2 + \omega^2\left(\omega^2 - \frac{c_{\text{sh}}^2}{d^2}\right) = 0 \quad (7.9.26)$$

where $c_{\text{sh}} = (\kappa G/\rho)^{1/2}$ is the *shear wave speed*. The characteristic equation indicates that for any ω, there are two corresponding values of μ^2, one positive and the other negative. We write these values as

$$\mu^2 = -\left(\frac{\alpha}{L}\right)^2, \left(\frac{\beta}{L}\right)^2 \quad \Rightarrow \quad \mu = \pm\frac{\alpha}{L}i, \pm\frac{\beta}{L} \quad (7.9.27)$$

where α and β are positive real quantities given by

$$\alpha = \left\{\left[\frac{1}{4}(\sigma^2 - 1)^2(kL)^4 + \frac{L^2}{d^2}(kL)^2\right]^{1/2} + \frac{1}{2}(\sigma^2 + 1)(kL)^2\right\}^{1/2}$$

$$\beta = \left\{\left[\frac{1}{4}(\sigma^2 - 1)^2(kL)^4 + \frac{L^2}{d^2}(kL)^2\right]^{1/2} - \frac{1}{2}(\sigma^2 + 1)(kL)^2\right\}^{1/2}$$

$$(7.9.28)$$

The parameters in these expressions are $k = \omega/c_{\text{bar}}$ and $\sigma = c_{\text{bar}}c_{\text{sh}} \equiv \sqrt{2(1 + \nu)/\kappa} > 1$.

Imaginary characteristic exponents in eqs. (7.9.27) lead to sinusoidal dependence on x, whereas real exponents represent evanescent waves that decay away from the boundaries. As was done for classical beam theory, the evanescent waves may be combined into hyperbolic functions. Hence, the general solution for transverse displacement is written as

$$w = \text{Re}[\psi_w(x)\exp(i\omega t)]$$

$$\psi_w = C_1\sin\left(\alpha\frac{x}{L}\right) + C_2\cos\left(\alpha\frac{x}{L}\right) + C_3\sinh\left(\beta\frac{x}{L}\right) + C_4\cosh\left(\beta\frac{x}{L}\right)$$

(7.9.29)

The easiest way to identify the corresponding general solution for χ is to employ the first of eqs. (7.9.9), which states that

$$\frac{\partial\chi}{\partial x} = \frac{\partial^2 w}{\partial x^2} + \sigma^2 k^2 w$$

We substitute the general solution for w, and integrate to find

$$\chi = \text{Re}[\psi_\chi(x)\exp(i\omega t)]$$

(7.9.30)

where the mode function ψ_χ for shear is

$$\psi_\chi = \frac{1}{L\alpha}(\sigma^2 k^2 L^2 - \alpha^2)\left[-C_1\cos\left(\alpha\frac{x}{L}\right) + C_2\sin\left(\alpha\frac{x}{L}\right)\right]$$

$$+ \frac{1}{L\beta}(\sigma^2 k^2 L^2 + \beta^2)\left[C_3\cosh\left(\beta\frac{x}{L}\right) + C_4\sinh\left(\beta\frac{x}{L}\right)\right]$$

(7.9.31)

The situation now is like it was at the stage in classical theory after the general solution for w had been derived. There are four coefficients C_n to determine. It still remains to satisfy the boundary conditions, which we take to be at $x = 0$ and $x = L$. There are a total of four homogeneous boundary conditions—two from each end according to the alternatives in eqs. (7.9.10). When we substitute eqs. (7.9.29) and (7.9.31) into these boundary conditions, the result will be four linear homogeneous equations for the mode coefficients. The general form of these equations will be

$$[D(\alpha, \beta, k)]\begin{bmatrix} C_1 & C_2 & C_3 & C_4 \end{bmatrix}^{\text{T}} = \{0\}$$

(7.9.32)

For a nontrivial solution, the coefficient matrix must be rank deficient,

$$|D(\alpha, \beta, k)| = 0$$

(7.9.33)

The primary difference between these boundary equations and the general conditions for classical beam theory, eq. (7.3.11), is the present dependence of the coefficients D_{jn} on α, β, and k. (The values of κ and σ are set for a specified beam.) The characteristic equation, eq. (7.9.33), represents one relation among these three quantities. The other two are eqs. (7.9.28). A modal solution requires determination of a trio of values α_n, β_n, and k_n satisfying these three equations. Each set of values must then be used to determine proportions of C_1, C_2, C_3, and C_4 by simultaneously solving three of the scalar equations contained in eqs. (7.9.32). Either C_1 or C_2 may be equated to unity in that analysis.

It is obvious that determining the modes for a Timoshenko beam with arbitrary boundary conditions is much more difficult than it is for classical beam theory. We will explore a general procedure for carrying out such an analysis in the next example. Fortunately, there is a case where the analysis may be performed fairly easily: a simply supported beam. Hinge supports require that $w = 0$ and $M = 0$ at each end. Correspondingly, we set $\psi_w = \partial\psi_\chi/\partial x$ at $x = 0$ and $x = L$. The conditions at $x = 0$ require that

$$
\left.
\begin{aligned}
C_2 + C_4 &= 0 \\
(\sigma^2 k^2 L^2 - \alpha^2)C_2 + (\sigma^2 k^2 L^2 + \beta^2)C_4 &= 0
\end{aligned}
\right\} \quad \Rightarrow \quad C_2 = C_4 = 0 \quad (7.9.34)
$$

With the elimination of C_2 and C_4, the boundary conditions at $x = L$ require

$$
C_1 \sin(\alpha) + C_3 \sinh(\beta) = 0
$$
$$
(\sigma^2 k^2 L^2 - \alpha^2)C_1 \sin(\alpha) + (\sigma^2 k^2 L^2 + \beta^2)C_3 \sinh(\beta) = 0
$$

Satisfying both conditions requires that $C_1 \sin(\alpha) = 0$ and $C_3 \sinh(\beta) = 0$. Because $\beta > 0$, it must be that $C_3 = 0$. This leads to $\sin(\alpha) = 0$, so that the eigenvalues are

$$
\alpha_n = n\pi \quad (7.9.35)
$$

These eigenvalues are the same as those for a simply supported beam according to classical theory. The difference is in the determination of the corresponding natural frequencies. In the present analysis, frequency is represented nondimensionally as $kL \equiv \omega L/c_{\text{bar}}$. Equation (7.9.28) defines α in terms of kL. The solution of this quadratic equation gives two frequencies for a specified α,

$$
kL(\alpha, L/d, \sigma) = \frac{1}{\sqrt{2}\sigma} \left\{ \left[\left(\frac{L}{d}\right)^2 + (\sigma^2 + 1)\alpha^2 \right] \right.
$$

$$
\left. \mp \left[\left(\frac{L}{d}\right)^4 + 2\left(\frac{L}{d}\right)^2 (\sigma^2 + 1)\alpha^2 + (\sigma^2 - 1)^2 \alpha^4 \right]^{1/2} \right\}^{1/2} \quad (7.9.36)
$$

The alternative values of kL obtained from these relations yield two branches, comparable to the two branches of the dispersion relation for wave propagation. As we did for wave propagation, we use subscripts a and b to denote respectively the lower and higher values of k obtained from this relation.

Because all coefficients except C_1 are zero, the modes of a simply supported beam are found from eqs. (7.9.29) and (7.9.31) to be

$$
\psi_w = C_{1n}\sin\left(n\pi\frac{x}{L}\right), \qquad \psi_\chi = -\frac{C_{1n}}{L}\left[\frac{\sigma^2(k_{a,b}L)^2}{n\pi} - n\pi\right]\cos\left(n\pi\frac{x}{L}\right) \quad (7.9.37)
$$

The relatively large value of $k_b L$ means that the cross-sectional rotations for the high-frequency branch will be much greater for a given transverse displacement.

Correspondingly, shear deformation is much greater for the modes of the high-frequency branch.

Classical beam theory gives only a single branch for each eigenvalue. The non-dimensional natural frequencies for a simply supported beam are

$$(kL)_{\text{classical}} = \frac{d}{L}(n\pi)^2 \tag{7.9.38}$$

Figure 7.17 compares the natural frequencies obtained from the Timoshenko approximation to those of classical theory. For both thickness ratios, classical theory is seen to overpredict the natural frequency increasingly with increasing mode number. To understand this trend, note that when $\alpha = n\pi$, eqs. (7.9.36) and (7.9.38) indicate that $(kL)_{\text{classical}}/(kL)_a$ depends only on nh/L. Thus, increasing the mode number has the same effect as increasing h. For the more slender beam in Figure 7.17, $h = L/20$, classical theory is in reasonably good agreement with Timoshenko results up to $n = 4$, at which point classical theory is 5.6% high. The same error is obtained at $n = 2$ when $h = L/10$. The nodal distance for the sinusoidal mode functions of a simply supported beam is L/n. Hence, the natural frequency results indicate that a 5.6% error occurs when the nodal distance is five times the thickness. Recall that the dispersion equation for flexural waves identified a guideline that classical theory will become inaccurate if the nodal distance is less than 10 times the thickness.

It is evident from Figure 7.17 that even for the deep beam, all natural frequencies for branch b are well above those for branch a. In all cases except ultrasonic applications, these frequencies are far greater than the highest frequency of the excitation. Consequently, those modes will not receive significant excitation, and may be ignored. As was the case for wave propagation, dynamic elasticity solutions indicate that this branch actually is quite inaccurate. Thus, it is fortunate that we need not employ this branch.

Regardless of the type of boundary conditions, solution of the eigenvalue problem in eq. (7.9.32) yields a sequence of eigenvalues α_j along the lower branch. Each eigenvalue gives a corresponding set of coefficients C_{1j}, C_{2j}, C_{3j}, and C_{4j}, with either the first or second coefficient equal to unity. This leads to a set of mode functions $\psi_{wj}(x)$ and $\psi_{\chi j}(x)$. These modes obey orthogonality conditions, which also serve to normalize the modes $\Psi_{wj}(x)$ and $\Psi_{\chi j}(x)$. To identify the orthogonality and

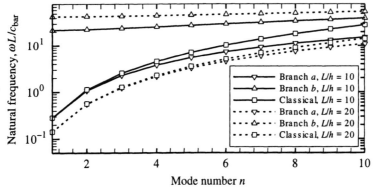

FIGURE 7.17 Natural frequencies of a rectangular beam according to Timoshenko and classical beam theory for two depth ratios, $\kappa = 5/6$, $\nu = 0.3$.

normalization conditions, we expand both displacements in a modal series. A modal coordinate η_j gives the contribution of a normal mode to each displacement variable, so the modal series are

$$w(x, t) = \sum_j \Psi_{wj}(x)\eta_j(t), \qquad \chi(x, t) = \sum_j \Psi_{\chi j}(x)\eta_j(t) \qquad (7.9.39)$$

Note that these series contain only the modes for the low-frequency branch, based on the decision to ignore the high-frequency branch.

The first orthogonality condition is such that substitution of the modal series into the kinetic energy leads to a sum of squares of the modal coordinates. The kinetic energy is given in eq. (7.9.11). Substitution of the displacement series leads to

$$T = \frac{1}{2}\int_0^L \left[\rho A \left(\sum_j \Psi_{wj}\dot{\eta}_j\right)\left(\sum_n \Psi_{wn}\dot{\eta}_n\right) + \rho I \left(\sum_j \Psi_{\chi j}\dot{\eta}_j\right)\left(\sum_n \Psi_{\chi n}\dot{\eta}_n\right) \right] dx$$

$$= \frac{1}{2}\sum_j \sum_n \left[\int_0^L (\rho A \Psi_{wj}\Psi_{wn}\dot{\eta}_n + \rho I \Psi_{\chi j}\Psi_{\chi n})\, dx \right] \dot{\eta}_j \dot{\eta}_n \qquad (7.9.40)$$

In order for this to reduce to a sum of squares,

$$T = \frac{1}{2}\sum_j \dot{\eta}_j^2 \qquad (7.9.41)$$

the first orthogonality condition must be

$$\int_0^L \left(\rho A \Psi_{wj}\Psi_{wn} + \rho I \Psi_{\chi j}\Psi_{\chi n}\right) dx = \delta_{jn} \qquad (7.9.42)$$

where the case $n = j$ yields the normalization rule for the mode functions. The modal series also diagonalizes the potential energy, such that

$$V = \frac{1}{2}\sum_j \omega_j^2 \eta_j^2 \qquad (7.9.43)$$

As we have seen for classical beam theory, there is no need to actually know the second orthogonality condition leading to the preceding.

The power input by the transverse distributed force is

$$\mathscr{P}_{in} = \int_0^L f_z \dot{w}\, dx = \int_0^L f_z \left(\sum_j \Psi_{wj}\dot{\eta}_j\right) dx \qquad (7.9.44)$$

The coefficient of each $\dot{\eta}_j$ is the corresponding modal generalized force,

$$Q_j = \int_0^L f_z \Psi_{wj}\, dx \qquad (7.9.45)$$

The modal equations of motion associated with these energy and power expressions is the standard set,

$$\ddot{\eta}_j + \omega_j^2 \eta_j = Q_j \qquad (7.9.46)$$

These modal equations may be solved for transient and steady-state harmonic response.

EXAMPLE 7.14

Develop an algorithm by which the eigenvalue problem, eq. (7.9.32), may be solved for arbitrary boundary conditions.

Solution This example will provide a procedure for determining the eigenvalues for boundary conditions other than simple supports. The problem confronting us is to identify conditions where eq. (7.9.33) is satisfied, when α, β, and k are related by eqs. (7.9.28). As evidenced by Figure 7.17, there is a value of k corresponding to every α, so we select β and k for elimination. For this operation, we limit our interest to the lower frequency branch for the reasons stated previously. The first of eqs. (7.9.28) may be rationalized to a biquadratic equation relating α and k. The solutions to this relation are given in eq. (7.9.36). We use the negative sign for the lower frequency branch, so we have

$$kL(\alpha, L\_d, \sigma) = \frac{1}{\sqrt{2}\sigma}\left\{\left[(L\_d)^2 + (\sigma^2 + 1)\alpha^2\right]\right.$$
$$\left. -\left[(L\_d)^4 + 2(L\_d)^2(\sigma^2 + 1)\alpha^2 + (\sigma^2 - 1)^2\alpha^4\right]^{1/2}\right\}^{1/2}$$

where L_d is used in lieu of L/d, in anticipation of programming steps that follow. To eliminate β we square both of eqs. (7.9.28) and take their difference, which leads to

$$\beta(\alpha, L\_d, \sigma) = \{\alpha^2 - (\sigma^2 + 1)[kL(\alpha, L\_d, \sigma)]^2\}^{1/2}$$

The parameters σ and L_d are properties of the beam, so the preceding expressions enable us to consider kL and β to be functions of α. In turn, that makes the coefficient matrix $[D]$ in eq. (7.9.33) a function of α. The task now is to find the values of α that make the determinant of $[D]$ vanish. In Mathcad one defines the functions for kL and β directly in the worksheet. The same steps are implemented in MATLAB by using function M-files to define each function. Correspondingly, in the definitions of the elements of $[D]$ the parameters kL and β are entered as functions. Also, because sinh (α) and cosh (α) grow rapidly with increasing α, it is a good idea to scale down the determinant by dividing both boundary conditions at $x = L$ by cosh (α).

With $[D]$ now defined as a 4×4 matrix function of α, L_D, and σ, one merely needs to use $|[D]|$ as the argument for the software equation solver. In Mathcad this is readily achieved by writing $root(|D(\alpha, L\_d, \sigma)|, \alpha)$. In MATLAB, we define $[D]$ in an M-file $D\_mat.m$, whose first line is $function f = D\_mat(alpha, L\_d, sigma)$, and we evaluate $|[D]|$ in an M-file $D\_det.m$ whose first line is $function f = D\_det(alpha, L\_d, sigma)$. The statement $fzero('D\_det', alpha, [], [], L\_d, sigma)$ then yields the eigenvalues. Good starting guesses for a numerical solver in either software are the α eigenvalues obtained from classical theory.

To find the coefficients C_n corresponding to each eigenvalue, whichever of C_1 or C_2 is nonzero for the classical beam functions should be set equal to 1. The first three rows of $[D]$, evaluated at an eigenvalue, then lead to three equations for the other coefficients.

EXAMPLE 7.15

Consider a concentrated harmonic force $F\cos(\omega t)$ acting at midspan of a simply supported rectangular beam. Determine the spatial dependence of the amplitudes of displacement $|w|$ and cross-section rotation $|\beta|$ at this frequency. Consider cases where the excitation is close to the fundamental and ninth natural frequency, $\omega = 1.05\omega_1$ and $\omega = 1.05\omega_9$. In both cases $L/h = 10$. Compare these results to those obtained from classical beam theory.

Solution This example demonstrates modal analysis using the mode functions of Timoshenko beam theory. Also, because $L/h = 10$ corresponds to a rather deep beam, the solution will provide an indication of the degree to which classical beam theory is erroneous in its response predictions. The eigenvalues for a simply supported beam are $\alpha_n = n\pi$, and the Timoshenko natural frequencies are $\omega_n = (c_{\text{bar}}/L)(k_a)_n L$, where $(k_a)_n L$ are obtained from eq. (7.9.36) with the negative sign corresponding to the lower branch. The mode functions are defined in eqs. (7.9.37). It is convenient to define the coefficient in ψ_χ as

$$R_n = \frac{\sigma^2[(k_a)_n L]^2}{n\pi} - n\pi$$

To determine the coefficient C_{1n} that normalizes the mode functions, we form eq. (7.9.42) for $j = n$, which requires that

$$C_{1n}^2 \int_0^L \left[\rho A \sin\left(n\pi\frac{x}{L}\right)^2 + \rho I \left(\frac{R_n}{L}\right)^2 \cos\left(n\pi\frac{x}{L}\right)^2 \right] dx = 1$$

This yields

$$C_{1n} = \frac{1}{(\rho AL)^{1/2}} \left[\frac{2}{1 + \left(\frac{d}{L}\right)^2 R_n^2} \right]^{1/2}$$

The stated excitation is $f_z = F\cos(\omega t)\delta(x - L/2)$, so the modal generalized forces obtained from eq. (7.9.45) are

$$Q_n = FC_{1n} \sin\left(\frac{n\pi}{2}\right) \cos(\omega t)$$

Only the odd-numbered terms are nonzero, because the even-numbered modes are antisymmetric relative to $x = L/2$, and therefore have a node at that location. Thus, the modal equations of motion are

$$\ddot{\eta}_n + \omega_n^2 \eta_n = FC_{1n} \sin\left(\frac{n\pi}{2}\right) \cos(\omega t), \quad n \text{ odd}$$

The corresponding steady-state responses are

$$\eta_n = \frac{FC_{1n} \sin\left(\frac{n\pi}{2}\right)}{(\omega_n)^2 - \omega^2} \cos(\omega t), \quad n \text{ odd}$$

When we substitute these solutions into the modal series for w and χ, we obtain

$$w = \sum_{n \text{ odd}} FC_{1n}^2 \frac{\sin\left(\frac{n\pi}{2}\right)}{(\omega_n)^2 - \omega^2} \cos(\omega t) \sin\left(n\pi\frac{x}{L}\right)$$

$$\chi = -\frac{1}{L}\sum_{n \text{ odd}} FC_{1n}^2 R_n \frac{\sin\left(\frac{n\pi}{2}\right)}{(\omega_n)^2 - \omega^2} \cos(\omega t) \cos\left(n\pi\frac{x}{L}\right)$$

It is desirable to write these expressions in a nondimensional form. Toward that end, we recall that $\omega_n = (c_{bar}/L)(k_a)_n L$, so $\omega = 1.05(c_{bar}/L)(k_a)_\ell L$, where $\ell = 1$ or 9. Further, we substitute the expression for C_1 and recall that $c_{bar}^2 = E/\rho$ and $I = Ad^2$. The result is

$$
w = \frac{FL^3}{EI} \sum_{n \text{ odd}} \left[\frac{2}{\left(\frac{L}{d}\right)^2 + R_n^2} \right] \frac{\sin\left(\frac{n\pi}{2}\right)}{[(k_a)_n]^2 - (1.05)^2[(k_a)_\ell]^2}
$$

$$
\times \cos\left[1.05((k_a)_\ell L)\frac{c_{bar}t}{L} \right] \sin\left(n\pi\frac{x}{L} \right)
$$

$$
\chi = -\frac{FL^2}{EI} \sum_{n \text{ odd}} \left[\frac{2R_n}{\left(\frac{L}{d}\right)^2 + R_n^2} \right] \frac{\sin\left(\frac{n\pi}{2}\right)}{\left[(k_a)_n\right]^2 - (1.05)^2\left[(k_a)_\ell\right]^2}
$$

$$
\times \cos\left[1.05((k_a)_\ell L)\frac{c_{bar}t}{L} \right] \cos\left(n\pi\frac{x}{L} \right)
$$

The respective amplitude functions are the quantities that remain when the harmonic time dependence is factored out.

The displacement according to classical theory obtained by the methods in Section 7.7.2 is

$$
w = \frac{FL^3}{EI} \sum_{n \text{ odd}} \frac{\sin\left(\frac{n\pi}{2}\right)}{(n\pi)^4 - (1.05)^2(\ell\pi)^4} \cos\left[1.05(\ell\pi)^2 \frac{c_{bar}d}{L^2} L \right] \sin\left(n\pi\frac{x}{L} \right)
$$

Note that the frequency for this expression is set to be 5% greater than the ℓ th natural frequency obtained from classical theory. The cross-sectional rotation in classical theory is $\partial w/\partial x$, whose series representation may be obtained by differentiating the series for w.

The results for $\ell = 1$ and $\ell = 9$ appear in the graphs. Given the proximity of ω to ω_ℓ, it is not surprising the ℓth mode is the dominant contributor in each case. What is surprising is the high level of agreement between Timoshenko and classical theories. Even at $\ell = 9$, in which case the nodal distance $L/9$ is almost equal to the depth, $h = L/10$, the results agree in terms of their shape and order of magnitude. These results suggest that Timoshenko theory is quite usable for deep beams, and equally importantly, that classical beam theory is quite good for beams that are reasonably slender.

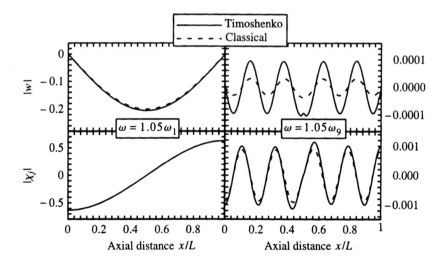

REFERENCES AND SELECTED READINGS

CRAIG, R. R., JR. 1981. *Structural Dynamics.* John Wiley & Sons, New York.

DOYLE, J. F. 1989. *Wave Propagation in Structures.* Springer-Verlag, New York.

GRAFF, K. F. 1975. *Wave Motion in Elastic Solids.* Ohio State University Press, Columbus.

MEIROVITCH, L. 1967. *Analytical Methods in Vibrations.* Macmillan, New York.

MEIROVITCH, L. 1997. *Principles and Techniques of Vibrations.* Prentice Hall, Englewood Cliffs, NJ.

MINDLIN, R. D., & GOODMAN, L. E. 1950. "Beam Vibrations with Time-Dependent Boundary Conditions." *ASME Journal of Applied Mechanics,* 17, 377–380.

NEWLAND, D. E. 1989. *Mechanical Vibration Analysis and Computation.* Longman Scientific and Technical, White Plains, NY.

PESTEL, E. C., & LECKIE, F. A. 1963. *Matrix Methods in Elastomechanics.* McGraw-Hill, New York.

RAYLEIGH, J. W. S. 1945. *Theory of Sound,* Vol. 1. Dover, New York.

SHAMES, I. H., & DYM, C. L. 1985. *Energy and Finite Element Methods in Structural Mechanics.* Hemisphere Publishing, Washington, DC.

TIMOSHENKO, S. P. 1922. "On the Transverse Vibrations of Bars of Uniform Cross-Section." *Philosophical Magazine.* VI, 43, 125–131.

TIMOSHENKO, S. P., & GOODIER, J. N. 1970. *Theory of Elasticity,* McGraw-Hill, New York.

WEAVER, W., TIMOSHENKO, S. P., & YOUNG, D. H. 1990. *Vibration Problems in Engineering.* 5th ed. John Wiley & Sons, New York.

YOUNG, D. 1962. "Continuous Systems." *Handbook of Engineering Mechanics.* W. Flügge, ed. McGraw-Hill, New York.

EXERCISES

7.1 Derive the boundary conditions corresponding to the case where Figure 7.5 corresponds to the end $x = 0$, with x increasing to the left.

7.2 Derive the boundary conditions corresponding to the case where Figure 7.6 depicts the end $x = 0$, with x increasing to the left.

7.3 Consider a beam to which a cube having mass m and side dimensions b is welded to the right end. A spring of stiffness k is attached to the cube. The left end of the bar is clamped. Determine the differential equation and boundary conditions governing flexural motion. Account for both the translational and rotatory inertia of the cube.

7.4 Consider a situation in which the bar depicted in the sketch undergoes both axial and flexural motion. What are the differential equations and boundary conditions to be satisfied by the displacement fields $u(x, t)$ and $w(x, t)$?

7.5 Consider a homogeneous, constant cross-section rod that is not supported at either end; that is, both ends are free. Determine the natural frequencies and mode functions for extensional vibration.

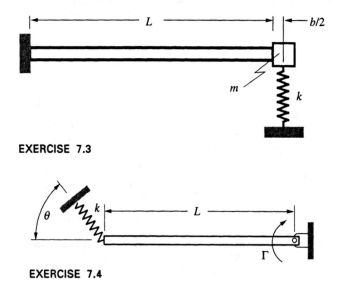

EXERCISE 7.3

EXERCISE 7.4

7.6 A uniform bar is suspended vertically by a spring whose stiffness is $\varepsilon\,EA/L$, where EA is the extensional rigidity of the bar. The lower end of the bar is free.

(a) Derive the characteristic equation and an expression for the extensional mode functions corresponding to an arbitrary value of ε. Solve the equations in the case of natural frequencies that are very high.

(b) Consider the limit where the spring stiffness is very small, $\varepsilon \ll 1$. Identify the limiting behavior of the natural frequencies, and of the fundamental and high-frequency mode functions in this case.

(c) Consider the limit where the spring stiffness is very large, so that $\varepsilon \gg 1$. Derive expressions for the natural frequencies and mode functions in this case. How do these results compare to the limiting behaviors in parts (a) and (b)?

(d) Consider the case of a 2 m long steel rod having a circular cross-section of 25 mm. Determine the lowest two natural frequencies corresponding to $\varepsilon = 0.01$ and $\varepsilon = 1$. Compare these values to the corresponding values in the limits $\varepsilon = 0$ and $\varepsilon \to \infty$, as well as the natural frequency when the system is modeled as a rigid body suspended by a spring of stiffness $\varepsilon\,EA/L$.

7.7 Consider the bar in Example 7.1. Derive the characteristic equation for torsional motion, and an expression for the mode function corresponding to a specific eigenvalue.

7.8 The cross-sectional area of a homogeneous bar increases linearly with axial distance, so that $A = A_1(x/L)$. The bar is unsupported at $x = 0$, while the end $x = L$ is a weld connection to a rigid wall. Derive expressions for the natural frequencies and mode functions of such a bar. (*Hint:* The differential equation for ψ is a specific case of Bessel's equation.)

7.9 Both ends of a uniform bar of length L are clamped. Derive the characteristic equation for flexural vibration and an expression for the mode functions. Evaluate the lowest three natural frequencies, and graph the corresponding mode functions.

7.10 Consider two uniform beams having identical length and cross-sectional properties. Both are supported at their left end by a pin. The right end of one is guided, whereas the right end of the second beam is free. Derive the characteristic equation for the natural frequencies of each system. Evaluate the lowest three nonzero natural frequencies and corresponding mode functions for each beam and compare them graphically. Then prove that the natural frequencies of one configuration are always lower than the corresponding frequencies of the other, and give a physical explanation for that property.

7.11 A torsional spring is attached to the left end of a pinned-pinned beam whose cross-section is constant. The torsional stiffness of the spring is $\kappa EI/L^2$. Derive the characteristic equation and an expression for the mode functions. Identify the behavior of high-frequency modes and identify at which mode such behavior is attained in the case where $\kappa = 20$.

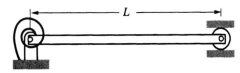

EXERCISE 7.11

7.12 A block whose mass is $\mu\rho AL$ is attached to one end of a uniform bar, where μ is a nondimensional parameter; the other end is clamped. The dimensions of the block are sufficiently small that its centroidal moment of inertia may be neglected. Derive the characteristic equation and an expression for the mode functions. Then examine two limiting behaviors: low natural frequencies with large μ, and high frequencies with small μ. For each limit, identify a simpler system that exhibits similar behavior.

7.13 The block attached to the right end of the pinned-pinned beam in the sketch has mass moment of inertia $I_B = \mu\rho AL^3$ about the right pin, where μ is a nondimensional parameter. Derive

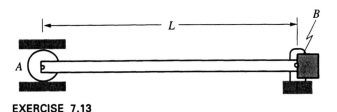

EXERCISE 7.13

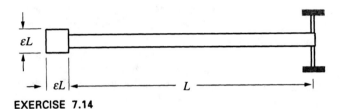

EXERCISE 7.14

the characteristic equation and an expression for the mode functions. Then compare the behavior of the eigensolution for small and large values of μ to the eigensolutions for pinned-pinned and pinned-clamped beams listed in Appendix C.

7.14 A block whose mass is $2\rho AL$ is attached to the left end of a uniform bar whose right end is guided. The block is a cube whose sides are εL, where ε is arbitrary. Derive the characteristic equation for flexural vibration, and an expression for the corresponding mode functions.

7.15 Show that flexural modal displacement of a uniform hinged-guided beam consists of only propagating waves.

7.16 Consider a uniform guided-free bar. Based on $x = 0$ being the guided end and $x = L$ being the free end, derive the characteristic equation and an expression for the mode functions. Then derive from those results a high-frequency asymptotic representation of the mode function. For the case of the second mode, evaluate the asymptotic representation for $0 \le x \le L$ and compare it to the result obtained by direct evaluation of the eigensolution. Then repeat the analysis for the eighth and sixteenth modes.

7.17 Perform the analysis called for in Exercise 7.16 for the case where $x = 0$ is the free end and $x = L$ is the guided end.

7.18 In Exercise 7.9 the characteristic equation of a uniform clamped-clamped beam was solved in order to determine the modal properties. Evaluate

modes 5, 10, and 15 according to that analysis, and compare the results to those obtained from a high-frequency asymptotic approximation.

7.19 The analytical solution for the modes of a free-free beam is described in Appendix C. Evaluate the tabulated entry for modes 5, 10, and 15, and compare the results to those obtained from a high-frequency asymptotic approximation.

7.20 Consider the beam in Example 7.5, in which one end is pinned and the other is supported by a transversely mounted spring. What are the amplitudes of the propagating and evanescent waves forming a high-frequency modal vibration?

7.21 What are the amplitudes of the propagating and evanescent waves forming a high-frequency modal vibration for the beam in Exercise 7.12?

7.22 A bar has identical springs $k = \varepsilon EA/L$ aligned axially at each end. Decompose the axial displacement into symmetric and antisymmetric parts. Derive the characteristic equation and an expression for the mode functions corresponding to each type of motion.

7.23 A bar has identical torsional springs $k = \varepsilon GJ/L$ aligned axially at each end. Decompose the torsional rotation into symmetric and antisymmetric parts. Derive the characteristic equation and an expression for the mode functions corresponding to each type of motion.

7.24 Consider torsional rotation of the circular shaft of radius R, which has identical gears

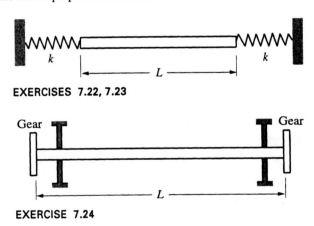

EXERCISES 7.22, 7.23

EXERCISE 7.24

attached at each end. The polar moment of inertia of each gear is $\sigma \rho A L R^2$. Decompose the rotation into symmetric and antisymmetric parts. Derive the characteristic equation and an expression for the mode functions corresponding to each type of motion. For the case where $\sigma = 0.25$, evaluate the modal properties for the first two deformational modes in each symmetry case.

7.25 Consider a free-free beam of length L. Decompose the extensional and flexural vibration into symmetric and antisymmetric modes. Derive the characteristic equation and an expression for the mode functions corresponding to each case: extensional and flexural, symmetric and antisymmetric.

7.26 Consider a beam in which both ends are clamped. Decompose the transverse displacement into symmetric and antisymmetric modes. Determine the lowest two natural frequencies for each type of motion, and graph the corresponding mode functions.

7.27 The collars guiding the ends of the beam in the figure each have mass $m = \varepsilon \rho A L$. Decompose the transverse displacement into symmetric and antisymmetric modes. Derive the characteristic equation and an expression for the mode functions associated with each type of motion.

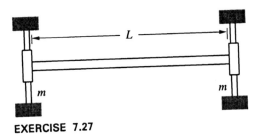

EXERCISE 7.27

7.28 Consider axial motion of the bars in Examples 7.3 and 7.6. What are the orthogonality conditions of the eigenmodes for each system?

7.29 Consider transverse motion of the bars in Example 7.5 and 7.12. What are the orthogonality conditions of the eigenmodes for each system?

7.30 Consider transverse motion of the bar in Exercise 7.14. What are the orthogonality conditions of the eigenmodes for this system?

7.31 A large cubic block of mass m, whose sides are length b, is attached to the left end of a uniform beam that is unsupported. The right end is supported by transverse spring k_S and torsional spring k_T. What are the orthogonality conditions of the eigenmodes for this system?

7.32 Solve Example 7.8 in the case where a spring $k = 5EA/L$, rather than a mass, is attached to the bar at the right end.

7.33 The shaft in Exercise 7.24 initially was undeformed and at rest. A torque $\Gamma h(t)$ is applied at $t = 0$. Use analytical modal analysis to derive an expression for the rotation of a cross-section. Evaluate and plot the rotations of the left and right ends as functions of time.

7.34 Use analytical modal analysis to solve Exercise 6.34.

7.35 Solve Example 7.9 in the case where the beam is simply supported.

7.36 A simply supported beam that is initially at rest is loaded by a concentrated transverse force $F \sin(\Omega t)$. This force is translating to the right at speed v, such that the point at which it is applied is at distance vt from the left end. Determine the resulting displacement, assuming that damping is negligible. Are there any combinations of v and Ω that should be avoided?

7.37 Consider a cantilevered beam that is initially at rest under a static transverse force F acting at its free end. At $t = 0$ this force is removed. Damping is negligible. Determine the oscillation of the beam. Plot the transverse displacement at the free end as a function of time.

7.38 Consider a simply supported beam. A block of mass $\rho A L$ is suspended from the middle by a cable. At $t = 0$, with the beam at rest, the cable is cut. Determine the resulting oscillation of the beam if the ratio of critical damping for each mode is $\zeta_j = 0.01$. Plot the transverse displacement at midspan as a function of time.

7.39 A uniform bar translating axially to the left at speed v strikes a wall at $t = 0$. If the wall is considered to be rigid, the axial displacement at the left end of the bar must be zero, until such time

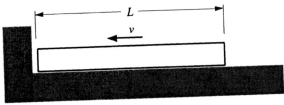

EXERCISE 7.39

that the bar begins to rebound. Derive an expression for the axial displacement field in the time interval during which there is contact between the bar and the wall, so that the displacement at the contact point is zero. How can this result be used to determine the rebound time? Also, prove that at the instant the bar rebounds, the bar is executing a translation to the right at speed v. (*Hint:* The modal series for the axial strain at the contact point corresponds to a recognizable function of time.)

7.40 A uniform bar pivoted at its left end and unsupported at its right end is rotating about its pivot clockwise at angular speed Ω when, at $t = 0$, an impulsive force $G\delta(t)$ is applied at the midpoint. Derive an expression for the ensuing transverse displacement.

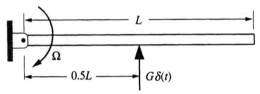

EXERCISE 7.40

7.41 A distributed load is applied to a clamped-guided beam. The load varies harmonically in time with an amplitude that is constant in x, so that $f_z = f_0 \sin(\omega t)$. Determine the steady-state displacement at the midpoint, $x = L/2$, when ω is 1% greater than the lowest natural frequency of a deformational mode. Give the answer in terms of magnitude and phase lag relative to a sine function for cases where damping is negligible, and where the structural damping loss factor is $\gamma = 0.01$.

7.42 The beam shown below is supported at its right end by a hinge, where a harmonically varying couple $\Gamma = \Gamma_0 \sin(\omega t)$ is applied. The spring stiffness is $k = 36EI/L^3$. The system is constructed such that the bar is horizontal when it is in static equilibrium. Derive a modal series for the displacement. Evaluate the steady-state displacement amplitude as a function of frequency at the loca-

tion of the spring and at midspan for the case of a steel beam having a 200 mm square cross-section with $L = 4$ m. Consider $0 < \omega < 1.2\omega_3$.

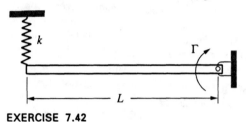

EXERCISE 7.42

7.43 A simply supported beam is initially at rest in its static equilibrium position. At $t = 0$, the right end is given a constant upward acceleration a_0. Derive an expression for the ensuing flexural displacement. Evaluate and graph this expression at $x = L/4$, $L/2$, and $3L/4$ for the interval $0 \le t \le 4\pi/\omega_1$, where ω_1 is the fundamental flexural natural frequency.

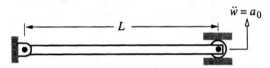

EXERCISE 7.43

7.44 Use analytically derived mode functions to solve Example 6.13.

7.45 A known harmonically varying axial displacement $B \sin(\omega t)$ is imposed on the left end by an unknown force $F(t)$ acting at that end. The right end of the bar is attached to an axial spring whose stiffness is $k = EA/L$. The system was initially at rest in the undeformed position at $t = 0$. Derive

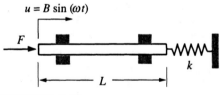

EXERCISE 7.45

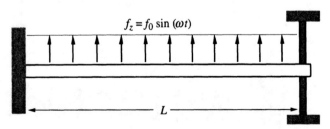

EXERCISE 7.41

expressions for the ensuing displacement field and the force F.

7.46 Solve Exercise 7.45 in the case where the spring is replaced by a block whose mass is $m = \rho AL$.

7.47 Consider eq. (6.5.17) for the γ_{jn} coefficients representing the elasticity effects associated with w_{bc}. Suppose that the Ritz basis functions ψ_j are the mode functions Ψ_j for a beam with clamped supports, and that the expansion functions g_n satisfy $d^4 g_n/dx^4 = 0$, as imposed in the Mindlin-Goodman formulation. Prove that in such circumstances, $\gamma_{jn} \equiv 0$.

7.48 A seismic shock wave propagating through the ground imparts a vertical displacement field $\chi(x, t) = f(t - x/c)$, where c is the propagation speed. When such a wave arrives at the clamped end of a horizontal cantilevered beam, the effect is to impart a vertical displacement to the end, and also to cause the end to rotate, such that

$$w\big|_{x=0} = \chi(0, t) = f(t)$$

$$\frac{\partial w}{\partial x}\bigg|_{x=0} = \frac{\partial}{\partial x}\chi(x, t)\bigg|_{x=0} = -\frac{1}{c}\dot{f}(t)$$

Suppose the wave constitutes a constant acceleration, such that $f(t) = \frac{1}{2}A_0 t^2$, and the beam was initially at rest when $t = 0$. Formulate an analysis whose completion would yield the displacement field $w(x, t)$ resulting from this excitation. Specify as completely as possible all equations required to perform the analysis. There is no need to actually solve these equations.

7.49 A uniform bar is pinned at its left end, and a small mass m is attached to its unsupported right end. Motion occurs in the horizontal plane. The bar is initially at rest in an undeformed position. It is desired to apply a torque $\Gamma(t)$ that will impart a constant angular acceleration $\ddot{\theta}$ to the pinned end.

(a) Determine the term w_{bc} accounting for the time-dependent boundary condition.

(b) Outline an analysis leading to expressions for the required moment and the corresponding transverse displacement field. Be as specific as possible.

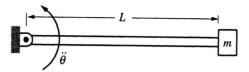

EXERCISE 7.49

7.50 A harmonically varying axial displacement $B \sin(\omega t)$ is imposed on the right end by an unknown force $F(t)$ acting at that end. The left end of the bar is attached to block $m = 0.5\rho AL$ and a dashpot $c = (\rho A^2 E)^{1/2}$. Derive expressions for the steady-state axial displacement field and the force F.

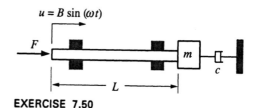

EXERCISE 7.50

7.51 A mass m is suspended from the ceiling by a spring K and dashpot c acting in parallel. A uniform bar is attached vertically to the mass. Dissipation within the bar is negligible. It is desired to apply an axial harmonic force at the lower end of the bar that produces an axial displacement $u_1 \sin(\omega t)$ at that end. Derive an expression for the steady-state force at the lower end required to attain this condition. Also derive an expression for the steady-state displacement at the upper end of the bar.

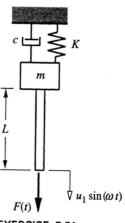

EXERCISE 7.51

7.52 A dashpot and block are attached at the end of the cantilevered beam, as shown on the next page. The clamped end of the beam is given a vertical displacement $w = 2 \sin(\omega t)$ mm. The beam is composed of steel, and its cross-section is a 20 mm square. The mass of the block is one-quarter of the mass of the beam, and the dashpot constant is $c = 1$ kN-s/m. Determine the amplitude of the displacement of the mass as a function of ω for the range $0 < \omega < 1000$ rad/s.

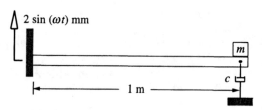

2 sin (ωt) mm

m

c

1 m

EXERCISE 7.52

7.53 Solve Exercise 7.52 for the case where the mass and dashpot are located at the midspan location.

7.54 The beam is supported by a spring-dashpot suspension at its left end and midpoint. A shaker pinned to the right end imparts a harmonic vertical displacement at that end given by 5 cos (ωt) mm. The bar is composed of steel, $E = 210(10^9)$ Pa, $\rho = 7800$ kg/m³, and material dissipation is negligible. The suspension parameters are $K = 100$ kN/m and $c = 2$ kN-s/m. The beam's cross-section is rectangular, with a height of 100 mm and a width of 50 mm. Derive an expression for the displacement amplitudes at the attachment points of the suspensions as a function of ω. Graph these amplitudes in the interval from zero to 1000 Hz.

7.55 A 100 N harmonic force is applied axially to a bar at its midpoint. One end of the bar is fixed and the other is free. The bar is composed of a vis-

coelastic polymer whose frequency-dependent Young's modulus is

$$E(\omega) = E_0 + (E_\infty - E_0)\frac{\omega^2\tau^2}{\omega^2\tau^2 + 1}\left(1 + \frac{i}{\omega\tau}\right)$$

In this expression E_0 is the static Young's modulus, corresponding to very low values of the frequency ω; E_∞ is the Young's modulus for very large ω, and $\tau > 0$ is the relaxation time. System parameters are $E_0 = 8(10^7)$ Pa, $E_\infty = 12(10^7)$ Pa, $\tau = 1.5$ ms, $\rho = 2000$ kg/m³, $A = 0.001$ m², and $L = 1$ m. Determine the axial displacement at the midpoint in a frequency range from 0 to 150 Hz.

7.56 Consider flexural motion of a bar whose material has a nonzero viscoelastic loss factor, such that $E = \hat{E}(\omega)[1 + i\gamma(\omega)]$. Derive the generalization of eq. (7.8.32). (*Hint:* The principal root of $(1 + i\gamma)^{1/4}$ has positive real and imaginary parts.)

7.57 Determine the boundary conditions for Timoshenko beam theory corresponding to the systems in the sketch.

7.58 Determine the boundary conditions for Timoshenko theory corresponding to the beam in the sketch.

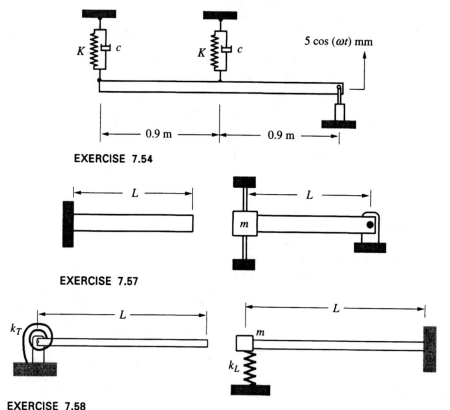

K c K c 5 cos (ωt) mm

0.9 m 0.9 m

EXERCISE 7.54

L m L

EXERCISE 7.57

k_T L m L k_L

EXERCISE 7.58

7.59 Identify Ritz series basis functions suitable for using Timoshenko theory to analyze the beams in Exercise 7.57.

7.60 Identify Ritz series basis functions suitable for using Timoshenko theory to analyze the beams in Exercise 7.58.

7.61 By taking appropriate limits it is possible to recover classical beam theory from eqs. (7.9.9). First show that letting $\kappa G \to \infty$ leads to $Y \to 0$. Then perform operations on both equations to eliminate terms containing $\partial w/\partial x - \chi$, and use the limit $Y \to 0$ to derive a single differential equation for w. As the last step, identify what other limit is necessary to recover the standard equation of motion, $EI\, \partial^4 w/\partial x^4 + \rho A \ddot{w} = f_z$.

7.62 Consider a simply supported beam having a rectangular cross-section with a span to depth ratio $L/h = 10$. At what mode number does the natural frequency for the lower branch exceed the natural frequency for $n = 1$ on the upper branch?

7.63 Use the algorithm developed in Example 7.14 to determine the lowest six natural frequencies for a cantilever beam whose cross-section is rectangular. The beam's depth is $h = 0.10L$. Compare these values to results obtained from classical theory.

7.64 Use the algorithm developed in Example 7.14 to determine the lowest six natural frequencies for a clamped-hinged beam whose cross-section is rectangular. The beam's depth is $h = 0.10L$. Compare these values to results obtained from classical theory.

7.65 A simply supported rectangular beam is loaded by a concentrated harmonic force at $x = L/4$. The depth of the beam is $h = L/12$, and the structural damping loss factor is 0.0001. Determine the steady-state amplitude of the nondimensional displacement EIw/FL^3 at midspan for excitation frequencies in the range $0 < \omega < 1.2\omega_6$, where ω_6 is the sixth natural frequency. Then compare that result to the displacement obtained from classical beam theory.

7.66 A concentrated force that varies in time as a step function, $F = F_0 h(t)$, is applied at the $L/4$ point of a simply supported beam that initially is at rest in its static equilibrium position. The beam's depth is $h = 0.10L$, and dissipation is negligible. Determine the vertical displacement as a function of time over the interval $0 \le t \le 4\pi/\omega_1$ at $x = L/4, L/2$, and $3L/4$. Compare the result to the displacement obtained from classical beam theory.

CHAPTER 8

THE FINITE ELEMENT METHOD

Ritz series and field equation formulations are powerful tools for evaluating the response of continuous systems, but they are difficult to implement when a system's configuration is irregular or intricate. There are two philosophies as to how structural complexity may be addressed. One, which is the scope of this chapter, decomposes a system in many small regular pieces. The idea is to obtain equations of motion for the individual pieces, and then assemble the system by fitting the pieces together. This is the basic notion underlying the finite element method (FEM, in common parlance). Finite difference methods also fit this description. However, their application in vibration studies is somewhat specialized, so we will not address them. The other philosophy for irregular systems decomposes a system into large individual components, whose equations of motion we presumably already have determined. For example, we might envision a structural framework or mechanical linkage as an assembly of beams. We will examine such a viewpoint in the next chapter. In practice, it often is useful to implement both philosophies.

Another difference between FEM and a Ritz series analysis is that the generalized coordinates in FEM are actual displacement variables for selected points, rather than amplitudes of basis functions covering the entire system, whose physical significance is more abstract. Nevertheless, the Ritz series method provides the technical foundation for finite elements. We will see that the essential difference is the manner in which the basis functions are defined.

8.1 INTERPOLATING FUNCTIONS

Many texts have been written explaining the derivation and application of FEM. Our treatment will adapt the general presentation provided by Shames and Dym (1985) to the special case of planar systems of interconnected bars undergoing axial and flexural displacement. The basic operations are typical of those required to treat any type of displacement field.

To set the stage for the technical development, let us consider a straight bar. We wish to approximate its displacement as a function of axial position by giving displacement values at a set of points whose location we denote as x_j. Let us consider these locations, which are called *mesh points* or *nodes*, to define segments of the bar. Each of these segments is a *finite element*. The overall procedure develops rules for a generic element, after which the finite elements are assembled to form the system. For simplicity in considering a specific element, we shall number the left and right mesh points as 1 and 2, respectively, and let L denote the length of the element.

We wish to fit the elements together in a manner that satisfies all geometric continuity conditions. Specifically, across the junction of elements, the bar's axial and transverse displacement must be continuous, as must the cross-sectional rotation. These continuity conditions will be met if we define as generalized coordinates the

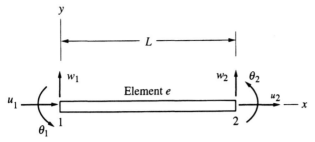

FIGURE 8.1 Mesh point displacements and rotations for a finite element.

displacement components and rotation at the mesh point where an element adjoins other elements. For a case of planar motion these variables are u_j, w_j, and θ_j at mesh point j, which we use to form the *mesh point displacement* vector,

$$\{q_j\} = [u_j \ w_j \ \theta_j]^T, \quad j = 1, 2 \tag{8.1.1}$$

The displacement variables for each mesh point and the coordinate system for an element are depicted in Figure 8.1.

Suppose we have determined the instantaneous values of the mesh point variables and we wish to draw a curve showing the dependence of u on x. If we connect the mesh point values u_j with a straight line, we would be using a linear interpolating function. Over one finite element (the segment between two mesh points), the curve u versus x would be a trapezoid, as shown in Figure 8.2(a). Figure 8.2(b) decomposes this trapezoid into two triangles, each of whose height is the value of u at the respective mesh points. Thus, the linear interpolating function is equivalent to representing the dependence of u over the element by a Ritz series in the form

$$u(x) = u_1 \psi_{u1}(x) + u_2 \psi_{u2}(x)$$

$$\psi_{u1}(x) = \frac{x}{L}, \qquad \psi_{u2}(x) = 1 - \frac{x}{L} \tag{8.1.2}$$

A Ritz series in this form uses basis functions that express the contribution of each mesh point. The basis functions for finite elements are referred to as *shape functions*.

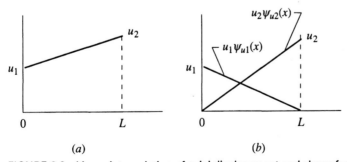

(a)   (b)

FIGURE 8.2 Linear interpolation of axial displacement and shape functions.

We now wish to identify shape functions for transverse displacement. Toward that end we note that there are two displacement variables that must be matched at each of the two mesh points, specifically

$$w(x = 0) = w_1 \qquad \frac{\partial w}{\partial x}(x = 0) = \theta_1$$

$$w(x = L) = w_2 \qquad \frac{\partial w}{\partial x}(x = L) = \theta_2$$

(8.1.3)

There are four conditions to satisfy, so we employ a cubic interpolating polynomial, which contains four coefficients,

$$w(x) = c_0 + c_1\frac{x}{L} + c_2\left(\frac{x}{L}\right)^2 + c_3\left(\frac{x}{L}\right)^3, \quad 0 \le x \le L$$

(8.1.4)

This function meets the criteria in eqs. (8.1.3) if

$$c_0 = w_1, \quad c_1 = L\theta_1, \quad c_2 = -3w_1 + 3w_2 - 2L\theta_1 - L\theta_2$$

$$c_3 = 2w_1 - 2w_2 + L\theta_1 + L\theta_2$$

(8.1.5)

The result of collecting the factors of each displacement variable, w_1, θ_1, w_2, θ_2, when the coefficients are used to form $w(x)$ is

$$w(x) = w_1\psi_{w1}(x) + \theta_1\psi_{w2}(x) + w_2\psi_{w3}(x) + \theta_2\psi_{w4}(x)$$

(8.1.6)

where the shape functions are

$$\psi_{w1}(x) = 1 - 3\frac{x^2}{L^2} + 2\frac{x^3}{L^3}, \quad \psi_{w2}(x) = L\left(\frac{x}{L} - 2\frac{x^2}{L^2} + \frac{x^3}{L^3}\right)$$

$$\psi_{w3}(x) = 3\frac{x^2}{L^2} - 2\frac{x^3}{L^3}, \quad \psi_{w4}(x) = L\left(-\frac{x^2}{L^2} + \frac{x^3}{L^3}\right)$$

(8.1.7)

Figure 8.3 depicts the displacement superposition in eq. (8.1.6). These interpolating functions, as well as those for axial displacement, are the static displacement a bar would undergo if the associated generalized coordinate were the only nonzero variable.

It is useful to describe both displacement components within the element in a single expression. Toward that goal we define an *elemental displacement vector*, which holds the mesh variables for the entire element. We use a superscript index e ranging from one to the number of elements to indicate which element is under consideration, so that

$$\boxed{\{q^e\} = [u_1 \ w_1 \ \theta_1 \ u_2 \ w_2 \ \theta_2]^T = \left[\{q_1\}^T \{q_2\}^T\right]^T}$$

(8.1.8)

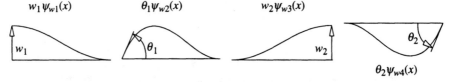

FIGURE 8.3 Cubic interpolating functions for flexural displacement.

The two displacement fields within the element are related to the element's mesh displacements by

$$\begin{Bmatrix} u(x, t) \\ w(x, t) \end{Bmatrix} = [N(x)]\{q^e\} \tag{8.1.9}$$

where $[N(x)]$ is formed from the shape functions,

$$[N] = \begin{bmatrix} \psi_{u1} & 0 & 0 & \psi_{u2} & 0 & 0 \\ 0 & \psi_{w1} & \psi_{w2} & 0 & \psi_{w3} & \psi_{w4} \end{bmatrix} \tag{8.1.10}$$

In essence, eq. (8.1.9) constitutes a Ritz series, with the shape functions serving as basis functions and the mesh displacement variables serving as the generalized coordinates. Thus, rather than using a single Ritz series to describe the response over the entire system, the finite element method uses different Ritz series for each element. Just as there may be several basis functions that are useful for a conventional Ritz series, shape functions other than the linear and cubic polynomials developed here are possible. Much work has gone into developing improved schemes. For example, mesh points interior to the element can be defined. Matching the mesh point displacements to interpolating functions then leads to higher order shape functions. From an overall perspective the development of such higher order schemes follows the general procedures we shall employ here.

8.2 ELEMENTAL MATRICES

The next step is to construct the mechanical energies and power expressions associated with the elemental displacement. For the purpose of demonstrating the finite element method, we shall focus on the bar itself. Attached rigid masses, springs, and dashpots would then be included as add-ons. Some finite element codes restrict cross-sectional properties to be constants. Instead, we shall use the shape functions to represent the axial dependence of the cross-sectional mass and stiffness parameters in terms of the respective values at the mesh points, according to

$$\begin{aligned} \rho A(x) &= (\rho A)_1 \psi_{u1} + (\rho A)_2 \psi_{u2} \\ EA(x) &= (EA)_1 \psi_{u1} + (EA)_2 \psi_{u2} \\ EI(x) &= (EI)_1 \psi_{u1} + (EI)_2 \psi_{u2} \end{aligned} \tag{8.2.1}$$

The kinetic energy of the element is obtained by using eq. (8.1.9) to characterize the velocity components, so that

$$\begin{aligned} T &= \frac{1}{2} \int_0^L \rho A(x)[\dot{u}^2 + \dot{w}^2]\, dx \\ &= \frac{1}{2} \int_0^L [(\rho A)_1 \psi_{u1}(\rho A)_2 \psi_{u2}][\dot{u}\ \dot{w}] \begin{Bmatrix} \dot{u} \\ \dot{w} \end{Bmatrix} dx \\ &= \frac{1}{2} \sum_{j=1}^2 (\rho A)_j \int_0^L \psi_{uj}\{\dot{q}^e\}^T [N]^T [N]\{\dot{q}^e\}\, dx \end{aligned} \tag{8.2.2}$$

The standard representation of kinetic energy is

$$T = \tfrac{1}{2}\{\dot{q}^e\}^T[M^e]\{\dot{q}^e\} \tag{8.2.3}$$

Comparison of these alternative descriptions of T leads to the *elemental inertia matrix*,

$$\boxed{[M^e] = \sum_{j=1}^{2} (\rho A)_j \int_0^L \psi_{uj}[N]^T[N]\,dx} \tag{8.2.4}$$

In the special case where the cross-section is constant over the element, the second of eqs. (8.12) indicates that $\psi_{u1} + \psi_{u2} = 1$, so that

$$\boxed{[M^e] = \rho A \int_0^L [N]^T[N]\,dx} \tag{8.2.5}$$

The description of potential energy proceeds in a similar fashion. We use an operator matrix $[D]$ to indicate the operations required to obtain the axial and bending strains from the displacement field, such that

$$\begin{Bmatrix} \partial u/\partial x \\ \partial^2 w/\partial x^2 \end{Bmatrix} = [D]\begin{Bmatrix} u(x,t) \\ w(x,t) \end{Bmatrix} = [D][N]\{q^e\}, \quad [D] = \begin{bmatrix} \partial/\partial x & 0 \\ 0 & \partial^2 \partial x^2 \end{bmatrix} \tag{8.2.6}$$

The sum of the extensional and flexural strain energies within the element is given by

$$V = \frac{1}{2}\int_0^L \left[EA(x)\left(\frac{\partial u}{\partial x}\right)^2 + EI(x)\left(\frac{\partial^2 w}{\partial x^2}\right)^2 \right] dx \tag{8.2.7}$$

The spatial variation of EA and EI is described by eqs. (8.2.1), and eq. (8.2.6) describes the strain field, from which we find that

$$V = \frac{1}{2}\int_0^L \{[D][N]\{q^e\}\}^T \begin{bmatrix} \displaystyle\sum_{j=1}^{2}\psi_{uj}(EI)_j & 0 \\ 0 & \displaystyle\sum_{j=1}^{2}\psi_{uj}(EI)_j \end{bmatrix} [D][N]\{q^e\}\,dx$$

$$= \frac{1}{2}\sum_{j=1}^{2}\int_0^L \psi_{uj}\{q^e\}^T[N]^T[D]^T \begin{bmatrix} (EA)_j & 0 \\ 0 & (EI)_j \end{bmatrix} [D][N]\{q^e\} \tag{8.2.8}$$

The standard form of V for an element is

$$V = \frac{1}{2}\{q^e\}^T[K^e]\{q^e\} \tag{8.2.9}$$

Matching the last two expressions leads to the *elemental stiffness matrix*,

$$[K^e] = \sum_{j=1}^{2} \int_0^L \psi_{uj} [[D][N]]^T \begin{bmatrix} (EA)_j & 0 \\ 0 & (EI)_j \end{bmatrix} [[D][N]]\, dx \tag{8.2.10}$$

If the cross-section is uniform along the element, this reduces to

$$[K^e] = \int_0^L [N]^T [D]^T \begin{bmatrix} EA & 0 \\ 0 & EI \end{bmatrix} [D][N]\, dx \tag{8.2.11}$$

It should be noted that the expressions for both $[M^e]$ and $[K^e]$ yield symmetric matrices. Formulas for these quantities in the case where the cross-sectional properties are constant are derived in Example 8.1.

Internal material damping may be accounted for with a structural damping loss factor, so we shall take $[C^e] = [0]$. Power may be input to the element by concentrated forces and couples, whose contribution we will consider later, and by distributed forces. To represent the distributed forces in terms of values at the mesh points we use the shape functions, in the manner of eq. (8.1.9),

$$\begin{Bmatrix} f_x(x) \\ f_z(x) \end{Bmatrix} = [N]\{f^e\} \tag{8.2.12}$$

where $\{f^e\}$ holds the distributed force values at the mesh points, arranged in correspondence to the definition of $\{q^e\}$, that is,

$$\{f^e\} = \begin{bmatrix} (f_x)_1 & (f_z)_1 & 0 & (f_x)_2 & (f_z)_2 & 0 \end{bmatrix}^T \tag{8.2.13}$$

The power input by the distributed load is then

$$(\mathcal{P}_{in})_{dist} = \int_0^L [f_x \dot{u} + f_z \dot{w}]\, dx = \int_0^L [\dot{u}\ \dot{w}]^T \begin{Bmatrix} f_x(x) \\ f_z(x) \end{Bmatrix} dx$$

$$\tag{8.2.14}$$

$$= \{\dot{q}^e\}^T \int_0^L [N]^T[N]\{f^e\}\, dx = \{\dot{q}^e\}^T [\gamma^e]\{f^e\}$$

where $[\gamma^e]$ is a matrix of weighting factors expressing the mesh point contributions of the distributed forces,

$$[\gamma^e] = \int_0^L [N]^T[N]\, dx \tag{8.2.15}$$

In situations where the bar's cross-sectional properties do not vary over the finite element, a comparison of the preceding with eq. (8.2.5) reveals that $[\gamma^e] = (1/\rho A)[M^e]$.

EXAMPLE 8.1

Consider the finite element in Figure 8.1 when the shape functions are those in eq. (8.1.7). Derive the elemental inertia and stiffness matrices for situations where the cross-sectional properties of the bar are constant.

Solution This example, which demonstrates the evaluation of elemental matrices, will provide fundamental values for further developments. Because the cross-sectional properties are independent of x, we evaluate the elemental matrices according to eqs. (8.2.5) and (8.2.11). For the purpose of evaluating the integrals, it is convenient to consider the shape functions to depend on the nondimensional distance $\xi = x/L$. We use a prime to denote differentiation with respect to ξ, so that

$$\frac{d\psi_{uj}}{dx} = \frac{1}{L}\psi'_{uj}, \qquad \frac{d^2\psi_{uj}}{dx^2} = \frac{1}{L^2}\psi''_{uj}, \ \dots$$

In view of eqs. (8.1.10) and (8.2.6), we have

$$[N]^{\mathrm{T}}[N] = \begin{bmatrix} \psi_{u1}^2 & 0 & 0 & \psi_{u1}\psi_{u2} & 0 & 0 \\ 0 & \psi_{w1}^2 & \psi_{w1}\psi_{w2} & 0 & \psi_{w1}\psi_{w3} & \psi_{w1}\psi_{w4} \\ 0 & \psi_{w1}\psi_{w2} & \psi_{w2}^2 & 0 & \psi_{w2}\psi_{w3} & \psi_{w2}\psi_{w4} \\ \psi_{u1}\psi_{u2} & 0 & 0 & \psi_{u2}^2 & 0 & 0 \\ 0 & \psi_{w1}\psi_{w3} & \psi_{w2}\psi_{w3} & 0 & \psi_{w3}^2 & \psi_{w3}\psi_{w4} \\ 0 & \psi_{w1}\psi_{w4} & \psi_{w2}\psi_{w4} & 0 & \psi_{w3}\psi_{w4} & \psi_{w4}^2 \end{bmatrix}$$

$$[D][N] = \begin{bmatrix} \dfrac{1}{L}\psi'_{u1} & 0 & 0 & \dfrac{1}{L}\psi'_{u2} & 0 & 0 \\ 0 & \dfrac{1}{L^2}\psi''_{w1} & \dfrac{1}{L^2}\psi''_{w2} & 0 & \dfrac{1}{L^2}\psi''_{w3} & \dfrac{1}{L^2}\psi''_{w4} \end{bmatrix}$$

$$[N]^{\mathrm{T}}[D]^{\mathrm{T}}\begin{bmatrix} EA & 0 \\ 0 & EI \end{bmatrix}[D][N] =$$

$$\begin{bmatrix} \dfrac{EA}{L^2}(\psi'_{u1})^2 & 0 & 0 & \dfrac{EA}{L^2}\psi'_{u1}\psi'_{u2} & 0 & 0 \\ 0 & \dfrac{EI}{L^4}(\psi''_{w1})^2 & \dfrac{EI}{L^4}\psi''_{w1}\,\psi''_{w2} & 0 & \dfrac{EI}{L^4}\psi''_{w1}\,\psi''_{w3} & \dfrac{EI}{L^4}\psi''_{w1}\psi''_{w4} \\ 0 & \dfrac{EI}{L^4}\psi''_{w1}\psi''_{w2} & \dfrac{EI}{L^4}(\psi''_{w2})^2 & 0 & \dfrac{EI}{L^4}\psi''_{w2}\,\psi''_{w3} & \dfrac{EI}{L^4}\psi''_{w2}\psi''_{w4} \\ \dfrac{EA}{L^2}\psi'_{u1}\psi'_{u2} & 0 & 0 & \dfrac{EA}{L^2}(\psi'_{u2})^2 & 0 & 0 \\ 0 & \dfrac{EI}{L^4}\psi''_{w1}\psi''_{w3} & \dfrac{EI}{L^4}\psi''_{w2}\psi''_{w3} & 0 & \dfrac{EI}{L^4}(\psi''_{w3})^2 & \dfrac{EI}{L^4}\psi''_{w3}\psi''_{w4} \\ 0 & \dfrac{EI}{L^4}\psi''_{w1}\psi''_{w4} & \dfrac{EI}{L^4}\psi''_{w2}\psi''_{w4} & 0 & \dfrac{EI}{L^4}\psi''_{w3}\,\psi''_{w4} & \dfrac{EI}{L^4}(\psi''_{w4})^2 \end{bmatrix}$$

As a result of changing variables, the integral of any function $f(x)$ over the interval $0 \le x \le L$ becomes

$$\int_0^L f(x)\,dx = L\int_0^1 f(\xi)\,d\xi$$

We use this property to integrate each term in the preceding matrix products. The resulting elemental inertia matrix is

$$[M^e] = \rho A L \begin{bmatrix} \dfrac{1}{3} & 0 & 0 & \dfrac{1}{6} & 0 & 0 \\[2mm] 0 & \dfrac{13}{35} & \dfrac{11}{210}L & 0 & \dfrac{9}{70} & -\dfrac{13}{420}L \\[2mm] 0 & \dfrac{11}{210}L & \dfrac{1}{105}L^2 & 0 & \dfrac{13}{420}L & -\dfrac{1}{140}L^2 \\[2mm] \dfrac{1}{6} & 0 & 0 & \dfrac{1}{3} & 0 & 0 \\[2mm] 0 & \dfrac{9}{70} & \dfrac{13}{420}L & 0 & \dfrac{13}{35} & -\dfrac{11}{210}L \\[2mm] 0 & -\dfrac{13}{420}L & -\dfrac{1}{140}L^2 & 0 & -\dfrac{11}{210}L & \dfrac{1}{105}L^2 \end{bmatrix}$$

Comparable operations applied to the terms for $[K^e]$ leads to

$$[K^e] = \begin{bmatrix} \dfrac{EA}{L} & 0 & 0 & -\dfrac{EA}{L} & 0 & 0 \\[2mm] 0 & 12\dfrac{EI}{L^3} & 6\dfrac{EI}{L^2} & 0 & -12\dfrac{EI}{L^3} & 6\dfrac{EI}{L^3} \\[2mm] 0 & 6\dfrac{EI}{L^2} & 4\dfrac{EI}{L} & 0 & -6\dfrac{EI}{L^3} & 2\dfrac{EI}{L} \\[2mm] -\dfrac{EA}{L} & 0 & 0 & \dfrac{EA}{L} & 0 & 0 \\[2mm] 0 & -12\dfrac{EI}{L^3} & -6\dfrac{EI}{L^3} & 0 & 12\dfrac{EI}{L^3} & -6\dfrac{EI}{L^2} \\[2mm] 0 & 6\dfrac{EI}{L^3} & 2\dfrac{EI}{L} & 0 & -6\dfrac{EI}{L^2} & 4\dfrac{EI}{L} \end{bmatrix}$$

In subsequent exercises we shall use the above expressions for $[M^e]$ and $[K^e]$ as standard formulas that depend solely on the cross-sectional properties and the length L of the element. A finite element program would have these values stored in a library of elements associated with different types of structures.

8.3 CONVERSION TO GLOBAL COORDINATES

According to the previous section, the properties of a single element are defined by the inertia matrix $[M^e]$, the stiffness matrix $[K^e]$, and the distributed force coefficients $[\gamma^e]$. The task now is to combine these terms for the various elements forming the system of interest. The combination process must account for two fundamental aspects associated with the fact that some mesh points are shared between elements. The first issue is that unless the elements are oriented in the same direction, the sense of their displacement components will differ. In addition, it is likely that the mesh points where the elements join will be numbered differently.

To address the differing orientations we define a *global coordinate system,* which we have labeled *XYZ* in Figure 8.4. In this terminology, *xyz* for an element is a *local coordinate system.* We wish to convert all displacement components to the global coordinate system. We define β_e as the angle between the X axis and the axial direction for the element, with positive β_e defined by the right hand rule relative to the Z axis. We denote displacement components relative to the global system with a caret. Projection of the global displacement components $\hat{u}_1$ and $\hat{w}_1$ onto the local coordinate axes gives

$$u_1 = \hat{u}_1 \cos\beta_e + \hat{w}_1 \sin\beta_e, \quad w_1 = -\hat{u}_1 \sin\beta_e + \hat{w}_1 \cos\beta_e \tag{8.3.1}$$

In addition, the z and Z axes are parallel, so no transformation is required for rotation. Similar relations obviously apply for the second mesh point of element e. Thus, we find that the local and global mesh variables for element e are related by

$$\begin{Bmatrix} u_1 \\ w_1 \\ \theta_1 \\ u_2 \\ w_2 \\ \theta_2 \end{Bmatrix} = [R^e] \begin{Bmatrix} \hat{u}_1 \\ \hat{w}_1 \\ \hat{\theta}_1 \\ \hat{u}_2 \\ \hat{w}_2 \\ \hat{\theta}_2 \end{Bmatrix} \tag{8.3.2}$$

A shorthand description of this relation is

$$\boxed{\{q^e\} = [R^e]\{\hat{q}^e\}} \tag{8.3.3}$$

where $\{\hat{q}^e\}$ is the global generalized coordinate vector for element e. The matrix $[R^e]$ is the *rotation transformation,*

$$[R^e] = \begin{bmatrix} \cos\beta_e & \sin\beta_e & 0 & 0 & 0 & 0 \\ -\sin\beta_e & \cos\beta_e & 0 & 0 & 0 & 0 \\ 0 & 0 & 1 & 0 & 0 & 0 \\ 0 & 0 & 0 & \cos\beta_e & \sin\beta_e & 0 \\ 0 & 0 & 0 & -\sin\beta_e & \cos\beta_e & 0 \\ 0 & 0 & 0 & 0 & 0 & 1 \end{bmatrix} \tag{8.3.4}$$

It is readily verified that $[R^e]^{-1} = [R^e]^T$, which is the *orthonormal property.* To employ the preceding definition it is important to note that β_e is measured counterclockwise from the global X axis to the local x axis, and the latter is oriented from local node 1 to local node 2.

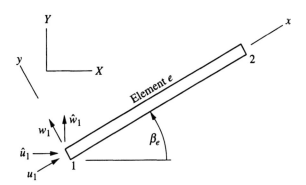

FIGURE 8.4 Rotation transformation between local and global coordinate systems.

The idea now is to treat eq. (8.3.3) as a redefinition of the generalized coordinates. To obtain the associated inertia matrix, we substitute this new definition into eq. (8.2.3), which gives

$$T = \tfrac{1}{2}\{\dot{\hat{q}}^e\}^{\mathrm{T}} [R^e]^{\mathrm{T}} [M^e][R^e]\{\dot{\hat{q}}^e\} \tag{8.3.5}$$

Thus, the inertia matrix for the element is transformed to

$$\boxed{[\hat{M}^e] = [R^e]^{\mathrm{T}}[M^e][R^e]} \tag{8.3.6}$$

which corresponds to the element's kinetic energy being given by

$$T = \tfrac{1}{2}\{\dot{\hat{q}}^e\}^{\mathrm{T}} [\hat{M}^e]\{\dot{\hat{q}}^e\} \tag{8.3.7}$$

The identification of the element's global stiffness applies similar steps to eq. (8.2.9), with the result that

$$\boxed{[\hat{K}^e] = [R^e]^{\mathrm{T}}[K^e][R^e]} \tag{8.3.8}$$

which corresponds to the strain energy being given by

$$V = \tfrac{1}{2}\{\hat{q}^e\}[\hat{K}^e]\{\hat{q}^e\} \tag{8.3.9}$$

EXAMPLE 8.2

One finite element for a system is isolated in the sketch. The cross-section is a 2 mm × 2 mm square. The global coordinate system is *XYZ*. Determine the elemental global inertia and stiffness matrices for this element.

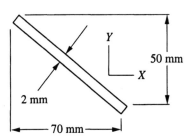

Solution The primary purpose of this example is to show how to formulate the rotation transformation consistently with the manner in which the nodes are numbered. The first step is to evaluate the elemental inertia and stiffness relative to the local coordinate system. For a square cross-section, we have $A = 0.002^2 = 4(10^{-6})$ m^2 and $I = \frac{1}{12}(0.002^4) = 1.3333(10^{-12})$ m^4. We also find that $L = (0.07^2 + 0.05^2)^{1/2} = 8.602(10^{-2})$ meter. Because the cross-sectional properties are constant, it does not matter for the elemental quantities which end we designate to be node 1. We substitute these values into the expressions for $[M^e]$ and $[K^e]$ in Example 8.1. The result is that

$$[M^e] = \rho(10^{-9}) \begin{bmatrix} 114.70 & 0 & 0 & 57.35 & 0 & 0 \\ 0 & 127.81 & 1.5505 & 0 & 44.24 & -0.9162 \\ 0 & 1.5505 & 0.02425 & 0 & 0.9162 & -0.018188 \\ 57.35 & 0 & 0 & 114.70 & 0 & 0 \\ 0 & 44.24 & 0.9162 & 0 & 127.81 & -1.5505 \\ 0 & -0.9162 & -0.018188 & 0 & -1.5505 & 0.02425 \end{bmatrix}$$

$$[K^e] = E(10^{-8}) \begin{bmatrix} 4650 & 0 & 0 & -4650 & 0 & 0 \\ 0 & 2.513 & 0.10811 & 0 & -2.513 & 0.10811 \\ 0 & 0.10811 & 0.00620 & 0 & -0.10811 & 0.00310 \\ -4650 & 0 & 0 & 4650 & 0 & 0 \\ 0 & -2.513 & -0.10811 & 0 & 2.513 & -0.10811 \\ 0 & 0.10811 & 0.00310 & 0 & -0.10811 & 0.00620 \end{bmatrix}$$

The wide range of magnitudes displayed by $[K^e]$ stems from the large value of the extensional rigidity terms EA relative to the flexural rigidity EI.

Our convention is to align the local xyz coordinate system such that node 1 is the origin. Thus, if we take node 1 to be the upper left end, and node 2 to be the lower right, then the x axis is at an angle $\tan^{-1}(50/70) = 0.6203$ rad below the X axis. Because positive β_e corresponds to a counterclockwise rotation from X to x, we set $\beta_e = -0.6203$. We find from eq. (8.3.4) that

$$[R^e] = \begin{bmatrix} 0.8137 & -0.5812 & 0 & 0 & 0 & 0 \\ 0.5812 & 0.8137 & 0 & 0 & 0 & 0 \\ 0 & 0 & 1 & 0 & 0 & 0 \\ 0 & 0 & 0 & 0.8137 & -0.5812 & 0 \\ 0 & 0 & 0 & 0.5812 & 0.8137 & 0 \\ 0 & 0 & 0 & 0 & 0 & 1 \end{bmatrix}$$

We then find from eqs. (8.3.6) and (8.3.8) that the elemental global inertia and stiffness matrices for this finite element are

$$[\hat{M}^e] = [R^e]^T[M^e][R^e]$$

$$= \rho(10^{-9}) \begin{bmatrix} 119.1 & 6.200 & -0.9012 & 52.92 & -6.200 & 0.5325 \\ 6.200 & 123.38 & -1.262 & -6.200 & 48.67 & 0.7455 \\ -0.9012 & -1.262 & 0.02425 & -0.5325 & -0.7455 & -0.01819 \\ 52.92 & -6.200 & -0.5325 & 119.13 & 6.200 & 0.9012 \\ -6.200 & 48.67 & -0.7455 & 6.200 & 123.38 & 1.262 \\ 0.5325 & 0.7455 & -0.01819 & 0.9012 & 1.262 & 0.02425 \end{bmatrix}$$

$$[\hat{K}^e] = [R^e]^T [K^e][R^e]$$

$$= E(10^{-8}) \begin{bmatrix} 3080 & -2198 & -0.0628 & -3080 & 2198 & -0.0628 \\ -2198 & 1572.6 & -0.0880 & 2198 & -1572.6 & -0.0880 \\ -0.0628 & -0.0880 & 0.0062 & 0.0628 & 0.0880 & 0.0031 \\ -3080 & 2198 & 0.0628 & 3080 & -2198 & 0.0628 \\ 2198 & -1572.6 & 0.0880 & -2198 & 1572.6 & 0.0880 \\ -0.0628 & -0.0880 & 0.0031 & 0.0628 & 0.0880 & 0.0062 \end{bmatrix}$$

Let us now consider what would have been the consequence of changing the node numbering such that the lower right end of the element is denoted as node 1. Placing the origin at node 1 leads to an x axis that runs up and to the left along the bar, so the angle $\beta_e = \pi - 0.6203$. This would reverse the signs of several terms in $[R^e]$. However, the resulting $[\hat{M}^e]$ and $[\hat{K}^e]$ would be unaltered because there is no physical difference in the properties associated with each mesh point.

8.4 ASSEMBLY OF THE ELEMENTS

Conversion of the inertia and stiffness matrices to global components is a necessary preliminary to combining them. We still must account for the fact that different elements have some mesh points in common. To avoid counting mesh points more than once, we assign a unique global number to each point; we shall use a prefix g to distinguish the global number from the local numbers 1 and 2 used within each element. We define a *global displacement vector* $\{\hat{q}\}$ holding the displacement variables for all mesh points, such that

$$\{\hat{q}\} = [\hat{u}_{g1} \ \ \hat{w}_{g1} \ \ \hat{\theta}_{g1} \ \ \hat{u}_{g2} \ \ \hat{w}_{g2} \ \ \hat{\theta}_{g2} \ldots]^T \tag{8.4.1}$$

This scheme is based on all bars being welded together, in which case all elements connected to a common mesh point must undergo the same rotation. If a connection is a pin, the elements must share the same global displacement, but they may undergo independent rotations. Such a situation would be described by allowing $\{\hat{q}\}$ to contain more than one $\hat{\theta}$ value at that mesh point. (We will consider a system featuring a pin connection in Example 8.3.)

To illustrate the process, consider the situation in Figure 8.5, where three elements form a triangle, so there are three mesh points. The element and mesh point numbers have been assigned in a counterclockwise manner in the figure, with the

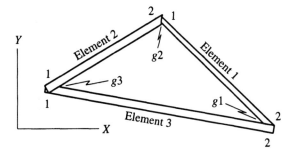

FIGURE 8.5 Local and global mesh point numbering.

latter numbers being preceded by the letter g. The local mesh point numbers are either 1 or 2. For this system, $\{\hat{q}\}$ consists of nine values,

$$\{\hat{q}\} = [\hat{u}_{g1} \ \hat{w}_{g1} \ \dots \ \hat{w}_{g3} \ \hat{\theta}_{g3}]^{\mathrm{T}} \tag{8.4.2}$$

We can select the six global displacements forming $\{\hat{q}^e\}$ for any element by multiplying $\{\hat{q}\}$ by a matrix of ones and zeroes, which serves as a sorting filter. For example, in the case of element 3 of the sample system, its mesh point 1 is $g3$, and mesh point 2 is $g1$. Such considerations for each generalized coordinate show that the global displacement $\{\hat{q}^3\}$ for element 3 is related to the full set of global displacements by

$$\{\hat{q}^3\} \equiv \begin{Bmatrix} \hat{u}_1 \\ \hat{w}_1 \\ \hat{\theta}_1 \\ \hat{u}_2 \\ \hat{w}_2 \\ \hat{\theta}_2 \end{Bmatrix} = \begin{bmatrix} 0 & 0 & 0 & 0 & 0 & 0 & 1 & 0 & 0 \\ 0 & 0 & 0 & 0 & 0 & 0 & 0 & 1 & 0 \\ 0 & 0 & 0 & 0 & 0 & 0 & 0 & 0 & 1 \\ 1 & 0 & 0 & 0 & 0 & 0 & 0 & 0 & 0 \\ 0 & 1 & 0 & 0 & 0 & 0 & 0 & 0 & 0 \\ 0 & 0 & 1 & 0 & 0 & 0 & 0 & 0 & 0 \end{bmatrix} \begin{Bmatrix} \hat{u}_{g1} \\ \hat{w}_{g1} \\ \hat{\theta}_{g1} \\ \hat{u}_{g2} \\ \hat{w}_{g2} \\ \hat{\theta}_{g2} \\ \hat{u}_{g3} \\ \hat{w}_{g3} \\ \hat{\theta}_{g3} \end{Bmatrix} \tag{8.4.3}$$

This idea may be applied in general. We let N be the number of generalized coordinates. If a system contains P mesh points, then $N \geq 3P$. For each element we define a $6 \times N$ array of ones and zeroes, which we denote as $[S^e]$. This matrix represents the *connectivity* for element e, because it tells us how an element is formed from the mesh points. The connectivity matrix has the property that

$$\boxed{\{\hat{q}^e\} = [S^e]\{\hat{q}\}} \tag{8.4.4}$$

We use this relation to describe the elemental kinetic and potential energies. From eqs. (8.3.7) and (8.3.9), we find that

$$T = \tfrac{1}{2}\{\dot{\hat{q}}\}^{\mathrm{T}}[S^e]^{\mathrm{T}}[\hat{M}^e][S^e]\{\dot{\hat{q}}\}$$

$$V = \tfrac{1}{2}\{\hat{q}\}^{\mathrm{T}}[S^e]^{\mathrm{T}}[\hat{K}^e][S^e]\{\hat{q}\} \tag{8.4.5}$$

These relations express the energy contributions of each element in terms of a common set of generalized coordinates. A summation over the index e gives the system energies. Factoring out the common premultiplying and postmultiplying factors leads to

$$T_{\text{total}} = \tfrac{1}{2}\{\dot{\hat{q}}\}^{\mathrm{T}}[\hat{M}]\{\dot{\hat{q}}\}, \qquad V_{\text{total}} = \tfrac{1}{2}\{\hat{q}\}^{\mathrm{T}}[\hat{K}]\{\dot{\hat{q}}\} \tag{8.4.6}$$

where the inertia and stiffness matrices associated with the mesh point global displacements are

$$
[\hat{M}] = \sum_e [S^e]^\mathrm{T}[\hat{M}^e][S^e] = \sum_e [S^e]^\mathrm{T}[R^e]^\mathrm{T}[M^e][R^e][S^e]
$$

$$
[\hat{K}] = \sum_e [S^e]^\mathrm{T}[\hat{K}^e][S^e] = \sum_e [S^e]^\mathrm{T}[R^e]^\mathrm{T}[K^e][R^e][S^e]
$$

(8.4.7)

It still remains to characterize the generalized forces associated with the mesh point displacements and rotations. The treatment of the distributed forces is different from the manner in which concentrated forces are addressed. For the distributed forces, we add eq. (8.2.14), which describes the contribution to the power input associated with each element,

$$
(\mathscr{P}_{\mathrm{in}})_{\mathrm{dist}} = \sum_e \{\dot{q}^e\}^\mathrm{T}[\gamma^e]\{f^e\}
\tag{8.4.8}
$$

As defined in eq. (8.2.13), $\{f^e\}$ consists of the values of the distributed forces relative to the local coordinate system for element e. Equations (8.3.3) and (8.4.4) map the elemental displacement variables onto the global set, according to

$$
\{q^e\} = [R^e][S^e]\{\hat{q}\}
\tag{8.4.9}
$$

When we substitute this expression into eq. (8.4.8), we find that the power input to the system by distributed forces is

$$
(\mathscr{P}_{\mathrm{in}})_{\mathrm{dist}} = \sum_e \{\dot{\hat{q}}\}^\mathrm{T}[S^e]^\mathrm{T}[R^e]^\mathrm{T}[\gamma^e]\{f^e\}
\tag{8.4.10}
$$

The description of concentrated force is more direct. It is implicit to the development that *all concentrated forces act at node points*. (The finite element mesh must be defined consistent with this requirement.) We represent the concentrated forces in terms of their components with respect to the global coordinate system. Thus, the force at global node gn has components $(F_x)_{gn}$ and $(F_y)_{gn}$. In addition, a concentrated torque Γ_{gn}, positive in the sense of the right-hand rule relative to the Z axis, may be applied at this node. We store these values in a column matrix that is arranged in the same manner as the definition of $\{\hat{q}\}$, specifically,

$$
\{\hat{F}\} = \left[(F_X)_{g1} \;\; (F_Y)_{g1} \;\; \Gamma_{g1} \;\; (F_X)_{g2} \;\; (F_Y)_{g2} \;\; \Gamma_{g2} \cdots \right]^\mathrm{T}
\tag{8.4.11}
$$

The similarity in the definition of $\{\hat{q}\}$ and $\{\hat{F}\}$ means that the contribution of concentrated forces to $\mathscr{P}_{\mathrm{in}}$ is merely

$$
(\mathscr{P}_{\mathrm{in}})_{\mathrm{conc}} = \{\dot{\hat{q}}\}^\mathrm{T}\{\hat{F}\}
\tag{8.4.12}
$$

The total power input is the sum of the contributions of the distributed forces, eq. (8.4.10), and concentrated forces. In general, the power input is related to the generalized force $\{Q\}$ by $\mathscr{P}_{\mathrm{in}} = \{\dot{\hat{q}}\}^\mathrm{T}\{\hat{Q}\}$. Because $\{\dot{\hat{q}}\}^\mathrm{T}$ is a common factor in $(\mathscr{P}_{\mathrm{in}})_{\mathrm{conc}}$ and $(\mathscr{P}_{\mathrm{in}})_{\mathrm{dist}}$, we find that the global generalized force is

$$
\{\hat{Q}\} = \{\hat{F}\} + \sum_e [S^e]^\mathrm{T}[R^e]^\mathrm{T}[\gamma^e]\{f^e\}
\tag{8.4.13}
$$

Once we have formed $[\hat{M}^e]$, $[\hat{K}^e]$, and $[\hat{Q}^e]$, we have completed the task of assembling the finite elements. However, these quantities cannot yet be used to determine response, primarily because some of the concentrated forces and couples are unknown reactions.

EXAMPLE 8.3

An inclined bar braces a cantilevered beam, with their junction being a pin. Both bars are composed of steel and their cross-sectional properties are $A = 0.0015 = m^2$, $I = 3.125(10^{-7})$ m². A very crude finite element model uses three elements, two of which are the segments of the cantilevered beam on either side of the pin connection; the third element is the entire inclined bar. Identify the global displacement vector and the connectivity matrix for each element, with XYZ taken to be the global coordinate system. Use these quantities to construct the global inertia, stiffness, and generalized force matrices.

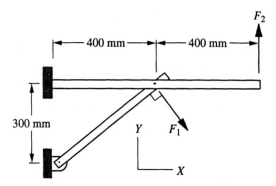

Solution The objective of this example is to demonstrate the construction of the connectivity matrices, and the subsequent assembly operations. We will go through the steps explicitly, but the process will make it evident that the finite element formulation is readily automated. The first step is to assign local and global numbers to the nodes. We show them in a simple line drawing, with a g prefix denoting the global numbers. We also show the element numbers by giving their e values.

To construct the global displacement we observe that the generalized coordinates for nodes $g1$, $g3$, and $g4$ are the global displacements in the X and Y directions and the flexural rotation at that location. In contrast, junction $g2$ is a pin connection, which permits the angled bar to undergo a flexural rotation that is different from that of the horizontal bar. Let us denote these rotation variables as θ_{g2b} for the angled bar and θ_{g2a} for the horizontal bar. Then the global displacement vector is

$$\{\hat{q}\} = \begin{bmatrix} \hat{u}_{g1} & \hat{w}_{g1} & \hat{\theta}_{g1} & \hat{u}_{g2} & \hat{w}_{g2} & \hat{\theta}_{g2a} & \hat{\theta}_{g2b} & \hat{u}_{g3} & \hat{w}_{g3} & \hat{\theta}_{g3} & \hat{u}_{g4} & \hat{w}_{g4} & \hat{\theta}_{g4} \end{bmatrix}^T$$

Note that $\{\hat{q}\}$ consists of 13 variables, so the global inertia and stiffness will be 13×13 matrices.

The idea now is to successively consider all aspects required to describe the contribution of a specific element to the global inertia and stiffness. For this we use the formulas in Example 8.1 to compute the local values $[M^e]$ and $[K^e]$. For each element we are given the values of A and I; $E = 210(10^9)$ Pa and $\rho = 7800$ kg/m^3 for steel. In a finite element program these operations are performed in a program loop; we will carry them out explicitly.

For element 1 we have $L = 0.4$ m. We compute $[M^1]$ and $[K^1]$ from the formulas. Because the x axis for an element runs from node 1 to node 2, this axis is parallel to the global X axis, so $\beta_1 = 0$. Correspondingly, the rotation transformation $[R^1]$ is the 6×6 identity matrix. To save space in writing the connectivity matrices, we observe that most of the elements of any $[S^e]$ will be zero. Thus, we shall only list the nonzero elements. In terms of the displacement variables for each of the two nodes, the global displacement vector for the first element is

$$\{\hat{q}^1\} = \begin{bmatrix} \hat{u}_1 & \hat{w}_1 & \hat{\theta}_1 & \hat{u}_2 & \hat{w}_2 & \hat{\theta}_2 \end{bmatrix}^{\mathrm{T}}$$

We replace the local node numbers with the corresponding global numbers, so that

$$\{\hat{q}^1\} = \begin{bmatrix} \hat{u}_{g1} & \hat{w}_{g1} & \hat{\theta}_{g1} & \hat{u}_{g2} & \hat{w}_{g2} & \hat{\theta}_{g2a} \end{bmatrix}^{\mathrm{T}}$$

The associated connectivity matrix is such that this expression is equivalent to $\{\hat{q}^1\} = [S^1]\{\hat{q}\}$, where $[S^1]$ is a 6×13 matrix. To filter $\hat{u}_{g1}$ from $\{\hat{q}\}$ we need a unit value in the $(1,1)$ position of $[S^1]$, while a unit value in the $(2,2)$ position of $[S^1]$ will give $\hat{w}_{g1}$, and so on. Thus, we identify

$$(S^1)_{1,1} = (S^1)_{2,2} = (S^1)_{3,3} = (S^1)_{4,4} = (S^1)_{5,5} = (S^1)_{6,6} = 1$$

Once again, the elements not listed are zero. In essence, the row number of nonzero $[S^e]$ elements matches the row number associated with the generalized coordinate in the local displacement $\{\hat{q}^e\}$. Similarly, the column number of nonzero elements matches the row number associated with the generalized coordinate in the global displacement $\{\hat{q}\}$.

With $[S^1]$ now known, we may evaluate the contribution of the first finite element to the global matrices. Note that the product $[R^e][S^e]$ arises in several locations of eqs. (8.4.7). To save operations we define this product to be $[\chi^e]$. Because $[R^1]$ is the identity matrix, we have $[\chi^1] = [S^1]$, which we use to compute $[\chi^1]^{\mathrm{T}}[M^1][\chi^1]$ and $[\chi^1]^{\mathrm{T}}[K^1][\chi^1]$. These two quantities represent the first contribution to $[\hat{M}]$ and $[\hat{K}]$, respectively.

Much of the work for evaluating the contribution of element 2 has been done already. Its local elemental matrices are such that $[M^2] = [M^1]$ and $[K^2] = [K^1]$, because elements 1 and 2 share the same cross-sectional properties and length. Furthermore, the x axis, which goes from the element's node 1 to node 2, is parallel to the X axis, so $[R^2]$ is also an identity matrix. What is different is the element's connectivity. As before, we explicitly write the element's generalized coordinates in terms of local node numbers, which we then translate to the global numbers. Specifically, we have

$$\{\hat{q}^1\} = [\hat{u}_1 \ \hat{w}_1 \ \hat{\theta}_1 \ \hat{u}_2 \ \hat{w}_2 \ \hat{\theta}_2]^{\mathrm{T}}$$
$$= [\hat{u}_{g2} \ \hat{w}_{g2} \ \hat{\theta}_{g2a} \ \hat{u}_{g3} \ \hat{w}_{g3} \ \hat{\theta}_{g3}]^{\mathrm{T}} = [S^2]\{\hat{q}\}$$

The nonzero elements of $[S^2]$ satisfying this description are

$$(S^2)_{1,4} = (S^2)_{2,5} = (S^2)_{3,6} = (S^2)_{4,8} = (S^2)_{5,9} = (S^2)_{6,10} = 1$$

In view of the fact that $[R^2]$ is also the identity matrix we have $[\chi^2] = [R^2][S^2] = [S^2]$. The contributions of the second element are $[\chi^2]^{\mathrm{T}}[M^2][\chi^2]$ and $[\chi^2]^{\mathrm{T}}[K^2][\chi^2]$, which add to the respective contributions of the first element.

For the third element we have $L = 0.5$ m, which, in combination with the constant values of ρA, EA, and EI, enables us to compute the local elemental matrices $[M^3]$ and $[K^3]$ from the general formulas. The x axis progressing from local node 1 to local node 2 of element 3 is at

36.87° counterclockwise from the X axis, so $\beta_3 = 36.87°$. We compute the corresponding rotation transformation $[R^3]$ according to eq. (8.3.4). To identify the connectivity matrix for this element we write

$$\{\hat{q}^3\} = [\hat{u}_1 \ \ \hat{w}_1 \ \ \hat{\theta}_1 \ \ \hat{u}_2 \ \ \hat{w}_2 \ \ \hat{\theta}_2]^\mathrm{T}$$

$$= [\hat{u}_{g4} \ \ \hat{w}_{g4} \ \ \hat{\theta}_{g4} \ \ \hat{u}_{g2} \ \ \hat{w}_{g2} \ \ \hat{\theta}_{g2b}]^\mathrm{T} = [S^3]\{\hat{q}\}$$

The nonzero elements of $[S^3]$ that satisfy the last equation are

$$(S^3)_{1,11} = (S^3)_{2,12} = (S^3)_{3,13} = (S^3)_{4,4} = (S^1)_{5,5} = (S^1)_{6,7} = 1$$

We next evaluate $[\chi^3] = [R^3] [S^3]$, and use that value to compute $[\chi^3]^\mathrm{T} [M^3] [\chi^3]$ and $[\chi^3]^\mathrm{T} [K^3] [\chi^3]$. Addition of these values to the accumulated values from elements 1 and 2 yields the assembled global inertia and stiffness matrices, $[\hat{M}]$ and $[\hat{K}]$, both of which are 13×13.

Many of the elements of $[\hat{M}]$ and $[\hat{K}]$ are zero, and the matrices are symmetric. For brevity, we shall list only the nonzero values in the upper right portion of each:

$$\hat{M}_{1,1} = 1.5600, \quad \hat{M}_{1,4} = 0.7800, \quad \hat{M}_{2,2} = 1.7383, \quad \hat{M}_{2,3} = 0.09806$$

$$\hat{M}_{2,5} = 0.6017, \quad \hat{M}_{2,6} = -0.05794, \quad \hat{M}_{3,3} = 0.007131, \quad \hat{M}_{3,5} = 0.05794$$

$$\hat{M}_{3,6} = -0.005349, \quad \hat{M}_{4,4} = 5.150, \quad \hat{M}_{4,5} = -0.1070, \quad \hat{M}_{4,7} = 0.09193$$

$$\hat{M}_{4,8} = 0.78, \quad \hat{M}_{4,11} = 0.8948, \quad \hat{M}_{4,12} = 0.10697, \quad \hat{M}_{4,13} = -0.05432$$

$$\hat{M}_{5,5} = 5.569, \quad \hat{M}_{5,7} = -0.1226, \quad \hat{M}_{5,9} = 0.6017, \quad \hat{M}_{5,10} = -0.05794$$

$$\hat{M}_{5,11} = 0.10697, \quad \hat{M}_{5,12} = 0.8324, \quad \hat{M}_{5,13} = 0.07243, \quad \hat{M}_{6,6} = 0.014263$$

$$\hat{M}_{6,9} = 0.05794, \quad \hat{M}_{6,10} = -0.005349, \quad \hat{M}_{7,7} = 0.013929, \quad \hat{M}_{7,11} = 0.05432$$

$$\hat{M}_{7,12} = -0.07243, \quad \hat{M}_{7,13} = -0.010446, \quad \hat{M}_{8,8} = 1.56, \quad \hat{M}_{9,9} = 1.7383$$

$$\hat{M}_{9,10} = -0.09806, \quad \hat{M}_{10,10} = 0.007131, \quad \hat{M}_{11,11} = 2.030, \quad \hat{M}_{11,12} = -0.10697$$

$$\hat{M}_{11,13} = -0.09193, \quad \hat{M}_{12,12} = 2.093, \quad \hat{M}_{12,13} = 0.12257, \quad \hat{M}_{13,13} = 0.013929$$

$$\hat{K}_{1,1} = 7.875(10^8), \quad \hat{K}_{1,4} = -7.875(10^8), \quad \hat{K}_{2,2} = 1.2305(10^7)$$

$$\hat{K}_{2,3} = 2.461(10^6), \quad \hat{K}_{2,5} = -1.2305(10^7), \quad \hat{K}_{2,6} = 2.461(10^6)$$

$$\hat{K}_{3,3} = 6.563(10^5), \quad \hat{K}_{3,5} = -2.461(10^6), \quad \hat{K}_{3,6} = 3.281(10^5)$$

$$\hat{K}_{4,4} = 1.9805(10^9), \quad \hat{K}_{4,5} = 2.994(10^8), \quad \hat{K}_{4,7} = 9.450(10^5)$$

$$\hat{K}_{4,8} = -7.875(10^8), \quad \hat{K}_{4,11} = -4.055(10^8), \quad \hat{K}_{4,12} = -2.994(10^8)$$

$$\hat{K}_{4,13} = 9.450(10^5), \quad \hat{K}_{5,5} = 2.554(10^8), \quad \hat{K}_{5,7} = -1.2600(10^6)$$

$$\hat{K}_{5,9} = -1.2305(10^7), \quad \hat{K}_{5,10} = 2.461(10^6), \quad \hat{K}_{5,11} = -2.994(10^8)$$

$$\hat{K}_{5,12} = -2.308(10^8), \quad \hat{K}_{5,13} = -1.2600(10^6), \quad \hat{K}_{6,6} = 1.3125(10^6)$$

$$\hat{K}_{6,9} = -2.461(10^6), \quad \hat{K}_{6,10} = 3.281(10^6), \quad \hat{K}_{7,7} = 5.250(10^5)$$

$$\hat{K}_{7,11} = -9.450(10^5), \quad \hat{K}_{7,12} = 1.2600(10^6), \quad \hat{K}_{7,13} = 2.625(10^5)$$

$$\hat{K}_{8,8} = 7.875(10^8), \quad \hat{K}_{9,9} = 1.2305(10^7), \quad \hat{K}_{9,10} = -2.461(10^6)$$

$$\hat{K}_{10,10} = 6.563(10^5), \quad \hat{K}_{11,11} = 4.055(10^8), \quad \hat{K}_{11,12} = 2.994(10^8)$$

$$\hat{K}_{11,13} = -9.450(10^5), \quad \hat{K}_{12,12} = 2.308(10^8), \quad \hat{K}_{12,13} = 1.2600(10^6)$$

$$\hat{K}_{13,13} = 5.250(10^5), \quad \hat{K}_{n,j} = \hat{K}_{j,n}$$

The last quantity to identify is the global generalized force $\{\hat{Q}\}$. Global node $g1$ is clamped, so there are constraint forces in the X and Y directions and a couple reaction at that point. Let us denote these as F_{X1}, F_{Y1}, and Γ_1. At mesh point $g2$, force F_1 is applied. Its X and Y components are $F_1 \cos(53.13°) = 0.6F_1$ and $-F_1 \sin(53.13°) = -0.8F$, respectively. No external torque is applied at this location. At mesh point $g3$, there is only F_2, which acts in the Y direction. Finally, mesh point $g4$ corresponds to a pin connection, so there are reactions forces in both global directions, F_{X4} and F_{Y4}. There are no distributed forces, so $\{\hat{Q}\}$ contains only the generalized forces and couples arranged in the same order as the way in which $\{\hat{q}\}$ is formed from the corresponding displacements and rotations. In particular, note that two zeroes will be placed in the positions corresponding to θ_{g2a} and θ_{g2b}. Thus, we form

$$\{\hat{Q}\} = [F_{X1} \;\; F_{Y1} \;\; \Gamma_1 \;\; 0.6F_1 \;\; -0.8F_1 \;\; 0 \;\; 0 \;\; 0 \;\; F_2 \;\; 0 \;\; F_{X4} \;\; F_{Y4} \;\; 0]^{\mathrm{T}}$$

8.5 REACTION FORCES AND CONSTRAINT CONDITIONS

At this stage in the development, each mesh point has three or more global generalized coordinates associated with it: displacement in the global X and Y directions, and one or more flexural rotations about the fixed Z axis. We now must characterize the manner in which the system is supported, which means that we need to account for motion constraints and the associated reaction forces.

Consider the simple situation in Figure 8.6, where the three finite elements we considered earlier are supported by a pin at mesh point $g2$ and a roller at mesh point $g3$. Suppose further that the base at $g2$ is executing a known vertical motion $G(t)$. Matching the motion of mesh point $g2$ to the base requires that $\hat{u}_{g2} = 0$ and $\hat{w}_{g2} = G(t)$. The roller at $g3$ prevents movement in the direction perpendicular to the guide. Let us denote displacement components parallel and perpendicular to the roller as s_3 and n_3, as shown in the figure. The constraint condition at $g3$ is that $n_3 = 0$, while s_3 is unrestricted. The conditions to be satisfied by the global displacement variables at mesh point 3 therefore are

$$\hat{u}_{g3} = s_3 \cos(\gamma) + n_3 \sin(\gamma), \quad \hat{w}_{g3} = -s_3 \sin(\gamma) + n_3 \cos(\gamma), \quad n_3 = 0 \quad (8.5.1)$$

In this example, we say that $\hat{u}_{g2}$, $\hat{w}_{g2}$, and n_3 are *constrained generalized coordinates*. When the system supports are stationary, the constrained generalized coordinates will be equated to zero. When the supports execute a specified motion, as is the case for mesh point 2 in the figure, the constrained generalized coordinates will be known functions of time.

The idea is to separate the constrained generalized coordinates from the unconstrained ones. This separation involves shifting the elements of $\{\hat{q}\}$ such that the unknown generalized coordinates come first, followed by the constrained variables. We write this rearrangement in partitioned form with $\{q_f\}$, where the subscript "f" indicates that these are the free, that is, unconstrained, variables stacked above $\{q_c\}$,

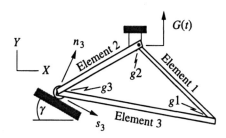

FIGURE 8.6 Example of constraints imposed on global displacements.

which contains the constrained generalized coordinates. The original and rearranged generalized coordinates are related by a *constraint coefficient matrix* [A], such that

$$\{\hat{q}\} = [A] \begin{Bmatrix} \{q_f\} \\ \{q_c\} \end{Bmatrix} \tag{8.5.2}$$

For example, in the case of Figure 8.6 we have

$$\begin{Bmatrix} \hat{u}_{g1} \\ \hat{w}_{g1} \\ \theta_{g1} \\ \hat{u}_{g2} \\ \hat{w}_{g2} \\ \theta_{g2} \\ \hat{u}_{g3} \\ \hat{w}_{g3} \\ \theta_{g3} \end{Bmatrix} = \begin{pmatrix} 1 & 0 & 0 & 0 & 0 & 0 & 0 & 0 & 0 \\ 0 & 1 & 0 & 0 & 0 & 0 & 0 & 0 & 0 \\ 0 & 0 & 1 & 0 & 0 & 0 & 0 & 0 & 0 \\ 0 & 0 & 0 & 0 & 0 & 0 & 1 & 0 & 0 \\ 0 & 0 & 0 & 0 & 0 & 0 & 0 & 1 & 0 \\ 0 & 0 & 0 & 1 & 0 & 0 & 0 & 0 & 0 \\ 0 & 0 & 0 & 0 & \cos\gamma & 0 & 0 & 0 & \sin\gamma \\ 0 & 0 & 0 & 0 & -\sin\gamma & 0 & 0 & 0 & \cos\gamma \\ 0 & 0 & 0 & 0 & 0 & 1 & 0 & 0 & 0 \end{pmatrix} \begin{Bmatrix} \hat{u}_1 \\ \hat{w}_1 \\ \theta_1 \\ \theta_2 \\ s_3 \\ \theta_3 \\ \hat{u}_2 \\ \hat{w}_2 \\ n_3 \end{Bmatrix} \tag{8.5.3}$$

We denote the constraint conditions as

$$\{q_c\} = \{G(t)\} \tag{8.5.4}$$

(If a displacement or rotation is not possible, the appropriate element of $\{G(t)\}$ is set to zero.) The fact that $\{q_c\}$ is known is used later in the development.

Each restriction on a displacement variable is imposed by an unknown constraint (or reaction) force that acts in the direction in which motion is prevented. Similarly, a restriction on rotation is imposed by an unknown couple reaction. For example, the restriction against flexural rotation at a clamped end is imposed by a bending moment. Thus, the generalized forces $\{\hat{Q}\}$ will contain two distinctly different effects. The external forces, which are known, are grouped as $\{Q_f\}$. These generalized forces are associated with mesh point displacement and rotations that are free. The forces that constrain generalized coordinates, and therefore are unknown, are grouped in $\{Q_c\}$. In the case of Figure 8.6, $\{Q_c\}$ consists of the X and Y components of the pin force at $g2$ and the roller force at $g3$ acting in the direction of the n_3 displacement.

We define the sequencing of $\{Q_f\}$ and $\{Q_c\}$ to match the definitions of $\{q_f\}$ and $\{q_c\}$, respectively. This enables us to employ eq. (8.5.2) to rearrange the generalized forces into the two types of forces, specifically,

$$\{\hat{Q}\} = [A] \begin{Bmatrix} \{Q_f\} \\ \{Q_c\} \end{Bmatrix} \tag{8.5.5}$$

The process of assembling the elements gives $\{\hat{Q}\}$. The task now is to decompose it into $\{Q_f\}$ and $\{Q_c\}$, which entails inverting [A]. A simple identity, which results from reconsidering the power input, is available for this purpose. When we use the original arrangement of global quantities to form $\mathcal{P}_{in}$, and then introduce eqs. (8.5.2) and (8.5.5), we find that

$$\mathcal{P}_{in} = \{\hat{q}\}^T\{\hat{Q}\} = [\{q_f\}^T\{q_c\}^T][A]^T[A]\begin{Bmatrix}\{Q_f\}\\\{Q_c\}\end{Bmatrix} \tag{8.5.6}$$

On the other hand, we may form $\mathcal{P}_{in}$ directly from the partitioned displacement and force vectors, which leads to

$$\mathcal{P}_{in} = \left[\{q_f\}^T \ \{q_c\}^T\right]\begin{Bmatrix}\{Q_f\}\\\{Q_c\}\end{Bmatrix} \tag{8.5.7}$$

This description of $\mathcal{P}_{in}$ must give the same result as eq. (8.5.6) for any set of forces and corresponding responses, which requires that

$$[A]^T[A] = [I] \quad \Rightarrow \quad [A]^{-1} = [A]^T \tag{8.5.8}$$

In other words [A] is an *orthonormal transformation*, which is a property shared by [R^e]. This property allows us to solve eq. (8.5.5), with the result

$$\boxed{\begin{Bmatrix}\{Q_f\}\\\{Q_c\}\end{Bmatrix} = [A]^T\{\hat{Q}\}} \tag{8.5.9}$$

The partitioned generalized forces form the right side of the equations of motion for the partitioned generalized coordinates. It also is necessary to rearrange the inertia and stiffness matrices consistent with the rearrangement of displacement variables. To do so, we use eq. (8.5.2) to adjust these system matrices to the new arrangement of generalized coordinates. As we did previously, we form T and V, then collect the coefficients of the quadratic sums, which leads to

$$T = \frac{1}{2}\left[\{\dot{q}_f\}^T \ \{\dot{q}_c\}^T\right][M]\begin{Bmatrix}\{\dot{q}_f\}\\\{\dot{q}_c\}\end{Bmatrix}, \quad V = \frac{1}{2}\left[\{q_f\}^T \ \{q_c\}^T\right][K]\begin{Bmatrix}\{q_f\}\\\{q_c\}\end{Bmatrix} \tag{8.5.10}$$

where the new inertia and stiffness matrices are given by

$$\boxed{[M] = [A]^T[\hat{M}][A], \quad [K] = [A]^T[\hat{K}][A]} \tag{8.5.11}$$

From these relations we can form the basic equations of motion, which are

$$[M]\begin{Bmatrix}\{\ddot{q}_f\}\\\{\ddot{q}_c\}\end{Bmatrix} + [K]\begin{Bmatrix}\{q_f\}\\\{q_c\}\end{Bmatrix} = \begin{Bmatrix}\{Q_f\}\\\{Q_c\}\end{Bmatrix} \tag{8.5.12}$$

The unknown quantities in the preceding are the free displacement variables in $\{q_f\}$ and the constraint forces in $\{Q_c\}$. We would like to have equations of motion whose only unknowns are the generalized coordinates. To arrive at that form we partition [M] and [K] in conformation with the partitioning of the displacement and forces. Hence, we write

$$[M] = \begin{bmatrix} [M_{ff}] & [M_{fc}] \\ [M_{cf}] & [M_{cc}] \end{bmatrix}, \quad [K] = \begin{bmatrix} [K_{ff}] & [K_{fc}] \\ [K_{cf}] & [K_{cc}] \end{bmatrix} \tag{8.5.13}$$

Symmetry of [M] and [K] leads to the conditions that $[M_{cf}] = [M_{fc}]^T$ and $[K_{cf}] = [K_{fc}]^T$.

We now impose the constraint conditions in eq. (8.5.4), $\{q_c\} = \{G(t)\}$. In combination with the partitioned form of the system matrices, eq. (8.5.12) becomes

$$\begin{bmatrix} [M_{ff}] & [M_{fc}] \\ [M_{cf}] & [M_{cc}] \end{bmatrix} \begin{Bmatrix} \{\ddot{q}_f\} \\ \{\ddot{G}(t)\} \end{Bmatrix} + \begin{bmatrix} [K_{ff}] & [K_{fc}] \\ [K_{cf}] & [K_{cc}] \end{bmatrix} \begin{Bmatrix} \{q_f\} \\ \{G(t)\} \end{Bmatrix} = \begin{Bmatrix} \{Q_f\} \\ \{Q_c\} \end{Bmatrix} \qquad (8.5.14)$$

The upper and lower partitions of this expression provide different information. We bring any term containing $\{G(t)\}$ to the right side, because it represents known functions. Thus, the upper partition gives

$$[M_{ff}]\{\ddot{q}_f\} + [K_{ff}]\{q_f\} = \{Q_f\} - [M_{fc}]\{\ddot{G}\} - [K_{fc}]\{G\} \qquad (8.5.15)$$

These are the equations of motion for the $\{q_f\}$ variables. The right side indicates that the generalized forces for this system are the actual mesh point forces associated with $\{q_f\}$, onto which are superposed inertial and stiffness forces associated with motion of the supports. The lower partition of eq. (8.5.14) also is useful. It states that

$$\{Q_c\} = [M_{cf}]\{\ddot{q}_f\} + [M_{cc}]\{\ddot{G}\} + [K_{cf}]\{q_f\} + [K_{cc}]\{G\} \qquad (8.5.16)$$

From this expression we may determine the reaction forces, after we have evaluated $\{q_f\}$ as a function of t.

The main concept to carry away from our development of the finite element method is that its theoretical foundation is the Ritz series method, with basis functions defined over small segments of the system. The segments are the finite elements, and the basis functions are the shape functions. The generalized coordinates associated with this Ritz series are displacements and rotations at mesh points.

Implementations of the finite element method usually consider each element sequentially. The physical properties and geometry are used to evaluate the elemental matrices, the rotation transformation, and the connectivity matrix for each element. Assembly of the elements consists of adding the contributions of each element's effects, and rearranging the results to account for motion constraints and associated constraint forces. The generalized forces are obtained by combining the distributed and mesh point forces. Once these operations have been performed, the finite element equations have the basic form of any other multi-degree-of-freedom system, and may be solved accordingly.

In this text, we perform the finite element operations manually in order to understand what is done at each stage. One feature that makes FEM so attractive is that every operation is readily automated into a computer program. In comparison to other analytical techniques, FEM codes place different burdens on the user. The models tend to have many mesh points, so there will be a large number of degrees of freedom. Faulty data input, such as system parameters, can lead to a great waste of computer resources. In addition, the great amount of output data associated with having many degrees of freedom requires auxiliary programs to assemble the output in a meaningful manner, as well as wisdom in assessing the data in the light of known vibrational phenomena.

EXAMPLE 8.4

Determine the natural frequencies and modes associated with the system in Example 8.3.

Solution This example completes the sequence of operations leading to a set of finite element equations that may be solved for the response. The global matrices were determined previously. The next step is to identify the constraint matrix.

The global displacement vector was defined to be

$$\{\hat{q}\} = \begin{bmatrix} \hat{u}_{g1} & \hat{w}_{g1} & \hat{\theta}_{g1} & \hat{u}_{g2} & \hat{w}_{g2} & \hat{\theta}_{g2a} & \hat{\theta}_{g2b} & \hat{u}_{g3} & \hat{w}_{g3} & \hat{\theta}_{g3} & \hat{u}_{g4} & \hat{w}_{g4} & \hat{\theta}_{g4} \end{bmatrix}^T$$

Mesh point $g1$ is a fixed end, so $\hat{u}_{g1} = \hat{w}_{g1} = \hat{\theta}_{g1} = 0$. Also, mesh point $g4$ is a pin, so $\hat{u}_{g4} = \hat{w}_{g4} = 0$. These are the constrained generalized coordinates, which correspond to

$$\{q_c\} = \begin{bmatrix} \hat{u}_{g1} & \hat{w}_{g1} & \hat{\theta}_{g1} & \hat{u}_{g4} & \hat{w}_{g4} \end{bmatrix}^T$$

The associated constraint equation is

$$\{q_c\} = \{G(t)\} = \begin{bmatrix} 0 & 0 & 0 & 0 & 0 \end{bmatrix}^T$$

This leaves eight generalized coordinates from which to form $\{q_f\}$. Our definition maintains these variables in the order in which they appear in $\{\hat{q}\}$,

$$\{q_f\} = \begin{bmatrix} \hat{u}_{g2} & \hat{w}_{g2} & \hat{\theta}_{g2a} & \hat{\theta}_{g2b} & \hat{u}_{g3} & \hat{w}_{g3} & \hat{\theta}_{g3} & \hat{\theta}_{g4} \end{bmatrix}$$

We now seek the 13×13 matrix $[A]$ that satisfies eq. (8.5.2) for the preceding definitions. We proceed row by row to identify the appropriate elements. For example, the first element of $\{\hat{q}\}$ is $\hat{u}_{g1}$, which is the first element of $\{q_c\}$, and therefore the ninth element of the vector formed by stacking $\{q_f\}$ above $\{q_c\}$. Correspondingly, we set $A_{1,9} = 1$, and all of the other elements in the first row of $[A]$ are zero. As another example, consider $\hat{\theta}_{g2b}$, which is the seventh element of $\{\hat{q}\}$. This variable is the fourth element of $\{q_f\}$, so we set $A_{7,4} = 1$, and all of the other elements in row seven of $[A]$ are zero. The full set of nonzero elements of $[A]$ is

$$A_{1,9} = A_{2,10} = A_{3,11} = A_{4,1} = A_{5,2} = A_{6,3} = A_{7,4} = 1$$

$$A_{8,5} = A_{9,6} = A_{10,7} = A_{11,8} = A_{12,12} = A_{13,13} = 1$$

We substitute $[A]$ and the $[\hat{M}]$ and $[\hat{K}]$ matrices found in Example 8.3 into eq. (8.5.11). These computations lead to the inertia $[M]$ and stiffness $[K]$, whose elements are sorted into partitions of free and constrained generalized coordinates. The constraint conditions at mesh points $g1$ and $g4$ correspond to $\{q_c\} = \{G\} = \{0\}$. Also, we set $\{Q_f\} = \{0\}$ for an evaluation of the modal solution. Thus, the equation of motion for $\{q_f\}$ reduces to

$$[M_{ff}]\{\ddot{q}\} + [K_{ff}]\{q\} = \{0\}$$

Because $\{q_f\}$ consists of eight generalized coordinates, setting $\{q_f\} = \text{Re}[\{\phi\}\exp(i\omega t)]$ leads to eight modes. The natural frequencies and normal modes that result are

$$\omega_n = 499.4, 2119, 5588, 8712, 9569, 18703, 25883, 34365 \text{ rad/s}$$

$$[\Phi] = \begin{bmatrix} 0.0002 & 0.006 & 0.087 & 0.057 & 0.274 & 0.077 & 0.024 & 0.403 \\ 0.197 & 0.332 & -0.028 & 0.002 & -0.166 & 0.142 & -0.312 & 0.008 \\ 0.904 & -0.477 & -0.019 & -6.679 & 1.983 & -2.877 & -7.373 & 0.084 \\ 0.735 & 1.173 & 6.666 & -1.402 & -6.510 & 3.383 & -4.117 & -1.656 \\ 0.0002 & 0.006 & 0.0956 & 0.073 & 0.365 & 0.339 & -0.122 & -0.654 \\ 0.631 & -0.639 & 0.026 & 0.768 & -0.051 & -0.448 & -1.131 & 0.012 \\ 1.149 & -3.246 & 0.260 & 9.268 & -0.499 & -9.440 & -27.476 & 0.326 \\ 0.148 & 0.261 & 0.137 & 0.082 & 0.200 & -0.561 & 0.257 & -0.244 \end{bmatrix}$$

Interpretation of the normal modes must be done in conjunction with the definition of $\{q_f\}$, whose elements are displacements in two directions and rotations. It is difficult to recognize by inspection the displacement pattern associated with a specific mode $\{\Phi_n\}$. This difficulty is amplified many times for realistic finite element models, in which there are many more mesh points. For this reason, an important aspect of a finite element software package is its postprocessors. These are programs that convert the computed mesh point displacements into user-friendly form. For example, a postprocessor could use the interpolating functions to evaluate the displacement pattern at any instant, which could be animated for video display. Another postprocessor could evaluate the axial and shear forces, and bending moment within each member, from which locations of maximum stress could be located.

REFERENCES AND SELECTED READINGS

BATHE, K. J. 1982. *Finite Element Procedures in Engineering Analysis.* Prentice-Hall, Englewood Cliffs, NJ.

COOK, R. D., MALKUS, D. S., & PLESHA, M. E. 1989. *Concepts and Applications of Finite Elements.* John Wiley & Sons, New York.

MEIROVITCH, L. 1997. *Principles and Techniques of Vibrations.* Prentice Hall, Englewood Cliffs, NJ.

SHAMES, I. H., & DYM, C. L. 1985. *Energy and Finite Element Methods in Structural Mechanics.* Hemisphere Publishing, Washington, DC.

THOMSON, W. T., & DAHLEH, M. D. 1993. *Theory of Vibration with Applications,* 5th ed. Prentice Hall, Englewood Cliffs, NJ.

ZIENKIEWICZ, O. C., & TAYLOR, R. I. 1991. *The Finite Element Method,* 4th ed. McGraw-Hill, New York.

EXERCISES

8.1 It is desired to develop a finite element describing displacement u of a bar solely in the axial direction, so that the elemental displacement is $\{q^e\} = \begin{bmatrix} u_1 & u_2 \end{bmatrix}^T$. The bar's cross-sectional properties are uniform. Derive expressions for $[M^e]$ and $[K^e]$.

8.2 Consider using the finite element in Figure 8.1 to describe only axial displacement, so that the elemental displacement is $\{q^e\} = \begin{bmatrix} u_1 & u_2 \end{bmatrix}^T$. Use eqs. (8.2.1) to describe the dependence of ρA and EA on distance along the bar; the shape functions are those in eq. (8.1.7). Evaluate the elemental inertia and stiffness matrices appropriate to this situation.

8.3 A finite element for torsion of circular bars uses the torsional rotation at each end of the element as generalized coordinates. Derive the interpolating functions for such an element, based on the cross-sectional properties being uniform. Then evaluate the corresponding elemental inertia and stiffness matrices.

8.4 Consider a torsional finite element in which the generalized coordinates are taken to be the rotations at the ends, θ_1 and θ_2 in the sketch, as well as rotation θ_3 at the midpoint. The corresponding elemental displacement vector is $\{q^e\} = \begin{bmatrix} \theta_1 & \theta_2 & \theta_3 \end{bmatrix}^T$. Derive interpolating functions based on matching $\theta(x)$ to the rotation at these mesh points.

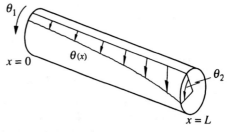

EXERCISES 8.3, 8.4

8.5 It is possible to define finite elements in which there are internal mesh points in addition to those at the ends of the elements. One possibility is that the interpolating functions are selected to match axial displacement u_3 and transverse displacement w_1 at the middle of the element, as well as both displacement components and flexural

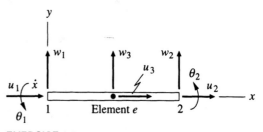

EXERCISE 8.5

rotation at the ends. Derive polynomial interpolating functions for such an element.

8.6 Finite elements can be constructed for other types of structural configurations. The sketch shows a triangular element describing plane strain in a thin plate. Nodes are the apexes of the triangle, and generalized coordinates are the displacement components relative to the local coordinate system xyz. Derive quadratic polynomial interpolating functions based on matching displacement functions $u(x, y)$ and $w(x, y)$ to the generalized coordinates. (*Hint:* Quadratic terms for both u and w consist of x^2, y^2, and xy.)

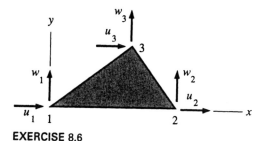

EXERCISE 8.6

8.7 Prove that the elemental rotation transformation $[R_e]$ has the *orthonormal property*, $[R_e]^T[R_e] = [I]$, from which it follows that $[R_e]^{-1} = [R_e]^T$.

8.8 Two local coordinate systems are $x_1 y_1 z_1$ at angle β_1 counterclockwise from the global XYZ system, and $x_2 y_2 z_2$ at counterclockwise angle $\beta_2 > \beta_1$ from XYZ.

(a) Use the fact that $\beta_2 - \beta_1$ is the angle from $x_1 y_1 z_1$ to determine the rotation transformation giving nodal displacement components u and w relative to $x_2 y_2 z_2$ in terms of corresponding $x_1 y_1 z_1$ components.

(b) Obtain the result in part (a) by applying matrix algebra and trigonometric identities to the transformations $[R_1]$ from XYZ to $x_1 y_1 z_1$ and $[R_2]$ from XYZ to $x_2 y_2 z_2$.

8.9 The sketch shows a triangular element describing plane strain in a thin plate. Mesh points are the apexes of the triangle, and generalized coordinates are the displacement components relative to the local coordinate system xyz. What is the elemental

rotation transformation that converts these generalized coordinates to global quantities with respect to XYZ?

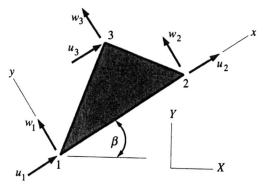

EXERCISE 8.9

8.10 Two circular bars are welded together collinearly, as shown. It is desired to develop a finite element model that solely accounts for axial displacement. The model should consist of three equal lengths. Identify the global displacement vector and connectivity matrix corresponding to this model. Then construct the corresponding global inertia and stiffness matrices.

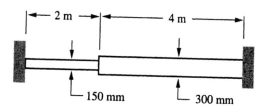

EXERCISE 8.10

8.11 A beam is supported at its midpoint by a roller. Because the bar is straight, flexural and axial displacements are uncoupled. Consequently, only flexural motion is induced by the transverse force excitation $F(t)$ at the right end. Consider using two equal-length finite elements to model the flexural motion. Identify the corresponding global displacement vector and connectivity matrix. Then construct the corresponding global inertia, stiffness, and generalized force matrices.

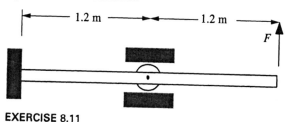

EXERCISE 8.11

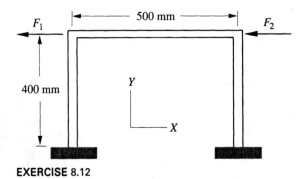

EXERCISE 8.12

8.12 The framework is composed of three aluminum bars having 30 mm × 30 mm square cross-sections that are welded at their junctions. Suppose each bar is defined to be a single finite element. Identify the global displacement vector and connectivity matrix corresponding to this model, with XYZ taken to be the global coordinate system. Then construct the corresponding global inertia, stiffness, and generalized force matrices.

8.13 The pair of bars depicted in the sketch are welded together. The distributed force f_z acts transverse to the inclined bar and grows linearly with distance from the roller. The global coordinate system is XYZ. Use two equal-length finite elements to describe the angled bar, and a single finite element to describe the vertical bar. Identify the global displacement vector and connectivity matrix corresponding to this model. Then evaluate the generalized forces.

8.14 Derive the global inertia and stiffness matrices for the finite element model described in Exercise 8.13. The bars are steel pipes with 40 mm outside diameter and 30 mm inside diameter. The length parameter is $L = 250$ mm.

8.15 Identify the constraint coefficient matrix [A] and the associated constraint conditions for the finite element model described in Exercise 8.13.

8.16 Bar AC in the linkage shown below is modeled as two finite elements sharing node B, which is a pin. Use a single finite element for bar CD, and two elements for bar AC. The global coordinate system is XYZ. Identify the connectivity matrices $[S^e]$, the constraint coefficient matrix [A], and the associated constraint conditions.

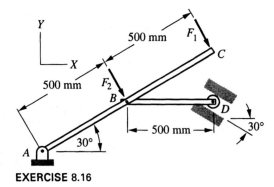

EXERCISE 8.16

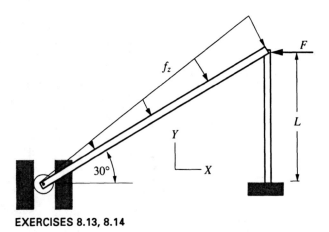

EXERCISES 8.13, 8.14

8.17 Determine the equations of motion for the finite element model described in Exercise 8.12. Solve those equations for the associated natural frequencies and mode shapes.

8.18 Determine the equations of motion for the finite element model described in Exercise 8.14.

Solve those equations for the associated natural frequencies and mode shapes.

8.19 Determine the natural frequencies and free vibration modes for the linkage model described in Exercise 8.16. Bar *AC* is composed of aluminum and bar *BD* is steel. Both bars have 50 mm × 50 mm square cross-sections.

CHAPTER 9

SUBSTRUCTURING CONCEPTS

Consider the following scenario. Two groups working for an automobile manufacturer have developed models for different aspects of a new prototype. One has an analytical model of the chassis, and the other has a model of the suspension system. It is now necessary to consider how the two work together. Rather than creating a new mathematical model that ignores previous efforts, it is possible to merge models, analogously to the way in which the vehicle is assembled from its components. Indeed, we shall use the term *component* to refer to the mathematical models of the subassemblies from which the system is synthesized. From a different viewpoint, we might have a very challenging system to model. Conceptualizing such a system as an assembly of components can simplify the modeling task.

The systems of interest are continuous media. The development will focus on situations where the component models have been formulated in terms of a finite number of degrees of freedom. For the sake of simplicity, only time-invariant systems will be addressed and Ritz series will be used as the framework for derivations. However, the results are equally valid for models that have been developed using the finite element method. Translation of the present results to a finite element formulation may be achieved by returning to the description of the displacement field as a locally defined Ritz series using the interpolation functions as the basis functions; see Section 8.1.

9.1 METHOD OF LAGRANGE MULTIPLIERS

When a machine or structure is physically assembled, the process of joining parts makes them act in unison. We shall refer to the locations where components are joined as *interfaces*. The specific kinematical conditions expressing the unity of motion at the interfaces are *constraint conditions*. The forces and/or moment couples exerted at the interfaces are *constraint forces*, which are more commonly called the *reactions*. The formulation requires that we identify what constraint conditions must be satisfied to make the system components move consistently. We also must account for the constraint forces that enforce these conditions. One of the main developments will be the fact that characterization of the constraint equations concurrently enables us to account for the constraint forces.

9.1.1 Constraint Equations and Reactions

A simple system consisting of a single component serves to set the stage for the main concepts. Suppose we have formulated the equations of motion for flexural vibration of a simply supported beam by using the Ritz series method. We wish to convert this model to one for a fixed-pinned beam. Figure 9.1(a) shows the displaced state of the simply-supported beam with some force excitation. The interface is the left end, which should be clamped. Figure 9.1(b) superposes a couple Γ at the interface. This couple induces a rotation at the end, opposite to that of the simply supported beam.

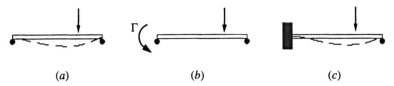

FIGURE 9.1 Conceptual aspects to treating a simply supported beam as a component of a fixed-pinned beam.

The role of Γ is to impose the condition that there can be no rotation at a clamped end. The formulation of the system equations essentially adjusts Γ to obtain this kinematical condition. The result is the fixed-pinned beam in Figure 9.1(c), where it is obvious that Γ is the moment reaction at a clamped end.

The Ritz series for the simply supported version of the beam uses basis functions that satisfy the geometric boundary conditions that there is no displacement at the ends,

$$w = \sum_{j=1}^{N} \psi_j(x)q_j(t), \quad \psi_j(0) = \psi_j(L) = 0 \tag{9.1.1}$$

There is no rotation at a clamped end. This condition is not identically satisfied by the Ritz series, so we must enforce it explicitly. We therefore require that

$$\frac{\partial w}{\partial x}\bigg|_{x=0} \equiv \sum_{j=1}^{N} \frac{\partial \psi_j}{\partial x}\bigg|_{x=0} q_j(t) = 0 \tag{9.1.2}$$

This is the constraint equation associated with clamping a simply supported beam. Because the basis functions have been set, the partial derivatives in the summation are constant coefficients. Let us denote these quantities as a_{1j}, where the unit value of the first subscript denotes that these coefficients are for the first constraint equation. Thus, the constraint equation is

$$\sum_{j=1}^{N} a_{1j}q_j = 0, \quad a_{1j} = \frac{\partial \psi_j}{\partial x}\bigg|_{x=0} \tag{9.1.3}$$

The next aspect to consider is the role of the moment reaction Γ. From the perspective of a simply supported beam, this is a torque load applied at the end. We therefore must account for the contribution of Γ in the generalized forces. The total generalized force may be decomposed into the portion Q_j attributable to known external forces and the contribution R_j of the unknown constraint force Γ. The quantity Q_j is the same as what it would be if the beam actually were simply supported. If we are using the power balance method, we identify R_j by formulating the power input by Γ, which is $\mathscr{P}_{in} = \Gamma(\partial \dot{w}/\partial x)|_{x=0}$. We use the series in eq. (9.1.2) to express the angular velocity term and equate the power to the standard relation between power input and generalized forces, which leads to

$$\sum_{j=1}^{N} R_j \dot{q}_j = \Gamma \sum_{j=1}^{N} \frac{\partial \psi_j}{\partial x}\bigg|_{x=0} \dot{q}_j \tag{9.1.4}$$

In view of eq. (9.1.3), matching like coefficients of generalized velocities in the preceding leads to

$$R_j = a_{1j}\Gamma \tag{9.1.5}$$

The appearance in R_j of the constraint equation coefficients a_{1j} is not a coincidence, as we shall soon see.

Lagrange's equations are more generally applicable, so let us examine how the preceding operation would evolve in that formulation. To identify R_j we would formulate the virtual work done by Γ. Equation (9.1.2) describes the rotation at $x = 0$. A virtual displacement increments each of the generalized coordinates by δq_j, so the virtual rotation at the end is

$$\delta\left(\left.\frac{\partial w}{\partial x}\right|_{x=0}\right) = \sum_{j=1}^{N} \left.\frac{\partial \psi_j}{\partial x}\right|_{x=0} \delta q_j \tag{9.1.6}$$

Equating the standard form of virtual work to the work done by Γ in the virtual rotation leads to

$$\sum_{j=1}^{N} R_j \delta q_j = \Gamma \delta\left(\left.\frac{\partial w}{\partial x}\right|_{x=0}\right) = \Gamma \sum_{j=1}^{N} \left.\frac{\partial \psi_j}{\partial x}\right|_{x=0} \delta q_j \tag{9.1.7}$$

Matching like coefficients of δq_j obviously leads to eq. (9.1.5). As shown in Appendix A, every operation in the power balance method has a parallel in the Lagrange equation formulation.

Let us assess the situation at this juncture. We started out with an N-term Ritz series for the simply supported beam. Thus, there are N unknown generalized coordinates to determine. The constraint force Γ is a new unknown introduced by clamping the interface. The available equations are the N ordinary differential equations of motion for which the inertia, stiffness, and dissipation coefficients are those for a Ritz series analysis of a simply supported beam. These are supplemented by the constraint equation. Because the number of equations to be satisfied equals the number of unknowns, we can proceed to solve these equations.

The insight we have gained from the simple system in Figure 9.1 may be generalized to fit any linear, time-invariant system. To do so consider the situation in Figure 9.2, where two component bars are welded at right angles. The weld connection

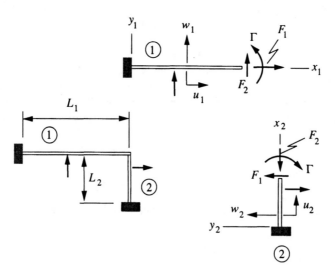

FIGURE 9.2 Assembly of two cantilevered bars as structural components.

requires that each bar undergoes the same displacement at that location. It also requires that the rotation be the same. Transverse displacement in one bar corresponds to axial displacement in the other, and the coordinate system for each component is different. This requires that we take special care to assign the appropriate direction and sense when we match the movement of each bar. As an aid in this task, we define a local coordinate system for each bar. The x axis for each bar is defined to coincide with the bar's axial direction, and all z axes are defined in the same sense normal to the plane. A superscript will denote which structural component is associated with a quantity, so the constraint conditions are

$$u^1(L_1) = -w^2(L_2), \qquad w^1(L_1) = u^2(L_2), \qquad \left.\frac{\partial w^1}{\partial x_1}\right|_{x_1=L_1} = \left.\frac{\partial w^2}{\partial x_2}\right|_{x_2=L_2} \tag{9.1.8}$$

Thus, there are three constraint conditions to be satisfied.

The Ritz series for each displacement field may have a different length,

$$u^\ell = \sum_{j=1}^{P_\ell} \psi_{uj}^\ell q_{uj}^\ell, \qquad w^\ell = \sum_{j=1}^{N_\ell} \psi_{wj}^\ell q_{wj}^\ell, \qquad \ell = 1, 2 \tag{9.1.9}$$

We use these series to quantify the constraint conditions in eq. (9.1.8), which leads to

$$\sum_{j=1}^{P_1} \psi_{uj}^1(L_1) q_{uj}^1 + \sum_{j=1}^{N_2} \psi_{wj}^2(L_2) q_{wj}^2 = 0$$

$$\sum_{j=1}^{N_1} \psi_{wj}^1(L_1) q_{wj}^1 - \sum_{j=1}^{P_2} \psi_{uj}^2(L_2) q_{uj}^2 = 0 \tag{9.1.10}$$

$$\sum_{j=1}^{N_1} \left.\frac{\partial \psi_{wj}^1}{\partial x_1}\right|_{x_1=L_1} q_{wj}^1 - \sum_{j=1}^{N_2} \left.\frac{\partial \psi_{wj}^2}{\partial x_2}\right|_{x_2=L_2} q_{wj}^2 = 0$$

The coefficient multiplying each generalized coordinate is a constant, so the preceding are three linear algebraic equations for the generalized coordinates that must be satisfied simultaneously with the equations of motion for each component.

To generalize the preceding to fit any situation involving linear time-invariant systems, we define a generalized coordinate vector $\{q\}$ such that the variables for each component are grouped together, specifically

$$\{q\} = \left[\{q^1\}^T \{q^2\}^T \{q^3\}^T \ldots\right]^T \tag{9.1.11}$$

The vector $\{q^n\}$ for each structural component may be formed from variables representing axial, transverse, or torsional displacement, in any combination. The constraint equations are linear, algebraic equations relating the generalized coordinates, so they may be written in matrix form as

$$[a]\{q\} = \{0\} \tag{9.1.12}$$

The coefficient matrix $[a]$ is called the *Jacobian constraint matrix*. The number of rows of $[a]$ will equal the number of constraint conditions to be imposed on the structural components, and the number of columns of $[a]$ will equal the total number of generalized coordinates associated with all components. We identify the specific coefficients by matching the general form to the actual constraint equations.

Now let us consider the generalized forces. The known external forces lead to a generalized force vector $\{Q\}$, whose elements are sequenced in the same order as the elements of $\{q\}$. In the case of Figure 9.2, each constraint condition is enforced by a constraint force: F_1 for horizontal displacement, F_2 for vertical displacement, and Γ for rotation. Let us denote as $\{R_k\}$, $k = 1$, 2, or 3, the contribution of each constraint force to the total generalized force. We could characterize these quantities by forming the power input or virtual work for each, as we did for the simply supported beam. Fortunately, there is a much simpler alternative.

When the components vibrate independently, the elements of $\{\dot{q}\}$ may have arbitrary values, which are determined solely by the external forces and the initial conditions. The velocity field obtained from the Ritz series in that case will be kinematically admissible for each component structure, but not for the assembly. A special class of generalized velocities $\{\dot{r}\}$ represents motions that satisfy all constraint equations, including those at the interface between components. In other words, $\{\dot{r}\}$ represents any velocity field that is kinematically admissible for the assembled system. According to this definition, $\{\dot{r}\}$ satisfies the velocity version of eq. (9.1.12),

$$[a]\{\dot{r}\} = \{0\} \qquad (9.1.13)$$

The number of different combinations of the individual elements of $\{\dot{r}\}$ is limited only by the number of its elements, because the particular mix depends on the forces and initial conditions for the system.

Any $\{\dot{r}\}$ satisfying eq. (9.1.13) will result in zero power input from the interface constraint forces, regardless of the specific values of its elements. To see why this is so in the case of Figure 9.2, we observe that when the constraint conditions are satisfied, the respective motions at the interface end of each cantilevered bar are the same. On the other hand, Newton's Third Law requires that the constraint forces acting at those ends have equal magnitude, but opposite sense. Thus, whatever power is input by a constraint force in the movement of one component is the negative of the power input by the corresponding force acting on the other component.

This property that constraint forces do not input power when the generalized velocities are consistent with the constraint conditions obviously applies for constraint forces that prevent movement, such as the couple reaction in Figure 9.1. Indeed, it is a fundamental property of constraints; see Ginsberg (1995). (It actually is more proper to consider virtual displacements $\{\delta q\} = \{\delta r\}$ and the associated virtual work. The present argument is sufficient for linear time-independent systems.) In other words, it must always be true that

$$\{\dot{r}\}^T\{R_k\} \equiv \{R_k\}^T\{\dot{r}\} = 0 \qquad (9.1.14)$$

This relation and eq. (9.1.13) will be satisfied simultaneously, regardless of the composition of $\{\dot{r}\}$, if $\{R_j\}^T$ is proportional to one of the rows of $[a]$. Without loss of generality, we associate row $[a_k]$ with $\{R_k\}$. The factor of proportionality is called a *Lagrange multiplier*, so we have

$$\boxed{\{R_k\} = [a_k]^T\lambda_k} \qquad (9.1.15)$$

For an arbitrary system, which has N_c constraint equations, there will be N_c constraint forces, and therefore N_c Lagrange multipliers. The total generalized force will be the sum of $\{Q\}$ due to known forces and each of the $\{R_k\}$ terms,

$$\{Q\}_{\text{total}} = \{Q\} + \sum_{k=1}^{N_c} \{R_k\} = \{Q\} + \sum_{k=1}^{N_c} [a_k]^{\mathrm{T}} \lambda_k \qquad (9.1.16)$$

In matrix notation, the preceding is

$$\boxed{\{Q\}_{\text{total}} = \{Q\} + [a]^{\mathrm{T}}\{\lambda\}} \qquad (9.1.17)$$

An important aspect of the development is that eqs. (9.1.12) and (9.1.17) are descriptive of all linearized vibration models. Thus, although we developed the relations using the Ritz series method for a continuum, it is equally valid for the finite element formulation, and for systems of particles and rigid bodies. It follows that the Jacobian constraint matrix, which we obtain from a description of the constraint conditions imposed in the generalized coordinates, provides all the information we need to characterize the role of the constraint forces in the assembly of structural components.

EXAMPLE 9.1

The sketch shows a bar bent at a right angle and lying in the horizontal plane. The force F at the tip acts in the vertical direction, so displacement in either of the horizontal directions may be ignored. Consider each straight segment of the bar to be a component. Select Ritz series appropriate to the displacement field of each component, then evaluate the corresponding Jacobian constraint matrix.

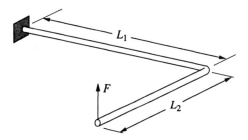

Solution The basic operations required to join structural components are the focus of this example. It also will give a hint of the expanded capability to model complicated systems provided by the Lagrange multiplier method. We define each straight bar segment as a component, and draw a sketch showing the coordinate system to be used for each.

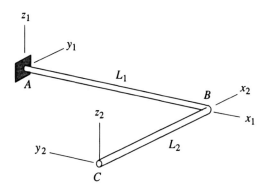

A static stress analysis shows that the vertical force F causes flexural deformation in bar BC. The static reactions where this segment is connected to bar AB consist of a shear force and bending moment. The reaction shear force applied by bar BC to bar AB induces a bending moment in bar AB, so it too undergoes flexural displacement. In contrast, the bending moment in bar BC at corner B represents a torsional load for bar AB. Similarly, the bending moment at the end of bar AB represents a torsional load for bar BC. Thus, both bars are subjected to a state of combined stress: flexure and torsion.

We define Ritz series for both types of displacement in each bar based on there being no connection at their common ends. End A is clamped to the wall, so the geometric boundary conditions for this segment are $w^1 = \partial w^1/\partial x_1 = \theta^1 = 0$ at $x_1 = 0$. Bar BC is unsupported at end C, so there are no geometric boundary conditions for the basis functions associated with that bar. We shall use power series terms as the basis functions. If we use the same number of terms for each series, we have

$$w^1 = \sum_{j=1}^{N} \psi_{wj}^1 q_{wj}^1, \qquad \psi_{wj}^1 = \left(\frac{x_1}{L_1}\right)^{j+1}$$

$$\theta^1 = \sum_{j=1}^{N} \psi_{\theta j}^1 q_{\theta j}^1, \qquad \psi_{\theta j}^1 = \left(\frac{x_1}{L_1}\right)^{j}$$

$$w^2 = \sum_{j=1}^{N} \psi_{wj}^2 q_{wj}^2, \qquad \psi_{wj}^2 = \left(\frac{x_2}{L_2}\right)^{j-1}$$

$$\theta^2 = \sum_{j=1}^{N} \psi_{\theta j}^2 q_{\theta j}^2, \qquad \psi_{\theta j}^2 = \left(\frac{x_2}{L_2}\right)^{j-1}$$

The generalized coordinate vectors for each structural component are

$$\{q^1\} = \left[\{q_w^1\}^{\mathrm{T}} \ \{q_\theta^1\}^{\mathrm{T}}\right]^{\mathrm{T}}, \qquad \{q^2\} = \left[\{q_w^2\}^{\mathrm{T}} \ \{q_\theta^2\}^{\mathrm{T}}\right]^{\mathrm{T}}$$

and the assembled set of generalized coordinates is

$$\{q\} = \left[\{q^1\} \ \{q^2\}\right]^{\mathrm{T}} \equiv \left[\{q_w^1\}^{\mathrm{T}} \ \{q_\theta^1\}^{\mathrm{T}} \ \{q_w^2\}^{\mathrm{T}} \ \{q_\theta^2\}^{\mathrm{T}}\right]^{\mathrm{T}}$$

Transverse displacements w^1 and w^2 are positive when in the positive z direction, and rotations are positive in the sense of the right-hand rule relative to the respective x axes. At joint B, it obviously must be that the transverse displacements in each bar are the same. Equating w^1 at $x_1 = L_1$ to w^2 at $x_2 = L_2$ yields the first constraint equation,

$$\sum_{j=1}^{N} q_{wj}^1 = \sum_{j=1}^{N} q_{wj}^2$$

The flexural rotation of bar AB at end B is $\partial w^1/\partial x_1$ at $x_1 = L_1$. Because w_1 is positive upward, a positive slope corresponds to rotation by the right-hand rule about the negative y_1 axis of bar AB. This is the same as the negative x_2 axis for bar BC. Thus, $\partial w^1/\partial x_1 = -\theta^2$ at corner B. Substituting the respective Ritz series leads to the second constraint equation,

$$\sum_{j=1}^{N} \left(\frac{j+1}{L_1}\right) q_{wj}^1 = -\sum_{j=1}^{N} q_{\theta j}^2$$

The third constraint equation is similar to the preceding. The flexural rotation of bar BC is $\partial w^2/\partial x_2$ about the negative y_2 axis for that bar. This corresponds to a rotation about the positive x_1 axis for bar AB, so

$$\sum_{j=1}^{N} \left(\frac{j-1}{L_2}\right) q_{wj}^2 = \sum_{j=1}^{N} q_{\theta j}^1$$

To identify $[a]$ we bring all terms in each of the constraint equations to the left side and compare each equation to a row of $[a]\{q\} = \{0\}$. Because there are three constraint equations and $\{q^1\}$ and $\{q^2\}$ are each $2N$ in size, $[a]$ is a $3 \times 4N$ matrix, whose nonzero elements are

$$a_{1j} = 1, \qquad a_{1(j+2N)} = -1, \qquad j = 1, ..., N$$

$$a_{2j} = \frac{j+1}{L_1}, \qquad a_{2(j+3N)} = -1, \qquad j = 1, ..., N$$

$$a_{3(j+N)} = -1, \qquad a_{3(j+2N)} = \frac{j-1}{L_2}, \qquad j = 1, ..., N$$

9.1.2 Synthesis of the Equations of Motion

The previous section addressed the issues related to assembling structural components. Here we return to considering the behavior of the individual parts. Let us designate as $\{q^\ell\}$ the vector of generalized coordinates associated with component ℓ. If appropriate, $\{q^\ell\}$ may itself be partitioned into terms associated with different types of displacement, as in the case of the preceding example. In any event, the prerequisite for using a subsystem as a structural component is that we have determined its equations of motion. This means that we have identified the inertia $[M^\ell]$, stiffness $[K^\ell]$, and damping $[C^\ell]$ corresponding to $\{q^\ell\}$. It also means that we have described the generalized forces $\{Q^\ell\}$ associated with the excitations. Thus, the equations of motion for component ℓ, isolated from the other components of the assembly, are

$$[M^\ell]\{\ddot{q}^\ell\} + [C^\ell]\{\dot{q}^\ell\} + [K^\ell]\{q^\ell\} = \{Q^\ell\} \qquad (9.1.18)$$

The generalized coordinates are defined individually for each component. Consequently, the set of system generalized coordinates $\{q\}$ may be defined by stacking the various $\{q^\ell\}$,

$$\{q\} = \left[\{q^1\}^T \ \{q^2\}^T \ \{q^3\}^T \ ... \right]^T \qquad (9.1.19)$$

We stack the equations of motion for the disjoint (that is, unconnected) structural components, which we write as

$$[M]\{\ddot{q}\} + [C]\{\dot{q}\} + [K]\{q\} = \{Q\} \qquad (9.1.20)$$

In view of the definition of $\{q\}$, the assembled coefficient matrices are

$$[M] = \begin{bmatrix} [M^1] & [0] & [0] & [0] \\ [0] & [M^2] & [0] & [0] \\ [0] & [0] & [M^3] & \cdots \\ \vdots & \vdots & \vdots & \ddots \end{bmatrix}, \qquad [C] = \begin{bmatrix} [C^1] & [0] & [0] & [0] \\ [0] & [C^2] & [0] & [0] \\ [0] & [0] & [C^3] & \cdots \\ \vdots & \vdots & \vdots & \ddots \end{bmatrix}$$

$$(9.1.21)$$

$$[K] = \begin{bmatrix} [K^1] & [0] & [0] & [0] \\ [0] & [K^2] & [0] & [0] \\ [0] & [0] & [K^3] & \cdots \\ \vdots & \vdots & \vdots & \ddots \end{bmatrix}, \qquad \{Q\} = \begin{Bmatrix} \{Q^1\} \\ \{Q^2\} \\ \{Q^3\} \\ \vdots \end{Bmatrix}$$

The generalized forces $\{Q\}$ account only for the external force acting on the individual components. To account for the fact that constraint forces act at the interface, we call on eq. (9.1.17). Thus, the equations of motion for the connected system are

$$[M]\{\ddot{q}\} + [C]\{\dot{q}\} + [K]\{q\} = \{Q\} + [a]^{\mathrm{T}}\{\lambda\}$$

$$[a]\{q\} = \{0\}$$

(9.1.22)

If the number of degrees of freedom for an unconnected component is N^{ℓ}, then the total number of generalized coordinates is $N = \sum N^{\ell}$. In addition, if N_c is the number of constraint equations, then there are N_c Lagrange multipliers to determine. The number of unknowns matches the N scalar equations contained in the first part of eqs. (9.1.22) and the N_c scalar constraint equations. Hence, we are ready to solve the coupled equations.

EXAMPLE 9.2

Synthesize the equations of motion for the bent bar in Example 9.1.

Solution We will see in this example that the steps leading to the coefficient matrices for the assembled system are essentially a repeated application of the Ritz method for a bar. The Ritz series for torsion and bending were selected in Example 9.1 such that

$$w^1 = \sum_{j=1}^{N} \psi^1_{wj} q^1_{wj}, \quad \psi^1_{wj} = \left(\frac{x}{L_1}\right)^{j+1}$$

$$\theta^1 = \sum_{j=1}^{N} \psi^1_{\theta j} q^1_{\theta j}, \quad \psi^1_{\theta j} = \left(\frac{x}{L_1}\right)^{j}$$

$$w^2 = \sum_{j=1}^{N} \psi^2_{wj} q^2_{wj}, \quad \psi^2_{wj} = \left(\frac{x}{L_2}\right)^{j-1}$$

$$\theta^2 = \sum_{j=1}^{N} \psi^2_{\theta j} q^2_{\theta j}, \quad \psi^2_{\theta j} = \left(\frac{x}{L_2}\right)^{j-1}$$

We consider each structural component to act individually at this stage, and the energies for each type of deformation are uncoupled. For the kinetic energy of component AB we have

$$T^1 = T_{\text{flex}} + T_{\text{torsion}} = \frac{1}{2}\int_0^{L_1} \rho A (\dot{w}^1)^2 \, dx + \frac{1}{2}\int_0^{L_1} \rho J (\dot{\theta}^1)^2 dx$$

$$= \frac{1}{2}\int_0^{L_1} \rho A \left(\sum_{j=1}^{N} \psi^1_{wj} \dot{q}^1_{wj}\right)\left(\sum_{n=1}^{N} \psi^1_{wn} \dot{q}^1_{wn}\right) dx$$

$$+ \frac{1}{2}\int_0^{L_1} \rho J \left(\sum_{j=1}^{N} \psi^1_{\theta j} \dot{q}^1_{\theta j}\right)\left(\sum_{n=1}^{N} \psi^1_{\theta n} \dot{q}^1_{\theta n}\right) dx$$

$$= \frac{1}{2}\{\dot{q}^1_w\}^{\mathrm{T}}[M^1_w]\{\dot{q}^1_w\} + \frac{1}{2}\{\dot{q}^1_\theta\}^{\mathrm{T}}[M^1_\theta]\{\dot{q}^1_\theta\}$$

The elements of the inertia matrices are

$$(M_w^1)_{jn} = \int_0^{L_1} \rho A \, \psi_{wj}^1 \psi_{wn}^1 \, dx = \frac{1}{j+n+3} \rho A L_1$$

$$(M_\theta^1)_{jn} = \int_0^{L_1} \rho J \psi_{\theta j}^1 \psi_{\theta n}^1 \, dx = \frac{1}{j+n+1} \rho J L_1$$

The treatment of the strain energy proceeds similarly, so that

$$V^1 = V_{\text{flex}} + V_{\text{torsion}} = \frac{1}{2} \int_0^{L_1} EI \left(\frac{\partial^2 w^1}{\partial x^2} \right)^2 dx + \frac{1}{2} \int_0^{L_1} GJ \left(\frac{\partial \theta^1}{\partial x} \right)^2 dx$$

$$= \frac{1}{2} \int_0^{L_1} EI \left(\sum_{j=1}^N \frac{\partial^2 \psi_{wj}^1}{\partial x^2} q_{wj}^1 \right) \left(\sum_{n=1}^N \frac{\partial^2 \psi_{wn}^1}{\partial x^2} q_{wn}^1 \right) dx$$

$$+ \frac{1}{2} \int_0^{L_1} GJ \left(\sum_{n=1}^N \frac{\partial \psi_{\theta j}^1}{\partial x^2} q_{\theta j}^1 \right) \left(\sum_{n=1}^N \frac{\partial \psi_{\theta n}^1}{\partial x} q_{\theta n}^1 \right) dx$$

$$= \frac{1}{2} \{q_w^1\}^{\mathrm{T}} [K_w^1]\{q_w^1\} + \frac{1}{2}\{q_\theta^1\}^{\mathrm{T}}[K_\theta^1]\{q_\theta^1\}$$

where the stiffness coefficients are

$$(K_w^1)_{jn} = \int_0^{L_1} EI \frac{\partial^2 \psi_{wj}^1}{\partial x^2} \frac{\partial^2 \psi_{wn}^1}{\partial x^2} dx = \frac{(j+1)j(n+1)n}{j+n-1} \frac{EI}{L_1^3}$$

$$(K_\theta^1)_{jn} = \int_0^{L_1} GJ \frac{\partial \psi_{\theta j}^1}{\partial x} \frac{\partial \psi_{\theta n}^1}{\partial x} dx = \frac{jn}{j+n-1} \frac{GJ}{L_1}$$

Aside from the difference in the basis functions, the evaluation for bar BC proceeds in the same manner. We find that

$$T^2 = \frac{1}{2}\{\dot{q}_w^2\}^{\mathrm{T}}[M_w^2]\{\dot{q}_w^2\} + \frac{1}{2}\{\dot{q}_\theta^2\}^{\mathrm{T}}[M_\theta^2]\{\dot{q}_\theta^2\}$$

$$V^2 = \frac{1}{2}\{q_w^2\}^{\mathrm{T}}[K_w^2]\{q_w^2\} + \frac{1}{2}\{q_\theta^2\}^{\mathrm{T}}[K_\theta^2]\{q_\theta^2\}$$

where the coefficients are

$$(M_w^2)_{jn} = \int_0^{L_2} \rho A \, \psi_{wj}^2 \psi_{wn}^2 \, dx = \frac{1}{j+n-1} \rho A L_2$$

$$(M_\theta^2)_{jn} = \int_0^{L_2} \rho J \psi_{\theta j}^2 \psi_{\theta n}^2 \, dx = \frac{1}{j+n-1} \rho J L_2$$

$$(K_w^2)_{jn} = \int_0^{L_2} EI \frac{\partial^2 \psi_{wj}^2}{\partial x^2} \frac{\partial^2 \psi_{wn}^2}{\partial x^2} dx = \begin{cases} 0 \text{ if } j \le 2 \text{ or } n \le 2 \\ \dfrac{(j-1)(j-2)(n-1)(n-2)}{j+n-5} \dfrac{EI}{L_2^3} \text{ otherwise} \end{cases}$$

$$(K_\theta^2)_{jn} = \int_0^{L_2} GJ \frac{\partial \psi_{\theta j}^1}{\partial x} \frac{\partial \psi_{\theta n}^1}{\partial x} dx = \begin{cases} 0 \text{ if } j = 1 \text{ or } n = 1 \\ \dfrac{(j-1)(n-1)}{j+n-1} \dfrac{GJ}{L_2} \text{ otherwise} \end{cases}$$

The only force that inputs power to the disjoint components is the force F at end C. The corresponding power input is

$$\mathcal{P}_{\text{in}}^2 = F\dot{w}^2(x_2 = 0) = F\sum_{j=1}^{N} \psi_{wj}^2(x_2 = 0)\dot{q}_{wj}^2 = F\dot{q}_{w1}^2$$

Thus, the only nonzero generalized force is

$$Q_{w1}^2 = F$$

We synthesize the system equations by partitioning the system matrices consistently with the generalized coordinates. The previous solution defined the component coordinates. Stacking them leads to

$$\{q\} = \left[\{q^1\}^T \ \{q^2\}^T\right]^T = \left[\{q_w^1\}^T \ \{q_\theta^1\}^T \ \{q_w^2\}^T \ \{q_\theta^2\}^T\right]^T$$

We arrange the inertia and stiffness matrices for the assembled system, as well as the Jacobian constraint matrix found previously, in the same manner. The resulting system equations matching eq. (9.1.22) are

$$\begin{bmatrix} [M_w^1] & [0] & [0] & [0] \\ [0] & [M_\theta^1] & [0] & [0] \\ [0] & [0] & [M_w^2] & [0] \\ [0] & [0] & [0] & [M_\theta^2] \end{bmatrix} \begin{Bmatrix} \{\ddot{q}_w^1\} \\ \{\ddot{q}_\theta^1\} \\ \{\ddot{q}_w^2\} \\ \{\ddot{q}_\theta^2\} \end{Bmatrix} + \begin{bmatrix} [K_w^1] & [0] & [0] & [0] \\ [0] & [K_\theta^1] & [0] & [0] \\ [0] & [0] & [K_w^2] & [0] \\ [0] & [0] & [0] & [K_\theta^2] \end{bmatrix} \begin{Bmatrix} \{q_w^1\} \\ \{q_\theta^1\} \\ \{q_w^2\} \\ \{q_\theta^2\} \end{Bmatrix}$$

$$= \begin{Bmatrix} \{0\} \\ \{0\} \\ \{Q_w^2\} \\ \{0\} \end{Bmatrix} + \begin{bmatrix} [a_w^1]^T \\ [a_\theta^1]^T \\ [a_w^2]^T \\ [a_\theta^2]^T \end{bmatrix} \begin{Bmatrix} \lambda_1 \\ \lambda_2 \\ \lambda_3 \end{Bmatrix}$$

$$\left[[a_w^1] \ [a_\theta^1] \ [a_w^2] \ [a_\theta^2]\right]\left[\{q_w^1\}^T \ \{q_\theta^1\}^T \ \{q_w^2\}^T \ \{q_\theta^2\}^T\right]^T = \{0\}$$

Seeing the assembled equations in partitioned form serves to emphasize an important aspect. The fact that all of the partitions of $[a]$ are nonzero means that the interface force and constraint conditions couple flexure and torsion in each bar, as well as coupling both bars.

9.1.3 Solution Methods

The form of the equations of motion for the synthesized system, eqs. (9.1.22), is somewhat different from any we have encountered thus far. The Lagrange multipliers occur as algebraic variables in the force equations, and not at all in the constraint equations. Equations such as these are called algebraic-differential equations. They cannot be solved directly by the standard methods we have used thus far. A number of solution strategies have been developed for solving equations of this type, which commonly arise in nonlinear dynamical systems; see for example Haug (1992) or Ginsberg (1995). One of these methods is especially suitable for linear vibratory systems. It involves a process of elimination that can be carried out algebraically if the number of equations is not too large, or numerically in any situation.

The constraint equations constitute a set of N_c algebraic relations that solely involve the generalized coordinates. We use these relations to solve for N_c generalized coordinates in terms of the other variables. To do so, we need to identify a set of N_c variables that occur in a linearly independent manner in one or more of the constraint equations. The remaining $N - N_c$ generalized coordinates may have any value without violating constraint conditions, so we refer to them as the *unconstrained* generalized coordinates, denoted $\{q_u\}$. The N_c generalized coordinates selected for elimination are the *constrained* variables, $\{q_c\}$.

The elimination process will be expedited if $\{q_c\}$ occurs at the top of $\{q\}$. In some cases, the first N_c generalized coordinates might meet the independence criterion, but it is quite probable that the elements of $\{q\}$ will need to be rearranged. Some individuals shift the variables manually, which requires reordering the columns of $[M]$, $[K]$, $[C]$, and $\{Q\}$ to maintain consistency. An alternative is to define a *sorting matrix*, which we will call $[P]$, which has the property that

$$\{q\} = [P]\begin{Bmatrix} \{q_c\} \\ \{q_u\} \end{Bmatrix} \qquad (9.1.23)$$

Fulfilling this definition will lead to a matrix $[P]$ filled with ones and zeroes. To demonstrate this operation, suppose $N = 5$, and we have selected $\{q_c\} = [q_3 \quad q_5]^T$, so that $\{q_u\} = [q_1 \quad q_2 \quad q_4]^T$. The term q_1 then occurs in the first row in the left side of eq. (9.1.23) and in the third row in the right side. Hence, the first row of $[P]$ contains zeroes, except in the third column where it is 1. Application of comparable considerations for each row of $\{q\}$ leads to the full $[P]$ matrix. The nonzero elements in this sample case are

$$P_{13} = P_{24} = P_{31} = P_{45} = P_{52} = 1 \qquad (9.1.24)$$

Substitution of eq. (9.1.23) rearranges the constraint equations to be

$$[\hat{a}]\begin{Bmatrix} \{q_c\} \\ \{q_u\} \end{Bmatrix} = \{0\}, \qquad [\hat{a}] = [a][P] \qquad (9.1.25)$$

The rearranged Jacobian constraint matrix $[\hat{a}]$ may be partitioned to conform to the partitioned generalized coordinates, such that

$$\begin{bmatrix} [\hat{a}_c] & [\hat{a}_u] \end{bmatrix}\begin{Bmatrix} \{q_c\} \\ \{q_u\} \end{Bmatrix} = \{0\} \qquad (9.1.26)$$

The expanded form of this expression is $[\hat{a}_c]\{q_c\} + [\hat{a}_u]\{q_u\} = \{0\}$. It is necessary that the $N_c \times N_c$ array $[\hat{a}_c]$ is full rank, that is, $\|[\hat{a}_c]\| \neq 0$. (If this condition is not met, it is necessary that we alter the selection of generalized coordinates to group into $\{q_c\}$.) In that case, it is possible to solve eq. (9.1.26) for $\{q_c\}$ corresponding to an arbitrary set of unconstrained generalized coordinates,

$$\{q_c\} = -[\hat{a}_c]^{-1}[\hat{a}_u]\{q_u\} \qquad (9.1.27)$$

From this, we find that the original generalized coordinate vector is

$$\{q\} = [B]\{q_u\}, \qquad [B] = [P]\begin{bmatrix} -[\hat{a}_c]^{-1}[\hat{a}_u] \\ [I] \end{bmatrix} \qquad (9.1.28)$$

Now consider the result of using this description of $\{q\}$ to form the original constraint equations,

$$[a]\{q\} = [a][B]\{q_u\} = \{0\} \qquad (9.1.29)$$

In view of eqs. (9.1.28) and (9.1.25), we have

$$[a][B] = [a][P] \begin{bmatrix} -[\hat{a}_c]^{-1}[\hat{a}_u] \\ [I] \end{bmatrix} = \begin{bmatrix} [\hat{a}_c] & [\hat{a}_u] \end{bmatrix} \begin{bmatrix} -[\hat{a}_c]^{-1}[\hat{a}_u] \\ [I] \end{bmatrix} \equiv [0] \quad (9.1.30)$$

The fact that $[a][B]$ vanishes merely demonstrates that eqs. (9.1.29) will be satisfied identically, regardless of $\{q_u\}$. In other words, any $\{q_u\}$ represents a displacement that is consistent with the constraint equations, and therefore kinematically admissible.

The equations governing $\{q_u\}$ are obtained by substituting eq. (9.1.28) into the force equations portion of eqs. (9.1.22). In order to maintain the symmetry of the coefficient matrices, we multiply the result of that substitution by $[B]^T$, which gives

$$[B]^T[M][B]\{\ddot{q}_u\} + [B]^T[C][B]\{\dot{q}_u\} + [B]^T[K][B]\{q_u\}$$
$$= [B]^T\{Q\} + [B^T][a]^T\{\lambda\} \quad (9.1.31)$$

In view of eq. (9.1.30), $[B]^T[a]^T \equiv [0]$, so the Lagrange multipliers drop out of consideration. Therefore the equations of motion reduce to

$$\boxed{[\hat{M}]\{\ddot{q}_u\} + [\hat{C}]\{\dot{q}_u\} + [\hat{K}]\{q_u\} = [B]^T\{Q\}} \quad (9.1.32)$$

where the matrices for the assembled system are

$$\boxed{[\hat{M}] = [B]^T[M][B], \qquad [\hat{C}] = [B]^T[C][B], \qquad [\hat{K}] = [B]^T[K][B]} \quad (9.1.33)$$

Some individuals find it surprising that the process of eliminating excess generalized coordinates from the governing equations simultaneously eliminates the Lagrange multipliers. However, it could not be otherwise. Any combination of $\{\dot{q}_u\}$ represents a kinematically admissible velocity field. The presence of Lagrange multipliers in equations of motion that solely contain $\{q_u\}$ would imply that the interface constraint forces input power. This would be contradictory, because the definition of constraint forces for time-invariant systems states that they are forces that do not input power in a kinematically admissible motion.

We began with a total of N combined equations of motion for the structural components and N_c constraint equations. The process of eliminating a set of N_c generalized coordinates entailed defining a sorting matrix $[P]$ according to eq. (9.1.23). This matrix led to a modified Jacobian constraint matrix $[\hat{a}]$ according to eq. (9.1.25). Inversion of a partition of $[\hat{a}]$ led to the transformation matrix $[B]$, defined in eq. (9.1.28), which enables us to determine the full set of generalized coordinates corresponding to the unconstrained variables $\{q_u\}$. Evaluation of $[B]$ is the central step, because it leads us to equations of motion for $\{q_u\}$, eq. (9.1.32), from which the Lagrange multipliers have been removed. These transformed equations of motion have the same form as those for any other time-independent system. Once we have determined the $\{q_u\}$ values, we find the full set of generalized coordinates, $\{q\}$, by evaluating eq. (9.1.28).

The elimination procedure enables us to perform modal analysis for transient and steady-state response. However, as we have seen on several occasions, direct frequency domain analysis can be quite useful for cases of harmonic excitation. The same is true when it comes to solving the assembled equations of motion. If all excitation forces acting on a system have frequency ω, the response and the Lagrange multipliers in the steady-state condition must also vary harmonically. As a result, the time-dependent terms in the assembled equations of motion, eqs. (9.1.22), will be

$$\begin{aligned}
\{Q\} &= \mathrm{Re}\{\{F\}\exp(i\omega t)\} \\
\{q\} &= \mathrm{Re}\{\{X\}\exp(i\omega t)\} \\
\{\lambda\} &= \mathrm{Re}\{\{\Lambda\}\exp(i\omega t)\}
\end{aligned} \tag{9.1.34}$$

When we substitute these descriptions into eqs. (9.1.22), we obtain a set of simultaneous algebraic equations,

$$\begin{bmatrix} [[K] + i\omega[C] - \omega^2[M]] & -[a]^{\mathrm{T}} \\ -[a] & [0] \end{bmatrix} \begin{Bmatrix} \{X\} \\ \{\Lambda\} \end{Bmatrix} = \begin{Bmatrix} \{F\} \\ \{0\} \end{Bmatrix} \tag{9.1.35}$$

The minus sign is introduced in the lower partition in order to make the assembled coefficient matrix symmetric.

Equation (9.1.35) offers a direct path to the complex response amplitudes, without the bother of performing the intermediate elimination step. A further benefit for some applications is that the approach concurrently yields the values of the Lagrange multipliers, which can be related to the amplitudes of the reaction forces.

EXAMPLE 9.3

A two-span beam has a torsional spring K_θ at its intermediate support. Use the Lagrange multiplier method to determine the natural frequencies and mode shapes for the system for three sets of system parameters: (a) $K_\theta L/EI = 0, b = 0.5L$; (b) $K_\theta L/EI = 500, b = 0.5L$; (c) $K_\theta L/EI = 500, b = 0.48L$.

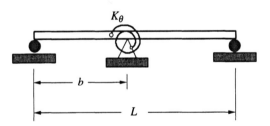

Solution This example is a straightforward illustration of modal analysis using the Lagrange multiplier method; what makes it especially interesting is the nature of the modes. The system has been the subject of numerous studies because it serves as a prototype for studying mode localization. The analysis we present is essentially the one offered by Pierre and Dowell (1987). We will formulate the system as a simply supported beam of length L subject to a constraint condition at $x = b$. Suitable basis functions are sines, which are the actual mode functions when the interior pin and torsional spring are not present. Thus, the Ritz series is

$$w = \sum_{j=1}^{N} q_j \sin\left(\frac{j\pi x}{L}\right)$$

The interior pin requires that the displacement at $x = b$ is zero. This leads to the constraint condition that

$$\sum_{j=1}^{N} q_j \sin\left(\frac{j\pi b}{L}\right) = 0$$

There are no other constraint conditions to satisfy, because slope continuity at $x = b$ is ensured by using a single Ritz series for the entire span. The constraint equation fits the general form $[a]\{q\} = \{0\}$, where $[a]$ is a $1 \times N$ array,

$$a_j = \sin\left(\frac{j\pi b}{L}\right)$$

To determine the inertia matrix we use the Ritz series to form the kinetic energy of the beam,

$$T = \frac{1}{2}\int_0^L \rho A(\dot{w})^2 \, dx = \frac{1}{2}\int_0^L \rho A\left[\sum_{j=1}^N \dot{q}_j \sin\left(\frac{j\pi x}{L}\right)\right]\left[\sum_{n=1}^N \dot{q}_n \sin\left(\frac{n\pi x}{L}\right)\right]dx$$

$$= \frac{1}{2}\{\dot{q}\}^T[M]\{\dot{q}\}, \qquad M_{jn} = \frac{\pi}{2}\rho AL\delta_{jn}$$

The potential energy in the torsional spring depends on the rotation of the beam at the pin, which is $\partial w/\partial x$ evaluated at $x = b$. Thus, the total potential energy is

$$V = \frac{1}{2}\int_0^L EI\left(\frac{\partial^2 w}{\partial x^2}\right)^2 dx + \frac{1}{2}K_\theta\left(\frac{\partial w}{\partial x}\bigg|_{x=b}\right)^2$$

$$= \frac{1}{2}\int_0^L EI\left[\sum_{j=1}^N \left(\frac{j\pi}{L}\right)^2 q_j^2 \sin\left(\frac{j\pi x}{L}\right)\right]\left[\sum_{n=1}^N \left(\frac{n\pi}{L}\right)^2 q_n\sin\left(\frac{n\pi x}{L}\right)\right]dx$$

$$+ \frac{1}{2}K_\theta\left[\sum_{j=1}^N \left(\frac{j\pi}{L}\right)q_j \cos\left(\frac{j\pi b}{L}\right)\right]\left[\sum_{n=1}^N \left(\frac{n\pi}{L}\right)^2 q_n\sin\left(\frac{n\pi x b}{L}\right)\right]$$

$$= \frac{1}{2}\{q\}^T[K]\{q\}, \qquad K_{jn} = j^4\frac{\pi^5 EI}{2L^3}\delta_{jn} + K_\theta\left(\frac{jn\pi^2}{L^2}\right)\cos\left(\frac{j\pi b}{L}\right)\cos\left(\frac{n\pi b}{L}\right)$$

Now that $[M]$, $[K]$, and $[a]$ have been determined, the next step is to eliminate constrained generalized coordinates. There is one constraint equation, so there is only one such variable. Any q_j for which the corresponding a_{1j} is nonzero may be selected for elimination; we shall use q_1. There then is no need to rearrange the elements of $\{q\}$, because $\{q_c\} = \{q_1\}$ and $\{q_u\} = [q_2 \ q_3 \ \cdots \ q_N]^T$. Hence, the sorting matrix in eq. (9.1.23) is $[P] = [I]$. For the notation used in eq. (9.1.28), we have

$$[a_c] = a_{11}, \qquad [a_u] = [a_{12} \ a_{13} \ \cdots \ a_{1N}]$$

The fact that $[a_c]$ consists of a single element leads to

$$\{q\} = [B]\{q_u\}, \qquad [B] = \begin{bmatrix} -\dfrac{a_{12}}{a_{11}} & -\dfrac{a_{13}}{a_{11}} & \cdots & -\dfrac{a_{1N}}{a_{11}} \\ 1 & 0 & \cdots & 0 \\ 0 & 1 & \cdots & 0 \\ \vdots & \vdots & \ddots & \vdots \\ 0 & 0 & \cdots & 1 \end{bmatrix}$$

We now have the quantities required to evaluate the $(N-1) \times (N-1)$ system matrices $[\hat{M}]$ and $[\hat{K}]$ according to eq. (9.1.33). Let us denote as $\{\hat{\Phi}_j\}$ the normal modes associated with these matrices, so that

$$[[\hat{K}] - \omega_j^2[\hat{M}]]\{\hat{\Phi}_j\} = \{0\}, \qquad \{\hat{\Phi}_j\}^T[\hat{M}]\{\hat{\Phi}_j\} = 1$$

The dimensional factors in $[\hat{M}]$ and $[\hat{K}]$ are removed by defining nondimensional natural frequencies Ω_j such that

$$\Omega_j \equiv \left(\frac{\rho A L^4}{EI}\right)^{1/2} \omega_j$$

Mathematical software yields $N - 1$ values of Ω_j and $N - 1$ corresponding modes $\{\hat{\Phi}_j\}$, which constitute the columns of the normal mode matrix $[\hat{\Phi}]$. This does not complete the analysis, because $\{\hat{\Phi}_j\}$ describes the proportions of the $N - 1$ unconstrained generalized coordinates. The mode vectors for the full set of N generalized coordinates are obtained by applying eq. (9.1.28),

$$[\Phi] = [B][\hat{\Phi}]$$

A mode function $\Psi_j(x)$ may then be obtained by using the elements of $\{\Phi_j\}$ as the Ritz series coefficients associated with the basis functions $\psi_n(x)$. The numerical algorithm we developed in Example 6.7 for carrying out this operation defines a rectangular array $[psi\_vals]$ whose column j holds the respective basis function values at a discrete set of points x_p, specifically, $psi\_vals_{pj} \equiv \psi_j(x_p)$. The mode functions at the x_p locations are then given by

$$\left[\{\Psi_1\} \ \{\Psi_2\} \ \cdots \ \{\Psi_{N-1}\}\right] = [psi\_vals][\Phi]$$

The first case to be considered is $b = 0.5L$ and $K_\theta = 0$. Because the intermediate support is at the midpoint, the system is symmetric, which means that the mode functions must be either symmetric or antisymmetric about the middle. The first four modes appear in the first graph. These results were obtained with $N = 15$, which is more than enough to attain convergence for the first four modes. The lowest natural frequencies are $\Omega_j = 39.48, 61.70, 157.91,$ and 200.12.

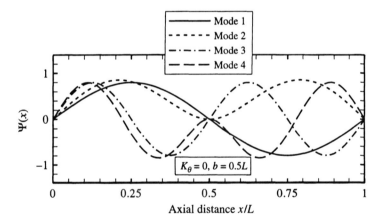

We see that the modes occur as pairs having the same general pattern on either span. Because a symmetric mode must have zero slope at the middle, it will have more curvature in the vicinity of $x = 0.5L$, and therefore more flexural strain energy, than the comparable antisymmetric mode. Consequently, the fundamental mode for this system features antisymmetric displacement.

In the second case, $b = 0.5L$ and $K_\theta L/EI = 500$, the torsional spring is quite stiff. The mode functions for this case are depicted in the second graph, and the lowest natural frequencies are $\Omega_j = 61.70, 61.90, 200.12,$ and 200.63. Here again $N = 15$ is sufficient to attain convergence for the first four mode functions.

The system is symmetric about $x = 0.5L$ in this case also, so the modes occur as symmetric and antisymmetric pairs. The difference here is that the natural frequencies for a pair are very close, with the symmetric mode having the lower value. A clue to the cause of this altered behavior lies in the fact that the first mode in this case is identical to the second mode in the

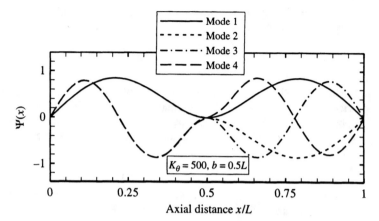

first case. Because there is no rotation at midspan in symmetric modes, the torsional spring has no effect for these modes. In contrast, the large value of K_θ limits the rotation in the antisymmetric modes, which is why the symmetric and antisymmetric modes have similar appearances over either span. Furthermore, although the rotation at the spring is relatively small, the large K_θ value leads to significant potential energy in the spring. This raises the natural frequencies of the antisymmetric modes. In the limit $K_\theta \to \infty$, the two spans would be uncoupled clamped-pinned beams having identical modal properties. The proximity of the pair of modes for large K_θ is a precursor of that limit.

In the third case, $b = 0.48L$, $K_\theta L/EI = 500$, the system is not symmetric. The lowest four natural frequencies in this case are $\Omega_j = 55.99, 65.86, 181.66,$ and 213.57, and the corresponding mode functions appear in the third graph. Obtaining these results requires $N > 40$ because there are small regions where the mode functions change rapidly. In comparison to the second case, changing the location of the pin support by only 2% of the total span changes the natural frequencies by 10%. In addition, the small change in the value of b drastically alters the mode functions. What makes these results especially interesting is the fact that each mode exhibits little displacement in one span. These modes are *localized*, just like the modes of the bladed disk system we encountered in Example 4.6. Chen and Ginsberg (1992) showed that this behavior arises because the asymmetric system's modes may be considered to be linear combinations of the modes for the $b = 0.5L$ case. This combination is such that the modal displacement on one side of the pin is reinforced, while it is canceled on the other side. The comparatively large change in the eigenvalues can be explained in terms of the properties of curves describing the eigenvalues as a function of b. Rather than the curves for different modes intersecting, the curves veer away sharply, resulting in large changes in the modal properties. Eigenvalue veering and model localization raise significant issues for design, because they suggest that the system parameters must be known quite accurately and the modal analysis must be done very carefully.

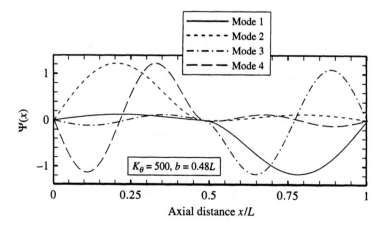

EXAMPLE 9.4

A frame consists of steel bars having a 100 mm $\times$ 100 mm square cross-section. Determine the lowest four natural frequencies and corresponding mode functions.

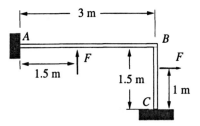

Solution This example, which addresses the coupling of axial and flexural displacements that occurs in frames, serves as a prototype for all operations entailed in implementing the Lagrange multiplier method. We identify bar AB as component 1 and bar BC as component 2. The disjoint bars are clamped at ends A and C, so power functions are suitable for the Ritz series for each displacement. Each bar undergoes axial and flexural displacement, so we have

$$u^1 = \sum_{j=1}^{N}\left(\frac{x}{L_1}\right)^j q_{uj}^1, \qquad w^1 = \sum_{j=1}^{N}\left(\frac{x}{L_1}\right)^{j+1} q_{wj}^1$$

$$u^2 = \sum_{j=1}^{N}\left(\frac{x}{L_2}\right)^j q_{uj}^2, \qquad w^2 = \sum_{j=1}^{N}\left(\frac{x}{L_2}\right)^{j+1} q_{wj}^2$$

Note that we have selected the same series length N for each displacement component.

The coordinate system for each component was defined in Figure 9.2. Equation (9.1.8) gives us the continuity conditions at the bend in terms of displacements, and eqs. (9.1.10) give the corresponding constraint conditions for arbitrary basis functions. For the power functions we have selected, these equations become

$$\sum_{j=1}^{N} q_{uj}^2 + \sum_{j=1}^{N} q_{wj}^2 = 0$$

$$\sum_{j=1}^{N} q_{wj}^1 - \sum_{j=1}^{N} q_{uj}^2 = 0$$

$$\sum_{j=1}^{N}\left(\frac{j+1}{L_1}\right) q_{wj}^1 - \sum_{j=1}^{N}\left(\frac{j+1}{L_2}\right) q_{wj}^2 = 0$$

The partitioned form of the assembled generalized coordinates is

$$\{q\} = [\{q_u^1\}^T \quad \{q_w^1\}^T \quad \{q_u^2\}^T \quad \{q_w^2\}^T]^T$$

and the corresponding form of the constraint equation is

$$[a]\{q\} = \{0\}, \qquad [a] = [[a_u^1][a_w^1][a_u^2][a_w^2]]$$

There are three constraint equations, so each partition of $[a]$ is a $3 \times N$ matrix. Comparing the standard form to the actual expressions shows that

$$[a_u^1] = \begin{bmatrix} 1 & 1 & \dots & 1 \\ 0 & 0 & \dots & 0 \\ 0 & 0 & \dots & 0 \end{bmatrix}, \quad [a_w^1] = \begin{bmatrix} 0 & 0 & \dots & 0 \\ 1 & 1 & \dots & 1 \\ 2/L_1 & 3/L_1 & \dots & (N+1)/L_1 \end{bmatrix}$$

$$[a_u^2] = \begin{bmatrix} 0 & 0 & \dots & 0 \\ -1 & -1 & \dots & -1 \\ 0 & 0 & \dots & 0 \end{bmatrix}, \quad [a_w^2] = \begin{bmatrix} 1 & 1 & \dots & 1 \\ 0 & 0 & \dots & 0 \\ -2/L_2 & -3/L_2 & \dots & -(N+1)/L_2 \end{bmatrix}$$

The inertia and stiffness matrices for each component are composed of $N \times N$ partitions consisting of the standard forms for extension and bending, which are uncoupled,

$$[M^1] = \begin{bmatrix} [M_u^1] & [0] \\ [0] & [M_w^1] \end{bmatrix}, \quad [M^2] = \begin{bmatrix} [M_u^2] & [0] \\ [0] & [M_w^2] \end{bmatrix}$$

$$[K^1] = \begin{bmatrix} [K_u^1] & [0] \\ [0] & [K_w^1] \end{bmatrix}, \quad [K^2] = \begin{bmatrix} [K_u^2] & [0] \\ [0] & [K_w^2] \end{bmatrix}$$

The inertia coefficients for component $\ell = 1$ or 2 are

$$(M_u^\ell)_{jn} = \int_0^{L_\ell} \rho A \left(\frac{x}{L_\ell}\right)^j \left(\frac{x}{L_\ell}\right)^n dx = \frac{1}{j+n+1} \rho A L_\ell$$

$$(M_w^\ell)_{jn} = \int_0^{L_\ell} \rho A \left(\frac{x}{L_\ell}\right)^{j+1} \left(\frac{x}{L_\ell}\right)^{n+1} dx = \frac{1}{j+n+3} \rho A L_\ell$$

and the stiffness coefficients are

$$(K_u^\ell)_{jn} = \int_0^{L_\ell} \frac{EA}{L_\ell^2} (jn) \left(\frac{x}{L_\ell}\right)^{j-1} \left(\frac{x}{L_\ell}\right)^{n-1} dx = \frac{jn}{j+n-1} \frac{EA}{L_\ell}$$

$$(K_w^\ell)_{jn} = \int_0^{L_\ell} \frac{EI}{L_\ell^4} (j+1)(n+1)(jn) \left(\frac{x}{L_\ell}\right)^{j-1} \left(\frac{x}{L_\ell}\right)^{n-1} dx$$

$$= \frac{(j+1)(n+1)(jn)}{j+n-1} \frac{EI}{L_\ell^3}$$

Upon substitution of the given parameters, $L_1 = 3$ m, $L_2 = 1.5$ m, $E = 210(10^9)$ Pa, $A = 0.01$ m^2, and $I = 8.333(10^{-6})$ m^4, we are ready to assemble the inertia and stiffness coefficients. In view of the uncoupled nature of extension and flexure in the disjoint components, the assembled inertia and stiffness matrices are $4N \times 4N$ arrays given by

$$[\hat{M}] = \begin{bmatrix} [M_u^1] & [0] & [0] & [0] \\ [0] & [M_w^1] & [0] & [0] \\ [0] & [0] & [M_u^2] & [0] \\ [0] & [0] & [0] & [M_w^2] \end{bmatrix}, \quad [\hat{K}] = \begin{bmatrix} [K_u^1] & [0] & [0] & [0] \\ [0] & [K_w^2] & [0] & [0] \\ [0] & [0] & [K_u^2] & [0] \\ [0] & [0] & [0] & [K_w^2] \end{bmatrix}$$

There are three constraint conditions, so we must select three generalized coordinates as the constrained variables. We cannot select any three variables because the columns of $[a]$ corresponding to the selected variables will form $[a_c]$, which must be nonsingular. Because $[a_u^1]$ has nonzero values in the first row, we select q_{u1}^1 as one constrained variable. For the second constrained variable, we observe that $[a_u^2]$ has nonzero values in the second row, so we select q_{u1}^2. The third variable we select is q_{w1}^2, which is acceptable because $[a_w^2]$ has nonzero elements in the third row. Thus, the scheme we select defines

$$\{q_c\} = \begin{bmatrix} q_{u1}^1 & q_{u1}^2 & q_{w1}^2 \end{bmatrix}^{\mathrm{T}}$$

$$\{q_u\} = \begin{bmatrix} q_{u2}^1 \dots q_{uN}^1 & q_{w1}^1 \dots q_{wN}^1 & q_{u2}^2 \dots q_{uN}^2 & q_{w2}^2 \dots q_{2N}^2 \end{bmatrix}^{\mathrm{T}}$$

The next step is to identify the $4N \times 4N$ sorting matrix $[P]$ in eq. (9.1.23). We achieve this by matching both sides of $\{q\} = [P]\big[\{q_c\}^{\mathrm{T}} \ \{q_u\}^{\mathrm{T}}\big]^{\mathrm{T}}$. The nonzero elements are found to be

$$P_{11} = P_{(2N+1)2} = P_{(3N+1)3} = 1$$

$$P_{j(j+2)} = 1, \quad j = 2, \dots, 2N$$

$$P_{j(j+1)} = 1, \quad j = 2N+2, \dots, 3N, \quad P_{jj} = 1, \quad j = 3N+2, \dots, 4N$$

The product $[\hat{a}] = [a][P]$ is the constraint matrix partitioned consistently with the definition of $\{q_c\}$. The left 3×3 partition of this product is $[a_c]$, such that

$$[[\hat{a}_c] \ [\hat{a}_u]] = [a][P]$$

We use the submatrix function on Mathcad or an index range in MATLAB to extract $[\hat{a}_c]$ and $[\hat{a}_u]$ from the product, and then compute $[B_1] = -[\hat{a}_c]^{-1}[\hat{a}_u]$, which is a $3 \times (4N-3)$ matrix. Stacking this above a $(4N-3) \times (4N-3)$ identity matrix produces the $4N \times (4N-3)$ transformation matrix $[B]$, as described by eq. (9.1.28).

The remainder of the computations are straightforward. The system matrices for the eigenvalue problem are obtained by using $[B]$ to transform the assembled inertia and stiffness matrices to $(4N-3) \times (4N-3)$ arrays associated with the unconstrained generalized coordinates,

$$[\hat{M}] = [B]^{\mathrm{T}}[M][B], \qquad [\hat{K}] = [B]^{\mathrm{T}}[K][B]$$

We input these matrices to our mathematical software in order to determine the $4N - 3$ solutions for the general eigenvalue problem, $[[\hat{K}] - \omega_j^2[\hat{M}]]\{\phi_j\} = \{0\}$. We normalize the eigenvectors with respect to $[\hat{M}]$, which yields the normal mode matrix,

$$[\hat{\Phi}] = [\{\hat{\Phi}_1\} \ \{\hat{\Phi}_2\} \ \cdots \ \{\hat{\Phi}_{4N-3}\}], \qquad \{\hat{\Phi}_j\}^{\mathrm{T}}[\hat{M}]\{\hat{\Phi}_j\} = 1$$

These mode vectors describe the proportions of the unconstrained generalized coordinates in a free vibration. The proportions of all generalized coordinates are described by

$$[\Phi] = [B][\hat{\Phi}]$$

Because the analysis yields $4N - 3$ eigensolutions, $[\Phi]$ has $4N$ rows and $4N - 3$ columns. The value of N used for the computations should be verified for convergence of the desired natural frequencies and modes. In the present case $N = 6$ is found to be adequate.

The physical displacements, rather than the proportions of the q_j variables, are the quantities of interest. These displacements may be determined by using each mode vector to synthesize the Ritz series. Each partition of $\{\Phi_j\}$ corresponds to a distinct displacement in one of the structural components. Because each partition has length N, we have

$$(u^1)_n = \sum_{j=1}^{N} \left(\frac{x}{L_1}\right)^j \Phi_{jn}, \qquad (w^1)_n = \sum_{j=1}^{N} \left(\frac{x}{L_1}\right)^{j+1} \Phi_{(j+N)n}$$

$$(u^2)_n = \sum_{j=1}^{N} \left(\frac{x}{L_2}\right)^j \Phi_{(j+2N)n}, \qquad (w^2)_n = \sum_{j=1}^{N} \left(\frac{x}{L_2}\right)^{j+1} \Phi_{(j+3N)n}$$

We now are faced with the question of how to display the results. We could create conventional graphs or u and w in each mode. A more meaningful picture emerges if we use the u and w values at a set of points to create displacement vectors. In view of the definition of the x and y directions in each component in Figure 9.2, u^2 is in the same direction as w^1 and w^2 is in the same direction as $-u^1$. Specialized graphical visualization software (Tecplot was used here) can convert a four-column array whose rows contain x and y coordinates of a point and x and y displacement components into a graphical picture, in which lines emanating from each point proportionally represent the displacement vectors. The results for the first four modes are graphed below.

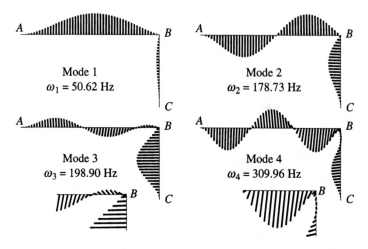

Only in the enlarged views in the vicinity of corner B is the axial displacement u evident. In essence, this behavior results because the extensional rigidity of a member is EA/L, which is much greater than the flexural rigidity EI/L^3. In the static analysis of frames it is standard to consider the structural members to be rigid in extension. Such an analysis would be successful for the lower natural frequencies. Because the axial displacements are very small in these modes, and u for bar BC is w for bar AB, and vice versa, both bars behave like clamped-hinged beams. In each of the first three modes, bar AB follows the successive modes of such a beam, while bar BC executes the first mode. The displacement in bar BC becomes larger relative to bar AB with increasing mode number as a consequence of the increasing rotation at corner B. The reason bar BC remains in the first mode pattern lies in the fact that the isolated bar's natural frequency is scaled by $(EI/\rho A L_2^4)^{1/2}$, which is four times higher than the comparable factor for bar AB. Thus, the vibration frequency must be much higher before bar BC shows evidence of the higher modal patterns. Such is the case for the fourth mode.

9.2 COMPONENT MODE SYNTHESIS

The method of Lagrange multipliers greatly enhances our capabilities. However, the astute reader might have recognized a limitation. Specifically, the method only addresses situations where the components are subjected to additional constraints in the process of assembling the system. In some situations it might be appropriate to build up the assembly from disjoint components that are more constrained than they are when joined. A simple example of such a situation may be found in Figure 9.2. The components treated in the context of that figure were cantilevered beams. Suppose we instead used two fixed-fixed beams as the components. Joining such bars would require modifying the clamped boundary conditions to allow for displacement and rotation at the junction.

Component mode synthesis builds on the Lagrange multiplier method by introducing additional basis functions that enable us to remove constraint conditions. The method's name conveys another modification from the preceding method: It is formulated in terms of the modal solutions associated with the component structures. The concept is that if we have a system with many degrees of freedom, we may perform a modal analysis of each component. If we use a subset of those component modes, truncated according to the guidelines for modal series, we may synthesize a system model that has far fewer degrees of freedom than the number of degrees of freedom contained in all of the components. Thus, component mode synthesis can be used to achieve model reduction.

Several versions of component mode synthesis have been formulated. The presentation here will focus on the rudimentary concepts. Hurty's paper (1965) provided the spark for much of the later development. The text by Craig (1981) provides further detail, and an overview of several versions may be found in the paper by Hintz (1975).

9.2.1 Relaxation of Constraints

Let us consider using a clamped-clamped beam as the component leading to a cantilevered beam. In Figure 9.3(a) we have the component clamped-clamped structure subjected to some excitation. The displacement and rotation at this end are zero, which are not geometric boundary conditions for the structure we wish to synthesize. Thus, we add basis functions that violate these constraints, which is the operation depicted in Figure 9.3(b). This is a process of relaxing constraint conditions.

An unlimited number of functions violate the constraint conditions to be relaxed. Hurty's development (1965) introduced the concept of *constraint modes*, which are defined as the *static* displacement that results when only one of the boundary conditions to be relaxed is violated. Synthesizing the cantilevered beam requires two constraint modes: One is the static displacement when the end is given a rotation θ_L with the displacement at that end held at zero, while the other constraint mode imparts a static displacement at that end with the rotation held at zero. The displacement field for the synthesized structure then is a series containing the terms representing the original clamped-clamped component beam and the constraint modes multiplied by coefficients to be determined. This series is used to enforce the relevant constraint equations, as well as to form the equations of motion for the coefficients of the Ritz series and the constraint modes.

Hurty's treatment of each structural component is a generalization of the preceding treatment of a beam. The individual components are defined such that they are immobilized against displacement and rotation at their interface with other components. For this reason, the Ritz series basis functions for the components are referred to as *fixed-interface modes*. In other words, the fixed-interface modes are kinematically admissible displacement patterns for a structural component in the case

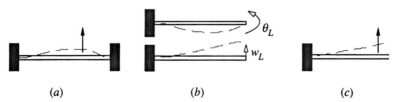

| (a) | (b) | (c) |

FIGURE 9.3 Addition of constraint modes to synthesize a cantilevered beam from a hinged-hinged beam.

where that component is immobilized at the interfaces. Hurty's development addresses situations where relaxation of the interface constraints for a component leads to the possibility of rigid body motion. For the sake of simplicity, we shall avoid such situations.

As noted in the preceding, transverse displacement has two constraint modes, corresponding to unit displacement and rotation at the interface. There also is a single constraint mode for axial displacement. Each of the constraint modes is obtained from a static displacement analysis. Hence, they satisfy $EA\, d^2u/dx^2 = 0$ or $EI\, d^4w/dx^4 = 0$. If L is the length of a component bar, integrating the static differential equation leads to

$$u = c_0 + c_1\frac{x}{L}, \quad w = d_0 + d_1\frac{x}{L} + d_2\left(\frac{x}{L}\right)^2 + d_3\left(\frac{x}{L}\right)^3$$

The constraint modes are defined to have a unit value for the geometric boundary condition they violate, and zero values for the other geometric boundary conditions. We shall use a superscript C to denote the constraint modes in order to distinguish them from the fixed-interface modes, which are denoted with superscript F. The table below lists the constraint modes and associated boundary conditions; these functions relax constraint conditions at $x = L$. (Situations where it is desired to relax conditions at $x = 0$ may be addressed by replacing x with $L - x$ in the tabulated functions. This transformation swaps the interface conditions between the ends, except that $d\psi_\chi^C/dx = -1$ at $x = 0$.)

Interface condition	$x = 0$	$x = L$
Relax axial displacement		
$\psi_u^C(x) = \dfrac{x}{L}$	$\psi_u^C = 0$	$\psi_u^C = 1$
Relax transverse displacement	$\psi_w^C = 0$	$\psi_w^C = 1$
$\psi_w^C(x) = 3\dfrac{x^2}{L^2} - 2\dfrac{x^3}{L^3}$	$\dfrac{d\psi_w^C}{dx} = 0$	$\dfrac{d\psi_w^C}{dx} = 0$
Relax rotation	$\psi_\chi^C = 0$	$\psi_\chi^C = 0$
$\psi_\chi^C(x) = L\left(-\dfrac{x^2}{L^2} + \dfrac{x^3}{L^3}\right)$	$\dfrac{d\psi_\chi^C}{dx} = 0$	$\dfrac{d\psi_\chi^C}{dx} = 1$

In principle, the fixed-interface modes could be any set of basis functions. However, using the true modes leads to the possibility of reducing the number of degrees of freedom, as we will see later. Also, developments subsequent to Hurty's paper, such as Craig and Brampton's (1968) work, have led to the ability to use any structural configuration as a basic component. In order to focus on the basic concepts, we will adhere to the original restriction to structural components that are fixed at their interfaces with other components.

The displacement field of each structural component in the assembled system is described by Ritz series composed of a combination of fixed-interface modes and constraint modes. These series are coerced through the Lagrange multiplier formulation to satisfy the interface constraint conditions for the assembled system. Both sets of functions shall be denoted as vectors to emphasize that the displacement field they represent may consist of axial, transverse, and/or torsional motions. The corresponding Ritz series for the displacement field in the ℓth structural component is

$$\vec{u}^{\ell} = \sum_{j=1}^{N_F} \bar{\psi}_j^{F\ell} q_j^{F\ell} + \sum_{j=1}^{N_C} \bar{\psi}_j^{C\ell} q_j^{C\ell} \qquad (9.2.1)$$

The number of terms N_F and N_C in the preceding may vary between component structures. Presumably, the number N_F of fixed-interface modes is such that the highest natural frequency of the fixed-interface modes is much higher than the upper limit of the frequency range of interest. The number N_C of constraint modes is set by the number of fixed constraints for member ℓ that must be relaxed. The generalized coordinate vector corresponding to eq. (9.2.1) is

$$\{q^{\ell}\} = \left[\{q^{F\ell}\}^{\mathrm{T}} \ \{q^{C\ell}\}^{\mathrm{T}} \right]^{\mathrm{T}} = \left[q_1^{F\ell} \ \dots \ q_{N_F}^{F\ell} \ q_1^{C\ell} \ \dots \ q_{N_C}^{C\ell} \right]^{\mathrm{T}} \qquad (9.2.2)$$

The general outline of the analysis from this juncture is much like the developments in Section 9.1. Thus, the vector $\{q\}$ holding all of the generalized coordinates is defined by stacking the variables for each component, as described by eq. (9.1.11). The Jacobian constraint matrix is then obtained by using each Ritz series, evaluated at the interface between structure components, to describe the interface constraint conditions. This yields a set of linear constraint equations in the form of eq. (9.1.12). By definition, the fixed-interface modes contribute to neither displacement nor rotation at interfaces. Also, the constraint modes yield a unit value for the constraint condition they relax. Consequently, the constraint equations will contain only some of the $\{q^{C\ell}\}$ variables.

EXAMPLE 9.5

The frame shown in the sketch is like the one in Example 9.4, except that end C is hinged. Describe the fixed-interface modes and constraint modes appropriate to an analysis of this structure using component mode synthesis. Then identify the Jacobian constraint matrix $[a]$, the set of generalized coordinates to be considered constrained, the corresponding sorting matrix $[P]$, and the transformation matrix $[B]$.

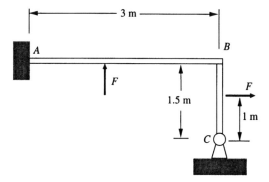

Solution By using a system similar to one we have already examined, this example will highlight the differences between component mode synthesis and generic Lagrange multiplier formulations. The structural components are bars AB and BC. The coordinate systems are illustrated in the first sketch.

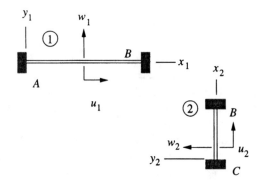

Note that bar BC is depicted as a clamped-clamped beam in order to illustrate how constraints at supports may be relaxed. A simpler analysis would consider bar BC to be pinned at end C. It then would only be necessary to relax the clamped constraints at ends B.

The fixed-interface modes are the normal modes for a clamped-clamped bar. For axial deformation, with $\ell = 1$ or 2 for the respective bars, we find from Appendix C that

$$\psi_{uj}^{F\ell} = \left(\frac{2}{\rho A L_\ell}\right)^{1/2} \sin\left(\frac{j\pi x}{L_\ell}\right), \quad \omega_j^\ell = \left(\frac{E}{\rho}\right)^{1/2}\frac{j\pi}{L_\ell}$$

The normalized clamped-clamped mode functions are given by

$$\psi_{wj}^{F\ell} = C_j^\ell \Psi\left(\frac{x}{L_\ell}, \alpha_j\right)$$

$$\psi\left(\frac{x}{L_\ell}, \alpha_j\right) = \left[\sin\left(\alpha_j\frac{x}{L_\ell}\right) - \sinh\left(\alpha_j\frac{x}{L_\ell}\right)\right] + B_j\left[\cos\left(\alpha_j\frac{x}{L_\ell}\right) - \cosh\left(\alpha_j\frac{x}{L_\ell}\right)\right]$$

$$C_j^\ell = \left[\rho A L_\ell \int_0^1 \psi(\xi, \alpha_j)^2 d\xi\right]^{-1/2}$$

where the characteristic equation for α_j and the expression for B_j^ℓ are listed in Appendix C.

The next step is to identify which fixed interface constraints should be relaxed. At end B, both bars are free to displace in the axial and transverse directions and to rotate. Hence, we must relax each of the clamped boundary conditions at end B. This requires three constraint modes for each bar, $\psi_1^{C\ell} = \psi_u^C(x_\ell)$, $\psi_2^{C\ell} = \psi_w^C(x_\ell)$, and $\psi_3^{C\ell} = \psi_\chi^C(x_\ell)$, $\ell = 1$ or 2, representing unit axial and transverse displacements, and a unit rotation at end B. Also, the fixed-interface modes for bar BC clamp end C, but the actual connection is a pin. Thus, we add an additional constraint mode, $\psi_4^{C2} = \psi_\chi^C(L_2 - x_2)$, which replaces x_2 with $L_2 - x_2$ in the tabulated function ψ_χ^C in order to relax the constraint condition at $x_2 = 0$.

The Ritz series for each bar is a sum of the fixed-interface and constraint modes. Let us use the same number N of axial and flexural modes. It is convenient to place the generalized coordinates for the constraint modes at the end of the generalized coordinate vector for each structural component, so that

$$\{q^1\} = \left[\{q_u^{F1}\}^T \; \{q_w^{F1}\}^T \; q_1^{C1} \; q_2^{C1} \; q_3^{C1}\right]^T$$

$$\{q^2\} = \left[\{q_u^{F2}\}^T \; \{q_w^{F2}\}^T \; q_1^{C2} \; q_2^{C2} \; q_3^{C2} \; q_4^{C2}\right]^T$$

The corresponding Ritz series for each displacement are

$$u^1(x) = \sum_{j=1}^{N} \psi_{uj}^{F1} q_{uj}^{F1} + \psi_1^{C1} q_1^{C1}$$

$$w^1(x) = \sum_{j=1}^{N} \psi_{wj}^{F1} q_{wj}^{F1} + \psi_2^{C1} q_2^{C1} + \psi_3^{C1} q_3^{C1}$$

$$u^2(x) = \sum_{j=1}^{N} \psi_{uj}^{F2} q_{uj}^{F2} + \psi_1^{C2} q_1^{C2}$$

$$w^2(x) = \sum_{j=1}^{N} \psi_{wj}^{F2} q_{wj}^{F2} + \psi_2^{C2} q_2^{C2} + \psi_3^{C2} q_3^{C2} + \psi_4^{C2} q_4^{C2}$$

The structure to be synthesized welds ends B of each bar. The displacement constraint conditions are the same as those in eqs. (9.1.10),

$$u^1(L_1) = -w^2(L_2), \quad w^1(L_1) = u^2(L_2), \quad \left.\frac{\partial w^1}{\partial x}\right|_{x=L_1} = \left.\frac{\partial w^2}{\partial x}\right|_{x=L_2}$$

The fixed-interface modes evaluate to zero at the ends, and each constraint mode gives either zero or one, depending on its type. Thus, the three constraint equations are

$$q_1^{C1} + q_2^{C2} = 0, \quad q_2^{C1} - q_1^{C2} = 0, \quad q_3^{C1} - q_3^{C2} = 0$$

Note that no constraint conditions are imposed on the fourth constraint mode because rotation about pin C is unrestricted.

The assembled set of generalized coordinates is

$$\{q\} \equiv \left[\{q^1\}^T \ \{q^2\}^T\right]^T$$

$$= \left[\{q_u^{F1}\}^T \ \{q_w^{F1}\}^T \ q_1^{C1} \ q_2^{C1} \ q_3^{C1} \ \{q_u^{F2}\}^T \ \{q_w^{F2}\}^T \ q_1^{C2} \ q_2^{C2} \ q_3^{C2} \ q_4^{C2}\right]^T$$

There are three constraint equations and $4N + 7$ generalized coordinates, so $[a]$ is a $3 \times (4N + 7)$ array. The nonzero elements $[a]$ obtained by matching $[a]\{q\} = \{0\}$ to the actual constraint equations are

$$a_{1(2N+1)} = a_{1(4N+5)} = a_{2(2N+2)} = -a_{2(4N+4)} = a_{3(2N+3)} = -a_{3(4N+6)} = 1$$

Because q_1^{C1}, q_1^{C2}, and q_3^{C2} appear in a linearly independent manner in the constraint equations, we select them as the three constrained generalized coordinates for $\{q_c\}$, and form $\{q_u\}$ from the remaining variables

$$\{q_c\} = \left[q_1^{C1} \ q_1^{C2} \ q_3^{C2}\right]^T$$

$$\{q_u\} = \left[\{q_u^{F1}\}^T \ \{q_w^{F1}\}^T \ q_2^{C1} \ q_3^{C1} \ \{q_u^{F2}\}^T \ \{q_w^{F2}\}^T \ q_2^{C2} \ q_4^{C2}\right]$$

To determine the sorting matrix $[P]$ we match the original and rearranged sets of generalized coordinate vectors,

$$\{q\} \equiv \left[\{q^1\}^T \ \{q^2\}^T\right]^T = [P]\left[\{q_c\}^T \ \{q_u\}^T\right]^T$$

Setting $P_{rs} = 1$ equates element r of $\{q\}$ to element s of $\left[\{q_c\}^T \ \{q_u\}^T\right]^T$, so matching the left and right sides of the preceding equation leads to

$$P_{j(j+3)} = P_{(j+2N+3)(j+2N+5)} = 1, \quad j = 1, 2, \ldots, 2N$$

$$P_{(2N+1)1} = P_{(2N+2)(2N+4)} = P_{(2N+3)(2N+5)} = 1$$

$$P_{(4N+4)2} = P_{(4N+5)(4N+6)} = P_{(4N+6)3} = P_{(4N+7)(4N+7)} = 1$$

The elements of $[P]$ not listed are zero.

The Jacobian constraint matrix corresponding to the altered sequence of generalized coordinates is

$$[\hat{a}] = \left[[\hat{a}_c] \ [\hat{a}_u]\right] = [a][P]$$

In the preceding, $[\hat{a}_c]$ is a 3×3 array. We compute $-[\hat{a}_c]^{-1}[\hat{a}_u]$ and stack the result above a $4N + 4$ identity matrix. Multiplying that matrix by $[P]$ yields the transformation matrix $[B]$ specified in eq. (9.1.28). For $N = 6$, the nonzero elements of $[B]$ are found to be

$$B_{jj} = 1, \quad j = 1, \ldots, 12, \quad B_{(13)(27)} = -1, \quad B_{(j+13)(j+12)} = 1, \quad j = 1, \ldots, 14$$

$$B_{(28)(13)} = B_{(29)(27)} = B_{(30)(14)} = B_{(31)(28)} = 1$$

9.2.2 Component Properties

Once the constraint modes have been defined, the procedure from that point onward follows the Lagrange multiplier method. We form the system matrices $[M^\ell]$, $[K^\ell]$, and $[C^\ell]$ for each component. We also form the generalized force $\{Q^\ell\}$ representing the external force system acting on each component. These quantities are obtained by the standard method of substituting the Ritz series, eq. (9.2.1), into the energy and power expressions appropriate to that component. In keeping with the partitioned definition of $\{q\}$, we partition the matrices in the same manner as $\{q^\ell\}$, so that

$$[M^\ell] = \begin{bmatrix} [M^\ell]^{FF} & [M^\ell]^{FC} \\ [M^\ell]^{CF} & [M^\ell]^{CC} \end{bmatrix}, \quad [K^\ell] = \begin{bmatrix} [K^\ell]^{FF} & [0] \\ [0] & [M^\ell]^{CC} \end{bmatrix}$$

$$[C^\ell] = \begin{bmatrix} [C^\ell]^{FF} & [C^\ell]^{FC} \\ [C^\ell]^{CF} & [C^\ell]^{CC} \end{bmatrix}, \quad \{Q^\ell\} = \begin{bmatrix} \{Q^\ell\}^F \\ \{Q^\ell\}^C \end{bmatrix} \tag{9.2.3}$$

The coefficient matrices are symmetric, so the following properties apply to the partitions:

$$[M^\ell]^{FF} = \left[[M^\ell]^{FF}\right]^T, \quad [M^\ell]^{CC} = \left[[M^\ell]^{CC}\right]^T, \quad [M^\ell]^{CF} = \left[[M^\ell]^{FC}\right]^T$$

$$[C^\ell]^{FF} = \left[[C^\ell]^{FF}\right]^T, \quad [C^\ell]^{CC} = \left[[C^\ell]^{CC}\right]^T, \quad [C^\ell]^{CF} = \left[[C^\ell]^{FC}\right]^T \tag{9.2.4}$$

$$[K^\ell]^{FF} = \left[[K^\ell]^{FF}\right]^T, \quad [K^\ell]^{CC} = \left[[K^\ell]^{CC}\right]^T$$

The off-diagonal partitions of $[K^\ell]$ are identically zero, which means that the constraint modes are elastically uncoupled from the fixed interface modes. To see why this is so, we observe that stiffness coefficients express the work done by conservative forces. A term K_{ij} is proportional to the work done by a conservative generalized force Q_i in a displacement in which the only nonzero generalized coordinate is q_j. The off-diagonal partitions of $[K^\ell]$ represent the work done by the interface (end) forces and couples associated with the constraint modes $\psi_i^{C\ell}$ when the component displaces proportionally to one of the fixed-interface modes $\psi_j^{F\ell}$. However, $\psi_j^{F\ell}$ is defined to have zero displacement and rotation at interfaces, so these stiffness coefficients are zero.

The other coefficients are determined by integration, in the usual manner for Ritz series. For example, the mass coefficients are

$$(M^\ell)_{jn}^{FF} = \int_0^{L_\ell} pA\ \bar{\psi}_j^{F\ell} \cdot \bar{\psi}_n^{F\ell}\ dx, \quad (M^\ell)_{jn}^{FC} = \int_0^{L_\ell} pA\ \bar{\psi}_j^{F\ell} \cdot \bar{\psi}_n^{C\ell}\ dx$$

$$(M^\ell)_{jn}^{CC} = \int_0^{L_\ell} pA\ \bar{\psi}_j^{C\ell} \cdot \bar{\psi}_n^{C\ell}\ dx$$

(9.2.5)

The vector notation for the basis functions and the dot product serve to emphasize that cross-terms involving functions for different types of displacement (axial, flexural, or torsional) will be zero because of the uncoupled nature of these displacements in a straight bar. Furthermore, when the fixed-interface modes are normal vibration modes, the orthogonality properties diagonalize the inertia and stiffness matrices for those modes. It follows that

$$[M^\ell]^{FF} = [I], \quad [K^\ell]^{FF} = [(\omega^\ell)^2]$$

(9.2.6)

At this stage, the analytical situation is the same as it was for the straightforward application of the Lagrange multiplier method. The matrices for the synthesized system are assembled according to eqs. (9.1.21) using the component values defined in eqs. (9.2.5). These matrices and the Jacobian constraint matrix $[a]$ are combined in accordance with eqs. (9.1.22). Transient and steady-state responses may then be determined by the techniques outlined in Section 9.1.3.

We have used the Ritz series method as the foundation for developing component mode synthesis. In practice, the technique is usually implemented in conjunction with finite element analysis. A finite element model of a structural component is likely to require many nodes, and therefore many degrees of freedom. Assembling the structural components directly would greatly raise the number of degrees of freedom. Instead, the normal modes of the individual components may be computed. Modes whose natural frequency is well above the upper limit of the frequency range of interest may be truncated from the series describing the structural component, thereby reducing the number of degrees of freedom for the synthesized structure.

As was mentioned earlier, versions of component mode synthesis that do not require fixed-interface modes have been developed. These provide greater freedom in selecting basis functions for the structural components. One advantage of the fixed-interface modes is that they are well suited to experimental techniques. Physical models of the components may be fabricated with all interface locations held stationary. Such models may be used to validate the finite element models for the components, rather than waiting to test the entire synthesized system.

EXAMPLE 9.6

Use component mode synthesis to determine the lowest four natural frequencies and corresponding modal displacements for the system in Example 9.5.

Solution The only difference between this system and the one in Example 9.4 is the manner in which end C is supported. By analyzing a system that is like one we studied previously using the generic Lagrange multiplier method, we will gain insight into the operational differences between the methods, as well as the influence of design changes. The basic work of setting up the Ritz series and generalized coordinates and identifying the Jacobian constraint matrix and the sorting matrix was performed in Example 9.5. For reference, the generalized coordinates for the components were arranged such that

$$\{q^1\} = \left[\{q_u^{F1}\}^{\mathrm{T}} \ \{q_w^{F1}\}^{\mathrm{T}} q_1^{C1} \ q_2^{C1} \ q_3^{C1}\right]^{\mathrm{T}}$$

$$\{q^2\} = \left[\{q_u^{F2}\}^{\mathrm{T}} \ \{q_w^{F2}\}^{\mathrm{T}} q_1^{C2} \ q_2^{C2} \ q_3^{C2} \ q_4^{C2}\right]^{\mathrm{T}}$$

The constraint modes corresponding to the preceding were defined to be $\psi_1^{C\ell} = \psi_u^C(x_\ell)$, $\psi_2^{C\ell} = \psi_w^C(x_\ell)$, $\psi_3^{C\ell} = \psi_{\chi}^C(x_\ell)$, $\ell = 1, 2$ and $\psi_4^{C2} = \psi_{\chi}^C(L_2 - x_2)$.

We begin by evaluating the inertia and stiffness matrices. There is no coupling between bending and extension in a straight bar. Furthermore, the fixed-interface modes in Example 9.5 are normal vibration modes. As a result, the nonzero coefficients in eqs. (9.2.5) are

$$(M^\ell)_{jj}^{FF} = 1, \quad j = 1, \ldots, 2N, \qquad (M^\ell)_{j1}^{FC} = \int_0^{L_\ell} pA \psi_{uj}^{F\ell} \psi_1^{C\ell} dx, \quad j = 1, \ldots, N$$

$$(M^\ell)_{jn}^{FC} = \int_0^{L_\ell} pA \psi_{wj}^{F\ell} \psi_n^{C\ell} dx, \quad \begin{cases} j = N+1, \ldots, 2N, \quad n = 2, 3 \quad \text{if } \ell = 1 \\ j = N+1, \ldots, 2N, \quad n = 2, 3, 4 \quad \text{if } \ell = 2 \end{cases}$$

$$(M^\ell)_{11}^{CC} = \int_0^{L_\ell} pA (\psi_1^{C\ell})^2 dx$$

$$(M^\ell)_{jn}^{CC} = \int_0^{L_\ell} pA \psi_j^{C\ell} \psi_n^{C\ell} dx, \quad \begin{cases} j, n = 2, 3 \quad \text{if } \ell = 1 \\ j, n = 2, 3, 4 \quad \text{if } \ell = 2 \end{cases}$$

Similar reasoning applied to the stiffness coefficients leads to

$$(K^\ell)_{jj}^{FF} = (\omega_j^\ell)^2, \quad j = 1, \ldots, 2N$$

$$(K^\ell)_{j1}^{FC} = \int_0^{L_\ell} EA \frac{d\psi_j^{F\ell}}{dx} \frac{d\psi_1^{C\ell}}{dx} dx, \quad j = 1, \ldots, N$$

$$(K^\ell)_{jn}^{FC} = \int_0^{L_\ell} EI \frac{d^2\psi_{wj}^{F\ell}}{dx^2} \frac{d^2\psi_n^{C\ell}}{dx^2} dx, \quad \begin{cases} j = N+1, \ldots, N, \quad n = 2, 3 \quad \text{if } \ell = 1 \\ j = N+1, \ldots, 2N, \quad n = 2, 3, 4 \quad \text{if } \ell = 2 \end{cases}$$

$$(K^\ell)_{11}^{CC} = \int_0^{L_\ell} EA \left(\frac{d\psi_1^{C\ell}}{dx}\right)^2 dx$$

$$(K^\ell)_{jn}^{CC} = \int_0^{L_\ell} EI \frac{d^2\psi_j^{C\ell}}{dx^2} \frac{d^2\psi_n^{C\ell}}{dx^2} dx, \quad \begin{cases} j, n = 2, 3 \quad \text{if } \ell = 1 \\ j, n = 2, 3, 4 \quad \text{if } \ell = 2 \end{cases}$$

As usual, all coefficients not listed are zero. Numerical methods are used to evaluate the integrals. Those values enable us to assemble $[M]$ and $[K]$ for the whole system according to eqs. (9.2.3) and (9.1.21). We set $N = 6$ for the computations, which is the value used in Example 9.4. Evaluation of integrals containing the fixed-interface mode functions is expedited by using the high-frequency asymptotic approximation of the flexural mode shape. The techniques described in Section 7.4 lead to

$$\Psi\left(\frac{x}{L_\ell}, \alpha_j\right) \approx \sin\left(\alpha_j \frac{x}{L_\ell}\right) - \cos\left(\alpha_j \frac{x}{L_\ell}\right) + \exp\left(-\alpha_j \frac{x}{L_\ell}\right) - (-1)^j \exp\left[-\alpha_j \left(1 - \frac{x}{L_\ell}\right)\right]$$

which is quite accurate for $j > 2$.

The next step is to eliminate the constrained generalized coordinates. We use the transformation matrix $[B]$ determined in Example 9.4 to evaluate the reduced $(4N + 4) \times (4N + 4)$ matrices $[\hat{M}]$ and $[\hat{K}]$ according to eq. (9.1.33). These matrices are used to compute the modal solution for the unconstrained generalized coordinates. The analysis yields $4N + 4$ modes, which are normalized with respect to $[\hat{M}]$,

$$[[\hat{K}] - \omega_j^2[\hat{M}]]\{\hat{\Phi}_j\} = \{0\}, \qquad \{\hat{\Phi}_j\}^T[\hat{M}]\{\hat{\Phi}_j\} = 1$$

The eigenvectors representing the modal proportions for all of the generalized coordinates are then obtained by computing

$$[\Phi] = [B][\{\hat{\Phi}_1\} \ \{\hat{\Phi}_2\} \ \cdots \ \{\hat{\Phi}_{4N+4}\}]$$

The process of evaluating the physical displacements in a mode is like that in Example 9.4. From Example 9.5, the Ritz series for displacements are

$$u^1(x) = \sum_{j=1}^{N} \psi_{uj}^{F1} q_{uj}^{F1} + \psi_1^{C1} q_1^{C1}, \qquad w^1(x) = \sum_{j=1}^{N} \psi_{wj}^{F1} q_{wj}^{F1} + \sum_{j=2}^{3} \psi_j^{C1} q_j^{C1}$$

$$u^2(x) = \sum_{j=1}^{N} \psi_{uj}^{F2} q_{uj}^{F2} + \psi_1^{C2} q_1^{C2}, \qquad w^2(x) = \sum_{j=1}^{N} \psi_{wj}^{F2} q_{wj}^{F2} + \sum_{j=2}^{4} \psi_j^{C2} q_j^{C2}$$

To obtain a modal displacement, we replace each generalized coordinate by its modal value in $\{\Phi_n\}$. This requires recognizing the sequence in which $\{q\}$ was defined. For mode n, the appropriate expressions are

$$(u^1(x))_n = \sum_{j=1}^{N} \psi_{uj}^{F1} \Phi_{jn} + \psi_1^{C1} \Phi_{(2N+1)n}$$

$$(w^1(x))_n = \sum_{j=1}^{N} \psi_{wj}^{F1} \Phi_{(j+N)n} + \sum_{j=2}^{3} \psi_1^{C1} \Phi_{(j+2N+1)n}$$

$$(u^2(x))_n = \sum_{j=1}^{N} \psi_{uj}^{F2} \Phi_{(j+2N+3)n} + \psi_1^{C2} \Phi_{(4N+4)n}$$

$$(w^2(x))_n = \sum_{j=1}^{N} \psi_{wj}^{F1} \Phi_{(j+3N+3)n} + \sum_{j=2}^{4} \psi_1^{C1} \Phi_{(j+4N+3)n}$$

The figure displays the displacement in each of the first four modes as a set of lines emanating from points along the frame.

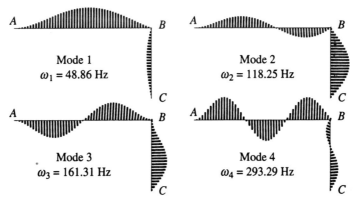

A comparison with the results in Example 9.4 shows that each of the present modes has a lower natural frequency than in the previous case. This is to be expected, because the sole difference between the systems is that end C here is hinged. Consequently, the system is less stiff than it is in the previous case, where end C is clamped. A comparison of the modal displacement patterns shows that the overall trends are alike, but bar BC now shows significant transverse displacement in each mode. The greater participation of bar BC relative to bar AB may be explained by observing that relaxing the clamped condition at end C lowers the stiffness of bar BC if it is isolated from the rest of the system. This offsets the greater stiffness of bar BC relative to bar AB resulting from the fact that L_2 is half of L_1. The isolated natural frequencies of the individual bars are comparable, so both bars exhibit comparable displacements when they are joined.

REFERENCES AND SELECTED READINGS

CHEN, P. T., & GINSBERG, J. H. 1992. "On the Relationship between Veering of Eigenvalue Loci and Parameter Sensitivity of Eigenfunctions." *ASME Journal of Vibration and Acoustics*, 114, 141–148.

CRAIG, R. R., JR. 1981. *Structural Dynamics*. John Wiley & Sons, New York.

CRAIG, R. R., JR., & BRAMPTON, M. C. C. 1968. "Coupling of Substructures Using Component Modes." *AIAA Journal*, 6, 1313–1319.

DOWELL, E. H. 1972. "Free Vibration of an Arbitrary Structure in Terms of Component Modes." *ASME Journal of Applied Mechanics*, 39, 727–732.

GINSBERG, J. H. 1995. *Advanced Engineering Dynamics*. Cambridge University Press, Cambridge, England.

HAUG, E. J. 1992. *Intermediate Dynamics*. Prentice Hall, Englewood Cliffs, NJ.

HINTZ, R. M. 1975. "Analytical Methods in Component Modal Analysis." *AIAA Journal*, 13, 1007–1016.

HURTY, W. C. 1965. "Dynamic Analysis of Coupled Systems Using Component Modes." *AIAA Journal*, 3, 678–685.

MEIROVITCH, L. 1997. *Principles and Techniques of Vibrations*. Prentice Hall, Englewood Cliffs, NJ.

PIERRE, C., & DOWELL, E.H. 1987. "Localization of Vibrations by Structural Irregularity." *Journal of Sound and Vibration*, 114, 549–564.

EXERCISES

9.1 The interior support for the two-span beam is a roller located midspan. Consider the following alternative decompositions:

(**a**) A single component consisting of a clamped-pinned beam.

(**b**) Two components consisting of a clamped-free (cantilevered) beam connected to a simply supported beam.

Identify Ritz series and the corresponding Jacobian constraint matrix for each formulation.

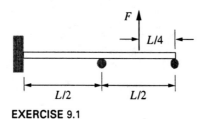

EXERCISE 9.1

9.2 Connection B in the frame is a pin, and the other connections are welded joints. Identify a set of structural components whose assembly forms the frame. Select Ritz series appropriate to the displacement field of each component, then evaluate the corresponding Jacobian constraint matrix.

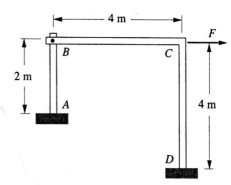

EXERCISE 9.2

9.3 The bent bar is loaded by forces in the vertical plane. Consider each straight segment of the bar to be a structural member. Select Ritz series appropriate to the displacement field of each component, then evaluate the corresponding Jacobian constraint matrix.

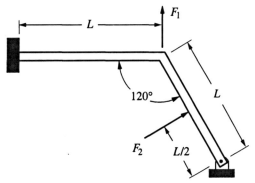

EXERCISE 9.3

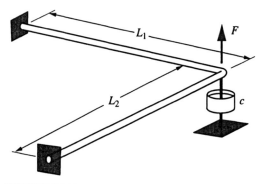

EXERCISE 9.5

9.4 A T-bar is clamped at ends A and B. A torque R is applied at end C, which is free to move. Consider a displacement field in which bar CD undergoes torsional motion only, while bar AB displaces in the vertical direction, with no displacement at point D. Identify a set of structural components whose assembly forms the T-bar. Select Ritz series appropriate to the displacement field of each component, then evaluate the corresponding Jacobian constraint matrix.

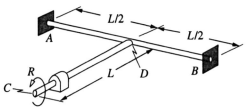

EXERCISE 9.4

9.5 The sketch shows a bar bent at a right angle and lying in the horizontal plane. Both ends are clamped to walls. The force F at the corner acts in the vertical direction. Consider each straight segment of the bar as a component. Select Ritz series appropriate to the displacement field of each component, then evaluate the corresponding Jacobian constraint matrix.

9.6 A beam spanning distance L is braced at both ends by shorter bars that are welded to it. Select Ritz series appropriate to the displacement field of each member, then evaluate the corresponding Jacobian constraint matrix.

9.7 Consider the component representation of the two-span beam described in Exercise 9.1(a). Find the corresponding equations of motion.

9.8 Consider the component representation of the two-span beam described in Exercise 9.1(b). Find the corresponding equations of motion.

9.9 The frame in Exercise 9.2 consists of three bars having the same composition and cross-section. Find the equations of motion. Express the answer in terms of E, A, I, and ρ for the bars.

9.10 Find the equations of motion for the bent bar in Exercise 9.3. Express the answer in terms of E, A, I, and ρ, which is the same for both bars.

9.11 Both members of the T-bar in Exercise 9.4 have identical cross-section and composition. Find the equations of motion. Express the answer in terms of E, A, I, and ρ.

9.12 The two-span beam in Exercise 9.1 is subjected to a harmonic excitation, $F = F_0 \cos(\omega t)$. The structural damping loss factor is $\gamma = 0.001$. The cross-section is square, with depth $h = L/20$. Find the steady-state amplitude of the displacement

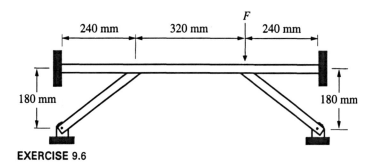

EXERCISE 9.6

at the location where F is applied. Graph the result as $w(EI/F_0 L^3)$ as a function of $\omega(\rho A L^4/EI)^{1/2}$.

9.13 Determine the first four natural frequencies and corresponding mode functions of the two-span beam in Exercise 9.1. The cross-section is square, with depth $h = L/20$.

9.14 Example 9.2 evaluated the equations of motion for a bent bar that undergoes flexural and torsional displacement. Consider the case where the bar is aluminum with a 250 mm diameter, and $L_1 = 5$ m, $L_2 = 3$ m. Determine the lowest four natural frequencies.

9.15 Both forces acting on the frame in Example 9.4 are $F = F_0 \sin(\omega t)$. The bars are steel, with a 100 mm $\times$ 100 mm square cross-section. Determine the amplitude of the horizontal and vertical displacements of the points at which the forces are applied. Plot these amplitudes as functions of frequency over a range covering the first four resonances.

9.16 Determine the lowest four natural frequencies and corresponding modal displacements when the frame in Example 9.4 is modeled as being rigid in extension. What conclusions regarding the significance of axial displacement in frames can be drawn from these results?

9.17 The forces acting on the bent bar in Exercise 9.3 are $F_1 = F_0 \cos(\omega t)$, $F_2 = F_0 \sin(\omega t)$. The bar has a circular cross-section, with radius $r = L/40$. Determine the steady-state amplitudes of the horizontal and vertical displacements at the corner where the bar is bent. Graph the result over a frequency range covering the first four resonances.

9.18 Determine the first three natural frequencies of the bent bar in Exercise 9.3, nondimensionalized relative to $(E/\rho L^2)^{1/2}$. Also determine the corresponding mode shapes. Graph the mode shapes for axial and transverse displacement as functions of distance along the bar, measured from the left end. The bar has a circular cross-section, with radius $r = L/40$.

9.19 The bent bar in Exercise 9.5 is subjected to a harmonically varying force $F \cos(\omega t)$. The bar is composed of aluminum, with an 80 mm cross-sectional diameter. The segment lengths are $L_1 = 2$ m, $L_2 = 0.9$ m. The dashpot constant is $c = 100$ N-s/m. Determine the steady-state amplitude of the vertical displacement at the point where the force is applied as a function of frequency, up to the fourth natural frequency.

9.20 Determine the lowest four (undamped) natural frequencies and corresponding modes of the system in Exercise 9.5. The bar is composed of aluminum, with an 80 mm cross-sectional diameter. The segment lengths are $L_1 = 2$ m, $L_2 = 0.9$ m. Plot the modal displacements as functions of arc length along the bar.

9.21 Consider each span of the beam in Exercise 9.1 to be a structural component. Identify fixed-interface modes and constraint modes for performing component mode synthesis. Then identify the corresponding Jacobian constraint matrix.

9.22 It is desired to use component mode synthesis to model the system in Exercise 9.4. Determine the constraint modes and Jacobian constraint matrix corresponding to using the flexural and torsional modes for a clamped-clamped beam as the fixed-interface modes.

9.23 It is desired to use component mode synthesis to model the system in Example 9.1. Determine the constraint modes and Jacobian constraint matrix corresponding to using the flexural and torsional modes for a clamped-clamped beam as the fixed-interface modes.

9.24 Identify the constraint modes and Jacobian constraint matrix corresponding to considering each straight segment of the frame in Exercise 9.2 to be a fixed-fixed structural component.

9.25 Consider component mode synthesis of the beam in Exercise 9.1 based on considering each span as a structural component. Determine the assembled inertia, stiffness, and generalized force matrices.

9.26 Consider using the flexural and torsional modes for a clamped-clamped beam as the fixed-interface modes for the T-bar in Exercise 9.4. Determine the assembled inertia, stiffness, and generalized force matrices. Use those results to determine the lowest four natural frequencies and corresponding mode shapes when $L = 100$ mm. The bars are steel and their cross-sectional diameter is 5 mm.

9.27 Use component mode synthesis to solve Exercise 9.19.

9.28 In Exercise 9.3 the members are aluminum circular bars, with 50 mm diameter and $L = 600$ mm. Use component mode synthesis to determine the lowest four natural frequencies of the system.

9.29 Use component mode synthesis to determine the lowest four natural frequencies and corresponding mode shapes of the frame in Exercise 9.2. The bars are steel, and have a 100 mm square cross-section.

CHAPTER 10

DAMPED MODAL ANALYSIS: THE STATE SPACE

The modal analysis techniques we have developed thus far are in wide use. However, they are not valid if the damping matrix $[C]$ is large and is not diagonalized by the mode matrix. We will develop here an alternative modal analysis approach that does not suffer from these limitations. Its modes describe the free vibration properties of the damped system, so we refer to them as *damped modes*. The eigenvalue problem associated with this formulation is like the one for undamped systems. The similarity extends to the fact that a modal expansion leads to uncoupled equations for the modal coordinates. However, there is a penalty associated with damped modal analysis, in that the size of the eigenvalue problem is doubled.

10.1 STATE-SPACE EQUATIONS

For many individuals the notion underlying damped modes seems to be paradoxical, in that it considers velocity to be independent of position, whereas knowledge of either variable as a function of time may be used to recover the other. We define a column of unknowns $\{x\}$ whose elements are the generalized coordinates q_j arranged sequentially, followed by the generalized velocities $\dot{q}_j$. Thus, $\{x\}$ for a system having N degrees of freedom would have $2N$ elements, whose partitioned form is

$$\{x\} = \begin{bmatrix} q_1 & \cdots & q_N & \dot{q}_1 & \cdots & \dot{q}_N \end{bmatrix}^{\mathrm{T}} = \begin{Bmatrix} \{q\} \\ \{\dot{q}\} \end{Bmatrix} \qquad (10.1.1)$$

We say that $\{x\}$ is the *state vector*, because we may consider the assembled values of the generalized coordinates and generalized velocities to represent the current state of the system. If we know the values of all elements of $\{x\}$ at the current instant, we may determine the generalized accelerations from the equations of motion, which then leads to the state of the system an infinitesimal instant later.

Because the velocity and position variables are treated as independent variables, we must explicitly enforce the property that each $\dot{q}_j$ variable is the derivative of the corresponding q_j. This operation, as well as others that follow, are greatly assisted by the properties of partitioned matrices. Let $[I]$ be the $N \times N$ identity matrix and $[0]$ be an $N \times N$ matrix of zeroes. To filter $\{q\}$ or $\{\dot{q}\}$ from $\{x\}$ we have

$$\{q\} = \begin{bmatrix} [I] & [0] \end{bmatrix} \begin{Bmatrix} \{q\} \\ \{\dot{q}\} \end{Bmatrix} \equiv \begin{bmatrix} [I] & [0] \end{bmatrix} \{x\}$$

$$\{\dot{q}\} = \begin{bmatrix} [0] & [I] \end{bmatrix} \begin{Bmatrix} \{q\} \\ \{\dot{q}\} \end{Bmatrix} \equiv \begin{bmatrix} [0] & [I] \end{bmatrix} \{x\}$$

(10.1.2)

This leads to the *transformation identity*,

$$\begin{bmatrix} [I] & [0] \end{bmatrix} \{\dot{x}\} - \begin{bmatrix} [0] & [I] \end{bmatrix} \{x\} = \{0\}$$

(10.1.3)

The preceding represents N differential equations relating the $2N$ variables contained in $\{x\}$. The equations of motion form the remaining N equations, but we must convert them to a form that depends on the x_j variables, rather than q_j and $\dot{q}_j$. To do so, we use eqs. (10.1.2) to eliminate $\{q\}$ and its derivatives from the standard equations of motion, such that

$$[M]\begin{bmatrix} [0] & [I] \end{bmatrix}\{\dot{x}\} + [C]\begin{bmatrix} [0] & [I] \end{bmatrix}\{x\} + [K]\begin{bmatrix} [I] & [0] \end{bmatrix}\{x\} = \{Q\} \qquad (10.1.4)$$

Premultiplication by an $N \times N$ matrix behaves like a scalar multiplication in the partitioning concept, so the equation of motion may be rewritten as

$$\begin{bmatrix} [0] & [M] \end{bmatrix}\{\dot{x}\} + \begin{bmatrix} [K] & [C] \end{bmatrix}\{x\} = \{Q\} \qquad (10.1.5)$$

Rather than dealing with two separate matrix equations, it is convenient to stack them into a single equation. If we stack eq. (10.1.3) above eq. (10.1.5), the form that results will have a coefficient matrix multiplying $\{\dot{x}\}$ that is symmetric, but the coefficient of $\{x\}$ will not be symmetric. A number of desirable properties result from having symmetry in both sets of matrices. For that reason, we observe that the coefficient matrix multiplying $\{x\}$ in eq. (10.1.5) has $+[K]$ in the lower left partition. Because $[K]$ is symmetric, multiplying eq. (10.1.3) by $-[K]$ leads to a symmetric appearance of the stacked equations. The result is a set of first-order differential equations,

$$[S]\frac{d}{dt}\{x\} - [R]\{x\} = \begin{Bmatrix} \{0\} \\ \{Q\} \end{Bmatrix} \qquad (10.1.6)$$

where the coefficient matrices are

$$[S] = \begin{bmatrix} -[K] & [0] \\ [0] & [M] \end{bmatrix}$$

$$[R] = -\begin{bmatrix} [0] & [K] \\ [K] & [C] \end{bmatrix}$$

(10.1.7)

Because the matrix contained in each partition is symmetric, and the off-diagonal partitions are alike, both $[R]$ and $[S]$ are symmetric. This symmetry has important implications for modal orthogonality, just as it did for the undamped modes.

10.2 EIGENVALUE PROBLEM

We begin our analysis of the state-space equations by considering free vibration, $\{Q\} = \{0\}$. Both $[R]$ and $[S]$ are constant, so eq. (10.1.6) in this case represents a set of $2N$ linear, homogeneous, ordinary differential equations with constant coefficients. The homogeneous solution for each variable will depend on t exponentially, such that

$$\{x\} = B\{\psi\} \exp(\lambda t) \tag{10.2.1}$$

The constant vector $\{\psi\}$ and characteristic coefficient λ will be determined by satisfying the differential equation, while the coefficient B is dictated by the initial conditions. We substitute eq. (10.2.1) into eq. (10.1.6) and cancel the common exponential factor, which gives

$$\boxed{[[R] - \lambda[S]]\{\psi\} = \{0\}} \tag{10.2.2}$$

This constitutes a general eigenvalue problem. The basic theory for solving the equation is the same as that for the undamped modes, with a few modifications. Because the size of $[A]$ is $2N \times 2N$, the characteristic polynomial is degree $2N$,

$$\boxed{|[R] - \lambda[S]| = a_{2N}\lambda^{2N} + a_{2N-1}\lambda^{2N-1} + \cdots + a_1\lambda + a_0 = 0} \tag{10.2.3}$$

The coefficients a_j are real. The roots of a real polynomial must either be real or else occur as pairs of complex conjugates. Because $[C]$ is nonzero, we expect that dissipation will cause a free vibration to decay with increasing t. If this is so, the real part of each eigenvalue will be negative. We shall number the eigenvalues based on their magnitudes, with complex conjugates numbered adjacently,

$$\boxed{|\lambda_n| \le |\lambda_{n+1}|, \quad \lambda_{n+1} = \lambda_n^* \text{ if } \text{Im}(\lambda_n) \ne 0} \tag{10.2.4}$$

In view of eq. (10.2.1), the imaginary part of each eigenvalue represents a damped natural frequency, and the real part is the decay rate.

For each eigenvalue λ_n, the corresponding eigenvector $\{\psi_n\}$ must be a nontrivial solution of

$$\boxed{[[R] - \lambda_n[S]]\{\psi_n\} = \{0\}} \tag{10.2.5}$$

Because the value of λ_n causes the determinant $|[R] - \lambda_n[S]|$ to vanish, the above eigenvector equation is rank deficient. In most situations, the reduction in rank matches the number of times the value of λ_n occurs as a root of the characteristic equation. For example, if all of the roots are distinct, then one of the equations is a linear combination of the others. A double root, which is analogous to the case of repeated natural frequencies for undamped systems, leads to two equations that are linear combinations of the others.

Exceptions to the preceding behavior correspond to critical damping of a mode, much like the behavior of one-degree-of-freedom systems, and to zero eigenvalues associated with rigid body motion. In both cases the rank reduction is less than the multiplicity of the root. Some researchers say that systems having such properties are *defective*, because there will be fewer modes than expected. Mathematical procedures for defective systems are available—see Kaplan (1958)—but simpler treatments are possible for engineering applications. Critical damping may be handled by changing any damping parameter, such as a dashpot constant, by a very small amount. The use of superposition to address systems having one or more rigid body modes is discussed separately in Section 10.5.4.

Hence, we shall assume here that the eigenvalues are distinct. Then the rank of $[[R] - \lambda_n[S]]$ is $2N - 1$, so only that number of equations is available to find the $2N$ elements of $\{\psi_n\}$. We may consider any nonzero element of $\{\psi_n\}$ to be arbitrary, or alternatively assign it a convenient numerical value, like unity. We find the other elements of $\{\psi_n\}$ in terms of the arbitrary one by requiring that the eigenvector satisfy eq. (10.2.5). The resulting set of eigenvectors are the *damped modes*.

If an eigenvalue λ_n is real, the coefficient matrix in eq. (10.2.5) is also real. In that case, the corresponding eigenvector $\{\psi_n\}$ obtained algebraically will be real. However, we will see in the next section that normalizing a real eigenvector might lead to a purely imaginary one. When an eigenvalue is complex, the same reasoning leads to the conclusion that the eigenvector must also be complex. The fact that complex eigenvalues occur as complex conjugate pairs has an important implication for the corresponding eigenvectors. Suppose we have determined $\{\psi_n\}$ corresponding to λ_n. The eigenvector associated with $\lambda_{n+1} = \lambda_n^*$ satisfies $[[R] - \lambda_n^*[S]]\{\psi_{n+1}\} = 0$. Because $[R]$ and $[S]$ are real, we may rewrite the eigenvector equation as $[[R^*] - \lambda_n^*[S^*]]\{\psi_{n+1}\} = \{0\}$. The coefficient matrix here is the complex conjugate of the coefficient matrix for $\{\psi_n\}$, which leads us to conclude that $\{\psi_{n+1}\}$ is the complex conjugate of $\{\psi_n\}$,

$$\lambda_{n+1} = \lambda_n^* \quad \Rightarrow \quad \{\psi_{n+1}\} = \{\psi_n^*\} \tag{10.2.6}$$

An important aspect of the eigenvalue problem in eq. (10.2.2) is that the same general eigenvalue solvers used to analyze undamped modes may be applied to damped modes, because the coefficient matrices are real and symmetric. In other words, we may input $[R]$ and $[S]$ to MATLAB's *eig* function, or Mathcad's *genvals* and *genvecs* functions, and thereby obtain the λ_n and $\{\psi_n\}$ values, even though the results might be complex. However, one of the troublesome aspects of mathematical software is that the numerical algorithms are not necessarily the same from program to program. This is an issue here because the eigenvectors returned by a program do not always have the mathematical properties we identified. Specifically, it can happen that they yield eigenvectors associated with a pair of complex conjugate roots that do not seem to satisfy eq. (10.2.6). Another possibility is that the eigenvector associated with a real eigenvalue is not real. Such behavior arises because any eigenvector may be multiplied by a nonzero number, real or complex, without violating the conditions for an eigensolution. Even if the eigenvectors returned by the software appear to be anomalous, the situation will be rectified when we normalize the eigenvectors, which is a topic in Section 10.4.

The significance of an eigenvector being complex may be recognized by returning to the modal response in eq. (10.2.1). Let us write the elements of $\{\psi_n\}$ in polar form as

$$\psi_{jn} = |\psi_{jn}| \exp(i\delta_{jn}) \tag{10.2.7}$$

To emphasize the complex nature of the eigenvalues, let us write them as $\lambda_n = -\alpha_n + i\beta_n$, where we take $\alpha_n > 0$ based on damping causing decay. Rather than considering the contribution of one eigensolution to $\{q\}$, let us combine a specific eigensolution with its complex conjugate mate. We will see later that the multiplicative factor B_n will be matched by its complex conjugate for mode $n + 1$. Adding the eigensolutions for n and $n + 1$ to form the elements of $\{q\}$ according to eq. (10.2.1) yields

$$q_j = B_n|\psi_{jn}| \exp(i\delta_{jn}) \exp[(-\alpha_n + i\beta_n)t]$$

$$+ B_n^*|\psi_{jn}| \exp(-i\delta_{jn}) \exp[(-\alpha_n + i\beta_n)t] + \cdots$$

$$\boxed{q_j = 2|B_n^*||\psi_{jn}| \exp(-\alpha_n t) \cos(\beta_n t + \delta_{jn} + \arg(B_n)) + \cdots} \qquad (10.2.8)$$

We see from the above that the pair of complex conjugate modes represents a free vibration in which all generalized coordinates oscillate at frequency $\beta_n = |\text{Im}(\lambda_n)|$. The vibration decays exponentially with increasing t at decay rate $\alpha_n = -\text{Re}(\lambda_n)$. The individual generalized coordinates have different phase angles δ_{jn} relative to a pure cosine variation, which means that the instant at which a peak value or zero occurs varies from one generalized coordinate to another. The overall amplitude of the contribution of this pair of damped modes is set by $|B_n|$, and $\arg(B_n)$ is an overall phase angle for the modal contribution. Hence, the two primary new features of the damped modes in comparison to the undamped ones are the decaying nature of a modal free vibration and the fact that the different generalized coordinates are not synchronous, although all generalized coordinates share the same decay rate and frequency.

10.3 RELATION TO PREVIOUS DEVELOPMENTS

In order to better understand the nature of the damped mode formulation, let us explore some situations that we have already investigated by other methods. As an aid to this study, we decompose any damped mode $\{\psi\}$ into upper and lower partitions $\{u\}$ and $\{v\}$, each having N elements,

$$\{\psi\} = \begin{Bmatrix} \{u\} \\ \{v\} \end{Bmatrix} \qquad (10.3.1)$$

The corresponding form of the eigenvalue problem in eq. (10.2.2) is

$$\left[\begin{bmatrix} [0] & -[K] \\ -[K] & -[C] \end{bmatrix} - \lambda \begin{bmatrix} -[K] & [0] \\ [0] & [M] \end{bmatrix} \right] \begin{Bmatrix} \{u\} \\ \{v\} \end{Bmatrix} = \begin{Bmatrix} \{0\} \\ \{0\} \end{Bmatrix} \qquad (10.3.2)$$

After we factor out the common factor $[K]$, the first partitioned set of equations gives

$$\{v\} = \lambda\{u\} \qquad (10.3.3)$$

In other words, the second set of N elements of damped mode $\{\psi_n\}$ are λ_n times the corresponding first set of N elements,

$$\psi_{(j+N)n} = \lambda_n \psi_{jn}, \quad j = 1, \ldots, N \tag{10.3.4}$$

(This relation can serve as a simple check for computed results.) The second partitioned set of equations described by eq. (10.3.2) is

$$[K]\{u\} + [[C] + \lambda[M]]\{v\} = \{0\} \tag{10.3.5}$$

When we use eq. (10.3.3) to eliminate $\{v\}$, we obtain

$$[[K] + \lambda[C] + \lambda^2[M]]\{u\} = \{0\} \tag{10.3.6}$$

Suppose rather than employing the state-space formulation to analyze cases where $[C] \neq [0]$, we had endeavored to solve the free vibration problem directly from the equations of motion for generalized coordinates, $[M]\{\ddot{q}\} + [C]\{\dot{q}\} + [K]\{q\} = \{0\}$. This would entail trying to construct a homogeneous solution in the form $\{q\} = \{\phi\}\exp(\lambda t)$. Aside from using a real exponent, such a trial solution is essentially the way we analyzed the undamped case. Substitution of the trial solution into the equation of motion would have produced eq. (10.3.6), with $\{\phi\}$ instead of $\{u\}$.

Equation (10.3.6) also represents an eigenvalue problem. However, because the eigenvalue occurs quadratically, we say that it constitutes a nonlinear eigenvalue problem. Its form does not match the general or standard eigenvalue problems that are addressed directly by most mathematical software, so we would have more difficulty solving the problem. Furthermore, the orthogonality properties of the modes obtained from eq. (10.3.6) would not be as useful as those we will soon establish for damped modes. The price we pay for using the state-space formulation is doubling the size of the eigenvectors, with the lower partition mirroring the upper partition according to eq. (10.3.4).

Let us use the above to assess the damped modes in the special case of an undamped system. Because $[C] = [0]$ in this case, eq. (10.3.6) becomes

$$[[K] + \lambda^2[M]]\{u\} = \{0\} \tag{10.3.7}$$

A comparison of this to the general eigenvalue problem for undamped modes, eq. (4.2.3), shows that $\lambda^2 = -\omega^2$ and $\{u\} = \{\phi\}$. Hence, when damping is not present, all state-space eigenvalues occur as purely imaginary conjugate pairs whose magnitude matches a natural frequency,

$$\lambda_{2n-1} = i\omega_n, \quad \lambda_{2n} = -i\omega_n, \quad n = 1, \ldots, N \tag{10.3.8}$$

Then, because the undamped modes are real, eqs. (10.3.3) indicate that the following relations exist between the damped mode partitions and the undamped properties,

$$\{u_{2n-1}\} = \{\phi_n\}, \quad \{v_{2n-1}\} = i\omega_n\{\phi_n\}$$
$$\{u_{2n}\} = \{\phi_n\}, \quad \{v_{2n}\} = -i\omega_n\{\phi_n\} \tag{10.3.9}$$

From these relations we recognize that the damped modes $\{\psi_j\}$ in the case where $[C] = [0]$ are merely the undamped modes, recast into a state-space form, in which the lower partition merely expresses the fact that it is the velocity version of the upper partition.

Another case that is instructive to consider is the damped modal behavior for a one-degree-of-freedom system. Because $[M]$, $[K]$, and $[C]$ then consist of only a single element, the corresponding state-space eigenvalue problem is

$$\begin{bmatrix} -\lambda K & K \\ -K & -\lambda M - C \end{bmatrix} \begin{Bmatrix} u \\ v \end{Bmatrix} = \begin{Bmatrix} 0 \\ 0 \end{Bmatrix} \tag{10.3.10}$$

The characteristic equation resulting from this is

$$(\lambda^2 M + C\lambda + K)K = 0 \tag{10.3.11}$$

which, aside from the common factor K, is identical to the quadratic characteristic equation derived in Section 2.2. The roots are either purely real if $C > 2(KM)^{1/2}$ for an overdamped system, or else they are complex conjugates if $C < 2(KM)^{1/2}$ for an underdamped system. (The case where $C = 2(KM)^{1/2}$ corresponds to critical damping, which yields a repeated root.)

In the overdamped case, the damped modes are real. When $u = 1$ is selected as the arbitrary coefficient, the first of the scalar equations described by eq. (10.3.10) yields $v = \lambda$. The state-space free vibration solution for each of the two modes is

$$\{x\} \equiv \begin{Bmatrix} q \\ \dot{q} \end{Bmatrix} = B_n \begin{Bmatrix} 1 \\ \lambda_n \end{Bmatrix} \exp(\lambda_n t), \quad n = 1, 2 \tag{10.3.12}$$

The first element constitutes one of the two terms forming the general solution for free vibration of an overdamped one-degree-of-freedom system, while the second element expresses the fact that velocity is the derivative of displacement.

In the underdamped case, the eigenvalues are complex conjugates. We know from our earlier study that the real part of λ_n is the decay constant, and the imaginary part is the damped natural frequency, such that

$$\lambda_1 = -\zeta \omega_{\text{nat}} + i\omega_{\text{d}}, \quad \lambda_2 = \lambda_1^* = -\zeta \omega_{\text{nat}} - i\omega_{\text{d}} \tag{10.3.13}$$

Without loss of generality, we take $u = 1$. Then the first scalar equation obtained from eq. (10.3.10) gives $v = \lambda_n$, $n = 1, 2$. When we combine the free vibration contributions of the two eigensolutions, in the manner of eq. (10.2.8), we find that

$$\{x\} \equiv \begin{Bmatrix} q \\ \dot{q} \end{Bmatrix} = B_1 \begin{Bmatrix} 1 \\ \lambda_1 \end{Bmatrix} \exp(\lambda_1 t) + B_2 \begin{Bmatrix} 1 \\ \lambda_2 \end{Bmatrix} \exp(\lambda_2 t) \tag{10.3.14}$$

The complex conjugate properties of the second mode, along with $B_2 = B_1^*$, then lead to

$$\begin{aligned} q &= 2\,\text{Re}[B_1 \exp(\lambda_1 t)] \\ &= 2|B_1| \exp(-\zeta \omega_{\text{nat}} t) \cos[\omega_{\text{d}} t + \arg(B_n)] \end{aligned} \tag{10.3.15}$$

The expression for $\dot{q}$ derived from the second row of eq. (10.3.14) merely reiterates that $\dot{q}$ is the time derivative of q.

EXAMPLE 10.1

The following properties are known for a certain three-degree-of-freedom system:

$$[M] = \begin{bmatrix} 200 & 0 & 400 \\ 0 & 600 & 0 \\ 400 & 0 & 1200 \end{bmatrix} \text{kg}, \qquad [K] = \begin{bmatrix} 50 & -20 & 0 \\ -20 & 40 & 0 \\ 0 & 0 & 30 \end{bmatrix} \text{kN/m}$$

$$[C] = \alpha \begin{bmatrix} 500 & 300 & -700 \\ 300 & 900 & 600 \\ -700 & 600 & 1400 \end{bmatrix} \text{N-s/m}$$

where α is a scaling factor for the damping. Determine the state-space eigenvalues and corresponding damped modes of the system when $\alpha = 1$ and $\alpha = 5$. For each case, compare the eigenvalues to those for equivalent one-degree-of-freedom systems. Also, compare the displacement partition of the damped modes to the corresponding undamped modes.

Solution This example provides a straightforward demonstration of the evaluation procedure for a damped eigenvalue problem. The purpose in seeking results for two damping scaling factors is to see cases where all eigenvalues are complex, and where some of the eigenvalues are real. In addition, comparing the results in both cases to those obtained for undamped modes will shed further light on the quality of the light modal approximation. The process begins by forming $[R]$ and $[S]$ according to eq. (10.1.7). The simplest approach is to first define $[M]$, $[K]$, and $[C]$ in our preferred software. Users of MATLAB may then form R and S by treating M, K, and C as elements of a matrix,

MATLAB: $\quad S = [[-K, zeros(3)], [zeros(3), M]]; \quad R = -[[zeros(3), K], [K, C]];$

where $zeros(3)$ is a function that returns a 3×3 array of zeroes. In Mathcad, these operations are performed using the *stack* and *augment* functions:

Mathcad: $\quad zero = identity(3) * 0; \quad S := stack(augment-K, zero),$
$\qquad\qquad (augment(zero, M)) \; R := -stack(augment(zero, K), augment(K, C))$

Note that the definition of *zero* postmultiplies by 0, because premultiplication by 0 merely yields a scalar zero.

Once the two state-space coefficient matrices have been defined, the eigenvalues and eigenvectors may be computed. In MATLAB, the syntax is the same as that used for undamped modes: $[psi, lambda] = eig(R, S)$. Sorting the eigensolutions in ascending order of the eigenvalue magnitudes requires modification of the algorithm we implemented in Example 4.4 for undamped modes. We use $|\lambda_j|$ to create an array for sorting. We wish to keep complex conjugate pairs together, with the one for which $\text{Im}(\lambda) > 0$ coming first. This may be achieved by adding a very small amount to any $|\lambda_j|$ for which $\text{Im}(\lambda_j) < 0$. The specific procedure proceeds as follows:

MATLAB: $\quad$ *for $j = 1 : length(lambda)$;*
$\qquad\qquad Mag(j) = abs(lambda(j)) + 0.0001 * (Im(lambda(j)) < 0);$
$\qquad\qquad$ *end*
$\qquad\qquad [Mag\_sort, n\_sort] = sort(Mag)$
$\qquad\qquad$ *for $j = 1 : N$; $lambda\_sort(j) = lambda(n\_sort(j))$;*
$\qquad\qquad psi\_sort(:, j) = psi(:, n\_sort(j))$; *end*

In Mathcad we employ the two-step procedure $\lambda := genvals(R, S)$, $\psi := genvecs(R, S)$. The *stack* function is used to create the array for sorting based on a row of adjusted $|\lambda_j|$ values, according to

Mathcad: $\quad j := 1; rows(\lambda) \quad Mag_j := |\lambda_j| + 0.0001 * (Im(\lambda_j) < 0)$
$Unsorted := stack(stack(Mag^\mathsf{T}, \lambda^\mathsf{T}), \psi)$
$Sorted := rsort(Unsorted, 1) \quad \lambda = (Sorted^\mathsf{T})^{<2>}$
$\psi = submatrix(Sorted, 3, rows(\lambda) + 2, 1, rows(\lambda))$

We obtain the undamped modal solution by using the same eigenvalue solvers, with K and M now set as the inputs. To compare the complex eigenvalues to the corresponding undamped properties, we recall eqs. (10.3.13). A standard modal analysis gives the undamped natural frequency ω_j, and the light damping approximation yields an estimate for the modal damping ratio ζ_j. If the modal coordinates were not coupled by damping, they would individually behave like one-degree-of-freedom systems having natural frequency ω_j and critical damping ratio ζ_j. The characteristic exponents for the homogeneous solution of such an oscillator are

$$\lambda_j = \begin{cases} -\zeta_j\omega_j \pm i\omega_j(1 - \zeta_j^2)^{1/2}, & \zeta_j < 1 \\ -\zeta_j\omega_j \pm \omega_j(\zeta_j^2 - 1)^{1/2}, & \zeta_j > 1 \end{cases}$$

In this manner, we can estimate two characteristic exponents associated with each undamped mode. We will compare each such pair to the λ_j values for the complex modes. Comparison of the respective modes is facilitated by standardizing the computed modes such that their first element is 1. This may be done in MATLAB by using a "for" loop over n, within which we have $psi(:, n) = psi(:, n)/psi(1, n)$. In Mathcad, we would define an index j and then write $\psi^{<j>} = \psi^{<j>}/\psi_{1,j}$. We apply the same standardization to the undamped modes.

The case where $\alpha = 1$ is the first we shall examine. The first table gives the complex damped eigenvalues, and the approximate complex eigenvalues corresponding to the undamped natural frequency and estimated modal damping ratio, as described above. It is evident that there is close agreement between the two formulations.

Light Damping: $\alpha = 1$						
Damped mode		**Undamped mode**				
Mode #j	λ_j	**Mode #j**	ω_j	ζ_j	λ_j	
1	$-0.47 + 4.79i$	1	4.77	0.098	$-0.47 + 4.74i$	
2	$-0.47 - 4.79i$				$-0.47 - 4.74i$	
3	$-0.89 + 7.32i$	2	7.35	0.122	$-0.90 + 7.29i$	
4	$-0.89 - 7.32i$				$-0.90 - 7.29i$	
5	$-8.38 + 26.93i$	3	28.55	0.294	$-8.39 + 27.29i$	
6	$-8.38 - 26.93i$				$-8.39 - 27.29i$	

There are three pairs of complex conjugate eigenvalues, so the corresponding damped modes also occur as complex conjugate pairs. The first set of graphs on the next page compare damped and undamped modes based on the correspondence in the first table. The damped mode selected for this comparison is the one for which the imaginary part of the eigenvalue is positive. There is close agreement between the undamped mode and the real part of the damped mode. We also see that the imaginary part of each damped mode is much smaller, in an average sense, than the real part. These features suggest that eq. (10.3.9), which describes the relation between the two types of modes when there is no dissipation, is satisfied approximately in the present case. This is quite surprising, given the fact that the estimate for the third undamped mode is that its damping ratio is approximately 30% of critical.

The situation changes drastically when $\alpha = 5$, which increases $[C]$ by a factor of 5. Changing the amount of damping does not affect the undamped natural frequencies ω_j, and the estimated modal damping ratios are increased proportionally, as shown in the second tabulation. This table indicates that the larger value of $[C]$ leads to two real eigenvalues and two pairs of complex conjugates. The damped eigenvalues are entered in the table in ascending order of their magnitude. The first and sixth damped eigenvalues are real, which correlates to the third

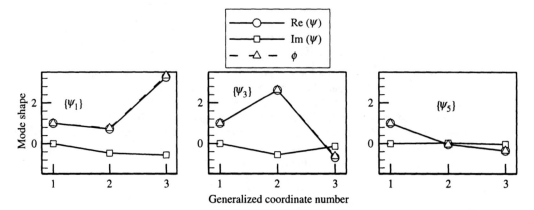

approximate eigenvalue obtained from undamped modal analysis. The sequencing for un-damped modes 1 and 2 in the table is based on the magnitude of the approximate λ_j. These es-timated characteristic values now differ significantly from the closest damped eigenvalues.

Heavy Damping: $\alpha = 5$					
Damped mode		Undamped mode			
Mode #j	λ_j	Mode #j	ω_j	ζ_j	λ_j
1	−5.87	3	28.55	1.469	−11.22
2	−4.03 + 4.38i	1	4.77	0.488	−2.32 + 4.16i
3	−4.03 − 4.38i				−2.32 − 4.16i
4	−1.92 + 7.52i	2	7.35	0.611	−4.49 + 5.82i
5	−1.92 − 7.52i				−4.49 − 5.82i
6	−79.73	3	28.55	1.469	−72.65

We use the correspondence in the second table to select which damped and undamped modes should be compared. As we did for $\alpha = 1$, we only compare one of each complex conjugate pair of modes to the undamped mode. The table indicates that the first and last damped modes corre-spond to the third undamped mode, so we use that as the comparison for both $\{\psi_1\}$ and $\{\psi_6\}$.

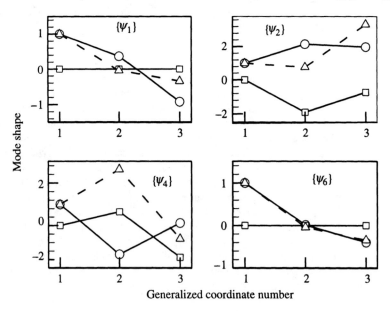

It is evident that the imaginary part of the complex mode pairs, $\{\psi_2\}$, $\{\psi_3\}$ and $\{\psi_4\}$, $\{\psi_5\}$, are not small compared to the real part. We also see that the undamped modes are poor approximations to these complex modes. Interestingly, the overdamped real modes $\{\psi_1\}$ and $\{\psi_6\}$ predicted by undamped modal analysis are better than the underdamped ones.

10.4 ORTHOGONALITY

Equation (10.2.2) constitutes a general symmetric eigenvalue problem. Its form is comparable to the general eigenvalue problem for undamped modes, with $[R]$ analogous to $[K]$ and $[S]$ analogous to $[M]$. As a consequence of this similarity, many of the steps in the analysis leading to uncoupled modal equations will strongly resemble those for undamped modes. For example, the modal orthogonality properties for the damped modes are like those for the undamped modes. Specifically, we have

$$\{\psi_j\}^{\mathrm{T}}[S]\{\psi_n\} = \nu_n \delta_{jn}$$
$$\{\psi_j\}^{\mathrm{T}}[R]\{\psi_n\} = \lambda_n \nu_n \delta_{jn}$$

(10.4.1)

where ν_n are *damped modal masses*. Note that these orthogonality properties apply even if mode n is the complex conjugate of mode j.

As we did for undamped modes, we normalize damped modes by dividing them by the square root of their modal mass. This will yield a set of *normalized damped modes* $\{\Psi_n\}$, whose corresponding modal masses are one. Hence, we define

$$\boxed{\{\Psi_n\} = \frac{\{\psi_n\}}{(\nu_n)^{1/2}}, \quad \nu_n = \{\psi_n\}^{\mathrm{T}}[S]\{\psi_n\}}$$

(10.4.2)

In undamped modal analysis, the modal masses ν_n are always positive values. When we compute ν_n and ν_{n+1} corresponding to a pair of complex conjugate eigensolutions, both values are complex conjugates, rather than being real. Another difference is that if $\{\psi_n\}$ is associated with a real eigenvalue, it might happen that the ν_n value is negative. The underlying reason for this difference between undamped and damped modal masses is the positive definiteness of $[M]$, which is not a property possessed by $[S]$. This fact may be demonstrated by using the definition of $[S]$ to form the quadratic sum associated with an arbitrary state vector $\{x\}$,

$$\{x\}^{\mathrm{T}}[S]\{x\} = \begin{bmatrix} \{u\}^{\mathrm{T}} & \{v\}^{\mathrm{T}} \end{bmatrix} \begin{bmatrix} -[K] & [0] \\ [0] & [M] \end{bmatrix} \begin{Bmatrix} u \\ v \end{Bmatrix} = -\{u\}^{\mathrm{T}}[K]\{u\} + \{v\}^{\mathrm{T}}[M]\{v\}$$

We know that $[M]$ is positive definite, and $[K]$ is at least positive semidefinite. Hence this quadratic sum may have either a positive or a negative value.

The only impact of the possibility that ν_n is not real and positive is that we shall use the square root value that is in the first or second quadrants of the complex plane, $0 \le \arg(\sqrt{\nu_n}) < \pi$. In the case of real λ_n, these observations suggest that the associated normalized eigenvector can be purely imaginary, rather than real, depending on whether the ν_n value is positive or negative.

Once we have evaluated the normal modes, the next step is to define the normal mode matrix. For undamped modes, this was an $N \times N$ matrix whose columns are the various modes. Now the normal mode matrix will be a $2N \times 2N$ matrix whose columns are the $2N$ damped modes,

$$[\Psi] = \left[\{\Psi_1\} \cdots \{\Psi_{2N}\} \right]$$
(10.4.3)

The orthogonality relations satisfied by $[\Psi]$ are

$$[\Psi]^T[S][\Psi] = [I], \qquad [\Psi]^T[R][\Psi] = [\lambda]$$
(10.4.4)

where $[\lambda]$ is a diagonal array of sequenced eigenvalues. Like the case of undamped modes, satisfaction of both orthogonality relations is a necessary and sufficient condition that λ_n and $\{\Psi_n\}$ constitute the normal mode solution. Thus, if we use the computed eigenvalues λ_n and eigenvectors $\{\Psi_n\}$ to check that eqs. (10.4.4) are actually satisfied, we will obtain a robust check on the correctness of the computation.

EXAMPLE 10.2

The system shown below has no mass at the junction of the spring and dashpot that are connected in series. Nevertheless, a generalized coordinate must be assigned to that location, so the system is sometimes said to have *one and one-half degrees of freedom.*

(a) Derive the state-space form of the equations of motion of this system.

(b) Determine the state-space eigenvalues and normalized damped modes when $\mu = (km)^{1/2}$.

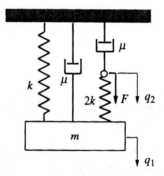

Solution The process of formulating the equations of motion will emphasize the specific form associated with damped modal analysis. In the process of carrying out the analysis, we will see how the mathematical software handles certain singularities, and the value of the orthogonality relations in checking that numerical software has performed correctly. We begin with the equations of motion, which we obtain by summing forces on the block, and on the spring-dashpot junction,

$$m\ddot{q}_1 = -kq_1 - \mu\dot{q}_1 - 2k(q_1 - q_2)$$
$$2k(q_1 - q_2) + F - \mu\dot{q}_2 = 0$$

We observe that $\ddot{q}_2$ does not occur in these equations, which implies that the value of $\dot{q}_2$ is defined by the second motion equation. This makes it tempting to define the state-space vector as the set of values q_1, $\dot{q}_1$, and q_2,

$$\{x\} = \left[q_1 \; q_2 \; \dot{q}_1 \right]^T$$

The transformation identity associated with this definition is

$$[1 \; 0 \; 0]\{\dot{x}\} = [0 \; 0 \; 1]\{x\}$$

When we stack this above the two equations of motion, the system of equations is

$$\begin{bmatrix} 1 & 0 & 0 \\ 0 & 0 & m \\ 0 & -\mu & 0 \end{bmatrix} \{\dot{x}\} - \begin{bmatrix} 0 & 0 & 1 \\ -3k & 2k & -\mu \\ -2k & 2k & 0 \end{bmatrix} \{x\} = \begin{Bmatrix} 0 \\ 0 \\ F \end{Bmatrix}$$

These equations could be used as the basis for numerical solution using a method such as Runge-Kutta. They also can be solved by the eigenvalue formulation developed in Chapter 11 for gyroscopic systems. However, because the coefficient matrices multiplying $\{x\}$ and $\{\dot{x}\}$ are not symmetric, the equations do not have the form appropriate to the damped modal formulation.

To use damped modes we must go back to the basic equations of motion. Suppose that there was a very small mass at the location whose displacement is q_2. (In reality, this would certainly be the case.) Rather than using the zero mass idealization, if we let the mass at the spring-mass junction be a small value m_2, the associated motion equation would be

$$m_2 \ddot{q}_2 = 2k(q_1 - q_2) + F - \mu \dot{q}_2$$

From this viewpoint, the equation of motion defines $\ddot{q}_2$, so the state-space variables are both generalized coordinates and their first-order derivatives. In other words, the state vector we should use is

$$\{x\} = \begin{bmatrix} q_1 & q_2 & \dot{q}_1 & \dot{q}_2 \end{bmatrix}^{\mathrm{T}}$$

This fits the definition we have employed thus far. The corresponding system matrices are

$$[M] = \begin{bmatrix} m & 0 \\ 0 & m_2 \end{bmatrix}, \qquad [K] = \begin{bmatrix} 3k & -2k \\ -2k & 2k \end{bmatrix}, \qquad [C] = \begin{bmatrix} \mu & 0 \\ 0 & \mu \end{bmatrix}, \qquad \{Q\} = \begin{Bmatrix} 0 \\ F \end{Bmatrix}$$

From this, we may construct the $[R]$ and $[S]$ matrices according to eqs. (10.1.7). In doing so, we set $m_2 = 0$ in recognition of the fact that we are interested in the idealization. Hence, we have

$$[S] = \begin{bmatrix} -3k & 2k & 0 & 0 \\ 2k & -2k & 0 & 0 \\ 0 & 0 & m & 0 \\ 0 & 0 & 0 & 0 \end{bmatrix}, \qquad [R] = \begin{bmatrix} 0 & 0 & -3k & 2k \\ 0 & 0 & 2k & -2k \\ -3k & 2k & -\mu & 0 \\ 2k & -2k & 0 & -\mu \end{bmatrix}$$

Because a row and column of $[S]$ are identically zero, $[S]$ is singular. In view of the fact that the eigenvalue problem $[[R] - \lambda[S]]\{\psi\} = \{0\}$ for damped modes is analogous to $[[K] - \omega^2[M]]\{\phi\} = \{0\}$ for undamped modes, singularity of $[S]$ is comparable to singularity of $[M]$. We know that a singular $[K]$ leads to zero eigenvalues. If we define a new eigenvalue parameter as $\rho = 1/\lambda$, the problem at hand becomes

$$[[S] - \rho[R]]\{\psi\} = \{0\}, \qquad \rho = 1/\lambda$$

From this we recognize that $\rho = 0$ is a possible eigenvalue satisfying $|[S] - \rho[R]| = 0$. However, $\rho = 0$ corresponds to λ being infinite, which we would expect to cause problems for mathematical software. The manner in which we address this possibility depends on the software we are using.

In order to convert the problem to a form suitable for numerical evaluation, we return to the original equations of motion and the transformation identity. In these equations, we substitute the given value $\mu = (km)^{1/2}$ and nondimensionalize time with a new variable

$\tau = t(k/m)^{1/2}$. Since $d/dt = (k/m)^{1/2}(d/d\tau)$, this change of variables converts the equations of motion to

$$\frac{d^2q_1}{d\tau^2} - q_1 - \frac{dq_1}{d\tau} - 2(q_1 - q_2)$$

$$0 = 2(q_1 - q_2) + \frac{F}{k} - \frac{dq_2}{d\tau} = 0$$

In other words considering q_1 and q_2 to be functions of τ has the effect of setting $m = 1$ and $k = 1$, with F scaled by k. We correspondingly redefine the state vector to be

$$\{x\} = \begin{bmatrix} q_1 & q_2 & \dfrac{dq_1}{d\tau} & \dfrac{dq_2}{d\tau} \end{bmatrix}^{\mathrm{T}}$$

The associated matrices for the state-space equations are

$$[S] = \begin{bmatrix} -3 & 2 & 0 & 0 \\ 2 & -2 & 0 & 0 \\ 0 & 0 & 1 & 0 \\ 0 & 0 & 0 & 0 \end{bmatrix}, \qquad [R] = \begin{bmatrix} 0 & 0 & -3 & 2 \\ 0 & 0 & 2 & -2 \\ -3 & 2 & -1 & 0 \\ 2 & -2 & 0 & -1 \end{bmatrix}$$

The homogeneous solution to the state-space equations is now taken to be $\{x\} = B\{\psi\}\exp(\hat{\lambda}\tau)$, where the dimensional eigenvalue is $\lambda = (k/m)^{1/2}\hat{\lambda}$.

In MATLAB, we find the eigensolution by writing $[psi, dum] = eig(R, S)$; and then extract a column of eigenvalues by writing $lambda = diag(dum)$;. The result output by MATLAB is

$$lambda = \begin{Bmatrix} \mathrm{NaN} - \mathrm{NaN}i \\ -1.2267 + 1.4677i \\ -1.2267 - 1.4677i \\ -0.5466 - 0.0000i \end{Bmatrix}$$

$$psi = \begin{bmatrix} 0 & 0.2876 + 0.0692i & -0.0694 - 0.2875i & 0.5158 + 0.0003i \\ 0 & 0.2354 - 0.2678i & 0.2677 - 0.2356i & 0.7098 + 0.0004i \\ 0 & -0.4544 + 0.3371i & -0.3369 + 0.4545i & -0.2820 - 0.0002i \\ -1 & 0.1043 + 0.6741i & -0.6741 - 0.1039i & -0.3880 - 0.0002i \end{bmatrix}$$

In these results NaN stands for "Not a Number," which means that a mathematically undefined operation like 0/0 was performed. This corresponds to the infinite eigenvalue caused by $[S]$ being singular.

Another anomaly is that the second and third eigenvalues are complex conjugates, yet the third eigenvector, that is, the third column of psi, is not the complex conjugate of the second eigenvector. Furthermore, the fourth eigenvalue is real, but the fourth eigenvector is complex. To check whether this is an error, we compute the orthogonality products, for which the MATLAB syntax is $prod1 = psi.' * S * psi$; $prod2 = psi.' * R * psi$;. (An often overlooked aspect of MATLAB is that the transpose operation is indicated with a prime. If the matrix on which this operation is performed is complex, MATLAB also applies the complex conjugate operation. In other words, the operation psi' returns the complex conjugate transpose of psi. This is not what is intended for the transpose operations written here. The way in which MATLAB is told not to apply the complex conjugate operation is to precede the prime with a period.) Both $prod1$ and $prod2$ are found to be diagonal, which confirms that the computations were correct. In essence, what has happened is that the eigenvectors computed by MATLAB are multiplied by different complex factors. This becomes evident when we use the diagonal elements of $prod1$, which are the modal masses ν_n, to normalize the eigenvectors. The value of $prod1(1, 1)$ is zero, which corresponds to a zero modal mass, so normalizing the first mode would be a meaningless operation. For the other three modes, we write

MATLAB: $for\ j = 2:4;\quad psi(:,j) = psi(:,j)/sqrt(prod1(j,j));\quad end$

The outcome is

$$psi = \begin{bmatrix} 0 & 0.3088 + 0.2954i & 0.3088 - 0.2954i & 0.0000 + 1.0080i \\ 0 & 0.4886 - 0.1633i & 0.4886 + 0.1633i & 0.0000 + 1.3872i \\ 0 & -0.8124 + 0.0908i & -0.8124 - 0.0908i & 0.0000 - 0.5510i \\ -1 & -0.3597 + 0.9176i & -0.3597 - 0.9176i & -0.0000 - 0.7582i \end{bmatrix}$$

We see that the normalized second and third eigenvectors are indeed complex conjugates. Note that the fourth eigenvector is imaginary, yet the corresponding eigenvalue is real. This aspect, which we anticipated earlier, arises because the value of $prod1(4,4)$ is negative.

Let us now consider how Mathcad analyzes this problem. The analogous steps to the procedure we implemented in MATLAB are $\lambda = genvals(R, S)$ and $\psi = genvecs(R, S)$. This leads to

$$\lambda = \begin{Bmatrix} -0.5466 \\ -1.2267 - 1.4677i \\ -1.2267 + 1.4677i \end{Bmatrix}$$

$$\psi = \begin{bmatrix} 0.5158 & -0.1809 + 0.2340i & -0.1809 - 0.2340i \\ 0.7098 & -0.3513 - 0.0614i & -0.3513 + 0.0614i \\ 0.2820 & 0.5654 - 0.0216i & 0.5654 + 0.0216i \\ 0.3880 & 0.3408 + 0.5909i & 0.3408 - 0.5909i \end{bmatrix}$$

In other words, the result obtained by Mathcad is the three finite eigenvalues, and their corresponding modes. Interestingly, Mathcad's modes corresponding to the complex conjugate eigenvalues are complex conjugates, whereas the raw results obtained from MATLAB did not show this property. Normalization requires division of each computed mode by the corresponding modal mass, according to

Mathcad: $j := 1;3\quad \nu = \psi^T * S * \psi\quad \Psi^{<j>} = \dfrac{\psi^{<j>}}{\sqrt{\nu_{j,j}}}$

Mathcad's failure to find the fourth mode is not a cause for concern. To see why, consider the nonlinear eigenvalue problem described by eq. (10.3.6) when λ is infinite. In that case, it must be that $\{u\}$, which is the upper partition of an eigenvector, is zero. Because the upper partition is associated with the generalized coordinates, this means that this mode will not contribute to the response. For some individuals, the failure of a program to find all the eigensolutions is troubling. A remedy is to use the reciprocal eigenvalue parameter $\rho = 1/\lambda$, as described earlier in the solution. The eigenvalue problem may then be solved in Mathcad with the operations $\rho = genvals(S, R)$, $\psi = genvecs(S, R)$, and $\lambda_j = if(\nu_j = 0; \infty, 1/\nu_j)$. This will yield $\rho_2 = 0$ and $\lambda_2 = 10^{307}$, which is the largest number allowed by the operating system. The eigenvectors associated with the finite eigenvalues would be the same as the Mathcad results listed above, and the eigenvector associated with the infinite eigenvalue would be $\psi_2 = \begin{bmatrix} 0 & 0 & 0 & -1 \end{bmatrix}^T$, as indicated by MATLAB.

10.5 MODAL COORDINATES

The parallelism between undamped and damped modes extends to the derivation of uncoupled equations of motion for forced vibration. The state-space modal transformation is

$$\{x\} = [\Psi]\{\xi\} = \sum_{n=1}^{2N} \{\Psi_n\}\xi_n \tag{10.5.1}$$

where ξ_n are a set of $2N$ *damped modal coordinates*. We substitute the modal transformation into the equations of motion, eq. (10.1.6), then premultiply by the transpose of the normal mode matrix. These operations yield

$$[\Psi]^T[S][\Psi]\{\dot{\xi}\} - [\Psi]^T[R][\Psi]\{\xi\} = [\Psi]^T \begin{Bmatrix} \{0\} \\ \{Q\} \end{Bmatrix} \tag{10.5.2}$$

In view of the orthogonality relations, eqs. (10.4.4), this reduces to

$$\{\dot{\xi}\} - [\lambda]\{\xi\} = [\Psi]^T \begin{Bmatrix} \{0\} \\ \{Q\} \end{Bmatrix} \tag{10.5.3}$$

The preceding represents a set of uncoupled differential equations, because $[\lambda]$ is diagonal. The scalar version is

$$\dot{\xi}_n - \lambda_n \xi_n = \{\Psi_n\}^T \begin{Bmatrix} \{0\} \\ \{Q\} \end{Bmatrix}, \quad n = 1, \ldots, 2N \tag{10.5.4}$$

These modal equations are first-order equations, so they are somewhat easier to solve than those for undamped modes. Unless we are solely interested in steady-state response, we will need to know the initial conditions. The modal differential equations, being first order, require only the values of each ξ_n at the initial instant, which we take to be $t = 0$. However, the initial conditions likely to be specified are the values of the generalized coordinates q_j and generalized velocities $\dot{q}_j$ at $t = 0$. To determine the modal initial values we recall the definitions of the state vector and the modal transformation, which state that

$$\{x(0)\} = \begin{Bmatrix} \{q(t = 0)\} \\ \{\dot{q}(t = 0)\} \end{Bmatrix} = [\Psi]\{\xi(t = 0)\} \tag{10.5.5}$$

Orthogonality of the modes expedites solution for $\{\xi(0)\}$. Postmultiplying the first of the modal orthogonality relations, eqs. (10.4.4), by $[\Psi]^{-1}$ leads to

$$[\Psi]^{-1} = [\Psi]^T[S] \tag{10.5.6}$$

from which we find

$$\{\xi(t = 0)\} = [\Psi]^T[S] \begin{Bmatrix} \{q(0)\} \\ \{\dot{q}(0)\} \end{Bmatrix} \tag{10.5.7}$$

The scalar form of this relation is

$$\xi_n(t = 0) = \{\Psi_n\}^{\mathrm{T}}[S]\begin{Bmatrix}\{q(0)\}\\\{\dot{q}(0)\}\end{Bmatrix} \tag{10.5.8}$$

Equations (10.5.4) and (10.5.8) provide some general insights regarding mathematical properties. Recall that a complex eigenvalue λ_n and corresponding normalized eigenvector $\{\Psi_n\}$ will be accompanied by complex conjugates $\lambda_{n+1} = \lambda_n^*$ and $\{\Psi_{n+1}\} = \{\Psi_n^*\}$. The differential equation and initial condition associated with the complex conjugate mode are found by introducing these complex conjugate properties into eqs. (10.5.4) and (10.5.8), which leads to

$$\dot{\xi}_{n+1} - \lambda_n^* \xi_{n+1} = \{\Psi_n^*\}^{\mathrm{T}}\begin{Bmatrix}\{0\}\\\{Q\}\end{Bmatrix}$$

$$\xi_{n+1}(t = 0) = \{\Psi_n^*\}^{\mathrm{T}}[S]\begin{Bmatrix}\{q(0)\}\\\{\dot{q}(0)\}\end{Bmatrix} \tag{10.5.9}$$

Each of these equations is the complex conjugate of the corresponding one for ξ_n. Therefore,

Regardless of the nature of the excitation, the modal coordinates associated with a pair of complex conjugate eigenvalues are complex conjugates.

$$\boxed{\lambda_{n+1} = \lambda_n^* \quad \Rightarrow \quad \xi_{n+1} = \xi_n^*} \tag{10.5.10}$$

Let us denote as $\{x^n\}$ the contributions of both modes to the state-space response. When we evaluate the modal transformation, eq. (10.5.1), the combined term will be

$$\{x^n\} = \{\Psi_n\}\xi_n + \{\Psi_{n+1}\}\xi_{n+1} \equiv 2\,\mathrm{Re}[\{\Psi_n\}\xi_n] \tag{10.5.11}$$

In other words, the complex conjugate modal property leads to a real $\{x\}$.

When λ_n is real and $\{\Psi_n\}$ is also real, it is clear that ξ_n will also be real. It also is possible that a real λ_n will lead to a purely imaginary $\{\Psi_n\}$. In that case, eqs. (10.5.4) and (10.5.8) indicate that ξ_n will also be imaginary. Multiplying the imaginary $\{\Psi_n\}$ by the imaginary ξ_n when we form eq. (10.5.1) will again lead to a purely real contribution to $\{x\}$.

Our computational procedure for determining response in general will be to solve the modal differential equations for each mode, regardless of whether there is a corresponding complex conjugate mode. The software will take care of canceling the imaginary parts. However, examining the special cases of free vibration, impulse response, and steady-state response to harmonic excitation will provide other perspectives for the way in which complex conjugate modal contributions combine.

10.5.1 Free Vibration

For free vibration, the homogeneous solution of eq. (10.5.4) starting with the initial value of ξ_n specified by eq. (10.5.8) is

$$\xi_n = \{\Psi_n\}^{\mathrm{T}}[S]\begin{Bmatrix}\{q(0)\}\\\{\dot{q}(0)\}\end{Bmatrix}\exp(\lambda_n t) \tag{10.5.12}$$

When λ_n is real, we let $\{x^n\}$ denote the contribution of this modal coordinate to the state vector. The modal expansion leads to

$$\text{Real } \lambda_n: \quad \{x^n\} = \{\Psi_n\}\{\Psi_n\}^T[S]\begin{Bmatrix}\{q(0)\}\\\{\dot{q}(0)\}\end{Bmatrix} \exp(\lambda_n t) \tag{10.5.13}$$

For a dissipative system, all real eigenvalues must be negative. By analogy with one-degree-of-freedom systems, we recognize that such modes are overdamped. The above shows that the contribution of overdamped modes decays without oscillation. Furthermore, the product $\{\Psi_n\}\{\Psi_n\}^T$ will be real, regardless of whether $\{\Psi_n\}$ is real or imaginary. This means that the various generalized coordinates respond in constant proportions independently of t.

To describe the contribution of a complex mode and its complex conjugate to a free vibration, we recall eq. (10.5.11), and use eq. (10.5.12) to represent the modal coordinate. The resulting contribution to the state space response is

$$\text{Complex } \lambda_n: \quad \{x^n\} = 2 \, \text{Re}\left[\{\Psi_n\}\xi_n\right]$$

$$= 2 \, \text{Re}\left[\{\Psi_n\}\{\Psi_n\}^T\exp(\lambda_n t)\right][S]\begin{Bmatrix}\{q(0)\}\\\{\dot{q}(0)\}\end{Bmatrix} \tag{10.5.14}$$

This is essentially the same as eq. (10.2.8), except that the amplitude coefficients B_n in the previous equation are now defined explicitly in terms of the initial conditions. This response is like that of an underdamped one-degree-of-freedom system, whose exponential decay factor is $\zeta\omega_{\text{nat}} = -\text{Re}(\lambda_n)$ and whose damped natural frequency is $\omega_{\text{d}} = \text{Im}(\lambda_n)$.

10.5.2 Impulsive Forces

The effect of impulsive forces is to change velocity suddenly, after which the response evolves as a free vibration. Impulsive excitations are represented by $\{Q\} = \{P\}\delta(t)$, where P_n is the impulse of the nth generalized force. Subsequent to the application of the impulse, the system executes a free vibration. To determine the initial conditions for that motion we integrate the modal equations of motion, eq. (10.5.4), from $t = 0^-$ to $t = 0^+$, which is the interval over which the impulse occurs. Because η_j is finite, its integral over the infinitesimal integration interval is zero. We also recognize that $\dot{\eta}_j \, dt$ is a perfect differential. Setting $\eta_j(0^-) = 0$, corresponding to the system being at rest in its equilibrium position prior to the impulse, then leads to

$$\int_{0^-}^{0^+} \eta_j \, dt = 0, \qquad \int_{0^-}^{0^+} \dot{\eta}_j \, dt = \eta_j(0^+) \tag{10.5.15}$$

Hence, integrating the equation of motion over this interval gives

$$\eta_j(0^+) = \{\Psi_n\}^T\begin{Bmatrix}\{0\}\\\{P\}\end{Bmatrix}\int_{0^-}^{0^+}\delta(t)\,dt = \{\Psi_n\}^T\begin{Bmatrix}\{0\}\\\{P\}\end{Bmatrix} \tag{10.5.16}$$

The corresponding free vibration solution for the modal coordinates is

$$\xi_n = \{\Psi_n\}^T\begin{Bmatrix}\{0\}\\\{P\}\end{Bmatrix}\exp(\lambda_n t) \tag{10.5.17}$$

This is equivalent to eq. (10.5.8) for a specific set of initial velocities $\{\dot{q}(0)\}$ with no initial displacement. An important aspect of this general impulse response solution is that it provides the basis for a number of experimental modal analysis techniques. The methods use time response measurements directly, rather than converting them to DFT values according to the concepts described in Section 5.4.2. The interested reader should refer to the papers by Ibraham and Mikulcik (1976) and by Leuridan, Brown, and Allemang (1986), and the text by Inman (1989).

10.5.3 Steady-State Harmonic Vibration

In the case of harmonic excitation, the modal coordinates are not necessarily real because their associated normal damped modes $\{\Psi_n\}$ may be real, imaginary, or complex. A straightforward way in which all cases may be considered is to use a complex exponential to represent the steady-state condition, but to require that only the state-space response $\{x\}$ be real,

$$\{x\} = \mathrm{Re}[\{X\}\exp(i\omega t)], \qquad \{\xi\} = \{Y\}\exp(i\omega t) \tag{10.5.18}$$

It follows from the modal transformation, $\{x\} = [\Psi]\{\xi\}$, that the two sets of complex amplitudes are related by

$$\{X\} = [\Psi]\{Y\} = \sum_{n=1}^{2N}\{\Psi_n\}Y_n \tag{10.5.19}$$

We represent the generalized forces as

$$\{Q\} = \mathrm{Re}[\{F\}\exp(i\omega t)] \tag{10.5.20}$$

When we substitute these representations of ξ_n and $\{Q\}$ into eq. (10.5.4) and match like coefficients of the exponential factor, we find that

$$Y_n = \frac{\{\Psi_n\}^{\mathrm{T}}\{\hat{F}\}}{(i\omega - \lambda_n)}, \qquad \{\hat{F}\} \equiv \begin{Bmatrix} \{0\} \\ \{F\} \end{Bmatrix} \tag{10.5.21}$$

These relations apply for both real and complex eigenvalues. Thus, they represent the fundamental relations for a computation of $\{X\}$. However, a more detailed examination of these relations will increase our understanding of the underlying physical phenomena. In the case of overdamped modes (real λ_n), the modal contribution to $\{X\}$ will be

$$\text{Real } \lambda_n: \quad \{X^n\} \equiv \{\Psi_n\}Y_n = \frac{\{\Psi_n\}\{\Psi_n\}^{\mathrm{T}}\{\hat{F}\}}{(i\omega - \lambda_n)} \tag{10.5.22}$$

Regardless of whether $\{\Psi_n\}$ is real or imaginary, the product $\{\Psi_n\}\{\Psi_n\}^{\mathrm{T}}$ will be real. Hence, the only complex aspect of an overdamped mode's contribution to $\{X\}$ is associated with the denominator $i\omega - \lambda_n$. When viewed in the complex plane, we see the real part, $-\lambda_n$, remains constant as the frequency increases from zero, while the imaginary part, $i\omega$, increases. Thus, $|i\omega - \lambda_n|$ increases monotonically with frequency, which means that the contribution to $\{X\}$ of overdamped modes becomes progressively smaller with increasing frequency—they cannot lead to resonance-type phenomena.

In the case of complex conjugate eigenvalues, it is evident from eq. (10.5.21) that $Y_{n+1} \neq Y_n^*$, even though $\lambda_{n+1} = \lambda_n^*$ and $\{\Psi_{n+1}\} = \{\Psi\}_n^*$. Let us consider the denominators of Y_n and Y_{n+1} as a function of ω. We take λ_n to be the root with the positive real part, so that $\lambda_n = -\alpha + \beta i$, $\lambda_{n+1} = -\alpha - \beta i$. Thus, the smallest value of $|i\omega - \lambda_n|$ occurs when $\omega = \beta$, while $|i\omega - \lambda_{n+1}|$ increases monotonically with increasing ω. In other words the resonance-like condition is manifested by ξ_n having a peak magnitude, while ξ_{n+1} decreases progressively with increasing ω. (This assumes that $\{\widehat{F}\}$ is independent of ω.)

We shall combine the contributions of each member of the pair Y_n and Y_{n+1} in order to compare the effect to the corresponding one for an undamped mode. Thus, we form

$$\{X^n\} \equiv \{\Psi_n\}Y_n + \{\Psi_{n+1}\}Y_{n+1}$$

$$= \{\Psi_n\}\frac{\{\Psi_n\}^T\{\widehat{F}\}}{(i\omega - \lambda_n)} + \{\Psi_{n+1}\}\frac{\{\Psi_{n+1}\}^T\{\widehat{F}\}}{(i\omega - \lambda_{n+1})} \qquad (10.5.23)$$

$$= \left[\frac{\{\Psi_n\}\{\Psi_n\}^T}{(i\omega - \lambda_n)} + \frac{\{\Psi_n^*\}\{\Psi_n^*\}^T}{(i\omega - \lambda_n^*)}\right]\{\widehat{F}\}$$

The last step is to place both terms over a common denominator, from which we find

Complex λ_n:

$$\{Y_n\} = \frac{2i\omega\,\text{Re}\,[\{\Psi_n\}\{\Psi_n\}^T] - 2\,\text{Re}\,[\lambda_n^*\{\Psi_n\}\{\Psi_n\}^T]}{\lambda_n\lambda_n^* - 2i\omega\,\text{Re}(\lambda_n) - \omega^2}\{\widehat{F}\} \qquad (10.5.24)$$

It is evident that this expression is more complicated than those we previously encountered for steady-state response. Nevertheless, certain similarities are apparent. For example, in the case of an underdamped one-degree-of-freedom system, eq. (10.3.13) gives $\lambda_1 = -\zeta\omega_{nat} + i\omega_d$. Because $\omega_d = \omega_{nat}(1 - \zeta^2)^{1/2}$ and $\lambda_1\lambda_1^* \equiv |\lambda_1|^2$, the denominator of the above becomes $\omega_{nat}^2 + 2i\zeta\omega\omega_{nat} - \omega^2$. If we replace ω_{nat} with ω_n, we find that this denominator matches the one in eq. (4.3.42), which describes the steady-state response of an N-degree-of-freedom system with modal damping.

The numerator of eq. (10.5.24) also is reminiscent of the analysis using undamped modes. The quantity $\{\Psi_n\}^T\{\widehat{F}\}$, which is a scalar, represents the modal force amplitude, while the leading factor $\{\Psi_n\}$ gives the weighted contribution of the modal coordinate to each element of $\{x\}$. The product $\{\Psi_n\}\{\Psi_n\}^T$ is obviously not real for underdamped modes, which means that the product introduces phase shifts to the steady-state response. These phase shifts have no counterpart when modal analysis is performed in terms of the undamped modes.

10.5.4 Rigid Body Motion

Systems that are capable of rigid body motion present a special case. Each type of motion, translation in orthogonal directions and rotation about orthogonal centroidal axes, will correspond to a zero eigenvalue. Each such eigenvalue will represent a double root of the characteristic equation. That is, if r is the number of rigid body modes, the total number of eigenvalues will be $2N - r$, with r of the eigenvalues being zero. Furthermore, when $\lambda = 0$, the characteristic equation is $|[R]| = 0$, and the rank of $[R]$ will be $2N - r$. In other words, only one rigid body mode will occur for each zero eigenvalue. This means that there will not be $2N$ eigenvectors. Thus, rigid body modes constitute a specific case where the multiplicity of the eigenvalues is greater than the reduction in rank of $[R] - \lambda[S]$. Such situations were noted in Section 10.2 to also occur in cases where a mode is critically damped.

The formal mathematical procedure for handling missing eigenvectors requires reduction of the eigenvalue problem to Jordan normal form. The details of this procedure were provided by Kaplan (1958); they are beyond the scope of the present treatment. Fortunately, a simple resolution is available for the case of rigid body motion. We may evaluate the rigid body contribution to the *generalized coordinates* by integrating the primary equations of motion, which are $\sum \bar{F} = m\,\bar{a}_G$ and $\sum \bar{M} = I\dot{\bar{\Omega}}$ for planar motion. The integration constants for such a solution should be left arbitrary. The deformational motion, which is associated with those modes for which $\lambda \neq 0$, is superimposed on this motion. Hence, we only solve eqs. (10.5.4) subject to the initial conditions in eq. (10.5.8) for the modes for which $\lambda \neq 0$. We form the modal transformation, eq. (10.5.1), by summing over only the deformational modes, and add that result to the rigid body motion. Then the integration constants for the rigid body displacement must be selected to offset initial values associated with the deformational response.

10.5.5 Closure

For systems in which $[C]$ is arbitrary and large, we now have an alternative to using the frequency domain formulation to determine steady-state response. Specifically, rather than employing the transfer function according to eq. (5.1.5), we may synthesize the damped modal responses in eqs. (10.5.24) and (10.5.22). The question of which is preferable to use depends on the situation. If we are interested in the response for a single excitation frequency ω, and a single set of forces $\{F\}$, the frequency domain approach is probably preferable. In contrast, when there are many loading cases to consider, the extra effort entailed in evaluating the damped modes is offset by the ability to perform subsequent operations algebraically without solving simultaneous equations. Furthermore, damped modal analysis may be applied directly to arbitrary excitations. In contrast, applying the frequency domain formulation to arbitrary excitations requires introduction of FFT concepts. Also, frequency domain analysis is incapable of addressing any situation where nonzero initial conditions must be satisfied.

EXAMPLE 10.3

The impulse response of a master system with attachments was evaluated in Example 5.3 by using FFT techniques. Determine whether there are significant discrepancies between that solution and the one obtained by using damped modal analysis.

Solution This problem will illustrate the ease with which the damped modal response may be ascertained after the eigenvalue problem has been solved. It also will verify the correctness of the previous FFT solution. We use the generalized coordinates defined in Example 5.3, which are q_1 to q_5 for the downward displacement of the attached masses, and q_6 for the master mass. The corresponding $[M]$, $[K]$, and $[C]$ were derived earlier, and the generalized forces are $Q_6 = \delta(t)$, $Q_j = 0$ otherwise.

The first step in the modal analysis is to form $[R]$ and $[S]$ according to eqs. (10.1.7). We next must solve the state-space eigenvalue problem, $[[R] - \lambda[S]]\{\psi\} = \{0\}$, for which we use our mathematical software as described in Example 10.1. Because the system has six degrees of freedom, the result is 12 eigenvalues λ_j and corresponding eigenvectors $\{\psi_j\}$. Not surprisingly in view of the small amount of damping within each attachment, the former are found to consist of six complex conjugate pairs. For the purpose of interpreting the results it is useful to identify the equivalent natural frequency and damping ratio as defined by eq. (10.3.13). Toward that end, we observe that $\omega_d = \omega_{nat}(1 - \zeta_{eq}^2)^{1/2}$, where ω_{nat} and ζ_{eq} are the natural frequency and damping ratio for the undamped mode equivalent to the pair of complex conjugate modes. Setting $\lambda_j = -\zeta_{eq}\omega_{nat} \pm \omega_d i$ shows that $\omega_{nat} = |\lambda_j|$ and $\zeta_{eq} = -\text{Re}(\lambda_j)/|\lambda_j|$. This leads to the tabulated eigenvalue properties.

Equivalent undamped modal parameters			
Mode #	λ_j	ω_{nat}	ζ_{eq}
1 & 2	$-2.038(10^{-3}) \pm 0.889i$	0.889	$2.292(10^{-3})$
3 & 4	$-4.643(10^{-3}) \pm 0.959i$	0.959	$4.844(10^{-3})$
5 & 6	$-4.791(10^{-3}) \pm 0.986i$	0.986	$4.859(10^{-3})$
7 & 8	$-4.940(10^{-3}) \pm 1.013i$	1.013	$4.879(10^{-3})$
9 & 10	$-5.093(10^{-3}) \pm 1.039i$	1.039	$4.902(10^{-3})$
11 & 12	$-3.729(10^{-3}) \pm 1.124i$	1.124	$3.317(10^{-3})$

The tabulated eigenvalues have been sorted in ascending order of their magnitude by invoking the algorithm in Example 10.1, which is an optional operation in the context of a response evaluation. The last step in the eigenvalue solution, which is not optional, is to normalize the modes according to eq. (10.4.2); see Example 10.2. Note that it is recommended that the off-diagonal elements of the orthogonality products $\{\psi\}^T[S]\{\psi\}$ and $\{\psi\}^T[R]\{\psi\}$, which are computed as an intermediate step, be checked to be certain that they are negligibly small.

Now that the eigenvalue problem has been solved, we are ready to proceed to the modal responses. We define the modal transformation, $\{x\} = [\Psi]\{\xi\}$, which leads us to the standard modal differential equations, eqs. (10.5.4). Only the last element of $\{Q\}$ is nonzero in the present case, so the differential equations to be solved are

$$\dot{\xi} - \lambda_n \xi_n = \Psi_{12,n} \delta(t), \quad n = 1, \ldots, 12$$

Integrating this differential equation over the interval of the impulse, as was done to obtain eq. (10.5.17), shows the initial value of the modal coordinate at $t = 0^+$ is $\xi_n(0^+) = \Psi_{12,n}$. Thus, the response is

$$\xi_n = \Psi_{12,n} \exp(\lambda_n t)$$

Once we have this solution we may evaluate the generalized coordinate response $\{q\}$ at any instant t by taking the upper partition of the modal transformation, $\{x\} = [\Psi]\{\xi(t)\}$. To compare the results of the present formulation to those obtained in Example 5.3, we set $t_k = (k - 1)\Delta$, with Δ equal to the time increment associated with the previous FFT, $\Delta = 1600/4096$. To use our software efficiently, we evaluate an array $[xi]$ whose kth column holds the modal coordinate values at t_k, that is, $[xi] = [\cdots \{\xi(t_k)\} \cdots]$. A corresponding array of the state variables is obtained from $[x] = [\Psi][xi]$. Row n of $[x]$ contains the values of generalized coordinate q_n at the succession of time instants, which is the data we would plot. We shall not show the result because the curve for each generalized coordinate would be sufficiently close to the one obtained in Example 5.3 that differences would be barely perceptible. For example, at $t_{2049} = 800$, the damped modal solution gives $q_6 = 0.09874$, while the FFT solution gave $q_6 = 0.10061$.

The foregoing points out one of the advantages of damped modes not noted previously: We may compute the response at only those instants of interest. Another advantage of the damped mode solution is that we can interrogate it for the underlying physical aspects leading to interesting response features. For example, we observe that each of the modal coordinates ξ_n decays at the rate $-\text{Re}(\lambda_n)$. The preceding table indicates that the slowest to decay will be ζ_1, $\xi_2 = \zeta_1^*$, ζ_{11}, and $\xi_{12} = \zeta_{11}^*$. Eventually, these will be the dominant contributors. In addition, the overall magnitude of each ξ_n is set by the associated value of $\Psi_{12,n}$. For this system, $|\Psi_{12,1}| = 0.443$, $|\Psi_{12,3}| = 0.118$, $|\Psi_{12,5}| = 0.117$, $|\Psi_{12,7}| = 0.113$, $|\Psi_{12,9}| = 0.106$, and $|\Psi_{12,11}| = 0.502$, so the elements $\Psi_{12,n}$ at $n = 1, 2, 11$, and 12 have the largest magnitudes. The damping ratios associated with these complex conjugate modes is indicated in the table to be $2.292(10^{-3})$ for the first two and $3.317(10^{-3})$ for the last two. The overall damping ratio measured from the envelope of the q_6 response in Example 5.3 was found to be 0.0025, which reflects the combination of values associated with each mode. These same considerations lead us to recognize the cause of the beats. The free oscillation of each modal coordinate occurs at the frequency ω_{nat} given in the table. These all are close to unity, so the oscillation within each envelope occurs at a period of $2\pi/\omega_{av} \approx 2\pi$. The difference frequency between mode pair 1,2 and mode pair 11,12 is 0.235. The corresponding beat period is $2\pi/0.235 \approx 27$, which is comparable to the pattern in the responses graphed previously.

EXAMPLE 10.4

A two-degree-of-freedom model representing an automobile was developed in Example 4.2. As stated there, the chassis has a mass of 1200 kg, and the radius of gyration about the center of mass G is 1.1 m. The spring stiffnesses are 15 kN/m for each of the two front springs, and 10 kN/m for the two rear springs. This model may be improved by considering dashpots to act parallel to each spring, with constant $c_A = 3100$ N-s/m for the front and $c_B = 2800$ N-s/m for the rear. It is desired to study the response of this vehicle when the ground elevation is given by $w(X) = 20\sin(\pi X/2)$ mm, where X is the horizontal distance from a reference location, measured in meters. As depicted in the sketch, the effect of variable ground elevation may be incorporated into the automobile model by applying vertical forces $2k_A w(X_A) + 2c_A \dot{w}(X_A)$ and $2k_B w(X_B) + 2c_B \dot{w}(X_B)$ at each spring attachment point. For a constant speed v of the automobile, the horizontal position of each wheel relative to some reference location will be $X_A = vt$ and $X_B = vt - 2.6$ m. Use damped modal analysis to determine as a function of the speed v the amplitudes of the center of mass's vertical displacement and of the pitching motion. Also determine the acceleration amplitude for each variable. The maximum speed for the evaluation is 20 m/s.

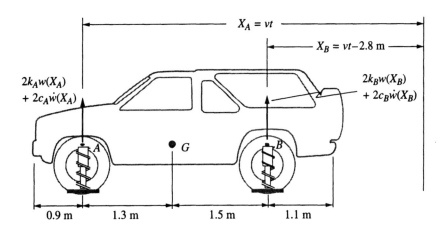

Solution The objective here is to illustrate most of the operations entailed in a damped mode analysis of steady-state response. Furthermore, the results will enable us to understand some of the issues encountered in treating a vehicle's ride quality. To expedite the solution we employ the same generalized coordinates as those in Example 4.2, specifically, $q_1 = y_G$, positive upward, and $q_2 = \theta$, positive in the counterclockwise sense relative to the side view of the automobile. The geometrical, inertial, and elastic properties are like those used previously, so we may directly use the previous descriptions of $[M]$ and $[K]$,

$$[M] = \begin{bmatrix} 1200 & 0 \\ 0 & 1452 \end{bmatrix}, \quad [K] = \begin{bmatrix} 50,000 & -9000 \\ -9000 & 95,700 \end{bmatrix}$$

Each spring is accompanied by a parallel dashpot, so $[C]$ has the same form as $[K]$,

$$[C] = \begin{bmatrix} 2(c_A + c_B) & 2(-1.3c_A + 1.5c_B) \\ 2(-1.3c_A + 1.5c_B) & 2(1.69c_A + 2.25c_B) \end{bmatrix} = \begin{bmatrix} 11,800 & 340 \\ 340 & 23,078 \end{bmatrix}$$

We shall derive the generalized forces from the power input associated with the specified suspension forces. The vertical velocities at the suspension points are $\dot{y}_A$ and $\dot{y}_B$. If F_A and F_B denote the respective forces associated with the elevation, we have $\mathcal{P}_{in} = F_A \dot{y}_A + F_B \dot{y}_B$. In the small displacement approximation, these velocity components are related to the upward displacement $\dot{y}_G$ and counterclockwise rotation $\dot{\theta}$ by

$$\dot{y}_A = \dot{y}_G - 1.3\dot{\theta}, \quad \dot{y}_B = \dot{y}_G + 1.5\dot{\theta}$$

From this, we find that

$$\mathcal{P}_{in} = F_A(\dot{y}_G - 1.3\dot{\theta}) + F_B(\dot{y}_G + 1.5\dot{\theta})$$

The generalized forces are the coefficients of $\dot{y}_G$ and $\dot{\theta}$ in $\mathcal{P}_{in}$, so we have

$$Q_1 = F_A + F_B = 2k_A w(X_A) + 2c_A \dot{w}(X_A) + 2k_B w(X_B) + 2c_B \dot{w}(X_B)$$

$$Q_2 = -1.3F_A + 1.5F_B = -2.6[k_A w(X_A) + c_A \dot{w}(X_A)] + 3.0[k_B w(X_B) + c_B \dot{w}(X_B)]$$

Furthermore, we know that $w(X) = 0.02 \sin(\pi X/2)$ m and $X_A = vt$, $X_B = vt - 2.8$ m. Let us define

$$\Omega = \frac{\pi v}{2} \text{ rad/s}$$

so that

$$w(X_A) = \text{Re}\left[\frac{0.02}{i}\exp(i\Omega t)\right] \quad \Rightarrow \quad \dot{w}(X_A) = \text{Re}\,[0.02\Omega \exp(i\Omega t)]$$

$$w(X_B) = \text{Re}\left[\frac{0.02}{i}\exp(i\Omega t - 1.4\pi i)\right] \quad \Rightarrow \quad \dot{w}(X_A) = \text{Re}\,[0.02\Omega \exp(i\Omega t - 1.4\pi i)]$$

It is obvious from these descriptions that the generalized forces vary harmonically at frequency Ω, such that $Q_j = \text{Re}\,[F_j \exp(i\Omega t)]$, where the complex amplitudes are functions of Ω given by

$$F_1(\Omega) = 0.04\left(\frac{k_A}{i} + \Omega c_A\right) + 0.04\left(\frac{k_B}{i} + \Omega c_B\right)\exp(-1.4\pi i)$$

$$F_2(\Omega) = -0.052\left(\frac{k_A}{i} + \Omega c_A\right) + 0.06\left(\frac{k_B}{i} + \Omega c_B\right)\exp(-1.4\pi i)$$

Now that we have determined the basic equations of motion, we proceed to analyze the damped modal properties. The state vector is

$$\{x\} = \begin{bmatrix} y_G & \theta & \dot{y}_G & \dot{\theta} \end{bmatrix}^{\text{T}}$$

We form $[R]$ and $[S]$ according to eq. (10.1.7), and use software to solve

$$[[R] - \lambda[S]]\{\psi\} = \{0\}$$

This yields four eigenvalues, two of which are complex conjugates and the other two are real. The eigenvalues are

$$\lambda_1 = -5.163 + 4.366i, \quad \lambda_2 = -5.163 - 4.366i, \quad \lambda_3 = -7.200, \quad \lambda_4 = -8.202$$

Correspondingly, we find that the first two modes are a complex conjugate pair, while $\{\psi_3\}$ and $\{\psi_4\}$ are real. We normalize these eigenvectors by following the procedures developed previously. The resulting normal damped mode matrix is

$$[\Psi] = (10^{-3}) \begin{bmatrix} -1.899 - 3.254i & 1.899 + 3.254i & 4.060i & 3.179 \\ -0.587 + 0.922i & -0.587 - 0.922i & 9.665i & 9.153 \\ 24.010 + 8.508i & 24.010 - 8.508i & -29.235i & -26.074 \\ -0.994 - 7.322i & -0.994 + 7.322i & -69.586i & -75.075 \end{bmatrix}$$

With these preliminaries completed, we are now ready to formulate the equations for the modal coordinates. Upon substitution of $\{x\} = [\Psi]\{\xi\}$, the state-space equations of motion reduce to

$$\dot{\xi}_n - \lambda_n \xi_n = \{\Psi_n\}^{\mathrm{T}} \begin{Bmatrix} \{0\} \\ \{Q\} \end{Bmatrix} = \{\Psi_n\}^{\mathrm{T}} \operatorname{Re} \left[\{\hat{F}\} \exp(i\Omega t) \right]$$

where the complex amplitudes of the excitation are

$$\hat{F}_1 = \hat{F}_2 = 0, \qquad \hat{F}_3 = F_1(\Omega), \qquad \hat{F}_4 = F_2(\Omega)$$

The complex amplitude of each ξ_n is given by eq. (10.5.21). Thus, for each mode, regardless of whether its eigenvalue is real or complex, we compute

$$Y_n = \frac{\{\Psi_n\}^{\mathrm{T}}\{\hat{F}\}}{(i\Omega - \lambda_n)}$$

We may evaluate these quantities at any specified value of Ω. From those values, we compute the complex amplitude of $\{X\}$ according to eq. (10.5.19),

$$\{X\} = \sum_{n=1}^{2N} \{\Psi_n\} Y_n = [\Psi]\{Y\}$$

In the present problem, we seek the values of $\{x\}$ when $0 < v < 20$ m/s. The manner in which we implement this computation is essentially the same in Mathcad and MATLAB, aside from syntactic differences. In Mathcad we discretize the range of v using subscript $p := 1, \ldots, P$, where $P = 201$ is more than adequate. The corresponding Ω_p values are $\pi v_p/2$. We then evaluate an array whose pth column consists of the Y_n values at Ω_p, after which multiplying Y by the state-space mode matrix gives an array X holding the $\{X\}$ values at each speed. Specifically, we write

Mathcad: $\quad Y_{n,p} = ((\Psi^{<n>})^{\mathrm{T}}*[0\ 0\ F_1(\Omega_p)\ F_2(\Omega_p)]^{\mathrm{T}})_{1,1} / (1i*\Omega_p - \lambda_n)$
$\qquad\qquad X = \Psi * Y$

where $F_1(\Omega)$ and $F_2(\Omega)$ are functions giving the spring attachment forces defined earlier. In view of the definition of $\{X\}$ for this system, the first row holds the complex amplitude of y_G and the second row holds the complex amplitude of θ. From those results we may evaluate the displacement amplitude by taking the magnitude of each $X_{1,p}$ or $X_{2,p}$. Because the complex acceleration amplitude is $-\Omega^2$ times the displacement amplitude in a harmonic vibration, the acceleration amplitudes of interest are $\Omega_j^2 |X_{1,j}|$ and $\Omega_j^2 |X_{2,j}|$. To employ MATLAB we would define a discrete set of v values, and compute a vector *omega* holding the corresponding Ω values. We then use a nested pair of *for* loops to carry out the main computation. The Y_n values for a speed v set by the outer loop are computed within the inner loop, such that

MATLAB: $v = 0 : 0.1 : 20$; $om = (pi/2) * v$; $for\ p = 1: length\ (v)$; $for\ n = 1 : 4$;
$Y(n) = psi(:,n).' * [0\ 0\ F1(om(p))\ F2(om(p))] .' / (i * om(p) - lambda(n))$;
end; $X(:,j) = psi * Y$; end.

The functions $F1(om)$ and $F2(om)$ would be defined in M-files.

The responses obtained from either of the above procedures are depicted graphically. Three cases are displayed. The curve marked "actual" describes the situation for the specified suspension parameters. To obtain the curves marked "actual × 1000," the spring and dashpot constants were multiplied by 1000, while "actual/10" corresponds to spring and dashpot constants that are one-tenth their specified values. For the stated set of parameters, a very mild peak in the vertical displacement amplitude occurs at 4 m/s, while the amplitude of the pitching rotation maximizes at approximately 5 m/s. Both peaks are associated with a maximum value of $|Y_1|$, which is the amplitude of the first complex conjugate mode. The denominator of the solution for this variable maximizes at $\Omega = \text{Im}(\lambda_1)$, which in view of the definition of Ω occurs when $v = (2/\pi)\,\text{Im}(\lambda_1) \approx 2.8$ m/s. The peak values of y_G and θ do not occur at that speed for two reasons. We have seen the excitation functions increase linearly with Ω, and these quantities appear in the numerator of X_1 and X_2. Another important factor is the contribution of the other modes. No resonance phenomena can be exhibited by these modes, but that does not mean that their effect is unimportant.

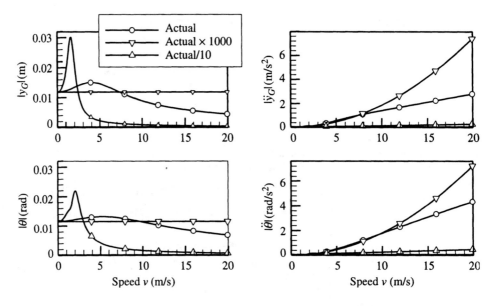

Dividing the suspension constants by 10 has the effect of making the suspension very soft. In that case, the eigenvalues are two complex conjugate pairs. The imaginary parts are within 25% of each other, so the resonant peaks of Y_1 and Y_3 merge into the single one displayed in the graph. Away from the peak, the displacement and rotation amplitudes for this case are very small. Hence, beyond a resonant peak, a soft suspension leads to displacement and rotation amplitudes that are quite small. This is contrasted by the case of a nearly rigid suspension, for which both amplitudes remain nearly constant at their zero speed values.

The acceleration amplitudes for the nearly rigid suspension increase almost quadratically with increasing speed. This trend stems from the Ω^2 factor that converts displacement to acceleration. The given suspension parameters have the effect of lowering the acceleration amplitudes at the higher values of v. For the parameters of the present system, the Ω^2 factor leading to acceleration is sufficiently strong that it suppresses the appearance of peaks in the acceleration graphs. The graphs indicate that a soft suspension greatly reduces the acceleration amplitudes.

The resonant displacement peak of the soft suspension is one important reason why automobile suspensions must be reasonably stiff. However, the penalty for that stiffness is greater acceleration imparted to the vehicle's occupants. Another reason for making a suspension stiff is vehicle handling (that is, maneuverability). If the suspension is very soft, it will be difficult to make the vehicle body follow sudden changes in the orientation of the wheels.

REFERENCES AND SELECTED READINGS

IBRAHAM, S. R., & MIKULCIK, E. C. 1976. "The Experimental Determination of Vibration Parameters from Time Responses." *Shock and Vibration Bulletin*, 46.5, 187–196.

INMAN, D. J. 1989. *Vibration*. Prentice Hall, Englewood Cliffs, NJ.

KAPLAN, W. 1958. *Ordinary Differential Equations*. Addison-Wesley, Reading, MA.

LEURIDAN, J. M., BROWN, D. L., & ALLEMANG, R. J. 1986. "Time Domain Parameter Identification Techniques for Linear Modal Analysis: A Unifying Approach." *ASME Journal of Vibration, Acoustics, Stress, and Reliability*, 108, 1–8.

MEIROVITCH, L. 1997. *Principles and Techniques of Vibrations*. Prentice Hall, Englewood Cliffs, NJ.

TONGUE, B. H. 1996. *Principles of Vibrations*. Oxford University Press, New York.

EXERCISES

10.1 System parameters are $m_1 = 1$ kg, $m_2 = 2$ kg, $k_1 = 100$ N/m, $k_2 = 50$ N/m, $k_3 = 250$ N/m, $c_1 = 6$ N-s/m, $c_2 = 3$ N-s/m, and $c_3 = 12$ N-s/m. Determine the state-space eigenvalues and corresponding damped modes. Compare the real and imaginary parts of the eigenvalues to the natural frequencies and modal damping ratios that are obtained from the light damping approximation.

10.2 In order to isolate a piece of electronic equipment ($m_1 = 2$ kg) from vertical vibration of the table, it is mounted on spring $k_1 = 100$ N/m. The table is rigid, and its mass is $m_2 = 100$ kg. Each of the two springs $k_2 = 400$ N/m surrounds a dashpot whose constant is $c_2 = 50$ N-s/m.

(a) Determine the state-space equations of motion for this system.

(b) Evaluate the eigenvalues and damped modes for the system.

(c) Compare the damped modal properties to corresponding quantities obtained from undamped modal analysis.

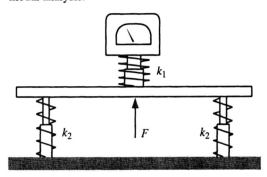

EXERCISE 10.2

10.3 The configuration shown in the sketch corresponds to the static equilibrium position of the system in the absence of the vertical force F. The masses

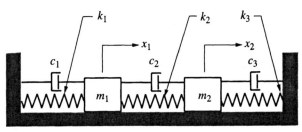

EXERCISE 10.1

are 4 kg for the upper bar and 2 kg for the lower bar. Also, $L = 800$ mm, $k_1 = k_2 = 1500$ N/m, and $c_1 = c_2 = 100$ N-s/m.

(a) Determine the state-space equations of motion for this system.

(b) Evaluate the eigenvalues and damped modes for the system.

(c) Compare the damped modal properties to corresponding quantities obtained from undamped modal analysis.

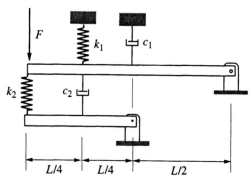

EXERCISE 10.3

10.4 Friction between the ground and each block is negligible. The dashpot constant is $c = (km)^{1/2}$. Determine the damped modal properties of this system. How many damped modes are there? Does this system possess a rigid mode according to damped modal analysis?

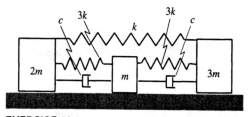

EXERCISE 10.4

10.5 In order to understand the effect of damping on mode localization, consider the 10-bladed disk assembly in Example 4.6 in the weakly coupled case, $R = 0.05$. The imperfection parameters for the blade stiffnesses, Δf_j, are as stated there. Suppose the effect of damping can be modeled as a dashpot acting in parallel to each blade stiffness, with $c_j = 2m_j\omega_{bj}$. Determine the damped modal properties. Are the modes still localized?

10.6 Solve Exercise 10.5 in the strongly coupled case, $R = 0.5$. Increase the number of blades to 16.

10.7 The equations of motion for a system are

$$4\ddot{y} + \ddot{z} + 20\dot{y} - 5\dot{z} + 200y - 50z = 0$$

$$\ddot{y} + 8\ddot{z} - 5\dot{y} + 10\dot{z} - 50y + 500z = 0$$

Determine the eigenvalues and normalized damped modes. Identify the natural frequencies and critical damping ratios for equivalent one-degree-of-freedom systems.

10.8 The following properties are known for a certain three-degree-of-freedom system:

$$[M] = \begin{bmatrix} 600 & 400 & 200 \\ 400 & 1200 & 0 \\ 200 & 0 & 800 \end{bmatrix} \text{kg},$$

$$[K] = \begin{bmatrix} 300 & 0 & -200 \\ 0 & 500 & 300 \\ -200 & 300 & 700 \end{bmatrix} \text{kN/m},$$

$$[C] = \begin{bmatrix} 500 & 300 & -400 \\ 300 & 900 & 600 \\ -400 & 600 & 1300 \end{bmatrix} \text{N-s/m}$$

Determine the eigenvalues and normalized damped modes.

10.9 In the triple pendulum depicted below, identical suspended bars having mass m are coupled by weak springs k and dashpots c. The system is constructed such that the springs are unstretched when the bars are vertical. Determine the eigenvalues and normalized damped modes for two cases: $k/m = 0.05g/L$, $c/m = 0.01(g/L)^{1/2}$ and $k/m = 0.05g/L$, $c/m = 2(g/L)^{1/2}$.

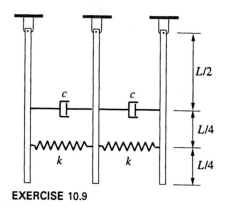

EXERCISE 10.9

10.10 Each floor of the four-story building modeled in the sketch is considered to translate horizontally as a slab, with the columns acting as massless beams. The inertia and flexibility matrices are

$$[M] = \begin{bmatrix} 50{,}000 & 0 & 0 & 0 \\ 0 & 50{,}000 & 0 & 0 \\ 0 & 0 & 40{,}000 & 0 \\ 0 & 0 & 0 & 35{,}000 \end{bmatrix} \text{kg}$$

$$[L] = \begin{bmatrix} 0.4 & 0.4 & 0.4 & 0.4 \\ 0.4 & 1.6 & 1.6 & 1.6 \\ 0.4 & 1.6 & 4 & 4 \\ 0.4 & 1.6 & 4 & 7 \end{bmatrix} \text{mm/kN}$$

The constant for each dashpot is $c = 200$ kN-s/m. Determine the natural frequencies, modal damping ratios, and normalized damped modes.

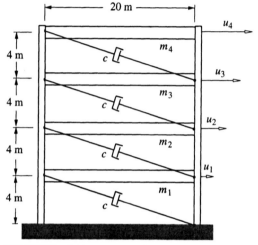

EXERCISE 10.10

10.11 Consider the damped two-degree-of-freedom system in Exercise 10.1. Initial conditions are $x_1 = 10$ mm, $\dot{x}_1 = 0$, $x_2 = 0$, and $\dot{x}_2 = 2$ m/s. Use damped modal analysis to determine the response.

10.12 The triple pendulum in Exercise 10.9 is released from rest at $t = 0$. The initial rotation of the middle bar is 0.08 rad counterclockwise, while the left and right bars are vertical at the instant of release. Determine and graph the free vibration of this system for both sets of parameters specified in Exercise 10.9.

10.13 The building in Exercise 10.10 is released from rest with an initial displacement that is proportional to the elevation above the ground, $u_1 = 20$ mm, $u_2 = 40$ mm, $u_3 = 60$ mm, and $u_4 = 80$ mm. Determine and graph the ensuing free vibration of each floor.

10.14 Consider the damped two-degree-of-freedom system in Exercise 10.1. Determine the response of the system when an exponentially decaying force $F(t) = 4 \exp(-2t)$ N is applied to the left block at $t = 0$. The system is initially at rest at its static equilibrium position.

10.15 Exercise 4.45 treated a three-degree-of-freedom model of a building. Use damped modal analysis to determine the displacement $x_j(t)$ of each floor. Graph the results.

10.16 The two-degree-of-freedom system undergoes free vibration, $F \equiv 0$. It is released from rest with mass m_1 at its static equilibrium position and mass m_2 at 100 mm left of its static equilibrium position. System parameters for the two-degree-of-freedom system are $m_1 = 1$, $m_2 = 4$ kg; $k_1 = 200$, $k_2 = 300$, $k_3 = 100$ N/m; and $c_1 = 40$, $c_2 = 10$ N-s/m. Use damped modal analysis to determine the resulting displacement of each mass as a function of time. Approximately how much time is required for the displacement of both masses from their equilibrium position not to exceed 1 mm?

10.17 At $t = 0$, both blocks in the two-degree-of-freedom system pass their static equilibrium positions with a velocity of 10 m/s to the left. At that instant m_2 is subjected to a step force $F = 50h(t)$ N acting to the right. System parameters are $m_1 = 1$, $m_2 = 4$ kg; $k_1 = 200$, $k_2 = 300$, $k_3 = 100$ N/m; and $c_1 = 40$, $c_2 = 10$ N-s/m. Use damped modal analysis to determine the resulting

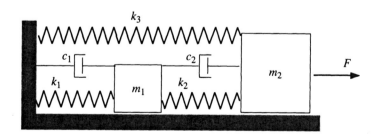

EXERCISES 10.16, 10.17, 10.18

displacement of each mass as a function of time. Graph the result.

10.18 Block m_2 in Exercise 10.16 is subjected to a harmonic force $F = 10 \cos (6t) + 5 \sin (6t - \pi/6)$ N. Use damped modal analysis to determine the steady-state response of each block. Give the amplitude and phase angle of each displacement relative to a pure cosine function.

10.19 The effect of a blast loading on the building in Exercise 10.10 may be modeled as a set of equal impulsive forces acting horizontally on each floor. The impulse of each such force is 40 kN-s, and the building was at rest in its equilibrium position prior to the impulse. Determine the response to this excitation. In particular, determine the maximum horizontal displacement each floor undergoes.

10.20 Suppose the blast loading in Exercise 10.19 is modeled as equal forces $F(t) = 10^{10} t \exp (-500t)$ N applied to each floor at $t = 0$. The impulse of each force is 40 kN-s, which matches the value for each impulsive force in Exercise 10.19. Use damped modal analysis to determine the ensuing response.

10.21 Modal decoupling associated with the light damping approximation was employed in Exercise 4.43. Use damped modal analysis to determine the response of that system. All parameters match those originally stated.

10.22 An impulsive force is applied to the right block of the system in Exercise 10.4. The system was at rest with the springs undeformed prior to application of the force. Use damped modal analysis to determine the displacement of each block. Graph each result as a function of $(k/m)^{1/2} t$.

10.23 Suppose the ground elevation in Example 10.4 is a speed bump described by $w(x) = 4A \times (Lx - x^2)[h(x) - h(x - L)]$, where $L = 500$ mm and $A = 80$ mm are, respectively, the width and height of the bump. Consider a case where the automobile was traveling at $v = 5$ m/s without ver-

tical motion when it first encountered the bump. Determine the maximum peak values of y_G and $\ddot{y}_G$ in the ensuing motion.

10.24 Consider the one-and-one-half-degree-of-freedom system in Example 10.2 with $\mu = (km)^{1/2}$. Use damped modal analysis to evaluate the steady-state response of the block when $F = F_0 \cos(\omega t)$. Graph the amplitude as a function of ω. Does the amplitude show any peak values? If so, at what frequencies do they occur?

10.25 Use damped modal analysis to determine the steady-state vertical motion of the vehicle in Exercise 4.56. In particular, graph the steady-state amplitudes $|y_1|$ and $|y_2|$ as a function of v for $0 < v < 20$ m/s.

10.26 The mass, stiffness, and damping matrices for a three-degree-of-freedom system are

$$[M] = \begin{bmatrix} 2 & 0 & 1 \\ 0 & 1 & 0 \\ 1 & 0 & 3 \end{bmatrix} \text{ kg}, [K] = \begin{bmatrix} 100 & 20 & -50 \\ 20 & 200 & 40 \\ -50 & 40 & 150 \end{bmatrix} \text{ N/m}$$

$$[C] = \begin{bmatrix} 12 & -3 & -6 \\ -3 & 9 & 3 \\ -6 & 3 & 6 \end{bmatrix} \text{ N-s/m}$$

The harmonic excitation is $Q_j = 3 \sin (5t - \pi/3) + 4 \sin (15t + \pi/6)$ N, $j = 1, 2, 3$. Correspondingly, the steady-state response must have the form $q_j = A_j \sin (5t - \chi_j) + B_j \sin (15t - \theta_j)$. Determine the amplitudes A_j and B_j, and the phase angles χ_j and θ_j by using each of the following approaches:

(a) Undamped modal analysis using the light damping approximation.

(b) Harmonic analysis using the system's transfer function.

(c) Damped modal analysis.

Based on the effort and the quality of the resulting solutions, discuss the merits of each approach.

CHAPTER 11

MODAL ANALYSIS OF GYROSCOPIC SYSTEMS

For the most part, the systems we have treated thus far had a static equilibrium position, which we used as the reference location for defining the generalized coordinates. In this chapter we will remove that restriction. The systems of particular interest will be those in which the reference state is a steady rotational motion. We will identify conditions in which the steady rotation is unstable, which would lead to catastrophic failure. We also will develop the ability to identify operational conditions that cause resonant-like conditions leading to excessively large motions.

Rotating machinery for propulsion, power generation, and pumps is of special concern. Related phenomena arise in treating linkages for robotic applications, as well as large flexible space structures. We will encounter two inertial effects that are unlike any we have considered thus far. Coriolis acceleration, which is associated with movement relative to a rotating reference frame, leads to the generation of forces that are perpendicular to the relative velocity. Similarly, rotation about an axis that is perpendicular to a nominal overall rotation leads to couples that are perpendicular to the nominal rotation axis. We will see that both phenomena are manifested by inertial terms in the equations of motion that are skew-symmetric. One of the consequences of this property is that the modal analysis techniques we developed for time-independent systems are not directly applicable.

Because the power balance method is not valid for time-dependent systems, we will employ Lagrange's equations, derived in Appendix A, to obtain equations of motion. Our work in this chapter begins by linearizing the equations of motion. The matrix eigenvalue problem for modal analysis will be seen to be somewhat different from those for undamped and damped modal analysis of time-invariant systems. In turn, this will require modifications of the method for analyzing forced response.

11.1 GENERAL LINEARIZED EQUATIONS OF MOTION

If position depends explicitly on time, as well as the generalized coordinates, $\bar{r} = \bar{r}(q_j, t)$, the velocity will contain a term $\partial \bar{r}/\partial t$ that does not arise in time-invariant systems; see eq. (A.4). Let us consider the implication of this difference for the kinetic energy of a single particle, for which we have

$$
\begin{aligned}
T &= \frac{1}{2}m\bar{v}\cdot\bar{v} = \frac{1}{2}m\left(\sum_{j=1}^{N}\frac{\partial\bar{r}}{\partial q_j}\dot{q}_j + \frac{\partial\bar{r}}{\partial t}\right)\cdot\left(\sum_{j=1}^{N}\frac{\partial\bar{r}}{\partial q_j}\dot{q}_j + \frac{\partial\bar{r}}{\partial t}\right) \\
&= \frac{1}{2}m\sum_{j=1}^{N}\sum_{n=1}^{N}\frac{\partial\bar{r}}{\partial q_j}\cdot\frac{\partial\bar{r}}{\partial q_n}\dot{q}_j\dot{q}_n + m\frac{\partial\bar{r}}{\partial t}\cdot\sum_{j=1}^{N}\frac{\partial\bar{r}}{\partial q_j}\dot{q}_j + \frac{1}{2}m\frac{\partial\bar{r}}{\partial t}\cdot\frac{\partial\bar{r}}{\partial t}
\end{aligned}
\tag{11.1.1}
$$

where the summation index for one of the double sums was changed in order to ensure that we account for all combinations of the dot product.

The generalized velocities $\dot{q}_j$ occur as quadratic products in the first term, linearly in the second term, and not at all in the third term. Furthermore, in each term the partial derivatives may depend on both the current values of q_j and t, because that is the dependence of position for a time-dependent system. As a consequence, we conclude that adding the kinetic energy of all particles in a system will lead to a kinetic energy that consists of a combination of terms that are quadratic and linear in the generalized velocities, as well as terms that are independent of the $\dot{q}_j$ variables. The coefficients of these terms may be functions of the generalized coordinates and time. We express the general form as

$$
T = \frac{1}{2}\sum_{j=1}^{N}\sum_{n=1}^{N}\hat{M}_{jn}(q_k, t)\dot{q}_j\dot{q}_n + \sum_{j=1}^{N}\hat{N}_j(q_k, t)\dot{q}_j + \hat{\Gamma}(q_k, t)
\tag{11.1.2}
$$

It is evident that this form will lead to numerous terms in the equations of motion that have no counterpart for time-invariant systems. There are two approaches to linearizing the equations of motion. One is to retain all terms in the energy expressions and delete nonlinear terms after Lagrange's equations have been evaluated. The alternative, which is like the procedure whereby the power balance method is implemented, entails truncating T and V to eliminate the occurrence of nonlinear terms in the equations of motion. For the formulation of specific equations, the former is often easier. However, by examining how the energy expressions are simplified, we will be able to recognize the general nature of the equations we will need to solve.

We shall use a zero subscript to denote quantities that are evaluated at the reference state of motion, in which the generalized coordinates are defined to be identically zero, $q_j = \dot{q}_j = 0$. Hence, the reference state is defined by

$$
\frac{d}{dt}\left(\frac{\partial T}{\partial\dot{q}_j}\right)_0 - \left(\frac{\partial T}{\partial q_j}\right)_0 + \left(\frac{\partial V}{\partial q_j}\right)_0 = (Q_j)_0
\tag{11.1.3}
$$

Note that the generalized forces might depend on the q_j variables. If so, they should also be evaluated at $q_j = \dot{q}_j = 0$.

Because the terms in Lagrange's equations are derivatives with respect to a generalized coordinate or velocity, linear terms in the equations of motion correspond to quadratic terms in the energies, as is the case for time-invariant systems. The Taylor series procedure we used previously to derive the quadratic representation of potential energy in eq. (1.5.17) may also be applied to the kinetic energy. The first term (the double sum) in eq. (11.1.2) already is quadratic in the generalized velocities, so linearization of this term is achieved by evaluating the coefficient functions $\hat{M}_{jn}$ at the reference state. The second term (the single sum) is linear in the generalized velocities, so limiting this term to quadratic terms is achieved by using a first-order Taylor series to approximate the coefficient functions N_j. By the same reasoning, we use a second-order Taylor series to approximate $\hat{\Gamma}$. Thus, the approximations we employ are

$$
M_{jn} \approx (\hat{M}_{jn})_0, \qquad \hat{N}_j \approx (\hat{N}_j)_0 + \sum_{n=1}^{N} \left(\frac{\partial \hat{N}_j}{\partial q_n} \right)_0 q_n
$$

$$
\hat{\Gamma} \approx (\hat{\Gamma})_0 + \sum_{n=1}^{N} \left(\frac{\partial \hat{\Gamma}}{\partial q_n} \right)_0 q_n + \frac{1}{2} \sum_{k=1}^{N} \sum_{n=1}^{N} \left(\frac{\partial^2 \hat{\Gamma}}{\partial q_k \partial q_n} \right)_0 q_k q_n
$$

(11.1.4)

The terms with subscript 0 are evaluated at the position where the generalized coordinates are zero, so they may depend only on t.

Substitution of these approximate representations into eq. (11.1.2) leads to the most general form of kinetic energy appropriate to a linearized analysis. It consists of terms that are quadratic and linear in the generalized coordinates and/or velocities, as well as terms that are independent of those variables. In order to expedite further operations, it is convenient to change the first summation index from j to k, so the most general linearized kinetic energy is

$$
T = \frac{1}{2} \sum_{k=1}^{N} \sum_{n=1}^{N} M_{kn} \dot{q}_k \dot{q}_n + \sum_{k=1}^{N} \sum_{n=1}^{N} B_{kn} \dot{q}_k q_n + \frac{1}{2} \sum_{k=1}^{N} \sum_{n=1}^{N} E_{kn} q_k q_n
$$
$$
+ \sum_{k=1}^{N} N_k \dot{q}_k + \sum_{j=1}^{N} J_k q_k + (\hat{\Gamma})_0
$$

(11.1.5)

where the coefficients are

$$
M_{kn} = M_{nk} = (\hat{M}_{kn})_0, \qquad B_{kn} = \left(\frac{\partial \hat{N}_k}{\partial q_n} \right)_0, \qquad E_{kn} = E_{nk} = \left(\frac{\partial^2 \hat{\Gamma}}{\partial q_k \partial q_n} \right)_0
$$
$$
N_k = (\hat{N}_k)_0, \qquad J_k = \left(\frac{\partial \hat{\Gamma}}{\partial q_k} \right)_0
$$

(11.1.6)

The same line of reasoning applies to the representation of potential energy. The result has the same form as the quadratic expression that was derived for the power balance method, eq. (1.5.17). Thus, the most general linearized potential energy is

$$
V = V_0 + \sum_{k=1}^{N} (F_0)_k q_k + \frac{1}{2} \sum_{k=1}^{N} \sum_{n=1}^{N} K_{kn} q_k q_n, \qquad K_{nk} = K_{kn}
$$

(11.1.7)

Note that the coefficients in the linearized kinetic and potential energies may be functions of time in the most general situation.

In order to recognize the basic nature of the equations of motion for time-dependent systems, let us apply Lagrange's equations to the general energy expressions. Equation (A.2.7) proves that a partial derivative with respect to $\dot{q}_j$ filters out the jth term from a summation. Thus, differentiating the terms in T that depend explicitly on a generalized velocity yields

$$\frac{\partial T}{\partial \dot{q}_j} = \sum_{n=1}^{N} M_{jn}\dot{q}_n + \sum_{n=1}^{N} B_{jn}q_n + N_j \tag{11.1.8}$$

The coefficients may be functions of time, so the first term in Lagrange's equations is

$$\frac{d}{dt}\left(\frac{\partial T}{\partial \dot{q}_j}\right) = \sum_{n=1}^{N} (M_{jn}\ddot{q}_n + \dot{M}_{jn}\dot{q}_n) + \sum_{n=1}^{N} (B_{jn}\dot{q}_n + \dot{B}_{jn}q_n) + \dot{N}_j \tag{11.1.9}$$

The partial derivative of T with respect to a generalized coordinate is

$$\frac{\partial T}{\partial q_j} = \sum_{k=1}^{N} B_{kj}\dot{q}_k + \sum_{n=1}^{N} E_{jn}q_n + J_j \tag{11.1.10}$$

We change the index for the first sum from k to n, which leads to

$$\frac{d}{dt}\left(\frac{\partial T}{\partial \dot{q}_j}\right) - \frac{\partial T}{\partial q_j} = \sum_{n=1}^{N} M_{jn}\ddot{q}_n + \sum_{n=1}^{N} [B_{jn} - B_{nj} + \dot{M}_{jn}]\dot{q}_n$$

$$+ \sum_{n=1}^{N} (\dot{B}_{jn} - E_{jn})q_n + \dot{N}_j - J_j \tag{11.1.11}$$

When the coefficients actually are functions of time, it is unlikely that we will be able to solve the equations of motion analytically, although we could use numerical methods. There are, however, important situations in which the coefficients are constants. We rewrite eq. (11.1.11) for such cases by equating to zero the time derivative of any coefficient. The inertial terms in Lagrange's equations in that case are

$$\frac{d}{dt}\left(\frac{\partial T}{\partial \dot{q}_j}\right) - \frac{\partial T}{\partial q_j} = \sum_{n=1}^{N} M_{jn}\ddot{q}_n + \sum_{n=1}^{N} G_{jn}\dot{q}_n - \sum_{n=1}^{N} E_{jn}q_n - J_j \tag{11.1.12}$$

where

$$\boxed{G_{jn} = B_{jn} - B_{nj}} \tag{11.1.13}$$

The coefficients M_{jn} multiplying $\ddot{q}_n$ are like the inertia coefficients for time-invariant systems. The E_{jn} terms, which are the coefficients of q_n terms, resemble stiffness effects, in that they represent accelerative loadings that are proportional to the displacement. Like stiffness coefficients, E_{jn} are symmetric. Centripetal acceleration effects typically are manifested in these terms. The J_j terms represent accelerative loadings that are independent of the displacement.

The coefficients G_{jn} multiplying $\dot{q}_n$ are unlike any effect that arises in a time-invariant system. The B_{jn} defining G_{jn} are associated with terms in T that are products of a generalized coordinate and a generalized velocity. An example of such an effect is Coriolis effects in polar coordinates, when the radial position and polar angle are used as generalized coordinates. The B_{jn} coefficients also arise in cases where a body rotates about several axes, thereby generating gyroscopic moments. It is standard to refer to $[G]$ as the *gyroscopic inertia matrix,* even when it actually represents Coriolis effects.

It is obvious from eq. (11.1.13) that $G_{jn} \equiv 0$ if $j = n$, and that swapping the index values changes the sign of the specific term. Thus, we find that

$$\boxed{G_{nj} = -G_{jn} \quad \text{and} \quad G_{jj} \equiv 0 \Leftrightarrow [G]^{\mathrm{T}} = -[G]} \qquad (11.1.14)$$

This is the property of *skew-symmetry.*

The process of differentiating V in eq. (11.1.7) to form Lagrange's equations is like that leading to eq. (11.1.8), so we have

$$\frac{\partial V}{\partial q_j} = \sum_{k=1}^{N} K_{jk} q_k + (F_0)_j \qquad (11.1.15)$$

It can be proven that the Rayleigh dissipation term $\partial \mathcal{D} / \partial \dot{q}_j$ inserted into eq. (A.2.15) in order to account for dissipation is also descriptive of time-dependent systems; see Ginsberg (1995). Because $\mathcal{D} \equiv \frac{1}{2} \mathcal{P}_{\mathrm{dis}}$ is formed by squaring velocity quantities, its most general linear form resembles eq. (11.1.5). In circumstances where the inertia coefficients in the equations of motion actually are constants, it is likely that the damping coefficients C_{jk} will also be constants, in which case the dissipation term to be added to the equations of motion is

$$\frac{\partial \mathcal{D}}{\partial \dot{q}_j} = \sum_{k=1}^{N} C_{jk} \dot{q}_k \qquad (11.1.16)$$

When we use eqs. (11.1.11) and (11.1.16) to form Lagrange's equations, then write the result in matrix form, we find that

$$[M]\{\ddot{q}\} + [[G] + [C]]\{\dot{q}\} + [[K] - [E]]\{q\} = \{Q\} + \{J\} - \{F_0\} \quad (11.1.17)$$

A system governed by equations of motion with nonzero $[G]$ is said to be *gyroscopic.* These linearized equations of motion contain two fundamentally different kinds of velocity-dependent terms, only one of which is associated with dissipation. Indeed, an important special case is that of a conservative gyroscopic system, for which $[C] = 0$ and $[G] \neq [0]$.

In some situations nonconservative forces contained in $\{Q\}$ are linear in the q_j and/or $\dot{q}_j$ variables. An important case is linear feedback in automatic control concepts. In position control, forces proportional to some or all generalized coordinates are applied to counter other dynamic effects. Velocity control exerts forces that are proportional to the generalized velocities. Exertion of such forces requires an external energy source, and therefore can appear in the equations of motion as nonsymmetric matrices multiplying $\{q\}$ and/or $\{\dot{q}\}$. Another situation giving rise to coefficient matrices that are neither symmetric nor skew-symmetric is *circulatory forces,* which are also known as *follower forces.* The orientation of circulatory forces tracks the position of the system. For example, a force that remains parallel to a bar's axis when the bar vibrates is a follower force. The effects of feedback control systems and circulatory forces may be incorporated into the equations of motion by decomposing the generalized forces into $\{Q_f\}$, which is known explicitly as a function of time, combined with general terms that are linear in the generalized coordinates and velocities. In other words,

$$\{Q\} = \{Q_f\} + [g_q]\{q\} + [g_v]\{\dot{q}\} \tag{11.1.18}$$

where $[g_q]$ and $[g_v]$ are constants that depend on the feedback gains or the properties of the circulatory force. These coefficient matrices may be asymmetric. This leads to a general equation for linear vibration,

$$
\boxed{
\begin{array}{c}
[M]\{\ddot{q}\} + [[G] + [C] - [g_v]]\{\dot{q}\} + [[K] - [E] - [g_q]]\{q\} \\
= \{Q\} + \{J\} - \{F_0\}
\end{array}
}
\tag{11.1.19}
$$

The modifications of the solution technique required to address the skew-symmetry of $[G]$ will also enable us to handle with equal ease systems for which $[g_q]$ or $[g_v]$ are nonzero.

In order to prepare the equations of motion for solution, we place them in state-space form. The $2N$ state vector is defined, as before, to be

$$\{x\} = \begin{Bmatrix} \{q\} \\ \{\dot{q}\} \end{Bmatrix} \tag{11.1.20}$$

The transformation identity, eq. (10.1.3), provides N state-space equations. To represent eq. (11.1.19) in state-space form, we modify eq. (10.1.5) for time-independent systems by replacing $[C]$ with $[C] + [G] - [g_v]$ and $[K]$ with $[K] - [E] - [g_q]$. It is desirable that the equations for the gyroscopic system resemble as much as possible those for damped modal analysis. For this reason, we multiply the transformation identity by $-[K] + [E]$ and stack it above the equation of motion. The resulting state-space equations have the same form as the previous version,

$$
\boxed{
[S]\frac{d}{dt}\{x\} - [R]\{x\} = \begin{Bmatrix} \{0\} \\ \{Q\} + \{J\} - \{F_0\} \end{Bmatrix}
}
\tag{11.1.21}
$$

where

$$
\boxed{
\begin{array}{l}
[S] = \begin{bmatrix} -[K] + [E] & [0] \\ [0] & [M] \end{bmatrix} \\[1.5em]
[R] = -\begin{bmatrix} [0] & [K] - [E] \\ [[K] - [E] - [g_q]] & [[C] + [G] - [g_v]] \end{bmatrix}
\end{array}
}
\tag{11.1.22}
$$

It is evident that $[S]$ is symmetric, as it was for damped modal analysis. However, $[R]$ will not be symmetric unless the gyroscopic inertia and feedback coefficients are all zero.

EXAMPLE 11.1

A small block is fastened to a turntable by four springs arranged symmetrically. The sketch shows the static equilibrium position of the block when the turntable is not rotating, $\omega = 0$. Generalized coordinates for the block are the displacement components x and y in the orthogonal directions of the springs. Derive the linearized equations of motion for the block corresponding to ω being a specified arbitrary function of time.

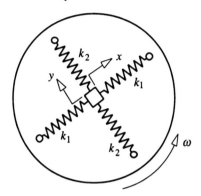

Solution This exercise demonstrates the derivation of linearized equations of motion for gyroscopic systems. In addition, the result will be a prototypical system that we will use for subsequent studies. We model the block as a particle whose location is defined by position coordinates $q_1 = x$ and $q_2 = y$ measured relative to an xyz reference frame attached to the turntable, with the z axis coincident with the axis of rotation. The rotational kinetic energy of the turntable is independent of the generalized coordinates, so it is irrelevant to the analysis. The kinematical relation for relative velocity states that the absolute velocity of a point P is given by

$$\bar{v}_P = \bar{v}_O + (\bar{v}_P)_{xyz} + \bar{\omega} \times \bar{r}_{P/O}$$

where point O is the origin of xyz, $(\bar{v}_P)_{xyz}$ is the velocity of point P as seen by an observer who moves in unison with xyz, and $\bar{\omega}$ is the angular velocity of xyz. In the present situation $\bar{v}_O = \bar{0}$ and $\bar{\omega} = \omega\bar{k}$. Also, because x and y are the position coordinates, $\bar{r}_{P/O} = x\bar{i} + y\bar{j}$. With respect to the turntable, $\bar{i}$ and $\bar{j}$ have constant orientations, so $(\bar{v}_P)_{xyz} = \dot{x}\bar{i} + \dot{y}\bar{j}$. It follows that

$$\bar{v}_{\text{block}} = \dot{x}\bar{i} + \dot{y}\bar{j} + \omega\bar{k} \times (x\bar{i} + y\bar{j}) = (\dot{x} - \omega y)\bar{i} + (\dot{y} + \omega x)\bar{j}$$

The corresponding kinetic energy is

$$T = \tfrac{1}{2}m[(\dot{x} - \omega y)^2 + (\dot{y} + \omega x)^2]$$

$$= \tfrac{1}{2}m(\dot{x}^2 + \dot{y}^2) + m\omega(-\dot{x}y + \dot{y}x) + \tfrac{1}{2}m\omega^2(x^2 + y^2)$$

A comparison of this expression to eq. (11.1.5) shows that

$$M_{11} = M_{22} = m, \qquad M_{12} = M_{21} = 0$$

$$B_{12} = -B_{21} = -m\omega, \qquad B_{11} = B_{22} = 0$$

$$E_{11} = E_{22} = m\omega^2, \qquad E_{12} = E_{21} = 0$$

$$N_1 = N_2 = J_1 = J_2 = 0$$

The corresponding nonzero gyroscopic inertia coefficients are

$$G_{12} = -G_{21} = B_{12} - B_{11} = -2m\omega$$

Potential energy is stored in the springs. Because of their orthogonal arrangement, small movement parallel to one set of springs does not alter the length of the other set, so

$$V = \tfrac{1}{2}(2k_1)x^2 + \tfrac{1}{2}(2k_2)y^2$$

Comparison of this expression to eq. (11.1.7) shows that

$$K_{11} = 2k_1, \qquad K_{22} = 2k_2, \qquad K_{12} = K_{21} = 0$$

$$(F_0)_1 = (F_0)_2 = 0$$

No mention is made in the problem statement regarding damping, so we take $[C] = [0]$. The only forces acting on the block are exerted by the springs, so we set $Q_1 = Q_2 = 0$. The resulting equations of motion obtained from eq. (11.1.17) are

$$\begin{bmatrix} m & 0 \\ 0 & m \end{bmatrix} \begin{Bmatrix} \ddot{x} \\ \ddot{y} \end{Bmatrix} + \begin{bmatrix} 0 & -2m\omega \\ 2m\omega & 0 \end{bmatrix} \begin{Bmatrix} \dot{x} \\ \dot{y} \end{Bmatrix} + \begin{bmatrix} 2k_1 - m\omega^2 & 0 \\ 0 & 2k_2 - m\omega^2 \end{bmatrix} \begin{Bmatrix} x \\ y \end{Bmatrix} = \begin{Bmatrix} 0 \\ 0 \end{Bmatrix}$$

The reader is invited to verify that forming Lagrange's equations by direct differentiation of the expressions for T and V leads to the same result.

EXAMPLE 11.2

The device, called a *spinning top,* consists of a rotor that spins at constant rate ω_1 about its axis of symmetry, as the gimbal rotates at constant speed ω_2 about the vertical axis. The angle θ of the axis of symmetry relative to the vertical axis is the generalized coordinate for the system. Only the inertia of the rotor is significant. There are three possible states of steady motion, in which θ is a constant value. Identify these values. Then derive a differential equation describing small vibrations relative to each steady motion. Based on the coefficients of these differential equations, assess the range of values of ω_1 and ω_2 for which each state represents a stable vibration.

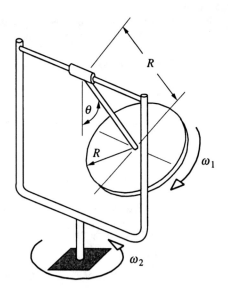

Solution   In this example, we will identify states of steady dynamic response and derive equations for vibration relative to such positions. The result will give a hint of some of the diverse phenomena introduced by the overall motion of a system. Formulating the kinetic energy requires some theorems from three-dimensional dynamics; see Ginsberg (1995). The ω_1 and ω_2 rotations are known, so the position of the disk is defined by the generalized coordinate $q_1 = \theta$. Under the assumption that the mass of the shaft supporting the disk is negligible, we only need to characterize the kinetic energy of the disk. To do this, we attach a moving reference frame to the disk. As shown in the sketch, we align the x axis with the shaft's axis and place the origin of xyz at the horizontal shaft.

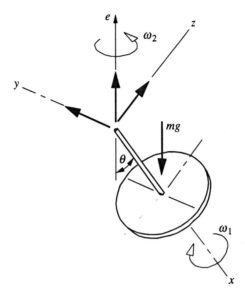

The disk is rotationally symmetric, so it is convenient to align the y axis with the horizontal shaft. The angular velocity $\bar{\omega}$ of the disk is the sum of the individual rotation rates defined as vectors according to the right-hand rule. The rotation ω_2 occurs about the fixed vertical axis, which is denoted as $\bar{e}$ in the free body diagram. The rotation $\dot{\theta}$ is about the y axis and ω_1 is about the x axis, so

$$\bar{\omega} = \omega_1 \bar{i} + \omega_2 \bar{e} - \dot{\theta}\bar{j}$$

Constructing the kinetic energy requires that $\bar{\omega}$ be described in terms of components relative to xyz. Because $\bar{e} = -\cos(\theta)\bar{i} + \sin(\theta)\bar{k}$, we have

$$\bar{\omega} = [\omega_1 - \omega_2\cos(\theta)]\bar{i} - \dot{\theta}\bar{j} + \omega_2\sin(\theta)\bar{k}$$

The x, y, and z axes are a set of principal axes of inertia. Standard tables of inertia properties and the parallel axis theorems give

$$I_{xx} = \tfrac{1}{2}mR^2, \qquad I_{yy} = I_{zz} = \tfrac{5}{4}mR^2$$

The corresponding expression for kinetic energy is

$$T = \tfrac{1}{2}(I_{xx}\omega_x^2 + I_{yy}\omega_y^2 + I_{zz}\omega_z^2)$$

$$= \tfrac{1}{2}mR^2\left\{\tfrac{1}{2}[\omega_1 - \omega_2\cos(\theta)]^2 + \tfrac{5}{4}\dot{\theta}^2 + \tfrac{5}{4}\omega_2^2\sin(\theta)^2\right\}$$

The potential energy of gravity is

$$V = -mgR\cos(\theta)$$

We assume that dissipation is negligible. Also, no nonconservative forces do work when θ is increased to $\theta + \delta\theta$, so we set $C_{11} = 0$ and $Q_1 = 0$.

The first task is to identify the steady motions at which θ is constant. We do this by following eq. (11.1.3) to form Lagrange's equation for θ, and then setting $\dot{\theta} = \ddot{\theta} = 0$. It is premature to carry out linearization, so we use the full expressions for T and V to form the derivatives in Lagrange's equation,

$$\frac{d}{dt}\left(\frac{\partial T}{\partial \dot{\theta}}\right) = \frac{d}{dt}\left(\frac{5}{4}mR^2\dot{\theta}\right) = \frac{5}{4}mR^2\ddot{\theta}$$

$$\frac{\partial T}{\partial \theta} = mR^2\left\{[\omega_1 - \omega_2\cos(\theta)][\omega_2\sin(\theta)] + \frac{5}{4}\omega_2^2\sin(\theta)\cos(\theta)\right\}$$

$$= mR^2\left[\omega_1\omega_2\sin(\theta) + \frac{1}{4}\omega_2^2\sin(\theta)\cos(\theta)\right]$$

$$\frac{\partial V}{\partial \theta} = mgR\sin(\theta)$$

Thus, Lagrange's equation is

$$\frac{5}{4}\ddot{\theta} - \omega_2^2\sin(\theta)\cos(\theta) - \frac{1}{4}\omega_1\omega_2\sin(\theta) + \frac{g}{R}\sin(\theta) = 0$$

Setting $\ddot{\theta} \equiv 0$ leads to

$$\omega_2^2\sin(\theta)\cos(\theta) + \frac{1}{4}\omega_1\omega_2\sin(\theta) = \frac{g}{R}\sin(\theta)$$

The roots of this equation are

$$\theta_1 = 0, \qquad \theta_2 = \pi, \qquad \theta_3 = \cos^{-1}\left(\frac{g}{R\omega_2^2} - \frac{1}{4}\frac{\omega_1}{\omega_2}\right)$$

The first two are obvious possibilities, for they constitute planar motions of the disk. One might think that the third root corresponds to a balance between the gravitational moment and centripetal acceleration resulting from rotation about the vertical axis. The incorrectness of such thinking is displayed by the facts that θ_3 depends on ω_1 and that it is greater than $\pi/2$ if $\omega_1\omega_2 > 4g/R$. In fact, θ_3 constitutes a situation where the gravitational moment matches the change of the disk's angular momentum vector as it rotates about the vertical axis.

Let us consider altering ω_1 with ω_2 fixed. The value of θ_3 ranges from 0 to π as ω_1 is increased from $(\omega_1)_{min}$ to $(\omega_1)_{max}$, where

$$(\omega_1)_{min} = 4\left(\frac{g}{R\omega_2} - \omega_2\right), \qquad (\omega_1)_{max} = 4\left(\frac{g}{R\omega_2} + \omega_2\right)$$

We wish to derive equations for small changes of θ relative to θ_1, θ_2, or θ_3. We could obtain such equations by linearizing T and V according to the procedures outlined in the preceding development. However, because we already have formed the general equation of motion, it is simpler to work directly with that equation by letting $\theta = \theta_n + \chi$, $n = 1, 2$, or 3. We use Taylor series to drop all terms that are quadratic or higher in χ. Thus, we have

$$\sin(\theta) = \sin(\theta_n + \chi) = \sin(\theta_n)\cos(\chi) + \cos(\theta_n)\sin(\chi)$$

$$\approx \sin(\theta_n) + \cos(\theta_n)\chi$$

$$\sin(\theta)\cos(\theta) = \frac{1}{2}\sin[2(\theta_n + \chi)] \approx \frac{1}{2}\sin(2\theta_n) + \cos(2\theta_n)\chi$$

We substitute these expressions into Lagrange's equation, and take account of the fact that the terms independent of χ constitute a solution when $\ddot{\chi} = 0$. This leads to

$$\frac{5}{4}\ddot{\chi} + \left[\left(\frac{g}{R} - \frac{1}{4}\omega_1\omega_2\right)\cos(\theta_n) - \omega_2^2\cos(2\theta_n)\right]\chi = 0$$

This is the equation of motion for a simple oscillator. If the coefficient of the χ term is positive, the disturbance results in an oscillation whose mean value is $\theta = \theta_n$, with the oscillation's frequency being

$$\omega_{nat} = \frac{2}{\sqrt{5}}\left[\left(\frac{g}{R} - \frac{1}{4}\omega_1\omega_2\right)\cos(\theta_n) - \omega_2^2\cos(2\theta_n)\right]^{1/2}$$

In contrast, if the coefficient of the χ term is negative, so that ω_{nat} is imaginary, χ grows exponentially. We interpret this to mean that a small disturbance will not lead to a stable oscillation centered about $\theta = \theta_n$.

In the case $\theta_1 = 0$, we have

$$\omega_{nat} = \frac{2}{\sqrt{5}}\left[\frac{g}{R} - \frac{1}{4}\omega_1\omega_2 - \omega_2^2\right]^{1/2}$$

Without loss of generality, let us consider ω_2 to be positive. Then $\theta_1 = 0$ gives $\omega_{nat} > 0$, and therefore represents a stable state of steady motion, if

$$\omega_1 < 4\left(\frac{g}{R\omega_2} - \omega_2\right) = (\omega_1)_{min}$$

Note that this condition is always satisfied if ω_1 is negative, corresponding to a rotation opposite the one shown in the sketch.

The case $\theta_2 = \pi$ leads to

$$\omega_{nat} = \frac{2}{\sqrt{5}}\left[-\frac{g}{R} + \frac{1}{4}\omega_1\omega_2 - \omega_2^2\right]^{1/2}$$

Thus, $\theta = \pi$ is a stable state if

$$\omega_1 > 4\left(\frac{g}{R\omega_2} + \omega_2\right) = (\omega_1)_{max}$$

Analysis of the case $\theta = \theta_3$ is slightly more complicated. We use the expression for θ_3, in conjunction with the identity that $\cos(2\theta_3) = [2\cos(\theta_3)^2 - 1]$, to simplify ω_{nat}. The simplest route is to substitute into ω_{nat} the relation

$$\left(\frac{g}{R} - \frac{1}{4}\omega_1\omega_2\right) = \omega_2^2\cos(\theta_3)$$

which eventually leads to

$$\omega_{nat} = \sqrt{\frac{2}{5}}\omega_2\sin(\theta_3)$$

In other words, if the value of θ_3 is a real value, it will represent a stable state of motion. We found earlier that θ_3 is a real value if $(\omega_1)_{min} \leq \omega_1 \leq (\omega_1)_{max}$.

In summary, we have found that $\theta = 0$ is a stable state of motion of $\omega_1 < (\omega_1)_{min}$. If ω_1 is increased from this minimum value, the tilted position $\theta = \theta_3$, which increases from zero, is the stable state. Continued increase of ω_1 increases θ_3 until $\omega = (\omega_1)_{max}$, at which value $\theta_3 = \pi$. Further increase of ω_1 beyond $(\omega_1)_{max}$ leads to a stable motion at $\theta = \pi$. In view of the fact that this condition is not possible in the absence of ω_1, we conclude that gyroscopic effects can stabilize a system.

11.2 RIGHT AND LEFT EIGENVALUE PROBLEMS

As we did for damped stationary systems, we begin by considering free vibration, so all inhomogeneous terms in the equations of motion, eq. (11.1.21), are set to zero. All solutions of the homogeneous differential equations must have the form

$$\{x\} = \{\psi\}\exp(\lambda t) \tag{11.2.1}$$

We substitute this representation into eq. (11.1.2) and require that the equation be satisfied for all t, which enables us to cancel the common factor $\exp(\lambda t)$. What remains is an eigenvalue problem whose appearance is like that for damped modes,

$$\boxed{[[R] - \lambda[S]]\{\psi\} = \{0\}} \tag{11.2.2}$$

The characteristic equation is

$$\boxed{|[R] - \lambda[S]| = a_{2N}\lambda^{2N} + a_{2N-1}\lambda^{2N-1} + \cdots + a_1\lambda + a_0 = 0} \tag{11.2.3}$$

Because $[R]$ and $[S]$ are both real, the a_n coefficients are real. It follows that the $2N$ eigenvalues are either real, or else they occur as complex conjugate pairs. As we did for damped modes, we number the eigenvalues based on their magnitudes, with the complex conjugates numbered adjacently,

$$\boxed{|\lambda_n| \le |\lambda_{n+1}|, \quad \lambda_{n+1} = \lambda_n^* \text{ if } \text{Im}(\lambda_n) \ne 0} \tag{11.2.4}$$

The eigenvector $\{\psi_n\}$ associated with eigenvalue λ_n is such that

$$\boxed{[[R] - \lambda_n[S]]\{\psi_n\} = \{0\}} \tag{11.2.5}$$

We assume that there are no repeated eigenvalues, in which case the rank of $[R] - \lambda_n[S]$ will be $2N - 1$. Thus, we may consider any nonzero element of $\{\psi_n\}$ to be arbitrary. Because $[R]$ and $[S]$ are real, a real λ_n value must be associated with a real or purely imaginary eigenvector. Similarly, if $\lambda_{n+1} = \lambda_n^*$, then $\{\psi_{n+1}\} = \{\psi_n^*\}$.

At this stage, the development seems to be the same as that leading to damped modes. The difference arises when we consider orthogonality conditions, because the previous orthogonality properties do not apply when $[R]$ is not symmetric. New orthogonality properties we will derive involve a related eigenvalue problem, which is stated as

$$\{\widetilde{\psi}\}^{\mathrm{T}}[[R] - \widetilde{\lambda}[S]] = \{0\} \tag{11.2.6}$$

The eigenvector $\{\widetilde{\psi}\}$ premultiplies the coefficient matrix. Thus, the preceding is referred to as a *left eigenvalue problem*, while eq. (11.2.2) is referred to as a *right eigenvalue problem*. The mathematical terminology that is often employed is to say that the left eigenvalue problem is the *adjoint* of the right one.

Equation (11.2.6) is an unconventional way of writing simultaneous algebraic equations, so let us take its transpose. In view of the fact that $[S]$ is symmetric, this operation leads to

$$[[R]^{\mathrm{T}} - \widetilde{\lambda}[S]]\{\widetilde{\psi}\} = \{0\} \tag{11.2.7}$$

The characteristic equation for the left eigenvalue problem is

$$\left|[R]^{\mathrm{T}} - \widetilde{\lambda}[S]\right| = 0 \tag{11.2.8}$$

The determinant of a matrix and of its transpose have the same value. Hence, the preceding is identical to $\left| [R] - \widetilde{\lambda}\,[S] \right| = 0$, because $[S]$ is symmetric. This is the same as the characteristic equation for the right eigenvalue problem, eq. (11.2.3), except that $\widetilde{\lambda}$ appears here, instead of λ. It follows that the left and right eigenvalues are identical,

$$\boxed{\widetilde{\lambda}_n = \lambda_n}$$
(11.2.9)

The left eigenvectors are different from the right, for they satisfy

$$\boxed{[[R]^T - \lambda_n[S]]\{\widetilde{\psi}_n\} = \{0\}}$$
(11.2.10)

Note that when $[G] = 0$ and there are no feedback forces, then eq. (11.1.22) indicates that $[R]^T = [R]$. In that case the left and right eigenvectors are identical, as expected for the case of damped modes.

Evaluation of the eigenvalues and eigenvectors for systems having more than three degrees of freedom is best done with the aid of a general linear eigenvalue solver contained in mathematical software. For $N = 2$ and $N = 3$, it is useful to work with the characteristic equation. In the most general situation, $[S]$ and $[R]$ are defined in eqs. (11.1.22). Substitution of those forms, along with partitioning $\{\psi\}$ as $\left[\{u\}^T \ \{v\}^T\right]^T$, gives

$$-\begin{bmatrix} [0] & [K] - [E] \\ [[K] - [E] - [g_q]] & [[C] + [G] - [g_v]] \end{bmatrix}\begin{Bmatrix} \{u\} \\ \{v\} \end{Bmatrix}$$

$$-\lambda\begin{bmatrix} -[K] + [E] & [0] \\ [0] & [M] \end{bmatrix}\begin{Bmatrix} \{u\} \\ \{v\} \end{Bmatrix} = \begin{Bmatrix} \{0\} \\ \{0\} \end{Bmatrix}$$
(11.2.11)

The upper half of eq. (11.2.2) leads to $v = \lambda\{u\}$, which is the same decomposition as eq. (10.3.1) for damped modes. Thus,

$$\{\psi\} = \begin{Bmatrix} \{u\} \\ \lambda\{u\} \end{Bmatrix}$$
(11.2.12)

The lower partition of eq. (11.2.11) then requires that

$$\boxed{[\lambda^2[M] + \lambda[[C] + [G] - [g_v]] + [K] - [E] - [g_q]]\{u\} = 0}$$
(11.2.13)

This is a nonlinear eigenvalue problem. The associated characteristic equation is

$$\left| \lambda^2[M] + \lambda[[C] + [G] - [g_v]] + [K] - [E] - [g_q] \right| = 0$$
(11.2.14)

This form is more suitable for manual work, because it only requires expansion of an $N \times N$ determinant. Once the $2n$ roots λ_n of this polynomial have been extracted, the corresponding right eigenvectors would be obtained by solving eq. (11.2.13) with one element of $\{u_n\}$ arbitrarily set. The lower partition of the eigenvector may then be filled in by applying eq. (11.2.12). A similar analysis, which is left for a homework exercise, reveals that the nonlinear left eigenvalue problem is defined by

$$\boxed{\{\widetilde{\psi}\} = \begin{Bmatrix} \{\widetilde{u}\} \\ \{\widetilde{v}\} \end{Bmatrix}}$$
(11.2.15)

where

$$\left[\lambda^2[M] + \lambda\Big[[C] - [G] - [g_v]^T\Big] + [K] - [E] - [g_q]^T\right]\{\tilde{v}\} = \{0\}$$

$$\lambda[[K] - [E]]\{\tilde{u}\} = \Big[[K] - [E] - [g_q]^T\Big]\{\tilde{v}\}$$

(11.2.16)

Evaluation of the left eigenvectors according to the preceding begins by using the first relation to evaluate the associated $\{\tilde{v}_n\}$ (with one element arbitrarily defined), after which the second relation yields the accompanying $\{\tilde{u}_n\}$.

An important class of problems involves conservative gyroscopic systems—that is, systems that do not feature dissipative, feedback, or circulatory forces, $[C] = [g_q] = [g_v] = [0]$. The nonlinear characteristic equation obtained from eq. (11.2.14) then reduces to

$$|\lambda^2[M] + \lambda[G] + [K] - [E]| = 0 \tag{11.2.17}$$

We know that $[M]$, $[K]$, and $[E]$ are symmetric, while $[G]$ is skew-symmetric. Because the determinant of a matrix equals the determinant of its transpose, it follows that the characteristic equation also is

$$\left|\big[\lambda^2[M] + \lambda[G] + [K] - [E]\big]^T\right| \equiv |\lambda^2[M] - \lambda[G] + [K] - [E]| = 0 \tag{11.2.18}$$

The significant aspect of the second form is that it is also the result of replacing λ by $-\lambda$ in eq. (11.2.17). Hence, any eigenvalue for a conservative gyroscopic system must be accompanied by a matching negative value. At the same time, we know that the coefficients of the characteristic equation are real, and the roots of a real polynomial must either be real, or else occur as complex conjugate pairs. There are two possible ways in which both conditions can be satisfied. If the eigenvalues are complex conjugate pairs, $\lambda_n = \alpha + \beta i$, $\lambda_{n+1} = \alpha - \beta i$, then the condition that $\lambda_{n+1} = -\lambda_n$ requires that $\alpha = 0$. The alternative possibility is that the eigenvalues are a pair of oppositely signed real values.

These alternatives could have been anticipated on physical grounds. Because the system is conservative, the mechanical energy $T + V$ must be conserved in any free vibration. If the dynamic equilibrium state $(q_j = 0)$ is stable, then displacement relative to that state must be oscillatory, without attenuation. This is the case of purely imaginary eigenvalues. In contrast, if the dynamic equilibrium is unstable, then a disturbance will cause the system to move away from that state. Such is the motion associated with a positive real eigenvalue. The issue of stability is explored in more detail in the next section.

When the eigenvalues of a conservative gyroscopic system are purely imaginary, the left and right eigenvectors have a simple relationship. The only complex quantity in the coefficient matrices of eq. (11.2.16) is the imaginary λ. Hence, substituting $\lambda = \pm i\omega_n$ into those relations with $[C] = [g_q] = [g_v] = [0]$ leads to

$$[-\omega_n^2[M] \mp i\omega_n[G] + [K] - [E]]\{\tilde{v}_n\} = \{0\}$$

$$\pm i\omega_n\{\tilde{u}_n\} = \{\tilde{v}_n\}$$

(11.2.19)

If $\omega_n \neq 0$, we may use the second relation to replace $\{\tilde{v}_n\}$ in the first equation. Then the fact that $\mp i\omega_n = (\pm i\omega_n)^*$ leads to

$$[-\omega_n^2[M] \pm i\omega_n[G] + [K] - [E]]^*\{\tilde{u}_n\} = \{0\} \tag{11.2.20}$$

In this special case the coefficient matrix is the complex conjugate of the one in eq. (11.2.14) for $\{u_n\}$, so it must be that $\{\tilde{u}_n\} = c\{u_n\}^*$, where c is an arbitrary constant. It follows that for conservative gyroscopic systems,

$$\lambda_n = \pm i\omega_n, \qquad \{\psi_n\} = \left\{\begin{array}{c} \{u_n\} \\ \lambda_n\{u_n\} \end{array}\right\}, \qquad \{\tilde{\psi}_n\} = c\left\{\begin{array}{c} \{u_n\}^* \\ \lambda_n\{u_n\}^* \end{array}\right\} \qquad (11.2.21)$$

The special properties of the eigensolution for conservative gyroscopic systems led Meirovitch (1975) to develop a modal analysis formulation that does not require the left eigenvector. However, considering both the right and left eigenvalue problems enables us to use the same procedures to treat any system.

EXAMPLE 11.3

Determine the eigenvalues and the right and left eigenvectors of the system in Example 11.1. Then for the case where $k_1 = 0.5k$, $k_2 = 1.5k$, and $\omega = 0.47(k/m)^{1/2}$, determine the path followed by the block when the system vibrates according to each mode. Describe this path as seen by an observer on the turntable, and as seen by a fixed observer.

Solution   In addition to illustrating basic operations, this example is intended to enhance physical understanding of gyroscopic modal properties. To employ the results of Example 11.1, we let $q_1 = x$ and $q_2 = y$. The coefficient matrices are

$$[M] = \begin{bmatrix} m & 0 \\ 0 & m \end{bmatrix}, \qquad [G] = \begin{bmatrix} 0 & -2m\omega \\ 2m\omega & 0 \end{bmatrix}$$

$$[K] = \begin{bmatrix} 2k_1 & 0 \\ 0 & 2k_2 \end{bmatrix}, \qquad [E] = \begin{bmatrix} m\omega^2 & 0 \\ 0 & m\omega^2 \end{bmatrix}$$

Because this is an $N = 2$ system, we shall employ eq. (11.2.13) and perform the operations algebraically. It is convenient to define two parameters

$$\Omega_x^2 = \frac{2k_1}{m}, \qquad \Omega_y^2 = \frac{2k_2}{m}$$

so the right eigenvalue problem becomes

$$\frac{1}{m}[\lambda^2[M] + \lambda[G] + [K] - [E]]\{u\} = \begin{bmatrix} \lambda^2 + \Omega_x^2 - \omega^2 & -2\omega\lambda \\ 2\omega\lambda & \lambda^2 + \Omega_y^2 - \omega^2 \end{bmatrix}\begin{Bmatrix} u_1 \\ u_2 \end{Bmatrix} = \begin{Bmatrix} 0 \\ 0 \end{Bmatrix}$$

The corresponding characteristic equation is

$$\lambda^4 + (\Omega_x^2 + \Omega_y^2 + 2\omega^2)\lambda^2 + (\Omega_x^2 - \omega^2)(\Omega_y^2 - \omega^2) = 0$$

The roots of this quadratic equation for λ^2 are

$$\lambda^2 = -\tfrac{1}{2}(\Omega_x^2 + \Omega_y^2 + 2\omega^2) \pm \tfrac{1}{2}[(\Omega_x^2 - \Omega_y^2)^2 + 8\omega^2(\Omega_x^2 + \Omega_y^2)]^{1/2}$$

If $\Omega_x < \omega < \Omega_y$, one of the λ values is positive. A positive value of λ indicates that the system is unstable, as we will see in the next section. The more interesting case is that in which $\omega < \Omega_x$ or $\omega > \Omega_y$, which leads to eigenvalues that are complex conjugate pairs,

$$\lambda_1 = i\alpha, \qquad \lambda_2 = -i\alpha, \qquad \lambda_3 = i\beta, \qquad \lambda_4 = -i\beta$$

$$\alpha = \frac{1}{\sqrt{2}}\{\Omega_x^2 + \Omega_y^2 + 2\omega^2 - [(\Omega_x^2 - \Omega_y^2)^2 + 8\omega^2(\Omega_x^2 + \Omega_y^2)]^{1/2}\}^{1/2}$$

$$\beta = \frac{1}{\sqrt{2}}\{\Omega_x^2 + \Omega_y^2 + 2\omega^2 + [(\Omega_x^2 - \Omega_y^2)^2 + 8\omega^2(\Omega_x^2 + \Omega_y^2)]^{1/2}\}^{1/2}$$

We obtain the right eigenvectors by substituting each λ_n into the matrix eigenvalue problem. The first of the scalar equations gives

$$u_{2n} = \frac{\lambda_n^2 + \Omega_x^2 - \omega^2}{2\omega\lambda_n} u_{1n}$$

It then follows from eq. (11.2.12) that the right eigenvector is

$$\{\psi_n\} = u_{1n}\left[1 \quad \frac{\lambda_n^2 + \Omega_x^2 - \omega^2}{2\omega\lambda_n} \quad \lambda_n \quad \frac{\lambda_n^2 + \Omega_x^2 - \omega^2}{2\omega}\right]^{\mathrm{T}}$$

Because this system is conservative, the left eigenvectors are found from eq. (11.2.21) to be

$$\{\widetilde{\psi}_n\} = \widetilde{u}_{1n}\left[1 \quad \frac{\lambda_n^2 + \Omega_x^2 - \omega^2}{2\omega\lambda_n^*} \quad \lambda_n \quad -\frac{\lambda_n^2 + \Omega_x^2 - \omega^2}{2\omega}\right]^{\mathrm{T}}$$

where the last element results from the fact that $\lambda_n/\lambda_n^* = -1$ when λ_n is imaginary.

For the stated values of k_1, k_2, and ω, we have $\Omega_x^2 = k/m$ and $\Omega_y^2 = 3k/m$. The corresponding eigenvalues are

$$\lambda = \left(\frac{k}{m}\right)^{1/2}\left[0.7466i \quad -0.7466i \quad 1.9709i \quad -1.9709i\right]^{\mathrm{T}}$$

The right eigenvectors are

$$\{\psi_1\} = u_{11}\begin{Bmatrix} 1 \\ -0.3159i \\ 0.7466i \\ 0.2358 \end{Bmatrix}, \qquad \{\psi_2\} = u_{12}\begin{Bmatrix} 1 \\ 0.3159i \\ -0.7466i \\ 0.2358 \end{Bmatrix}$$

$$\{\psi_3\} = u_{13}\begin{Bmatrix} 1 \\ 1.6762i \\ 1.9709i \\ -3.3035 \end{Bmatrix}, \qquad \{\psi_4\} = u_{14}\begin{Bmatrix} 1 \\ -1.6762i \\ -1.9709i \\ -3.3035 \end{Bmatrix}$$

The left eigenvectors differ from these by having opposite signs in the second and fourth rows.

It is instructive to consider the result of using mathematical software to solve the eigenvalue problem at the specified ω. Toward that end, we set $k/m = 1$ and form

$$[R] = -\begin{bmatrix} [0] & [[K] - [E]] \\ [[K] - [E]] & [G] \end{bmatrix}, \qquad [S] = \begin{bmatrix} [-[K] + [E]] & [0] \\ [0] & [M] \end{bmatrix}$$

with $\omega = 0.47$. In MATLAB, we write [psi_r, lambda] = eig(R, S), which yields the right eigenvector matrix psi_r, while [psi_l, lambda] = eig(R', S) yields the left eigenvector matrix. Comparable operations in Mathcad are carried out with λ = genvals(R, S), ψ_r = genvecs(R, S), ψ_l = genvecs(R^T, S). The eigenvalues obtained from either program match the values obtained here. To compare the eigenvectors, it is necessary to divide each column of the computed result by its first element, because the present eigenvectors are based on setting the first element to unity. When that operation is carried out, both sets of eigenvectors match the values obtained here.

A modal free vibration corresponding to a pair of complex conjugate modes is a real quantity obtained by summing eq. (11.2.1) for each pair. Hence, the time dependence of x and y in a free vibration involving mode n and its complex conjugate mode $n + 1$ is

$$x = \psi_{1n}\exp(\lambda_n t) + \psi_{1(n+1)}\exp(\lambda_{n+1}t) = 2\,\mathrm{Re}[\psi_{1n}\exp(\lambda_n t)]$$

$$y = \psi_{2n}\exp(\lambda_n t) + \psi_{2(n+1)}\exp(\lambda_{n+1}t) = 2\,\mathrm{Re}[\psi_{2n}\exp(\lambda_n t)]$$

Because $\psi_{1n} = u_{1n}$, we express this factor scaling the respective eigenvectors by writing it in polar form as $u_{1n} = |r_n|\exp(-i\chi_n)$. We define a nondimensional time $\tau = (k/m)^{1/2}t$, which leads to

Modes 1 and 2: $\qquad x = |r_1|\cos(0.7466\tau - \chi_1), \qquad y = 0.3159|r_1|\sin(0.7466\tau - \chi_1)$

Modes 3 and 4: $\qquad x = |r_3|\cos(1.9709\tau - \chi_3), \qquad y = -1.6762|r_3|\sin(1.9709\tau - \chi_3)$

The values of each $|r_n|$ and χ_n are defined by the initial conditions.

Each modal motion represents an elliptical path, which may be verified by eliminating τ to find

$$\text{Modes 1 and 2:} \quad x^2 + \left(\frac{y}{0.3159}\right)^2 = |r_1|^2$$

$$\text{Modes 3 and 4:} \quad x^2 + \left(\frac{y}{1.6762}\right)^2 = |r_3|^2$$

These paths are shown in the plots. To identify the sense of the movement, indicated by arrows, we consider the polar angle θ from the radial line to the x axis, which is defined by $\tan(\theta) = y/x$, which leads to

$$\text{Modes 1 and 2: } \tan(\theta) = 0.3519\tan(0.7466\tau - \chi_1)$$

$$\text{Modes 3 and 4: } \tan(\theta) = 1.6762\tan(-1.9709\tau + \chi_3)$$

The first relation shows that θ increases with increasing τ, so the motion in modes 1 and 2 proceeds in the general direction of the turntable's rotation. In modes 3 and 4, θ decreases with increasing τ, so the elliptical motion proceeds in the opposite sense from the rotation.

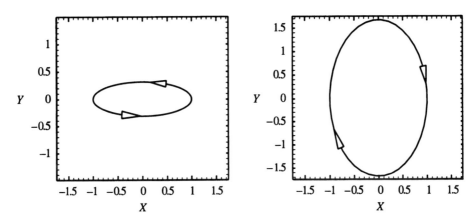

Note that in neither case does θ change at a constant rate. The period to complete each orbit is $\Delta\tau = 2\pi/0.7466 = 8.4157$ for modes 1 and 2, and $\Delta\tau = 2\pi/1.9709 = 3.188$ for modes 3 and 4.

These elliptical paths are the manner in which an observer rotating with the turntable would perceive the motion. The position relative to a fixed observer is represented by the position coordinates relative to a fixed reference frame, which we denote as XYZ. If we take $t = 0$ to be the instant at which the x and X axes coincide, then $\omega t = 0.47\tau$ is the angle from X to x, and from Y to y, measured counterclockwise. The rotation transformation between coordinates in that case is

$$\begin{Bmatrix} X \\ Y \end{Bmatrix} = \begin{bmatrix} \cos(0.47\tau) & -\sin(0.47\tau) \\ \sin(0.47\tau) & \cos(0.47\tau) \end{bmatrix} \begin{Bmatrix} x \\ y \end{Bmatrix}$$

For modes 1 and 2 we have

$$X = |x_1|[\cos(0.47\tau)\cos(0.7466\tau - \chi_1) - 0.3159\sin(0.47\tau)\sin(0.7466\tau - \chi_1)]$$
$$Y = |x_1|[\sin(0.47\tau)\cos(0.7466\tau - \chi_1) + 0.3159\cos(0.47\tau)\sin(0.7466\tau - \chi_1)]$$

Identities for products of harmonic functions convert these expressions to

Modes 1 and 2: $\begin{cases} X = |x_1|[0.3421\cos(0.2766\tau - \chi_1) + 0.6579\cos(1.2166\tau - \chi_1)] \\ Y = |x_1|[-0.3421\sin(0.2766\tau - \chi_1) + 0.6579\sin(1.2166\tau - \chi_1)] \end{cases}$

Following the same process for mode 3 leads to

Modes 3 and 4: $\begin{cases} X = |x_3|[1.3381\cos(1.5009\tau - \chi_3) - 0.3381\cos(2.4409\tau - \chi_3)] \\ Y = |x_3|[-1.3381\sin(1.5009\tau - \chi_3) - 0.3381\cos(2.4409\tau - \chi_3)] \end{cases}$

It is substantially more difficult to find an equation for the path by eliminating τ from these expressions. Instead, we may graph the path by considering τ to be a parameter, and evaluating $X(\tau)$ and $Y(\tau)$ over a range of τ. Plotting Y versus X for each τ leads to the path, which is known as a *Lissajous pattern*. (The elliptical paths of x versus y also are Lissajous patterns.) The results are the second set of graphs. The dots mark the position of the block at intervals of one-quarter the rotation period, which corresponds to $\Delta\tau = (\pi/2)/0.7466$ and $(\pi/2)/1.9709$ for the respective cases. The starting point, $t = 0$, corresponds to setting $|x_1| = |x_3| = 1, \chi_1 = \chi_3 = \pi/6$.

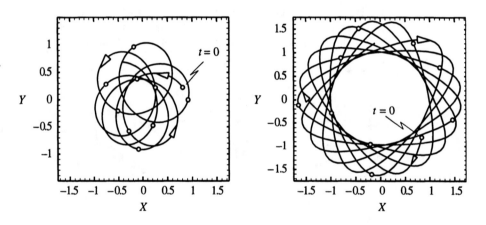

Seeing the paths as a function of Y versus X provides a different view of the motion. The block oscillates between a minimum radial distance from the origin that is half the minor diameter of the elliptical path relative to xy, and a maximum radial distance that is half the major diameter of that ellipse. In modes 1 and 2 this radial oscillation is accompanied by an overall rotation in the sense of ω; this is called *forward precession*. In that motion the rotation of the turntable adds to the rotation of the block in the elliptical path. For modes 3 and 4, the counterclockwise rotation of the turntable is slower than the clockwise average rotation rate in the elliptical path for modes 3 and 4. Hence, the overall motion is opposite the sense of ω; this is *retrograde precession*. The occurrence of modes featuring forward and retrograde precession is a common feature of rotating systems.

It is instructive to consider the outcome of this analysis in the special case $k_2 = k_1$, so that $\Omega_x = \Omega_y \equiv \Omega$. The eigensolutions in that case are

$$\lambda_1 = (\Omega - \omega)i, \qquad \lambda_2 = -(\Omega - \omega)i, \qquad \lambda_3 = (\Omega + \omega)i, \qquad \lambda_4 = -(\Omega + \omega)i$$

$$u_{21} = -iu_{11}, \qquad u_{22} = iu_{12}, \qquad u_{23} = iu_{13}, \qquad u_{24} = -iu_{14}$$

The path in each modal motion then is a circle, with $\Omega - \omega$ being the constant forward (counterclockwise) rotation rate relative to the turntable in modes 1 and 2, while $\Omega + \omega$ is the constant backward rotation rate relative to the turntable in modes 3 and 4. The turntable is rotating forward at angular speed ω. Hence, relative to the fixed XY reference frame, such motions are seen to be circular motions in the forward and retrograde directions at rotation rate Ω, which is the natural frequency of the system in the orthogonal directions of the springs in the absence of turntable rotation.

11.3 DYNAMIC STABILITY

We have seen that the eigenvalues λ_j are real or complex conjugates. If the real part of any value is positive, the corresponding homogeneous solution given by eq. (11.2.1) will increase exponentially with t. Such a solution represents a *dynamic instability*, which refers to a condition in which the displacement following a disturbance increasingly differs from the dynamic equilibrium state as time elapses. The possibilities for a system are

- *Asymptotically stable*: Real negative eigenvalues, or complex eigenvalues with all real parts being negative. In this case any set of initial conditions will lead to a free vibration in which all generalized coordinates ultimately decay to zero.

- *Stable*: Purely imaginary eigenvalues. This condition can only be attained if energy dissipation is not present. Also, because other effects, notably nonlinear forces, have been ignored, it might happen that the system actually is unstable. For this reason, some individuals refer to purely imaginary eigenvalues as being a case of *marginal stability*.

- *Divergence instability*: Purely real eigenvalues, with at least one being positive. In this case the homogeneous solution grows monotonically. Euler buckling of a beam is a typical divergence instability. It is possible for gyroscopic and centripetal acceleration effects to cause divergence. When the divergent response becomes sufficiently large, nonlinear effects might take over to keep the response finite. However, it is likely that a divergent response will become sufficiently large to damage an engineering system.

- *Flutter instability*: Complex eigenvalues with at least one positive real part. The homogeneous solution associated with these eigenvalues will have an envelope that grows exponentially, and the response within this envelope will be oscillatory. One way of picturing this is to consider the response of an underdamped system in the case where the damping ratio ζ is negative. For this reason systems that flutter are often said to have *negative damping*. Aerodynamic and circulatory forces are the most common causes of flutter.

Exponential growth is far more rapid than the linear growth in amplitude associated with resonant excitation of an undamped system. Thus, it is imperative that the system parameters be such that the homogeneous solutions representing free vibration be stable. Also, observe that if the free vibration of a system is unstable, it cannot be stabilized by applying a conventional excitation, for such forces are manifested solely by particular solutions. Stabilization of an unstable system requires modification of $[G]$, $[E]$, $[g_q]$, $[g_v]$, $[C]$, $[K]$, and/or $[M]$.

Design studies typically consider a few system parameters to be adjustable. Computing the eigenvalues for each parameter combination in a discretized range enables us to identify situations where stability is an issue. An alternative way of performing the same investigation is to use the *Routh-Hurwitz criteria,* which indicate whether a particular set of parameters will lead to instability, without actually computing the eigenvalues. The Routh-Hurwitz criteria are formulated in terms of the coefficients of the characteristic equation, eq. (11.2.3). It is easier to recognize the manner in which the criteria are formed if the coefficient indices are reversed, so that

$$b_j = a_{2N-j}, \quad j = 2N, 2N - 1, \ldots, 0 \tag{11.3.1}$$

Thus, the standard form of the characteristic equation for applying the Routh-Hurwitz criteria is

$$b_0 \lambda^{2N} + b_1 \lambda^{2N-1} + \cdots + b_{2N-1} \lambda + b_{2N} = 0 \tag{11.3.2}$$

We consider b_0 to be positive, without loss of generality. To make the pattern of the criteria more apparent, we zero pad the b_n coefficients with $2N$ zeroes, such that

$$b_{2N+1} = b_{2N+2} = \cdots = b_{4N} = 0 \tag{11.3.3}$$

The b_n coefficients are used to form a $2N \times 2N$ array, which we shall denote as $[\Delta]$, according to

$$[\Delta] = \begin{bmatrix} b_1 & b_0 & 0 & 0 & 0 & 0 & \cdots & 0 & 0 & 0 \\ b_3 & b_2 & b_1 & b_0 & 0 & 0 & \cdots & 0 & 0 & 0 \\ b_5 & b_4 & b_3 & b_2 & b_1 & b_0 & \cdots & 0 & 0 & 0 \\ \vdots & \vdots & \vdots & \vdots & \vdots & \vdots & \vdots\vdots\vdots & \vdots & \vdots & \vdots \\ b_{4N-3} & b_{4N-4} & b_{4N-5} & b_{4N-6} & b_{4N-7} & b_{4N-8} & \cdots & b_{2N} & b_{2N-1} & b_{2N-2} \\ b_{4N-1} & b_{4N-2} & b_{4N-3} & b_{4N-4} & b_{4N-5} & b_{4N-6} & \cdots & b_{2N+2} & b_{2N+1} & b_{2N} \end{bmatrix} \tag{11.3.4}$$

The Routh-Hurwitz criteria pertain to the $2N$ partial determinants formed from the upper left $n \times n$ partitions $[\Delta_n]$ of this array. The criteria state that all eigenvalues will have negative real parts corresponding to the asymptotic stability of the system if the coefficients of the characteristic equation are all positive and if all sub-determinants are positive,

$$\boxed{\begin{array}{ll} b_n > 0, & n = 0, 1, \ldots, 2N \\ |[\Delta_n]| > 0, & n = 1, 2, \ldots, 2N \end{array}} \tag{11.3.5}$$

Note that as a consequence of zero padding the b_n coefficients, the only nonzero element in the last row of $[\Delta]$ is b_{2N}. Because b_{2N} must be positive according to the first set of criteria, the condition $|[\Delta_{2N}]| > 0$ will be satisfied identically if $|[\Delta_{2N-1}]| > 0$.

In practice, obtaining the coefficients of the characteristic equation is quite tedious for systems having more than three degrees of freedom. For the case $N = 2$ the criteria reduce to

$$N = 2: \begin{cases} b_1, b_2, b_3, b_4 > 0 \\ |[\Delta_2]| = b_1 b_2 - b_0 b_3 > 0 \\ |[\Delta_3]| = b_1 b_2 b_3 - b_0 b_3^2 - b_1 b_4^2 > 0 \end{cases} \tag{11.3.6}$$

The usefulness of the Routh-Hurwitz criteria lies in the possibility of using them to identify algebraic inequality constraints that must be satisfied by a system's parameters.

EXAMPLE 11.4

The linkage consists of identical bars having mass m, with torsional springs k at each joint. The springs are undeformed when the bars are in the upright position. The compressive force F is a follower force whose magnitude is constant and whose orientation is always parallel to the upper bar, as shown in the sketch. Gravity is negligible compared to F. Identify the range of values of F for which the system is stable. Compare this range to the case where F remains vertical as the system moves. Then consider the situation where a small torsional dashpot c acts in parallel to each spring. Examine whether c alters the range of F for which the system is stable.

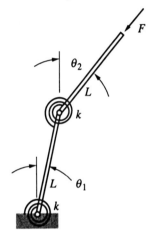

Solution The main objective of this example is to demonstrate most of the operations required for a stability assessment. In addition, the results will show that damping can have surprising effects. This is a time-invariant system. However, because of the manner in which the force $\bar{F}$ varies, we shall use virtual work to formulate the generalized forces. We use the rotation angles θ_1 and θ_2 as generalized coordinates. The inertia and stiffness matrices are

$$[M] = mL^2 \begin{bmatrix} 4/3 & 1/2 \\ 1/2 & 1/3 \end{bmatrix}, \qquad [K] = k \begin{bmatrix} 2 & -1 \\ -1 & 1 \end{bmatrix}$$

The standard linearization techniques would oversimplify the effect of $\bar{F}$ because coefficients of the force that depend linearly on θ_1 or θ_2 would be dropped. We therefore describe the virtual work corresponding to the system being in an arbitrary position, and then perform linearization operations. Because the force intersects the joint connecting the bars, the simplest evaluation is $\delta W = \bar{F} \cdot \delta \bar{r}_{B/A}$. In terms of horizontal and vertical components we have

$$\bar{F} = -F \sin(\theta_2)\bar{\imath} - F \cos(\theta_2)\bar{\jmath}, \qquad \bar{r}_{B/A} = L \sin(\theta_1)\bar{\imath} + L \cos(\theta_1)\bar{\jmath}$$

$$\delta \bar{r}_{B/A} = \frac{\partial \bar{r}_{B/A}}{\partial \theta_1}\delta\theta_1 + \frac{\partial \bar{r}_{B/A}}{\partial \theta_2}\delta\theta_2 = L\delta\theta_1 \cos(\theta_1)\bar{\imath} - L\delta\theta_1 \sin(\theta_1)\bar{\jmath}$$

$$\delta W = FL\delta\theta_1[-\sin(\theta_2)\cos(\theta_1) + \cos(\theta_2)\sin(\theta_1)] \equiv FL\delta\theta_1 \sin(\theta_1 - \theta_2)$$

The generalized forces are such that virtual work is $\delta W = Q_1\delta\theta_1 + Q_2\delta\theta_2$, so the preceding expression indicates that $Q_1 = FL \sin(\theta_1 - \theta_2)$, $Q_2 = 0$. Our interest lies in situations where θ_1 and θ_2 are small, which leads us to the approximation that

$$Q_1 = FL(\theta_1 - \theta_2), \qquad Q_2 = 0$$

The reason we had to be more careful than usual in describing the effect of an external force is that here, the follower force effect is a moment that depends on the angles not being zero. For a conventional force the primary effect is a force or moment that is independent of the generalized coordinates.

The generalized force is linear in θ_1 and θ_2, but the effect directly excites only one generalized coordinate. This is typical of circulatory forces, and it gives rise to nonsymmetric terms in the equations of motion, specifically,

$$mL^2 \begin{bmatrix} 4/3 & 1/2 \\ 1/2 & 1/3 \end{bmatrix} \begin{Bmatrix} \ddot{\theta}_1 \\ \ddot{\theta}_2 \end{Bmatrix} + k \begin{bmatrix} 2 & -1 \\ -2 & 1 \end{bmatrix} \begin{Bmatrix} \theta_1 \\ \theta_2 \end{Bmatrix} = FL \begin{bmatrix} 1 & -1 \\ 0 & 0 \end{bmatrix} \begin{Bmatrix} \theta_1 \\ \theta_2 \end{Bmatrix}$$

Relative to the standard form of the equation of motion, the term in the right side of the preceding is $[g_q]\{q\}$.

Because this is a second-order system we shall derive the characteristic equation by forming the nonlinear eigenvalue problem, eq. (11.2.13). For free vibration we set each $\theta_j = u_j \exp(\lambda t)$, which leads to

$$\begin{bmatrix} (\frac{4}{3}mL^2\lambda^2 + 2k - FL) & (\frac{1}{2}mL^2\lambda^2 - k + FL) \\ (\frac{1}{2}mL^2\lambda^2 - k) & (\frac{1}{3}mL^2\lambda^2 + k) \end{bmatrix} \begin{Bmatrix} u_1 \\ u_2 \end{Bmatrix} = \begin{Bmatrix} 0 \\ 0 \end{Bmatrix}$$

The characteristic equation is

$$\frac{7}{36}\nu^4 + \left(3 - \frac{5}{6}\frac{FL}{k}\right)\nu^2 + 1 = 0$$

where ν is a nondimensional eigenvalue parameter defined as

$$\nu = \left(\frac{mL^2}{k}\right)^{1/2} \lambda$$

Let us first consider applying the Routh-Hurwitz criteria, eq. (11.3.5). The coefficients for the $[\Delta]$ array are $b_0 = \frac{7}{36}$, $b_1 = 0$, $b_2 = 3 - \frac{5}{6}(FL/k)$, $b_3 = 0$, $b_4 = 1$. Because some of these coefficients are not positive, the criteria cannot be met, which means that the system is not asymptotically stable. This is not surprising, for there is no damping to dissipate energy. To determine values of F that lead to instability, we evaluate the roots of the biquadratic characteristic equation for ν,

$$\nu^2 = -\frac{1}{7}\left(54 - 15\frac{FL}{k}\right) \pm \frac{3}{7}\sqrt{296 - 180\frac{FL}{k} + 25\left(\frac{FL}{k}\right)^2}$$

A critical condition corresponds to the discriminant being negative. In such a situation, the values of λ^2 will be complex conjugates. By deMoivre's theorem, the square root of a complex number has two complex roots, one of which has a negative real part. Thus, a negative discriminant leads to two pairs of complex conjugate roots, with one pair having a negative real part. We conclude that a flutter instability arises if $296 - 180k(FL/k) + 25(FL/k)^2 < 0$. Solving this inequality leads to

$$\text{Flutter instability:} \quad \frac{18 - 2\sqrt{7}}{5} < \frac{FL}{k} < \frac{18 + 2\sqrt{7}}{5}$$

When FL/k is outside the preceding range, the roots ν^2 are real. This leads to another critical condition. For $FL/k < (18 - 2\sqrt{7})/5 = 2.542$, both values of ν^2 are negative, and all eigenvalues are purely imaginary. This is the case of marginal stability. In contrast, $FL/k > (18 + 2\sqrt{7})/5 = 4.658$ leads to two positive values of ν^2, so two of the eigenvalues will be real negative values. This is the case of divergence instability. From the standpoint of usability, it does not matter if the system is unstable in divergence or flutter. Thus, $FL/k < 2.542$ is the safe operational range. The first set of graphs displays the eigenvalues $\nu = (mL^2/k)^{1/2}\lambda$ as a function of FL/k when there is no dissipation.

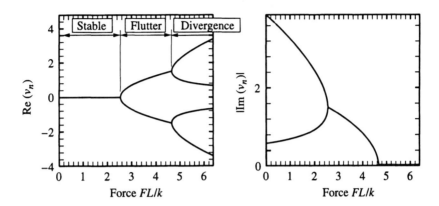

We now turn our attention to the case where the force is always vertical, $\bar{F} = -F\bar{j}$. The virtual work in this case is $\delta W = \bar{F} \cdot \delta\bar{r}_{C/A}$, where

$$\bar{r}_{C/A} = [L\sin(\theta_1) + L\sin(\theta_2)]\bar{i} + [L\cos(\theta_2) + L\cos(\theta_2)]\bar{j}$$

$$\delta\bar{r}_{C/A} = \frac{\partial\bar{r}_{C/A}}{\partial\theta_1}\delta\theta_1 + \frac{\partial\bar{r}_{C/A}}{\partial\theta_2}\delta\theta_2$$

$$= L\delta\theta_1[\cos(\theta_1)\bar{i} - \sin(\theta_1)\bar{j}] + L\delta\theta_2[\cos(\theta_2)\bar{i} - \sin(\theta_2)\bar{j}]$$

It follows that $\delta W = FL[\sin(\theta_1)\delta\theta_1 + \sin(\theta_2)\delta\theta_2] = Q_1\delta\theta_1 + Q_2\delta\theta_2$. Introducing the small angle approximation leads to

$$Q_1 = FL\theta_1, \qquad Q_2 = FL\theta_2$$

The corresponding equations of motion are

$$mL^2\begin{bmatrix} 4/3 & 1/2 \\ 1/2 & 1/3 \end{bmatrix}\begin{Bmatrix} \ddot{\theta}_1 \\ \ddot{\theta}_2 \end{Bmatrix} + k\begin{bmatrix} 2 & -1 \\ -2 & 1 \end{bmatrix}\begin{Bmatrix} \theta_1 \\ \theta_2 \end{Bmatrix} = FL\begin{bmatrix} 1 & 0 \\ 0 & 1 \end{bmatrix}\begin{Bmatrix} \theta_1 \\ \theta_2 \end{Bmatrix}$$

Here the effect of $\bar{F}$ is manifested in the equations of motion as symmetric terms. This force is conservative, for its effect is equivalent to that of a gravitational force acting at point C.

The characteristic equation resulting from substituting $\theta_j = u_j\exp(\lambda t)$ is

$$\frac{7}{36}\nu^4 + \left(3 - \frac{5}{3}\frac{FL}{K}\right)\nu^2 + \left[1 - 3\frac{FL}{K} + \left(\frac{FL}{K}\right)^2\right] = 0$$

The roots of this quadratic characteristic equation are

$$\nu^2 = -\frac{1}{7}\left(54 + 30\frac{FL}{K}\right) \pm \frac{6}{7}\sqrt{74 - 69\frac{FL}{K} + 18\left(\frac{FL}{k}\right)^2}$$

The discriminant in this expression is always positive. As a result, the values of λ^2 are either negative or positive real values. The positive case corresponds to a divergence instability. The largest value of F for which the system is stable gives $\lambda^2 = 0$ as one root. The result is

$$\text{Divergence instability: } \frac{FL}{k} > \frac{3 - \sqrt{5}}{2} = 0.38197$$

The critical value of F in this case is much smaller than the flutter value for the follower force.

Now let us consider the effect of damping in the follower force case. The damping matrix is proportional to the stiffness, because each spring is accompanied by a parallel torsional dashpot, so

$$[C] = c \begin{bmatrix} 2 & -1 \\ -1 & 1 \end{bmatrix}$$

This leads to the eigenvalue problem

$$\begin{bmatrix} \left(\frac{4}{3}\nu^2 + 2\hat{c}\nu + 2 - \frac{FL}{k}\right) & \left(\frac{1}{2}\nu^2 - \hat{c}\nu - 1 + \frac{FL}{k}\right) \\ \left(\frac{1}{2}\nu^2 - \hat{c}\lambda - 1\right) & \left(\frac{1}{3}\nu^2 + \hat{c}\lambda + 1\right) \end{bmatrix} \begin{Bmatrix} u_1 \\ u_2 \end{Bmatrix} = \begin{Bmatrix} 0 \\ 0 \end{Bmatrix}$$

where $\hat{c} = c/(mL^2k)^{1/2}$. The characteristic equation is

$$\frac{7}{36}\nu^4 + 3\hat{c}\nu^3 + \left(3 + \hat{c}^2 - \frac{5}{6}\frac{FL}{k}\right)\nu^2 + 2\hat{c}\nu + 1 = 0$$

Finding the roots of this quartic equation requires a numerical solution, which is best implemented using the *polyroots* function in Mathcad, or the *roots* function in MATLAB. (Both functions find all roots of a polynomial without setting starting values.) The next set of graphs, for which $\hat{c} = 0.01$, shows that the eigenvalues ν_n are two pairs of complex conjugates until $FL/k \approx 4.6$, at which value there is a divergence instability.

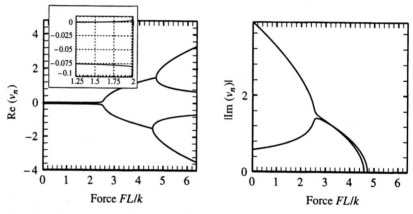

Superficially, it seems as though the flutter instability range is essentially what it was for $\hat{c} = 0$, but a careful examination of the enlarged view of the real parts shows a significant change. The real parts of the ν_n values are quite small for $FL/k < 2.5$, but one becomes positive around $FL/k = 1.6$. In other words, addition of damping lowers the critical force for a flutter instability relative to the undamped case. Many individuals find this to be unexpected, for conventional thinking is that damping stabilizes a system by draining energy. The explanation lies in the Routh-Hurwitz criteria. The first set of criteria, that all b_n coefficients should be positive, is satisfied if $FL/k < (6/5)(3 + \hat{c}^2)$. The Routh-Hurwitz array $[\Delta]$ described eq. (11.3.4) is

$$[\Delta] = \begin{bmatrix} 3\hat{c} & \frac{7}{36} & 0 & 0 \\ 2\hat{c} & \left(3 + \hat{c}^2 - \frac{5}{6}\frac{FL}{k}\right) & 3\hat{c} & \frac{7}{36} \\ 0 & 1 & 2\hat{c} & \left(3 + \hat{c}^2 - \frac{5}{6}\frac{FL}{k}\right) \\ 0 & 0 & 0 & 1 \end{bmatrix}$$

The nontrivial subdeterminants are

$$|[\Delta_2]| = \frac{1}{18}\hat{c}\left(155 + 54\hat{c}^2 - 45\frac{FL}{k}\right) > 0$$

$$|[\Delta_3]| = \frac{1}{9}\hat{c}^2\left(74 + 54\hat{c}^2 - 45\frac{FL}{k}\right) > 0$$

The smallest value of the force for which either requirement is not met is $FL/k = 74/45 + (6/5)\hat{c}^2$. Thus we conclude that

$$\text{Flutter instability:} \frac{FL}{k} > 1.6444 + 1.20\hat{c}^2 \text{ if } \hat{c} > 0$$

Thus, even in the limit as $\hat{c}$ becomes infinitesimal, the system will flutter at a smaller value of FL/k than it would if $\hat{c}$ were identically zero. Note that raising the value of $\hat{c}$ from a very small value does indeed stabilize the system by raising the critical value of FL/k.

The divergence instability arising when $\bar{F}$ is always oriented in the vertical direction is predictable by a static buckling analysis. In contrast, the flutter instability for the follower force case can only be identified from a dynamic analysis. Failure to recognize the role of follower forces was the source of some spectacular failures in the early United States space program. An extensive treatment of these issues is provided by Leipholz (1980).

11.4 BI-ORTHOGONALITY AND MODAL EQUATIONS OF MOTION

Once we have ascertained that a system is stable, we may proceed to evaluate its response. The presence of gyroscopic and/or circulatory forces alters the method by which we obtain uncoupled equations for modal coordinates. As was mentioned earlier, the left and right eigenvectors are both needed to form an orthogonality property. We consider the jth right eigenvector and the nth left eigenvector, where j and n are arbitrarily selected. By definition, we have

$$[[R] - \lambda_j[S]]\{\psi_j\} = \{0\}$$
$$[[R]^T - \lambda_n[S]]\{\widetilde{\psi}_n\} = \{0\}$$
(11.4.1)

We move the terms containing $[S]$ to the right side, then multiply the first equation by $\{\widetilde{\psi}_n\}^T$ and the second by $\{\psi_j\}^T$,

$$\{\widetilde{\psi}_n\}^T[R]\{\psi_j\} = \lambda_j\{\widetilde{\psi}_n\}^T[S]\{\psi_j\}$$
$$\{\psi_j\}^T[R]^T\{\widetilde{\psi}_n\} = \lambda_n\{\psi_j\}^T[S]\{\widetilde{\psi}_n\}$$
(11.4.2)

In view of the symmetry of $[S]$, taking the transpose of the second equation gives a result that is like the first equation, except that λ_n replaces λ_j. Thus, the difference of the two equations is

$$(\lambda_j - \lambda_n)\{\widetilde{\psi}_n\}^T[S]\{\psi_j\} = 0$$
(11.4.3)

If $\lambda_n \neq \lambda_j$, which is always the case if j and n are distinct and the eigenvalues are not repeated, then the preceding reduces to the *first bi-orthogonality condition*,

$$\{\widetilde{\psi}_n\}^T[S]\{\psi_j\} = 0 \text{ if } \lambda_n \neq \lambda_j$$
(11.4.4)

The *second bi-orthogonality condition* follows immediately from the first of eqs. (11.4.2),

$$\{\widetilde{\psi}_n\}^T[R]\{\psi_j\} = 0 \text{ if } \lambda_n \neq \lambda_j$$
(11.4.5)

We are working under the assumption that the eigenvalues are not repeated, so both bi-orthogonality conditions apply for $n \neq j$.

It is useful to define the scaling factor for normal modes such that the first bi-orthogonality condition gives a unit value when $n = j$. We denote *normal right* and *normal left eigenvectors* as $\{\Psi_j\}$ and $\{\widetilde{\Psi}_j\}$, respectively. Because the orthogonality properties involve both $\{\Psi_j\}$ and $\{\widetilde{\Psi}_j\}$, we shall use the same normalizing factor c_j for both. Thus, we set

$$\{\widetilde{\Psi}_j\}^{\mathrm{T}}[S]\{\Psi_j\} = c_j^2\{\widetilde{\psi}_j\}^{\mathrm{T}}[S]\{\psi_j\} = 1 \tag{11.4.6}$$

It follows that the normal eigenvectors may be obtained from computed values by

$$\{\Psi_j\} = \frac{\{\psi_j\}}{[\{\widetilde{\psi}_j\}^{\mathrm{T}}[S]\{\psi_j\}]^{1/2}}, \qquad \{\widetilde{\Psi}_j\} = \frac{\{\widetilde{\psi}_j\}}{[\{\widetilde{\psi}_j\}^{\mathrm{T}}[S]\{\psi_j\}]^{1/2}} \tag{11.4.7}$$

When we use the normal eigenvectors to form the first of eqs. (11.4.2) for the case $n = j$, we find that

$$\{\widetilde{\Psi}_j\}^{\mathrm{T}}[R]\{\Psi_j\} = \lambda_j \tag{11.4.8}$$

The right and left normal eigenvector matrices are $2N \times 2N$ arrays formed from the respective eigenvectors,

$$[\Psi] = \left[\{\Psi_1\}\ \{\Psi_2\}\ \cdots\ \{\Psi_{2N}\}\right], \qquad [\widetilde{\Psi}] = \left[\{\widetilde{\Psi}_1\}\ \{\widetilde{\Psi}_2\}\ \cdots\ \{\widetilde{\Psi}_{2N}\}\right] \tag{11.4.9}$$

The orthogonality and normalization properties for the individual eigenvectors lead to

$$[\widetilde{\Psi}]^{\mathrm{T}}[S][\Psi] = [I], \qquad [\widetilde{\Psi}]^{\mathrm{T}}[R][\Psi] = [\lambda] \tag{11.4.10}$$

where $[\lambda]$ is a diagonal array of sequenced eigenvalues.

One should be aware that software will not necessarily return the left and right eigensolutions in the same sequence. This would be evident if the left eigenvalues are not listed in the same order as the right eigenvalues. If they are not, eqs. (11.4.10) will not be satisfied. For that reason it is good practice to apply the sorting procedures in Example 10.1 individually to the left and right eigensolutions, prior to implementing the step of normalizing the modes.

The bi-orthogonality relations lead to uncoupled modal equations. The modal transformation is defined in terms of the right eigenvectors,

$$\{x\} = [\Psi]\{\xi\} \tag{11.4.11}$$

We substitute this transformation into eq. (11.1.21), and premultiply by the transpose of the left eigenvector matrix. In view of eqs. (11.4.10), these operations lead to

$$\{\dot{\xi}\} - [\lambda]\{\xi\} = [\widetilde{\Psi}]^{\mathrm{T}}\begin{Bmatrix} \{0\} \\ \{Q\} \end{Bmatrix} \tag{11.4.12}$$

The scalar version of the preceding is a set of uncoupled equations that are like those for damped modal coordinates, except that the modal forces are defined in terms of the left eigenvectors,

$$\dot{\xi}_n - \lambda_n\xi_n = [\widetilde{\Psi}_n]^{\mathrm{T}}\begin{Bmatrix} \{0\} \\ \{Q\} \end{Bmatrix}, \qquad n = 1, \ldots, 2N \tag{11.4.13}$$

Solution of these equations for other than steady-state harmonic vibration requires initial values of the ξ_n variables. Given the initial value of the x_j variables we find from eq. (11.4.11) that

$$\{\xi\} = [\Psi]^{-1}\{x\} \tag{11.4.14}$$

The inverse of the normal mode matrix may be obtained from the first bi-orthogonality relation, which yields

$$\boxed{\{\xi(t = 0)\} = [\widetilde{\Psi}][S]\{x(t = 0)\}} \tag{11.4.15}$$

It is evident that the modal coordinate equations for systems with gyroscopic inertia and/or circulatory forces are essentially the same as those for damped modal analysis. In both formulations, the modes are purely real, or else occur as complex conjugates, and the modal differential equations have the same forms. Thus, the solution techniques discussed in Section 10.5 for free, transient, and steady-state harmonic excitation are equally suitable to gyroscopic modal analysis. For example, if the excitation is harmonic, we have

$$\boxed{\begin{Bmatrix}\{0\} \\ \{Q\}\end{Bmatrix} = \text{Re}\{\{F\}\exp(i\omega t)\} = \tfrac{1}{2}\{F\}\exp(i\omega t) + \tfrac{1}{2}\{F^*\}\exp(-i\omega t)} \tag{11.4.16}$$

The corresponding steady-state solution of eq. (11.4.13) is

$$\boxed{\xi_n = \frac{1}{2}\frac{\{\widetilde{\Psi}_n\}\{F\}}{(i\omega - \lambda_n)}\exp(i\omega t) + \frac{1}{2}\frac{\{\widetilde{\Psi}_n\}\{F^*\}}{(-i\omega - \lambda_n)}\exp(-i\omega t)} \tag{11.4.17}$$

This expression sheds light on resonance phenomena, which we see correspond to a minimum value of $i\omega - \lambda_n$, that is, $\omega = \text{Im}(\lambda_n)$. However, it often is the case that a direct frequency domain evaluation of steady-state amplitudes based on substituting $\{q\} = \text{Re}[\{Y\}\exp(i\omega t)]$ into eq. (11.1.19) is more convenient for computations.

An overview of the developments thus far shows that the presence of gyroscopic or circulatory effects leads to one major change in the modal analysis of response: the need to determine left, as well as right, eigenvectors. The left eigenvectors are used to evaluate the modal forces in the right side of the modal coordinate differential equations, eqs. (11.4.13), as well as to evaluate initial conditions for the modal coordinates according to eq. (11.4.15).

EXAMPLE 11.5

A bar having rectangular cross-section spins about its base at constant angular speed. The hub at the base serves as a fixed end, so the bar is effectively a rotating cantilevered beam. The sketch represents a crude model in which two lumped masses serve to capture the inertia effects of the beam. Each mass equals half the total mass, $m_1 = m_2 = \frac{1}{2}\rho A \ell$. The bar is considered to be rigid in extension, so the displacement components y_j and z_j of each lumped mass serve as generalized coordinates. The xyz reference frame rotates with the base. The principal axes of the bar's cross-section are y and z, so that flexural displacements in the xy

and xz planes are elastically uncoupled. A static deflection analysis indicates that the flexibility of a cantilevered beam is

$$L(x, \chi) = \begin{cases} \dfrac{\chi^2(3x - \chi)}{6EI}, & x \geq \chi \\ L(\chi, x), & x \leq \chi \end{cases}$$

where L denotes the displacement at location x resulting from a static unit force at location χ. This relation describes the displacement in the xy plane if $I = I_{zz}$ and in the xz plane if $I = I_{yy}$. Consider a situation in which the rotation rate ω is 95% of the fundamental frequency of the beam in the absence of rotation, with $I_{yy} = \frac{2}{3}I_{zz}$. Determine the response of the system due to a unit impulsive force acting on the upper mass m_2 in the y direction.

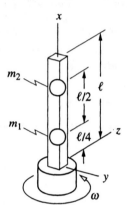

Solution In addition to illustrating all steps required to implement modal analysis in the presence of gyroscopic inertia effects, this example will serve as a precursor for our investigation of rotordynamics in the next chapter. The formulation of the inertia terms in the equations of motion is essentially the same as that for the system in Example 11.1. We define the generalized coordinates such that

$$\{q\} = \begin{bmatrix} y_1 & z_1 & y_2 & z_2 \end{bmatrix}^T$$

The velocity of each mass is the sum of the velocity relative to xyz and the contribution due to the rotation of xyz,

$$\bar{v}_n = (\dot{y}_n \bar{j} + \dot{z}_n \bar{k}) + \bar{\omega} \times \bar{r}_{n/O} = (\dot{y}_n - \omega z_n)\bar{j} + (\dot{z}_n + \omega y_n)\bar{k}$$

The similarity of this expression to the velocity of the block in Example 11.1 enables us to write the inertia terms in the equations of motion as

$$\frac{d}{dt}\left(\frac{\partial T}{\partial \dot{y}_n}\right) - \frac{\partial T}{\partial y_n} = m_n(\ddot{y}_n - \omega^2 y_n - 2\omega \dot{z}_n)$$

$$\frac{d}{dt}\left(\frac{\partial T}{\partial \dot{z}_n}\right) - \frac{\partial T}{\partial z_n} = m_n(\ddot{z}_n - \omega^2 z_n + 2\omega \dot{y}_n)$$

The corresponding inertia and gyroscopic matrices are

$$[M] = \frac{1}{2}\rho A \ell [I], \qquad [E] = \frac{1}{2}\rho A \ell \omega^2 [I], \qquad [G] = \frac{1}{2}\rho A L \omega \begin{bmatrix} 0 & -2 & 0 & 0 \\ 2 & 0 & 0 & 0 \\ 0 & 0 & 0 & -2 \\ 0 & 0 & 2 & 0 \end{bmatrix}$$

The elastic effects for this system are described in terms of flexibility properties, which were discussed in Section 4.2.4. Bending in the xy plane and xz plane are not coupled elastically, so the flexibility coefficients couple y_1 and y_2, and z_1 and z_2. Because of the manner in which the generalized coordinates have been sequenced, the coefficients for bending in the xy plane are L_{jn}, $j, n = 1, 3$, while the coefficients for bending in the xz plane are $L_{jn}, j, n = 2, 4$. The former use I_{zz}, while the latter use $I_{yy} = \frac{2}{3}I_{zz}$. Thus, the nonzero flexibility coefficients are

$$L_{1,1} = L(0.25\ell, 0.25\ell)|_{I_{zz}}, \qquad L_{3,3} = L(0.75\ell, 0.75\ell)|_{I_{zz}}$$

$$L_{3,1} = L(0.75\ell, 0.25\ell)|_{I_{zz}}, \qquad L_{1,3} = L_{3,1}$$

$$L_{2,2} = L(0.25\ell, 0.25\ell)|_{I_{yy}}, \qquad L_{4,4} = L(0.75\ell, 0.75\ell)|_{I_{yy}}$$

$$L_{4,2} = L(0.75\ell, 0.25\ell)|_{I_{yy}}, \qquad L_{2,4} = L_{4,2}$$

The resulting flexibility matrix is

$$[L] = \frac{\ell^3}{EI_{zz}}(10^{-2})\begin{bmatrix} 0.5208 & 0 & 2.083 & 0 \\ 0 & 0.7813 & 0 & 3.125 \\ 2.083 & 0 & 14.063 & 0 \\ 0 & 3.125 & 0 & 21.094 \end{bmatrix}$$

We need the stiffness matrix to form the state-space equations, so we invert $[L]$,

$$[K] = [L]^{-1} = \frac{EI_{zz}}{\ell^3}\begin{bmatrix} 471.27 & 0 & -69.82 & 0 \\ 0 & 314.18 & 0 & -46.55 \\ -69.82 & 0 & 17.45 & 0 \\ 0 & -46.55 & 0 & 11.64 \end{bmatrix}$$

The impulse is applied to y_2, so $Q_3 = \delta(t)$ is the only nonzero generalized force. In order to eliminate the system parameters from the analysis, let us define a nondimensional time τ and corresponding nondimensional rotation rate $\hat{\omega}$,

$$\tau = \left(\frac{EI_{zz}}{\rho A \ell^4}\right)^{1/2} t, \qquad \hat{\omega} = \left(\frac{\rho A \ell^4}{EI_{zz}}\right)^{1/2} \omega$$

The nondimensional form of the equations of motion is

$$[M']\frac{d^2}{d\tau^2}\{q\} + [G']\frac{d}{d\tau}\{q\} + [[K'] - [E']]\{q\} = \{Q'\}$$

where $[K']$ consists of the numerical factors in the $[K]$ array, and

$$[M'] = \tfrac{1}{2}[I], \qquad [E'] = \tfrac{1}{2}\hat{\omega}^2[I], \qquad [G'] = \begin{bmatrix} 0 & 0 & -\hat{\omega} & 0 \\ \hat{\omega} & 0 & 0 & 0 \\ 0 & 0 & 0 & -\hat{\omega} \\ 0 & 0 & \hat{\omega} & 0 \end{bmatrix}$$

The rotation rate is specified to be 95% of the fundamental frequency of the nonrotating system. We find that quantity by solving $[[K'] - (\omega_{nat})^2[M']]\{\phi\} = \{0\}$. This yields

$$(\omega_{nat})_n = 3.043, 3.739, 25.345, 31.041$$

The (nonnormalized) modes of the nonrotating system are

$$\{\phi_1\} = \begin{bmatrix} 0 & 1 & 0 & 6.65 \end{bmatrix}^T$$

$$\{\phi_2\} = \begin{bmatrix} 1 & 0 & 6.65 & 0 \end{bmatrix}^T$$

$$\{\phi_3\} = \begin{bmatrix} 0 & 1 & 0 & -0.150 \end{bmatrix}^T$$

$$\{\phi_4\} = \begin{bmatrix} 1 & 0 & -0.150 & 0 \end{bmatrix}^T$$

The corresponding value of the rotation rate is

$$\hat{\omega} = 0.95(\omega_{\text{nat}})_1 = 2.8909$$

We next form $[R]$ and $[S]$ according to eqs. (11.1.22), which then are input to our math software to solve the right and left eigenvalue problems. We apply the sorting procedures in Example 10.1 individually to the left and right eigensolutions in order that the left and right eigensolutions be in the correct correspondence. After the eigensolutions are sorted, we normalize both sets of eigenvectors according to eqs. (11.4.7). The result, sorted in the order of increasing $|\lambda_j|$, is

$$\lambda_n = \pm 0.355i, \pm 6.309i, \pm 24.121i, \pm 32.262i$$

The right and left eigenvectors occur as four complex conjugate pairs, arranged adjacently. For the sake of brevity, we only list the upper half of each eigenvector, because the lower half is a multiple of the upper partition; see eq. (11.2.12). The results are

$$\{\Psi_1\} = \{\Psi_2\}^* = \begin{bmatrix} -0.0587i & -0.1548 & -0.3903i & -1.0296 & \cdots \end{bmatrix}^T$$

$$\{\Psi_3\} = \{\Psi_4\}^* = \begin{bmatrix} -0.0233i & 0.0219 & -0.1552i & 0.1457 & \cdots \end{bmatrix}^T$$

$$\{\Psi_5\} = \{\Psi_6\}^* = \begin{bmatrix} 0.0138 & -0.0370i & -0.0021 & 0.0056i & \cdots \end{bmatrix}^T$$

$$\{\Psi_7\} = \{\Psi_8\}^* = \begin{bmatrix} -0.0289i & 0.0132 & 0.0043i & -0.0020 & \cdots \end{bmatrix}^T$$

$$\{\widetilde{\Psi}_1\} = \{\widetilde{\Psi}_2\}^* = \begin{bmatrix} -0.0587i & 0.1548 & -0.3903i & 1.0296 & \cdots \end{bmatrix}^T$$

$$\{\widetilde{\Psi}_3\} = \{\widetilde{\Psi}_4\}^* = \begin{bmatrix} -0.0233i & -0.0219 & -0.1552i & -0.1457 & \cdots \end{bmatrix}$$

$$\{\widetilde{\Psi}_5\} = \{\widetilde{\Psi}_6\}^* = \begin{bmatrix} -0.0138 & -0.0370i & 0.0021 & 0.0056i & \cdots \end{bmatrix}^T$$

$$\{\widetilde{\Psi}_7\} = \{\widetilde{\Psi}_8\}^* = \begin{bmatrix} -0.0289i & -0.0132 & 0.0043i & 0.0020 & \cdots \end{bmatrix}^T$$

The interesting aspect of these eigenvectors is that their proportions match the modes $\{\phi_n\}$ of the nonrotating system, specifically,

$$\frac{\phi_{4,1}}{\phi_{2,1}} = \frac{\phi_{3,2}}{\phi_{1,2}} = \frac{\psi_{3,1}}{\psi_{1,1}} = \frac{\psi_{4,1}}{\psi_{2,1}} = \frac{\psi_{3,3}}{\psi_{1,3}} = \frac{\psi_{4,3}}{\psi_{2,3}} = 6.65$$

$$\frac{\phi_{4,3}}{\phi_{2,3}} = \frac{\phi_{3,4}}{\phi_{1,4}} = \frac{\psi_{3,5}}{\psi_{1,5}} = \frac{\psi_{4,5}}{\psi_{2,5}} = \frac{\psi_{3,7}}{\psi_{1,7}} = \frac{\psi_{4,7}}{\psi_{2,7}} = -0.150$$

We further note that, in the stationary modes, either the second and fourth elements or the first and third elements are zero, which means that these modes represent free vibration that displaces the beam in the xy or xz planes, respectively. In the gyroscopic modes, $\{\Psi_n\}$, all elements are nonzero, with the second and fourth elements being 90° out of phase from the first and third elements. In other words, the base rotation splits each planar mode of the stationary system into two pairs of rotating modes. We will see that the first of each pair represents forward precession, while the second corresponds to retrograde precession.

We are now ready to evaluate the response. It was noted earlier that $Q_3 = \delta(t)$ is the only nonzero generalized force. We introduce the modal transformation, $\{x\} = [\Psi]\{\xi\}$, which leads to the modal equations of motion,

$$\dot{\xi}_n - \lambda_n \xi_n = \{\widetilde{\Psi}_n\}^T \begin{Bmatrix} \{0\} \\ \{Q\} \end{Bmatrix} = \widetilde{\Psi}_{7n}\delta(t), \quad n = 1, \ldots, 2N$$

At $t = 0$, the beam was at its equilibrium position and stationary relative to xyz, so we have $\xi_n = \dot{\xi}_n = 0$ prior to application of the impulse. According to eq. (10.5.16), the coefficient of

$\delta(t)$ in a first-order modal differential equation for ξ_n is the increment imparted to ξ_n by the impulse. After the impulse, each modal coordinate is a free vibration, so the modal response is

$$\xi_n = \widetilde{\Psi}_{7n} \exp(\lambda_n t)$$

Because the λ_n values are all imaginary, these responses consist of four pairs of harmonic vibration at frequencies $|\lambda_1|$, $|\lambda_3|$, $|\lambda_5|$, and $|\lambda_7|$. If we were to evaluate $\{x\} = [\Psi]\{\xi\}$ using all the modal coordinates, we would see a highly irregular pattern of displacement. Instead, let us decompose the response into contributions from each complex conjugate mode pair. Specifically, we define

$$\{x_{1\_2}(t)\} = \sum_{n=1}^{2} \{\Psi_n\}\xi_n(t), \qquad \{x_{3\_4}(t)\} = \sum_{n=3}^{4} \{\Psi_n\}\xi_n(t)$$

$$\{x_{5\_6}(t)\} = \sum_{n=5}^{6} \{\Psi_n\}\xi_n(t), \qquad \{x_{7\_8}(t)\} = \sum_{n=7}^{8} \{\Psi_n\}\xi_n(t)$$

The first four elements of each vector are the instantaneous values of y_1, z_1, y_2, and z_2 associated with each mode. To develop a time history we carry out the evaluation over an interval of t. The result is four sets of graphs.

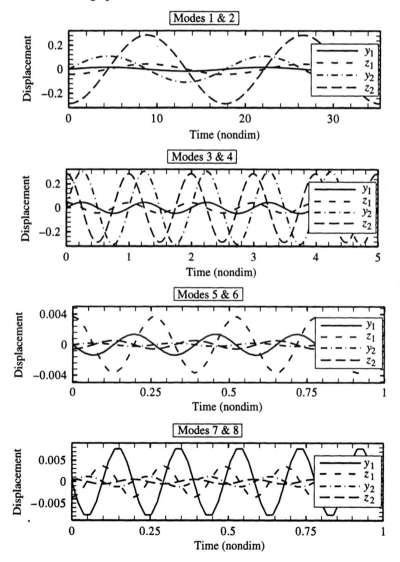

The frequencies of each are $\mathrm{Im}(\lambda_1) = 0.355$ for modes 1 and 2, $\mathrm{Im}(\lambda_3) = 6.309$ for modes 3 and 4, $\mathrm{Im}(\lambda_5) = 24.121$ for modes 5 and 6, and $\mathrm{Im}(\lambda_1) = 32.262$ for modes 7 and 8. An interesting aspect of the graphed responses is that some of them have nonzero initial values. However, the sum of their contributions to each generalized coordinate does satisfy the initial conditions that $q_j = 0$ at $t = 0$.

A different picture of the response may be obtained by drawing Lissajous diagrams for the lower mass m_1, whose displacement relative to xyz is $y_1 \bar{j} + z_1 \bar{k}$, and the upper mass m_2, whose displacement is $y_2 \bar{j} + z_2 \bar{k}$. Because each pair of displacements occurs as harmonic motions at the same frequency, but 90° out of phase, each displacement forms an elliptical path relative to the rotating base. The properties of $\{\Psi_n\}$ correspond to the upper and lower masses following elliptical paths that have the same aspect ratio. Furthermore, the diametral ratios for the ellipses match the proportions of the planar motions executed by the lower and upper masses in a stationary mode.

The initial position of each mass is shown as a dot in the Lissajous patterns, and the direction of the precession is indicated by an arrow. The latter may be identified by referring to the time response graphs. For example, for modes 1 and 2 we observe that at $t = 0$, $y_1 = 0$, $z_1 < 0$, $\dot{y}_1 > 0$, and $\dot{z}_1 > 0$. Thus, the initial point must displace to the right, which corresponds to a forward precession. This consideration reveals that modes 5 and 6 also represent forward precession, while modes 3 and 4 and modes 7 and 8 represent retrograde precession.

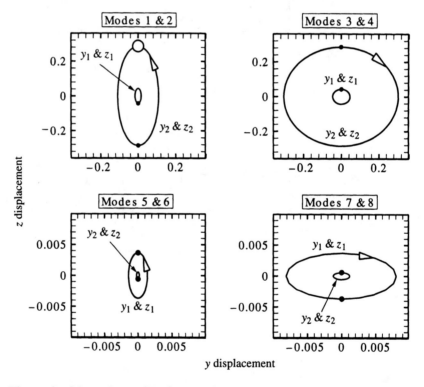

The result of these observations is recognition that the impulse force excites four fundamental responses. The first pair of gyroscopic modes is associated with a motion in which the beam is bent in a plane in proportions matching the first and second modes of the stationary beam. This plane precesses forward at an average angular speed $\mathrm{Im}(\lambda_1)$ relative to the base, which rotates at angular speed $\hat{\omega}$. The Lissajous patterns show that as the plane of the beam rotates, the beam goes from a minimum deflection when it aligns with the xy plane to a maximum deflection when it aligns with the xz plane. The other mode pairs behave similarly, except that the plane of the beam rotates backward relative to xyz in modes 3 and 4 and modes 7 and 8. Interestingly, both $\mathrm{Im}(\lambda_3)$ and $\mathrm{Im}(\lambda_7)$ are greater than $\hat{\omega}$, which means that, in the backward precessing modes, a fixed observer would see the plane of the beam rotating opposite the over-all rotation. The forward precessing modes obviously represent rotations of the beam's plane in the same sense as the rotation, but faster. Each type of motion is known as *shaft whirl*.

11.5 MOVING CONTINUOUS SYSTEMS

The methods we have developed to analyze the vibratory response of moving systems may be readily adapted to address the vibration of bars that undergo overall motion. The technique is a generalization of the Ritz method in Section 6.5, which treated a bar whose geometrical boundary conditions are time dependent. The situation we consider here is one in which the objects supporting the bar move in unison as a rigid body. We define a moving reference frame *xyz* that is attached to these objects, and therefore undergoes the specified motion. A typical case is the cantilevered beam in Example 11.5, where the base of the bar is mounted on a turntable that executes a planar rotation. It follows from this definition that the displacement $\bar{u}$ of the beam relative to *xyz* constitutes the bar's deformation. We denote the origin of *xyz* as point O. Basic theorems of kinematics state that the position and velocity of any point P at position $\bar{r}_{P/O}$ relative to *xyz* are given by

$$\bar{r}_P = \bar{r}_O + \bar{r}_{P/O}$$

$$\bar{v}_P = \bar{v}_O + (\bar{v}_P)_{xyz} + \bar{\Omega} \times \bar{r}_{P/O} \tag{11.5.1}$$

where $\bar{\Omega}$ is the angular velocity of *xyz* and $(\bar{v}_P)_{xyz}$ is the relative velocity, that is, the velocity of point P as seen by an observer who moves in unison with *xyz*.

It is useful to define the x axis to coincide with the undeformed axis of the bar. In the most general situation, the displacement $\bar{u}$ may consist of axial deformation in the x direction, and flexural displacements in each of the principal axis directions of the cross-section. Let us define y and z to be these directions. Then the displacement field may be represented as

$$\bar{u} = u_1 \bar{i} + u_2 \bar{j} + u_3 \bar{k} \tag{11.5.2}$$

(Using a tensorial-type notation to denote the displacement components expedites the analysis of deformation that is neither purely axial or flexural in a single direction. In special cases where we are solely concerned with one type of deformation, we could let the nonzero displacement be either u or w.) The relative position of a point on the beam's axis is the sum of deformational displacement and its position in the undeformed state, so

$$\bar{r}_{P/O} = \bar{u} + x\bar{i} \tag{11.5.3}$$

An observer on *xyz* is unaware that the directions of the x, y, and z axes might not be constant, so the relative velocity is

$$(\bar{v}_P)_{xyz} = \frac{\partial \bar{u}}{\partial t} = \dot{u}_1 \bar{i} + \dot{u}_2 \bar{j} + \dot{u}_3 \bar{k} \tag{11.5.4}$$

where the partial derivative is intended to denote that the unit vectors are held constant. We use Ritz series to represent each deformational displacement,

$$u_1 = \sum_j \psi_j^{(1)}(x) q_j^{(1)}(t), \quad u_2 = \sum_j \psi_j^{(2)}(x) q_j^{(2)}(t), \quad u_3 = \sum_{j=1}^N \psi_j^{(3)}(x) q_j^{(3)}(t) \tag{11.5.5}$$

The terms $\psi_j^{(n)}(x)$ are the basis functions for the displacement component u_n, and the series coefficients $q_j^{(n)}$ are the generalized coordinates for that displacement. (We may address situations where only one displacement component is important by setting the generalized coordinates for the other displacements identically to zero.) Because the generalized coordinates describe displacement relative to xyz, the geometric boundary conditions to be satisfied by each set of basis functions are those that would apply if xyz did not move. Consider for example the system in Example 11.5, with x defined as the axis of the beam, which coincides with the rotation axis. Axial vibration is not excited by the rotation, and flexural displacement consists of components in both transverse directions. Consequently, we need two sets of basis functions, $\psi_j^{(2)}$ and $\psi_j^{(3)}$. To examine the relative motion, we arrest the turntable, from which viewpoint we see that the bar behaves like a cantilevered beam flexing in the xy and xz planes. If we place the origin O at the base of the bar, the geometrical boundary conditions for the basis functions are $\psi_j^{(2)} = \psi_j^{(3)} = d\psi_j^{(2)}/dx = d\psi_j^{(3)}/dx = 0$ at $x = 0$.

The procedure from this juncture is the procedure used in Section 6.5, where kinetic and potential energies associated with the Ritz series were formulated and then input to Lagrange's equations. The kinetic energy of the bar is

$$T = \frac{1}{2}\int_0^L \rho A\left[\bar{v}_O + \frac{\partial \bar{u}}{\partial t} + \bar{\Omega} \times (\bar{u} + x\bar{i})\right] \cdot \left[\bar{v}_O + \frac{\partial \bar{u}}{\partial t} + \bar{\Omega} \times (\bar{u} + x\bar{i})\right] dx \quad (11.5.6)$$

We collect terms according to whether they are quadratic, linear, or independent of $\bar{u}$, with the result that

$$T = \frac{1}{2}\int_0^L \rho A\left[\frac{\partial \bar{u}}{\partial t} \cdot \frac{\partial \bar{u}}{\partial t} + 2\frac{\partial \bar{u}}{\partial t} \cdot (\bar{\Omega} \times \bar{u}) + (\bar{\Omega} \times \bar{u}) \cdot (\bar{\Omega} \times \bar{u})\right] dx$$

$$+ \int_0^L \rho A(\bar{v}_O + \bar{\Omega} \times x\bar{i}) \cdot \left(\frac{\partial \bar{u}}{\partial t} + \bar{\Omega} \times \bar{u}\right) dx \quad (11.5.7)$$

$$+ \frac{1}{2}\int_0^L \rho A(\bar{v}_O + \bar{\Omega} \times x\bar{i}) \cdot (\bar{v}_O + \bar{\Omega} \times x\bar{i}) \, dx$$

To carry out the operations required to process the preceding, let us define a column of generalized coordinates $\{s\}$ in partitioned form as

$$\{s\} = \begin{Bmatrix} \{q^{(1)}\} \\ \{q^{(2)}\} \\ \{q^{(3)}\} \end{Bmatrix} \quad (11.5.8)$$

Corresponding to this is a column array holding the basis functions, with each partition multiplied by the unit vector for the direction associated with that set,

$$\{\bar{\psi}\} = \begin{Bmatrix} \{\psi^{(1)}\}\bar{i} \\ \{\psi^{(2)}\}\bar{j} \\ \{\psi^{(3)}\}\bar{k} \end{Bmatrix} \quad (11.5.9)$$

where

$$\{\psi^{(n)}\} = \left[\psi_1^{(n)} \; \psi_2^{(n)} \; \psi_3^{(n)} \; \dots \right]^{\mathrm{T}} \tag{11.5.10}$$

These definitions enable us to write the displacement as

$$\bar{u} = \{\bar{\psi}\}^{\mathrm{T}}\{s\} = \{s\}^{\mathrm{T}}\{\bar{\psi}\} \tag{11.5.11}$$

The other displacement terms appearing in the kinetic energy may be written as

$$\frac{\partial \bar{u}}{\partial t} = \{\bar{\psi}\}^{\mathrm{T}}\{\dot{s}\} = \{\dot{s}\}^{\mathrm{T}}\{\bar{\psi}\}$$

$$\bar{\Omega} \times \bar{u} = (\bar{\Omega} \times \{\bar{\psi}\})^{\mathrm{T}}\{s\} = \{s\}^{\mathrm{T}}(\bar{\Omega} \times \{\bar{\psi}\}) \tag{11.5.12}$$

Note that the cross product is commutative in terms of matrix operations, so that

$$\bar{\Omega} \times \{\bar{\psi}\} = \begin{Bmatrix} \{\psi^{(1)}\}\bar{\Omega} \times \bar{i} \\ \{\psi^{(2)}\}\bar{\Omega} \times \bar{j} \\ \{\psi^{(3)}\}\bar{\Omega} \times \bar{k} \end{Bmatrix} \tag{11.5.13}$$

We next substitute eqs. (11.5.11) and (11.5.12) into eq. (11.5.7). The result is the matrix equivalent of eq. (11.1.5),

$$T = \tfrac{1}{2}\{\dot{s}\}^{\mathrm{T}}[M]\{\dot{s}\} + \{\dot{s}\}^{\mathrm{T}}[B]\{s\} + \tfrac{1}{2}\{s\}^{\mathrm{T}}[E]\{s\} + \{\dot{s}\}^{\mathrm{T}}\{N\} + \{s\}^{\mathrm{T}}\{J\} + T_{\mathrm{rb}} \tag{11.5.14}$$

In this expression, T_{rb} denotes the kinetic energy of rigid body motion, that is, the kinetic energy the bar would have if it did not deform. The coefficient matrices are

$$[M] = \int_0^L \rho A\{\bar{\psi}\} \cdot \{\bar{\psi}\}^{\mathrm{T}} dx$$

$$[B] = \int_0^L \rho A\{\bar{\psi}\} \cdot (\bar{\Omega} \times \{\bar{\psi}\})^{\mathrm{T}} dx$$

$$[E] = \int_0^L \rho A(\bar{\Omega} \times \{\bar{\psi}\}) \cdot (\bar{\Omega} \times \{\bar{\psi}\})^{\mathrm{T}} dx \tag{11.5.15}$$

$$\{N\} = \int_0^L \rho A(\bar{v}_0 + \bar{\Omega} \times x\bar{i}) \cdot \{\bar{\psi}\} \, dx$$

$$\{J\} = \int_0^L \rho A(\bar{v}_0 + \bar{\Omega} \times x\bar{i}) \cdot (\bar{\Omega} \times \{\bar{\psi}\}) \, dx$$

Each integrand in the preceding contains the product of column and row arrays, so its elements are the products of respective elements of each array. For example, orthogonality of the unit vectors leads to

$$\{\bar{\psi}\} \cdot \{\bar{\psi}\}^{\mathrm{T}} = \begin{Bmatrix} \{\psi^{(1)}\}\bar{i} \\ \{\psi^{(2)}\}\bar{j} \\ \{\psi^{(3)}\}\bar{k} \end{Bmatrix} \cdot \begin{bmatrix} \{\psi^{(1)}\}^{\mathrm{T}}\bar{i} & \{\psi^{(2)}\}^{\mathrm{T}}\bar{j} & \{\psi^{(3)}\}^{\mathrm{T}}\bar{k} \end{bmatrix}$$

$$= \begin{bmatrix} \{\psi^{(1)}\} \cdot \{\psi^{(1)}\}^{\mathrm{T}} & [0] & [0] \\ [0] & \{\psi^{(2)}\} \cdot \{\psi^{(2)}\}^{\mathrm{T}} & [0] \\ [0] & [0] & \{\psi^{(3)}\} \cdot \{\psi^{(3)}\}^{\mathrm{T}} \end{bmatrix}$$

The square submatrices lying on the diagonal are

$$\{\psi^{(n)}\} \cdot \{\psi^{(n)}\}^{\mathrm{T}} = \begin{bmatrix} (\psi_1^{(n)})^2 & \psi_1^{(n)}\psi_2^{(n)} & \cdots \\ \psi_1^{(n)}\psi_2^{(n)} & (\psi_2^{(n)})^2 & \cdots \\ \vdots & \vdots & \end{bmatrix}, \quad n = 1, 2, \text{ or } 3$$

It is clear from this development that $[M]$ is the same as the result that would be obtained if the bar was not rotating.

The description of the potential energy is substantially simpler. We split up V into the elastic strain energy and gravitational potential energy. A mass element of the bar is $\rho A \, dx$. From eqs. (11.5.1) and (11.5.3), the instantaneous position of the element is

$$\bar{r}_P = \bar{R}_O + x\bar{i} + \bar{u} \tag{11.5.16}$$

where $\bar{r}_O$ is the position of the origin of xyz relative to a point at the datum. The vertical component of this position, which is obtained from a dot product with the vertical unit vector, is Z for the mass element. Integrating the contribution of each element leads to

$$V_{\mathrm{grav}} = \int_0^L \rho g A (\bar{u} \cdot \bar{K}) \, dx + V_{\mathrm{rb}} \tag{11.5.17}$$

where V_{rb} denotes the gravitational potential energy that would be obtained if the bar did not deform. Because this term is independent of $\bar{u}$, it will play no role in the equations of motion governing $\bar{u}$. When we use eq. (11.5.11) to represent $\bar{u}$, we find that

$$V_{\mathrm{grav}} = \{s\}^{\mathrm{T}}\{F_{\mathrm{grav}}\} + V_{\mathrm{rb}} \tag{11.5.18}$$

where

$$\{F_{\mathrm{grav}}\} = \int_0^L \rho g A \{\bar{\psi}\} \cdot \bar{K} \, dx \equiv \begin{Bmatrix} (\bar{i} \cdot \bar{K}) \int_0^L \rho g A \{\psi^{(1)}\} \, dx \\ (\bar{j} \cdot \bar{K}) \int_0^L \rho g A \{\psi^{(2)}\} \, dx \\ (\bar{k} \cdot \bar{K}) \int_0^L \rho g A \{\psi^{(3)}\} \, dx \end{Bmatrix} \tag{11.5.19}$$

It should be noted some or all of the $\{F_{\text{grav}}\}$ elements might be functions of time because the vertical components of $\bar{i}$, $\bar{j}$, and $\bar{k}$ depend on the instantaneous orientation of xyz.

The elastic strain energy obviously depends on $\bar{u}$, but it is independent of the motion of xyz. Therefore, the elastic strain energy is the same as it would be if the reference frame were fixed. Consequently,

The contribution of strain energy to the stiffness coefficients is the same [K] matrix as what is obtained when the Ritz series is used to analyze the stationary system.

We thereby find that the potential energy of the bar is

$$V = \tfrac{1}{2}\{s\}^{\text{T}}[K]\{s\} + \{s\}^{\text{T}}\{F_{\text{grav}}\} + V_{\text{rb}} \tag{11.5.20}$$

Not much can be stated about the nature of the generalized forces $\{Q\}$ because the virtual work δW depends on the specific features of the external forces. One should bear in mind, however, that the virtual displacement of a point on the bar is produced by incrementing the generalized coordinates with time held constant. Thus, the only quantity in eq. (11.5.16) to be given a virtual increment is $\bar{u}$, that is, $\delta \bar{r}_p = \delta \bar{u}$. Because the basis functions are set, we obtain $\delta \bar{u}$ by incrementing the generalized coordinates. In terms of the matrix representation in eq. (11.5.11), these considerations lead to

$$\delta \bar{u} = \{\delta s\}^{\text{T}}\{\bar{\psi}\} \tag{11.5.21}$$

The generalized forces are obtained by using the preceding to describe the virtual displacement of each point where a nonconservative force is applied. We then sum $\bar{F} \cdot \delta \bar{u}$ to compute the virtual δW. Matching the result to $\delta W = \{\delta s\}^{\text{T}}\{Q\}$ leads to identification of $\{Q\}$.

The development has focused on the energy terms associated with a beam. The presence of attachments would lead to additional terms. The kinematical formulas employed for displacement and velocity are equally useful for the description of the attachments.

In essence, we have shown in this section that using Ritz series to represent deformational displacement of a bar relative to a moving reference frame leads to energy terms that precisely match those of discrete systems. The generalized coordinates $\{q\}$ for a discrete system become the series coefficients, which are assembled into $\{s\}$. Analysis of these equations for cases of free, transient, and steady-state response follow the procedures already established for gyroscopic modal analysis. The only alteration required to employ eq. (11.1.17) is to replace $\{Q\}$ appearing there with $\{Q\} - \{F_{\text{grav}}\}$. It follows that any of the phenomena encountered for discrete systems, including dynamic instability, can also arise in continuous systems.

EXAMPLE 11.6

Transverse vibration of axially moving cables is important for a variety of applications, such as conveyor belts, cable-pulley systems, and computer tape drives. Consider a cable that is taut between two rollers, such that the cable would translate horizontally at constant speed v if there were no transverse displacement. The cable has mass per unit length m and its tension F is large compared to the tension required to suspend it statically between the rollers. Determine the natural frequencies and mode shapes. Identify the maximum v for safe operation.

Solution This example will illustrate the manner in which the Ritz series formulation may be extended to treat time-dependent systems. By setting $v = 0$, we will see that the vibrational properties of stationary cables are directly analogous to those for axial displacement of a bar. A corollary of that observation will be evidence that overall motion of a system can have a drastic influence on its modal properties. Wickert and Mote (1990) analyzed the response of this system by solving the field equations. The present procedures represent a simpler alternative.

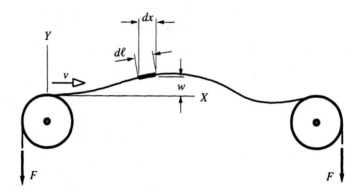

We begin by considering the kinetic energy of the cable segment highlighted in the sketch, whose arc length in the displaced position is $d\ell$. The transverse displacement is $w(x, t)$, where x is the horizontal position. Because the cable is translating at speed v in the horizontal direction, at time $t + \Delta t$ this segment will be at $x + v\Delta t$, so its vertical displacement at the later instant will be $w(x + v\Delta t, t + \Delta t)$. Thus, the velocity of this segment is

$$\bar{v}_s = v\bar{i} + \lim_{\Delta t \to 0}\left[\frac{w(x + v\Delta t, t + \Delta t) - w(x, t)}{\Delta t}\right]\bar{j} = v\bar{i} + \left[\frac{\partial w}{\partial t} + v\frac{\partial w}{\partial x}\right]\bar{j}$$

By the Pythagorean theorem, the arc length is

$$d\ell = \left[1 + \left(\frac{\partial w}{\partial x}\right)^2\right]^{1/2} dx$$

For the purpose of evaluating the mass of the segment, we invoke the linearization argument that $\partial w/\partial x \ll 1$, so the segment's mass is $m\, dx$. It follows that the kinetic energy of the cable spanning the rollers is

$$T = \frac{1}{2}\int_0^L [\bar{v}_s \cdot \bar{v}_s]m\, dx = \frac{1}{2}m\int_0^L\left[\left(\frac{\partial w}{\partial t}\right)^2 + 2v\frac{\partial w}{\partial t}\frac{\partial w}{\partial x} + v^2\left(\frac{\partial w}{\partial x}\right)^2\right]dx + \frac{1}{2}mLv^2$$

The middle term in the integrand represents an interaction between transverse velocity and rotation. In the field equations this effect would be manifested as a Coriolis acceleration effect. From that perspective, the last term represents a centripetal acceleration effect. The term $\frac{1}{2}mLv^2$ is the translational kinetic energy the cable would have if it did not vibrate. It may be ignored because it does not depend on w.

Potential energy is stored as a result of stretching the cable, as well as gravitational potential energy due to displacing the cable vertically. We take the datum of segment $d\ell$ to be the line between the rollers, so

$$V_{\text{grav}} = \int_0^L wmg\, dx$$

To construct the potential energy resulting from stretching the cable, we introduce the assumption that, because the cable tension F is very high, it changes little due to transverse displacement, and therefore is constant along the cable. We determine the strain by constructing the arc length of the cable in the displaced position. This is merely the integral of $d\ell$ described previously. The fact that $\partial w / \partial x$ is small enables us to use a binomial series to simplify the square root. Dropping terms that have the magnitude of $(\partial w / \partial x)^4$ leads to

$$\ell = \int_0^L d\ell = \int_0^L \left[1 + \left(\frac{\partial w}{\partial x} \right)^2 \right]^{1/2} dx \approx \int_0^L \left[1 + \frac{1}{2} \left(\frac{\partial w}{\partial x} \right)^2 \right] dx$$

The arc length in the undisplaced position is L, so the axial strain is

$$\varepsilon = \frac{\ell - L}{L} = \frac{1}{2L} \int_0^L \left(\frac{\partial w}{\partial w} \right)^2 dx$$

The potential energy per unit cable length is $F \varepsilon$, so the total potential energy is the sum of $F \varepsilon L$ and V_{grav},

$$V = \frac{1}{2} F \int_0^L \left(\frac{\partial w}{\partial x} \right)^2 dx + \int_0^L wmg \, dx$$

Examination of this equation shows that the last term is linear in w, which means that it is a static effect. If we redefine w to be measured from the static position the cable would have if v was zero or, alternatively, consider gravity to be negligible because $F \gg mgL$, we may ignore the second integral. In that case V has the mathematical form of the strain energy associated with axial displacement of a bar, with u replaced by w, and EA replaced by F. Furthermore, in situations where the conveyor belt is stationary, $v = 0$, we find that the analogy extends to the kinetic energy if ρA is replaced by m. Thus, any questions that arise regarding vibrations of stationary cables may be solved by the same techniques as those used for axial displacement of bars. This is much like the manner in which we exploited the analogous behavior of torsion and axial displacement.

Now let us implement the Ritz series. We assume that the rollers are mounted on rigid shafts, and therefore do not permit the cable to displace transversely. The corresponding geometric boundary conditions are $w = 0$ at $x = 0$ and $x = L$, so a suitable series expansion is

$$w = \sum_{j=1}^N \psi_j q_j, \qquad \psi_j = \sin\left(\frac{j\pi x}{L} \right)$$

Substitution of the series into the expression for T yields a result in the form of eq. (11.1.5) with

$$M_{kn} = m \int_0^L \psi_k \psi_n \, dx = \frac{1}{2} mL \delta_{kn}$$

$$B_{kn} = mv \int_0^L \psi_k \frac{\partial \psi_n}{\partial x} \, dx = \begin{cases} 2mv \dfrac{kn}{k^2 - n^2} & \text{if } k + n \text{ is odd} \\ 0 \text{ if } k + n \text{ is even} \end{cases}$$

$$E_{kn} = mv^2 \int_0^L \frac{\partial \psi_k}{\partial x} \frac{\partial \psi_n}{\partial x} \, dx = \frac{1}{2} \frac{m}{L} v^2 k^2 \pi^2 \delta_{kn}$$

According to eq. (11.1.13), the gyroscopic inertia matrix is

$$G_{kn} = B_{kn} - B_{nk} = \begin{cases} 4mv \dfrac{kn}{k^2 - n^2} & \text{if } k + n \text{ is odd} \\ 0 \text{ if } k + n \text{ is even} \end{cases}$$

Next, we substitute the Ritz series into the expression for V (with the gravitational term equated to zero), which leads to the stiffness coefficients,

$$K_{kn} = F \int_0^L \frac{\partial \psi_k}{\partial x} \frac{\partial \psi_n}{\partial x} \, dx = \frac{F}{2L} \pi^2 k^2 \delta_{kn}$$

The quantity $(F/m)^{1/2}$ is analogous to the bar speed for extensional waves, $(E/\rho)^{1/2}$. We use this speed to nondimensionalize v, and divide it by L to obtain a reference frequency that nondimensionalizes t, such that

$$\sigma = \left(\frac{m}{F}\right)^{1/2} v, \qquad \tau = \left(\frac{F}{m}\right)^{1/2} \frac{t}{L}$$

Introducing these quantities into the matrix coefficients and the equations of motion, eq. (11.1.17) leads to

$$[M']\frac{d^2}{d\tau^2}\{q\} + \sigma[G']\frac{d}{d\tau}\{q\} + [[K'] - \sigma^2[E']]\{q\} = \{0\}$$

where the matrices with a prime are the numerical values obtained by setting $F = m = L = v = 1$ in the respective matrices described above.

We use our software to form the state space coefficient matrices $[R]$ and $[S]$ described by eqs. (11.1.22), which depend on the value of σ,

$$[S(\sigma)] = \begin{bmatrix} [-[K'] + \sigma^2[E']] & [0] \\ [0] & [M'] \end{bmatrix}$$

$$[R(\sigma)] = -\begin{bmatrix} [0] & [[K'] - \sigma^2[E']] \\ [[K'] - \sigma^2[E']] & [G] \end{bmatrix}$$

The right eigenvalue problem to be solved is

$$[[R(\sigma)] - \lambda[S(\sigma)]]\{\psi\} = \{0\}$$

The easiest way to identify critical values of σ is to solve for the eigenvalues over a discretized range of σ, rather than to apply the Nyquist criteria. To carry out such an evaluation in our software, we define a discretized set of σ values, solve the eigenvalue problem for each σ, and store the sorted real part of the $2N$ eigenvalues as a column in a rectangular array. The imaginary part is treated in a similar fashion. A program fragment for performing these operations is

MATLAB: $sig = 0.02 : 0.04 : 1.5; for\ j = 1 : length\ (U);$
$eigval = eig(S, R(sig(j))); Re\_lam(:, j) = sort(real(eigval));$
$Im\_lam(:, j) = sort(real(eigval)); end$

It is implicit to this program fragment that $R(sig(j))$ is evaluated with a function M-file. There is no algorithmic difference in implementing the same procedure in Mathcad,

Mathcad: $j := 1; 75 \quad \sigma_j := j * 0.02 \quad \lambda^{\langle j \rangle} := genvals(R(\sigma_j), S)$
$Re\_\lambda^{\langle j \rangle} = sort(\text{Re}(\lambda^{\langle j \rangle})) \quad Im\_\lambda^{\langle j \rangle} = sort(\text{Im}(\lambda^{\langle j \rangle}))$

The result of plotting a row of either part against the corresponding value of σ is the first graph. Only the imaginary part is shown, because the real parts of all λ_n are zero if $\sigma < 1$.

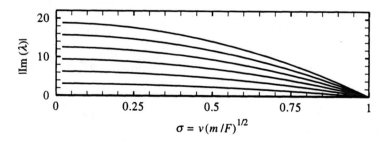

The eigenvalues vanish at $\sigma = 1$, that is, $v = (F/m)^{1/2}$, at which value a divergence instability occurs. Because $(F/m)^{1/2}$ is the speed at which flexural waves propagate along the cable, we interpret the divergence speed to be a requirement that the cable must translate *subsonically*.

An interesting aspect of the analysis comes when we examine the convergence properties of the solution as N is increased. For $\sigma \ll 1$, all except the last two λ_n are correctly determined, just like the behavior of natural frequencies identified in the discussion of the Rayleigh ratio for undamped systems. As σ approaches unity, fewer eigenvalues are correctly obtained if N is held fixed. For example, computations for a variety of N indicate that the λ_n values at $\sigma = 1$ are quite large for $n > N$. In addition, Wickert and Mote (1990) showed by solving the field equations that for any mode, $\lambda_n = \pm n\pi(1 - \sigma^2)i$ for $\sigma < 1$, which very closely matches the computed dependence of λ_n only for $n \leq N$.

The eigenfunction associated with a specified σ may be obtained from the associated eigenvector $\{\psi_n\}$. We normalize this quantity according to eq. (11.4.7), which requires that we also determine the left eigenvector $\{\tilde{\psi}_n\}$. The time dependence of a modal coordinate in free vibration is proportional to $\exp(i\lambda_n\tau)$, so the coefficients of the Ritz series vary as $\{\Phi_n\}\exp(i\lambda_n\tau)$ in a vibration at mode n, where $\{\Phi_n\}$ is the upper half of the normalized version of $\{\psi_n\}$. To obtain the normal mode function $w_n(x)$ we multiply each Ritz coefficient Φ_{jn} by the corresponding basis function $\psi_j(x)$, which leads to

$$w_n(x) = \begin{bmatrix} \psi_1(x) & \psi_2(x) & \ldots \end{bmatrix}\{\Phi_n\}\exp(i\lambda_n\tau)$$

A mode function may be graphed by replacing the row of basis function values in the preceding by an array whose rows are the function values at specific values of x. The result of such a computation when $\sigma = 0.9$ forms the next set of graphs. Each mode corresponds to the eigenvalue of the conjugate pair for which $\mathrm{Im}(\lambda_n) > 0$.

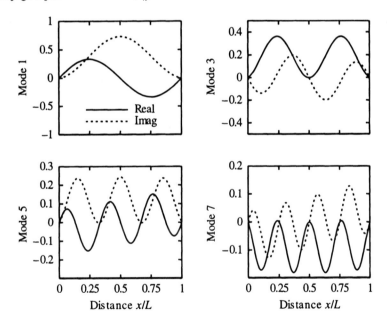

Several features of the damped mode functions are worth noting. The system is symmetric, so responses can be decomposed into symmetric and antisymmetric parts. The real and imaginary parts of the complex modes have opposite symmetries. The number of peaks and minima increases with mode number, like undamped mode functions, but the symmetric part here does not show any sign reversals. A node, at which $w_n(x) = 0$, occurs only if $\mathrm{Re}(w_n) = 0$ and $\mathrm{Im}(w_n) = 0$ at the same x. Inspection of the graphs shows that the nodes of a pair of complex conjugate modes for $v \neq 0$ match the corresponding stationary mode's zeroes. However, away from the nodes, the motion does not occur synchronously at different x.

EXAMPLE 11.7

A pipe transporting fluid flowing at speed U is clamped at its left end, and its right end is free. The cross-section of the pipe is circular, with inner diameter R_i and outer diameter R_o. The free end is terminated by a nozzle having cross-sectional area A_N. Fluid momentum change across the nozzle requires that the nozzle exert a compressive follower force tangent to the axis of the pipe at the free end. As shown in the sketch, the magnitude of this force is $\varepsilon \rho_f A_f U^2$, where ρ_f is the density of the pipe, $A_f = \pi R_i^2$, and $\varepsilon = A_f/A_N - 1$. At sufficiently high flow rates, the pipe will become dynamically unstable. Use the Ritz series method to formulate equations of motion whose solution would give the free vibrational properties of the pipe at a specified value of U. Then, for the case of water flowing through a steel pipe, with $R_0 = 75$ mm, $R_i = 50$ mm, $A_f/A_N = 4$, and $L = 2$ m, determine the highest speed U for which the pipe is stable. Damping of the pipe in the absence of fluid flow is estimated to be 2% of critical for each mode.

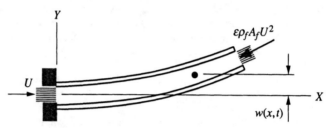

Solution   The effect of fluid flow is to generate gyroscopic terms in the equations of motion as a result of Coriolis acceleration. Thus, this example will give us experience with the analysis of bar vibration in the presence of gyroscopic, damping, and circulatory force effects. The analysis begins by deriving expressions for T, V, and δW appropriate to the system.

The fluid enters at two phases of the Ritz series formulation. The presence of fluid within the pipe affects the kinetic energy, while the follower force at the end is incorporated in the virtual work. Although this force compresses the bar, the static extensional rigidity is much larger than the flexural rigidity. We therefore shall ignore the effects of axial deformation. We assume that the cross-sectional velocity profile of the fluid is constant, so we focus on a fluid particle that is at a specific axial position x at instant t. To describe the position of this particle we define a fixed XYZ reference frame having its origin O at the clamped end. The instantaneous position of the fluid particle is $\bar{r}_{f/O}(t) = x\bar{I} + w(x, t)\bar{J}$. The flow speed U is the rate at which the axial distance of the fluid particle changes. Thus, an instant dt later will place the fluid particle at axial distance $x + U\,dt$. The corresponding position at this instant will be $\bar{r}_{P/O}(t + dt) = (x + U\,dt)\bar{I} + w(x + U\,dt, t + dt), \bar{J}$. From this, we deduce that the particle's velocity is

$$\bar{v}_f = \frac{\bar{r}_{f/O}(t + dt) - \bar{r}_{f/O}(t)}{dt} = U\bar{I} + \left[\frac{w(x + U\,dt, t + dt) - w(x, t)}{dt}\right]\bar{J}$$

$$= U\bar{I} + \left(\dot{w} + U\frac{\partial w}{\partial x}\right)\bar{J}$$

Note that this expression is the same as that for the velocity of a point on the cable in the previous example. A differential mass element for the fluid is $\rho_f A_f\,dx$, so the kinetic energy of this element is

$$dT_f = \frac{1}{2}(\rho_f A_f\,dx)\bar{v}_f \cdot \bar{v}_f = \frac{1}{2}(\rho_f A_f\,dx)\left[\dot{w}^2 + 2U\dot{w}\frac{\partial w}{\partial x} + U^2\left(\frac{\partial w}{\partial x}\right)^2 + U^2\right]$$

We combine this with the kinetic energy of a differential beam element, which leads to

$$T = \frac{1}{2}(\rho A + \rho_f A_f)\int_0^L \dot{w}^2\,dx + \rho_f A_f U\int_0^L \dot{w}\frac{\partial w}{\partial x}\,dx$$

$$+ \frac{1}{2}\rho_f A_f U^2\int_0^L \left(\frac{\partial w}{\partial x}\right)^2 dx + \frac{1}{2}\rho_f A_f U^2 L$$

The strain energy is the standard expression for flexure,

$$V = \frac{1}{2}EI \int_0^L \left(\frac{\partial w}{\partial x}\right)^2 dx$$

Proper evaluation of the virtual work requires careful consideration of the end displacement. We decompose the follower force into its X and Y components. In the deformed position, the angle of the tangent to the beam axis relative to the fixed X axis is $\beta = \tan^{-1}(\partial w/\partial x) \approx \partial w/\partial x \ll 1$. The tangent direction is $\bar{e}_t = \cos(\beta)\bar{I} + \sin(\beta)\bar{J}$, which leads to a description of the follower end thrust as

$$\bar{F} = -\varepsilon\rho_f A_f U^2 \bar{e}_t|_{x=L} = -\varepsilon\rho_f A_f U^2 [\cos(\beta)\bar{I} + \sin(\beta)\bar{J}]_{x=L}$$

The virtual work done by the X component of the thrust depends on the displacement of the end in that direction. We have made the assumption that axial deformation is negligible. Thus, the arc length measured along the deformed axis of the bar should be L. The sketch shows a segment of the bar whose length along the displaced axis of the bar is dx because we are considering the bar to be inextensible axially.

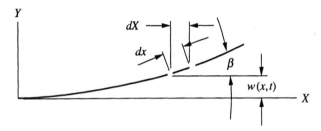

The projection of this distance onto the X axis is $dX = \cos(\beta)\,dx$. It would be an oversimplification to take $\cos(\beta) = 1$. Instead, we use the identity that

$$\cos(\beta) = \frac{1}{[1 + \tan(\beta)^2]^{1/2}} = \frac{1}{\left[1 + \left(\frac{\partial w}{\partial x}\right)^2\right]^{1/2}} \approx 1 - \frac{1}{2}\left(\frac{\partial w}{\partial x}\right)^2$$

From this, we may evaluate the projection of the bar's length onto the X axis in the deformed state according to

$$X_L = \int_0^L \cos(\beta)\,dx = \int_0^L \left[1 - \frac{1}{2}\left(\frac{\partial w}{\partial x}\right)^2\right] dx = L - \frac{1}{2}\int_0^L \left(\frac{\partial w}{\partial x}\right)^2 dx$$

The virtual displacement at the end is then found by evaluating

$$\delta\bar{r}(x = L) = \delta X_L \bar{I} + \delta w(x = L)\bar{J}$$

In conjunction with the component description of the end thrust, this leads to the virtual work being

$$\delta W = \bar{F} \cdot \delta\bar{r}(x = L) = -\varepsilon\rho_f A_f U^2 [\cos(\beta)\delta X_L + \sin(\beta)\delta w]_{x=L}$$

Because $X_L - L$ depends on the square of displacement, while δw is linear in displacement, it is consistent to take $\cos(\beta) = 1$ and $\sin(\beta) = \partial w/\partial x$ in the preceding. The virtual work then becomes

$$\delta W = -\varepsilon\rho_f A_f U^2 \left[\delta X_L + \frac{\partial w}{\partial x}\,\delta w\right]_{x=L}$$

We now proceed to formulate the Ritz series,

$$w = \sum_{j=1}^N \psi_j q_j$$

A suitable set of basis functions is the normal modes of the cantilevered pipe containing the fluid at rest. Thus, we take the mass per unit length to be $\rho A + \rho_f A_f$, where A is the area of the steel cross-section, $A = \pi(R_o^2 - R_i^2)$. According to Appendix C, the normal modes of a cantilevered beam are given by

$$\psi_j = C_j\left\{\left[\sin\left(\alpha_j\frac{x}{L}\right) - \sinh\left(\alpha_j\frac{x}{L}\right)\right] - R_j\left[\cos\left(\alpha_j\frac{x}{L}\right) - \cosh\left(\alpha_j\frac{x}{L}\right)\right]\right\}$$

where the modal parameters are given by

$$\cos(\alpha)\cosh(\alpha) = -1$$

$$R_j = \frac{\sin(\alpha_j) + \sinh(\alpha_j)}{\cos(\alpha_j) + \cosh(\alpha_j)}$$

(The Ritz series will be sufficiently short that the high-frequency asymptotic representation of a mode function will not be required.) Consistent with the definition of mass per unit length, the C_j coefficients are defined according to the normalization

$$(\rho A + \rho_f A_f)\int_0^L \psi_j^2\, dx = 1$$

The natural frequencies of the pipe with fluid in a quiescent state are

$$\omega_j = \left[\frac{EI}{(\rho A + \rho_f A_f)L^4}\right]^{1/2}\alpha_j^2$$

where $I = (\pi/4)(R_o^4 - R_i^4)$.

We substitute the Ritz series into the expression for T. In view of the orthogonality properties of the modes, we find that

$$T = \frac{1}{2}\sum_{j=1}^N \dot{q}_j^2 + \rho_f A_f U\sum_{j=1}^N\sum_{n=1}^N b_{jn}\dot{q}_j q_n + \frac{1}{2}\rho_f A_f U^2\sum_{j=1}^N\sum_{n=1}^N e_{jn}q_j q_n + \frac{1}{2}\rho_f A_f U^2 L$$

where

$$b_{jn} = \int_0^L \psi_j\frac{\partial\psi_n}{\partial x}\,dx, \qquad e_{jn} = \int_0^L \frac{\partial\psi_n}{\partial x}\frac{\partial\psi_n}{\partial x}\,dx = e_{nj}$$

A comparison of this description of T to the standard form in eq. (11.1.5) shows that

$$M_{jn} = \delta_{jn}, \qquad B_{jn} = \rho_f A_f U b_{jn}, \qquad E_{jn} = \rho_f A_f U^2 e_{jn}, \qquad N_k = J_k = 0$$

The strain energy is easy to obtain because of the second orthogonality property of the mode functions, specifically,

$$V = \frac{1}{2}\sum_{j=1}^N \omega_j^2 q_j^2$$

Thus, the stiffness coefficients are

$$K_{jn} = \omega_j^2\delta_{jn}$$

We determine the generalized forces by using the Ritz series to describe δW. Substitution of the series into the expression for X_L yields

$$X_L = L - \frac{1}{2}\sum_{j=1}^N\sum_{n=1}^N e_{jn}q_j q_n$$

The chain rule for differentiation, along with eq. (A.2.7), then leads to

$$\delta X_L = \sum_{k=1}^N\frac{\partial X_L}{\partial q_k}\delta q_k = -\sum_{k=1}^N\sum_{n=1}^N e_{kn}q_n\delta q_k$$

where the last step follows from symmetry of the e_{kn} coefficients. From the Ritz series itself, we find that

$$\delta w(x = L) = \sum_{k=1}^{N} \psi_k \big|_{x=L} \delta q_k$$

The virtual work corresponding to these expressions is

$$\delta W = \varepsilon \rho_f A_f U^2 \sum_{k=1}^{N} \sum_{n=1}^{N} e_{kn} q_n \delta q_k - \varepsilon \rho_f A_f U^2 \sum_{k=1}^{N} \sum_{n=1}^{N} \left(\frac{\partial \psi_n}{\partial x} \psi_k \right) \bigg|_{x=L} q_n \delta q_k$$

Matching this expression to the standard form for virtual work leads to

$$Q_j = \varepsilon \rho_f A_f U^2 \sum_{n=1}^{N} (e_{jn} - f_{jn}) q_n$$

$$f_{jn} = \left(\frac{\partial \psi_n}{\partial x} \psi_k \right) \bigg|_{x=L}$$

With this, all terms required to form Lagrange's equations have been obtained. The gyroscopic inertia coefficients are

$$G_{jn} = \rho_f A_f U g_{jn}, \qquad g_{jn} = (b_{jn} - b_{nj})$$

We substitute the matrices identified here into eq. (11.1.17), thereby obtaining

$$\{\ddot{q}\} + \rho_f A_f U \, [g]\{\dot{q}\} + [[\omega^2] - \rho_f A_f U^2 [e]]\{q\} = \varepsilon \rho_f A_f U^2 \, [[e] - [f]]\{q\}$$

where $[I]$ is an $N \times N$ identity matrix and $[\omega^2]$ is a diagonal array of the natural frequencies in the absence of flow. It was specified that each flexural mode has a critical damping ratio $\zeta_j = 0.005$. In view of the diagonal nature of $[M]$ and $[K]$, we correspondingly take the damping matrix to be diagonal, according to

$$C_{jn} = 2 \zeta_j \omega_j \delta_{jn}$$

Bringing all terms that depend on $\{q\}$ to the left side yields the final form of the equations of motion,

$$\{\ddot{q}\} + [2\zeta\omega]\{\dot{q}\} + \rho_f A_f U [g]\{\dot{q}\}$$
$$+ [[\omega^2] - (1 + \varepsilon)\rho_f A_f U^2 [e] + \varepsilon \rho_f A_f U^2 [f]]\{q\} = 0$$

The term $(1 + \varepsilon)\rho_f A_f U^2 [e]$ corresponds to $[E]$ in eq. (11.1.17). The ε portion represents the role of the $\overline{I}$ component of the end thrust, which acts like a buckling force. The portion that is independent of ε arose in the kinetic energy. It can be shown to represent the distributed force required to make the fluid follow the curved axis of the pipe in the deformed state. The last term, $\rho_f A_f U^2 [f]\{q\}$, represents the circulatory follower force effect, as is evident from the fact that $[f]$ is asymmetric.

The state-space equations of motion have the standard form

$$[S] \frac{d}{dt}\{x\} - [R(U)]\{x\} = \{0\}$$

where $\{x\} = [\{q\}^T \{\dot{q}\}^T]^T$. The matrices $[R]$ and $[S]$ in eqs. (11.1.22) are

$$[S] = \begin{bmatrix} -[[\omega^2] - \rho_f A_f U^2 (1 + \varepsilon)[e]] & [0] \\ [0] & [I] \end{bmatrix}$$

$$[R(U)] = - \begin{bmatrix} [0] & [[\omega^2] - \rho_f A_f U^2 (1 + \varepsilon)[e]] \\ [[\omega^2] - \rho_f A_f U^2 [(1 + \varepsilon)[e] - \varepsilon[f]]] & [[2\zeta\omega] + \rho_f A_f U [g]] \end{bmatrix}$$

We are interested in determining the maximum value of U for which the system is stable. This depends on the eigenvalues resulting from substituting $\{x\} = \chi \exp(\lambda t)$ into the state-space equations, which leads to

$$[[R(U)] - \lambda[S]]\{\chi\} = \{0\}$$

We will compute and graph the eigenvalues as a function of U. We begin with a guess that instability will occur for $U < 50$ m/s. An unstable response corresponds to a situation where any eigenvalue λ_j has a positive real part.

The first step is to evaluate the coefficient matrices corresponding to the given pipe parameters. (The integrations may be performed analytically or numerically using our math software.) We shall use $N = 6$, which leads to 12 eigenvalues. This number is more than sufficient, because we will see that the stability is governed by the smallest eigenvalues. The computational procedures discussed in the previous example also are applicable here, with the incremented parameter for the computations now set as the speed U.

For the lowest value $U = 0.05$ m/s, the eigenvalues are found to be

$$\lambda_j = -0.97 \pm 194i, \quad -6.09 \pm 1218i, \quad -17.05 \pm 3411i$$

$$-33.42 \pm 6683i, \quad -55.24 \pm 11048i, \quad -82.52 \pm 16504i$$

A scan of the data for different U values shows that the real parts of the larger eigenvalues remain negative throughout the range. Consequently, we only graph the real and imaginary data closest to zero. The subscripts used to describe the graphed curves sequence the eigenvalues by magnitude.

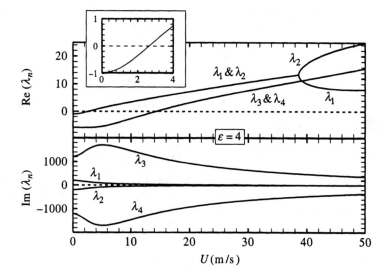

These graphs show that up to $U = 36.6$ m/s, λ_1 and λ_2 are complex conjugates, whereas λ_3 and λ_4 are complex conjugates throughout the plotted range. For $U < 2.64$ m/s, the real parts of all eigenvalues are negative, so the pipe is asymptotically stable up to that flow rate. Because Re (λ_1) and Re (λ_2) become positive for $U > 2.64$ m/s with nonzero imaginary parts, we conclude that a flutter instability occurs at this speed. The imaginary parts of the eigenvalues show that the frequency of the first two complex modes decreases steadily with increasing U, which we interpret as a loss of stiffness due to the flow. Interestingly, the third and fourth modes seem to become more stiff until $U = 5$ m/s. Other flow rates of interest are $U \approx 14$ m/s, at which a flutter instability occurs in the third and fourth complex modes, and $U = 36.6$ m/s, where λ_1 and λ_2 become real and positive, which means that there is a divergence instability beyond that value of U.

It is interesting to consider the effect of the end thrust relative to the effects arising within the span of the pipe. Setting $\varepsilon = 0$, corresponding to $A_N = A_f$, removes the effects of the follower force. The eigenvalue dependence on U in this case appears in the second set of graphs.

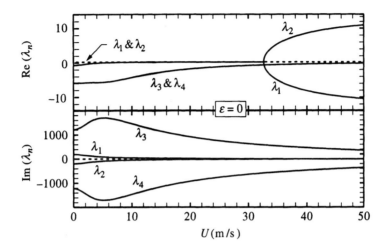

The topology of the graphs is recognizable from the previous result, but the real parts of all eigenvalues remain negative until $U = 32.8$ m/s, at which flow rate λ_1 and λ_2 become real. Thus, the modified pipe is asymptotically stable up to $U = 32.8$, after which it diverges. From this observation, we conclude that the compressive and circulatory effects of the end force, which are represented by the $[f]$ term and half the $[e]$ term in the equations of motion, have a stronger destabilizing influence than the wall interaction between the flowing fluid and the pipe.

Pipeline stability is especially important for the nuclear power industry, where large volumes of coolant flow at high rates. The various issues associated with this subject are discussed in detail in the recent book by Païdoussis (1998).

REFERENCES AND SELECTED READINGS

BLEVINS, R. D. 1986. *Flow-Induced Vibration,* Robert E. Krieger Publishing, New York.

DIMAROGONAS, A. 1996. *Vibration for Engineers,* 2nd ed. Prentice Hall, Englewood Cliffs, NJ.

GINSBERG, J. H. 1973. "Dynamic Stability of a Pipe Conveying a Pulsatile Flow." *International Journal of Engineering Sciences,* 11, 1012–1024.

GINSBERG, J. H. 1995. *Advanced Engineering Dynamics.* Cambridge University Press, Cambridge, England.

HUSEYN, K. 1978. *Vibration and Stability of Multiple Parameter Systems.* Sijthoff & Noordhoff, Germantown, MD.

INMAN, D. J. 1989. *Vibration.* Prentice Hall, Englewood Cliffs, NJ.

LEIPHOLZ, H. H. E. 1980. *Stability of Elastic Systems.* Sijthoff & Noordhoff, Germantown, MD.

MEIROVITCH, L., 1975. "A Modal Analysis for the Response of Linear Gyroscopic Systems." *Journal of Applied Mechanics,* 42, 446–450.

MEIROVITCH, L. 1997. *Principles and Techniques of Vibrations.* Prentice Hall. Englewood Cliffs, NJ.

NEMAT-NASSER, S., PRASAD, S. N., & HERMANN, G. 1967. "Destabilizing Effect of Velocity-Dependent Forces in Nonconservative Continuous Systems." *AIAA Journal,* 4, 1276–1280.

PAÏDOUSSIS, M. P. 1998. *Fluid-Structure Interactions: Slender Structures and Axial Flow, Volume 1.* Academic Press, San Diego.

WICKERT, J. A., & MOTE, C. D., JR. 1990. "Classical Vibration Analysis of Axially Moving Continua." *Journal of Applied Mechanics,* 57, 738–745.

EXERCISES

11.1 The turntable rotates in the horizontal plane at constant rate ω. When $\omega = 0$, the distance from the rotation center to the slider is L_0, and R_0 is the constant radial distance that is a possible state of steady motion. Derive the linearized equation of motion governing vibration of the block relative to $R = R_0$. Is there a maximum rotation rate beyond which the dynamic equilibrium state is unstable?

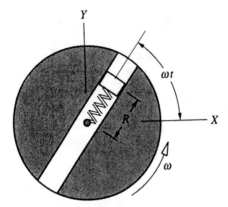

EXERCISE 11.1

11.2 The collar supporting the bar is given an oscillatory displacement $y(t) = \varepsilon \sin(\Omega t)$, where $\varepsilon/L \ll 1$. The system lies in the vertical plane. Derive the equation of motion for the system, assuming that the bar undergoes small rotations θ from the vertical orientation.

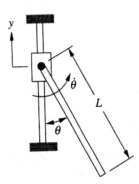

EXERCISE 11.2

11.3 In an earthquake, the ground displacement for an inverted double pendulum is $w(t)$ upward. Derive linearized equations of motion governing small rotations of each bar from the vertical position, at which the torsional springs are unstrained.

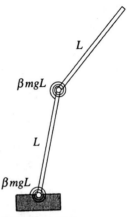

EXERCISE 11.3

11.4 Angle θ measures the position of bar AB relative to the T-bar, to which it is pinned. The T-bar rotates about the vertical axis at constant angular speed ω. The kinetic energy of the system is

$$T = \tfrac{1}{2}mL^2\Big[\tfrac{1}{3}(\dot{\theta}d\,)^2\Big(1 + \cos(\theta)$$

$$+ \tfrac{1}{3}\cos(\theta)^2\Big)\omega^2\Big] + \tfrac{1}{2}I_T\omega^2$$

where m is the mass of bar AB and I_T is the moment of inertia of the T-bar about its rotation axis.

(a) Derive an algebraic equation whose solution will give the constant angle θ that is a possible state of steady motion. Find the two roots θ_1 and θ_2 for the case where $\omega = 2\sqrt{g/L}$.

(b) Derive a linearized equation for small amplitude vibrations about each possible steady motion when $\omega = 2\sqrt{g/L}$. Thus, consider $\theta = \theta_1 + \chi$ and $\theta = \theta_2 + \chi$. Based on the coefficients of the differential equation for χ in each case, assess the stability of each steady motion.

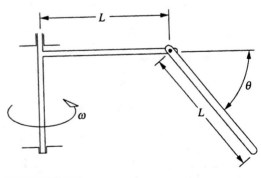

EXERCISE 11.4

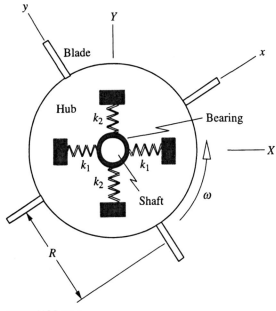

EXERCISE 11.6

11.5 A block is suspended from a rotating T-bar by a spring that has radial stiffness k_x, vertical stiffness k_z, and circumferential stiffness k_y. The sketch shows the position of the block when the rotation rate $\omega = 0$. Let x, y, and z be small displacements of the center of the block from this position, measured relative to the T-bar. Derive equations of motion for these variables.

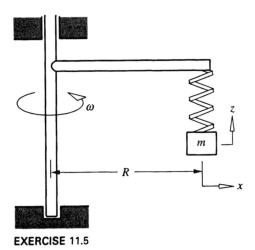

EXERCISE 11.5

11.6 The sketch depicts a model to be used to study the effect of elasticity at the bearings of a turbine. The bearing is supported by stationary springs, and the rigid hub-disk assembly rotates at constant speed ω relative to the bearing. The mass of the assembly is m and its centroidal moment of

inertia is I. Derive the equations of motion corresponding to using as generalized coordinates the hub's displacement components x_O, y_O measured parallel to the axes of the body-fixed xyz reference frame.

11.7 A MEMS sensor intended to measure rotation is modeled as a rigid bar of length L that is mounted on a turntable by four equal springs k. The sketch shows the position of the bar when the springs are undeformed and the turntable is not rotating. Use the position coordinates of the center C and the rotation θ of the bar, all measured relative to the turntable, as generalized coordinates. Derive equations of motion for this system.

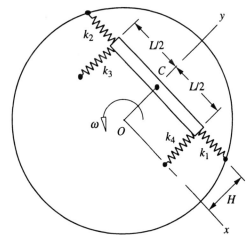

EXERCISE 11.7

11.8 The diagram depicts a crude model for a turbine engine's fan blades. The outer ring rotates at constant angular speed ω. A blade is represented as a thin rigid bar having mass m that is attached to the ring by a spring whose torsional stiffness is κ. The spring is undeformed when the attached blade is aligned radially. A blade's weight has negligible effect. Derive the equation of motion for small rotation of a blade relative to its radial orientation. Assume that the torsional spring only permits the blade to rotate about the attachment point O.

11.9 It is desired to refine the model of the fan blade in Exercise 11.8 in order to account for radial and transverse displacement at the spring attachment point O, as well as torsional rotation. Let xyz be a rotating reference frame having origin at the center with its x axis collinear with the blade in the undeformed position. The displacement of point O relative to xyz is $x_O\vec{i} + y_O\vec{j}$, and the torsional rotation of the blade relative to the hub is $\theta\vec{k}$. The corresponding forces opposing these displacements are $k_1 x_O$ and $k_2 y_O$, and the torsional moment opposing θ is $k_3\theta$. Derive the equations of motion governing x_O, y_O, and θ when $\omega \neq 0$. Identify the corresponding $[M]$, $[G]$, $[E]$, and $[K]$ matrices.

11.10 The sketch shows a prototype of a motorcycle wheel. The wheel is attached to the hub by eight spokes, which are represented as springs that form an angle β with the radial line when the wheel is constructed. The spokes are tensioned to a high value, so the stiff-spring approximation may be employed. Each spoke is supposed to have stiffness k_1, but one of the spokes is defective, with stiffness k_2. The tire-rim combination has mass m, and the radius of gyration about the center of the hub is $1.1R$. The center hub and the xyz coordinate system rotate at constant rate ω about the z axis. When the system is not rotating, the static equilibrium position of the outer rim is such that its center coincides with the origin of xyz and the spokes are symmetrically arranged, as shown. The position of the tire relative to xyz is defined by the

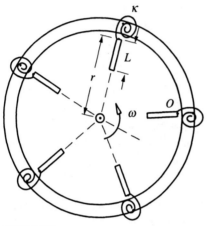

EXERCISE 11.8

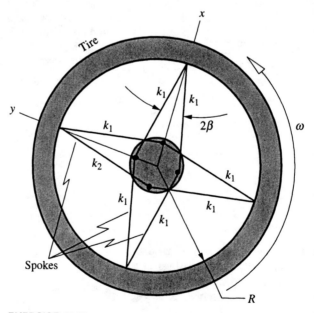

EXERCISE 11.10

position coordinates x_O and y_O of the center relative to this reference frame, and its rotation angle ϕ about the z axis. (The total angle of rotation of the rim about the z axis, as seen by a fixed observer, is $\phi + \omega t$.) Find linearized equations of motion for the rim corresponding to using x_O, y_O, and ϕ as generalized coordinates. Gravity effects are negligible.

11.11 Consider the system in Exercise 11.5 in the case where the spring constants are $k_x = k_z = k_y = k$ and the rotation rate is $\omega = 0.4(k/m)^{1/2}$. Determine the natural frequencies and the corresponding right and left eigenvectors. Draw Lissajous diagrams depicting the path followed by the block in each modal vibration, as it would be seen by an observer who rotates with the T-bar.

11.12 Solve Exercise 11.11 for the case where $k_x = k$, $k_z = 2k$, and $k_y = 0.5k$.

11.13 The parameters for the MEMS sensor in Exercise 11.7 are $k_1 = k_2 = 5$ mN/m, $k_3 = 10$ mN/m, $k_4 = 15$ mN/m, $m = 10^{-8}$ kg, $L = 5$ mm, and $H = 2$ mm. To account for dissipation within each spring, consider there to be proportional damping, with $[C] = 10^{-3}[K]$. Determine the eigenvalues, and the right and left eigenvectors of this system when $\omega = 20$ rad/s.

11.14 In Exercise 11.9 the stiffnesses are $k_1 = k_2 = k$ and $k_3 = 1.5kL^2$. The radial distance to a spring is $r = 5L$. Consider damping to be proportional, with $[C] = 0.01(m/k)^{1/2}[K]$. Determine the eigenvalues, and the right and left eigenvectors of this system when the rotation rate is $\omega = 0.8(k/m)^{1/2}$.

11.15 In Exercise 11.10 stiffness $k_2 = 0$, corresponding to the spoke being broken. The angle for a spoke is $\beta = 10°$. Determine the natural frequencies and right and left eigenvectors for this system. The rotation rate is $\omega = 0.6(k_1/m)^{1/2}$.

11.16 Prove that eqs. (11.2.15) and (11.2.16) govern the left eigenvector. Also prove that the nonlinear characteristic equation obtained from these equations is the same as eq. (11.2.14).

11.17 Consider the system in Exercise 11.5 in the case where the spring constants are $k_x = k$, $k_z = 2k$, and $k_y = 0.5k$. Is there a maximum rotation rate ω above which the suspended position of the block is not dynamically stable?

11.18 Consider the system in Example 11.1 when $k_1 = 0.5k$ and $k_2 = 2k$.

(a) Assess the stability of the system for $0 < \omega < 10(k/m)^{1/2}$.

(b) Repeat the analysis in part **(a)** for the case where the springs are identical, $k_1 = k_2 = 0.5k$.

11.19 Consider the system in Example 11.1 when $k_1 = 0.5k$ and $k_2 = 2k$. Suppose each spring is accompanied by a parallel dashpot giving proportional damping, with $c_1/k_1 = c_2/k_2 = 0.01(m/k)^{1/2}$. (This is said to be an *internal damping force*, whose effects are addressed in detail in Chapter 12.) Assess the stability of the system for $0 < \omega < 10(k/m)^{1/2}$.

11.20 The parameters for the MEMS sensor in Exercise 11.7 are $k_1 = k_2 = 5$ mN/m, $k_3 = 10$ mN/m, $k_4 = 15$ mN/m, $m = 10^{-8}$ kg, $L = 5$ mm, and $H = 2$ mm. To account for dissipation within each spring, consider there to be proportional damping, with $[C] = 10^{-3}[K]$. Determine the lowest rotation rate ω at which the dynamic equilibrium state is unstable.

11.21 In Exercise 11.9 the stiffnesses are $k_1 = k_2 = k$ and $k_3 = 1.5kL^2$. The radial distance to a spring is $r = 5L$. Consider damping to be proportional, with $[C] = 0.01(m/k)^{1/2}[K]$. Determine the maximum rotation rate ω for which stability of the radially aligned position of the blade is ensured. What is the type of instability when ω exceeds this value?

11.22 Repeat the analysis in Example 11.5 for the case where the beam has a square cross-section, so that $I_{yy} = I_{zz}$.

11.23 A shaker is mounted at a height of $2L/3$ above the base of the beam in Example 11.5. The shaker force is $\varepsilon\hat{\Omega}^2 \sin(\hat{\Omega}\tau)$ and it acts in the z direction relative to the turntable. The rotation rate is $\hat{\omega} = 2.5$.

(a) Determine how the amplitude of each displacement component depends on the excitation frequency $\hat{\Omega}$. What are the resonance frequencies?
(b) Determine the steady-state response of each mass if $\hat{\Omega}$ is 2% larger than the lowest resonant frequency. Draw Lissajous diagrams showing the path of each lumped mass as seen by an observer following the moving xyz coordinate system.

11.24 A force F having constant magnitude and acting in the fixed horizontal direction Y is applied to the beam in Example 11.5. The force acts at a height of $3L/4$ above the base of the beam. In terms of body-fixed components, this force is $F[\cos(\hat{\omega}\tau)\bar{j} - \sin(\hat{\omega}\tau)\bar{k}]$, where $\hat{\omega}$ is the rotation rate. Determine and graph the frequency dependence of the steady-state y and z displacement amplitudes of each lumped mass. Explain why the number of $\hat{\omega}$ values at which resonance occurs is fewer than the number of pairs of complex conjugate eigenvalues.

11.25 Consider the system in Example 11.5 when $\hat{\omega} = 2.0$. At $t = 0$ the beam is in a deformed position, such that the initial displacements are $y_1 = 0.001L$, $y_2 = 0.004L$, $z_1 = 0.002L$, and $z_2 = -0.002L$. The beam is stationary relative to the turntable at this initial instant. Determine the resulting free vibration response. Draw Lissajous figures showing the path of each lumped mass as seen by an observer following the moving *xyz* coordinate system.

11.26 Consider the system in Exercise 11.5 when $\omega = 0.9(k/m)^{1/2}$. The spring's stiffness parameters are $k_x = k$, $k_z = 2k$, and $k_y = 0.5k$. Initially, the block is stationary relative to the T-bar with the spring undeformed, and therefore oriented vertically. At $t = 0$, the block is released. Determine the ensuing free vibration. Draw a Lissajous diagram showing the path followed by the block in the horizontal plane, as seen by an observer who rotates with the T-bar.

11.27 Spoke k_2 of the motorcycle wheel in Exercise 11.10 is broken, so $k_2 = 0$. The spoke angle is $\beta = 10°$. At $t = 0$, an impulsive force $F = \delta(t)$ acting radially inward is applied to the wheel at the location where k_2 should be attached. For $t < 0$, the center of the wheel coincided with the hub's center and the rotation of the wheel relative to the hub was zero. Determine the response to the impulsive force.

11.28 Consider a pipe having the same parameters as the one in Example 11.7, except that it is simply supported. The nozzle is situated at the end where there is a roller. Because this support prevents transverse displacement, the force at the end may be considered to be $\varepsilon \rho_f A_f U^2$ acting horizontally. In other words, it does not act like a follower force. Determine the maximum value of U for which this pipe is stable.

11.29 A bar is clamped to the inside of a drum that rotates at constant rate ω. If the rotation rate is too high the beam will buckle due to centripetal loading. Develop a Ritz series analysis of this system using three basis functions to represent the transverse displacement w. (The bar

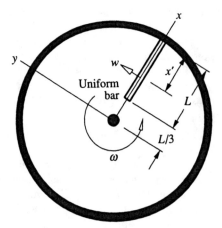

EXERCISE 11.29

may be assumed to be rigid in extension.) Determine the eigenvalues of this system when the rotation rate $\omega = (EI/\rho AL^4)^{1/2}$. Then consider ω to be a parameter and graph the eigenvalues as a function of ω. From that graph, ascertain the maximum rotation rate for which the bar will not buckle.

11.30 Rather than using a lumped mass model to represent the bar in Example 11.5, perform a Ritz series analysis of the system. Use $\psi_j = (x/L)^{j+1}$ as basis functions representing transverse displacement in the y and z directions. Compare the eigenvalues and modal displacement patterns obtained from a two-term Ritz series for each displacement component to the results obtained previously.

11.31 Consider the moving cable in Example 11.6. Wickert and Mote (1990) also analyzed this system in the case where the cable is sufficiently thick to represent it as an elastic beam with negligible axial deformation. Consider the ends $x = 0$ and $x = L$ to be hinged. The cable has a circular cross-section with radius $r = L/40$. Use the Ritz series formulation to construct a flexural vibration model for the translating cable. Identify the maximum speed v at which the belt may be safely operated according to this model. Then evaluate the first four mode shapes when v is 90% of critical.

CHAPTER 12

INTRODUCTION TO ROTORDYNAMICS

Water and gas turbines, computer disk drives, compressors, and pumps typify important engineering applications in which a nominally axisymmetric set of parts rotates about a common axis. Efficient operation often requires high rotational rates, which gives rise to concern that the system will resonate at unacceptably large amplitudes or that it will be unstable. Our objective here will be to survey the basic dynamic phenomena captured by a few canonical models. The Jeffcott model in Section 3.5 is fundamental. Many important results may be derived by improving that model with refined descriptions of elastic and dissipation effects. Other models incorporate gyroscopic effects. A comprehensive study of rotordynamics is multidisciplinary, for the dynamic behavior depends on the behavior of bearings, which requires understanding of fluid mechanics and tribology. The interested reader may find detailed discussions of numerous dynamical and tribological issues in the texts by Vance (1988) and Childs (1992).

12.1 ELEMENTS OF A ROTORDYNAMIC MODEL

Transverse forces leading to flexural displacement of shafts can arise from numerous sources, especially imbalance. Flexural loads act in addition to the torsional loads that the shaft transfers between various system components. We saw in our earlier studies that the fundamental frequency for torsional deformation of a slender bar is much higher than it is for flexure. We therefore focus on flexure, and consider the shaft to be rigid in torsion.

A simple rotordynamic system appears in Figure 12.1. It consists of a single rotor, which might be a bladed disk in a turbine, a flywheel, a propeller, or any other large component fastened to a shaft. When the shaft is undeformed, the rotor's geometric center C is considered to be situated on the bearing axis Z, and the distance

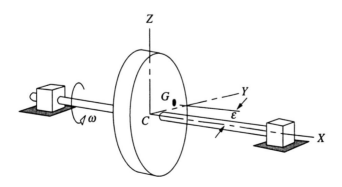

FIGURE 12.1 A rotor mounted on a shaft that is supported by bearings.

from the bearing axis to the center of mass G is the eccentricity ε. The ends of the shaft are supported by journal, roller, or ball bearings, or pressurized seals. The Jeffcott model described in Section 3.5, which accounts for translational inertia of a rotor and elasticity of the shaft, is fundamental to further exploration. In that model, the shaft is restricted to a circular cross-section, the rotor is considered to translate in directions transverse to the shaft's centerline, and the bearings are rigid. Let us consider some of the ways in which the Jeffcott model might be improved.

When a bar has a circular cross-section, flexure may be considered to occur in any two convenient planes containing the bearing axis, without coupling between the planes. This leads to *isotropic* stiffness coefficients, by which we mean that the coefficients relating shear forces and bending moments to the displacements are the same, regardless of which two orthogonal planes are used to view the displacement field. Isotropy made it possible to formulate the equations of motion for the Jeffcott model in terms of displacement Y_C and Z_C of the rotor's center relative to a fixed reference frame. The free body diagram for the Jeffcott model appears in Figure 12.2. The stiffness coefficient k is the same for displacement in each direction.

Suppose that, either because of its design, such as a noncircular cross-section, or because of a defect, such as a large crack, the shaft is not isotropic. This would be comparable to the situation in Example 11.3. The principal axes of the cross-section, for which the area products of inertia are zero, represent preferred directions because the displacement fields in each principal direction are uncoupled elastically, with constant stiffness coefficients representing the relationship between displacement and elastic force. Thus, a modified Jeffcott model, in which the shaft is not isotropic, is more readily formulated in terms of generalized coordinates that are the position coordinates of the center relative to a reference frame that rotates with the shaft.

Another elasticity effect that can be extremely important is associated with bearings, which are considered to be rigid in the Jeffcott model. Forces exerted by the bearings on the shaft require reactions that are applied to the supporting structure, thereby inducing displacement. Stiffness coefficients describe the relation between the displacement and the forces, both of which are defined relative to a fixed reference frame appropriate to the supporting structure. If flexibility of the supports and of the shaft are comparable in magnitude, and the shaft is not isotropic, we are faced with a dilemma because measurement of displacement with respect to either a fixed or rotating reference frame will only be suitable for one aspect of elasticity. In contrast, if either bearing or shaft elasticity is isotropic, we are free to use whichever set of

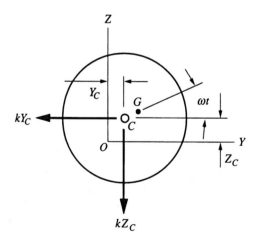

FIGURE 12.2 Jeffcott model of an unbalanced rotor on a circular shaft.

displacements expedites describing the other factors. A third source of elasticity lies within the bearings themselves. Ball and roller bearings and some lubricated bearings feature elastic effects that are unrelated to the behavior of the supporting structure. Such effects are usually taken to be isotropic.

Dissipation has significant impact on some phenomena. An obvious source of dissipation is external damping, in which power is dissipated into the surrounding fluid medium, either as heat due to viscosity or as acoustic energy that radiates away from the system. The Jeffcott model represented this effect as a linear damping force proportional to the velocity of the geometric center C of the rotor. Another source of dissipation is *internal damping*, which can arise from internal inelastic effects, such as viscoelasticity and microfractures. Since internal damping depends on strain rates within the rotating parts, the damping force will depend on the velocity of center C measured relative to a reference frame that rotates with the shaft. Thus, the effects of damping may be represented by either or both of the following:

$$\bar{f}_{ext} = -c_E \bar{v}_C, \qquad \bar{f}_{int} = -c_I(\bar{v}_C)_{xyz}$$

The position of point C may be described in terms of coordinates Y_C and Z_C relative to the fixed reference frame or, alternatively, in terms of coordinates y_C and z_C measured relative to reference frame xyz that rotates in unison with the shaft. Thus, the motion variables are

$$\bar{r}_{C/O} = Y_C \bar{I} + Z_C \bar{J}, \qquad \bar{v}_C = \dot{Y}_C \bar{I} + \dot{Z}_C \bar{J}$$

$$\bar{r}_{C/O} = y_C \bar{i} + z_C \bar{j}, \qquad (\bar{v}_C)_{xyz} = \dot{y}_C \bar{i} + \dot{z}_C \bar{j} \tag{12.1.1}$$

It is possible to represent either damping model in terms of components relative to either coordinate system using

$$\bar{v}_C = (\bar{v}_C)_{xyz} + \bar{\omega} \times \bar{r}_{C/O} \tag{12.1.2}$$

where $\bar{\omega} = \omega \bar{I} = \omega \bar{i}$. With the aid of this relation, the alternative descriptions of the damping forces are

$$\bar{f}_{ext} = -c_E \dot{Y}_C \bar{J} - c_E \dot{Z}_C \bar{K} = -c_E(\dot{y}_C - \omega z_C)\bar{j} - c_E(\dot{z}_C + \omega y_C)\bar{k}$$

$$\bar{f}_{int} = -c_I(\dot{Y}_C + \omega Z_C)\bar{J} - c_I(\dot{Z}_C - \omega Y_C)\bar{K} = -c_I \dot{y}_C \bar{j} - c_I \dot{z}_C \bar{k} \tag{12.1.3}$$

In terms of components relative to fixed directions, the matrix representation of the combined dissipation effect is

$$\begin{Bmatrix} (f_d)_Y \\ (f_d)_Z \end{Bmatrix} = -\begin{bmatrix} c_E + c_I & 0 \\ 0 & c_E + c_I \end{bmatrix} \begin{Bmatrix} \dot{Y}_C \\ \dot{Z}_C \end{Bmatrix} - \begin{bmatrix} 0 & c_I \omega \\ -c_I \omega & 0 \end{bmatrix} \begin{Bmatrix} Y_C \\ Z_C \end{Bmatrix} \tag{12.1.4}$$

and the corresponding description in terms of components relative to a rotating reference frame is

$$\begin{Bmatrix} (f_d)_y \\ (f_d)_z \end{Bmatrix} = -\begin{bmatrix} c_E + c_I & 0 \\ 0 & c_E + c_I \end{bmatrix} \begin{Bmatrix} \dot{y}_C \\ \dot{z}_C \end{Bmatrix} - \begin{bmatrix} 0 & -c_E \omega \\ c_E \omega & 0 \end{bmatrix} \begin{Bmatrix} y_C \\ z_C \end{Bmatrix} \tag{12.1.5}$$

In eq. (12.1.4) the velocity coefficient matrix is the usual damping matrix. The second term, which is proportional to c_I and depends on the instantaneous position, is a follower force. To see why, observe that the term $\bar{\omega} \times \bar{r}_{C/O}$ in eq. (12.1.2) is always perpendicular to $\bar{r}_{C/O}$. Hence, a portion of the internal damping force is always perpendicular to the plane formed by the bearing axis and center C. Such forces can destabilize a system, as we saw in Example 11.7.

The coefficient matrix multiplying the displacement terms in eq. (12.1.4) constitutes a set of *cross-coupling coefficients*. Cross-coupling coefficients may arise from other sources, such as bearing forces and fluid forces acting on bladed disks. In some situations the cross-coupling coefficients are not equal in magnitude, in which case their matrix is not skew-symmetric.

In addition to the translational inertia included in the Jeffcott model, rotational inertia effects might be significant. These arise from the fact that the rotor will rotate about axes perpendicular to the shaft's centerline when the slope of that line changes. Such rotation occurs in addition to the spin of the shaft. The result is gyroscopic effects, which exist whenever the angular momentum of a body is not collinear with the body's angular velocity. In the vast majority of applications, the rotor is axisymmetric and the shaft is circular. Such situations are isotropic, so generalized coordinates may be defined relative to a fixed reference frame. If the rotor is not axisymmetric or the shaft is not circular, then it would be preferable to employ generalized coordinates defined relative to a rotating reference frame.

It should be obvious from the discussion thus far that it is quite difficult to construct a single model that describes all effects. Even if we were to construct such a model, its analysis would present a formidable task. Our approach here will be to construct a few models that highlight some basic issues affecting rotating machinery.

12.2 EFFECTS OF INTERNAL AND EXTERNAL DAMPING

The different ways in which internal and external damping affect a rotating system may be demonstrated by modifying the Jeffcott model, whose free body diagram appeared in Figure 12.2. When we add the internal damping part of eq. (12.1.4) to eqs. (3.5.3), which already contain external damping, we obtain

$$\left\{ \begin{matrix} \ddot{Y}_C \\ \ddot{Z}_C \end{matrix} \right\} + 2\zeta\omega_{\text{nat}} \left\{ \begin{matrix} \dot{Y}_C \\ \dot{Z}_C \end{matrix} \right\} + \begin{bmatrix} \omega_{\text{nat}}^2 & 2\zeta_I\omega\omega_{\text{nat}} \\ -2\zeta_I\omega\omega_{\text{nat}} & \omega_{\text{nat}}^2 \end{bmatrix} \left\{ \begin{matrix} Y_C \\ Z_C \end{matrix} \right\} = \left\{ \begin{matrix} \varepsilon\omega^2\cos(\omega t) \\ \varepsilon\omega^2\sin(\omega t) \end{matrix} \right\} \quad (12.2.1)$$

where ω_{nat} is the natural frequency for displacement in either the X or Y directions when $\omega = 0$, ζ is a damping parameter accounting for both internal and external damping, and ζ_I is associated with internal damping only,

$$\omega_{\text{nat}} = \left(\frac{k}{m} \right)^{1/2}, \qquad \zeta_I = \frac{c_I}{2(km)^{1/2}}, \qquad \zeta_E = \frac{c_E}{2(km)^{1/2}}, \qquad \zeta = \zeta_I + \zeta_E \quad (12.2.2)$$

It is convenient to define nondimensional variables τ for time and Ω for rotation rate, such that

$$\tau = \omega_{\text{nat}}t, \qquad \Omega = \frac{\omega}{\omega_{\text{nat}}} \quad (12.2.3)$$

which transforms the equations of motion to

$$\frac{d^2}{d\tau^2}\begin{Bmatrix} Y_C \\ Z_C \end{Bmatrix} + 2\zeta\frac{d}{d\tau}\begin{Bmatrix} Y_C \\ Z_C \end{Bmatrix} + \begin{bmatrix} 1 & 2\zeta_I\Omega \\ -2\zeta_I\Omega & 1 \end{bmatrix}\begin{Bmatrix} Y_C \\ Z_C \end{Bmatrix} = \begin{Bmatrix} \varepsilon\Omega^2\cos(\Omega\tau) \\ \varepsilon\Omega^2\sin(\Omega\tau) \end{Bmatrix} \qquad (12.2.4)$$

Because this system has two degrees of freedom, we may obtain an analytical eigensolution by solving the nonlinear eigenvalue problem described by eq. (11.2.13). Thus, we let

$$\begin{Bmatrix} Y_C \\ Z_C \end{Bmatrix} = \begin{Bmatrix} u_1 \\ u_2 \end{Bmatrix}\exp(\lambda\tau) \qquad (12.2.5)$$

Substitution of this expression into eq. (12.2.4) leads to

$$\begin{bmatrix} \lambda^2 + 2\zeta\lambda + 1 & 2\zeta_I\Omega \\ -2\zeta_I\Omega & \lambda^2 + 2\zeta\lambda + 1 \end{bmatrix}\begin{Bmatrix} u_1 \\ u_2 \end{Bmatrix} = \begin{Bmatrix} 0 \\ 0 \end{Bmatrix} \qquad (12.2.6)$$

The characteristic equation is the sum of two squares, so that

$$\lambda^2 + 2\zeta\lambda + 1 = \pm 2i\zeta_I\Omega \qquad (12.2.7)$$

Either sign leads to two roots of a quadratic equation. Each root may be simplified by using the leading terms in a binomial series for the square root, which is valid if damping is light. In order to avoid any ambiguity, let us use $\nu = \pm 1$ to denote the alternative sign coefficient in eq. (12.2.7). The result is

$$\lambda = -\zeta \pm [i(1 - \zeta^2)^{1/2} + \nu\zeta_I\Omega] \qquad (12.2.8)$$

Substitution of eq. (12.2.7) into eq. (12.2.6) leads to

$$u_2 = -\nu i u_1 \qquad (12.2.9)$$

The alternative values of ε result in two pairs of complex conjugate modal solutions,

$$\begin{aligned}
\lambda_1 &= -\zeta + \zeta_I\Omega + i(1 - \zeta^2)^{1/2}, & \{\phi_1\} &= \begin{bmatrix} 1 & -i \end{bmatrix}^T \\[6pt]
\lambda_2 &= -\zeta + \zeta_I\Omega - i(1 - \zeta^2)^{1/2}, & \{\phi_2\} &= \begin{bmatrix} 1 & i \end{bmatrix}^T \\[6pt]
\lambda_3 &= -\zeta - \zeta_I\Omega + i(1 - \zeta^2)^{1/2}, & \{\phi_3\} &= \begin{bmatrix} 1 & i \end{bmatrix}^T \\[6pt]
\lambda_4 &= -\zeta - \zeta_I\Omega - i(1 - \zeta^2)^{1/2}, & \{\phi_4\} &= \begin{bmatrix} 1 & -i \end{bmatrix}^T
\end{aligned} \qquad (12.2.10)$$

The second pair of eigenvalues always have a negative real part. The first pair will have a positive real part, corresponding to a flutter instability, if the rotation rate is too large, specifically,

$$\text{Flutter:} \quad \Omega > \frac{\zeta}{\zeta_I} \quad \Rightarrow \quad \omega > \left(\frac{\zeta_E}{\zeta_I} + 1\right)\omega_{\text{nat}} \qquad (12.2.11)$$

We see that this instability arises only when $\zeta_I \neq 0$. Furthermore, for a specified value of ζ_I, increasing ζ_E raises the instability threshold. It follows that the follower force portion of the internal damping can destabilize a system, whereas external damping is beneficial. Note that when $\zeta_E = 0$, the condition $\omega = \omega_{nat}$ marks a maximum allowable value of ω, beyond which the system is unstable. In contrast, in the original Jeffcott model, $\omega = \omega_{nat}$ merely marks a critical speed at which a resonance occurs.

It is interesting to consider the nature of the modal free vibration when the system is stable. To do so we combine the complex conjugate pairs of solutions, multiplied by an arbitrary complex amplitude factor. For the first pair of eigensolutions this leads to

$$
\begin{Bmatrix} Y_C \\ Z_C \end{Bmatrix}_{1\&2} = \frac{1}{2} A_1 \{\phi_1\} \exp(\lambda_1 \tau) + \frac{1}{2} A_1^* \{\phi_1\}^* \exp(\lambda_1^* \tau)
$$

$$
= |A_1| \exp[-(\zeta\omega_{nat} - \zeta_I\omega)t] \begin{Bmatrix} \cos[\omega_{nat}(1 - \zeta^2)^{1/2}t + \arg(A_1)] \\ \sin[\omega_{nat}(1 - \zeta^2)^{1/2}t + \arg(A_1)] \end{Bmatrix}
$$

$$(12.2.12)$$

The radial distance from the axis of rotation to the rotor center is given by

$$
R_C = (Y_C^2 + Z_C^2)^{1/2} = |A_1| \exp[-(\zeta\omega_{nat} - \zeta_I\omega)t] \tag{12.2.13}
$$

and the polar angle relative to the Y axis is

$$
\theta_C = \tan^{-1}\left(\frac{Z_C}{Y_C}\right) = \omega_{nat}(1 - \zeta^2)^{1/2}t + \arg(A_1) \tag{12.2.14}
$$

It follows that point C follows an inward exponential spiral, with the radial line rotating in the counterclockwise sense at angular speed $\omega_{nat}(1 - \zeta^2)^{1/2}$. The shaft rotation $\bar{\omega}$ also is counterclockwise, so this is a forward whirling mode.

Similar operations may be applied to the second pair of modes, with the result that

$$
\begin{Bmatrix} Y_C \\ Z_C \end{Bmatrix}_{3\&4} = |A_3| \exp[-(\zeta\omega_{nat} + \zeta_I\omega)t] \begin{Bmatrix} \cos[\omega(1 - \zeta^2)^{1/2}t + \arg(A_1)] \\ -\sin[\omega(1 - \zeta^2)^{1/2}t + \arg(A_1)] \end{Bmatrix}
$$

$$(12.2.15)$$

This motion constitutes an inward spiral, defined by

$$
R_C = |A_3| \exp[-(\zeta\omega_{nat} + \zeta_I\omega)t]
$$

$$
\theta_C = -\omega_{nat}(1 - \zeta^2)^{1/2}t + \arg(A_3) \tag{12.2.16}
$$

Because θ_C decreases with time, this is a clockwise motion, so the second pair of eigensolutions form a backward whirling mode. For given values of the parameters, the backward whirling mode decays more quickly. Internal damping can lead to instability of the forward whirling mode, but not the one that whirls backward.

Our analysis of the Jeffcott model in Section 3.5 focused on identifying the critical rotation rate, at which the system is nearly resonant as a consequence of imbalance. Here, we shall use harmonic analysis of eq. (12.2.4) to determine how internal damping affects the critical speed, and the steady-state response in general. We set

$$
\begin{Bmatrix} Y_C \\ Z_C \end{Bmatrix} = \text{Re}\{\{U\}\exp(i\Omega\tau)\} \tag{12.2.17}
$$

which when substituted into eq. (12.2.4) leads to

$$\begin{bmatrix} 1 + i2\zeta\Omega - \Omega^2 & 2\zeta_I\Omega \\ -2\zeta_I\Omega & 1 + i2\zeta\Omega - \Omega^2 \end{bmatrix} \begin{Bmatrix} U_1 \\ U_2 \end{Bmatrix} = \begin{Bmatrix} \varepsilon\Omega^2 \\ -i\varepsilon\Omega^2 \end{Bmatrix} \qquad (12.2.18)$$

A solution using Cramer's rule requires the determinant Δ of the coefficient matrix, but such a determinant is identical to the characteristic equation, eq. (12.2.7), with λ replaced by $i\Omega$. The eigenvalues are the roots of the characteristic equation, so we have

$$\Delta = (i\Omega - \lambda_1)(i\Omega - \lambda_2)(i\Omega - \lambda_3)(i\Omega - \lambda_4) \qquad (12.2.19)$$

The amplitude factors described by eq. (12.2.18) are then found to be

$$U_1 = iU_2 = R, \quad R = \varepsilon \frac{\Omega^2[1 - \Omega^2 + 2i(\zeta_I + \zeta)\Omega]}{(i\Omega - \lambda_1)(i\Omega - \lambda_2)(i\Omega - \lambda_3)(i\Omega - \lambda_4)} \qquad (12.2.20)$$

The response obtained from eq. (12.2.17) is

$$Y_C = |R|\cos[\Omega t + \arg(R)], \qquad Z_C = |R|\sin[\Omega t + \arg(R)] \qquad (12.2.21)$$

This indicates that the rotor center C follows a circular path of radius $|R|$. The radial line to point C from the bearing axis rotates at angular speed ω in the same sense as the shaft rotation. In other words, the plane formed by the bearing axis and point C, which is the plane in which the bent shaft lies, rotates at the same rate and in the same direction as the shaft itself. This is a case of *synchronous whirl*.

The factor $|R|$ represents the amplitude of the whirl, and $\arg(R)$ is the angle by which the plane of the bent shaft leads the rotation of the shaft. Equation (12.2.20) gives the amplitude as a function of ω. Critical speeds correspond to resonant peaks. According to eq. (12.2.20), peak values of $|R|$ will occur whenever $\Omega \approx \mathrm{Im}(\lambda_j)$, $j = 1$, 2, 3, 4. However, because $\mathrm{Im}(\lambda_2)$ and $\mathrm{Im}(\lambda_4) < 0$, and $\mathrm{Im}(\lambda_1) = \mathrm{Im}(\lambda_3) \approx \omega_{\text{nat}}$ when damping is light, there is only one critical speed,

$$\omega_{\text{cr}} \approx \omega_{\text{nat}} \qquad (12.2.22)$$

It is interesting to note that if $\zeta_E = 0$, so that $\zeta = \zeta_I$, the determinant of the coefficient matrix in eqs. (12.2.18) actually vanishes when $\omega = \omega_{\text{nat}}$. In other words, there is a true resonance at this rotation rate, even though there is internal damping. The explanation for this apparent anomaly is that the shaft deformation is unaltered when the rotor center follows a circular path that is synchronized with the shaft. Consequently, no internal damping forces are generated in such a motion. Figure 12.3 displays the

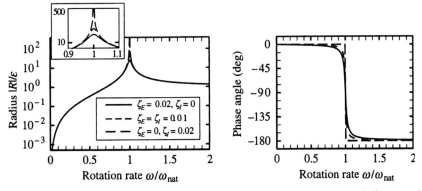

FIGURE 12.3 Whirl radius amplitude and phase angle according to the Jeffcott model with external and internal damping.

amplitude and phase angle dependence for three combinations of ζ_I and ζ_E for which $\zeta = 0.02$. The peak amplitude decreases with increasing ζ_I, but the other features are essentially independent of ζ_I. The maximum ω for the graph corresponds to the lowest flutter instability for the three damping cases. According to eq. (12.2.11), the case $\zeta_I = \zeta_E = 0.01$ leads to $\omega_{\text{flutter}} = 2\omega_{\text{nat}}$.

EXAMPLE 12.1

The rotor of the Jeffcott model with internal damping is turning at constant rate ω when, at $t = 0$, it is hit with an impulse force applied at its center in the Y direction. The impulse of the force is $100m\varepsilon\omega_{\text{nat}}$, and the damping ratios are $\zeta_E = 0.02$, $\zeta_I = 0.04$. Determine the ensuing motion of the center when $\omega = \omega_{\text{nat}}$.

Solution   The intent here is to shed light on how rotating systems evolve to steady-state conditions. The example is also intended to bring together the main concepts required for rotordynamic analyses. The given rotation rate is the critical speed, which is in the stable range given by eq. (12.2.11). Evaluation of transient response calls for modal analysis. We use a nondimensional time $\tau = \omega_{\text{nat}}t$, and correspondingly define $\Omega = \omega/\omega_{\text{nat}}$. By definition $[K]$ is the symmetric part of the coefficient matrix multiplying $\{q\}$. Hence, the coefficient matrix $[S]$ corresponding to eq. (12.2.4) is a 4×4 diagonal matrix whose nonzero elements are

$$S_{1,1} = S_{2,2} = -1, \qquad S_{3,3} = S_{4,4} = 1$$

while the other coefficient matrix is

$$[R] = \begin{bmatrix} 0 & 0 & -1 & 0 \\ 0 & 0 & 0 & -1 \\ -1 & -2\zeta_I\Omega & -2\zeta & 0 \\ 2\zeta_I\Omega & -1 & 0 & -2\zeta \end{bmatrix}$$

The impulsive force is applied at $t = 0$, and the imbalance forces are always present. After nondimensionalization, the excitation vector for the state-space equations is

$$\begin{Bmatrix} \{0\} \\ \{Q\} \end{Bmatrix} = \begin{Bmatrix} 0 \\ 0 \\ 100\varepsilon\delta(\tau) + \varepsilon\Omega^2\cos(\Omega\tau) \\ \varepsilon\Omega^2\sin(\Omega\tau) \end{Bmatrix}$$

For a specified value of Ω, we solve the right and left eigenvalue problems, as described by eqs. (11.2.3) and (11.2.6), and then normalize the eigenvectors according to eqs. (11.4.7). The nondimensional eigenvalues for $\Omega = 1$, $\zeta_E = 0.02$, $\zeta_I = 0.04$, are found to be

$$\{\hat{\lambda}\} = \begin{bmatrix} -0.020 + 0.999i & -0.020 - 0.999i & -0.100 + 0.999i & -0.100 - 0.999i \end{bmatrix}^{\mathsf{T}}$$

The upper half of the right and left normalized eigenvector matrices are

$$[\Psi] = \begin{bmatrix} 0.304 + 0.398i & 0.304 - 0.398i & -0.449 - 0.220i & -0.449 + 0.220i \\ 0.398 - 0.304i & 0.398 + 0.304i & 0.220 - 0.449i & 0.220 + 0.449i \end{bmatrix}$$

$$[\widetilde{\Psi}] = \begin{bmatrix} -0.280 + 0.416i & -0.280 - 0.416i & 0.435 - 0.246i & 0.435 + 0.246i \\ -0.416 - 0.280i & -0.416 + 0.280i & -0.246 - 0.435i & -0.246 + 0.435i \end{bmatrix}$$

The modal transformation $\{x\} = [\Psi]\{\xi\}$ uncouples the modal coordinates. The differential equations obtained from eqs. (11.4.13) for the present case are

$$\frac{d\xi_n}{d\tau} - \hat{\lambda}_n\xi_n = \{\tilde{\Psi}_n\}^{\mathrm{T}}\begin{Bmatrix} \{0\} \\ \{Q\} \end{Bmatrix} = 100\varepsilon\tilde{\Psi}_{3n}\delta(t) + \varepsilon\Omega^2[\tilde{\Psi}_{3n}\cos(\Omega\tau) + \tilde{\Psi}_{4n}\sin(\Omega\tau)]$$

It is given that at the instant that the impulse was applied the rotor was executing the steady-state response corresponding to the harmonic portion of the excitation. It follows that the response for $t > 0$ consists of a superposition of that steady-state response, and the impulse response corresponding to zero initial conditions.

The impulse response solution of state-space modal equations was discussed in Example 11.5. The modal mass is unity, so the impulse of the excitation, which is $100\varepsilon\tilde{\Psi}_{3n}$, equals the jump in the value of ξ_n at $\tau = 0^+$. Thus, this part of the response is given by

$$(\xi_n)_1 = 100\varepsilon\tilde{\Psi}_{3n}\exp(\hat{\lambda}_n\tau)$$

To solve for the harmonic portion of the response we replace the cosine and sine excitations with their complex representations. In doing so we do not use real parts, because of the complex nature of the differential equations. Instead, we use Euler's equation to write

$$\varepsilon\Omega^2[\tilde{\Psi}_{3n}\cos(\Omega\tau) + \tilde{\Psi}_{4n}\sin(\Omega\tau)] = \tfrac{1}{2}\varepsilon\Omega^2[(\tilde{\Psi}_{3n} - \tilde{\Psi}_{4n}i)\exp(i\Omega\tau)$$
$$+ (\tilde{\Psi}_{3n} + \tilde{\Psi}_{4n}i)\exp(-i\Omega\tau)]$$

The steady-state response of the modal coordinate equations corresponding to this excitation is

$$(\xi_n)_2 = \tfrac{1}{2}\varepsilon[A_n\exp(i\Omega\tau) + B_n\exp(-i\Omega\tau)]$$

where the complex amplitudes are

$$A_n = \Omega^2\frac{\tilde{\Psi}_{3n} - \tilde{\Psi}_{4n}i}{i\Omega - \hat{\lambda}_n}, \qquad B_n = \Omega^2\frac{\tilde{\Psi}_{3n} + \tilde{\Psi}_{4n}i}{-i\Omega - \hat{\lambda}_n}$$

For the parameters of the present case, these amplitudes are

$$\{A\} = \begin{Bmatrix} -40.003 - 29.967i \\ 0 \\ 0 \\ -0.428 - 0.259i \end{Bmatrix}, \qquad \{B\} = \begin{Bmatrix} 0 \\ -40.003 + 29.967i \\ -0.428 + 0.259i \\ 0 \end{Bmatrix}$$

We now may superpose the two solutions. Rather than bothering with the complex conjugate properties of the left and right eigenvectors, as well as of $\{A\}$ and $\{B\}$, we let our mathematical software handle the calculations. The decay rate of each modal impulse response is set by the real part of the corresponding λ_n, so we set $\tau_{\max} = 4/\mathrm{Im}(-\lambda_1)$. We discretize this interval into time steps $\Delta = 2\pi/20$, which gives 20 points per cycle of either the impulse or harmonic portion of the first modal response. For each value τ_p, we compute a column of modal coordinates, such that

$$\zeta\_vals_{jp} = [(\xi_j)_1 + (\xi_j)_2]_{\tau=\tau_p}$$

A matrix of state-space responses then results from computing

$$[x\_vals] = [\Psi][\zeta\_vals]$$

The first row of $[x\_vals]$ contains Y_C at the respective τ instants, while the second row holds the corresponding Z_C values.

The first set of graphs displays an early interval of the total responses, as well as the portions attributable to the impulsive excitation.

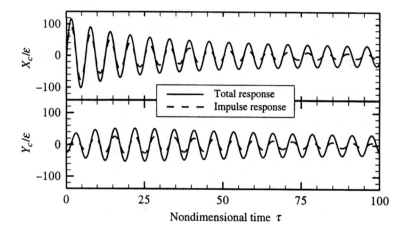

We see that the impulse response is almost in phase with the harmonic part. The Lissajous pattern in the next graph shows the path of the rotor center. It is evident that the impulse initially pushes the centerpoint away from its circular synchronous orbit.

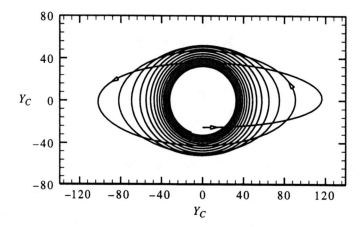

Because the harmonic response prior to the impulse force is a circular path, with $Y_C = 0$ and $Z_C = -R$ at $t = 0$, an impulse force in the Y direction at that instant pushes the rotor tangentially to the circular path. The resulting path seems to be nearly elliptical in each cycle, with the diameter and eccentricity of the ellipse decreasing progressively. Ultimately, the center will return to its original circular steady-state path, corresponding to synchronous whirl.

The inward spiraling nature of the impulse response serves to emphasize the significance of system stability. If we had carried out the same analysis for an Ω value in the flutter range, $\Omega > \zeta_I/\zeta_E$, the spiral would proceed outward at a rapid rate. In an actual system this would quickly lead to catastrophic failure.

12.3 FLEXIBLE ORTHOTROPIC BEARINGS

In the Jeffcott model the elasticity of the system is considered to be the result of flexure of the shaft. An alternative considers the shaft to be rigid, and the supports to be flexible. The simplest model of this type restricts the shaft to translate in either of two orthogonal directions. (Without such a restriction to translation, we would need to

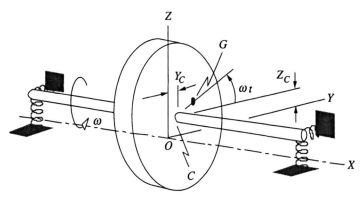

FIGURE 12.4 Modified Jeffcott model allowing for orthotropic bearing stiffnesses.

consider gyroscopic effects associated with the rotor's angular momentum, which is the topic of the next section.) From a practical standpoint, this type of model might be suitable for situations where a large massive rotor is mounted on a shaft that is short relative to the rotor's radius of gyration. The system we shall consider appears in Figure 12.4. The fixed X axis marks the bearing axis in the undeformed state. Because the shaft is only permitted to translate in the Y and Z directions, fixed reference frame coordinates Y_C and Z_C of the rotor center serve as generalized coordinates for the system.

Let us denote the transverse stiffness of each bearing as k_Y and k_Z, so that $-2k_Y Y_C$ and $-2k_Z Z_C$ are the forces acting on the system and shaft. These forces replace the elastic forces exerted by the shaft in the original Jeffcott model, eqs. (12.2.4). The shaft is rigid here, but some bearings, such as pressurized seals, can give rise to an internal damping effect. Therefore the only alteration in the aforementioned equations of motion is replacement of the elastic terms, which leads to

$$m\ddot{Y}_C + (c_I + c_E)\dot{Y}_C + 2k_Y Y_C + c_I \omega Z_C = \varepsilon \omega^2 \cos(\omega t)$$
$$m\ddot{Z}_C + (c_I + c_E)\dot{Z}_C - c_I \omega Y_C + 2k_Z Z_C = \varepsilon \omega^2 \sin(\omega t)$$

(12.3.1)

It is convenient to define a reference frequency ω_{ref} based on the average stiffness. In addition, we define an orthotropy parameter σ, such that

$$\omega_{\text{ref}} = \left[\frac{(k_Z + k_Y)}{m}\right]^{1/2}, \qquad \sigma = \frac{k_Z - k_Y}{k_Z + k_Y}$$

(12.3.2)

We use ω_{ref} to define nondimensional time τ, nondimensional rotation rate Ω, and nondimensional damping ratios, according to

$$\tau = \omega_{\text{ref}} t, \qquad \Omega = \frac{\omega}{\omega_{\text{ref}}}, \qquad \zeta_E = \frac{c_E}{2m\omega_{\text{ref}}}, \qquad \zeta_I = \frac{c_I}{2m\omega_{\text{ref}}}, \qquad \zeta = \zeta_E + \zeta_I$$

(12.3.3)

The resulting equations of motion are

$$\frac{d^2}{d\tau^2}\begin{Bmatrix} Y_C \\ Z_C \end{Bmatrix} + 2\zeta \frac{d}{d\tau}\begin{Bmatrix} Y_C \\ Z_C \end{Bmatrix} + \begin{bmatrix} (1-\sigma) & 2\zeta_I\Omega \\ -2\zeta_I\Omega & (1+\sigma) \end{bmatrix}\begin{Bmatrix} Y_C \\ Z_C \end{Bmatrix} = \begin{Bmatrix} \varepsilon\Omega^2\cos(\Omega\tau) \\ \varepsilon\Omega^2\sin(\Omega\tau) \end{Bmatrix}$$

(12.3.4)

To explore the modal properties of this model we set

$$
\left\{ \begin{array}{c} Y_C \\ Z_C \end{array} \right\} = \left\{ \begin{array}{c} u_1 \\ u_2 \end{array} \right\} \exp(\lambda t)
\tag{12.3.5}
$$

which leads to

$$
\begin{bmatrix} (\lambda^2 + 2\zeta\lambda + 1 - \sigma) & 2\zeta_I\Omega \\ -2\zeta_I\Omega & (\lambda^2 + 2\zeta\lambda + 1 + \sigma) \end{bmatrix} \left\{ \begin{array}{c} u_1 \\ u_2 \end{array} \right\} = \left\{ \begin{array}{c} 0 \\ 0 \end{array} \right\}
\tag{12.3.6}
$$

The characteristic equation is

$$
\lambda^4 + 4\zeta\lambda^3 + (2 + 4\zeta^2)\lambda^2 + 4\zeta\lambda + (1 - \sigma^2 + 4\zeta_I^2\Omega^2) = 0
\tag{12.3.7}
$$

This characteristic equation is not readily factorized, so we use the Routh-Hurwitz criteria to assess stability. A comparison of the present characteristic equation to the standard form in eq. (11.3.2) shows that $b_0 = 1$, $b_1 = b_3 = 4\zeta$, $b_2 = 2 + 4\zeta^2$, and $b_4 = 1 - \sigma^2 + 4\zeta_I^2\Omega_2$. Because $|\sigma| < 1$, all coefficients are positive, which is the first criterion. The only condition stated by eqs. (11.3.6) that is not identically satisfied is

$$
[\Delta_3]| = b_1 b_2 b_3 - b_0 b_3^2 - b_1^2 b_4 = 16\zeta^2(4\zeta^2 + \sigma^2 - 4\zeta_I^2\Omega^2) > 0
\tag{12.3.8}
$$

Thus, the system is asymptotically stable if $\omega < \omega_{\text{flutter}}$, where

$$
\omega_{\text{flutter}} = \omega_{\text{ref}}\left(\frac{\zeta^2}{\zeta_I^2} + \frac{\sigma^2}{4\zeta_I^2} \right)^{1/2}
\tag{12.3.9}
$$

In view of the fact that ζ_I is small, the preceding shows that even a small deviation from isotropic bearing stiffnesses can greatly extend the range of rotation rates for which the system is stable.

State-space modal analysis provides an alternative method for identifying stability limits, while concurrently disclosing critical speeds. A modal solution proceeds in the manner of Example 12.1. The coefficient matrices corresponding to eq. (12.3.1) are given by

$$
S_{1,1} = -(1 - \sigma), \qquad S_{2,2} = -(1 + \sigma), \qquad S_{3,3} = S_{4,4} = 1
$$

$$
[R] = \begin{bmatrix} 0 & 0 & -(1 - \sigma) & 0 \\ 0 & 0 & 0 & -(1 + \sigma) \\ -(1 - \sigma) & -2\zeta_I\Omega & -2\zeta & 0 \\ 2\zeta_I\Omega & -(1 + \sigma) & 0 & -2\zeta \end{bmatrix}
\tag{12.3.10}
$$

The state-space excitation in complex form is

$$
\left\{ \begin{array}{c} \{0\} \\ \{Q\} \end{array} \right\} = \frac{1}{2}\varepsilon\Omega^2 \left\{ \begin{array}{c} 0 \\ 0 \\ \exp(i\Omega t) + \exp(-i\Omega t) \\ -i\exp(i\Omega t) + i\exp(-i\Omega t) \end{array} \right\}
\tag{12.3.11}
$$

We solve the right and left eigenvalue problems, as described by eqs. (11.2.3) and (11.2.6), and then normalize the eigenvectors according to eqs. (11.4.7). The results of that analysis are the eigenvalues λ_n and right and left normalized eigenvector matrices, $[\Psi]$ and $[\widetilde{\Psi}]$. The modal transform $\{x\} = [\Psi]\{\xi\}$ leads to first-order, uncoupled differential equations like eqs. (11.4.13). The steady-state solution for the modal coordinates corresponding to the present excitation is

$$\xi_n = \frac{1}{2}\varepsilon\Omega^2\left[\frac{\widetilde{\Psi}_{3n} - \widetilde{\Psi}_{4n}i}{i\Omega - \lambda_n}\exp(i\Omega t) + \frac{\widetilde{\Psi}_{3n} + \widetilde{\Psi}_{4n}i}{-i\Omega - \lambda_n}\exp(-i\Omega t)\right] \qquad (12.3.12)$$

The physical response is then obtained from the upper half of $\{x\} = [\Psi]\{\xi\}$.

Critical speeds correspond to resonances, which essentially occur when one of the denominators has a minimum value. When Ω is in the stable range, we know that the eigenvalues occur as complex conjugate pairs, $\lambda_{n+1} = \lambda_n^*$, $n = 1, 3$. Thus, ξ_n and ξ_{n+1} have their maximum magnitude when

$$\boxed{|\mathrm{Im}(\lambda_n)| = \Omega} \qquad (12.3.13)$$

This condition is descriptive of other systems, because the state-space solution for steady-state response at frequency Ω always contains $i\Omega - \lambda_n$ or $-i\Omega - \lambda_n$ in a denominator; see eq. (11.4.17). This leads to a general procedure for identifying critical rotation rates. We compute and graph $|\mathrm{Im}(\lambda_n)|$ as a function of Ω. For a stable N-degree-of-freedom system, there are N pairs of complex conjugate roots, so there will be N curves to plot. In the same set of graphs, we plot the straight line, $|\mathrm{Im}(\lambda_n)| = \Omega$, which is called the *synchronous line*. The intersection of the synchronous line with any of the eigenvalue curves marks a critical speed. This graphical construction is called a *Campbell diagram*.

Figure 12.5 shows a typical Campbell diagram for orthotropic bearings. The critical speeds at which the synchronous line intersects $|\mathrm{Im}(\lambda)|$ are $\Omega_1 = 0.984$ and $\Omega_2 = 1.014$. The figure also shows a graph of $\mathrm{Re}(\lambda_n)$ as a function of Ω. Such a graph serves to identify the stable range of rotation rates, within which $\mathrm{Re}(\lambda_n) < 0$. The value of Ω at which the real part of one pair of eigenvalues becomes positive matches eq. (12.3.9).

When a Campbell diagram is constructed in practice, it is conventional to plot the rotation rate in units of rev/min, while rad/s is used for the eigenvalues. Also,

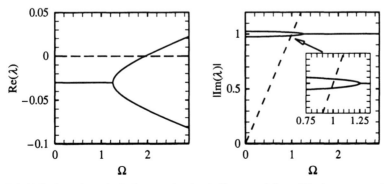

FIGURE 12.5 Campbell diagram for the Jeffcott model modified to account for orthotropic bearing stiffness, $\sigma = 0.05$, $\zeta_E = 0.01$, $\zeta_I = 0.02$.

many texts on rotordynamics have not made use of the complex variable approach we have used. The critical speeds that result are for the undamped system. The eigenvalues obtained from such analyses are $\chi_n = i\lambda_n$, which are positive and negative real numbers because λ_n are purely imaginary for gyroscopic conservative systems. Hence, a Campbell diagram associated with such a formulation needs to consider two synchronous lines, $\chi_n = \Omega$ and $\chi_n = -\Omega$.

In principle, if we wish to know the actual displacement amplitudes as a function of the rotation rate we could perform the computations associated with eq. (12.3.12). However, this would require an eigensolution for each value of the rotation rate. It is far simpler to compute the steady-state amplitudes directly using frequency domain analysis. Thus, to determine the steady-state solution of eqs. (12.3.4) over a range of Ω, we set

$$
\left\{\begin{array}{c} Y_C \\ Z_C \end{array}\right\} = \mathrm{Re}\left\{\left\{\begin{array}{c} U_Y \\ U_Z \end{array}\right\}\exp(i\Omega\tau)\right\} \tag{12.3.14}
$$

The complex amplitudes are found by solving

$$
\begin{bmatrix} (1-\sigma) + 2i\zeta\Omega - \Omega^2 & 2\zeta_I\Omega \\ -2\zeta_I\Omega & (1+\sigma) + 2i\zeta\Omega - \Omega^2 \end{bmatrix}\left\{\begin{array}{c} U_Y \\ U_Z \end{array}\right\} = \varepsilon\Omega^2\left\{\begin{array}{c} 1 \\ -i \end{array}\right\} \tag{12.3.15}
$$

whose solution is

$$
\left\{\begin{array}{c} U_Y \\ U_Z \end{array}\right\} = \frac{\varepsilon\Omega^2}{\mu(\Omega)}\left\{\begin{array}{c} (1+\sigma-\Omega^2) + 2i(\zeta+\zeta_1)\Omega \\ -i(1-\sigma-\Omega^2) + 2(\zeta+\zeta_1)\Omega \end{array}\right\} \tag{12.3.16}
$$

$$
\mu(\Omega) = (1 + 2i\zeta\Omega + \Omega^2)^2 - \sigma^2 + 4\zeta_I^2\Omega^2
$$

Typical amplitude-frequency diagrams are displayed in Figure 12.6. For small damping values, the peak amplitudes occur when $\mu(\Omega)$ has a minimum. If $\zeta_I = \zeta = 0$, this leads to $\Omega_1 = 1 - \sigma$ and $\Omega_2 = 1 + \sigma$ as the critical speeds, which explains why the peaks for the higher level of isotropy are separated farther. The graph shows that the first critical speed corresponds to peak $|Y_C|$, while the second critical speed maximizes $|Z_C|$. (In more general systems, depending on the nature of

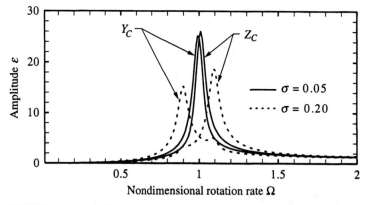

FIGURE 12.6 Amplitude-frequency diagram for the orthotropic Jeffcott model, $\zeta_E = 0.01$, $\zeta_I = 0.02$.

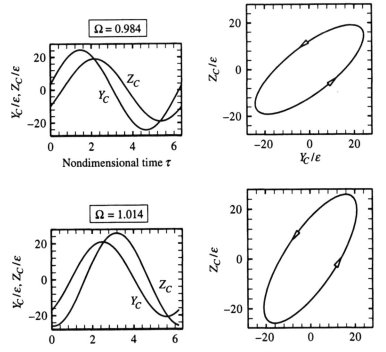

FIGURE 12.7 Orbits of the rotor center at the critical speeds of the Jeffcott model with orthotropic bearing stiffnesses, $\zeta_E = 0.01$, $\zeta_I = 0.02$.

the excitation, critical speeds are likely to produce peak amplitudes for more than one generalized coordinate.)

To close the discussion it is interesting to consider the nature of the path followed by the rotor center. Recall that isotropic bearings led to a circular path at any rotation rate. Figure 12.7 displays the paths at the two critical speeds identified in Figure 12.6. Orthotropy of the bearing stiffnesses leads to highly elliptical orbits, both of which correspond to synchronous forward whirl. When Ω is well away from the critical speeds, the amplitudes $|U_1|$ and $|U_2|$ are much closer in magnitude, so the path becomes more circular.

12.4 GYROSCOPIC EFFECTS

The models we have considered thus far restrict the rotor's movement to a plane that is perpendicular to the shaft. However, flexibility of the shaft and bearings allows the rotor to rotate out of this plane. Here we shall consider a relatively simple model that discloses some consequences of such motion. In Figure 12.8(a) a rotor is mounted perpendicularly on a flexible shaft whose bearings are rigid. When the shaft flexes, the rotor undergoes small rotations about axes that are transverse to the shaft because it must maintain its perpendicular orientation relative to the shaft's centerline. This system is referred to as the *Stodola-Green model* in the rotordynamics literature, in recognition of the contributions of A. Stodola (1924) and R. Green (1948).

The dashed line in Figure 12.8(a) indicates the deflected shape of the shaft's centerline. We define the tangent to that line as the *x* axis of a moving reference frame, as shown in Figure 12.8(b). To locate the orientation of this axis we define a pair of

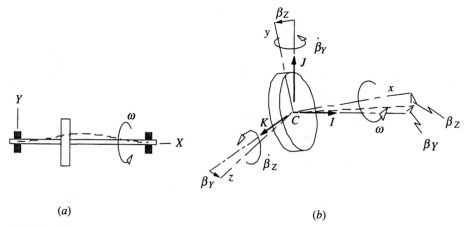

FIGURE 12.8 Possible rotations of a rotor that is aligned with a flexible shaft rotating at angular speed ω.

rotations that take us from the fixed XYZ reference frame directions to directions defined relative to the rotor. For discussion purposes it is convenient to refer to Y and Z as vertical and horizontal, but the actual orientation of these axes is arbitrary. The first rotation is β_Y, which occurs about the Y axis. This rotates the axis of the rotor, as well as the horizontal diameter, in the horizontal plane. The new position of this horizontal diameter is marked as the z axis, which is the axis for the second rotation, β_Z. This rotates the axis of the rotor and the vertical diameter in a vertical plane. These angles define the tangent to the deformed shaft, so they must be very small. As a result, when we look down the Y axis we essentially see β_Y as the angle between the X axis and the projection of the x axis onto the XZ plane. Similarly, β_Z is the angle between the X axis and the projection of the x axis onto the XY plane.

Because the shaft is substantially stiffer in torsion than it is in bending, the rotation rate ω occurs about the deformed centerline, which is the x axis. Thus, $\omega \bar{i}$ is the rotation of the rotor relative to the xyz reference frame. The total angular velocity of the rotor consists of contributions of this rotation, $\dot{\beta}_Y$ about the Y axis, and $\dot{\beta}_Z$ about the z axis, so

$$\bar{\omega}_r = \omega \bar{i} + \dot{\beta}_Y \bar{J} + \dot{\beta}_Z \bar{k} \qquad (12.4.1)$$

The principal axes of inertia for the disk are xyz, so we resolve $\bar{J}$ into xyz components. The angle between the Y and y axes is β_Z. Because this angle is limited to very small values, we may approximate $\cos(\beta_Z) \approx 1$, $\sin(\beta_Z) \approx \beta_Z$, from which we find that

$$\bar{J} \approx \bar{j} + \beta_Z \bar{i} \qquad (12.4.2)$$

In turn, this leads to

$$\bar{\omega}_r = (\omega + \dot{\beta}_Y \beta_Z) \bar{i} + \dot{\beta}_Y \bar{j} + \dot{\beta}_Z \bar{k} \qquad (12.4.3)$$

The kinetic energy is the sum of rotational and translational contributions referenced to the center of mass. We consider the center of mass G to be situated at distance ε from the rotor's center as in the Jeffcott model. If we take $t = 0$ to be the instant when point G is at the top-dead-center position, the positions of the points are defined by

$$\bar{r}_C = Y_C \bar{J} + Z_C \bar{K}, \qquad \bar{r}_{G/C} = \varepsilon \cos(\omega t) \bar{J} + \varepsilon \sin(\omega t) \bar{K} \qquad (12.4.4)$$

from which we find that

$$\bar{v}_G = \frac{d}{dt}(\bar{r}_G + \bar{r}_{G/C}) = [\dot{Y}_C - \varepsilon\omega\cos(\omega t)]\bar{J} + [\dot{Z}_C + \varepsilon\omega\sin(\omega t)]\bar{K} \quad (12.4.5)$$

From the expressions for $\bar{v}_G$ and $\bar{\omega}$, we find that

$$T = \tfrac{1}{2}mv_G^2 + \tfrac{1}{2}\left[I_{xx}(\omega_r)_x^2 + I_{yy}(\omega_r)_y^2 + I_{yy}(\omega_r)_z^2\right]$$

$$= \tfrac{1}{2}m\left[\dot{Y}_C^2 + \dot{Z}_C^2 - 2\varepsilon\omega\dot{Y}_C\cos(\omega t) + 2\varepsilon\omega\dot{Z}_C\sin(\omega t) + \varepsilon^2\omega^2\right] \quad (12.4.6)$$

$$+ \tfrac{1}{2}\left[I_{xx}(\omega^2 + 2\omega\dot{\beta}_Y\beta_Z) + I_{yy}(\dot{\beta}_Y^2 + \dot{\beta}_z^2)\right]$$

where I_{xx} and I_{yy} are moments of inertia of the rotor about axes having origin G parallel and perpendicular to the x axis, respectively. (The parallel axis theorems may be used to relate these moments of inertia to those relative to the geometric center C, but the differences may be ignored if ε is small.) It should be noted that products of β_Y or β_Z higher than quadratic have been dropped from the preceding expression for T, consistent with the smallness approximation for a linearized analysis. We may obtain the inertial terms in the equations of motion by explicitly evaluating Lagrange's equations, or by matching T to the standard form in eq. (11.1.5).

As a consequence of the shaft being circular, flexural displacement may be pictured as occurring independently in any two orthogonal planes. The displacement Y_C and rotation β_Z elastically couple to form the flexural displacement in the vertical XY plane, while the displacement Z_C and rotation β_Y arise from the flexural displacement in the horizontal plane. These displacement fields are uncoupled elastically. Specifically, a force F_Y in the Y direction at point C would produce a displacement Y_C and a rotation β_Z. For linear elasticity, we have $F_Y = K_{ww}Y_C + K_{w\beta}\beta_Z$. A couple Γ_Z about the Z axis also produces both displacement variables, so $\Gamma_Z = K_{\beta w}Y_C + K_{\beta\beta}\beta_Z$. The principle of reciprocity requires that $K_{\beta w} = K_{w\beta}$. Axisymmetry of the shaft requires that the stiffness coefficients be alike for the horizontal and vertical planes. However, $-\beta_Z$ is like β_Y because positive β_Z displaces the tip of $\bar{i}$ in the positive Y direction, while positive β_Y displaces the tip of $\bar{i}$ in the negative Z direction. In view of this similarity, the companion relations for flexure in the XZ plane are $F_Z = K_{ww}Z_C - K_{w\beta}\beta_Y$ and $\Gamma_Y = -K_{\beta w}Z_C + K_{\beta\beta}\beta_Y$. If the generalized coordinate vector is defined as

$$\{q\} = \begin{bmatrix} Y_C & Z_C & \beta_Y & \beta_Z \end{bmatrix}^\mathrm{T} \quad (12.4.7)$$

the forces and moments are given by

$$\begin{Bmatrix} F_Y \\ F_Z \\ \Gamma_Y \\ \Gamma_Z \end{Bmatrix} = \begin{bmatrix} K_{ww} & 0 & 0 & K_{w\beta} \\ 0 & K_{ww} & (-K_{w\beta}) & 0 \\ 0 & (-K_{w\beta}) & K_{\beta\beta} & 0 \\ K_{w\beta} & 0 & 0 & K_{\beta\beta} \end{bmatrix} \{q\} \quad (12.4.8)$$

The coefficient matrix multiplying $\{q\}$ is the stiffness $[K]$. Because the coefficients are derived from a static analysis the structural properties are often stated as a set of flexibility coefficients for bending in a plane. The relationship between the two sets of coefficients is

$$\begin{bmatrix} K_{ww} & K_{w\beta} \\ K_{w\beta} & K_{\beta\beta} \end{bmatrix} = \begin{bmatrix} L_{ww} & L_{w\beta} \\ L_{w\beta} & L_{\beta\beta} \end{bmatrix}^{-1} \qquad (12.4.9)$$

Expressions for the stiffness coefficients K_{ww}, $K_{w\beta}$, and $K_{\beta\beta}$ may be obtained from a static structural analysis.

External damping will generate a resistance force in opposition to the motion associated with each generalized coordinate. In order to highlight the gyroscopic effects, let us neglect internal damping. The resulting equations of motion have the standard form

$$[M]\{\ddot{q}\} + [[G] + [C]]\{\dot{q}\} + [K]\{q\} = \{Q\} \qquad (12.4.10)$$

The nonzero elements of the matrices other than $[K]$ are

$$M_{11} = M_{22} = m, \qquad M_{33} = M_{44} \doteq I_{yy}$$

$$G_{34} = -G_{43} = 2I_{xx}\omega \qquad (12.4.11)$$

$$C_{11} = C_{22} = c_w, \qquad C_{33} = C_{44} = c_\beta$$

The excitation consists of the centripetal effect of the imbalance,

$$\{Q\} = \varepsilon m\omega^2 \begin{bmatrix} \cos(\omega t) & \sin(\omega t) & 0 & 0 \end{bmatrix}^T \qquad (12.4.12)$$

If the rotor were to be mounted on the shaft obliquely, so that the axis of the rotor is not parallel to the shaft, the rotor would be said to be dynamically unbalanced. Dynamic imbalance, which results when the shaft's centerline is not a principal axis of inertia of the rotor, can be represented as a couple of constant magnitude that rotates at $\omega\bar{i}$; see Ginsberg (1995) or Hibbeler (1998).

The addition of gyroscopic effects has led us to an $N = 4$ model. Algebraic computations are difficult to perform on this model, so we must invoke numerical procedures. This requires that we evaluate the stiffness coefficients and moments of inertia appropriate to the specific system. Two types of analyses may be followed. The eigenvalue problem may be solved for a range of ω. Toward that end we would form the state-space equations of motion according to eqs. (11.1.21) and (11.1.22). Plotting the real parts of the eigenvalues as a function of ω enables us to identify ranges of rotation rates within which the system is unstable. Intersections of a plot of the imaginary part of the eigenvalues with the synchronous line, $\text{Im}(\lambda) = \omega$, enables us to identify critical speeds. Critical speeds, as well as the vibration amplitudes at any ω, may also be obtained from a steady-state analysis, in which all displacements are taken to vary at frequency ω. Peaks in a graph of each generalized coordinate's magnitude as a function of ω mark the critical speeds. Both analyses are implemented in the next example. The results are typical of those obtained from the Stodola-Green model.

EXAMPLE 12.2

A lightweight elastic shaft supports a thin disk of radius r that is situated at $L/3$ from the left end, where L is the distance between bearings. The ends are supported by roller bearings that are considered to act as simple supports.

(a) Identify any ranges of rotation rates at which the system is unstable.

(b) Determine the critical speeds.

(c) Determine the amplitudes of the flexural displacements of the center of the disk and the flexural rotations of the disk as a function of the rotation rate.

Solution The primary objective here is to demonstrate how to interpret the results obtained from the Stodola-Green model. The first step in the analysis is to determine the stiffness coefficients. A general procedure is to use Castigliano's theorems (Shames and Dym, 1985) to evaluate the static displacement and rotation of the shaft resulting from application of unit transverse force at the $L/3$ location. These quantities are the influence coefficients L_{ww} and $L_{\beta w}$. Similarly, application of a unit bending moment at the location of the rotor in conjunction with Castigliano's theorems yields the displacement $L_{w\beta}$ and the rotation $L_{\beta\beta}$. If we have performed these analyses correctly, we will find that $L_{\beta w} = L_{w\beta}$. The stiffness coefficients are then found from eq. (12.4.9). The resulting values for the case of a simply supported beam with forces and couples at $L/3$ are

$$K_{ww} = \frac{729}{8}\frac{EI}{L^3}, \qquad K_{w\beta} = -\frac{81}{4}\frac{EI}{L^2}, \qquad K_{\beta\beta} = \frac{27}{2}\frac{EI}{L}$$

We shall nondimensionalize the equations of motion in order to focus on the relevant parametric combinations. Toward that end, we define the following parameters

$$q_1' = \frac{X_C}{L}, \qquad q_2' = \frac{Y_C}{L}, \qquad q_3' = \beta_Y, \qquad q_4' = \beta_Z$$

$$\omega_{\text{ref}} = \left(\frac{EI}{mL^3}\right)^{1/2}, \qquad \Omega = \frac{\omega}{\omega_{\text{ref}}}, \qquad \tau = \omega_{\text{ref}}t$$

$$\zeta_w = \frac{c_{Ew}}{2m\omega_{\text{ref}}}, \qquad \zeta_\beta = \frac{c_{E\beta}}{2mL^2\omega_{\text{ref}}}$$

It is reasonable to take the damping ratios to be equal, $\zeta_\beta = \zeta_w = \zeta$. Substitution of these quantities, as well as the stiffness coefficients, converts eq. (12.4.10) to

$$[M']\left\{\frac{d^2q'}{d\tau^2}\right\} + [[G'] + [C']]\left\{\frac{dq'}{d\tau}\right\} + [K']\{q'\} = \{Q'\}$$

where

$$[K'] = \begin{bmatrix} 729/8 & 0 & 0 & -81/4 \\ 0 & 729/8 & 81/4 & 0 \\ 0 & 81/4 & 27/2 & 0 \\ -81/4 & 0 & 0 & 27/2 \end{bmatrix}$$

$$M_{11}' = M_{22}' = 1, \qquad M_{33}' = M_{44}' = \kappa_x^2$$

$$G_{34}' = -G_{43}' = 2\kappa_y^2\Omega$$

$$C_{11}' = C_{22}' = \zeta_w, \qquad C_{33}' = C_{44}' = \zeta_\beta$$

$$Q_1' = \frac{\varepsilon}{L}\Omega^2\cos(\Omega\tau), \qquad Q_2' = \frac{\varepsilon}{L}\Omega^2\sin(\Omega\tau)$$

The parameters κ_x and κ_y are nondimensional radii of gyration for the rotor, $\kappa_x^2 = I_{xx}/mL^2$, $\kappa_y^2 = I_{yy}/mL^2$. For a thin disk, $\kappa_x = r/\sqrt{2}L$ and $\kappa_y = r/2L$. It should be noted that altering the boundary conditions at the bearings solely changes $[K']$.

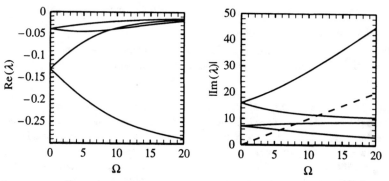

Campbell diagram for the Stodola-Green model of a thin disk at one-third the span between simple bearing supports, $r/L = 0.5$, $\kappa_x = 0.707 r/L$, $\kappa_y = 0.50 r/L$, $\zeta = 0.01$.

The first issue we shall explore is identification of rotation rates at which the system response becomes unstable, and of critical speeds corresponding to resonance. We place the equations of motion in state-space form according to eq. (11.1.17), substitute $\{q\} = \{\psi\}\exp(\lambda t)$, and solve the eigenvalue $[[R(\Omega)] - \lambda[S]]\{\psi\} = \{0\}$ for a range of Ω. At each value of Ω, there will be eight eigenvalues, which occur as four pairs of complex conjugates, unless a divergence instability occurs. To identify unstable situations we plot the real parts of the eigenvalues as a function of Ω. The result for a thin disk at one-third span appears as the first graph. All values are negative in the plotted range, and the indication is that the curves approach a constant value with increasing Ω. We conclude that the system is asymptotically stable for all rotation rates.

The second graph is the Campbell diagram. The imaginary parts of the eigenvalues split into four branches with increasing Ω. The synchronous (dashed) line intersects three branches, so there are three critical speeds. These values and the associated modal properties are tabulated below. (Only the first of the conjugate pair of modes is listed.)

Critical Ω	Mode no.	Eigenvalue	Mode (displacement part)
5.82	1 & 2	$-0.045 + 5.82i$	$\begin{Bmatrix} -0.0377 + 0.0132i \\ -0.0132 - 0.0377i \\ 0.0380 + 0.1064i \\ -0.1064 + 0.0380i \end{Bmatrix}$
8.25	3 & 4	$-0.023 + 8.25i$	$\begin{Bmatrix} -0.0406 + 0.0387i \\ 0.0387 + 0.0406i \\ -0.0446 - 0.0459i \\ -0.0459 + 0.0446i \end{Bmatrix}$
11.37	5 & 6	$-0.032 + 11.37i$	$\begin{Bmatrix} -0.0284 - 0.0062i \\ 0.0062 - 0.0284i \\ 0.0123 - 0.0533i \\ 0.0533 + 0.0123i \end{Bmatrix}$

Let Ω_k denote a critical speed and $\{u_k\}$ denote the corresponding mode vector in the tabulation. For $k = 1$ and $k = 3$, the table indicates that $u_{2k} = iu_{1k}$ and $u_{3k} = -iu_{4k}$, whereas $u_{22} = -iu_{12}$ and $u_{32} = iu_{42}$. The modal vibration response is obtained by computing

$$\begin{bmatrix} Y_C & Z_C & \beta_Y & \beta_Z \end{bmatrix}^T = \text{Re}\{\{u_k\}\exp(i\lambda_k \tau)\}$$

If we omit the exponential decay factor associated with the real part of the eigenvalue and replace $\text{Im}(\lambda_k)$ with the critical rotation rate, the preceding leads to

$$k = 2: \begin{cases} Y_C = |u_{1k}| \cos[\Omega_k \tau + \arg(u_{1k})] \\ Z_C = |u_{1k}| \sin[\Omega_k \tau + \arg(u_{1k})] \\ \beta_Z = |u_{4k}| \cos[\Omega_k \tau + \arg(u_{4k})] \\ -\beta_Y = |u_{4k}| \sin[\Omega_k \tau + \arg(u_{4k})] \end{cases}$$

$$k = 1 \text{ and } k = 3: \begin{cases} Y_C = |u_{1k}| \cos[\Omega_k \tau + \arg(u_{1k})] \\ Z_C = -|u_{1k}| \sin[\Omega_k \tau + \arg(u_{1k})] \\ \beta_Z = |u_{4k}| \cos[\Omega_k \tau + \arg(u_{4k})] \\ -\beta_Y = -|u_{4k}| \sin[\Omega_k \tau + \arg(u_{4k})] \end{cases}$$

The second critical modal vibration, which occurs at $\omega = \Omega_2$, is such that $|Z_C| = |Y_C|$ and $|\beta_Y| = |\beta_Z|$, with Z_C and $-\beta_Y$ respectively *lagging* Y_C and β_Z by 90°. Because $Z_C/Y_C = \tan[+\Omega_k t + \arg(u_{1k})]$ and $\beta_Z/(-\beta_Y) = \tan[+\Omega_k t + \arg(u_{4k})]$, we conclude that the deformed shape of the shaft rotates at $+\Omega_k \bar{i}$, which matches the shaft's rotation. This critical mode executes a forward precession. The deflected shape of the centerline remains constant in this motion, such that center C follows a circular path. In contrast, critical modes 1 and 3 are such that Z_C and $-\beta_Y$ respectively *lead* Y_C and β_Z by 90°, with $|Z_C| = |Y_C|$ and $|\beta_Y| = |\beta_Z|$. The deformed shape of the shaft in this case rotates at $-\Omega_k \bar{i}$, which is exactly opposite the shaft's rotation. Hence, the first and third critical modes execute backward precession. In either type of precession, cross-sections of the shaft rotate about the centerline at $+\Omega \bar{k}$. The modal pattern at each critical speed is displayed with displacement and rotations greatly exaggerated.

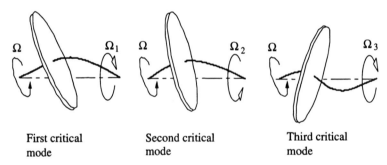

First critical mode   Second critical mode   Third critical mode

Mode shapes of the Stodola-Green model at the critical speeds, simple supports, $r/L = 0.5$.

The alternative to identifying critical speeds from a Campbell diagram is to compute the steady-state displacement amplitudes as a function of Ω. Toward that end we substitute $\{q\} = \text{Re}\{\{U\} \exp(i\Omega t)\}$ into eq. (12.4.10) and solve for $\{U\}$ at a set of incremental values of Ω, that is,

$$[[K'] + i\Omega[G'] + i\Omega[C'] - \Omega^2[M']]\{U\} = \frac{\varepsilon}{L}\Omega^2[1 \ -i \ 0 \ 0]^T$$

The result of this computation appears in the last graph.

In general, peak amplitudes occur at critical speeds. The graph indicates only one peak, which occurs at Ω_2. Although the amplitudes show no peak at Ω_1 and Ω_3, there is no disagreement with the Campbell diagram. The situation merely reflects the fact that the effective force due to imbalance rotates at the angular speed of the shaft. This is a synchronous force, which only excites synchronous modes. The first and third modes, which

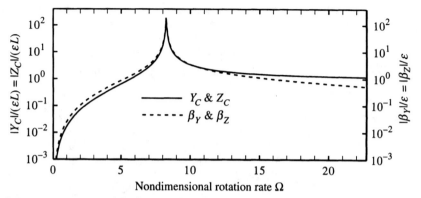

Frequency response of the Stodola-Green model of a thin disk at one third the span between simple bearing supports, $r/L = 0.5$, $\kappa_x = 0.707 r/L$, $\kappa_y = 0.50 r/L$, $\zeta = 0.01$.

precess backward, are orthogonal to the synchronous excitation. Whether the absence of resonant excitation actually makes the first and third critical speeds safe operational speeds depends on other circumstances. For example, excitation of the base due to ground vibration and small levels of orthotropy in the bearings can excite all critical modes (Childs, 1992).

This example illustrates a number of general phenomena encountered in the Stodola-Green model. In comparison to the Jeffcott model, inclusion of gyroscopic effects leads to the possibility of multiple critical speeds in forward or backward whirl, with the further possibility that some potentially resonant modes might not be excited. In both models, the isotropic nature of the system properties leads to whirling motion in which the rotor center follows a circular orbit. The difference is that gyroscopic effects couple flexural rotation in the horizontal and vertical planes. As a result, the deflected centerline of the shaft does not lie in a plane, whereas it does in the Jeffcott model.

12.5 FLEXIBLE ORTHOTROPIC SHAFTS

This section examines the implications of orthotropy of the shaft, which can arise either because the shaft has a noncircular cross-section or because of defects, such as stress fractures, in circular shafts. It is natural in such circumstances to define motion relative to a reference frame that rotates at the nominal shaft speed. The model we will develop is analogous to the Jeffcott model, in that gyroscopic effects will be ignored. However, internal and external damping effects will be included. The free body diagram for the system appears in Figure 12.9, where the generalized coordinates y_C and

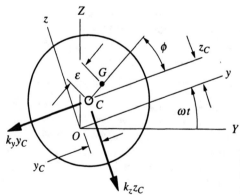

FIGURE 12.9 Free body diagram showing forces exerted on a rotor by a flexible orthotropic shaft.

z_C locate the rotor center relative to the xyz reference frame that rotates at angular speed $\omega \overline{k}$ with its origin on the bearing axis.

We assume that the shaft is symmetric with respect to either the xz or xy plane, so that bending in the two planes is uncoupled elastically; the stiffness coefficients are k_y and k_z for flexure in the respective planes. The $[K]$ matrix corresponding to using y_C and z_C as generalized coordinates is diagonal, with $K_{11} = k_y$ and $K_{22} = k_z$. The angle ϕ locates the center of mass relative to the y axis, so $\overline{r}_{G/C} = \varepsilon \cos(\phi)\overline{j} + \varepsilon \sin(\phi)\overline{k}$ and $\overline{r}_{C/O} = y_C\overline{j} + z_C\overline{k}$ at any instant. Because $\overline{r}_{G/C}$ does not change when viewed from xyz, the relative velocity is $(\overline{v}_G)_{xyz} = (\overline{v}_C)_{xyz} = \dot{y}_C\overline{j} + \dot{z}_C\overline{k}$. The absolute velocity of the center of mass therefore is

$$\overline{v}_G = \overline{v}_O + (\overline{v}_G)_{xyz} + \omega\overline{k} \times (\overline{r}_{G/C} + \overline{r}_{C/O})$$

$$= [\dot{y}_C - \omega z_C - \varepsilon\omega\sin(\phi)]\overline{j} + [\dot{z}_C + \omega y_C + \varepsilon\omega\cos(\phi)]\overline{k} \tag{12.5.1}$$

Using this expression to describe $\overline{v}_G \cdot \overline{v}_G$ leads to

$$T = \tfrac{1}{2}m[\dot{y}_C^2 + \dot{z}_C^2 - 2\omega\dot{y}_C z_C + 2\omega y_C \dot{z}_C + \omega^2 y_C^2 + \omega^2 z_C^2 + 2\varepsilon\omega\dot{y}_C\sin(\phi)$$

$$+ 2\varepsilon\omega\dot{z}_C\cos(\phi) + 2\varepsilon\omega^2 z_C\sin(\phi) + 2\varepsilon\omega^2 y_C\cos(\phi) + \varepsilon^2\omega^2] + \tfrac{1}{2}I_G\omega^2 \tag{12.5.2}$$

The effects of internal and external damping are incorporated into the model by using eq. (12.1.5). Based on considering ω to be constant, the equations of motion that result are

$$m\ddot{y}_C + (c_I + c_E)\dot{y}_C - 2m\omega\dot{z}_C + (k_y - m\omega^2)y_C - c_E\omega z_C = \varepsilon m\omega^2 \cos(\phi)$$

$$m\ddot{z}_C + (c_I + c_E)\dot{z}_C + 2m\omega\dot{y}_C + (k_y - m\omega^2)z_C + c_E\omega y_C = \varepsilon m\omega^2 \sin(\phi) \tag{12.5.3}$$

These equations in the balanced, conservative case, $\varepsilon = c_E = c_I = 0$, are the same as those in Example 11.1.

Before we address the solution of the preceding equations, it is instructive to consider the equivalent set of equations that would be obtained if we had used as generalized coordinates displacement components relative to the fixed Y and Z directions. All terms except the elastic ones would be like those in eqs. (12.3.1). We may obtain the stiffness coefficients from the strain energy, for which we need the rotation transformation,

$$y_C = Y_C \cos(\omega t) + Z_C \sin(\omega t)$$

$$z_C = -Y_C \sin(\omega t) + Z_C \cos(\omega t) \tag{12.5.4}$$

We substitute these expressions into V, which gives

$$V = \tfrac{1}{2}k_y y_C^2 + \tfrac{1}{2}k_z z_C^2$$

$$= \tfrac{1}{2}\{[k_y\cos(\omega t)^2 + k_z\sin(\omega t)^2]Y_C^2 + [k_y\sin(\omega t)^2 + k_z\cos(\omega t)^2]Z_C^2 \tag{12.5.5}$$

$$+ 2[(k_y - k_z)\sin(\omega t)\cos(\omega t)]Y_C Z_C\}$$

We match the coefficients of the preceding to the standard linearized form of V, and invoke the double-angle formulas for sine and cosine functions, which leads to the following equations of motion in the dissipationless case:

$$m\ddot{Y}_C + \tfrac{1}{2}[(k_y + k_z) + (k_y - k_z)\cos(2\omega t)]Y_C$$

$$+ \tfrac{1}{2}[(k_y - k_z)\sin(2\omega t)]Z_C = \varepsilon\omega^2\cos(\omega t + \phi)$$

$$m\ddot{Z}_C + \tfrac{1}{2}[(k_y + k_z) - (k_y - k_z)\cos(2\omega t)]Z_C \tag{12.5.6}$$

$$+ \tfrac{1}{2}[(k_y - k_z)\sin(2\omega t)]Y_C = \varepsilon\omega^2\sin(\omega t + \phi)$$

In other words, the differential equations of motion for fixed reference frame displacements have periodic coefficients. They fall in the class of Hill's equations, which are quite difficult to solve analytically for arbitrary ω (Ince, 1956). At certain values of ω, the solutions are periodic, but not harmonic. These parameter combinations mark the boundary between stable and unstable responses; the response in the latter case is said to be a *parametric resonance*. By using the rotating reference frame displacements y_C and z_C, we have obtained a set of differential equations of motion with constant coefficients, which are much easier to solve.

It is convenient to nondimensionalize the equations of motion in a manner similar to the analysis of orthotropic bearing effects, Section 12.3. Hence, we define

$$\omega_{\text{ref}} = \left[\frac{(k_z + k_y)}{2m}\right]^{1/2}, \qquad \sigma = \frac{k_z - k_y}{k_z + k_y}, \qquad \tau = \omega_{\text{ref}}t, \qquad \Omega = \frac{\omega}{\omega_{\text{ref}}}$$

$$\zeta_E = \frac{c_E}{2m\omega_{\text{ref}}}, \qquad \zeta_I = \frac{c_I}{2m\omega_{\text{ref}}}, \qquad \zeta = \zeta_E + \zeta_I \tag{12.5.7}$$

Without loss of generality, we may consider $k_z \geq k_y$, so $0 \leq \sigma < 1$. The corresponding nondimensionalized equations of motion are

$$\frac{d^2}{d\tau^2}\begin{Bmatrix} y_C \\ z_C \end{Bmatrix} + \begin{bmatrix} 2\zeta & -2\Omega \\ 2\Omega & 2\zeta \end{bmatrix}\frac{d}{d\tau}\begin{Bmatrix} y_C \\ z_C \end{Bmatrix}$$

$$+ \begin{bmatrix} (1 - \sigma - \Omega^2) & -2\zeta_E\Omega \\ 2\zeta_E\Omega & (1 + \sigma - \Omega^2) \end{bmatrix}\begin{Bmatrix} y_C \\ z_C \end{Bmatrix} = \begin{Bmatrix} \varepsilon\Omega^2\cos(\phi) \\ \varepsilon\Omega^2\sin(\phi) \end{Bmatrix} \tag{12.5.8}$$

We begin by searching for conditions leading to instability and resonances. Toward that end we shall formulate the nonlinear eigenvalue problem according to eqs. (12.2.12) and (12.2.13). Thus, we set

$$\begin{Bmatrix} y_C \\ z_C \end{Bmatrix} = \{u\}\exp(\lambda t) \tag{12.5.9}$$

which leads to

$$\begin{bmatrix} (\lambda^2 + 2\zeta\lambda + 1 - \sigma - \Omega^2) & -2\Omega\lambda - 2\zeta_E\Omega \\ 2\Omega\lambda + 2\zeta_E\Omega & (\lambda^2 + 2\zeta\lambda + 1 + \sigma - \Omega^2) \end{bmatrix}\{u\} = \begin{Bmatrix} 0 \\ 0 \end{Bmatrix} \tag{12.5.10}$$

The corresponding characteristic equation, written in a form suitable for the Routh-Hurwitz criteria, is

$$b_0 \lambda^{2N} + b_1 \lambda^{2N-1} + \cdots + b_{2N-1} \lambda + b_{2N} = 0$$

$$b_0 = 1, \qquad b_1 = 4\zeta, \qquad b_2 = 2(\Omega^2 + 1 + 4\zeta^2) \tag{12.5.11}$$

$$b_3 = 4\Omega^2(2\zeta_E - \zeta) + 4\zeta, \qquad b_4 = (\Omega^2 - 1)^2 - \sigma^2 + 4\zeta_E^2 \Omega^2$$

Let us begin with the dissipationless case, in which case the eigenvalues satisfy

$$\lambda^2 = -(1 + \Omega^2) \pm (4\Omega^2 + \sigma^2)^{1/2} \tag{12.5.12}$$

Both λ^2 values obtained from this equation are real, which means that the system will be marginally stable (purely imaginary eigenvalues) or unstable (one or more positive real eigenvalues). The range in which the latter condition occurs is $1 - \sigma < \Omega^2 < 1 + \sigma$. The dimensional condition for stability therefore is

$$\omega < \left(\frac{k_y}{m}\right)^{1/2} \quad \text{or} \quad \omega > \left(\frac{k_z}{m}\right)^{1/2} \tag{12.5.13}$$

The system exhibits a parametric instability if the rotation rate lies between the natural frequencies of the nonrotating system. This instability does not occur in a circular shaft, for which $k_z = k_y$.

The characteristic equation when damping is present is truly quartic, rather than biquadratic, so we shall employ the Routh-Hurwitz criteria to ascertain the range of rotation rates where the system is unstable. Some of the conditions for stability listed in eqs. (11.3.5) are identically satisfied. The ones that are not are

$$b_3 > 0, \qquad b_4 > 0, \qquad |[\Delta_3(\Omega)]| = b_1 b_2 b_3 - b_1^2 b_4 - b_0 b_3^2 > 0 \tag{12.5.14}$$

A symbolic math program is particularly useful for simplifying the last criterion. The resulting conditions required for the system to be stable are

$$\text{I:} \quad \zeta - (\zeta - 2\zeta_E)\Omega^2 > 0$$

$$\text{II:} \quad \Omega^4 - 2(1 - 2\zeta_E^2)\Omega^2 + 1 - \sigma^2 > 0 \tag{12.5.15}$$

$$\text{III:} \quad -(\zeta - \zeta_E)^2\Omega^4 + \zeta^2[1 - (\zeta - \zeta_E)^2]\Omega^2 + \zeta^2(\zeta^2 + \tfrac{1}{4}\sigma^2) > 0$$

These conditions are linear or biquadratic in Ω^2, which allows us to solve the inequalities. Because $\zeta - \zeta_E = \zeta_I$, the response will be stable if the following conditions are all met:

$$\text{I:} \quad \begin{cases} \zeta_E \geq \zeta_I \quad \text{or} \\[2mm] \Omega^2 < \dfrac{\zeta}{\zeta_I - \zeta_E} \quad \text{if } \zeta_E < \zeta_I \end{cases}$$

$$\text{II:} \quad \begin{cases} \Omega^2 < 1 - 2\zeta_E^2 - (\sigma^2 - 4\zeta_E^2 + 4\zeta_E^4)^{1/2} \quad \text{or} \\[2mm] \Omega^2 > 1 - 2\zeta_E^2 + (\sigma^2 - 4\zeta_E^2 + 4\zeta_E^4)^{1/2} \end{cases} \tag{12.5.16}$$

$$\text{III:} \quad \Omega^2 < \frac{\zeta^2}{2\zeta_I^2}\left\{(1 - \zeta_I^2) + \frac{1}{2}\left[(1 + \zeta_I^2)^2 + \sigma^2\frac{\zeta_I^2}{\zeta^2}\right]^{1/2}\right\}$$

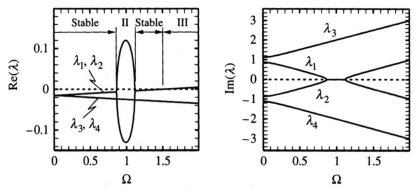

FIGURE 12.10 Stable and unstable regions for a rotor on an orthotropic shaft, $\sigma = 0.25$, $\phi = 30°$, $\zeta_E = 0.005$, $\zeta_I = 0.010$.

Because $\zeta \geq \zeta_I$, it can be shown that condition III gives a lower value for Ω^2 than does condition I. When $\sigma = 0$, this condition reduces to the flutter stability condition for isotropic shafts, eq. (12.2.11). It follows that orthotropy, $\sigma > 0$, raises the rotational rate at which internal damping causes a flutter instability. Condition II reduces to eq. (12.5.13) when $\zeta_E = 0$, and it is independent of internal damping. This shows that orthotropy of the shaft leads to a range of rotation rates at which the system has a parametric instability, regardless of the amount and type of damping. Typical results for the eigenvalues are depicted in Figure 12.10, which shows graphically that there are two ranges of safe operation.

Now let us turn our attention to the steady-state response and the associated issue of critical speeds. Suppose we have determined the state-space modal properties. The state-space excitation corresponding to eq. (12.5.8) is

$$\left\{ \begin{array}{c} \{0\} \\ \{Q\} \end{array} \right\} = \varepsilon\Omega^2 \begin{bmatrix} 0 & 0 & \cos(\phi) & \sin(\phi) \end{bmatrix}^{\mathrm{T}} \tag{12.5.17}$$

The modal transformation $\{x\} = [\Psi]\{\xi\}$ leads to

$$\dot{\xi}_n + \lambda_n \xi_n = [\widetilde{\Psi}_n]^{\mathrm{T}} \left\{ \begin{array}{c} \{0\} \\ \{Q\} \end{array} \right\} = \varepsilon\Omega^2 [\widetilde{\Psi}_{3n} \cos(\phi) + \widetilde{\Psi}_{4n} \sin(\phi)] \tag{12.5.18}$$

Because the modal excitation terms are constants, the steady-state modal response will be

$$\xi_n = \varepsilon\Omega^2 \frac{\widetilde{\Psi}_{3n} \cos(\phi) + \widetilde{\Psi}_{4n} \sin(\phi)}{\lambda_n} \tag{12.5.19}$$

This shows that the steady-state modal coordinates are independent of time, so the physical displacement will feature constant values of y_C and z_C. Critical speeds correspond to peak steady-state amplitudes, which occurs in this case when one of the values $|\lambda_n|$ has a minimum. If ζ_E and ζ_I are reasonably small, this condition may be approximated as $|\text{Im}(\lambda_n)| = 0$. This condition is the analogue for rotating reference frame formulations of the synchronous line, $|\text{Im}(\lambda_n)| = \omega$, for a Campbell diagram. For orthtropic shafts the condition $|\text{Im}(\lambda_n)| = 0$ for orthotropic shafts also approximates the condition II parametric stability limits when ζ_E is small.

We may determine the constant steady-state values of y_C and z_C directly. Setting the time derivatives in eq. (12.5.8) to zero yields

$$\begin{Bmatrix} y_C \\ z_C \end{Bmatrix} = \begin{bmatrix} (1 - \sigma - \Omega^2) & -2\zeta_E \Omega \\ 2\zeta_E \Omega & (1 + \sigma - \Omega^2) \end{bmatrix}^{-1} \begin{Bmatrix} \varepsilon\Omega^2 \cos(\phi) \\ \varepsilon\Omega^2 \sin(\phi) \end{Bmatrix}$$

$$= \frac{\varepsilon\Omega^2}{[(1 - \Omega^2)^2 - \sigma^2 + 4\zeta_E^2 \Omega^2]} \begin{Bmatrix} (1 + \sigma - \Omega^2)\cos(\phi) + 2\zeta_E \Omega \sin(\phi) \\ (1 - \sigma - \Omega^2)\sin(\phi) - 2\zeta_E \Omega \cos(\phi) \end{Bmatrix}$$

(12.5.20)

The radial distance from the bearing axis to the center C is $R = (x_C^2 + y_C^2)^{1/2}$, which gives

$$R_C = \varepsilon\Omega^2 \frac{(1 - \Omega^2)^2 + \sigma^2 + 4\zeta_E^2 \Omega^2 + 2(1 - \Omega^2)\sigma\cos(2\phi) + 4\sigma\zeta_E\Omega\sin(2\phi)}{(1 - \Omega^2)^2 - \sigma^2 + 4\zeta_E^2 \Omega^2}$$

(12.5.21)

This constant distance is the radius of the circular path that the center of the rotor follows, so the system undergoes synchronous forward whirl.

The value of ζ_I does not affect the steady-state amplitudes, because the deformation of the shaft is constant when center C is at a constant radial distance. The denominator in eq. (12.5.20) is identical to the left side of the condition II stability criterion, eqs. (12.5.15). This confirms that critical Ω values leading to maximum y_C and z_C are the same as the Ω values bounding regions of parametric instability. A typical plot of amplitude as a function of rotation rate appears in Figure 12.11. Only the response in regions of stability is depicted.

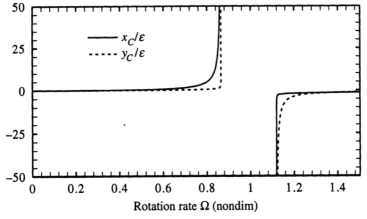

FIGURE 12.11 Steady-state displacement coordinates for a rotor on an orthotropic shaft, $\sigma = 0.25$, $\phi = 30°$, $\zeta_E = 0.005$, $\zeta_I = 0.010$.

REFERENCES AND SELECTED READINGS

CHILDS, D. 1992. *Rotordynamics of Turbomachinery.* John Wiley & Sons, New York.

DIMARGONAS, A. 1996. *Vibration for Engineers*, 2nd ed. Prentice Hall, Englewood Cliffs, NJ.

GINSBERG, J. H. 1995. *Advanced Engineering Dynamics*, 2nd ed. Cambridge University Press, Cambridge, England.

GREEN, R. 1948. "Gyroscopic Effects of the Critical Speeds of Flexible Rotors," *Journal of Applied Mechanics*, 15, 369–376.

HIBBELER, R. C. 1998. *Engineering Mechanics-Dynamics*, 8th ed. Prentice Hall, Englewood Cliffs, NJ.

INCE, E. L. 1956. *Ordinary Differential Equations.* Dover, New York.

LEE, C. W. 1993. *Vibration Analysis of Rotors.* Kluwer Academic Publishers, Norwell, MA.

SHAMES, I. H., & DYM, C. L. 1985. *Energy and Finite Element Methods in Structural Mechanics.* Hemisphere, New York.

STODOLA, A. 1924. *Dampf- und Gasturbinen.* Springer-Verlag, Berlin.

VANCE, J. M. 1988. *Rotordynamics of Turbomachinery.* John Wiley & Sons, New York.

WEAVER, W., TIMOSHENKO, S., & YOUNG, D. H. 1990. *Vibration Problems in Engineering*, 5th ed. John Wiley & Sons, New York.

EXERCISES

12.1 Laser measurements of a whirling rotor indicate that the peak radius for synchronous whirl occurs when the system rotates at 1200 rev/min, with maximum vertical displacement of the rotor center being 10 mm. Impulse response measurements suggest that the system will flutter if the rotation rate exceeds 3600 rev/min. It also is observed that the amplitude of the vertical displacement slightly below 3600 rev/min is 0.5 mm. Determine the corresponding values of ω_{nat}, ε, ζ_I, and ζ_E.

12.2 A steel circular shaft having a diameter of 50 mm supports a thin 20 kg disk at the middle of its 750 mm span. Both bearings are roller-type, which may be modeled as applying clamped boundary conditions to the shaft. External damping is estimated to be 1% of critical.

(a) What is the maximum amount of internal damping for which the system may be safely operated at a speed of 15,000 rev/min?

(b) If internal damping is at the level identified in part (a), what is the maximum eccentricity for which the rotor displacement will not exceed 10 mm when the system rotates at 15,000 rev/min.?

12.3 Determine the response of the system in Example 12.1 in the case where $\omega = 1.5\omega_{nat}$. Plot the corresponding path followed by the rotor center.

12.4 Determine the response of the system in Example 12.1 in the case where the rotation rate is 1% greater than the rate for onset of the flutter instability. Plot the corresponding path followed by the rotor center.

12.5 A 100 kg rotor is mounted on a short shaft that is much less deformable than the bearing sup-

ports. It is observed that when the shaft is not turning the natural frequencies for vibration in the horizontal and vertical direction are 100 Hz and 150 Hz, respectively. Internal and external damping factors are estimated to be $\zeta_E = \zeta_I = 0.03$. Determine the maximum permissible rotation rate, and the critical rotation rates for this system.

12.6 The bearings of a short shaft are mounted on an elastic structure. The rotor mass is 50 kg. It is observed that a 1 kN force applied statically to either bearing in the vertical Y direction causes the bearing to displace 2 mm in that direction. It also is observed that application of a 1 kN force in the transverse horizontal Z direction causes the bearing to displace 4 mm in that direction. Internal and external damping factors are estimated to be $\zeta_E = \zeta_I = 0.02$.

(a) Determine the maximum permissible rotation rate and the critical rotation rates for this system.

(b) Determine the ratio of the displacement amplitudes of the centroid C to its eccentricity when the shaft rotates at the critical speeds.

12.7 Consider the orthotropic Jeffcott model in a situation where $\sigma = -0.10$. Determine and graph the steady-state amplitude $|X_C|/\varepsilon$ and $|Y_C|/\varepsilon$ for three combinations of damping ratios that give $\zeta = 0.02$: (a) $\zeta_E = 0.02$, $\zeta_I = 0$; (b) $\zeta_E = 0$, $\zeta_I = 0.02$; and (c) $\zeta_E = \zeta_I = 0.01$. What conclusions can one derive from a comparison of the results regarding the role of internal and external damping?

12.8 A 200 kg rotor is mounted midspan on a short shaft. The bearings at both ends are mounted on a shock absorbing system. Measurements made prior to installation of the shaft indicate that the relationship between forces F_Y and F_Z(N) applied to the bearings in the horizontal and vertical directions and displacements Y and Z(m) in those directions is

$$\begin{Bmatrix} F_Y \\ F_Z \end{Bmatrix} = \begin{bmatrix} 5 & -3 \\ -3 & 11 \end{bmatrix}\left\{ 10^6 \begin{Bmatrix} Y \\ Z \end{Bmatrix} + 2(10^3) \begin{Bmatrix} \dot{Y} \\ \dot{Z} \end{Bmatrix} \right\}$$

Measurements of the shaft behavior indicate that the internal damping coefficient is $c_I = 8(10^3)$ N-s/m. Construct a Campbell diagram for the system, and from it determine the critical speeds and the largest rotation at which the system is stable.

12.9 The sketch shows a four-bladed turboprop propeller at the end of a short shaft. Although the propeller is not axisymmetric with respect to the shaft, its mass distribution is such that it behaves dynamically as though it were, with $I_{xx} = 2I_{yy} = 80$ kg-m^2 and $m = 100$ kg. Consider the situation where the propellers are rigid, and the shaft acts like a cantilevered beam. The flexibility coefficients for the shaft are $L_{ww} = L^3/3EI$, $L_{\beta w} = L^2/2EI$, and $L_{\beta\beta} = L/EI$. The shaft is steel with a 100 mm cross-sectional diameter. The eccentricity of the center of mass is $\varepsilon = 2$ mm and the damping ratios are $\zeta_w = \zeta_\beta = 0.005$. Use frequency domain analysis to determine the steady-state amplitudes of the motion of the propeller's center as a function of rotation rate. What are the critical rotation rates identified by this analysis?

12.10 Consider the system in Exercise 12.9. What are the critical rotation rates according to the Campbell diagram for the system?

12.11 The sketch shows a model that can be used to account for gyroscopic effects in combination with elasticity of the bearings. The shaft is taken to be rigid for this model. If the springs at each end are unequal, or the rotor is not situated at the mid-span, that is $b \neq L/2$, then Y_C and β_Z are elastically coupled, as are Z_C and β_Y. The expression for kinetic energy is the same as for the Stodola-Green model, so β_Y and β_Z are coupled by gyroscopic effects. Consider a case where the rotor is a thin disk with $m = 20$ kg and $r = 300$ mm. The shaft is steel with a 50 mm diameter and its span is $L = 900$ mm. The rotor is situated at $b = 150$ mm. The bearings are isotropically elastic, with $k_{YA} = k_{ZA} = 800$ kN/m, $k_{YB} = k_{ZB} = 600$ kN/m, and $\varepsilon = 1$ mm.

(a) Derive the equations of motion for this system.

(b) Construct the Campbell diagram for the system, and identify the critical rotation rates.

(c) Describe the spatial motion of the shaft's centerline in each critical mode. Explain why these patterns cause the critically excited modes to be referred to as cylindrical and conical modes.

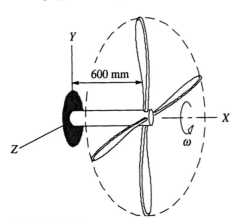

EXERCISE 12.9

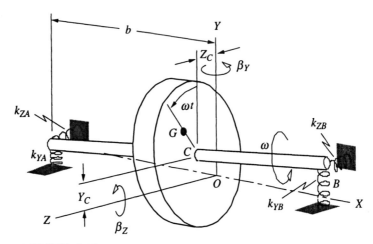

EXERCISE 12.11

12.12 If k_Y and k_Z for a bearing are allowed to be different, the system in Exercise 12.11 can be used to describe the combined effects of gyroscopic action and orthotropic bearing stiffness. Perform the analysis called for in the aforementioned exercise when all parameters are as stated there, except $k_{ZA} = 1600$ kN/m.

12.13 The model in Exercise 12.11 considers the role of gyroscopic effects with isotropic bearings, and Exercise 12.12 modifies the system by allowing for orthotropic bearings. Suppose that dissipation is well represented by a structural damping model, so that the complex stiffness is $[K][1 + i\gamma]$, where $\gamma = 0.01$ is the loss factor. Use frequency domain analysis to determine the steady-state amplitudes of Y_C and Z_C at the critical speed for each model. What does a comparison of the results suggest regarding the effect of orthotropy on displacement amplitudes?

12.14 In order to assess the relative significance of orthotropic bearings and shafts, consider the modified Jeffcott models in Figures 12.9 and 12.4. Suppose the stiffnesses $2k_X$ and $2k_Y$ and mass m for the bearing model respectively equal k_y, k_z, and m for the shaft model. Determine the stable ranges of Ω and the critical values of Ω predicted by each model when $\sigma = 0.05$ and $\sigma = 0.50$. The damping ratios are $\zeta_E = 0.005$ and $\zeta_I = 0.01$.

12.15 In order to assess the relative significance of the loss of symmetry for bearings and shafts, consider the modified Jeffcott models in Figures 12.9 and 12.4. Suppose the stiffnesses $2k_X$ and $2k_Y$ and mass m for the bearing model respectively equal k_y, k_z, and m for the shaft model. For the case where $\sigma = 0.5$, $\Omega = 0.65$, determine the fixed reference frame displacement components X_C and Y_C predicted by each model. Then use those results to graph the path followed by center-point C according to each model.

12.16 As shown in the sketch, the cam exerts a constant force $F = \beta m \omega_{ref}^2$ on the perimeter of the rotor, where ω_{ref} is defined in eqs. (12.5.7). The line of action of this force always intersects the center of the rotor. Determine the steady-state amplitudes $|y_C|/\varepsilon$ and $|z_C|/\varepsilon$ as functions of the nondimensional rotation rate. Parameters for the analysis are $\beta = \varepsilon$, $\sigma = 0.8$, $\phi = 0$, $\zeta_E = \zeta_I = 0.01$.

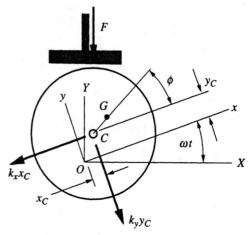

EXERCISE 12.16

12.17 An orthotropic shaft, for which $\sigma = 0.4$, $\phi = 0$, $\zeta_E = 0.004$, and $\zeta_I = 0.01$, is spinning at a constant rate with the center of the rotor constrained to the bearing axis, $y_C = z_C = 0$. At $t = 0$, this constraint is removed. Determine the ensuing responses y_C and z_C as functions of time. Then use the rotation transformation to determine the fixed reference frame components Y_C and Z_C at each instant, and graph the corresponding path. The rotation rate is 1% less than the lowest rate at which a condition II instability occurs.

12.18 Perform the analysis in Exercise 12.17 for the case where the rotation rate is 1% less than the lowest rate for a condition III instability.

LAGRANGE'S EQUATIONS

If we define a system's generalized coordinates relative to a reference state of motion, we say that the system is *time dependent*. A simple example is the spring-mass system mounted on a horizontal turntable, shown in Figure A.1(a). The groove restricts the block to slide radially relative to the turntable. The rotation rate ω is constant, so the angle of rotation ωt is considered to be known. The position of the block is therefore specified by the radial distance R, which means that the system has one degree of freedom. If the turntable does not rotate, the radial distance to the mass at static equilibrium equals the unstretched length of the spring, $R = L_0$. A steady rotation rate ω leads to a state of dynamic equilibrium at which the radial distance is constant at a value R_0 larger than L_0. (We could determine R_0 from Newton's Second Law.) If a small disturbance, such as an initial position that does not quite coincide with the dynamic equilibrium position, is applied, the result will be an oscillatory response in which R varies relative to R_0 by a small amount. We have two basic alternatives as to how we could define the generalized coordinate describing such a response. We could use a position coordinate measured relative to the fixed reference XYZ. However, in that case we would find that the generalized coordinate is not small because the angle of rotation ωt is large. The power balance method cannot describe situations where the generalized coordinates are large, because doing so requires retention of nonlinear effects.

We could instead use a generalized coordinate that locates the mass relative to the current position of the turntable. For example, a suitable generalized coordinate for a vibratory response would be $q_1 = R - R_0$, which could be considered to be sufficiently small to perform linearization. Even then, we cannot use the power balance method because of the action of constraint forces. To understand this issue, consider the free body diagram of the block in Figure A.1(b). The normal force $\overline{N}$ is

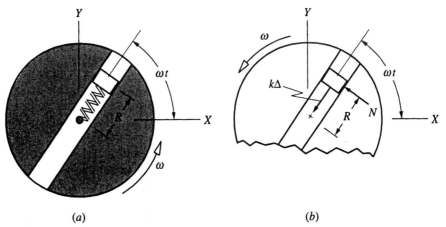

(a) (b)

FIGURE A.1 Example of a time-dependent system. (*a*) Rotating turntable. (*b*) Free body diagram of the sliding block.

a constraint force exerted by the groove walls; it imposes the kinematical restriction that radial motion is the only possible movement of the block relative to the turntable. In terms of polar coordinates the velocity components of the block are $v_x = \dot{R}$, $v_y = R\omega$. Because potential energy accounts for the effect of the spring, the power input is $\mathcal{P}_{in} = Nv_y = NR\omega$. Recall that we are using $R = q_1 + R_0$, so $\mathcal{P}_{in} = N\omega(q_1 + R_0)$. Two aspects of this expression illustrate the basic difficulty with the power balance method. First, note that $\mathcal{P}_{in}$ does not contain a $\dot{q}_1$ factor, so it does not fit the standard form for power input, eq. (1.5.23). Equally troublesome is the fact that the reaction force appears in $\mathcal{P}_{in}$. In contrast, a primary reason for using power balance for time-invariant systems is its ability to avoid consideration of reaction forces. The occurrence of constraint forces in the power input is typical of time-dependent systems. It stems from the fact that the constraint forces can impose specified states of motion, whereas the constraint forces for time-invariant systems prevent motion. We seek to derive equations of motion that do not contain the (unknown) constraint forces.

A.1 DERIVATION

For the sake of simplicity, the derivation of Lagrange's equations presented here assumes that there are no auxiliary kinematical relations to be satisfied by the generalized coordinates. Given the definition of generalized coordinates as variables that locate a system, it might seem as though this would always be the case. However, there are systems, such as bodies that roll, in which velocity variables must satisfy certain conditions. Because velocity is the derivative of position, such *velocity constraints* represent additional differential equations that must be satisfied by the general coordinates. If these differential equations governing velocity cannot be solved independently of the equations of motion, we say that the system is *nonholonomic* (the Greek word for "not integrable"). A general derivation of Lagrange's equations for nonholonomic systems may be found in the text by Ginsberg (1995), but that development introduces a number of concepts that we may avoid by restricting our attention to holonomic systems.

If N is the number of degrees of freedom for a holonomic system, a set of N generalized coordinates, unrelated by auxiliary kinematical conditions, defines the system's position. Any combination of these variables represents a *kinematically admissible state*; which of these states actually occurs depends on the forces that are applied to the system. For this reason, the generalized coordinates are said to be *unconstrained*.

The desire to derive equations of motion that do not contain constraint forces led d'Alembert to consider the notion of *virtual displacement*. In essence, a virtual displacement is an infinitesimal displacement that would be produced if the generalized coordinates at an arbitrary instant were given infinitesimal increments, *without changing the value of the time variable t*. The word "virtual" indicates the fact that such a displacement does not describe the actual motion of the system. When a virtual displacement is imparted to a system, there is no motion in the direction of the constraint forces, because time does not change. For example, in the case of the turntable in Figure A.1, a virtual displacement entails incrementing $q_1 = R - R_0$, with the rotation angle ωt held constant. As a result, the block moves radially in the vertical displacement, so the normal force $\bar{N}$ does no work. This feature is quite general. In fact, it may be used to express a consistent definition of a constraint force, specifically, *a con-*

straint force is any force that imposes a restriction on motion, and does no work in any movement that does not violate that restriction.

The work done by the forces when a system undergoes a virtual displacement is the *virtual work*, which we shall denote as δW. (The symbol δ is the standard notation used to indicate that a quantity is associated with a virtual displacement.) To characterize the virtual displacement of the particle, we recall that its position dependence is such that $\bar{r} = \bar{r}(q_j, t)$. The virtual displacement is produced by incrementing the particle's generalized coordinates by infinitesimal amounts δq_j. It is important to later developments to recognize that these increments may be assigned arbitrarily. Because t is held constant, the corresponding virtual displacement $\delta \bar{r}$ is found by the chain rule for differentials to be

$$\delta \bar{r} = \frac{\partial \bar{r}}{\partial q_1} q_1 + \frac{\partial \bar{r}}{\partial q_2} q_2 + \cdots = \sum_{j=1}^{N} \frac{\partial \bar{r}}{\partial q_j} \delta q_j \qquad (A.1.1)$$

In the foregoing, N represents the number of degrees of freedom. (For a single particle, N can be no greater than 3.)

Newton's Second Law obviously applies to the single particle under consideration, so the virtual work is given by

$$\delta W \equiv \bar{F} \cdot \delta \bar{r} = m\bar{a} \cdot \delta \bar{r} \qquad (A.1.2)$$

D'Alembert's idea was to use this relation as the foundation for extending the method of virtual work for static systems to dynamics. He reasoned that by suitably selecting the virtual displacement, one could obtain relations between the externally applied forces, excluding reactions, and the acceleration components. This perspective can be quite useful in some circumstances.

However, Lagrange used d'Alembert's development as the starting point for a formulation in terms of the kinetic energy. To do so, we substitute eq. (A.1.1) into the right side of the above expression for δW, and recall that $\bar{a} = d\bar{v}/dt$, so that,

$$\delta W = m\bar{a} \cdot \sum_{j=1}^{N} \frac{\partial \bar{r}}{\partial q_j} \delta q_j \equiv \sum_{j=1}^{N} m \frac{d\bar{v}}{dt} \cdot \frac{\partial \bar{r}}{\partial q_j} \delta q_j$$

$$\equiv \sum_{j=1}^{N} \left[\frac{d}{dt} \left(m\bar{v} \cdot \frac{\partial \bar{r}}{\partial q_j} \right) - m\bar{v} \cdot \frac{d}{dt} \left(\frac{\partial \bar{r}}{\partial q_j} \right) \right] \delta q_j \qquad (A.1.3)$$

The next step is to introduce an identity derived from eq. (A.1.1). Because $\bar{r} = \bar{r}(q_j, t)$, the velocity is found by the chain rule to be

$$\bar{v} = \sum_{n=1}^{N} \frac{\partial \bar{r}}{\partial q_n} \dot{q}_n + \frac{\partial \bar{r}}{\partial t} \qquad (A.1.4)$$

Note that the coefficients of $\dot{q}_n$ in the above are the same as the coefficients of δq_n in eq. (A.1.1). We may obtain the coefficient associated with a specific index j by differentiating $\bar{v}$ with respect to q_j, which shows that

$$\frac{\partial \bar{v}}{\partial \dot{q}_j} \equiv \frac{\partial \bar{r}}{\partial q_j} \qquad (A.1.5)$$

We will substitute this identity into the first term in eq. (A.1.3). For the second term, we observe that derivatives with respect to different variables may be taken in any sequence, so that

$$\frac{d}{dt}\left(\frac{\partial \bar{r}}{\partial q_j}\right) \equiv \frac{\partial}{\partial q_j}\left(\frac{d\bar{r}}{dt}\right) \equiv \frac{\partial \bar{v}}{\partial q_j} \tag{A.1.6}$$

These two identities lead to

$$\delta W = \sum_{j=1}^{N}\left[\frac{d}{dt}\left(m\bar{v}\cdot\frac{\partial \bar{v}}{\partial \dot{q}_j}\right) - m\bar{v}\cdot\frac{\partial \bar{v}}{\partial q_j}\right]\delta q_j \tag{A.1.7}$$

The kinetic energy of a particle is $T = \frac{1}{2}m\bar{v}\cdot\bar{v}$, and the rules for differentiation indicate that $\partial(\bar{v}\cdot\bar{v})/\partial\beta = 2\bar{v}\cdot\partial\bar{v}/\partial\beta$, where β is arbitrary. Hence, the virtual work for a single particle is related to the particle's kinetic energy by

$$\delta W = \sum_{j=1}^{N}\left[\frac{d}{dt}\left(\frac{\partial T}{\partial \dot{q}_j}\right) - \frac{\partial T}{\partial q_j}\right]\delta q_j \tag{A.1.8}$$

The above was derived by considering a single particle, but it also applies to any system of particles, including systems of rigid bodies. To prove this we observe that the kinetic energy and total virtual work of a system are the sums of the respective contributions for the constituent particles. Furthermore, the preceding expression is linear in both T and respective δW, so summing that equation for each particle merely reproduces the relation for the entire system. The crucial feature of this summation is that the constraint forces will still not contribute to the virtual work. Hence, eq. (A.1.8) relates the known external forces to the generalized coordinates.

It might seem as though eq. (A.1.8) is a single relation, whereas we need to find N equations of motion governing the generalized coordinates. However, we have not yet exploited a basic aspect of virtual displacement, whose definition states that the increments δq_j given to the generalized coordinates at any instant are independent and arbitrary. In order to make use of this arbitrariness, we use eqs. (A.1.1) and (A.1.2) to represent the virtual work. We use f as the index indicating which external force is under consideration, and $\bar{r}_f$ is the position where the force is applied. Thus,

$$\delta W = \sum_{f}\bar{F}_f\cdot\delta\bar{r}_f = \sum_{f}\bar{F}_f\cdot\sum_{j=1}^{N}\frac{\partial\bar{r}_f}{\partial q_j}\delta q_j \tag{A.1.9}$$

We define the coefficient of each virtual increment δq_j to be the generalized force Q_j. Thus, we have

$$\delta W = \sum_{j=1}^{N}Q_j\delta q_j, \quad Q_j = \sum_{f}\bar{F}_f\cdot\frac{\partial\bar{r}_f}{\partial q_j} \tag{A.1.10}$$

When we use this expression to describe the virtual work in eq. (A.1.8), we may write the result as

$$\sum_{j=1}^{N}\left[\frac{d}{dt}\left(\frac{\partial T}{\partial \dot{q}_j}\right) - \frac{\partial T}{\partial q_j} - Q_j\right]\delta q_j = 0 \tag{A.1.11}$$

This relation must be satisfied for any set of values of the δq_j quantities. We may select N different combinations, in which only one of these virtual increments is nonzero. In such a circumstance, the only possibility for satisfying the relation is that the coefficient of each δq_j vanishes identically,

$$\frac{d}{dt}\left(\frac{\partial T}{\partial \dot{q}_j}\right) - \frac{\partial T}{\partial q_j} = Q_j, \quad j = 1, 2, \ldots, N \tag{A.1.12}$$

These are *Lagrange's equations*, whose evaluation for a specific system yields N differential equations of motion.

Lagrange's equations usually are implemented in a slightly altered form. The generalized forces appearing above describe all external forces, regardless of whether or not they are conservative. However, our efforts with the power balance method showed that it is much more convenient to represent the effect of conservative forces in terms of the potential energy. The definition of a conservative force is that the work it does in the transition between any initial state 1 and final state 2 of a system is related to its potential energy at those states by

$$W_{1 \to 2} = V_1 - V_2 \tag{A.1.13}$$

Let us take these two states to be the position of the system before and after the virtual displacement. Because potential energy can only depend on the instantaneous position of the system, we know that V can depend only on the generalized coordinates and time. Time is held constant in a virtual displacement, so we have $V_1 = V(q_j, t)$ and $V_2 = V(q_j + \delta q_j, t)$. Thus the virtual work done by conservative forces is described by

$$\delta W_{\text{cons}} = V(q_j, t) - V(q_j + \delta q_j, t) \tag{A.1.14}$$

The virtual increments are infinitesimal in magnitude, which enables us to use differential calculus to represent the above difference as

$$\delta W_{\text{cons}} = -\sum_{j=1}^{N} \frac{\partial V}{\partial q_j} \delta q_j \tag{A.1.15}$$

We use this relation to partition the virtual work into contributions from nonconservative and conservative forces. Henceforth, we shall consider the generalized force parameters Q_j to consist of only forces that have not been included in the potential energy, so we have

$$\delta W = \delta W_{\text{cons}} + \delta W_{\text{nc}} = \sum_{j=1}^{N} Q_j \delta q_j - \sum_{j=1}^{N} \frac{\partial V}{\partial q_j} \delta q_j \tag{A.1.16}$$

A comparison of this representation of virtual work with eq. (A.1.12) shows that the Q_j parameters appearing previously should be replaced by $Q_j - \partial V/\partial q_j$, where the only forces now described by Q_j are those not already contained in V. Correspondingly, Lagrange's equations become

$$\boxed{\frac{d}{dt}\left(\frac{\partial T}{\partial \dot{q}_j}\right) - \frac{\partial T}{\partial q_j} + \frac{\partial V}{\partial q_j} = Q_j, \quad j = 1, 2, \ldots, N} \tag{A.1.17}$$

The preceding relations may be employed to represent the motion of any N-degree-of-freedom holonomic system. To evaluate the resulting differential equations for a specific system, one must describe the system's kinetic and potential energies, as well as the generalized forces, corresponding to the specific selection of generalized coordinates. The process of carrying out the operations prescribed by eq. (A.1.17) requires that one be cognizant of the difference between partial and total derivatives. In particular, for the partial derivatives, the generalized coordinates q_j are considered to be independent of the generalized velocities $\dot{q}_j$, so only terms that depend explicitly on the differentiation variable will contribute to the respective derivative. After evaluation of $\partial T/\partial \dot{q}_j$, the derivative with respect to t must account for the fact that the generalized coordinates and generalized velocities are implicitly functions of time.

A.2 POWER BALANCE VERSUS LAGRANGE'S EQUATIONS

We shall consider here a system whose position is located by a set of generalized coordinates measured relative to a static equilibrium. In this circumstance, the position of any point in the system depends solely on the current values of the generalized coordinates, $\bar{r} = \bar{r}(q_j)$. We further limit the discussion to a linearized description based on smallness of the generalized coordinates.

The first issue we shall consider is the method we developed in Chapter 1 for deriving the kinetic energy function. Because the position of points in a time-invariant system does not explicitly depend on t, the term $\partial \bar{r}/\partial t$ in eq. (A.1.14) is identically zero. The velocity of the point described by $\bar{r}$ then reduces to

$$\bar{v} = \sum_{n=1}^{N} \frac{\partial \bar{r}}{\partial q_n} \dot{q}_n \tag{A.2.1}$$

The coefficient $\partial \bar{r}/\partial q_n$ of each generalized velocity will usually depend on the instantaneous values of the generalized coordinates. However, the kinetic energy for a linearized analysis requires that these coefficients be constant, so that forming $\bar{v} \cdot \bar{v}$ leads to terms whose only variables are quadratic products of generalized velocities. In order to simplify the coefficients, we use the fact that the generalized coordinates are very small. Thus, the error introduced by evaluating the coefficients at the equilibrium position, where all $q_n = 0$, should be negligible. This leads to

$$\bar{v} = \sum_{n=1}^{N} \left(\frac{\partial \bar{r}}{\partial q_n} \right)_0 \dot{q}_n \tag{A.2.2}$$

As always, a zero subscript on a quantity denotes that the quantity is evaluated at the reference position. This description of the velocity is the mathematical equivalent to specifying, as we did in Chapter 1, that all geometrical factors affecting the velocity should be evaluated at the static equilibrium position.

When eq. (A.2.2) is used, the kinetic energy will be a quadratic sum. Also, as we saw in Chapter 1, truncation of a Taylor series for V leads to a quadratic sum for the potential energy. There is no need to consider the static force contributions $(F_0)_j$ if we take the external forces causing the motion to be disturbance values. Thus the linearized energies are

$$T = \frac{1}{2} \sum_{j=1}^{N} \sum_{n=1}^{N} M_{jn} \dot{q}_j \dot{q}_n, \qquad V = \frac{1}{2} \sum_{j=1}^{N} \sum_{n=1}^{N} K_{jn} q_j q_n \tag{A.2.3}$$

Because Lagrange's equations do not explicitly address dissipation effects, we shall defer such considerations until later.

When the generalized coordinates are limited to being very small, and the system is time invariant, the generalized forces defined by eqs. (A.1.10) are the same as the quantities defined by eq. (1.5.23) for the power balance method. To prove this, we use the smallness restriction to simplify the virtual displacement in eq. (A.1.1), which enables us to evaluate the partial derivative terms at the static equilibrium position,

$$\delta \bar{r}_f = \sum_{j=1}^{N} \left(\frac{\partial \bar{r}_f}{\partial q_{f}} \right)_0 \delta q_j \tag{A.2.4}$$

Virtual work consists of the sum of $\bar{F}_f \cdot \delta \bar{r}_f$ terms, whereas the instantaneous power input is a sum of $\bar{F}_f \cdot \bar{v}_f$ terms. A comparison of eqs. (A.2.4) and (A.2.2) reveals that the coefficient of each δq_j in the linearized virtual displacement is identical to the coefficient of $\dot{q}_j$ in the linearized velocity. In that case the generalized force derived from $\mathcal{P}_{in}$ will be the same as the quantity derived from δW.

The next issue is evaluation of the derivative terms in Lagrange's equations. We wish to carry out the required operations associated with a selected generalized coordinate q_j. In order to avoid confusion with the summation index used to form the energy expressions in eqs. (A.2.3), we change the first summation index in those expressions from j to k. The product rule for differentiation then leads to

$$\frac{\partial T}{\partial \dot{q}_j} = \frac{1}{2} \frac{\partial}{\partial \dot{q}_j} \left(\sum_{k=1}^{N} \sum_{n=1}^{N} M_{kn} \dot{q}_k \dot{q}_n \right) = \frac{1}{2} \sum_{k=1}^{N} \sum_{n=1}^{N} M_{kn} \left(\frac{\partial \dot{q}_k}{\partial \dot{q}_j} \dot{q}_n + \dot{q}_k \frac{\partial \dot{q}_n}{\partial \dot{q}_j} \right) \tag{A.2.5}$$

Recall that each variable is considered to be independent of the others in the process of partial differentiation, so $\partial \dot{q}_k / \partial \dot{q}_j = 0$ unless $k = j$, in which case the derivative is 1. Thus, the first partial derivative term filters out of the double summation all terms whose index k does not match j. In the same manner, the second term filters out all terms whose index n does not match j. The result is that

$$\frac{\partial T}{\partial \dot{q}_j} = \frac{1}{2} \sum_{n=1}^{N} M_{jn} \dot{q}_n + \frac{1}{2} \sum_{k=1}^{N} M_{kj} \dot{q}_k \tag{A.2.6}$$

We now observe that we may change the symbol used as the summation index without altering the result. Furthermore, the M_{jn} coefficients are symmetric, $M_{nj} = M_{jn}$, so we have

$$\frac{\partial T}{\partial \dot{q}_j} = \frac{1}{2} \sum_{n=1}^{N} M_{jn} \dot{q}_n + \frac{1}{2} \sum_{n=1}^{N} M_{nj} \dot{q}_n = \sum_{n=1}^{N} M_{jn} \dot{q}_n \tag{A.2.7}$$

The only time-dependent terms appearing in this expression are the generalized velocities, so the first term appearing in Lagrange's equations is

$$\frac{d}{dt} \left(\frac{\partial T}{\partial \dot{q}_j} \right) = \sum_{n=1}^{N} M_{jn} \ddot{q}_n \tag{A.2.8}$$

The remaining terms in Lagrange's equations feature derivatives with respect to a generalized coordinate. We observe that the kinetic energy for a time-invariant system, which is the first of eqs. (A.2.3), does not depend explicitly on the generalized coordinates, so that $\partial T / \partial q_j = 0$. To represent the potential energy term in Lagrange's equations, we note that the form of V in eqs. (A.2.3) is the same as T, except that generalized coordinates replace generalized velocities. It follows that the process of forming $\partial V/\partial q_j$ will be comparable to those that led to eq. (A.2.7), from which we deduce that

$$\frac{\partial V}{\partial q_j} = \sum_{n=1}^{N} K_{jn} q_n \tag{A.2.9}$$

The Lagrange's equations corresponding to these descriptions of the individual terms are

$$\sum_{n=1}^{N} M_{jn} \ddot{q}_n + \sum_{n=1}^{N} K_{jn} q_n = Q_j \tag{A.2.10}$$

This is the summation form of the matrix equation

$$[M]\{\ddot{q}\} + [K]\{q\} = \{Q\} \tag{A.2.11}$$

which we obtained previously by applying power balance considerations.

One difference between the two formulations is that the damping terms arise naturally in the power balance formulation. For Lagrange's equations in the case of nonlinear systems, dissipative effects must be incorporated into the generalized forces. However, it is possible to modify Lagrange's equations if one limits their attention to linear systems. In that case dissipation effects require insertion of the term $[C]\{\dot{q}\}$ into eq. (A.2.11). Note the similarity of this term to $[K]\{q\}$, which stems from the potential energy term in Lagrange's equations. This similarity led Rayleigh to define a *dissipation function* $\mathcal{D}$ that is similar in form to V, except that the damping constants replace the stiffness constants and the generalized velocities replace the generalized coordinates,

$$\mathcal{D} = \frac{1}{2} \sum_{j=1}^{N} \sum_{n=1}^{N} C_{jn} \dot{q}_j \dot{q}_n \tag{A.2.12}$$

Clearly, $\mathcal{D}$ is half the instantaneous power dissipation,

$$\boxed{\mathcal{D} = \tfrac{1}{2} \mathcal{P}_{\text{dis}}} \tag{A.2.13}$$

so forming this quantity requires no new considerations. The corresponding form of Lagrange's equations is

$$\frac{d}{dt}\left(\frac{\partial T}{\partial \dot{q}_j}\right) - \frac{\partial T}{\partial q_j} + \frac{\partial \mathcal{D}}{\partial \dot{q}_j} + \frac{\partial V}{\partial q_j} = Q_j, \quad j = 1, 2, \ldots, N \tag{A.2.14}$$

By analogy to the potential energy term, the Rayleigh dissipation function leads to the insertion of a term $[C]\{\dot{q}\}$ in the matrix equation of motion.

Lagrange's equations do not provide any additional insight or capabilities in regard to the linearized equations of motion governing time-invariant systems. The utility of Lagrange's equations is that they are valid for all situations: The generalized coordinates are not limited to being small, and the system is not limited to being time invariant.

REFERENCES AND SELECTED READINGS

GINSBERG, J. H. 1995. *Advanced Engineering Dynamics*, 2nd ed. Cambridge University Press, Cambridge, England.

TRANSIENT RESPONSES FOR UNDERDAMPED ONE-DEGREE-OF-FREEDOM SYSTEMS

$$\ddot{q} + 2\zeta\omega_{nat}\dot{q} + \omega_{nat}^2 q = \frac{F(t)}{M} \quad \zeta < 1, \quad \omega_d = \omega_{nat}\sqrt{1 - \zeta^2}$$

- Free vibration: $F(t) = 0$

$$q = \exp(-\zeta\omega_{nat}t)[q(0)\cos(\omega_d t) + \frac{\dot{q}(0) + \zeta\omega_{nat}q(0)}{\omega_d}\sin(\omega_d t)]$$

- Impulse excitation: $F(t) = \delta(t)$

$$q = \frac{1}{M\omega_d}\exp(-\zeta\omega_{nat}t)\sin(\omega_d t)h(t)$$

- Step excitation: $F(t) = h(t)$

$$q = \frac{1}{M\omega_{nat}^2}\{1 - \exp(-\zeta\omega_{nat}t)[\cos(\omega_d t) + \frac{\zeta\omega_{nat}}{\omega_d}\sin(\omega_d t)]\}h(t)$$

- Ramp excitation: $F(t) = th(t)$

$$q = \frac{1}{M\omega_{nat}^3}\{(\omega_{nat}t) - 2\zeta + \exp(-\zeta\omega_{nat}t)[2\zeta\cos(\omega_d t) - (1 - 2\zeta^2)\frac{\omega_{nat}}{\omega_d}\sin(\omega_d t)]\}h(t)$$

- Quadratic excitation: $F(t) = t^2 h(t)$

$$q = \frac{1}{M\omega_{nat}^4}\{(\omega_{nat}t)^2 - 4\zeta(\omega_{nat}t) - 2(1 - 4\zeta^2) + \exp(-\zeta\omega_{nat}t) \times [2(1 - 4\zeta^2)\cos(\omega_d t) + (6\zeta - 8\zeta^3\frac{\omega_{nat}}{\omega_d}\sin(\omega_d t)]\}h(t)$$

- Exponential excitation:
$$F(t) = \exp(-\beta t) h(t)$$

$$q = \frac{1}{M(\omega_{nat}^2 - 2\zeta\omega_{nat}\beta + \beta^2)}\{\exp(-\beta t) - \exp(-\zeta\omega_{nat}t)[\cos(\omega_d t) + \frac{\zeta\omega_{nat} - \beta}{\omega_d}\sin(\omega_d t)]\}h(t)$$

- Transient sinusoidal excitation:
$$F(t) = \sin(\omega t)h(t), \quad \omega \neq \omega_{nat} \text{ if } \zeta \neq 0$$

$$q = \frac{1}{M[(\omega_{nat}^2 - \omega^2)^2 + 4\zeta^2\omega_{nat}^2\omega^2]}$$

$$\times \{(\omega_{nat}^2 - \omega^2)\sin(\omega t) - 2\zeta\omega_{nat}\omega\cos(\omega t) + \omega\exp(-\zeta\omega_{nat}t)[2\zeta\omega_{nat}\cos(\omega_d t) - \frac{(1 - 2\zeta^2)\omega_{nat}^2 - \omega^2}{\omega_d}\sin(\omega_d t)]\}h(t)$$

- Transient co-sinusoidal excitation:
$$F(t) = \cos(\omega t)h(t), \quad \omega \neq \omega_{nat} \text{ if } \zeta \neq 0$$

$$q = \frac{1}{M[(\omega_{nat}^2 - \omega^2)^2 + 4\zeta^2\omega_{nat}^2\omega^2]}$$

$$\times \{(\omega_{nat}^2 - \omega^2)\cos(\omega t) + 2\zeta\omega_{nat}\omega\sin(\omega t) - \exp(-\zeta\omega_{nat}t)[(\omega_{nat}^2 - \omega^2)\cos\omega_d t + \frac{\zeta\omega_{nat}(\omega_{nat} + \omega^2)}{\omega_d}\sin(\omega_d t)]\}h(t)$$

DEFORMATION MODES FOR UNIFORM BARS

SEE BACK INSIDE COVER

Index

DEFORMATION MODES FOR UNIFORM BARS

Axial Motion: $\omega_n = \left(\dfrac{E}{p}\right)^{1/2} \alpha_n$

- Fixed ends at $x = 0$ and $x = L$:

$$\psi_n = \sin\left(\alpha_n \frac{x}{L}\right), \quad \alpha_n = n\pi, \quad n = 1, 2, \ldots.$$

- Fixed end at $x = 0$ and free end at $x = L$:

$$\psi_n = \sin\left(\alpha_n \frac{x}{L}\right), \quad \alpha_n = \left(\frac{2n-1}{2}\right)\pi, \quad n = 1, 2, \ldots$$

- Free ends at $x = 0$ and $x = L$:

$$\psi_n = \cos\left(\alpha_n \frac{x}{L}\right), \quad \alpha_n = (n-1)\pi, \quad n = 2, 3, \ldots.$$

Flexural Motion: $\omega_n = \left(\dfrac{EI}{\rho A L^4}\right)^{1/2} \alpha_n^2$

- Hinged end at $x = 0$:

$$\psi_n = \sin\left(\alpha_n \frac{x}{L}\right) + R_n \sinh\left(\alpha_n \frac{x}{L}\right)$$

- Hinged end at $x = L$:

$$\alpha_n = n\pi, \quad n = 1, 2, \ldots, \qquad R_n = 0$$

- Clamped end at $x = L$:

$$\tan(\alpha_n) - \tanh(\alpha_n) = 0, \quad \alpha_1 = 3.9266, \quad \alpha_n \approx \frac{4n+1}{4}\pi \ \text{ if } \ n \geq 2$$

$$R_n = -\left[\frac{\sin(\alpha_n)}{\sinh(\alpha_n)}\right]$$

- Guided end at $x = L$:

$$\alpha_n = \left(\frac{2n-1}{2}\right)\pi, \quad n = 1, 2, \ldots, \qquad R_n = 0$$

- Free end at $x = L$:

$$\tan(\alpha_n) - \tanh(\alpha_n) = 0, \quad \alpha_2 = 3.9266; \quad \alpha_n \approx \frac{4n-3}{4}\pi \ \text{ if } \ n \geq 3$$

$$R_n = \frac{\sin(\alpha_n)}{\sinh(\alpha_n)}$$

Printed in the United States
100973LV00002B/14/A

9 780471 370840